ECOLOGIGAL PROCESSES IN COASTAL ENVIRONMENTS

ECOLOGICAL PROCESSES IN COASTAL ENVIRONMENTS

THE FIRST EUROPEAN ECOLOGICAL SYMPOSIUM AND THE 19TH SYMPOSIUM OF THE BRITISH ECOLOGICAL SOCIETY

NORWICH, 12–16 SEPTEMBER 1977

EDITED BY

R. L. JEFFERIES

Department of Botany, University of Toronto, Canada

AND

A. J. DAVY

School of Biological Sciences,
University of East Anglia, Norwich

SPONSORING SOCIETIES

The British Ecological Society

The Ecological and Ethological Section of the Royal Zoological Society of Belgium

Gesellschaft für Ökologie

Oecologische Kring

Scandinavian Society Oikos

Societa Italiana di Ecologia

Société d'Écologie

BLACKWELL SCIENTIFIC PUBLICATIONS

OXFORD LONDON EDINBURGH MELBOURNE

Osney Mead, Oxford, OX2 0EL
8 John Street, London, WC1N 2ES
9 Forrest Road, Edinburgh, EH1 2QH
P.O. Box 9, North Balwyn,
Victoria, Australia

First published 1979
Distributed in U.S.A. and Canada by
Halsted Press, a division of
John Wiley & Sons Inc
New York

Printed and bound in Great Britain by
William Clowes & Sons Ltd
Beccles, Gt. Yarmouth and London

British Library
Cataloguing in Publication Data

Ecological processes in coastal environments.
—(British Ecological Society. Symposia; 19).
1. Coastal ecology — Congresses
I. Jefferies, R L II. Davy, A J
III. Series
574.5′2636 QH541.5.C65

ISBN 0-632-00472-X

CONTENTS

VII · Abstracts of Poster Papers

PLANNING COMMITTEE

P. J. M. VAN DER AART *Instituut voor Oecologisch Onderzoek, 'Weevers' Duin', Oostvoorne, The Netherlands*

A. J. DAVY *School of Biological Sciences, University of East Anglia, Norwich, U.K.*

E. A. G. DUFFEY *Institute of Terrestrial Ecology, Monks Wood Experimental Station, U.K.*

A. J. GRAY *Institute of Terrestrial Ecology, Furzebrook Research Station, U.K.*

C. HEIP *Laboratorium voor Morfologie en Systematik, Museum voor Dierkunde, Gent, Belgium*

B. HEYDEMANN *Zoologisches Institut der Universität, Kiel, B.R.D.*

R. L. JEFFERIES *Department of Botany, University of Toronto, Canada*

J. A. LEE *Department of Botany, University of Manchester, U.K.*

D. S. RANWELL *Institute of Terrestrial Ecology, Colney Research Station, Norwich, U.K.*

J. A. RIOUX *Ecologie Médicale et Pathologie Parasitaire, Université de Montpellier, France*

T. R. E. SOUTHWOOD *Imperial College Field Station, University of London, U.K.*

P. VESTERGAARD *Institute of Plant Ecology, University of Copenhagen, Denmark*

FOREWORD

THE EUROPEAN ECOLOGICAL SYMPOSIA

Ecology is like a seedling forest tree, after many years of unspectacular but sound growth, it has in the last decade suddenly come into the sunlight. Increasing concern with environmental problems has led politicians and planners to turn to ecologists for advice and predictions. The problems are international, indeed global, yet ecology has again been like tropical forest saplings, to some extent isolated spatially and with different developmental facies. The International Biological Programme brought ecologists together on a worldwide scale; the co-operation proved profitable and the International Ecological Congresses have tended to follow up the theme of IBP. Additionally, there is a place for international meetings, more limited in geographical scope and exploring other facets of ecology. Nowhere is there greater potential for such meetings than in Europe where linguistic and political divisions have tended to exaggerate and compound the differences between various 'schools of ecology'.

I was therefore delighted when, during my Presidency, the British Ecological Society (as the oldest European Ecological Society) decided to invite its sister Societies and the Ecological Institutes of Academies to participate in a series of symposia. The present symposium therefore represents the first of what I hope will be a long line to be held in different locations, probably every three years. (The next will be in Germany in 1980 and the third in France.) Additionally the European Societies have arranged to circulate details of their meetings, to welcome each other's members at these meetings and to seek co-operation in other ways so that the symposia will be a part of a continuing dialogue.

The sponsoring Societies hope to be able to welcome further Societies or Academies from other parts of Europe to the group of sponsors. There is no intention to limit co-operation to the founder Societies, nor as the lists of speakers and participants at the first symposium show, will individual attendance be restricted to those resident in Europe.

All those who attended and enjoyed (the two are synonymous) the first symposium will, I know, wish to join me in thanking its organizers and in wishing the Standing Committee and future organizers every success.

T. R. E. Southwood

PREFACE

This book is the outcome of the first symposium to be sponsored jointly by the Ecological Societies and Academies of Europe. Consequently, we have striven to maintain a balance in the numbers of contributors from different European countries so as to obtain as wide a representation as possible. The volume comprises a nucleus of 22 invited papers, together with a further 15 offered papers; because of limitations of space, 37 papers which were contributed to the symposium as posters are represented only by brief abstracts. We hope that the references appended to many of these abstracts will serve to lead the reader to more detailed accounts of this work. The inclusion of offered papers and a poster display certainly enabled many more European ecologists to contribute to the meeting than would otherwise have been the case. In particular, there was a gratifyingly large number of relatively young workers, which is an encouraging sign for the future of European ecology.

When it was decided that the 19th symposium of the British Ecological Society should also be the first European symposium, a planning committee composed of members of the different European societies was formed and the scope of the meeting was broadened from its originally intended theme of salt marsh ecology. Throughout the planning, emphasis has been placed on the ecological processes within coastal environments rather than on discussion of the different coastal habitats. However, most of the papers are concerned with salt marshes, dunes lagoons or estuaries as these environments represent much of the European coastline and are particularly vulnerable to misuse by man. In spite of these constraints, as well as those of time, a large number of topics was discussed; some may say too many! Many of the papers contain significant new material; indeed rather few papers are synthetic in approach and this reflects the present need for relevant quantitative data on many kinds of ecological process. The relative paucity of information concerning the roles of animals and microorganisms in coastal environments is particularly conspicuous.

The introductory paper is concerned with the geographical distribution and phytosociological relationships of the plant communities of European sand dunes and salt marshes. In spite of considerable progress, the authors indicate that, in some areas, knowledge of plant communities is still fragmentary and that relationships between communities are not always apparent. Several contributors emphasize the predictability of both sand dune and salt marsh environments, and the spatial and temporal heterogeneity which characterizes them. Likewise, the life cycle strategies of plants and animals in relation to this environmental heterogeneity are examined in a number of papers.

Another group of papers reflects the current interest in osmotic, ionic and water regulation in plants and animals living in a saline environment and, in particular, the role of soluble organic compounds which may act as compatible osmotic solutes. However, the concentrations of these compounds within the cellular compartments are unknown and it is evident that such solutes may have other roles during growth and development. Furthermore, changes in turgor in plants may be all that are necessary to accomodate small transient changes in external salinity.

The significance of particular species in regulating the flow of energy and materials in coastal ecosystems is considered by a large group of contributors. The detailed discussions range from the ecophysiology of photosynthetic adaptations, through primary production and secondary production to decomposition and nitrogen cycling. The models concerned with energy and nutrient flow within and between coastal ecosystems are based on studies in North America, as there are no comparable European studies. How far the results are applicable to western European marshes is uncertain, especially as the directions and magnitudes of exchanges appear to vary greatly from marsh to marsh.

Papers associated with the final theme of the volume are directly or indirectly concerned with the conservation of coastal ecosystems. They do not set out to provide answers to specific management problems but examine the progress and extent of research applicable to an understanding of the conservation of important coastal sites in Europe. The nature of the various problems emphasizes the need for close co-operation between scientists and the governments of Europe, a point which was frequently and strongly expressed at the meeting. We hope that this volume represents a step in the direction of closer collaboration between European ecologists—something that is long overdue.

ACKNOWLEDGMENTS

There are unusual difficulties in organizing a symposium of this complexity when the two organizers are separated by 3,000 miles for much of the time. One of us (R.L.J.) gratefully acknowledges the very careful planning and execution of the local arrangements by Dr A.J. Davy. We are very grateful to Miss C.E.P. Andrews for an enormous amount of secretarial assistance throughout both the organization of the symposium and the editing of this volume; Mr P.G. Mallott organized the assembly of the poster display; Mrs L.M. Davy contributed much to the organization of the meeting. Mrs J.C. Mallott, Mr N. Perkins, Mr T. Rudmik, Mr D. Walls and Dr and Mrs A.R. Watkinson gave us a considerable amount of help and support, especially in the period immediately prior to and during the meeting. We are fully appreciative of their efforts.

We thank the University of East Anglia for the provision of many facilities and in particular, the Pro Vice-Chancellor, Professor M.J. Frazer who opened the meeting and also was host at the University reception; the Librarian, Mr W. L. Guttsman, who kindly allowed the use of the library for the reception and the poster display; and the Assistant Estates Officer, Mr R. Lloyd, who put at our disposal the full resources of the University conference office. Our thanks are due to the School of Biological Sciences of the University of East Anglia and to the Department of Botany of the University of Toronto for much logistic support.

One of the highlights of the meeting was the civic reception given by the Lord Mayor and Lady Mayoress, Councillor and Mrs Roe at Norwich Castle; we deeply appreciate the generous hospitality extended by the City of Norwich. We are most impressed by the enthusiasm with which members of the Norwich Society, led by Mrs J. Ogden, ensured that accompanying persons had an interesting and varied programme.

We thank Dr L.A. Boorman, Dr A.J. Gray, Mr R. Scott and Dr S.M. Coles from the Institute of Terrestrial Ecology and Dr C. Beale and Mr C. Johnstone of the Nature Conservancy Council's East Anglian Region for leading excursions to the North Norfolk coast and the Wash. It is a pleasure to acknowledge the services of our eleven chairmen, including Dr E.A.G. Duffey who led the discussion of European conservation.

In conclusion we extend our thanks to the members of the planning committee, listed on p. xii, for all their assistance; especially we are grateful to Professor T.R.E. Southwood FRS for his help and encouragement throughout the planning of this symposium. It was largely as a result of his thinking of the need for closer co-operation between the Ecological Societies of Europe that this meeting took place.

I

INTRODUCTION

1. THE DIVERSITY OF EUROPEAN COASTAL ECOSYSTEMS

V. WESTHOFF AND M.G.C. SCHOUTEN

Botanisch Laboratorium, Faculteit der Wiskunde en Natuurwetenschappen, Katholieke Universiteit, Nijmegen, The Netherlands

Dedicated to Professor Heinrich Walter on the occasion of his eightieth birthday

SUMMARY

This paper considers the geographical variation in the plant communities of both the xerosere of sand dunes and the halosere of salt marshes along European coasts. The communities are discussed in relation to their distribution within these ecosystems.

Differences in the plant communities of sand dunes can be related to edaphic factors, such as calcareous soils or soils deficient in lime, and the climatic conditions associated with the Mediterranean, Atlantic and Baltic coasts. The communities of stabilized dunes show much geographical variation, as in these habitats edaphic and climatic factors exert considerable influence. The communities of the foredunes, in contrast, are relatively similar to one another, which reflects the marked influence of mobile sand on the vegetation of these dunes.

The geographical variation in coastal salt marsh communities of Europe is determined largely by differences in climatic conditions. The communities of the Mediterranean coast differ from those of coasts of western and northern Europe. These differences are associated with dissimilarities in the precipitation/evapotranspiration balance between these coasts. In general terms, the impoverishment of the floristic and phytosociological diversity of salt marshes can be related to increasing latitude and decreasing temperature.

RÉSUMÉ

Dans cette publication, nous examinons la variation géographique des communautés végétales de la xérosérie des dunes de sable et de la halosérie des marais salés, le long des côtes européennes. Nous discutons la distribution des communautés dans ces écosystèmes. On peut relier les différences des communautés végétales aux facteurs édaphiques (sol calcaire ou déficient en chaux) et aux conditions climatiques associées aux côtes meditérranéenne, atlantique et baltique. Les communautés des dunes stabilisées montrent beaucoup de variations géographiques; dans ces sites les facteurs édaphiques et climatiques exercent une influence considérable. Au contraire, les communautés des dunes jeunes sont relativement similaires les unes aux autres, ce qui reflète l'influence prédominante du sable mobile sur la végétation de ces dunes. La variation géographique des communautés des marais salés des côtes européennes est largement déterminée par les différences de conditions climatiques. Les communautés de la côte meditérranéenne diffèrent de celles de l'Europe de l'ouest et du nord. Ces différences sont associées aux dissimilarités dans le rapport précipitation/évapo-transpiration entre ces côtes. De facon générale l'appauvrissement de la diversité floristique autant que phytosociologique des marais salés peut être relié à l'augmentation de la latitude et la diminution de la température.

INTRODUCTION

Traditionally plant and animal communities of sand dunes and salt marshes have been described and zoned in relation to environmental gradients. The vegetation present within these different zones in these ecosystems has been studied extensively and will not be discussed here; the reader is referred to van Dieren (1934), Tansley (1949), Salisbury (1952), Gimingham (1964), van der Maarel (1966), Ranwell (1972) and Beeftink (1977).

In this paper geographical variation of coastal plant communities of Europe is examined for the circumboreal, atlantic and mediterranean domains. The communities of the xerosere of sand dunes and the halosere of mud flats and salt marshes are discussed initially. The plant communities of the strand line are described separately. The reader is referred to Appendix 1 and Westhoff and van der Maarel (1973) for the meaning of technical terms used in the classification of vegetation.

THE VEGETATION OF COASTAL DUNES OF EUROPE

A summary of the plant communities of sand dunes is given in Fig. 1.1.

Plant communities of the foredunes

In the sub-Arctic and Boreal regions coastal dunes are not as developed as those of the Atlantic and Mediterranean coasts. The areas of these flat northern dunes which show the greatest accretion of wind-blown sand are

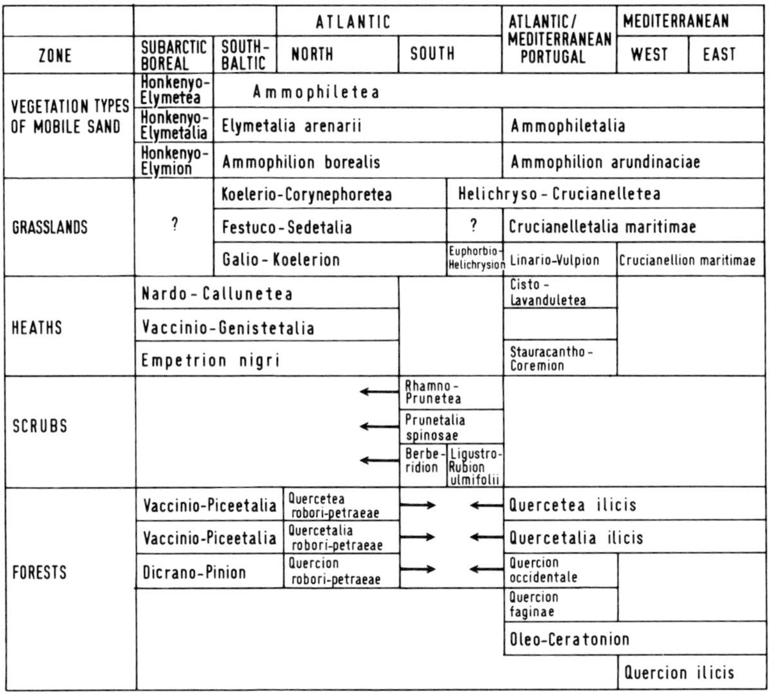

			ATLANTIC		ATLANTIC/ MEDITERRANEAN	MEDITERRANEAN	
ZONE	SUBARCTIC BOREAL	SOUTH-BALTIC	NORTH	SOUTH	PORTUGAL	WEST	EAST
VEGETATION TYPES OF MOBILE SAND	Honkenyo-Elymetea	Ammophiletea					
	Honkenyo-Elymetalia	Elymetalia arenarii			Ammophiletalia		
	Honkenyo-Elymion	Ammophilion borealis			Ammophilion arundinaciae		
GRASSLANDS	?	Koelerio-Corynephoretea		Helichryso-Crucianelletea			
		Festuco-Sedetalia		?	Crucianelletalia maritimae		
		Galio-Koelerion		Euphorbio-Helichrysion	Linario-Vulpion	Crucianellion maritimae	
HEATHS	Nardo-Callunetea				Cisto-Lavanduletea		
	Vaccinio-Genistetalia						
	Empetrion nigri				Stauracantho-Coremion		
SCRUBS			←	Rhamno-Prunetea			
			←	Prunetalia spinosae			
			←	Berberidion / Ligustro-Rubion ulmifolii			
FORESTS	Vaccinio-Piceetalia		Quercetea robori-petraeae	→ ←	Quercetea ilicis		
	Vaccinio-Piceetalia		Quercetalia robori-petraeae	→ ←	Quercetalia ilicis		
	Dicrano-Pinion		Quercion robori-petraeae	→ ←	Quercion occidentale		
					Quercion faginae		
					Oleo-Ceratonion		
						Quercion ilicis	

FIG. 1.1. Geographical distribution of the higher dune-xerosere syntaxa.

occupied by plant communities in which the Arctic–Boreal grass *Elymus arenarius* is dominant, although *Honkenya peploides* ssp. *peploides* and *Mertensia maritima* also are present. These communities belong to the class Honkenyo-Elymetea Tüxen 1966 but in general the *Elymus* communities of these regions have not been well studied.

The class Honkenyo-Elymetea characteristic of the northern dunes of Europe may be considered a northern counterpart of the class Ammophiletea Braun-Blanquet et Tüxen 1943 of the Atlantic, southern Baltic and Mediterranean dunes. This class is characterized by *Eryngium maritimum*, *Calystegia soldanella* and *Euphorbia paralias*, which are distributed along the Atlantic and Mediterranean coasts of Europe. In contrast to the dunes of the sub-Arctic and Boreal regions, elsewhere in Europe the plant communities of the low transient coastal foredunes are different from those of the young permanent dune ridges. Communities of *Ammophila arenaria* are present on the latter whereas *Agropyron junceiforme* grows on the foredunes. On the southern shores of the Baltic sea, however, *Agropyron junceiforme* has a local distribution and is largely confined to the west where the salinity is higher. On some exposed beaches in northern Europe, although *Agropyron junceiforme* is present, *Ammophila arenaria* initiates dune formation (van Dieren 1934;

Gimingham 1964). At other sites, as for example on the west coast of Ireland, *Ammophila arenaria* is scarce and *Agropyron junceiforme* is responsible for the stabilization of sand.

The class Ammophiletea is subdivided into an Atlantic order, Elymetalia arenarii Br.-Bl. et Tx. 1943, which is characterized by *Honkenya peploides* agg., and a Mediterranean order, Ammophiletalia Br.-Bl. et Tx. 1943, that has *Matthiola sinuata, Diotis maritima, Vulpia fasciculata, Medicago litoralis, Echinophora spinosa, Orlaya maritima, Sporobolus maritimus* and *Pancratium maritimum* as character-taxa. The geographical separation of these two orders is based on the distributions of two sub-species of *Ammophila arenaria. A.arenaria* ssp. *arenaria* is present on Atlantic coasts whereas *A.arenaria* ssp. *arundinacea* grows on the dunes of the Mediterranean coast. The geographical boundary between the distribution areas of these two sub-species as well as of *Agropyron junceiforme* ssp. *junceiforme* and *A.junceiforme* ssp. *mediterraneum* is the coast of northern Portugal.

Although *Elymus arenarius* is mainly distributed in the sub-Arctic and Boreal regions of Europe it occurs on the coasts of north-west Europe and it serves as a differential taxon to separate north Atlantic associations. On the foredunes the association Agropyretum boreo-atlanticum (Warming 1909) Br.-Bl. et de Leeuw 1936 em. Tx. 1952 has been recognized while on young dunes the association Elymo-Ammophiletum id. is present. The two associations on the Atlantic coasts of south-west Europe which correspond to those of the northern Atlantic coasts are Euphorbio-Agropyretum juncei Tx. (1945) 1952 and Euphorbio-Ammophiletum id. respectively. The frequency of *Euphorbia paralias* and *Calystegia soldanella* within the latter communities is used to separate them from the associations of the dunes of the northern Atlantic coast. These species occur also in dunes of north-west Europe but there they are restricted to the sheltered lee slope of young dunes. The geographical boundary which separates these Atlantic communities are the coasts of the Netherlands and England. The communities on the Irish coasts for the most part belong to the associations characteristic of south-west Europe.

Westhoff (1947), Fukarek (1961) and Géhu (1971) have indicated that the Elymo-Ammophiletum communities of the eastern Baltic coast, excluding the Gulf of Bothnia, are different from those of the western Baltic and North Sea coasts. In the former *Petasites spurius, Linaria odora, Tragopogon heterospermum* and *Coryspermum hyssopifolium* occur, whereas the latter are characterized by *Lathyrus maritimus* and *Sonchus arvensis* var. *maritimus*.

Géhu (1969, 1971) has stated that on the Atlantic coast of southern France where the floristic composition of Euphorbio-Ammophiletum communities is particularly rich, a further subdivision of this association may be possible as some communities include species which have an Atlantic–Mediterranean distribution. These include *Matthiola sinuata, Diotis maritima, Silene thorei* and endemics such as *Linaria thymifolia* and *Galium arenarium*.

The vegetation of the Portuguese coast shows a strong Mediterranean influence. For example, the species *Pancratium maritimum* which is common on the coasts of the Mediterranean is present in the psammophilous communities here. Braun-Blanquet (1972) classified the sand dune communities of the Portuguese coast into a special order, the Artemisietalia crithmifoliae, which contains one alliance, the Linario-Vulpion Br.-Bl., Rozeira et Pinto da Silva 1972. This order consists of communities of both the mobile sand and the stabilized dunes. In our opinion, however, this view is at variance with present day general concepts. The alliance Linario-Vulpion should only comprise the vegetation of the stabilized dunes and it should be assigned to the class Helichryso-Crucianelletea (see below). The communities of mobile sand should be classified as a separate alliance belonging to the order Ammophiletalia and comparable with the Ammophilion alliance of Mediterranean coasts. The psammophilous communities of the latter region have not been studied fully but it appears likely that the recognition of several associations which are geographically separated will be possible.

The plant communities of stabilized dunes

Differences in the floristic composition of the plant communities of the stabilized coastal dunes of Europe are considerable and they reflect the different edaphic and climatic conditions which prevail along the coastlines of Europe.

On the continent there is a marked difference between the dunes north of Bergen in the Netherlands, which are deficient in lime content, and those south of Bergen which are calcareous or partially calcareous. The relatively stable dunes of the northern area are characterized by grasslands which mainly belong to the alliance Galio-Koelerion and by heaths belonging to the class Nardo-Callunetea. The stabilization of the calcareous dunes in contrast is mainly the result of the establishment of shrub communities of the class Rhamno-Prunetea on dunes. Most dune grasslands in this area are of secondary origin and result from the action of rabbits or else from agricultural use such as cutting and grazing.

The dune grassland alliance Galio-Koelerion (Tx. 1937) den Held et Westhoff 1969 has an Atlantic distribution; it occurs on the coasts of France, the British Isles and the Netherlands and extends as far north as the coasts of Scania and the Baltic. The alliance comprises several associations.

Perhaps the best known of these communities, which was the first one described, is the pioneer association Tortulo-Phleetum arenarii (Massart 1908) Br.-Bl. et de Leeuw 1936. It is characterized by the occurrence of a number of winter annuals and by a carpet of mainly acrocarpic mosses such as *Tortula ruraliformis*. This association is widespread and its distribution is similar to that of the alliance Galio-Koelerion; it occurs on the coasts of western Europe as far north as Denmark. The association Tortulo-Phleetum thrives on soils which contain some lime but which are slightly overblown with sand. This community is also dependent on relatively warm and dry microclimatic conditions, so that it is mainly represented on the south and south-west facing slopes of the dunes.

The association Violo-Corynephoretum Westhoff (1943) 1947, which is characterized by *Viola canina* var. *dunensis* and by a carpet of lichens occurs mainly on non-calcareous dunes, where often it represents a structural regression phase after ephemeral *Hippophäe* scrub has died. This community is present on coastlines from France to southern Sweden but is most common on dunes north of Bergen, Holland. On the coasts of the south-west Baltic the association Violo-Corynephoretum is replaced by the association Helichryso-Jasionetum Libbert 1940 which is characterized by the abundance of *Helichrysum arenarium* and *Jasione montana* var. *litoralis*. The comparable association on the south-eastern Baltic coast is the association Helichryso-Artemisietum balticum Fukarek 1961, of which *Cardaminopsis arenosa*, *Artemisia campestris*, *Astragalus arenarius*, *Tragopogon heterospermum* and *Epipactis atrorubens* are character-taxa.

Under the influence of rabbit and cattle grazing both the associations Tortulo-Phleetum and Violo-Corynephoretum may be replaced by the association Festuco-Galietum maritimi (Onno 1933) Br.-Bl. et de Leeuw 1936. The less calcareous the substrate the greater the grazing pressure must be in order to maintain this association. The association occurs on the north European coast from Belgium to the south-west Baltic region.

The closed dune grassland of the alliance Galio-Koelerion on the eastern coasts of Britain is mainly represented by the association Astragalo-Festucetum arenariae Westhoff et Tx. apud Birse and Robertson 1976 of which *Astragalus danicus* is one of the character-taxa. On the western coasts of Britain the association Euphrasio-Festucetum Géhu et Tx. apud Birse and Robertson 1976 occurs which is characterized by a number of different orchid species (Gimingham 1964, Birse and Robertson 1976). The northernmost dune systems of Scotland contain a considerable number of Boreal and sub-Arctic species such as *Dryas octopetala*, *Oxytropis halleri*, *Draba incana*, *Carex maritima* and *Saxifraga aizoides* (Gimingham 1964).

Tüxen and Westhoff (1969) have described the association Gentiano-Pimpinelletum saxifragae which is characterized by *Draba incana*, *Rhytidium rugosum*, *Gentianella campestris* ssp. *baltica*, *Gentianella amarella* ssp. *uliginosa* and *Antennaria hibernica* and which occurs on calcareous dune systems in southern Norway. Tüxen (1970) also has recognized the *Silene maritima-Festuca cryophila* association on the Icelandic coast, which is composed of such species as *Festuca rubra* ssp. *cryophila*, *Silene maritima* and *Armeria maritima*. Since most character-taxa of the alliance Galio-Koelerion do not extend that far north, the syntaxonomical position of this association is far from clear. Along the Gulf of Bothnia the Galio-Koelerion probably is replaced by the continental alliance Koelerion glaucae of which *Koeleria glauca* and *Festuca polesica* are character-taxa.

In dune areas which have a stabilized soil rich in lime, plant communities may occur belonging to the alliance Mesobromion Br.-Bl. et Moor 1938 em. Oberd. 1957. This higher syntaxon is not confined to coastal ecosystems but comprises the mesic dry calcareous grasslands of western and central Europe. An example of Mesobromion in dunes is the association Camptothecio-Asperuletum cynanchicae Br.-Bl. et Tx. 1952, which occurs on the west coast of Ireland and is characterized by the following species; *Gentiana verna*, *Neotinea intacta* and *Asperula cynanchica*. The association Anthyllidi-Silenetum nutantis (de Leeuw 1938) Boerboom 1957 on the north facing slopes of Dutch calcareous dunes represents a transitional community between the Galio-Koelerion and the Mesobromion. From Brittany southwards the communities of the alliance Galio-Koelerion are replaced by communities in which chamaephytes such as *Ephedra distachya*, *Helichrysum stoechas* and *Rosa pimpinellifolia* are present indicating the sub-Mediterranean character of the vegetation. These communities belong to the alliance Euphorbio-Helichrysion stoechadis (Géhu et Tx. 1972 n.n.) Sissingh 1974.

The lime deficient dunes of the Portuguese coast contain plant communities of which *Vulpia alopecurus*, *Helichrysum angustifolium*, *Jasione lusitanica* and *Scrophularia frutescens* are character-taxa. The communities have been placed in the alliance Linario-Vulpion Br.-Bl. et Pinto da Silva 1972, within which the association Scrophulario-Vulpietum id. is the most widely distributed. Another alliance, the Crucianellion maritimae (Rivas-Goday et Rivas-Martinez 1963), is represented on the stabilized dunes of the Mediterranean coast. This alliance together with Linario-Vulpion can be placed in the order Crucianelletalia maritimae (Sissingh 1974) of the class Helichryso-Crucianelletea Géhu, Rivas-Mart. et R. Tx. 1975 em. Sissingh 1974. Character-taxa of the Crucianellion include *Teucrium polium* var. *maritimum*, *Scabiosa maritima*, *Scleropoa hemipoa*.

As mentioned earlier most northern European dunes are poor in lime. The plant communities characteristic of these dunes include the Galio-Koelerion communities discussed above and heath communities in which *Empetrum nigrum*, *Calluna vulgaris*, *Erica cinerea*, *Polypodium vulgare* and *Salix repens* ssp. *argentea* are abundant. In moist habitats *Erica tetralix* is frequent. On dry dune soils the heath communities belong to the alliance Carici-Empetrion nigri Böcher 1943 em. Schubert 1960 of the class Nardo-Callunetea Preising 1949. The most widespread association is the Polypodio-Empetretum (Meltzer 1941) Westhoff 1947, which occurs on coasts from Dutch Westfrisian Islands northwards to southern Norway and Sweden. These *Empetrum* heaths often develop on north and north-east facing slopes of stabilized dunes and the communities form, together with those of the association Violo-Corynephoretum of south facing slopes, a 'climax swarm' as defined by Tüxen and Diemont (1938). In dune systems where there are large differences in relief, the development of woodland appears to be retarded and the dry heath communities remain. In contrast woodlands have developed on the old landward dunes of low relief. Oak woodlands, such as the association Melampyro-Quercetum roboris Tx. (1930) 1947 in southern Sweden (Olsson 1972), and the association Querco-Betuletum polypodietosum Br.-Bl. et de Leeuw 1936 near Bergen, Holland, are representative. The pine forests on the Baltic shores belong to the association Leucobryo-Pinetum Matuszkiewicz 1962.

On the calcareous dunes south of Bergen in the Netherlands, shrub communities are characteristic of the stabilized part of the dune system, whereas grasslands appear to be secondary origin. The heath communities are restricted to some of the inner dune systems which date from older Holocene periods. The shrubs belong to the class Rhamno-Prunetea Rivas-Goday et Borja Carbonell 1961. From the southern Netherlands to Brittany this class is represented by the alliance Berberidion Br.-Bl. (1947) 1950. Two associations can be distinguished here; the Hippophao-Ligustretum Meltzer 1941 em. Boerboom 1960, and the Hippophao-Sambucetum Boerboom 1960. The latter association is characteristic of dune soils that are rich in nitrogen. From Brittany southward Hippophäe is replaced by a number of other shrubs that include *Ulex europaeus*, *Rubia peregrina* and *Cistus salviaefolius*. The shrub communities which contain these species belong to the alliance Ligustro-Rubion ulmifolii Géhu et Delelis 1972. Along the Atlantic coast of south-west France because of the effects of human disturbance, these communities at many sites have been replaced by grasslands and they are restricted largely to the edges of coastal pine plantations.

On the landward parts of the dune systems of the Portuguese coast which have a low lime content, heath communities are common. Species such as *Corema album*, *Thymus capitellatus*, *Lavandula stoechas* ssp. *lusitanica*, *Halimium halimifolium* and *Staurocanthus genistoides* characterize these communities. The latter belong to the order Ulicino-Cistetalia Br.-Bl., Pinto da Silva et Rozeira 1964 of the class Cisto-Lavanduletea Br.-Bl. 1940 which comprises the macchia of soils poor in lime. Along the Portuguese coast there are numerous plantations of *Pinus pinaster*. The natural littoral forest, however, appears to belong to one of the following alliances: the Quercion occidentale Br.-Bl., Pinto da Silva et Rozeira 1959 (north Portuguese coast), the Quercion fagineae id. (central Portuguese coast) and the Oleo-Ceratonion Br.-Bl. 1936 (southern Portuguese coast).

Finally along the Mediterranean coasts of southern Europe, the dune grasslands give way to the garrigue communities at the landward edge of the dunes. In the western Mediterranean region these communities belong to the order Lavanduletalia stoechidis Br.-Bl. (1931) 1940, class Cisto-Lavanduletea.

The corresponding communities of the eastern coasts of the Mediterranean have been placed in the order Cisto-Ericetalia Horvatic 1965 (class Quercetea ilicis Br.-Bl. 1936). Further inland beyond the garrigue communities are the evergreen climax forests of the alliance Quercion ilicis Br.-Bl. (1931) 1936.

THE VEGETATION OF COASTAL SALT MARSHES OF EUROPE

A summary of the plant communities of European salt marshes is shown in Fig. 1.2.

Communities of the sublittoral

The *Zostera* communities of the alliance Zosterion Christiansen 1934 (Zosteretea Pignatti 1953) have a very wide distribution along the coastlines of Europe. *Zostera marina* extends from the Arctic coast to southern Spain. It occurs on the Baltic shores as far north as the Gulf of Bothnia. In southern Europe, although present at some northern sites along the Mediterranean coast, this species is only common in the Black Sea. *Zostera noltii* is distributed from southern Norway and the British Isles southwards to West Africa; it grows along the south-western coasts of the Baltic Sea and is present locally on the shores of the Mediterranean and Black Sea (Den Hartog 1970). In the sublittoral of the European coasts, three associations of the Zosterion can be distinguished based on the abundance of either *Zostera noltii* or *Zostera marina* var. *marina* or *Zostera marina* var. *stenophylla*. The last-mentioned is an annual plant in northern Europe as frost kills most rhizomes each winter.

Communities of the eulittoral

The communities of the annual *Salicornia* species belong to the alliance Thero-Salicornion Br.-Bl. 1933 em. Tx. 1950 (Thero-Salicornietea Tx. 1954 apud Tx. et Oberd. 1958). They extend from the north Atlantic and west Baltic coasts to the west coast of France. On the Portuguese coasts and the northern coasts of the Mediterranean Sea the communities are distributed locally. The character-taxon of class, order and alliance is *Salicornia europaea* agg., which consists of several species or sub-species which grow in different habitats.

The communities of the alliance Spartinion Conard 1952 (Spartinetea Tx. 1961), characterized by the genus *Spartina* are represented along the Atlantic coasts from Denmark southwards to the west coast of Morocco. Although three *Spartina* species occur on European coasts, rarely do they grow together. The distribution of *Spartina maritima* extends from the coasts of southern

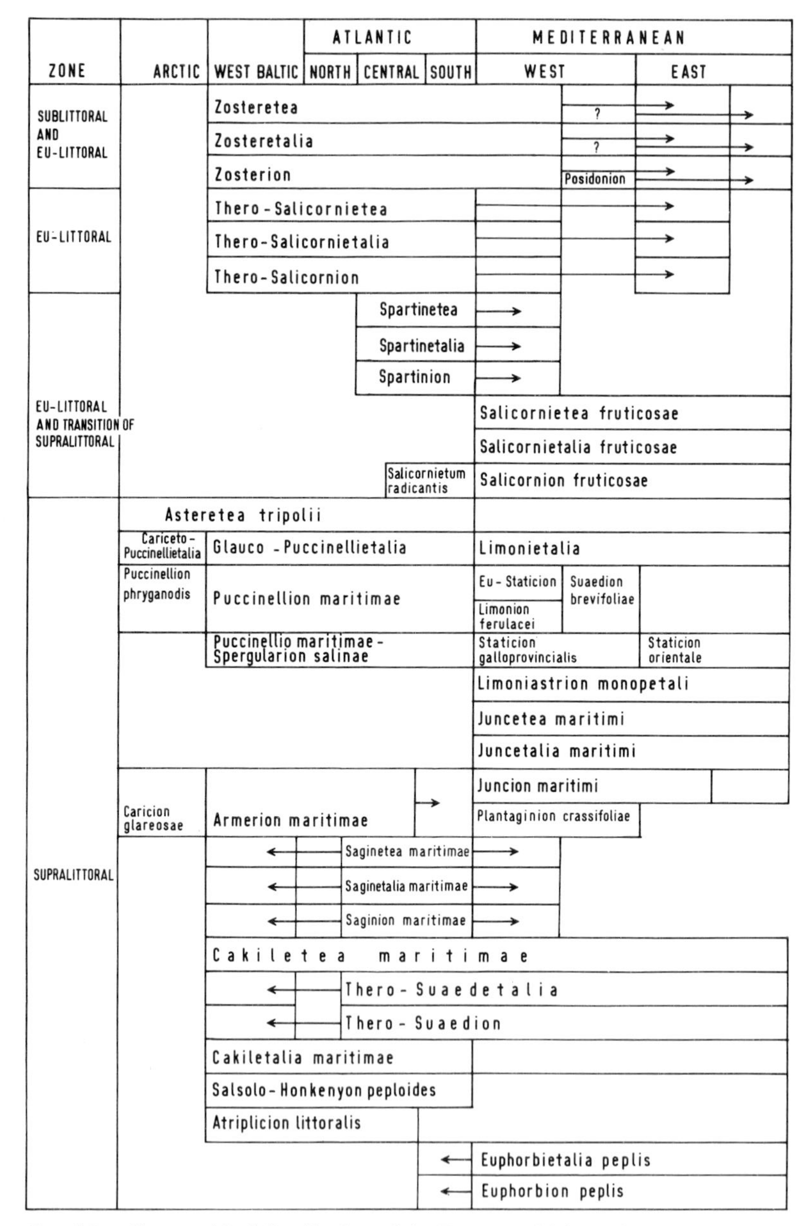

FIG. 1.2. Geographical distribution of the European higher salt marsh syntaxa (after Beeftink 1965).

England and the south-west Netherlands to the west coast of Morocco, but it also grows in the Mediterranean region on the shores of the Gulf of Venice. *Spartina alterniflora* which was introduced from North American grows locally along the Atlantic coasts. Lastly *Spartina townsendii* agg., a hybrid between *S.alterniflora* and *S.maritima* which has been introduced into many parts of the world because of its excellent ability to accrete silt, occurs in Europe from west Denmark, east and west Scotland and Ireland to south-west France and north-west Spain. It is still colonizing marshes along the coasts of western Europe often at the expense of the indigenous salt marsh pioneer communities.

The alliances Zosterion, Thero-Salicornion and Spartinion are not represented along the coasts of extreme northern Europe which may reflect the susceptibility of *Zostera* and *Spartina* to the action of frost and the freezing of coastal waters. In the case of *Salicornia europaea* agg., the summer is probably too short for this therophyte to set seed satisfactorily.

The perennial species *Salicornia fruticosa* is a characteristic plant of the communities which occur along the Mediterranean coastline. The latter belong to the order Salicornietalia fruticosae (Br.-Bl. et Tx. 1943) Tx. et Oberd. 1958 of the class Salicornietea fruticosae Br.-Bl. et Tx. 1943.

Communities of the supralittoral

Usually the supralittoral zone of the salt-marshes of west European coasts is occupied by communities of the class Astererea tripolii Westhoff et Beeftink 1962, of which *Aster tripolium*, *Plantago maritima* and *Triglochin maritima* are character-taxa. In north-western Europe two orders can be distinguished. The Carici-Puccinellietalia Beeftink et Westhoff 1965 occurs in the Arctic and sub-Arctic domain and the character-taxa of this order include *Plantago maritima* ssp. *borealis*, *Stellaria humifusa*, *Potentilla anserina* ssp. *egedii*. On the Atlantic coasts the order Glauco-Puccinellietalia Beeftink et Westhoff 1962 has been described and *Limonium vulgare* ssp. *vulgare* and *Spergularia media* are character-taxa of this order.

The Carici-Puccinellietalia can be further subdivided into two alliances; the Puccinellion phryganodis Hadač 1946 (character-taxon: *Puccinellia phryganodes*) and the Caricion glareosae Nordhagen 1954 (character-taxon *Carex glareosa*) for the lower and upper levels of Arctic salt-marshes respectively. The order Glauco-Puccinellietalia comprises three alliances of which Puccinellion maritimae Christiansen 1927 em. R. Tx. 1937 is restricted to the lower levels of salt-marshes. Character-taxa include *Puccinellia maritima*, *Cochlearia anglica* and *Bostrichia scorpioides*. The alliance Armerion maritimae Br.-Bl. et de Leeuw 1936 occurs in the upper levels of salt marshes and character-taxa include *Armeria maritima*, *Juncus gerardii* and *Glaux maritima*. Lastly the

Puccinellio-Spergularion salinae Beeftink 1965 (character-taxon: *Spergularia salina*) comprises the ephemeral communities of cattle tracks and enclosed areas where the habitat is characterized by a degree of instability as a result, for example, of trampling and seepage of water.

Grazing strongly influences the type of plant community which is present in the lower marsh. In the ungrazed salt marshes of the lower levels of Atlantic coasts, especially in the north Atlantic domain, *Limonium vulgare* ssp. *vulgare* and *Plantago maritima* are abundant (association Plantagini-Limonietum Westhoff et Segal 1961). In grazed marshes these species are replaced by species such as *Puccinellia maritima* which characterize the association Puccinellietum maritimae (Warming 1890) Christiansen 1927. The upper levels of both grazed and ungrazed salt marshes of the Atlantic coast often are occupied by communities of the association Juncetum gerardii Warming 1906, of which *Armeria maritima* var. *maritima* is a character-taxon. In sandy habitats the association Junco-Caricetum extensae Br.-Bl. et de Leeuw 1936 may be represented. Character-taxa of this association include *Carex extensa* and *Odontites litoralis*.

Communities of *Artemisia maritima* often are present on sandy elevated areas of salt marshes. The association Artemisietum maritimae (Hocquette 1927) Br.-Bl. et de Leeuw 1936 often replaces the association Halimionetum portulacoides (character-taxon *Halimione portulacoides*) when the accumulation of sediment blocks creeks. The synsystematic position of the latter community is not fully understood yet. Most northern European authors classify this association, which is represented as far north as Denmark, in the class Asteretea tripolii (Beeftink 1965; Westhoff & den Held 1969). Beeftink (1965) however has emphasized the close ecological relationship between this association and the Mediterranean communities with *Salicornia fruticosa*.

Where a seepage of fresh water occurs in the supralittoral zone communities may be present in which *Scirpus maritimus* var. *compactus* is dominant. These communities which belong to the association Halo-Scirpetum maritimi (Van Langendonck 1931) Dahl et Hadač 1941, occur on western European coasts from southern Norway to Brittany. The syntaxonomic position of this association is still under discussion.

Environmental conditions of salt marshes of the Mediterranean region are different from those of western Europe. One of the characteristic features of these Mediterranean marshes is the marked seasonal fluctuation in salinity associated with precipitation/evapotranspiration balance. The flora of salt marshes in the two regions is different, there being only a few species which are common to both areas. Of the class Salicornietea fruticosae the order Salicornietalia fruticosae which is characteristic of habitats that are hypersaline for much of the year has been mentioned already. Communities of the order Limonietalia Br.-Bl. et De Bolós 1957 have a distribution within salt marshes comparable to the communities characteristic of the lower levels of western European marshes. Character-taxa of the order include *Frankenia intermedia*, *Artemisia gallica*, *Atriplex hastata* var. *salina*, *Arthrocnemum glaucum*, *Suaeda fruticosa*, *Inula crithmoides* and *Limonium ferulaceum*.

Communities of the Juncetea maritimi Br.-Bl. 1931 em. Beeftink (1964) 1965 occur in the salt marshes of the Mediterranean region in less saline habitats than the above-mentioned vegetation types. These communities can be considered Mediterranean counterparts of those of the alliance Armerion maritimae, which are characteristic of the upper levels of west European salt marshes. Character-taxa include *Juncus maritimus*, *J.acutus*, *Sonchus maritimus*, *Carex extensa* and *Tetragonolobus siliquosus*. Some of these species such as *Juncus maritimus* and *Tetragonolobus siliquosus* are also present on the Atlantic coasts but only as facultative halophytes and, in general, as members of communities of the alliance Agropyro-Rumicion crispi Nordh. 1940 em. Tx. 1959 (Plantaginetea majoris Tx. et Preising 1950). However, the vegetation of Mediterranean salt marshes has not been studied intensively, so that a division into lower syntaxa is not yet possible.

Plant communities of the storm-flood zone, which are only inundated by the sea-water at storm tides

The communities of the Saginion maritimae Westhoff, Van Leeuwen et Adriani 1962 (Saginetea maritimae, Westhoff, Van Leeuwen et Adriani 1962) occur along west European coasts from southwest Sweden and the British Isles to western France. In the Mediterranean region they are present on the coast of south-east France and possibly also elsewhere. Character-taxa include *Sagina maritima*, *Cochlearia danica*, *Plantago coronopus*, *Catapodium marinum*, *Bupleurum tenuissimum* and *Parapholis strigosa*. An important association of the west European salt marshes is the Sagino maritimae-Cochlearietum danicae (Tx. 1937) Tx. et Gillner 1957, but another association, the Sagino maritimae-Catapodietum maritimae Tx. 1963 occurs on the rocky coasts of Brittany. *Hordeum marinum*, *Frankenia laevis*, *Limonium bellidifolium*, *L.occidentale*, *Hutchinsia procumbens* and *Tortella flavovirens* are abundant in the Saginion communities of southern Portugal compared with their frequency in similar communities further north in western France and southern England. In contrast *Cochlearia danica* is rare in southern communities (Beeftink 1965).

The Saginion communities of the Mediterranean coast have not been studies in detail although Beeftink, Tüxen and Westhoff (1963) have described an association Sagino maritimae-Tortelletum flavovirentis on the south-east coast of France. On the Atlantic and Baltic coasts the alliance Saginion maritimae occurs in the transitional zone between the xero- and halosere, whereas in the Mediterranean region these communities form a vegetational mosaic with the communities of the order Limonietalia.

Strand line plant communities

The vegetation which grows on tidal litter is not characteristic of the halosere or the xerosere. These halo-nitrophilous communities belong to the class Cakiletea maritimae Tx. et Preising 1950, and are represented on all European coasts from the sub-Arctic to the Mediterranean regions. Important taxa include *Salsola kali*, *Cakile maritima*, *Suaeda maritima*, *Bassia hirsuta*, *Atriplex littoralis*, *A.hastata*, *Matricaria maritima* ssp. *inodora* var. *salina*, *Suaeda fruticosa* and *Lepidium latifolium*. Often these communities are transient and they depend for survival on the frequency of litter deposition and the quantity of plant litter washed ashore.

Where the remains of filamentous algae and eelgrass are deposited in the lower salt marsh on sites which remain permanently moist, communities with *Suaeda maritima* and *Bassia hirsuta* may develop. These communities which are placed in the order Thero-Suaedetalia Br.-Bl. et De Bolnós 1957 em. Beeftink 1962 occur on coastlines from south-west Sweden to Spain, as well as along the northern coasts of the Mediterranean and the Black Sea.

Large quantities of the litter of phanerogams and fucoid brown algae often are deposited by storm tides along the high water mark in the upper salt marsh or at the foot of the main coastal dune ridge. Communities consisting of species of *Atriplex* may develop in this litter. On the Atlantic and Baltic coasts they belong to the order Cakiletalia maritimae Tx. apud Oberd. 1949 of which *Cakile maritima* and *Salsola kali* var. *polysarca* are included as character-taxa. In the Mediterranean region, however, this type of community is represented by the order Euphorbietalia peplis Tx. 1950, of which *Euphorbia peplis* and *Polygonum maritimum* are character-taxa. The order Cakiletalia maritimae is subdivided into two alliances. The first is Atriplicion littoralis (Nordh. 1940 p.p.) Tx. 1959, which is characterized by *Atriplex littoralis* and *Matricaria maritima* ssp. *inodora* var. *salina*. Communities of strand lines which are not covered by wind-blown sand generally belong to this alliance. The other alliance, Salsola-Honkenyion peploides Tx. 1950, occurs on strand lines which are subject to blown sand. Important taxa in this alliance include *Glaucium flavium*, *Atriplex sabulosa*, *Cakile maritima* and probably *Atriplex glabriuscula*. At present in the order Euphorbietalia peplis only one alliance has been described which is the Euphorbion peplis Tx. 1950. Tüxen (1975), however, distinguished different groups of communities for the western and eastern Mediterranean shores and for the coastline of the Black Sea on the basis of the distribution of certain species. *Calystegia soldanella* and *Agropyron junceiforme* are found in the west, *Cynodon dactylon* occurs in the east and *Cakile maritima* ssp. *euxiria* and *Salsola kali* ssp. *ruthenica* are members of communities on the shores of the Black Sea.

The syntaxonomic position of the association Atriplici-Agropyretum pungentis Beeftink et Westhoff 1962, comprising the communities which contain

Agropyron pungens is not fully understood. This association occurs on the highest parts of creek banks along western European coasts from Denmark to Portugal.

APPENDIX I

Scheme of Syntaxa used in this paper

Syntaxon	Suffix	Example	Denominating taxa
Class	-etea	Ammophiletea	*Ammophila arenaria*
Order	-etalia	Elymetalia	*Elymus arenarius*
Alliance	-ion	Galio-Koelerion	*Galium verum and Koeleria gracilis*
Association	-etum	Euphorbio-Agropyretum	*Euphorbia paralias and Agropyron junceiforme*

List of abbreviations used in the nomenclature of vegetation classification

em. = emendavit
apud = in the paper by
n.n. = nomen nudum
p.p. = pro parte

These terms are used in vegetation classification in the same way as they are used in taxonomy.

REFERENCES

Adriani M.J. & van der Maarel E. (1968) Voorne in de branding (Summ.: Breakers on Voorne). Stichting Wetenschappelijk Duinonderzoek, Oostvoorne.

Baudière A., Simoneau P. & Voelckel Ch. (1976) Les Sagnes de l'étang de Salses (Pyrénées-Orientales). *La Végétation des Vases Salées*, pp. 1–34. Colloques Phytosociologiques IV. J. Cramer, Vaduz.

Beeftink W.G. (1964) Die Systematik der europäischen Salzpflanzengesellschaften. *Ber. Int. Symp. Pflanzensoz. Syst. Stolzenau/Weser*, 239–72.

Beeftink W.G. (1965) De zoutvegetatie van ZW-Nederland beschouwd in Europees verband (Summ.: Salt marsh communities of the SW-Netherlands in relation to the European halophytic vegetation). *Meded. Landb. hogesch. Wageningen* **65**, 1–167.

Beeftink W.G. (1977) The coastal salt marshes of western and northern Europe: an ecological and phytosociological approach. *Ecosystems of the World* (Ed. by V.J. Chapman), pp. 109–55. Elsevier Scientific Publishing Company, Amsterdam.

Benzing L. (1969) Pflanzengesellschaften auf rezenten Küstendünen in Mittel-Portugal. *Vegetatio* **16**, 376–83.

Berghen C. Vanden (1958) Etude sur la végétation des dunes et des landes de la Bretagne. *Vegetatio* **8**, 193–208.

Birse E.L. & Robertson J.S. (1976) *Plant communities and soils of the lowland and southern upland regions of Scotland.* The Macaulay Institute for Soil Research, Craigiebuckler, Aberdeen.

Boerboom J.H.A. (1957) Les pelouses sèches des dunes de la côte néerlandaise. *Act. Bot. Neerl.* **6**, 642–80.

Boerboom J.H.A. (1958) De vegetatie van Meyendel. *Beplanting en Recreatie in de Haagse Duinen*, pp. 17–46. Meded. Instituut voor toegepast Biologisch onderzoek in de natuur (ITBON) 39, Arnhem.

Boerboom J.H.A. (1960) De plantengemeenschappen van de Wassenaarse duinen (Summ.: The plant communities of the Wassenaar dunes near the Hague). *Meded. Landb. hogesch. Wageningen* **60**, 1–135.

Braun-Blanquet J. (1967) La chênaie acidophile ibéro-atlantique (Quercion occidentale) en Sologne. *Comm. Stat. Intern. Géobot. Médit. et Alp. Montpell.* **178**, 53–87.

Braun-Blanquet J., Braun-Blanquet G., Rozeira A. & Pinto Da Silva A.R. (1972) Résultats de trois excursions géobotaniques à travers le Portugal septentrional et moyen IV, esquisse sur la végétation dunale. *Agronomia Lusit.* **33**, 217–34.

Braun-Blanquet J., Pinto da Silva A.R. & Rozeira A. (1956) Résultats de deux excursions géobotaniques à travers le Portugal septentrional et moyen II, chênaies à feuilles caduques (Quercion occidentale) et chênaies à feuilles persistantes (Quercion fagineae) au Portugal. *Agronomia Lusit.* **18**, 167–234.

Braun-Blanquet J., Roussine N. & Nègre R. (1951) *Les groupements végétaux de la France méditerranéenne.* Centre National des Recherches Scientifiques (Service de la Carte de Groupements Végétaux), Montpellier.

Braun-Blanquet J. & Tüxen R. (1952) Irische Pflanzengesellschaften. *Die Pflanzenwelt Irlands* (Ed. by W. Lüdi), pp. 224–415. Veröff. Geobot. Instit. Rübel 25. H. Huber, Bern.

Celiński F. & Piotrowska H. (1964) Trifolio-Anthyllidetum maritimae, a new psammophilous association on the southern Baltic coast. *Bull. Amis. Scienc. et lettr. Poznań série D*, **6**, 147–55.

Dieren J.W. van (1934) *Organogene Dünenbildung.* Thesis. Martinus Nijhoff, 's-Gravenhage.

Dierschke H. (1975) Beobachtungen zur Küstenvegetation Korsikas. *Anal. Inst. Bot. Cavanilles*, **32**, 967–91.

Fukarek F. (1961) Die Vegetation des Darss und ihre Geschichte. *Pflanzensoziologie* 12, pp. 1–321. G. Fischer, Jena.

Géhu J.-M. (1968a) Essai sur la position systématique des végétations vivaces halonitrophiles des côtes atlantiques françaises (Agropyretea pungentis Cl. Nov.) *Bull. Soc. Bot. Nord. France* **21**, 71–77.

Géhu J.-M. (1968b) Sur la vicariance géographique des associations végétales des dunes mobiles de la côte atlantique française. *C. R. Acad. Sci. Paris* **266**, 2422–25.

Géhu J.-M. (1969) Essai synthétique sur la végétation des dunes armoricaines. *Penn ar Bed* 7, 81–104.

Géhu J.-M. (1975) Essai systématique et chorologique sur les principales associations végétales du littoral atlantique français. *Anal. Real Acad. Farmac.* **41**, 207–27.

Géhu J.-M. (1976) Approche phytosociologique synthétique de la végétation des vases salées du littoral atlantique français (synsystématique et synchorologie). *La Végétation des Vases Salées*, pp. 395–462. Colloques Phytosociologiques IV. J. Cramer, Vaduz.

Géhu J.-M., Caron B. & Bon M. (1976) Données sur la végétation des prés salés de la Baie de Somme. *La Végétation des Vases Salées*, pp. 197–226. Colloques Phytosociologiques IV. J. Cramer, Vaduz.

Géhu J.-M. & Delzenne Ch. (1976) Apport à la connaissance phytosociologique des prairies salées de l'Angleterre. *La Végétation des Vases Salées*, pp. 227–48. Colloques Phytosociologiques IV. J. Cramer, Vaduz.

Géhu J.-M. & Géhu J. (1975) Les fourrés des sables littoraux du Sud-Ouest de la France. *Beitr. naturk. Forsch. Südw.-Dtl.* **34**, 79–94.

Géhu J.-M. & Petit M. (1965) Notes sur la végétation des dunes littorales de Charente et de Vendée. *Bull. Soc. Bot. Nord France* **18**, 69–88.

Géhu J.-M. & Tüxen R. (1975) Essai de synthése phytosociologique des dunes atlantique-européennes. *La Végétation des Dunes Maritimes*, pp. 61–70. Colloques Phytosociologiques I. J. Cramer, Vaduz.

Géhu-Franck J. & Géhu J.-M. (1975) Données écosystémiques et évaluation de la phytomasse dans le transect dunaire de Wimereux-Ambleteuse (Pas-de-Calais). *La Végétation des Dunes Maritimes*, pp. 253–81. Colloques Phytosociologiques I. J. Cramer, Vaduz.

Gimingham C.H. (1964) Maritime and sub-maritime communities. *The Vegetation of Scotland* (Ed. by J.H. Burnett), pp. 67–142. Oliver & Boyd, Edinburgh.

Hallberg H.P. (1971) Vegetation auf den Schalenablagerungen in Bohuslän, Schweden. *Act. phytogeogr. Suec.* **56**, 1–133.

Hallberg H.P. & Ivarsson R. (1965) Vegetation of coastal Bohuslän. *The Plant Cover of Sweden*, pp. 111–22. *Act. phytogeogr. Suec.* **50**.

Hartog C. den (1970) The Sea-grasses of the world. *Verh. K. Ned. Akad. Wet. Afd. Nat.* **59**, 1–275.

Heykena A. (1965) Vegetationstypen der Küstendünen an der östlichen und südlichen Nordsee. *Mitt. Arbeitsgem. Florist. Schlesw.-Holst. und Hamb.* **13**, 1–135.

Hueck K. (1932) Erläuterung zur vegetationskundlichen Karte der Lebanehrung (Ostpommern). *Beitr. Naturdenkmalpfl.* **15**, 99–133.

Ivimey-Cook R.B. & Proctor M.C.F. (1966) The plant communities of the Burren, Co. Clare. *Proc. R.I.A. 64, sect. B, No. 15*, 211–301.

Lavrentiades G.J. (1964) The ammophilous vegetation of the western Peloponnesos coasts. *Vegetatio* **12**, 223–87.

Lavrentiades G.J. (1975) On the vegetation of sand dunes of Greek coasts. *La Végétation des Dunes Maritimes*, pp. 91–98. Colloques Phytosociologiques I. J. Cramer, Vaduz.

Lemberg B. (1933) Ueber die Vegetation der Flugsandgebiete an den Küsten Finnlands. *Acta Bot. Fennica* **12**, 1–135.

Loriente E. (1974) Vegetación y flora de las playas y dunas de la provincia de Santander. Institución Cultural de Cantabria, Instituto de Ciencias Fisico-Quimicas y Naturales, Santander.

Loriente E. (1975) Nueva associación psamofila para las dunas muertas de la costa santanderina. *Anal. Inst. Bot. Cavanilles* **32**, 441–52.

Maarel E. van der (1966) Dutch studies on coastal sand dune vegetation, especially in the Delta region. *Wentia* **15**, 47–82.

Maarel E. van der (1975) Observations sur la structure et la dynamique des dunes de Voorne (Pays-Bas). *La Végétation des Dunes Maritimes*, pp. 167–84. Colloques Phytosociologiques. J. Cramer, Vaduz.

Maarel E. van der & Westhoff V. (1964) The vegetation of the dunes near Oostvoorne. *Wentia* **12**, 1–61.

Massart J. (1907) Essai de géographie botanique des districts littoraux et alluviaux de la Belgique. *Bull. Soc. roy. bot. belg.* **43**, 167–584.

Massart J. (1908) Essai de géographie botanique des districts littoraux et alluviaux de la Belgique: Annexe. Rec. de l'Inst. bot. Leo Errera 7, Bruxelles.

Matuszkiewicz W. (1962) Zur Systematik der natürlichen Kiefernwälder des mittel-und osteuropäischen Flachlandes. *Mitt. flor.-soz. Arbeitsgem. N.F.* **9**, 145–86.

Olsson H. (1974) Studies on south Swedish sand vegetation. *Act. Phytogeogr. Suec.* **60**, 1–170.

Onno M. (1933) Die Strandformationen an der mittleren Lübecker Bucht. *Bericht. Deuts. Bot. Gesells.* **51**, 232–66.

Pawlowski B., Medwecka A. & Kornaś J. (1966) Review of terrestrial and fresh-water plant communities. *The Vegetation of Poland* (Ed. by W. Szafer), pp. 241–534. PWN-Polish Scientific Publishers, Warszawa.

Pettersson B. (1965) Maritime sands. *The Plant Cover of Sweden*, pp. 105–10. Act. Phytogeogr. Suec. 50.

Piotrowska H. & Celiński F. (1965) Zespoły psammofilne wysp Wolina i po udniowo-wschodniego Uznamu (Summ.: Psammophilous vegetation of the Wolin island and that of south-eastern Uznam). *Bad. Fizjogr. nad Polska Zach.* **16**, 123–70.

Praeger R.LL. (1934) *The botanist in Ireland.* Hodges, Figgis, & Co., Dublin.

Ranwell D.S. (1972) *Ecology of salt marshes and sand dunes.* Chapman and Hall, London.

Rivas-Martinez S. (1976) Esquema sintaxonómico de la classe Juncetea maritimi en España. *La Végétation des Vases Salées*, pp. 193–96. Colloques Phytosociologiques IV. J. Cramer, Vaduz.

Rivas-Martinez S. & Costa M. (1976) Datos sobre la vegetación halófila de la Mancha (España). *La Végétation des Vases Salées*, pp. 81–98. Colloques Phytosociologiques IV. J. Cramer, Vaduz.

Salisbury E. J. (1952) *Downs & Dunes, their plant life and its environment.* Bell & Sons, London.

Sissingh G. (1974) Comparaison du Roso-Ephedretum de Bretagne avec des unités de végétation analogues. *Docum. phytosociol.* **7–8**, 95–106.

Skogen A. (1972) The Hippophaë rhamnoides alluvial forest at Leinöra, Central Norway. A phytosociological study. *K. norske Vidensk. Selsk. Skr.* **4**, 1–115.

Skye E. (1965) Glimpses of the Bothnian coast. *The Plant Cover of Sweden*, pp. 176–79. Act. Phytogeogr. Suec. 50.

Sloet van Oldruitenborgh C.J.M. (1976) Duinstruwelen in het Deltagebied (with a summary). *Meded. landb. hogesch. Wageningen* **76**, 1–112.

Steffen H. (1931) Vegetationskunde von Ostpreussen. *Pflanzensoziologie* 1, pp. 1–406. Gustav Fischer, Jena.

Tansley A.G. (1949) *The British Islands and their Vegetation.* Cambridge University Press.

Tüxen R. (1950) Grundriss einer Systematik der nitrophilen Unkrautgesellschaften in der Eurosibirischen Region Europas. *Mitt. flor.-soz. Arbeitsgem. N.F.* **2**, 94–175.

Tüxen R. (1951) Eindrücke während der pflanzengeographischen Exkursionen durch Süd-Schweden. *Vegetatio* **3**, 149–72.

Tüxen R. (1955) Das System der nordwestdeutschen Pflanzengesellschaften. *Mitt. flor. soz. Arbeitsgem. N.F.* **5**, 155–76.

Tüxen R. (1966) Ueber nitrophile Elymus-Gesellschaften an nordeuropäischen, nord-japanischen und nordamerikanischen Küsten. *Ann. Bot. Fenn.* **3**, 358–67.

Tüxen R. (1969) Pflanzensoziologische Beobachtungen an südwestnorwegischen Küsten-Dünengebieten. *Aquilo, Ser. Botanica* **6**, 241–72.

Tüxen R. (1970) Pflanzensoziologische Beobachtungen an isländischen Dünengesell-schaften. *Vegetatio* **20**, 251–78.

Tüxen R. (1975) Sobre las communidades del orden Euphorbietalia (Cakiletea maritimae). *Anal. Inst. Bot. Cavanilles* **32**, 453–64.

Tüxen R. (1976) La côte européenne occidentale, Domaine de lutte et de vie. *La Végétation des Vases Salées*, pp. 503–16. Colloques Phytosociologiques IV. J. Cramer, Vaduz.

Tüxen R., Böttcher H. & Dierssen K. (1971) Bolboschoenetea maritimi. *Bibliogr. Phytosoc. Syntax.* **1**, 1–25.

Tüxen R. & Oberdorfer E. (1958) Eurosibirische Phanerogamen-Gesellschaften Spaniens. *Die Pflanzenwelt Spaniens II*, pp. 1–328. Veröff. Geobot. Inst. Rübel. H. Huber, Bern.

Tüxen R. & Westhoff V. (1963) Saginetea maritimae, eine Gesellschaftsgruppe im wechselhalinen Grenzbereich der europäischen Meeresküsten. *Mitt. flor.-soz. Arbeitsgem. N.F.* **10**, 116–29.

Westhoff V. (1947) The vegetation of dunes and salt marshes on the Dutch islands of Terschelling, Vlieland and Texel. Thesis. C.J. van der Horst, 's-Gravenhage.

Westhoff V. (1952) Gezelschappen met houtige gewassen in de duinen en langs de binnenduinrand (Summ.: Plant communities with woody species ('lignosa') found in the Dutch dune area and its inner border). *Jaarb. Nederl. Dendr. Ver.*, 9–49.

Westhoff V. (1975) La végétation des dunes pauvres en calcaire des îles frisonnes hollandaises. *La Végétation des Dunes Maritimes*, pp. 71–78. Colloques Phytosociologiques I. J. Cramer, Vaduz.

Westhoff V. & den Held A.J. (1969) *Plantengemeenschappen in Nederland.* Thieme & Cie. Zutphen.

Westhoff V., van Leeuwen C.G., Adrinai M.J. & van der Voo, E. (1961) Enkele aspecten van vegetatie en bodem der duinen van Goeree, in het bijzonder de contactgordels tussen zout en zoet milieu (Summ.: Some aspects of the relation between vegetation and soil in the dunes of the island of Goeree (Holland), with special attention to the salt-fresh ecotones in xero- and hygrosere). *Jaarb. Wetensch. Genootsch. v. Goeree-Overflakkee*, 46–92.

Westhoff V. & van der Maarel E. (1973) The Braun-Blanquet approach. *Handbook of Vegetation Science Part V, Ordination and Classification of Vegetation* (Ed. by R.H. Whittaker), pp. 617–726. Dr. W. Junk B.V., Publishers, The Hague.

Wieman P. & Domke W. (1967) Pflanzengesellschaften der ostfriesischen Insel Spiekeroog. *Mitt. Staatsinst. Allg. Bot. Hamburg* **12**, 191–353.

Wojterski T. (1963) Bory bagienne na pobrzeżu Zachodniokaszubskim (Summ.: Wet pine-woods in the west Kashubian coastal region). *Bad. Fizjogr. nad Polska Zach.* **14**, 139–91.

Wojterski T. (1964) Bory sosnowe na wydmach nadmorskich na Polskim wybrzeżu (Summ.: Pine forests on sand dunes at the Polish Baltic coast). *Pozn. Tow. Przyj. Nauk prace komisji biolog.* **28**, 1–217.

Wojterski T. (1964) Schematy strefowego układu roślinności nadmorskiej na południowym wybrezeżu Baltyku (Summ.: Schemata of the zonal vegetation system on the southern coast of the Baltic sea). *Bad. Fizjogr. nad Polska Zach.* **14**, 87–105.

II

ESTABLISHMENT AND DIFFERENTIATION OF POPULATIONS IN RELATION TO ENVIRONMENTAL HETEROGENEITY

2. BENTHIC MICROALGAL POPULATIONS ON INTERTIDAL SEDIMENTS AND THEIR ROLE AS PRECURSORS TO SALT MARSH DEVELOPMENT

S.M. COLES

Institute of Terrestrial Ecology,
Colney Research Station,
Colney Lane, Norwich, U.K.

SUMMARY

On soft sediment shores the overriding factors controlling the type of sediment are the source of sediment supply and the degree of exposure. Biological factors can only be considered as mere embellishments over most of the range of possible degrees of exposure, however evidence from investigations in the Wash (E. England) indicate that in sheltered areas, organisms can strongly influence the type of sediment present on the upper intertidal zones.

The salt marshes and upper mud flats of the Wash are characterized by more or less continual net accretion of fine sediments, whereas no appreciable net gains occur on the sand flats.

Populations of benthic microalgae are high on the salt marshes and upper mud flats, but low on the sand flats. The low numbers on the sand flats appear to be largely the result of the grazing of numerous macroinvertebrates in the area.

The most abundant microalgae are epipelic diatoms, these are motile and produce copious mucus, which traps and binds fine sediments. Accretion of fine sediment on the salt marshes and mud flats can be stopped by the removal of the microalgae. Conversely the accretion of mud on the sand flats can be induced by the removal of the macroinvertebrates which allows development of large microalgal populations. Long term accretion of mud, however, depends on the compaction of the mud so it can withstand erosion during storm conditions.

RÉSUMÉ

Sur les côtes à sédiments mous, les facteurs les plus importants contrôlant le type de sédiment sont la source de l'apport de sédiment et le degré d'exposition. Les facteurs biologiques peuvent seulement être considérés comme un élément supplémentaire sur presque toute la série d'expositions possibles. Cependant des évidences provenant d'études dans le Wash indiquent que dans les surfaces protégées, les organismes peuvent fortement influencer le type de sédiment présent dans les zones supérieures, soumises au balancement des marées.

Les marais salés et les surfaces boueuses supérieures du Wash sont caractérisés par une accumulation nette de sédiments fins, plus ou moins continuelle, alors qu'il n'y a pas de gain net appréciable sur les surfaces sableuses.

Les populations de microalgues benthiques sont importantes dans les marais salés et les surfaces boueuses supérieures, mais faibles dans les surfaces sableuses. Ce faible nombre semble fortement résulter de la pâture de cette région par de nombreux macroinvertébrés.

Les microalgues les plus nombreuses sont les diatomées épipéliques; celles ci sont mobiles et produisent un abondant mucus qui piège et lie les fines particules de sédiments. On peut stopper l'apport de sédiment fin dans les marais salés et les surfaces boueuses en enlevant les microalgues. Réciproquement on peut induire l'augmentation de boues sur les surfaces sableuses en retirant les macroinvertébrés, ce qui permet le développement de larges populations de microalgues. L'augmentation à long terme de boue dépend de sa compacité, lui permettant de résister à l'érosion pendant des conditions de tempête.

INTRODUCTION

'Physically, a mud flat is uninviting, if not repellent, but it has much to offer the intellect' (Eltringham 1971). Mud is a very general term relating to any sediment which contains a significant proportion of mineral particles which are smaller than sand (i.e. less than 20 μm). Studies of processes occurring in intertidal mud have been neglected by sedimentologists, oceanographers and biologists alike.

The properties of individual particles of sand are mostly retained during the sedimentation or erosion of sand. In contrast, the fine particles of clay and silt are cohesive particles which form particle complexes by means of chemical bonds. The presence of inorganic salts or organic compounds in sea water may result in the flocculation of these cohesive sediments when in suspension. Similarly these particles often become agglomerated in the presence of organisms. Critical studies to evaluate the importance of these different effects on the rate of sedimentation have never been made. Even the importance of the flocculation of dispersed river-borne solids in the salt water of estuaries, as it affects the rate of sedimentation, is still contested (Meade 1972).

Deposition of intertidal mud and the consequential development of salt marshes occur in sheltered areas such as in estuaries or behind spits or in sites fronted by wide tidal flats, over which the energy of waves and tidal currents is dissipated. Although there have been numerous descriptions of tidal deposits recently (Ginsburg 1975) little is known of the processes involved in the deposition of cohesive particles. On a tidal flat the sediment decreases in size in a landward direction as a result of the decreasing velocity of the water. The accumulation of fine particles may also be aided by the effects of settling and scour lag as postulated by Postma (see Straaten & Kuennen 1957). Straaten and Kuennen (1958) also point out that several biological factors influence the trapping of cohesive sediments, especially at sites above the low water mark of the tides. For example, filter feeders like cockles and mussels filter fine particles from the water and produce comparatively coherent faecal pellets, the vegetation of salt marshes traps fine sediment, and diatoms produce slime which binds clay particles.

Meade (1972) has stated that the forces affecting the movement of water are so high that biological factors merely embellish the effects of these forces on the distribution of sediments. However, as already indicated organisms appear to influence the deposition of cohesive particles.

In this study an attempt has been made to determine the effects of biological factors on the rate of accretion of sediment in the Wash, England. This investigation was part of a larger study on the ecological implications of building bunded reservoirs for the storage of fresh water on part of the intertidal flats of the Wash. One objective was to determine the rôle of algae in stabilizing fine sediments on the tidal flats, thereby providing conditions suitable for the development of salt marshes.

THE WASH

The Wash is a large embayment on the east coast of England (Fig. 2.1), and is more exposed than most estuarine situations (N.E.R.C. 1976). Evans (1965) recognized the four main zones of deposition of sediment, on the west shore of the Wash above mean tide level, which are given below.

1. Salt marshes, which are areas with a slow rate of sedimentation. Mainly very fine sediments are deposited, and these are not reworked by waves or organisms.
2. Upper mud flats, which are areas where the rate of sedimentation is fairly rapid. A high proportion of the sediment which is deposited is fine, and is only reworked to a limited extent by waves and organisms.
3. Inner sand flats, which are areas where the rate of sedimentation is slow. Most of the sediment is fine sand but at some sites the deposit contains a high proportion of fine sediments which are reworked extensively by organisms and to a limited extent by waves.

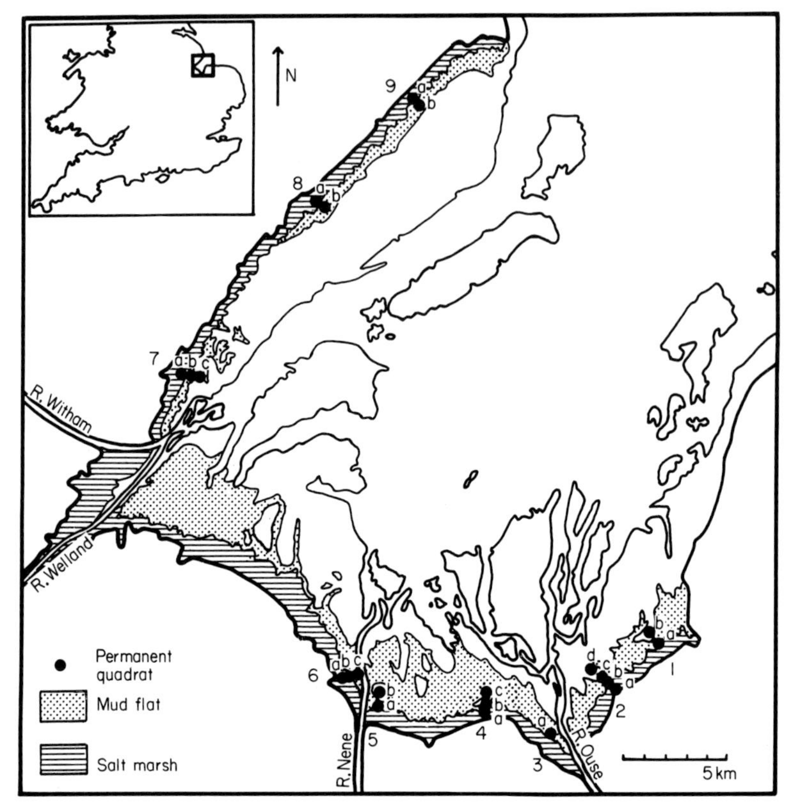

FIG. 2.1. Map of the Wash showing the locations of the permanent quadrats, which were monitored in 1973–74. The quadrats were placed along transects (1–9) and are labelled (a–d) in a seaward direction.

4. *Arenicola* sand flats which are composed of slightly larger sand particles than those of the inner sand flats and these particles are reworked extensively by waves and organisms.

On most of the shores of the Wash there is forward growth of the seaward edge of salt marshes, largely as a result of land reclamation and other engineering projects (Inglis & Kestner 1958; Kestner 1962, 1975). Land reclamation often alters the tidal forces so that accretion rather than erosion is favoured. Mud flats appear to be essential precursors for the growth of salt marshes in the Wash. The surface of the younger mud flats is relatively smooth, but they become progressively dissected by numerous creeks as they age, which leads to the formation of mud mounds between adjacent creeks. Where the salt marshes are eroding there are no upper mud flats and the inner sand flats abut onto the salt marsh.

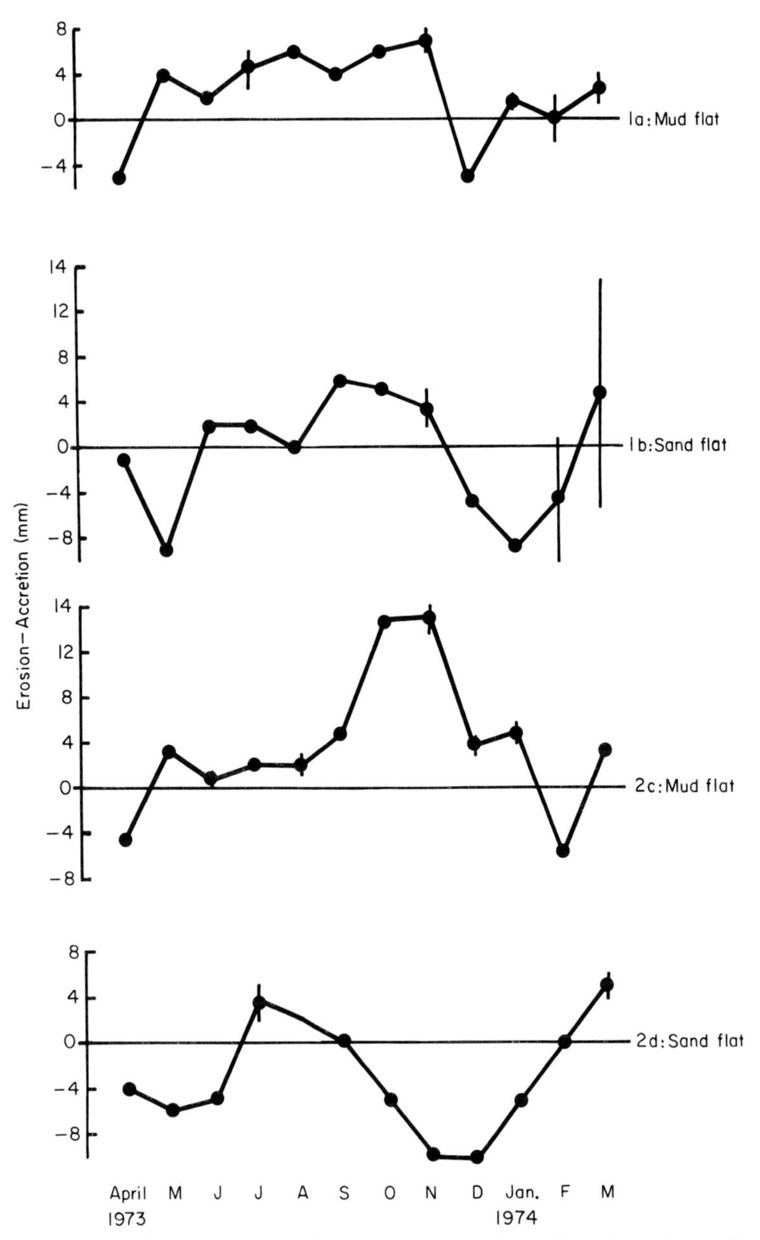

FIG. 2.2. Sediment accretion in permanent quadrats (locations shown in Fig. 2.1) on upper mud flats compared with the accretion in quadrats on adjacent inner sand flats. Where there was noticeable variation within a site, the range of variation is shown by a vertical bar.

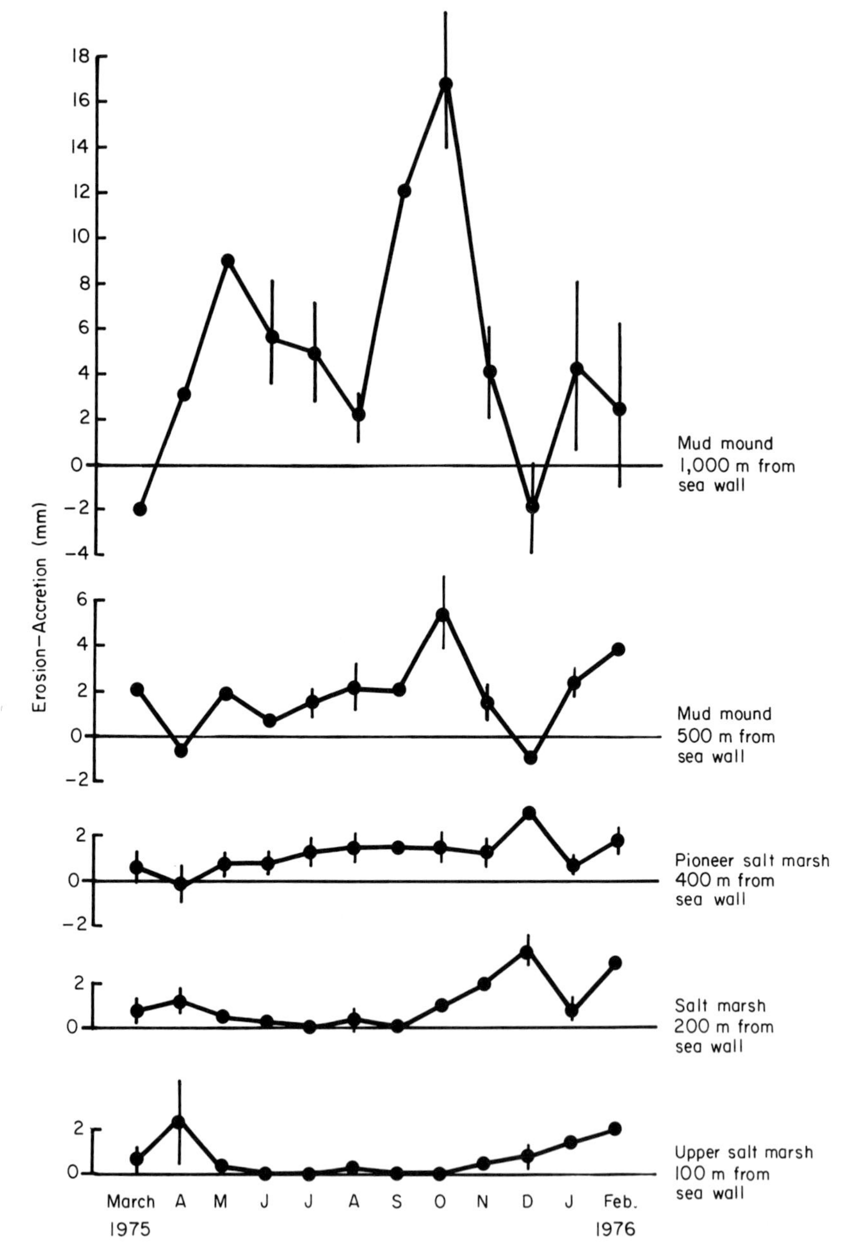

FIG. 2.3. Sediment accretion, monitored in 1975–76, along transect 4, from upper mud flat to upper salt marsh. Where there was noticeable variation within a site, the range of variation is shown by a vertical bar.

Measurement of rates of sediment accretion

Before the effect of algae on the deposition of sediment was examined, rates of sedimentation were measured at different sites in the upper intertidal zones of the Wash. These sites were mainly positioned in the upper mud flats and in the pioneer zones of the salt marshes (see Fig. 2.1), areas in which it was initially thought that the algae might be of most significance in sediment stabilization.

Sedimentation was measured at monthly intervals by means of marker layers, such as white silica flour, which were spread over areas of about

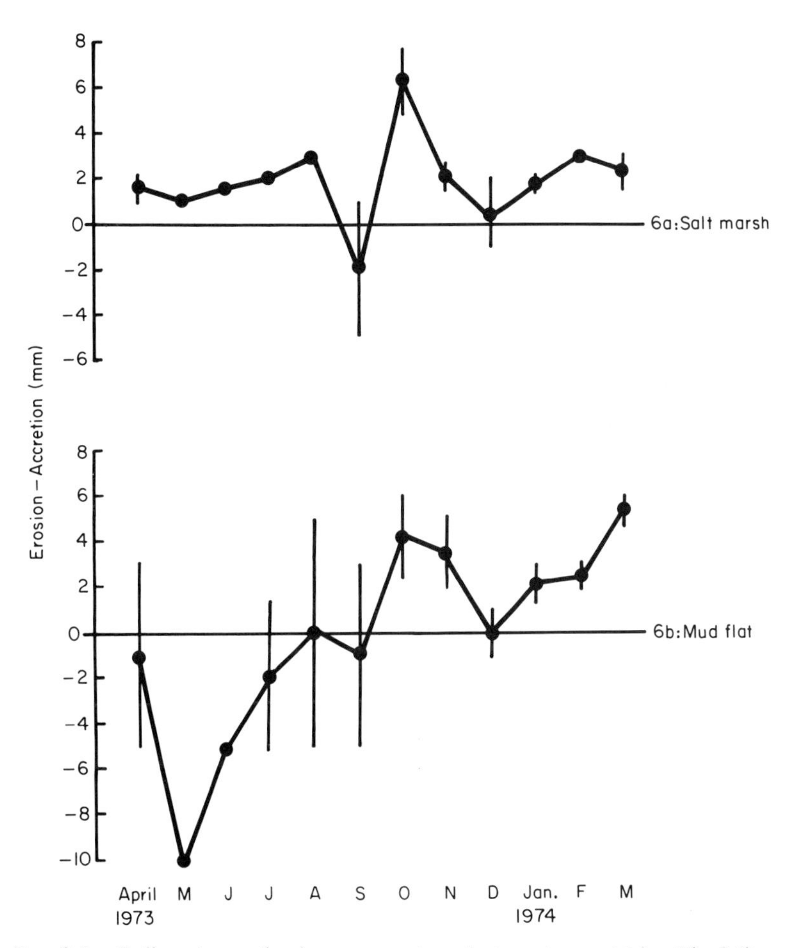

FIG. 2.4. Sediment accretion in permanent quadrats on transect 6 (see Fig. 2.1), comparing the accretion on the salt marsh with that on the upper mud flat which had ceased forward growth. Where there was noticeable variation within a site, the range of variation is shown by a vertical bar.

0·25 m^2 of the surface of the sediment. These measurements, due to the subsequent compaction of freshly deposited sediment, are not equivalent to height changes of the sediment surface. On the inner sand flats, erosion of the sediments was the dominant process during the winter months but in summer there was a net rate of accretion (Fig. 2.2). Amounts of sediment which accreted or were eroded were relatively large but at the end of a year the net gains and losses approximately balanced. On the pioneer salt marshes and the upper mud flats the highest rates of accretion were recorded in the autumn but accretion was continuous throughout most of the year (Fig. 2.2). On the upper salt marshes net accretion was greatest during winter when storms caused some erosion of sediment on the upper mud flats (Fig. 2.3). Erosion of sediments on the upper mud flats was generally rather limited and only rarely was sediment which lay 5 mm or more below the surface removed between consecutive readings. However, at some sites on the upper mud flats where the seaward advance of the mud flats and salt marshes appeared to be slowing or stopping, erosion of sediment was often greater and accretion of sediment only just managed to keep pace with the rates of erosion and compaction, with no resultant increase in height of the mud flats at the end of the year (Fig. 2.4).

In some respects the net accretion or erosion of sediments is possibly a better criterion for distinguishing between the inner sand flats and upper mud flats, than is the type of sediment present. The type of sediment which is deposited on the upper mud flats depends primarily on the source of the supply. For example the percentage of sand in the sediment increases if the

Table 2.1. *The percentage of sand (particles >0·02 mm) in the top 10 mm of sediment at the sites of the permanent quadrats (measured by means of a hydrometer method). Locations of sites shown in Fig. 2.1.*

	Permanent quadrat	% of sand March 1974	% of sand August 1974
Upper sand flats	1b	91	81
	2d	85	77
Upper mud flats near sand flats	1a	83	39
	2c	43	15
	8a	68	37
	8b	76	36
	9a	61	24
	9b	69	43
Upper mud flats some distance away from sand flats	3a	33	31
	4b	43	46
	4c	52	52
	5a	37	31
	6b	32	41
	6c	34	35

site is close to sand flats. In winter the quantity of sand deposited on the upper mud flats from nearby sand flats increases compared with the amount in summer because of a greater frequency of storms. The sediment which is deposited on the lower zones of the upper mud flats may be very similar to that found on the upper sand flats during the winter, but usually in the summer the percentage of fine sediments on the upper mud flats is much greater than on the upper sand flats (Table 2.1). Sites on upper mud flats distant from sand flats show little difference in the type of sediment deposited in winter and summer.

EFFECTS OF ALGAE ON SEDIMENTATION

Macroalgae

Several authors have suggested that macroalgae affect the rate of accretion of sediment. For example Carey and Oliver (1918) state that salt marsh forms of *Fucus* and *Vaucheria* influence the rate of accretion. Scoffin (1970) found that in the Bahamas sediment covered by dense *Enteromorpha*, was not eroded by water moving at a velocity which was five times greater than that which eroded unbound sand grains. In the Wash the effects of macroalgae on the rate of accretion are very localized and often short-lived. Macroalgae are rare in areas where the rate of mud accretion is rapid as they cannot tolerate being smothered by fine sediments.

Microalgae

Attached forms of microalgae, such as episammic diatoms, are rare where fine sediment is accreting. In contrast epipelic (motile, free-living) microalgae are abundant. A common characteristic of all epipelic algae, which occur in any abundance on mud, is that of leaving a trail of mucus when moving through sediment. The mechanism of the motility of these algae is controversial (Hopkins 1966, 1967; Walsby 1968) and few investigations of the chemical and physical properties of the mucus have been made.

Observations on the binding of sediment by microalgae have been primarily concerned with algal sediments and stromatolites formed largely by blue-green algae. Scoffin (1970) and Neumann *et al.* (1970), using an underwater flume, found that intact blue-green algal mats could withstand current velocities of at least twice that of unbound sediment before breaking up. Sediments of this type are of geographically widespread occurrence in the geological record, but present day examples are more or less limited to a few localities in the tropics and subtropics. In areas where there is normal sea water dense mats of blue-green algae are restricted to the upper levels of the intertidal zone where there are few grazers and burrowers (Garrett 1970). If conditions are hypersaline they occur over the whole intertidal zone as invertebrate organisms are virtually absent (Logan *et al.* 1970).

In temperate areas blue-green algae rarely occur as dense mats: their growth is more seasonal and intertidal conditions which are too dry or too hypersaline for animal grazers are rare. Epipelic diatoms, however, are often abundant on tidal flats. Many authors have suggested that the mucus which is produced by these epipelic diatoms, is involved in the trapping and stabilization of fine sediments on mud flats, and they have noted that the mud content is much greater in diatom-rich areas than would correspond to the normal deposition in the prevailing hydrological conditions.

Until recently experimental work to substantiate that diatoms can affect the rate of sedimentation appears to be totally lacking. Holland *et al.* (1974) tested the stabilizing powers of diatoms by agitating flasks containing diatom cultures to which various sediments were added and demonstrated that diatoms which produced significant quantities of mucus reduced the amount of sediment in suspension and retarded the laminar flow of the sediments.

Density of epipelic microalgae in the Wash

Initially a study was made of the seasonal abundance of benthic microalgae in a number of permanent quadrats (Fig. 2.1). Most of the quadrats were positioned on the upper mud flats and on the pioneer zones of the salt marshes, areas in which it was initially thought that the algae might be of most significance in sediment stabilization. The algae were scraped off a known area of sediment to a depth of 1·5 mm, and their abundance was assessed by means of haemocytometer counts. Although the presence of sediment interferes with counting, this was found to be the most efficient method of estimating the algal population. The material was initially only identified into the major algal groups.

Only the epipelic microalgae were abundant on areas composed of fine sediments. The most abundant group were the diatoms, and often these were the only groups of microalgae which occurred in significant numbers. The only other abundant epipelic microalgae were blue-green algae and members of the Euglenaceae. Numbers of epipelic diatoms varied from less than 1×10^3 cm^{-2} of sediment surface on some sand flats to over $5{\cdot}5 \times 10^5$ cm^{-2} on sheltered creek banks within salt marshes; values which are similar to those given by Brockman (1935), Linke (1939) and Grøntved (1949). The number of diatoms on the permanent quadrats mostly ranged from 5×10^4 cm^{-2} to 1×10^6 cm^{-2}, although at sites on the mud flats the estimates were between 1–5 $\times 10^5$ cm^{-2} on most occasions.

Low numbers of microalgae were recorded on the inner sand flats compared with the persistently high numbers of the upper mud flats and pioneer salt marshes (Figs. 2.5 and 2.6). Although few quadrats were placed on the sand flats, observations showed that blooms of diatoms rarely appeared and then only in very localized areas. This difference appears to be primarily the result of grazing by large numbers of deposit feeding macroinvertebrates,

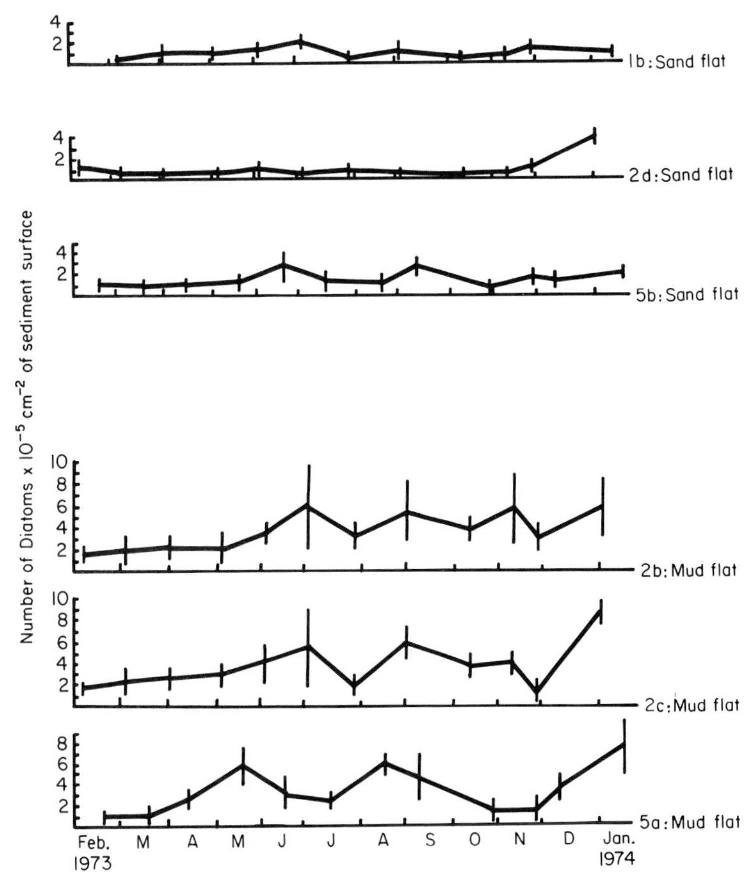

FIG. 2.5. A comparison of the seasonal variation in the number of epipelic diatoms in permanent quadrats (locations shown in Fig. 2.1) on upper and sand flats and upper mud flats. The graphs are based on the means of two samples, the range between the samples is shown as a vertical bar.

which occur on the inner sand flats, and not the result of differences in the nature of the sediment. The landward distribution of the macroinvertebrates on to higher intertidal zones is restricted probably by the effects of drought. On the muddy areas the size of diatom populations appears to be controlled mainly by animal grazing (meio- and microfauna) and the water content of the sediment. The highest peaks in the size of populations were found often on sites which had dried out during the previous summer, with a consequential decrease in the fauna. At these sites a large influx of fine sediment in autumn occurred, which produced a soft wet layer of mud. A similar influx of sediment

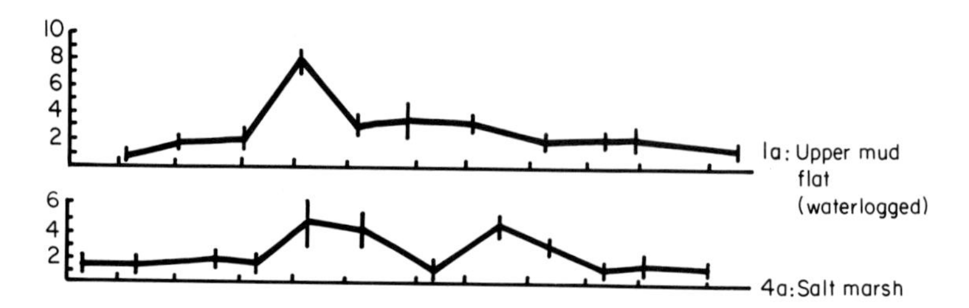

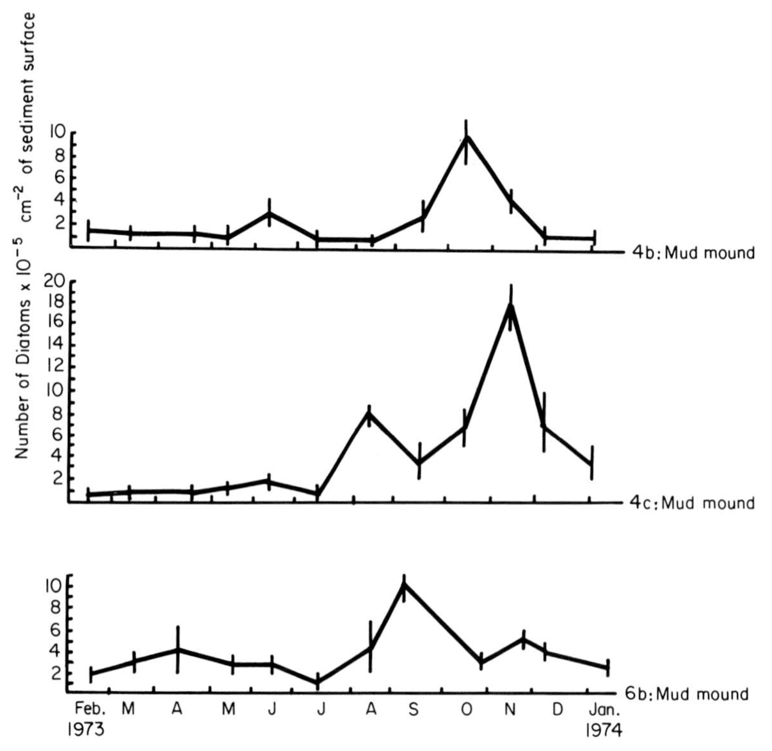

FIG. 2.6. A comparison of the seasonal variation in the number of epipelic diatoms in permanent quadrats (locations shown in Fig. 2.1) on an upper mud flat and a salt marsh which remains wet all year, with those on upper mud flats which are in the form of mud mounds and which dried out in early summer. The graphs are based on the means of two samples, the range between the samples is shown as a vertical bar.

to sites which remained wet all summer, did not produce marked algal blooms.

The blue-green algae showed more marked seasonal peaks in number than the diatoms. Low numbers of blue-green algae were found on many of the quadrats for most of the year, but the largest populations occurred in summer and autumn, and then only at sites which had suffered a drought in early summer. Blue-green algae appear to be more susceptible to grazing than diatoms. Members of the Euglenaceae were most abundant on areas of soft mud, especially the sides of creeks.

Experimental studies on the effect of epipelic diatoms on mud accretion

The occurrence of high densities of epipelic microalgae on mud and the much lower densities on sand flats suggest a causal relationship between the density of microalgae and the type of sediment. Laboratory studies have shown that particles of fine sediment readily adhere to the mucus produced by diatoms (Drum & Hopkins 1966), and that the mucus can stabilize sediment against erosion by currents (Holland *et al.* 1974). However it is uncertain whether these effects occur under natural conditions.

In order to test the effect of diatoms on mud accretion under natural conditions, diatoms were cleared from mud, after which the net accretion of sediments at these sites was compared with that at adjacent sites which had a normal microflora complement. The diatoms were removed either by treatment with chemicals (bleach or formalin) or by inverting shallow cores of sediment (4 cm in depth), so that the diatoms were buried. The treatments were arranged randomly in blocks, with 0·5 m allowed between each site in case of edge effects. The experimental sites were coated with a marker layer of white silica flour in order to help assess the amount of accretion. Zones in which these experiments were carried out included mud mounds some 500 m seaward of the salt marsh edge, mud mounds near the salt marsh edge, an area on the marsh of dense *Puccinellia maritima*, and one dominated by *Halimione portulacoides*; this last zone was about half way up the marsh, 200 m from its seaward edge.

On all of the control sites the layer of silica flour remained and fresh sediment was found above this. On the treated areas no freshly accreted sediment occurred, and in addition most of the silica flour had been washed away on the mud flats, but in the salt marshes a layer of silica flour remained. These results indicate that the deposition of mud on the upper intertidal zones can only in exceptional cases occur under the control of physical forces alone. It seems that normally mucus produced by the epipelic microflora aids the process of accretion of mud. The presence of mucus on the surface of the mud probably helps to 'trap' sediment. In addition the migration of algae through the freshly deposited sediment to the surface, releases mucus which stabilizes the sediment before it can be remobilized by the ebb tide.

THE EFFECT OF HIGHER PLANTS ON SEDIMENT DEPOSITION

It is quite often said that the salt marsh plants cause accretion. The statement made by Straaten and Kuennen (1957) that 'much sedimentation of mud takes place in consequence of the trapping effect of marsh plants', is typical of the literature on salt marshes.

Hummocks of sediment build up around the pioneer plants where the sediments are predominantly sandy (Chapman 1960). At sites where the sediments are mostly fine grained, it is not so much that the plants aid accretion but that they prevent erosion (Kestner 1975). In areas where mud flats are actively developing seaward, the rate of accretion can be much greater than it is in the salt marshes (Fig. 2.3). In areas where the physical forces are such that salt marshes and mud flats are being eroded, a net gain of sediment may occur still in the shelter of the plants on the salt marshes but away from the shelter of the plants a net loss of sediment takes place (Fig. 2.4).

EFFECT OF INVERTEBRATES ON SEDIMENT DEPOSITION

Only the direct effects of invertebrates on sedimentation have been considered previously and the indirect effects, such as the grazing of the microflora, mostly have been ignored. Of the more direct effects, local stabilization of sediment has been reported as a result of the activity of tube building species such as some polychaete worms (Gray 1974). Of much greater effect is the disturbance of the sediment by the deposit feeding species. Rhoads and Young (1970) have shown that the stability of the sea floor in Buzzards Bay, Massachusetts, is controlled largely by the feeding activities of the mud-dwelling deposit-feeders. Feeding produces an uncompacted surface which is easily resuspended by tidal scour. It is not yet known how such action affects the stability of the sediment compared with that of the filter-feeders which produce compacted faecal pellets. Verwey (1952) calculated that the mussels in the Dutch Wadden Sea deposit between 25,000 and 175,000 tonnes of detritus annually. Cadée (1976) has estimated that lugworms reworked a quantity of sediment which was equivalent of that to a depth of 6–7 cm each year over the same area.

Most deposit-feeding macroinvertebrates are known to ingest microalgae along with the sediment and detritus (Green 1968). Bacteria are generally a more important food source for most meio- and microinvertebrates, but benthic microalgae do form a substantial part of the diet of many species and are fed on more or less exclusively by some species of most groups of these invertebrates (Green 1968; Fenchel 1969). The distribution of the macroinvertebrates in the Wash may explain the differences in the densities of epipelic diatoms between the sand flat and the mud flat areas.

Among the most abundant macroinvertebrates on the inner sand flats of the Wash are the snail *Hydrobia ulvae* (Pennant) and the crustacean *Corophium volutator* (Pallas). When *Hydrobia* and *Corophium* were introduced to diatom rich sediment (approximately 5×10^5 diatoms cm^{-2} of sediment surface) the feeding of *Hydrobia* was patchy, but on the areas grazed by them the microalgae were eliminated. *Corophium* also drastically reduced the diatom population. At densities of four *Corophium* cm^{-2}, the diatom population was reduced by 50% in 12 hours, each *Corophium* having removed approximately 50,000 diatoms. Examination of the guts of the *Corophium* showed that some of the diatoms were the largest ingested particles. Experiments were carried out to see if *Hydrobia* and *Corophium* select sediments with large epipelic microalgal populations. The sediment which was used in these experiments came from the area where the animals were collected. It was presented to the animals either in its natural state or with the surface inoculated with epipelic diatoms at a density of about 2×10^5 diatoms cm^{-2}. In each experiment four boxes of sediment, each divided into four sections (each section 25 cm^2) were placed in a tray of sea water: two sections, randomly chosen, of each box were inoculated with diatoms. When *Hydrobia* was tested, 100 animals were placed on a small watch glass in the centre of each box. When *Corophium* was tested about 1,000 animals were distributed as evenly as possible in between the boxes. *Hydrobia* showed a distinct preference for sediment enriched with epipelic diatoms. After only 30 minutes 72% of the *Hydrobia* which had moved onto sediment, had selected sediment enriched by diatoms and after 16 hours this percentage had increased to 89%. The *Corophium* showed no such preference, and tests confirmed that they burrowed rapidly into the sediment which they first encountered.

Effect of the removal of animal grazing from the inner sand flats

Experiments to find the effect of removing animal grazing from the inner sand flats were carried out on a small area of sand flat which was surrounded by mud mounds. This area was chosen as it was considered to be a critical area in terms of sediment deposition, being difficult to comprehend why such a small, remote area should persist as a sand flat. The animals were killed in areas of about 0·5 m^2 by the application of concentrated domestic bleach. The effect of bleach in removing the animals did not last long and in most experiments the *Corophium* recolonized the sites within a few days. On one occasion this did not happen, and there was a dramatic explosion of diatom numbers, which reached an average of 95×10^4 diatoms cm^{-2} within a week compared with only 5×10^4 diatoms cm^{-2} on the surrounding sand flats. Accompanying this change was an increase in the rate of sedimentation, which resulted in a square patch of mud, slightly raised above the level of the sand flat. This 'mud' had a clay plus silt content of 30% compared with 11% in the surrounding sediment.

DISCUSSION

It appears that the accretion of intertidal mud usually only occurs when mucus, which is produced by the epipelic microflora, is present. On intertidal flats in temperate areas the most important group of microflora are epipelic diatoms. The experiments in the salt marshes of the Wash showed that even where the vegetation was dense the plants were unable to provide sufficient shelter to allow mud deposition to occur under the control of physical forces alone. But permanent mud deposition of sediment in any area must depend in part on the mud compacting rapidly enough to withstand the scour of subsequent storms. An important feature in the Wash, especially on the more exposed west and east shores, is that much of the upper mud flat is above high water neap tide level. As a result the mud flats frequently dry out, which greatly aids the compaction of the mud.

The difference in patterns of accretion at the various sites in the Wash, with the sand flats fluctuating and mud flats more or less continuously gaining material, may have been because compacted clay is far less easily eroded than fine sand. In sheltered estuaries, such as on the River Stour in Essex, where extensive mud flats occur at the lower tidal zones, the mud is not so compacted as on the mud flats of the Wash, and annual patterns of sediment deposition are more similar to those of the sand flats of the Wash. Even on this type of mud flat most of the short-term accretion is dependent on the trapping and binding effects of mucus produced by the diatom populations.

The complex interplay of the various biological and physical factors involved in the accretion or erosion of fine sediment on developing mud flats and salt marshes are only just beginning to be understood.

It may be that on most of the sand flats of the Wash, it is not merely the animal grazers that prevent the development of large diatom populations and the subsequent accretion of mud, but that the sand flats themselves may be too exposed to waves and strong currents for large surface populations of diatoms to develop. The transition between sand and mud flats has been examined in several areas of the Wash, at sites where the mud flats are moving seaward. Initially the inner sand flats in two of these sites had dense populations of *Corophium*. In the spring of 1976 these populations of *Corophium* crashed and at one site the mud flat moved dramatically seaward for several hundred metres over the previous sand flat. At the other site there was no apparent change.

ACKNOWLEDGMENTS

Most of the work described in this paper was funded by the Water Resources Board as part of the Wash Feasibility Study into fresh water storage in the

Wash. The Central Water Planning Unit became responsible for the completion of the study, and I am grateful to its Director for allowing me to publish this information.

I am particularly indebted to Mr M.G. Curry for his considerable assistance and I wish to thank my colleagues, especially Drs O.W. Heal, D.S. Ranwell and A.J. Gray for their help and encouragement.

REFERENCES

Brockman C. (1935) Diatomen under Schlick im Jade-Gebiet. *Abh. senckenb. naturforsch. Ges.* No. 430, 64 pp.

Cadée G.C. (1976) Sediment reworking by *Arenicola marina* on tidal flats in the Dutch Wadden Sea. *Neth. J. Sea Research* **10**, 440–60.

Carey A.E. & Oliver F.W. (1918) *Tidal lands, a study of shore problems.* Blackie, London.

Chapman V.J. (1960) *Salt marshes and salt deserts of the world.* Leon Hill, London.

Drum R.W. & Hopkins J.T. (1966) Diatom locomotion: an explanation. *Protoplasma* **62**, 1–33.

Eltringham S.K. (1971) *Life in mud and sand.* The English Universities Press, London.

Evans G. (1965) Intertidal flat sediments and their environments of deposition in the Wash *Q. Jl geol. Soc. Lond.* **121**, 209–45.

Fenchel T. (1969) The ecology of marine meiobenthos IV. Structure and function of the benthic ecosystem, its chemical and physical factors and the micro-fauna communities with special reference to the ciliated protozoa. *Ophelia* **6**, 1–182.

Garrett P. (1970) Phanerozoic stromatolites: non-competitive ecologic restriction by grazing and burrowing animals. *Science, N.Y.* **169**, 171–73.

Ginsburg R.N. (1975) Preface in *Tidal deposits: a case book of recent examples and fossil counterparts* (Ed. by R.N. Ginsburg), pp. 1–12. Springer-Verlag, Berlin.

Gouleau D. (1976) Le rôle des diatomées benthiques dans l'engraissement rapide des vasières atlantiques découvrantes. *C. r. hebd. Séanc. Acad. Sci., Paris, Sér. D.* **283**, 21–23.

Gray J.S. (1974) Animal-sediment relationships. *Oceanogr. Mar. Biol. Ann. Rev.* **12**, 223–61.

Green J. (1968) *The biology of estuarine animals.* Sidgwick and Jackson, London.

Grøntved J. (1949) Investigations on the phytoplankton in the Danish Waddensea in July 1941. *Meddr. Komm Danm. Fisk. -og Havunders. Serie: Plankton* **5**, 1–56.

Holland A.F., Zingmark R.G. & Dean J.M. (1974) Quantitative evidence concerning the stabilization of sediments by marine benthic diatoms. *Mar. Biol.* **27**, 191–96.

Hopkins J.T. (1966) The role of water in the behaviour of an estuarine mud-flat diatom. *J. mar. biol. Ass. U.K.* **46**, 617–26.

Hopkins J.T. (1967) The diatom trail. *J. Quekett microsc. Club* **30**, 209–17.

Inglis C.C. & Kestner F.J.T. (1958) Changes in the Wash as affected by training walls and reclamation works. *Proc. Instn. civ. Engrs.* **11**, 435–66.

Kamps L.F. (1962) Mud distribution and land reclamation in the eastern Wadden Shallows. *Rijkswat St. Commun.* Nr. 4.

Kestner F.J.T. (1962) The old coastline of the Wash. *Geogrl. J.* **128**, 457–78.

Kestner F.J.T. (1975) The loose-boundary regime on the Wash. *Geogrl. J.* **141**, 388–414.

Linke O. (1939) Die Biota des Jadebusenwalles. *Helgoländer wiss. Meeresunters.* **1**, 201–48.

Logan B.W., Davies G.R., Read J.F. & Cebulski, D.E. (1970) *Carbonate sedimentation and environments, Shark Bay, Western Australia.* Am. Assoc. Petroleum Geologists Mem. **13.**

Meade R.H. (1972) Transport and deposition of sediments in estuaries. *Environmental framework of coastal plain estuaries* (Ed. by B.W. Nelson), pp. 91–120. Geol. Soc. Amer. Mem. **133**.

Natural Environment Research Council (1976) The Wash water storage scheme feasibility study, a report on the ecological studies. N.E.R.C. Publications Series C No. 15.

Neuman A.C., Gebelein C.D. & Scoffin T.P. (1970) The composition structure and erodability of subtidal mats, Abaco, Bahamas. *J. sedim. Petrol.* **40**, 274–97.

Reinhold Th. (1949) Over het mechanisme der sedimentatie op de Wadden. *Meded. geol. Stricht., Nw. Ser.* **3**, 75–81.

Rhoads D.C. & Young D.K. (1970) The influence of deposit-feeding organisms on sediment stability and community trophic structure. *J. mar. Res.* **28**, 150–78.

Scoffin T.P. (1970) The trapping and binding of subtidal carbonate sediments by marine vegetation in Bimini Lagoon, Bahamas. *J. sedim. Petrol.* **40**, 249–73.

Straaten L.M.J.U. van (1951) Texture and genesis of Dutch Wadden Sea sediments. *Proc. 3rd Int. Congr. Sedim., Netherlands*, 225–44.

Straaten L.M.J.U. van & Kuennen Ph.H. (1957) Accumulation of fine grained sediments in the Dutch Wadden Sea. *Geologie en Munbouw (Nw. Ser.)* **19c**, 329–54.

Straaten L.M.J.U. van & Kuennen Ph.H. (1958) Tidal action as a cause of clay accumulation. *J. Sedim. Petrol.* **28**, 406–13.

Verwey J. (1952) On the ecology and the distribution of cockle and mussel in the Dutch Wadden Sea, their rôle in the sedimentation and the source of their food supply. *Arch. Néerl. Zologie* **10**, 171–239.

Walsby A.E. (1968) Mucilage secretion and the movements of blue-green algae. *Protoplasma* **65**, 223–38.

3. THE GENETIC STRUCTURE OF PLANT POPULATIONS IN RELATION TO THE DEVELOPMENT OF SALT MARSHES

A.J. GRAY*, R.J. PARSELL AND R. SCOTT
Institute of Terrestrial Ecology,
Colney Research Station,
Colney Lane, Norwich, U.K.

SUMMARY

Genetic differentiation within and between populations of plant species occupying heterogeneous environments is now a widely-observed phenomenon. Salt marshes provide both gradients and patches of spatial and temporal variation in habitat factors; notably in edaphic conditions related to the height of the marsh surface and the frequency of tidal submergence. Previous work on intraspecific variation in salt marsh species indicates that heritable morphological and physiological differences frequently occur between individuals in different parts of a marsh.

In *Aster tripolium* L., which has a high ecological amplitude, population differentiation occurs in response to the major gradient in marsh height. Plants from high marsh populations show significant differences from those in low marsh populations in a range of characters including life cycle, flowering time and fruit size, number and germination properties.

For perennial species with high ecological amplitudes the question arises: Do mature, high-level, marshes contain survivor populations consisting of individuals which have been selected during marsh development from the open mudflat pioneer phase? Measurement of heritable morphological characters indicated that pioneer populations of *Puccinellia maritima* contain a wider range of biotypes than mature populations. This suggests that genotypes closely adapted to the conditions of increased plant density on the upper marsh are selected from highly variable colonizing populations. However studies of clone structure on adjacent grazed and ungrazed marshes showed that this process does not necessarily involve reduction in the number of genetically different individuals (genets) per unit area. Using estimates of

* Present address: I.T.E., Furzebrook Research Station, Wareham, Dorset.

individual plant size based on the distribution of different isoenzyme phenotypes within 250 cm² grids, about twice the number of genets per unit area were found in *Puccinellia*-dominated swards which were ungrazed.

It is suggested that this contrast between grazed and ungrazed populations may be accounted for by differences in the reproductive biology of the survivors, differences imposed by the contrasting types of selection.

RÉSUMÉ

La différenciation génétique dans et entre les populations d'espèces végétales dans des environnements hétérogènes est maintenant un phénomène largement observé. Les marais salés fournissent à la fois des gradients et des mosaïques de variations temporelles et spaciales, dans les facteurs de l'habitat, notamment dans les conditions édaphiques reliées à la hauteur de la surface des marais et à la fréquence de la submersion par la marée. Un travail antérieur sur la variation spécifique des espèces dans le marais salant, indique que les différences héréditaires morphologiques et physiologiques ont lieu fréquemment pour différentes parties du marais.

La différenciation de la population, pour *Aster tripolium* L. qui présente une grande étendue écologique, répond au principal gradient de hauteur du marais. Les plantes des populations des hauts marais sont significativement différentes de celles des marais inférieurs pour une gamme de caractères comprenant le cycle de la vie, l'époque de la floraison et la taille du fruit, le nombre et les propriétés de la germination.

Pour les espèces vivaces avec une large distributipn écologique, une question se pose: Les hauts marais contiennent-ils des populations survivantes comprenant des individus selectionnés durant le développement du marais depuis la phase de colonisation des surfaces boueuses dénudées? La mesure des caractères morphologiques héréditaires indique que les populations pionnières de *Puccinellia maritima* contiennent une plus large série de biotypes que les populations matures. Ceci suggère que les génotypes fortement adaptés aux conditions de densité végétale importante du marais superieur, sont selectionnés à partir de colonies très variables. Cependant des études de structure de clones sur les marais adjacents, patures ou non, montrent que ce processus ne concerne pas nécessairement la réduction du nombre d'individus génétiquement différents (génettes) par unité de surface. Par des estimations de la taille des individus, basées sur la distribution des différents phénotypes d'isoenzymes, dans des grilles de 250 cm², nous trouvons environ deux fois le nombre de génettes par unité de surface dans les pelouses, dominées par *Puccinellia*, qui ne sont pas en pature.

Nous suggérons que ce contraste entre les populations, patures ou non, soit dû aux différences de biologie de reproduction des survivants, différences imposées par les types contrastants de sélection.

INTRODUCTION

Salt marsh ecology is easy to teach and difficult to study. The apparently clear zonation and floristic simplicity which makes the salt marsh an ideal habitat in which to demonstrate to the student the simpler principles of plant succession are features which evaporate rapidly upon detailed investigation. Almost all of the few plant species present on British marshes turn up, often in large numbers, in almost all parts of the zonation so that zones become blurred and the recognition of communities often becomes an arbitrary exercise in mathematics. The gradual increase in the height of the marsh surface and the concomitant decrease in the frequency of tidal submergence produces overall gradients up the marsh in habitat factors such as water table height, drainage characteristics, soil aeration and salinity (e.g. Hinde 1954; Chapman 1960; Adams 1963). However on inspection these gradients are found to be disturbed by local point-to-point variation in topography and soil type and by interaction with cyclical and seasonal variation (Brereton 1971; Gray 1972; Gray & Bunce 1972; Gray & Scott 1977a; Jefferies 1972). For example patches of hypersaline soils may be found in midsummer at high levels on a marsh (Tyler 1971; Ranwell 1972; Jefferies 1977).

However it is precisely the combination of these two features, the apparently high ecological amplitude of the common species and the underlying environmental heterogeneity, which makes salt marshes interesting places to study plant evolution. There is now considerable evidence that the populations of plant species distributed over heterogeneous environments frequently display genetic differences resulting from the effects of disruptive (diversifying) selection. Examples of such population differentiation include responses to metal contaminated soils (Bradshaw, McNeilly & Gregory 1965; McNeilly 1968), to variation in soil nutrient levels, either natural (Smith 1965; Antonovics, Lovett & Bradshaw 1967) or induced by management (Snaydon 1970; Snaydon & Davies 1972), to differences in soil moisture or flooding regimes (Hamrick & Allard 1972; Linhart & Baker 1973; Brown, Marshall & Albrecht 1974; Quinn 1975) and to different degrees of exposure to wind (Mogford 1974; Abbott 1976) and spray (Aston & Bradshaw 1966). Patterns of differentiation may be measured as inter-population differences in the gross morphology or physiological tolerances of individuals involving characters either known to be under polygenic control (Gartside & McNeilly 1974a, b) or presumed to be so (Aston & Bradshaw 1966) or as contrasting allelic frequencies in simpler genic systems (Hamrick & Allard 1972; Linhart & Baker 1973; Brown *et al.* 1974; Schaal 1975). The patterns of differentiation may be abrupt (Jain & Bradshaw 1966; Snaydon & Davies 1976) or clinal (Smith 1965; Antonovics & Bradshaw 1970) and may evolve rapidly (Snaydon 1970; Lin Wu & Bradshaw 1972) under extremely high coefficients of selection (Bradshaw 1971; Cook, Léfèvbre & McNeilly 1972; Davies & Snaydon 1976).

In the light of these results it would be surprising if salt marsh species, particularly those with relatively high amplitudes, did not show some form of genetic differentiation. Indirect evidence that differentiation may be widespread can be obtained from the literature on common British salt marsh species and this is briefly reviewed in the first part of this paper. We then discuss differentiation in response to the major environmental gradient of marsh height and tidal relations, referring to the results of a study of *Aster tripolium*. The first half of the paper concludes with a general summary of the factors which may be expected to affect population structure and in the second half these are considered in relation to the results of a study of *Puccinellia maritima*. In particular, we examine the proposition that populations of perennial species occurring at the higher levels of marshes may consist of individuals which have been selected during marsh development.

INTRASPECIFIC VARIATION IN SALT MARSH PLANTS

Coastal, particularly salt marsh, plants have had a special role in the development of our knowledge about environmentally induced variation in natural populations. As his early papers reveal Turesson's observations of variation in coastal plants played a large part in his formulation of the ecotype concept (Turesson 1920, 1922) and the development of this concept by Gregor, in a series of classic papers (e.g. Gregor 1930, 1938, 1944, 1946) was based on studies of differentiation in populations of *Plantago maritima*.

Since this early work intraspecific variation has been detected in many salt marsh plants. Sharrock (1967) has shown that the differences in leaf morphology in the varieties *parvifolia* and *angustifolia* of the species *Halimione portulacoides* has a genetic basis, although in the variety *latifolia* it may be modified by the environment. Boucaud (1962, 1970, 1972) describes differences in the physiological tolerances and germination requirements of three varieties of *Suaeda maritima*. Extensive variation within *Aster tripolium* includes both morphological and physiological characters with widely differing degrees of genetic control (Gray 1971, 1974). Significant differences in growth rates in response to inorganic nitrogen between populations from high and low marsh sites in *Salicornia europaea* agg. and *Triglochin maritima* (and in *Aster tripolium* and *Plantago maritima*) are reported by Jefferies (1977). The wide range of morphological variation in *Limonium vulgare*, within which a number of varieties have been described, appears from garden transplants to have a genetic component (Boorman 1967).

Several species found on British salt marshes also occur in inland habitats and distinct morphological and/or physiological differences between the inland and coastal taxa have been described in *Agrostis stolonifera* (Tiku & Snaydon 1971), *Festuca rubra* (Hannon & Bradshaw 1968), *Armeria maritima* (Léfèbvre 1974), and *Rumex crispus* (Cavers & Harper 1967).

POPULATION DIFFERENTIATION IN RELATION TO THE GRADIENT IN MARSH HEIGHT

The above examples do no more than suggest the possibility of widespread interaction between ecological amplitude of populations and environmental heterogeneity. However they do have a persistent theme. This is that differences are frequently found between plants of the same species growing in low and high marsh sites (or coastal and inland ones). It is perhaps to be expected that the response which populations of a species make to conditions prevailing at different heights on a marsh should be easily detected as these conditions reflect the predominating environmental gradient.

Aster tripolium is an appropriate species in which to study this response. It has an exceptionally high ecological amplitude on British salt marshes, being found in at least three types of habitat: low level marshes on pioneer sites (the Asteretum tripolii of Tansley 1953 and Chapman 1960), the middle levels (the 'general salt marsh' community of Chapman 1960) and brackish high level marshes and saline seepage areas above the limits of tidal submergence and often landward of sea walls (Gray 1977).

As part of a genecological investigation of this species (Gray 1971) samples of plants and seed were collected from 24 sites in Britain and the sites were subsequently divided into three groups (roughly corresponding to the three habitat types above). The grouping was based on two methods, association-analysis of the species lists recorded from each sampling area (Williams & Lambert 1959) and a division based on the frequency of tidal submergence and associated physical factors. There was good agreement between the two methods and the three groups were designated as 'high', 'mid' and 'low' marsh sites. Ten morphological and physiological characters of *Aster* plants were measured quantitatively in collateral cultivation conditions on sufficient population samples to allow a comparison between those from different marsh types (Table 3.1).

The results indicate that large differences exist between populations of *A.tripolium* at the different marsh levels. These differences appear in uniform cultivation and may have a large genetic component. Eight out of ten characters showed significant differences, usually between population samples from 'high' and 'low' marshes. On average a third of the plants from high marsh sites flowered in a year from the time of seed germination and most behaved as annuals, dying after flowering. In contrast, all the low marsh plants were perennial and some flowered for the first time after three years. High marsh plants were late flowering, producing in each capitulum many small fruits, a large percentage of which required a period of cold treatment to break seed dormancy. Low marsh plants flowered relatively earlier producing fewer but larger fruits in each capitulum, most of the seeds of which germinated precociously.

Table 3.1. *Summary of variation in relation to marsh type in* Aster tripolium (*data from Gray 1971*).

Character	High marsh sites		Mid marsh sites		Low marsh sites		
	$\bar{X}$ of population sample means	Number of populations tested and total plants ( )	$\bar{X}$ of population sample means	Number of populations tested and total plants ( )	$\bar{X}$ of population sample means	Number of populations tested and total plants ( )	*P*†
Per cent of flowering in first year	33·8	5 (220)	1·6	5 (150)	0	5 (198)	***
Flowering date (days after 1 August)	64·2	8 (103)	40·9	4 (40)	37·8	4 (58)	***
Erectness of young plant (ht/width)	0·82	5 (151)	0·84	4 (195)	1·48	4 (182)	**
Inflorescence ht. (cm)	57·5	5 (65)	38·7	4 (47)	57·5	3 (37)	n.s.
Number of flowering heads per plant	36·9	4 (57)	64·2	2 (27)	51·4	4 (45)	n.s.
Number disc fruits/head	35·5	3 (51)	29·9	1 (9)	24·3	3 (34)	**
Weight 10 disc fruits (mg)	6·8	3 (51)	11·0	1 (9)	15·6	3 (34)	***
Per cent fruit requiring prechilling to germinate	91·4	3 (600)	42·5	2 (400)	14·4	3 (600)	***
Per cent germination in 40% sea water after 28 days	5·3	3 (900)	17·7	2 (600)	32·4	3 (600)	***
Survival rate (%) of seedlings planted out in garden soil	83·0	7 (352)	63·5	5 (107)	38·4	5 (91)	***

† Significance tests were performed using either χ^2, F, or t tests of the mean difference between paired marsh types. The asterisks indicate that there are significant differences for that character at the 0·001 level of *P* (***) or the 0·01 level of *P* (**) between the means of populations belonging to at least two marsh types (i.e. between high and low, high and mid, or low and mid). n.s. = no significant difference between any of the marsh types.

By grouping populations into those from ecologically similar sites in this way and comparing them, using as error item the variance between their sample means, we are less likely to misinterpret as evidence of adaptive habitat-correlated variation a low estimate of within-sample variance which could arise from repeatedly sampling the same genotype (as may happen in an area where extensive vegetative spread has occurred (e.g. Wilkins 1959; Harberd 1961a).

FACTORS AFFECTING POPULATION STRUCTURE

The magnitude of the differences between plants of *A.tripolium* from different marsh levels, including differences in flowering dates, suggests that the low and high marsh populations in any geographical area may be effectively isolated. However, the pattern we obtain of population differences is often predetermined by the pattern of sampling. Sampling at widely separated points, say the top, middle and bottom of a marsh several hundred metres wide, gives a useful indication of overall trends in variation in relation to a major environmental gradient. However it fails to detect a precise relationship between the pattern of differentiation and small scale environmental differences such as that demonstrated, for example, in *Agrostis stolonifera* on sea cliffs (Aston & Bradshaw 1966; Bradshaw 1971). In this species both abrupt and clinal differentiation patterns occur in response to abrupt and gradual environmental changes.

Environmental factors

As indicated earlier small scale change is a feature of the salt marsh environment. The local, often small, differences in marsh height associated with topographical features such as hummocks, depressions, pans, and the sides and edges of creeks, are apparently sufficient to produce differences in the composition of species found at these sites (e.g. Gray & Scott 1977a). In some cases, for example pans, the selection pressures associated with the small-scale differences are sufficiently intense to eliminate all but a few species (such as *Salicornia* species). In others, such as the slopes of large hummocks or those bordering drainage creeks the changes are more gradual and selection may act indirectly through, for example, the increasing height and competitive vigour of a species such as *Halimione portulacoides* as the creek edge is approached. The creek edges themselves provide ribbon-like habitats intersecting the marsh and extending the area available to many species, often bringing them into contact with populations of conspecifics at all levels of the gradient from low to high marsh. For example plants in the creek bank populations of *Aster*, which have similar characters to those in the low marsh populations and share similar soil and tidal submergence conditions, may be found within a few metres of plants in high marsh populations of this species growing in very different conditions on the flat areas between the creeks. In such an outbreeding species which is insect pollinated gene flow between the two populations is a clear possibility.

Salt marsh plant populations are therefore presented with a heterogeneous environment in which there is complex variation in the size and shape of the patches and in the intensities and direction of selection. The effects of a similarly patchy environment on the genetic structure of a population is discussed by Dickinson and Antonovics (1973) using computer simulation. Their model demonstrates the complexity of effects at just a single locus on the differentiation or adaptedness of the population (measured as a genotype/environment correlation) to be expected from variation in the number and size of patches, the density of distribution of offspring and the intensities of selection. Fluctuations in the intensity of selection are likely to be an important aspect of the salt marsh environment where, for example, the interaction between rainfall and the tidal cycle may produce variations in the water potential and salinity of sediments. The potential influence of the tide in distributing seed or vegetative fragments over a wide area may also affect the genetic structure of populations.

Genetic factors

The extent to which a particular species can respond by population differentiation to the spatial and temporal heterogeneity of the environment depends on its intrinsic genetic properties. Among these we can distinguish the properties of the species' breeding system and those of its reproductive strategy, which together will affect the degree to which a species can exploit the spatial and temporal variation in its environment. Since they both influence the pattern and amounts of gene flow and genetic recombination within and between populations, they will also affect the extent to which differentiation, having arisen, can be maintained. The magnitude of the response to selection will depend further on the number of genes controlling the differentiating character, their mode of inheritance, their selective value, and the level of genetic variability in the original population.

In the second part of this paper we report preliminary results of a study of genetic and environmental factors affecting the population structure of *Puccinellia maritima.*

POPULATION STRUCTURE IN *PUCCINELLIA MARITIMA*

Puccinellia maritima is a common grass widespread on salt marshes throughout the British Isles and Europe. It is found as a pioneer on open mudflats and grows in the middle and upper parts of marshes. It may occur as scattered plants or as dense stands and frequently it is a major component of the vegetation, particularly in grazed marshes (Gray & Scott 1977b).

Several aspects of the species' reproductive biology and genetic system enable it to survive in a heterogeneous environment and buffer it against the exigencies of a fluctuating one. It is perennial and can be very long-lived (Gray & Scott 1977b). The plant spreads extensively by means of stolons and tiller fragments which are uprooted by grazing animals and become established when partly buried by accreting sediments. It is also capable of self-fertilization, although experiments using mutual bagging suggest that it is predominately outcrossing in nature (the species is wind-pollinated). In addition it is highly polyploid; counts from over forty populations in Britain give the somatic chromosome number of $2n = 56$, which is the octoploid level for the genus (Scott & Gray 1976). The high ploidy level will tend to slow down the rate at which genetic fixation (i.e. complete homozygosity) is reached as a result of repeated inbreeding (McConnell & Fyfe 1975). The combination of vegetative reproduction, self-compatibility and high ploidy level, enable the species to survive, and ensure the maintenance of some genetic variability, in populations colonizing isolated sites (Antonovics (1968) has discussed the comparative effects of gene flow on populations of annuals and perennials during the early stages of colonization). Plants of some populations of *P.maritima* flower freely and produce large numbers of viable fruits in the field and in garden cultivation. In these situations frequent outcrossing will enable genetic recombination to occur, maintaining high levels of heterozygosity, and ensuring a wide spectrum of genetic variability. This facilitates the evolution of genotypes adapted to local conditions.

The components of intraspecific variation

P.maritima exhibits a wide range of phenotypic variation in the field. In a preliminary assessment of variation in British populations 56 plants (clones) taken at random from six geographical regions were cultivated in pots in a gravel bed and a single-tiller ramet from each clone was grown in each of seven independent completely randomized blocks (Gray & Scott, unpublished). Twenty characters which included rates of tillering, vegetative and flowering tiller heights and widths, leaf dimensions, dates of ear emergence and anthesis and seed weights were measured on all plants. The within- and between-clone components of variance were used to estimate the genetic coefficient of variation and the heritability of each character (for a discussion of these parameters see Burton & DeVane 1953). Although the results obtained so far must be interpreted with caution as no breeding tests have been completed many of the provisional estimates of heritability were high indicating that much of the variation is genetically controlled.

Examination of the first three principal components extracted from a correlation matrix of the clone mean values for all 20 characters distinguished three major aspects of variation. The first component reflected variation in vegetative biotype, and contrasted plants which produced many small short-leaved tillers with those which produced a few large long-leaved tillers. Variation in growth habit was associated with this component; the rapid tillering biotypes were more prostrate than those which tillered slowly. This is a familiar pattern of biotype variation in grasses (Edwards & Cooper 1968). The second component provided a measure of yield, and plants were separated along an axis from high to low above-ground dry weight. The third component defined the variation in reproductive strategy and allowed a separation of plants on the basis of the proportion of their resources allocated to the production of flowering, as opposed to vegetative, tillers.

Only those characters which provided the best estimates of the variation in these three components of variation, vegetative biotype, yield, and reproductive strategy, were measured in later trials. One such trial was concerned with the variation in local populations on pioneer and mature marshes.

Pioneer and mature populations

In attempting to compare the structure of populations from different habitats we are confronted with two questions. Firstly, what constitutes a population and secondly where do different habitats begin and end? We have adopted here what is probably the plant ecologist's most common use of the word 'population' when referring to conspecifics: a group of individuals occurring within an area, the boundaries of which are not strictly drawn but within which it would be reasonable to assume that interbreeding can occur. This definition usefully holds in abeyance for the time being such questions as the levels of gene flow and of migration of individuals. In the study below we have sampled the populations in two ways. Individuals have been sampled as single-tiller units from within a 15 m radius of a randomly defined point within the habitat; for studies of clone structure tillers were collected within a 250 cm square grid. Defining the limits of a habitat is extremely subjective because we have no prior knowledge of the pattern of spatial or temporal variation in the environment or the type and magnitude of the selective factors present in the different patches. However on accreting marshes *P.maritima* is found in at least two recognizably distinct situations; those in which the individual plants are scattered across areas of bare mud and where annual colonists such as *Salicornia* spp. also may be present and those in which they form a more or less continuous sward or are members of a vegetated sward of different perennial species. Populations from these situations we have called 'pioneer' and 'mature' respectively.

Ten populations were sampled by the first method, six from Morecambe Bay in north-west England and four from Wells, Norfolk, in East Anglia. In Morecambe Bay pioneer and mature populations were sampled from two sheep-grazed marshes, at Meathop and Silverdale, and one ungrazed marsh near Grange-over-Sands. At Wells, pioneer populations were sampled at two stages of an accreting marsh, and the two mature populations were from ecologically contrasting areas of the middle marsh. All were believed to be from actively accreting marshes.

The 15 plants from each population, after six months' growth in John Innes compost, were each divided into five ramets and the resulting 750 single-tiller units grown in the same type of compost in pots arranged in rows in a single completely randomized block. Characters were measured on the basis of variation in the preliminary trial and some of the results, which are to be reported in full elsewhere, are discussed below. All character values referred to are the clone mean values (i.e. the mean value of the five ramets).

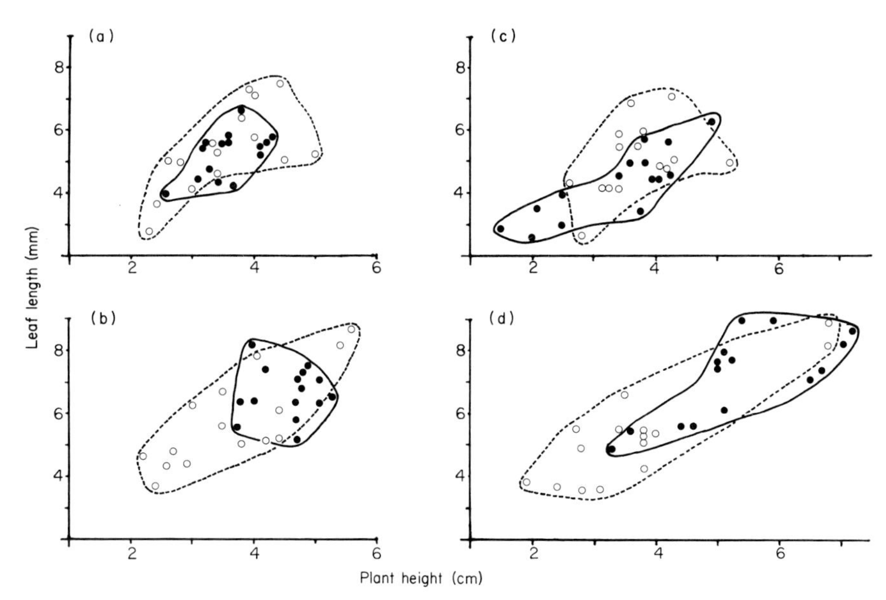

FIG. 3.1. Vegetative biotype in pioneer and mature populations of *Puccinellia maritima* (○ = plants from pioneer population, ● = plants from mature population on same marsh). (a) Meathop marsh (grazed), (b) Grange marsh (ungrazed), (c) Silverdale marsh (grazed), (d) Wells marsh (ungrazed). The values for each plant are the means of 5 ramets (leaf length = length 3rd flag leaf from ligule to tip. Plant height = height above soil level of plant at centre). See text for interpretation.

Fig. 3.1, which provides a measure of the vegetative biotype of the plants, illustrates a consistent feature of the data. With the exception of the Silverdale populations, which are discussed later, for most characters the variation is greater in the pioneer population than in the mature population from the same marsh. For example, the pioneer populations at Meathop and Grange both consisted of a range of vegetative biotypes whereas the Meathop mature population contained mainly small, prostrate, short-leaved plants and the Grange mature population mainly large, erect, long-leaved ones. Meathop marsh is heavily grazed by sheep whereas Grange marsh is ungrazed and the marshes are adjacent. As the plant has a perennial growth habit the results prompt the following question: Are the populations in the mature marsh composed of genotypes which have survived selection during marsh development?

The Meathop and Grange populations

The Meathop and Grange marshes are separated only by the outfall of a narrow river (Fig. 3.2). When the embankment carrying the Ulverston-Lancaster railway was constructed in 1857 the river Winster was canalized and a tide-operated gate built near Meathop crag (Gray & Adam 1974). This

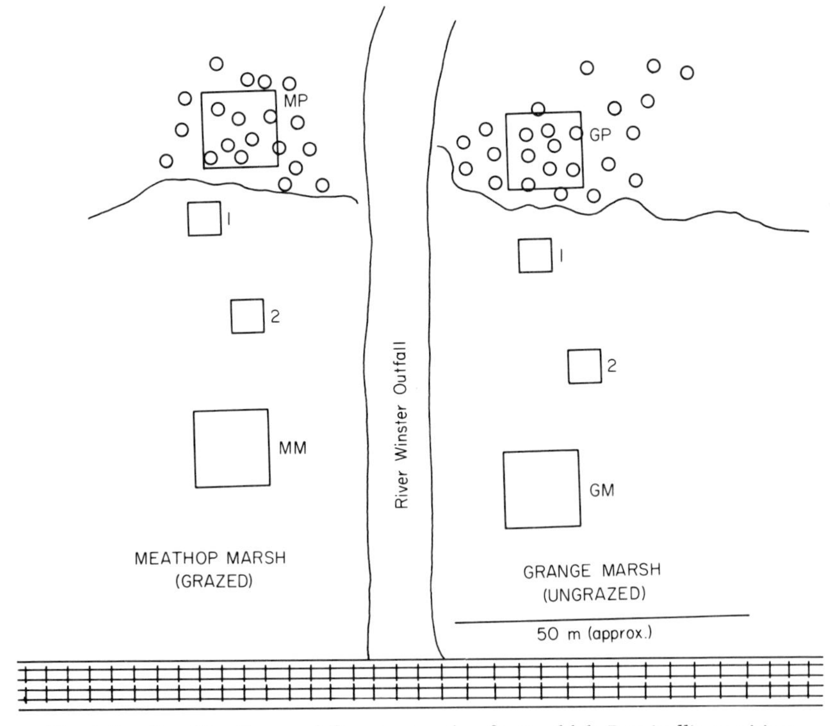

FIG. 3.2. The Meathop and Grange marshes from which *Puccinellia maritima* was sampled from pioneer (MP, GP) and mature (MM, GM) zones and from grids (numbered 1 and 2 on each marsh).

trapped a mudflat between the river outfall in the east and a road bank to Holme Island to the West on which salt marsh developed. The marsh which has not been sheep grazed, covered an area of 10·8 ha in 1969 (Gray 1972). The grazing intensity on Meathop marsh is high (about 6 sheep/ha) and the short compact turf, dominated by *P.maritima*, *Festuca rubra* and *Agrostis stolonifera*, rarely exceeds 3 cm in height. By contrast the vegetation on the ungrazed Grange marsh averages 30–40 cm in height and includes tall plants of *Halimione portulacoides* and *Aster tripolium*.

The pioneer and mature populations were sampled within 50 m of the river at sites which represented equivalent stages of the presumed succession on both marshes. The results of variation in the length of the flowering tiller (measured from ground level to the tip of the panicle) and the above-ground dry weight (divided into flowering and vegetative tillers) in a collateral cultivation trial of these populations is shown in Figs. 3.3 and 3.4 respectively.

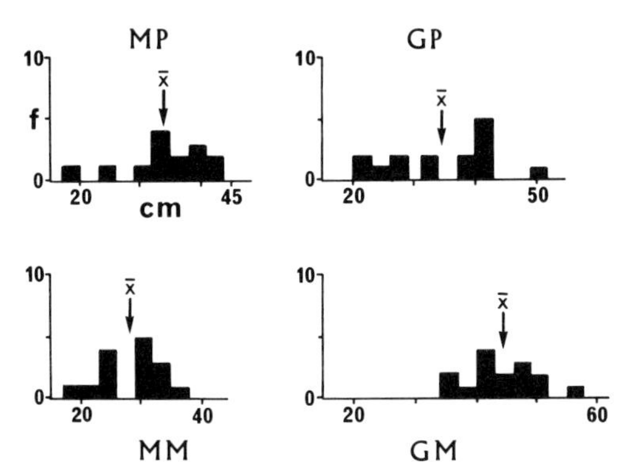

FIG. 3.3. Length of flowering tillers (from ground level to panicle tip) in plants of *Puccinellia maritima* from Meathop and Grange pioneer and mature populations (MP, MM, GP, GM) grown in collateral cultivation. The pioneer populations have similar means (Meathop 34·1 cm, Grange 34·8 cm) but the mature populations have significantly different means (Meathop 28·4 cm, Grange 44·6 cm, $P = 0{\cdot}001$ in Students *t* test). The values are the means of 5 ramets/plant.

Although the means of the lengths of the flowering tillers are similar in the two pioneer populations the corresponding means for the two mature populations are significantly different from one another. The data of tiller length from each of the pioneer populations also show a greater variance compared with similar data from the mature populations. The dry weights of plants from the Grange mature populations are significantly higher than those from the Meathop mature population. Grange mature plants also allocate significantly more (33% as opposed to 23%) of their resources to the production of flowering tillers (Fig. 3.4).

These results are consistent with the hypothesis that, as marsh development proceeds, grazing favours the selection of small, prostrate, short-leaved biotypes which allocate a relatively small proportion of their resources to flowering (see Thomas 1973; Mahmoud, Grime & Furness 1975), whereas in ungrazed marshes tall, erect, long-leaved biotypes with a high proportion of flowering tillers are selected. The results of populations originally collected

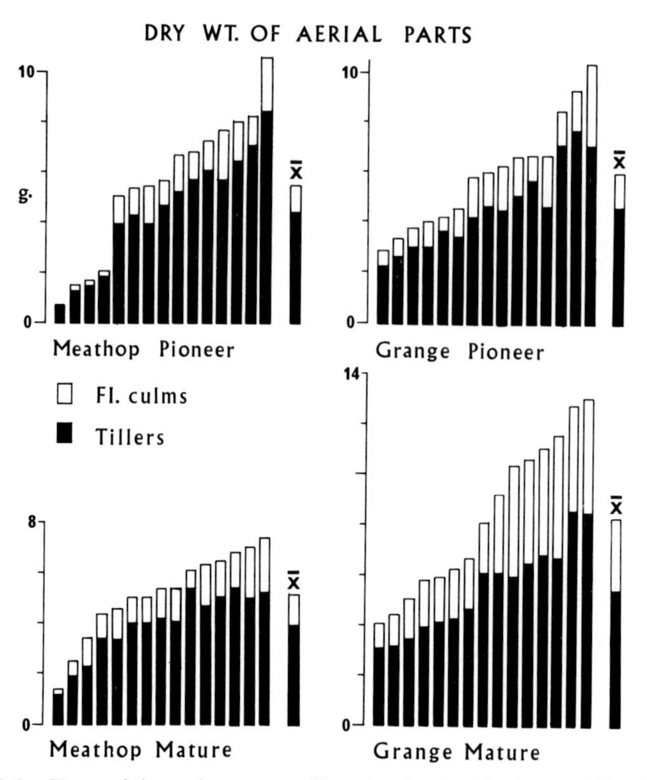

FIG. 3.4. Dry weight and resource allocation in the 15 plants of *Puccinellia maritima* from each of the Meathop and Grange populations in collateral cultivation. The values for each plant are the means of 5 ramets. Note the higher variance for both the yield and proportion of dry weight allocated to flowering tiller production in the pioneer populations compared to the mature populations. The Grange mature population is significantly higher yielding and the plants allocate significantly more of their dry weight to flowering tiller production than the Meathop population. (Grange mature, mean dry weight = 8·37 gm, % dry weight to flowering tillers = 33·2: Meathop mature, mean dry wt. = 5·25 gm, % dry wt. to flowering tillers = 23·4—Dry wts. significantly different at 0·002 level of *P* and % flowering tillers significantly different at 0·001 level of *P*.)

from Silverdale and Wells are also consistent with the hypothesis. The Wells mature population is composed of tall long-leaved individuals (Fig. 3.1) and the Silverdale mature population contains a number of plants which, although outside the range shown by the pioneer population, are smaller and shorter-leaved. There is also apparently a smaller range of variation in the pioneer population at Silverdale compared with elsewhere on which selection is able to act. Two facts could help to explain this difference. Firstly, although the plants from grazed marshes flowered in the experimental trial, it is often difficult to find flower heads in the field on such marshes. Unlike those of *F.rubra* and *A.stolonifera* the flowering heads of *P.maritima* apparently are grazed by sheep. Secondly, the Silverdale population is some distance (c. 3 km) from the nearest source of propagules from an ungrazed population. If, as we might predict, the major source of new genetic variation is from recombination and segregation during sexual reproduction, the smaller variation of the Silverdale population, with its very long history of grazing, and the restricted gene flow from ungrazed, flowering populations is to be expected.

Clone structure and individual plant size

The narrow range of biotypes on mature marshes may be the result of selection acting in at least two ways. During the development of the marsh individuals which are unable to tolerate the effects of high plant densities and, on grazed marshes, the effects of grazing, particularly repeated defoliation, may be eliminated. If this happens the mature marshes should on average contain fewer individuals of *Puccinellia maritima* per unit area. Alternatively individuals in the pioneer marsh may be replaced in the mature marsh by new individuals, in which case the number of individuals per unit area on pioneer and mature marshes may be similar or show no consistent relationship. There is a third possibility, namely that pioneer and mature populations are not developmentally related and are zones rather than spatial equivalents of stages in a temporal process. This cannot be discounted (although we believe it unlikely in these populations) and an explanation of the wider range of variation in pioneer populations might be sought in the greater environmental heterogeneity (more microhabitats, greater edaphic variation) of the pioneer marsh zone. From a superficial appearance of the two zones, we again believe this to be unlikely.

We have attempted to estimate the number of individuals per unit area in the Meathop and Grange populations from the pattern of variation in certain isoenzyme polymorphisms. Two 250 cm square grids were sampled on each marsh, one as near as possible to the point where the scattered colonist clumps coalesced to form a pure *P.maritima* sward and the second above this in the mature sward (Fig. 3.2). Each grid was divided by string into 100, 25 cm squares and the plant at the grid intersects was sampled (as a single tiller). The other species present in the Meathop upper grid were *Festuca rubra*, *Agrostis stolonifera*, *Armeria maritima*, *Glaux maritima*, *Spergularia media* and *Plantago maritima* and in the Grange upper grid *Halimione portulacoides*, *Aster tripolium*, *Suaeda maritima*, *Plantago maritima* and *Festuca rubra*. Where one of these species covered the grid intersect point the nearest plant of *Puccinellia maritima* to that point was sampled.

The 400 tillers thus obtained were grown in culture solution and crude extracts for electrophoresis were prepared from the terminal centimetre of freshly produced roots. Electrophoresis was conducted on horizontal acrylamide gels. The details of the methods used and genetic analysis of the polymorphic systems are to be reported elsewhere. In this paper we refer to variation in non-specific esterases, peroxidases and acid phosphatases. We

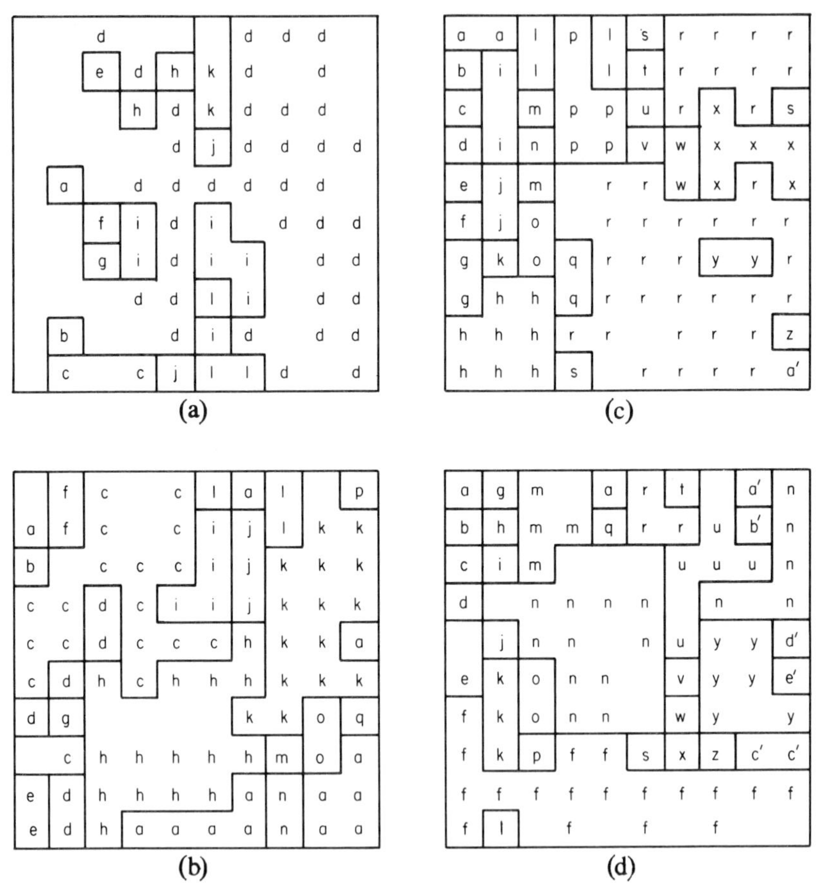

FIG. 3.5. The 2·5 m grids from Meathop and Grange marshes (a = Meathop lower; b = Meathop upper, c = Grange lower, d = Grange upper). The individuals of *Puccinellia maritima* within the grid which shared all alleles in three polymorphic enzyme systems (tentatively identified at 9 loci in total) have been given the same code letter. The letters are different for each grid and do not correspond to those in any other grid (although, for example, 'd' of Meathop lower is identical to 'h' of Meathop upper). Spaces indicate that plants from that intersect point died during cultivation or that the bands were too poorly resolved to calculate migration distances. Lines are drawn around areas containing identical adjacent plants.

have tentatively identified 9 loci in total; 4 esterase, 2 peroxidase and 3 acid phosphatase. For each grid the plants having the same alleles at all loci have been given a common code letter. Each grid has been coded separately for reasons discussed below—thus the letters are different for each grid and do not correspond to those in any other grid.

The results, set out in grid form (Fig. 3.5), provide an interesting insight into the possible structure of the populations. The Meathop grids contain only 12 and 17 different 'types' in the upper and lower grids respectively whereas those on Grange contained 26 and 31 different 'types' respectively, about twice the number. If we assume that the sampling of identical types from neighbouring points is equivalent to sampling the same genotype at two different points and that the intervening unsampled space is occupied exclusively by that genotype, the contrasting clone structure on the two marshes is evident (Fig. 3.6).

There are of course dangers in these assumptions. First, in this exercise we can easily establish that two samples are genetically different but in practice, given the total number of all possible loci, we can never finally establish that they are identical. Second, we have no information about the genotype of plants in the space surrounding sample points. However, the greater the number of identical enzyme phenotypes from plants at adjacent points, the greater is our confidence that we are sampling a single clone. Correspondingly we can have less confidence that plants with the same enzyme phenotype but in different grids are part of the same clone and have therefore deliberately used a different code for each grid. Despite these problems an examination of the spatial distribution of plants with different enzyme phenotypes (see Wu, Bradshaw & Thurman 1975; Schaal & Levin 1976) provides valuable data on the structure of plant populations in addition to that obtained from a comparison of morphological differences, self-incompatibility patterns or cytological variation (e.g. Harberd 1961b, 1967; Smith 1972).

The results shown in Fig. 3.5 and Fig. 3.6 (which is derived from the data of Fig. 3.5) indicate that, although the grids on both marshes contain 3 or 4 large clones of comparable size, in the Meathop grids the space between the large clones is apparently occupied by a small number of medium-sized clones whereas in the Grange grids it is occupied by a larger number of small clones. Because of lack of intensive sampling the number of clones shown in Fig. 3.6 is probably underestimated.

This contrast in clone structure on the two marshes might be predicted if the space between colonists on grazed marshes is largely filled by the vegetative spread of the colonists whereas that on ungrazed marshes is more often filled by seedlings. As the marsh develops, grazing is likely to reduce the numbers of seed produced both by physical removal of the flower heads and as a result of the differential survival of those genotypes which allocate resources to clonal reproduction and produce few flowering tillers. The prostrate rapidly tillering

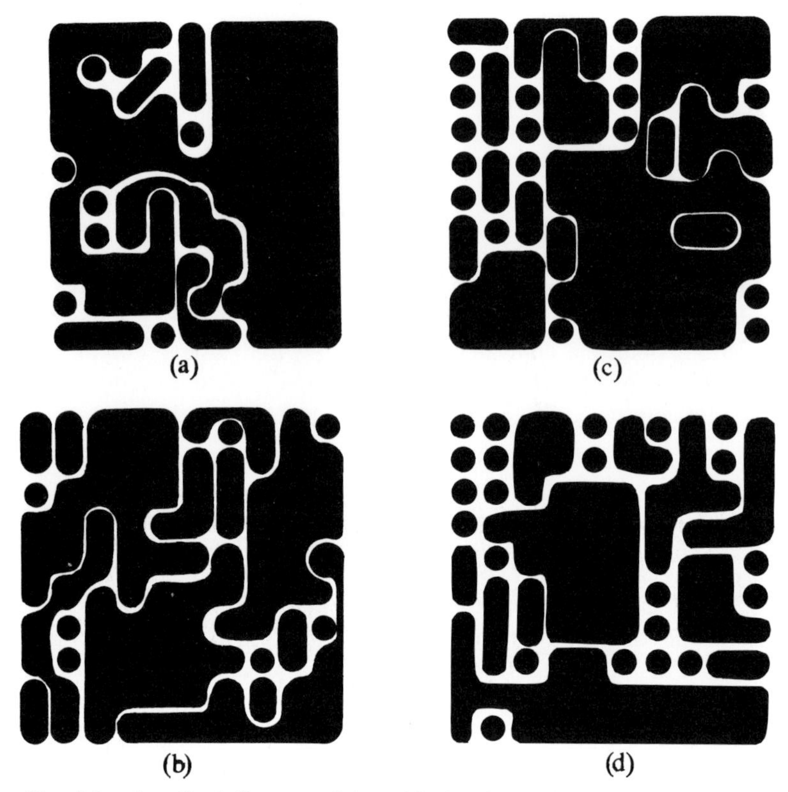

FIG. 3.6. A stylized diagram of the grids showing a clone structure of *Puccinellia maritima* which would give the results in Fig. 3.5 assuming that the samples from each intersect were from plants which occupied the whole of the 25 cm square surrounding them and that adjacent samples with identical enzyme phenotypes were from different parts of the same individual (genet). The black shapes thus represent the physical limits of one individual. (a = Meathop lower; b = Meathop upper; c = Grange lower; d = Grange upper.) For discussion of the differences between grids see text.

plants favoured by grazing probably quickly reduce the amount of space available for seedling recruitment. Spaces are likely to arise more often in the mature ungrazed marsh, particularly under *Halimione* bushes and other tall species which fall over or die back after flowering, than in the dense turf of the grazed marsh. Selection on the ungrazed marsh, for tall erect high-yielding biotypes, may also favour those genotypes which allocate a relatively high proportion of their resources to flowering and seed production. In a situation where selection favoured high seed production the species' high rate of out-breeding would ensure a constant rain of seeds carrying new genetic combinations. The effect of these phenomena on gene frequencies and the average heterozygosity of the populations will be discussed elsewhere.

CONCLUSION

We have argued that the variability between different mature populations in heritable morphological characters is adaptive and has been developed and maintained in response to diversifying selection as the marsh develops. This response although reducing the overall range of variation, need not necessarily involve a reduction in the number of genetically different individuals (genets) per unit area. We have suggested that the contrast in this respect between grazed and ungrazed mature populations may be accounted for by differences in the reproductive strategy of the survivors. These differences arise from contrasting types of selection and, since they affect the amounts of genetic recombination, are self-reinforcing, maintaining the differences in the substructure of the two populations.

There are several ways in which we might test these hypotheses. It would be of interest, for example, to attempt to measure the physical limits of individuals at both a smaller grid sampling distance and over a larger area. We are also currently investigating the competitive interaction of short and tall biotypes under the contrasting regimes of repeated defoliation or no defoliation.

In conclusion we commend to the reader the value of salt marsh systems as areas to study population biology. For example, it will not have escaped the notice of some that the pioneer and mature marshes respectively have many features of predominately *r*-selected and *K*-selected environments. In this paper an attempt has been made to focus down from the wider patterns of infra-specific variation and population differentiation through the local population to the factors which affect the individual and its survival and spread within the population. We cannot improve on a recent timely reminder (Mack & Harper 1977) that 'It would seem prudent for both ecologists and evolutionary biologists to evaluate their experiments at the level of the individual, particularly as biological phenomena demand an evolutionary interpretation and natural selection acts on individuals'.

ACKNOWLEDGMENTS

We thank Dr P. Mathews for technical advice and support during the development of the electrophoresis programme and many others, mostly vacational students, who have assisted in the experiments summarized in this paper. The work on *Aster tripolium* was done during the tenure of a Demonstratorship in Biology at the University of Keele and A.J.G. wishes to acknowledge the support there of Dr K.M. Goodway.

REFERENCES

Abbott R.J. (1976) Variation within common Groundsel *Senecio vulgaris* L. II. Local differences within cliff populations on Puffin Island. *New Phytol.* **76**, 165–72.

Adams D.A. (1963) Factors influencing vascular plant zonation in North Carolina salt marshes. *Ecology* **44**, 445–56.

Antonovics J. (1968) Evolution in closely adjacent plant populations. VI. Manifold effects of gene flow. *Heredity* **23**, 507–24.

Antonovics J. & Bradshaw A.D. (1970) Evolution in closely adjacent plant populations. VIII. Clinal patterns at a mine boundary. *Heredity* **25**, 349–62.

Antonovics J., Lovett J. & Bradshaw A.D. (1967) The evolution of adaptation to nutritional factors in populations of herbage plants. *Isotopes in plant nutrition and physiology*, pp. 549–67. I.A.E.A. Vienna.

Aston J.L. & Bradshaw A.D. (1966) Evolution in closely adjacent plant populations. II. *Agrostis stolonifera* in maritime habitats. *Heredity* **21**, 649–64.

Boorman L.A. (1967) Biological Flora of the British Isles. *Limonium vulgare* Mill. and *L.humile* Mill. *J. Ecol.* **55**, 221–32.

Boucaud J. (1962) Etude morphologique et écophysiologique de la germination de trois variétés de *Suaeda maritima*. Dum. *Bull. Soc. Linn. Norm.* 10 série, **3**, 63–74.

Boucaud J. (1970) Action de la salinité, de la composition du milieu et du prétraitement des semences sur la croissance de *Suaeda maritima* var. *flexilis* Focke en cultures sans sol. Comparaison avec *Suaeda maritima* var. *macrocarpa* Mog. *Bull. Soc. Linn. Norm.* **101**, 135–48.

Boucaud J. (1972) Auto-écologie et étude expérimentale des exigences écophysiologiques de *Suaeda maritima* (L) Dum. var *macrocarpa* Moq. et var. *flexilis* Focke. *Oecol. Plant.* **7**, 99–123.

Bradshaw A.D. (1971) Plant evolution in extreme environments. *Ecological Genetics and Evolution* (Ed. by R. Creed), pp. 20–50. Blackwell Scientific Publications, Oxford.

Bradshaw A.D., McNeilly T.S. & Gregory R.P.G. (1965) Industrialization, evolution and the development of heavy metal tolerance in plants. *Ecology and the Industrial Society* (Ed. by G.T. Goodman, R.W. Edwards & J.M. Lambert), pp. 327–43. Blackwell Scientific Publications, Oxford.

Brereton A.J. (1971) The structure of the species populations in the initial stages of salt marsh succession. *J. Ecol.* **59**, 321–38.

Brown A.H.D., Marshall D.R. & Albrecht L. (1974) The maintenance of alcohol dehydrogenase polymorphism in *Bromus mollis* L. *Aust. J. Biol. Sci.* **27**, 545–59.

Burton G.W. & DeVane E.E. (1953) Estimating heritability in tall fescue (*Festuca arundinacea*) from replicated clonal material. *Agron. J.* **45**, 478–81.

Cavers P.B. & Harper J.L. (1967) The comparative biology of closely related species living in the same area. IX. *Rumex*: the nature of adaptation to a sea-shore habitat. *J. Ecol.* **55**, 73–82.

Chapman V.J. (1960) *Salt marshes and Salt Deserts of the World*. Leonard Hill, London.

Cook S.C.A., Léfèbvre C. & McNeilly T. (1972) Competition between metal tolerant and normal plant populations on normal soil. *Evolution* **26**, 366–72.

Davies M.S. & Snaydon R.W. (1976) Rapid population differentiation in a mosaic environment. III. Measures of selection pressures. *Heredity* **36**, 59–66.

Dickinson H. & Antonovics J. (1973) The effects of environmental heterogeneity on the genetics of finite populations. *Genetics* **73**, 713–35.

Edwards K.J.R. & Cooper J.P. (1963) The genetic control of leaf development in *Lolium*. II. Response to selection. *Heredity* **18**, 307–17.

Gartside D.W. & McNeilly T. (1974a) The potential for evolution of heavy metal tolerance in plants. II. Copper tolerance in normal populations of different plant species. *Heredity* **32**, 335–48.

Gartside D.W. & McNeilly T. (1974b) Genetic studies in heavy metal tolerant plants. II. Zinc tolerance in *Agrostis ternuis*. *Heredity* **33**, 303–8.

Gray A.J. (1971) *Variation in* Aster tripolium *L. with particular reference to some British populations*. Ph.D. thesis. University of Keele.
Gray A.J. (1972) The Ecology of Morecambe Bay. V. The salt marshes of Morecambe Bay. *J. appl. Ecol.* **9**, 207–20.
Gray A.J. (1974) The genecology of salt marsh plants. *Hydro. Bull.* **8**, 152–65.
Gray A.J. (1977) Reclaimed Land. *The Coastline* (Ed. by R.S.K. Barnes), pp. 253–70. John Wiley & Sons, Chichester.
Gray A.J. & Adam P. (1974) The reclamation history of Morecambe Bay. *Nature Lancs* **4**, 13–20.
Gray A.J. & Bunce R.G.H. (1972) The Ecology of Morecambe Bay. VI. Soils and vegetation of the salt marshes: a multivariate approach. *J. appl. Ecol.* **9**, 221–34.
Gray A.J. & Scott R. (1977a) The Ecology of Morecambe Bay. VII. The distribution of *Puccinellia maritima*, *Festuca rubra* and *Agrostis stolonifera* in the salt marshes. *J. appl. Ecol.* **14**, 229–41.
Gray A.J. & Scott R. (1977b) Biological Flora of the British Isles. *Puccinellia maritima* (Huds.) Parl. *J. Ecol.* **65**, 699–716.
Gregor J.W. (1930) Experiments on the genetics of wild populations. I. *Plantago maritima*. *J. Genet.* **22**, 15–25.
Gregor J.W. (1938) Experimental taxonomy. II. Initial population differentiation in *Plantago maritima* L. of Britain. *New Phytol.* **45**, 254–70.
Gregor J.W. (1944) The ecotype. *Biol. Rev.* **19**, 20–30.
Gregor J.W. (1946) Ecotypic differentiation. *New Phytol.* **45**, 254–70.
Hamrick J.L. & Allard R.W. (1972) Microgeographical variation in allozyme frequencies in *Avena barbata*. *Proc. Natl. Acad. Sci. U.S.A.* **69**, 2100–4.
Hannon N. & Bradshaw A.D. (1968) Evolution of salt tolerance in two coexisting species of grass. *Nature, Lond.* **220**, 1342–43.
Harberd D.J. (1961a) The case for extensive rather than intensive sampling in genecology. *New Phytol.* **56**, 269–80.
Harberd D.J. (1961b) Observations on population structure and longevity of *Festuca rubra* L. *New Phytol.* **60**, 184–206.
Harberd D.J. (1967) Observations on natural clones in *Holcus mollis*. *New Phytol.* **66**, 401–8.
Hinde H.P. (1954) The vertical distribution of salt marsh phanerogams in relation to tide levels. *Ecol. Monogr.* **24**, 209–25.
Jain S.K. & Bradshaw A.D. (1966) Evolutionary divergence among adjacent plant populations. I. The evidence and its theoretical analysis. *Heredity* **21**, 407–41.
Jefferies R.L. (1972) Aspects of salt marsh ecology with particular reference to inorganic plant nutrition. *The Estuarine Environment* (Ed. by R.S.K. Barnes & J. Green), pp. 61–85. Applied Science, London.
Jefferies R.L. (1977) Growth responses of coastal halophytes to inorganic nitrogen. *J. Ecol.* **65**, 847–65.
Lévèbvre C. (1974) Population variation and taxonomy in *Armeria maritima* with special reference to heavy-metal-tolerant populations. *New Phytol.* **73**, 209–19.
Linhart Y.B. & Baker I. (1973) Intra-population differentiation of physiological response to flooding in a population of *Veronica peregrina* L. *Nature* **242**, 275–76.
Mack R.N. & Harper J.L. (1977) Interference in dune annuals: spatial pattern and neighbourhood effects. *J. Ecol.* **65**, 345–64.
Mahmoud A., Grime J.P. & Furness S.B. (1975) Polymorphism in *Arrhenatherum elatius* (L.). Beauv. ex J. & C. Presl. *New Phytol.* **75**, 269–76.
McConnell G. & Fyfe J.L. (1975) Mixed selfing and random mating with polysomic inheritance. *Heredity* **34**, 271–72.

McNeilly T. (1968) Evolution in closely adjacent plant populations. III. *Agrostis tenuis* on a small copper mine. *Heredity* **23**, 99–108.

Mogford D.J. (1974) Flower colour polymorphism in *Cirsium palustre* 1. *Heredity* **33**, 241–56.

Quinn J.A. (1975) Variability among *Danthonia sericea* (Gramineae) populations in responses to substrate moisture levels. *Amer. J. Bot.* **62**, 884–91.

Ranwell D.S. (1972) *Ecology of salt marshes and sand dunes.* Chapman & Hall, London.

Schaal B.A. (1975) Population structure and local differentiation in *Liatris cylindracea. Amer. Natur.* **109**, 511–28.

Schaal B.A. & Levin D.A. (1976) The demographic genetics of *Liatris cylindracea* Michx. (Compositae). *Amer. Natur.* **110**, 191–206.

Scott R. & Gray A.J. (1976) Chromosome number of *Puccinellia maritima* (Huds.) Parl. in the British Isles. *Watsonia* **17**, 53–57.

Sharrock J.T.R. (1967) *A study of morphological variation in* Halimione portulacoides *(L.) Aell. in relation to variations in the habitat.* Ph.D. thesis. University of Southampton.

Smith A. (1965) The assessment of patterns of variation in *Festuca rubra* L. in relation to environmental gradients. *Scott. Pl. Breed. Sta. Rep.* 1965, 163–95.

Smith A. (1972) The pattern of distribution of *Agrostis* and *Festuca* plants of various genotypes in a sward. *New Phytol.* **71**, 937–45.

Snaydon R.W. (1970) Rapid population differentiation in a mosaic environment. I. The response of *Anthoxanthum odoratum* populations to soils. *Evolution* **24**, 257–69.

Snaydon R.W. & Davies M.S. (1972) Rapid population differentiation in a mosaic environment. II. Morphological variation in *Anthoxanthum odoratum. Evolution* **26**, 390–405.

Snaydon R.W. & Davies M.S. (1976) Rapid population differentiation in a mosaic environment. IV. Populations of *Anthoxanthum odoratum* at sharp boundaries. *Heredity* **37**, 9–26.

Tansley A.G. (1953) *The British Isles and their Vegetation*, Vol. II. Cambridge University Press.

Thomas R.L. (1973) Analysis of variation in growth habit of *Lolium* populations. *Ann. Bot.* **37**, 481–86.

Tiku B.L. & Snaydon R.W. (1971) Salinity tolerance within the grass species *Agrostis stolonifera* L. *Pl. Soil.* **35** 421–31.

Turesson G. (1920) The cause of plagiotrophy in maritime shore plants. *Lunds. Univ. Arsskr. Adv.* **16**, 1–33.

Turesson G. (1922) The genotypical response of the plant species to the habitat. *Hereditas* **3**, 211–350.

Tyler G. (1971) Studies in the ecology of Baltic sea-shore meadows. III. Hydrology and Salinity of Baltic sea-shore meadows. *Oikos* **22**, 1–20.

Wilkins D.A. (1959) Sampling for genecology. *Scott. Pl. Breed. Sta. Rep.* 1959, 92–6.

Williams W.T. & Lambert J.M. (1959) Multivariate methods in plant ecology. I. Association-analysis in plant communities. *J. Ecol.* **47**, 83–101.

Wu Lin & Bradshaw A.D. (1972) Aerial pollution and the rapid evolution of copper tolerance. *Nature, Lond.* **238**, 167–69.

Wu Lin, Bradshaw A.D. & Thurman D.A. (1975) The potential for evolution of heavy metal tolerance in plants. III. The rapid evolution of copper tolerance in *Agrostis stolonifera. Heredity* **34**, 165–87.

4. L'ÉQUILIBRE DES BIOCÉNOSES VÉGÉTALES SALÉES EN BASSE CAMARGUE

J.J. CORRÉ

Laboratoire de systématique et géobotanique méditerranéennes,
Institut de Botanique, 163, rue Auguste Broussonnet,
34 000 Montpellier, France

SUMMARY

The jointed glassworts (*Salicornia* spp.) dominate much of the vegetation of the lower Camargue. In a nature reserve more than 45 years old, we have tried to determine the degree of stability that such communities attain in the absence of human interference and grazing.

We have observed the zonation around the shore of a temporary lagoon over a period of 10 years. Five types of vegetation, which included a total of 8 species (*Aeluropus littoralis, Arthrocnemum glaucum, Aster tripolium, Limonium vulgare, Ruppia brachypus, Salicornia fruticosa, Salicornia radicans, Salsola soda*), were recognized.

These groupings of species reflected the following:

1 The overall effect of environmental change during this period.

2 The fact that after a change in environmental conditions the plant communities are often transient. Depending on the severity of the environmental change, fluctuations in the abundance of different species may continue for a long time until a new community is established.

The dependance of the communities upon edaphic factors, which are strongly influenced by seasonal changes in climatic conditions, suggests that we are not observing the development of communities towards a hypothetical climax. Rather, development is halted by the severity of environmental conditions, in hot and dry seasons because of the high salinity, and in cold and rainy seasons because of flooding.

These results confirm the poor homeostasis characteristic of this type of biocenosis. Periodical disturbance in environmental conditions keeps these biocenoses in a juvenile phase.

RÉSUMÉ

En Basse Camargue, la végétation dominée par les Salicornes reste soumise aux variations du milieu provoquées par l'irrégularité du climat d'une année à l'autre. Elle évolue constamment à la recherche d'un équilibre précaire que des bouleversements périodiques dans les conditions du milieu remettent en cause. Il s'ensuit que ces communautés végétales sont maintenues en permanence dans un état juvénile.

INTRODUCTION

Le contraste des milieux en Basse Camargue tient à la présence d'un système de nappes phréatiques plus ou moins salées situées à faible profondeur (Heurteaux 1969; Astier *et al.* 1970); c'est pourquoi, le moindre dénivellé, de l'ordre du cm ou du dm et l'hétérogénéité sédimentaire modifiant les caractéristiques hydrodynamiques du sol ont une influence primordiale sur la structure de la végétation.

Ces nappes présentent des fluctuations saisonnières de niveau et de salinité auxquelles les espèces végétales se sont adaptées. Le climat est le facteur principal qui en détermine le rythme. Celui-ci, de type méditerranéen subhumide (Emberger 1955) est très irrégulier d'une année sur l'autre (Fig. 4.1), aussi avons nous essayé de voir comment la végétation pouvait réagir à l'instabilité du milieu qui en résulte.

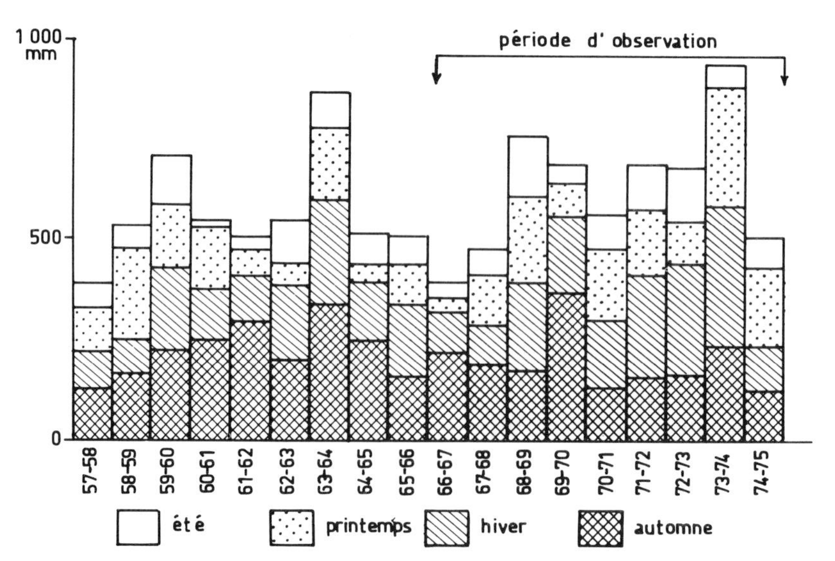

Fig. 4.1. Pluviosité saisonnière et annuelle (mm) à la Tour du Valat (Camargue) 1957–1975. Le poste météorologique est à moins de 2 km, en terrain plat, de la station étudiée.

Seasonal and annual rainfall (mm) at Tour du Valat (Camargue) 1957–1975. The meteorological station is less than 2 km across flat land from the study site.

MATÉRIEL ET MÉTHODES

Nous avons choisi à Salin de Badon dans la Réserve de Camargue un bord d'étang (Fig. 4.2) où le tapis végétal, dominé par des Salicornes vivaces (*Salicornia radicans*, *Salicornia fruticosa*, *Arthrocnemum glaucum*) y est assez représentatif des paysages de cette région (Corré 1975a, b; Molinier et Tallon 1970). Les mésures de protection qui s'y exercent depuis plus de 45 ans permettaient d'éliminer les facteurs liés à la présence de l'homme.

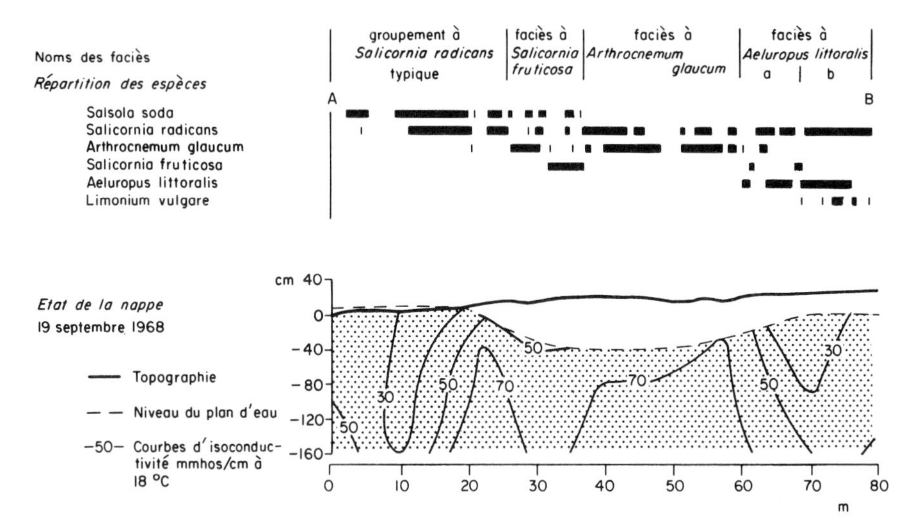

FIG. 4.2. Coupe de la bordure d'étang à Salin de Badon montrant la répartition des espèces végétales, la conductivité des eaux de la nappe aquifère (mmhos cm^{-1} à 18°C) à différentes profondeurs (zéro arbitraire) le long d'un transect de 78 m. Les observations correspondent au 19 septembre 1968.

Section of shoreline of the lagoon at Salin de Badon showing the occurrence of plant species, the conductivity of interstitial water (mmhos cm^{-1} at 18°C) in sediments at different depths along a transect line 78 m in length. Recordings were made on 18th september 1968.

Les observations ont été faites le long d'une ligne permanente, sans épaisseur, tendue perpendiculairement à la berge de l'étang entre des repères fixes. Cette ligne a été divisée en 780 segments contigus de 10 cm. De septembre 1966 à septembre 1975, deux fois par an, à la fin du printemps et pendant l'automne, on a relevé la position de chaque espèce en repérant les segments interceptés. Le nombre d'interceptions correspond pour chaque espèce à une fréquence absolue d'observations. Ses variations d'une lecture à l'autre donnent une mésure de l'évolution du couvert de l'espèce et par intégration du couvert de la végétation. Outre les 3 Salicornes précédemment citées, on a rencontré : *Aeluropus littoralis*, *Aster tripolium*, *Limonium vulgare*, *Ruppia*

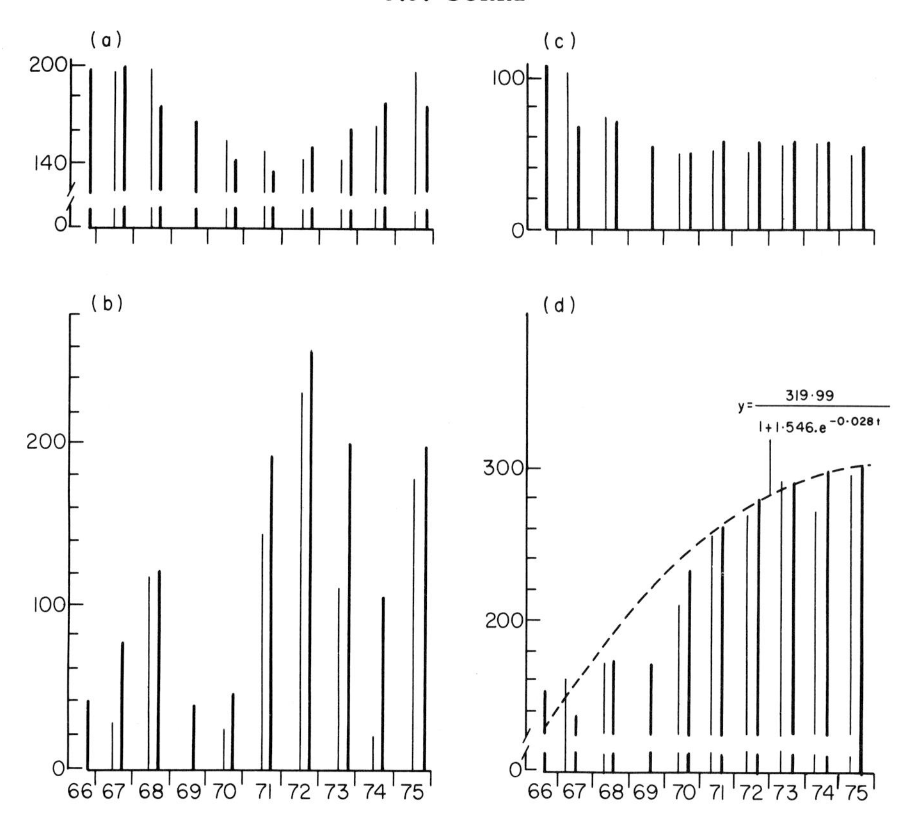

FIG. 4.3. Variation de la fréquence d'observations, à la fin du printemps (trait fin) et en automne (trait gros) d'*Arthrocnemum glaucum* (a), *Salsola soda* (b), *Salicornia radicans*, entre 0–20 m (c), *Salicornia radicans*, entre 20–60 m (d) le long d'un transect de 78 m de long observé entre 1966 et 1975. Le tireté—correspond à la courbe théorique d'extension de *S. radicans* selon l'équation indiquée sur le graphique.

Annual and seasonal variations in the number of occurrences at the end of spring (thin line) and during autumn (thick line) of *Arthrocnemum glaucum* (a), *Salsola soda* (b), *Salicornia radicans* between 0–20 m (c), *Salicornia radicans* between 20–60 m (d) along a transect 78 m in length, between 1966 and 1975. The fitted curve (—) predicts the rise in the size of the population of *Salicornia* according to the equation indicated on the graph.

brachypus, *Salsola soda*. Cet ensemble floristique se répartit en 5 biocénoses végétales.

RÉSULTATS

Les résultats bruts donnant les variations de fréquence de chaque population en fonction du temps permettent de classer les espèces en deux catégories: celles qui présentent des variations cycliques (Fig. 4.3a et b) et celles dont le caractère monotone paraît évident (Fig. 4.3c et d).

Les variations cycliques

Avec *Arthrocnemum glaucum*, la relation qui rend le mieux compte des variations de fréquence s'établit avec la pluviosité estivale. Elle est de la forme:

$$y = \frac{316{\cdot}31}{x^{0{\cdot}13636}}$$

y = fréquence absolue de l'espèce
x = pluie en mm pour juin, juillet et août
r = 0·9285 pour 4 degrés de libertés

ce qui signifie que l'extension latérale de l'espèce est réduite durant les années à été humide. Ce n'est cependant pas le seul facteur actif. En comparant (Tableau 4.1) données calculées et observées, on constate pour certaines

Tableau 4.1. *Fréquence absolue observée et calculée d'*Arthrocnemum glaucum *pour chaque année, entre 1966 et 1975, le long d'un transect de 78 m. Le taille de la population est calculée selon l'équation donnée ici. Pour des raisons expliquées dans le texte, les données sur les populations de 1966, 70, 71 et 72 n'ont pas été prises en considération dans l'établissement de l'équation.*

Observed and predicted occurrences of Arthrocnemum glaucum *for each year between 1966 and 1975 along a transect of 78 m long. The population size is predicted by the equation given above. For reasons given in the text, data of the population numbers in 1966, 70, 71 and 72 were eliminated in fitting the equation.*

Année	1966	1967	1968	1969	1970	1971	1972	1973	1974	1975
Valeurs observées (fréquences absolues)	198	199	174	165	141	135	149	161	178	173
Valeurs calculées	176	192	178	160	188	171	166	162	181	176
Écart	+22	+7	−4	+5	−47	−36	−17	−1	−3	−3

années des divergences importantes. En 1970, l'écart négatif est consécutif à une période de submersion anormalement longue qui s'est produite en 1969 et a détruit une partie de la population (Fig. 4.4). Par la suite, cet écart n'a cessé de se réduire.

La valeur très élevée observée en 1966 correspond au terme d'une période sèche pendant laquelle *A.glaucum* s'était avancé vers le centre de l'étang.

Ruppia brachypus est une espèce aquatique normalement vivace mais qui peut avoir un comportement d'annuelle puisqu'elle fructifie dès la première année. Cette particularité lui permet de coloniser les étangs temporaires, ce qui est le cas pour l'étang qui nous concerne. Elle germe à la fin de l'hiver et fleurit au printemps. Comme on peut s'y attendre, la meilleure expression de

```
               25                                                        31 m
               +        +         +         +         +        +          +
Septembre  66        IIIIIIIIIIII  IIIIIIIIIIII IIIIIIIIIIIIII        I
Juin       67           0I0IIIIII   II IIIIIIII  I0    IIIII0II     0
Septembre  67        IIIIIIIIIIIII IIIIIIIIIIIII II I IIIIIII           I
Juillet    68        IIIIIIIIIIII IIIIIIIIIIIIII IIIIIIIIIIIIII0
Septembre  68        IIIIIIIIIIII  IIIIIIIIIIIIII IIIIIIIIIIII
Septembre  69       III 00I I0III  IIIIIIIIIIIII  IIII0IIII
Juillet    70            0   0000  0   000000000  0000IIIII
Septembre  70                0000  IIII0000II0   III  IIIIII
Juillet    71                     IIIIII0I000    I I  IIIIIIII        I
Septembre  71                     IIIII 00  II  III   IIIIII          I
Juillet    72               I     I0IIIIIIIII   II IIIIIIIIIIII
Octobre    72                    IIIIIII II II  I IIIIIII IIIII III   I
Juillet    73                     IIIIII   III     III IIIIIIII   I   0
Octobre    73                    IIIIII  I IIIIII IIIIIIIIIIIII   I
Juin       74            00    IIIIIII0I I II      II IIIIIIIIII
Octobre    74            0   I 0IIIIIIIIIIIIII 00  IIIIIIIIIIIII
Juillet    75               00 II IIII IIIIIIII    IIIIIIIIIIII     II
Octobre    75                  IIIIIIII III II0  IIIIIIIIIIIIIII      II I
```

FIG. 4.4. Evolution de la répartition d'*Arthrocnemum glaucum* entre 25 et 35 m le long de la coupe représentée sur la Fig. 4.2. Les observations sont faites le long de segments contigus de 10 cm. (1) Intersection de la ligne par une partie d'un individu (0) Idem, mais le fragment est mort.
Remarquer la réduction de la zone occupée à partir de juillet 70 jusqu'en juillet 1971.

Changes in the presence of *Arthrocnemum glaucum* between 25 and 35 m along the transect line shown in Fig. 4.2. Observations were made at intervals of 10 cm along the transect. (1) presence of the plant within an interval of 10 cm. (0) As above but fragment of plant dead.
Note the mortality of plants between July '70 and July '71.

ses variations de fréquence est donnée par la durée des submersions durant sa période végétative.

$$y = 4{\cdot}14.10^{-4}.z^{3{\cdot}821}$$

y = fréquence absolue
z = durée en semaines de la submersion entre le ler mars et le 31 août
$r = 0{\cdot}9994$ pour 2 degrés de liberté

En 1970, la durée de submersion étant de 16 semaines, la fréquence attendue est de 16 à 17. La valeur observée est de 119 (Fig. 4.5). La différence tient à un changement du type biologique de l'espèce. En 1969, année très pluvieuse, l'étang ne s'est pas asséché. Bien naturellement *Ruppia* a pu alors se comporter en vivace, ce qui a transformé les conditions initiales de son développement saisonnier et lui a permis de prendre cette extension inattendue.

Salsola soda est une halophyte annuelle, hydrochore et nitrophile. Elle se développe en bordure d'étang là où l'eau a abandonné ses semences. La croissance de cette annuelle est rapide entre le printemps et l'automne, ce qui explique les variations de fréquence entre ces deux saisons.

L'année 1969, particulièrement humide, a été favorable au développement des biocénoses aquatiques. En 1970, le retour aux conditions habituelles, a entraîné une hécatombe des espèces non adaptées à une phase d'émersion.

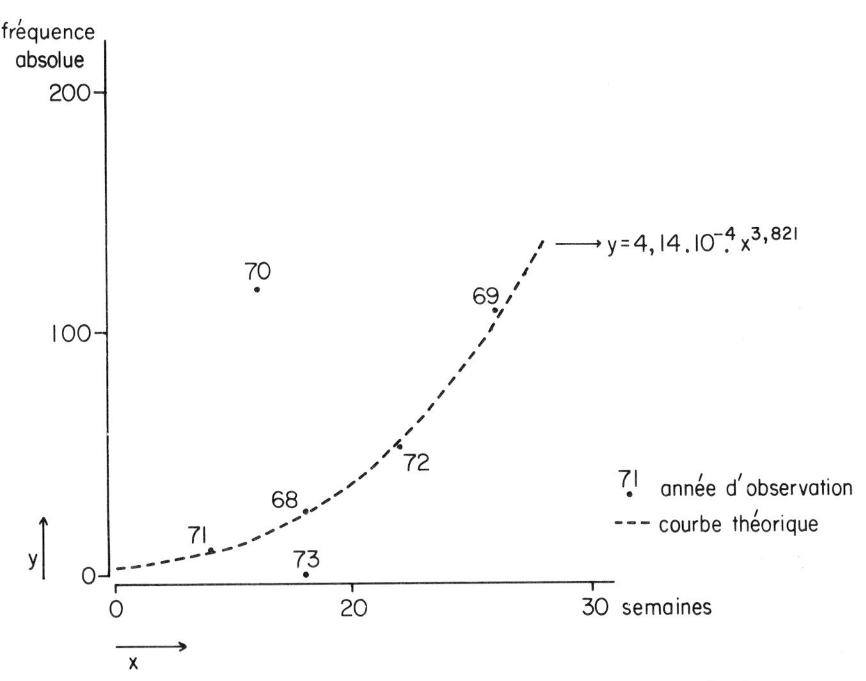

FIG. 4.5. Relation entre la fréquence absolue des observations de *Ruppia brachypus* et la durée de submersion, en semaines, entre mars et août, le long d'un transect de 78 m de long, pour la période comprise entre 1968 et 1973. La courbe en pointillé donne le nombre d'observations prévisibles d'après l'équation indiquée sur le graphique.

Relationship between the number of occurrences of *Ruppia brachypus* and the time in weeks the site was submerged between March and August from 1968 to 1973. The fitted curve of population size is predicted from the equation shown on the graph.

Les analyses d'Helwani (1971) portant sur cette période ont montré alors un accroissement de l'activité microbiologique et un enrichissement en substances azotées auquels correspond un maximum de fréquence de *Salsola soda* (Fig. 4.3b).

Les variations monotones

Le type en est représenté par *Salicornia radicans*. Cette espèce prend depuis quelques années en Camargue un développement important dans des milieux où elle passait inaperçue. A Salin de Badon, l'espèce est connue de longue date mais nous y avons remarqué ces dernières années un nouvel arrangement de son aire de répartition.

C'est une espèce de stations saumâtres périodiquement inondées où la hauteur d'eau doit être telle qu'au printemps les bourgeons végétatifs sont exondés.

Dans les parties basses de la coupe, entre 0 et 20 m, ces conditions n'ont pas été remplies en 1969 et vraisemblablement en 1967, ce qui explique la régression par paliers que les Fig. 4.3c et 4.6 mettent en évidence.

Entre 25 et 60 m, à une altitude légèrement supérieure ce facteur limitant n'intervient plus. Au contraire le cycle de plus forte pluviosité, qui a entraîné les mises en eau prolongées de l'étang, provoque un abaissement de salinité qui se ressent sur les nappes phréatiques. Les nouvelles conditions de biotope qui en résultent déterminent une poussée de croissance de *Salicornia radicans* (Fig. 4.3d). Elle se traduit par une courbe de variation de fréquence d'allure logistique :

$$y = \frac{319{\cdot}99}{1 + 1{\cdot}546 . e^{-0{\cdot}028t}}$$

y = fréquence absolue
r = 0·9549 pour 16 degrés de liberté
t = temps exprimé en mois

Au-delà de 60 m, les conditions de milieu physique et biotique changent. La salinité, en particulier, est moins-élevée. N'étant plus un facteur limitant pour *Salicornia radicans* son abaissement n'a plus la même influence sur sa population qui ici demeure stable.

Le groupement que notre coupe intercepte au-delà de 60 m comprend 5 espèces : *Salicornia radicans* déjà cité, *Aeluropus littoralis*, *Aster tripolium* *Limonium vulgare*, *Salicornia fruticosa*. Cette dernière espèce est citée pour mémoire car sa fréquence demeure faible et ses variations sont difficilement interprétables.

Hormis *Salicornia radicans*, chaque espèce présente plus ou moins tardivement une phase où sa fréquence augmente (Fig. 4.7). Pour *Aeluropus littoralis*, elle commence précocément et revêt une allure exponentielle jusqu'en 1972.

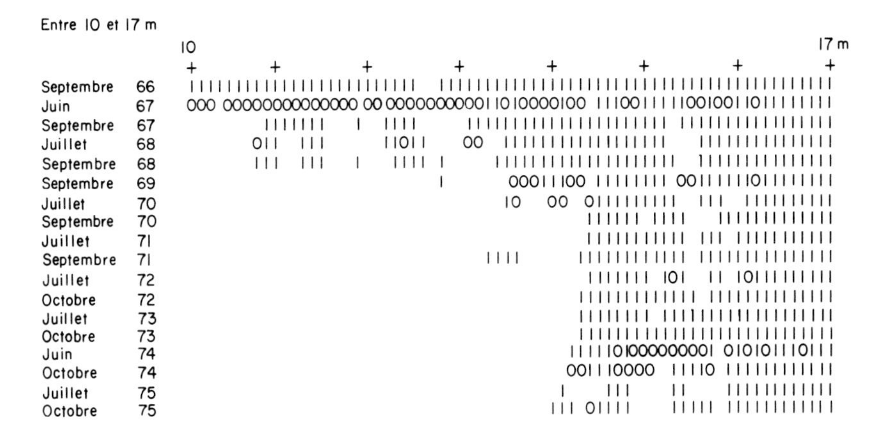

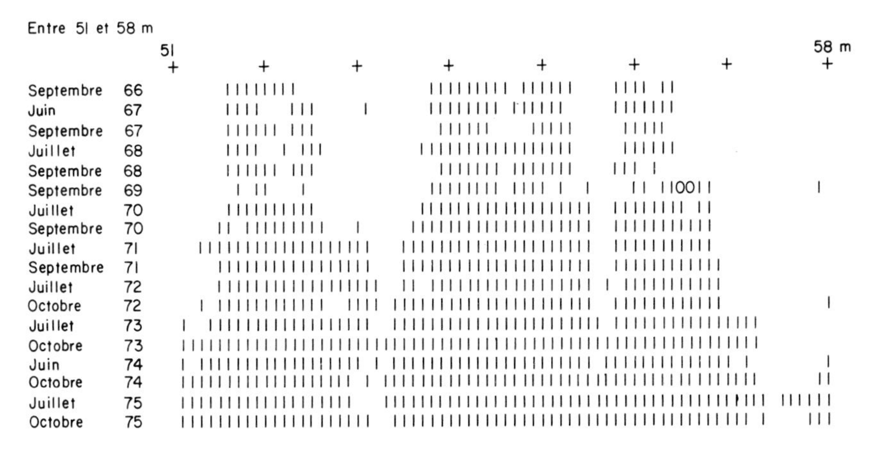

FIG. 4.6. Evolution de la répartition de *Salicornia radicans* entre 10 et 17 m et entre 51 et 58 m le long de la coupe représentée sur la Fig. 2.
Les observations sont faites le long de segments contigus de 10 cm. (1) Intersection de la ligne par une partie d'un individu. (0) Idem, mais le fragment est mort.
Remarquer entre 10 et 15 m la réduction de la zone occupée par *Salicornia radicans* et par contre son extension entre 51 et 58 m.

Changes in the presence of *Salicornia radicans* between 10 and 17 m and between 51 and 58 m along a transect line shown in Fig. 4.2. Observations were made at intervals of 10 cm along the transect. (1) Presence of the plant within an interval of 10 cm. (0) As above but fragment of plant dead.
Note that between 10 and 15 m there is a reduction in the abundance of *Salicornia radicans* compared with an increase in abundance between 51 and 58 m.

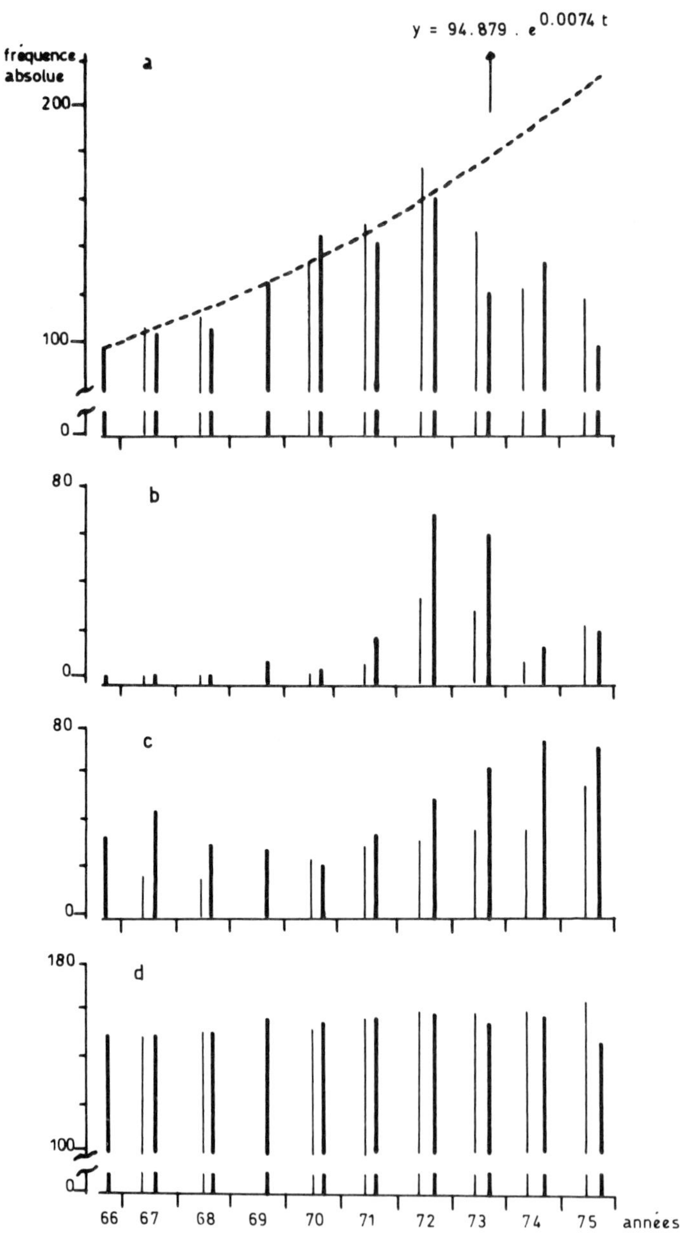

FIG. 4.7. Variations de fréquence absolue en fonction du temps des observations d'*Aeluropus littoralis* (a), *Aster tripolium* (b), *Limonium vulgare* (c) et *Salicornia radicans* au-delà de 60 m (d), entre 1966 et 1975 le long d'un transect de 78 m. La courbe en pointillé donne le nombre d'observations prévisibles d'*Aeluropus littoralis* d'après l'équation indiquée sur la figure. (Trait fin), observations de fin de printemps, (trait gros) observations courant automne.

Variations in the number of occurrences of *Aeluropus littoralis* (a), *Aster tripolium* (b), *Limonium vulgare* (c) and *Salicornia radicans* beyond 60 m (d), between 1966 and 1975 along a transect 78 m in length. The fitted curve of population size is predicted from the equation shown on the graph; (thin line) number of occurrences at the end of spring, (thick line) number of occurrences in autumn.

$$y = 94{\cdot}879 . e^{0{\cdot}0074t}$$

y = fréquence absolue
t = temps exprimé en mois
r = 0·9705 pour 10 degrés de liberté

Aster tripolium et *Limonium vulgare*, après une phase de latence qui dure jusqu'en 1970 voient également leur fréquence s'élever : *Aster tripolium* jusqu'en 1972, *Limonium vulgare* de façon apparémment logistique, compte non tenu des fluctuations saisonnières.

Le groupement a un recouvrement de 100% et il est bien évident que la concurrence interspécifique doit être vive. La baisse de salinité due à l'augmentation de pluviosité a, dans un premier temps, favorisé de façon plus ou moins précoce la croissance des différentes espèces du groupement. Par la suite, les phénomènes de concurrence ont été une contrainte imposant une limite, voir une phase de régression à certaines d'entre elles (*Aeluropus littoralis*, *Aster tripolium* après 1972).

CONCLUSIONS

Il ressort de cette étude que chaque espèce réagit à des facteurs limitants multiples selon des modalités qui lui sont propres.

Deux types de dynamique peuvent être mis en évidence. Le premier type est déclenché par une variation du milieu. L'évolution se poursuivra jusqu'à ce qu'un nouvel équilibre défini par les conditions de concurrence soit atteint. On peut penser qu'il s'agit d'oscillations à long terme même si nous n'avons pas pu, par suite de conditions climatiques défavorables, observer le retour au stade initial. Le deuxième type correspond à un contrôle beaucoup plus étroit des populations par le milieu naturel puisqu'il s'établit sur chaque cycle saisonnier. Il s'agit là d'oscillations à court terme.

Le point de départ des modifications d'équilibre du tapis végétal est bien souvent lié à l'année 1969. Elle est remarquable par sa pluviosité élevée d'autant qu'elle fait suite à une période sèche. Vue sur une plus longue période de temps, une telle situation n'est pas unique. La remise en cause périodique de l'équilibre de ces biocénoses est donc un fait habituel qui les caractérise.

En définitive, il est remarquable de constater le rôle important que continue à jouer le milieu extérieur sur des communautés d'âge pourtant relativement ancien, ce qui confirme le faible degré d'homéostasie de ce type de biocénose.

Soumises à un milieu extérieur très contraignant et toujours instable, elles conservent indéfiniment des caractères juvéniles.

REMERCIEMENTS

Ce travail a été possible grâce au soutien financier de la Délégation générale à la recherche scientifique et technique et à l'aide qu'ont bien voulu nous apporter le professeur Rioux et son équipe du laboratoire d'écologie médicale de Montpellier, la réserve nationale de Camargue, le centre d'écologie de Camargue, le C.E.P.E. Louis Emberger, M. Betaille du laboratoire d'informatique de l'Institut universitaire de technologie (Montpellier). Monsieur le professeur R.L. Jefferies a bien voulu relire le manuscrit et me faire part de ses remarques qui m'ont été très utiles dans la mise au point du texte. L'illustration a été réalisée avec l'aide de Mme Passama et de M. Barbry. Que tous veuillent bien trouver ici l'expression de ma gratitude.

REFERENCES

Astier A. et al. (1970) *Camargue. Etude hydrogéologique, pédologique et de salinité. Rapport général.* Direction départementale de l'agriculture (B. du Rhône).

Corré J.J. (1975a) *Etude phyto-écologique des milieux littoraux salés en Languedoc et en Camargue.* T 1 : texte et tabl., T 2 : figures et annexes. Thèse Doctorat, Montpellier, C.N.R.S. A0 3131.

Corré J.J. (1975b) Flore et végétation de la Réserve de Camargue. *Le Courrier de la Nature* **35**, 18–27.

Emberger L. (1955) Une classification biogéographique des climats. *Recueil des Travaux des Laboratoires de Botanique, Géologie, Zoologie de la Faculté des Sciences de l'Université de Montpellier*, Bot. **7**, 3–43.

Helwani H. (1971) *Contribution à l'étude de l'activité biologique des sols salés littoraux.* Thèse de Docteur-Ingénieur, Montpellier, C.N.R.S. A0 5930.

Heurteaux P. (1969) *Recherches sur les rapports des eaux souterraines avec les eaux de surface (étangs, marais, rizières), les sols halomorphes et la végétation en Camargue.* Thèse de Doctorat d'Etat, Montpellier.

Molinier R. et Tallon G. (1970) Prodrome des unités phytosociologiques observées en Camargue. *Bull. du Mus. d'Hist. nat. de Marseille*, **30**, 5–110.

5. THE STRUCTURE OF SALT MARSH COMMUNITIES IN RELATION TO ENVIRONMENTAL DISTURBANCES

W.G. BEEFTINK

Delta Institute for Hydrobiological Research,
*Yerseke, The Netherlands**

SUMMARY

The results of three types of disturbance on salt marsh vegetation in the south-west Netherlands are discussed. These disturbances are sudden and extreme changes in weather conditions (e.g. winter frost), and a sudden increase or a decrease in tidal influence (e.g. barrage schemes). Their effects on the vegetation of permanent plots were studied each year, using the 14-point scale estimation method.

Die-back in *Halimione portulacoides* is followed by sequential maxima in the densities of *Suaeda maritima, Aster tripolium* and *Puccinellia maritima.* Recovery of the original vegetation takes 4 to 15 years, and at least 10 to 50 years are required when the soil has been involved in the disturbance. An increase in the frequency of tidal immersion causes a regression in the development of the plant communities which is proportional to the increase in tidal height. Intermediate changes in the species composition of the vegetation similar to those mentioned above may occur. Sudden cessation of tidal cover results in the disappearance of halophytes within 1 to 7 years, and the colonization of the site by glycophytes. These are either tall herbs, or else pasture plants when the marsh is grazed. Spatial and temporal patterns in the decline of halophytes and increase in glycophytes are described. Vegetational changes in salt marshes as a consequence of disturbances can be predicted with some accuracy.

RÉSUMÉ

L'auteur discute les résultats de trois sortes de perturbation sur la végétation des salés au prés sud-ouest des Pays-Bas. Ils ont été causés par des changements du temps extrêmes et brusques (gelée en hiver) et des augmentations ou des diminutions de l'influence de la marée soudaines (système du barrage).

* Communication No. 166.

Chaque année nous avons étudié leurs effets sur la végétation par étudier des carrés permanents, en utilisant une méthode d'estimer la dominance et l'abundance des espèces conformément une échelle plus raffinée que celle de Braun-Blanquet.

Le déclin de *Halimione portulacoides* est suivie par des maxima de densité séquentiels de *Suaeda maritima, Aster tripolium* et *Puccinellia maritima.* Quand la perturbation est temporelle le rétablissement de la végétation originelle prend 4 à 15 ans, mais au moins 10 à 50 ans lorsque le sol est mêlé dans la perturbation.

Une augmentation durable de la fréquence de l'immersion par la marée cause une régression du développement des communautés végétales proportionnelle à l'augmentation de hauteur de la marée. Des changements intermédiaires de la composition des espèces de la végétation similaires à ceux mentionnés ci-dessus peuvent intervenir. Un arrêt brutal de l'immersion par la marée provoque la disparition des halophytes pendant 1 à 7 ans et la colonisation du site par des glycophytes. Le sont des herbes asser hautes, ou autrement des plantes de pré quand le marais est paturé. Les patrons spatiaux et temporels concernant le déclin des halophytes et l'angmentation des glycophytes ont été deċrits. Les changements végétals aux préssalés à cause de perturbations peuvent être prédits assez précisemment.

INTRODUCTION

Three approaches may be used to study long-term changes in the composition of vegetation (Beeftink 1977a). These are, firstly, examining the effect of sudden and extreme changes of weather on the vegetation; secondly, studying the effect on vegetation of planned technical interference, such as drainage improvements, and lastly the use of experimental methods to investigate the dynamics of vegetational changes. These three approaches have been used in this study in order to examine the responses of different plants in salt marsh communities to environmental heterogeneity. Results of such studies provide essential information for the preparation of management plans (Beeftink 1977b).

MATERIALS AND METHODS

From 1961 onwards about 500 permanent plots were established in the marshes of the estuaries of Rhine, Meuse and Scheldt in order to study vegetational changes (Beeftink 1975a). In some of these plots the effects of the different processes outlined above on the vegetation were evident. For example, the severe winter frost of 1962–63 damaged patches of *Halimione portulacoides* which were not snow-covered in Springersgors salt marsh (see Fig. 5.1).

On other plots the effects on vegetation of different agricultural practices,

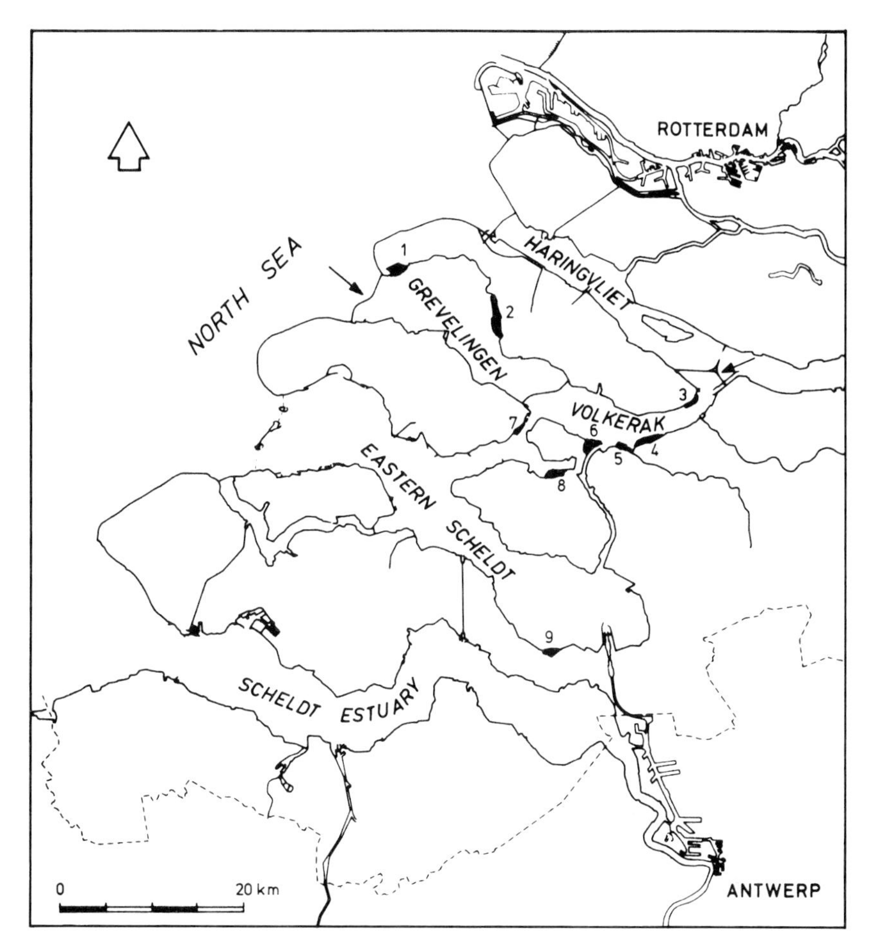

FIG. 5.1. Survey of the localities investigated. 1. Springersgors, 2. Slikken van Flakkee, 3. Weipolderse gors, 4. Drievriendenpolder, 5. Steenbergse Sas, 6. Slikken van de Heen, 7. Zijpe, 8. St. Annaland, and 9. Stroodorpepolder salt marsh.

similar to the traditional practices of marshmen have been examined (see Beeftink 1975b). These plots were established in 1964 and 1971 in the Stroodorpepolder salt marsh (Fig. 5.1).

Finally, the Delta Plan (Hydro Delft 1971) for coastal protection against flood disasters in the Rhine-Meuse estuary gave the opportunity to study the effect of barrage constructions on vegetation. A barrage built in the Volkerak in 1969 resulted in an increase in the mean high-water level of the tides of about 50 cm near the barrage to a few centimetres in the Eastern Scheldt. In 1971 a barrage was built in the Grevelingen excluding the tides from the area.

The permanent plots vary from 10 to 40 m² in size depending on the vegetation pattern. Species composition, expressed as cover and number of individuals, was estimated each year according to the 14-point scale method of Doing Kraft (1954). The complete list of values for all species at any time in one permanent plot is called a relevé following the method of Braun-Blanquet.

RESULTS

The effects of sporadic disturbances on vegetation

Plant succession following die-back of plants of *Halimione portulacoides* has been examined by Beeftink *et al.* (1978). It appeared that an orderly and predictable pattern of succession occurs, independent of the environmental factor(s) responsible for die-back, which include waterlogged conditions, severe frost, application of herbicides, and the destruction and removal of the upper layers of sediment. After these disturbances *Suaeda maritima, Aster tripolium* and *Puccinellia maritima* successively show high densities (Fig. 5.2). Minor environmental changes, such as winter frost, produce only small successional changes. The less severe the conditions, the fewer successional

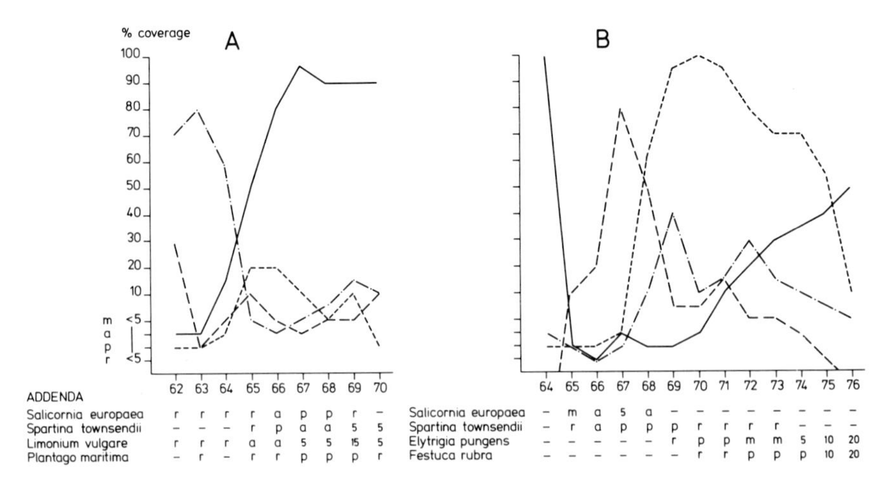

	62	63	64	65	66	67	68	69	70
Salicornia europaea	r	r	r	r	a	p	p	r	–
Spartina townsendii	–	–	–	r	p	a	a	5	5
Limonium vulgare	r	r	r	a	a	5	5	15	5
Plantago maritima	–	r	–	r	r	p	p	p	r

	64	65	66	67	68	69	70	71	72	73	74	75	76
Salicornia europaea	–	m	a	5	a	–	–	–	–	–	–	–	–
Spartina townsendii	–	r	a	p	p	p	r	r	r	r	–	–	–
Elytrigia pungens	–	–	–	–	–	r	p	p	m	m	5	10	20
Festuca rubra	–	–	–	–	–	–	r	r	p	p	p	10	20

FIG. 5.2. Population dynamics in communities of *Halimione portulacoides*. A. Changes in the composition of species on the Springersgors salt marsh after the extremely waterlogged conditions of 1961; size of plot 4·5 × 5 m. B. Changes in the composition of species in the Stroodorpepolder salt marsh after herbicide spraying in 1964, with paraquat and diquat (0·1 g/m² active ingredient of both); size of plot 2 × 6 m. Number of plants when cover is <5%: r = 1–3; p = 4–15; a = 16–40; and m = >40 individuals. (————), *Halimione portulacoides*; (— — — —), *Suaeda maritima*; (—·—·—·—), *Aster tripolium*; (-----), *Puccinellia maritima.*

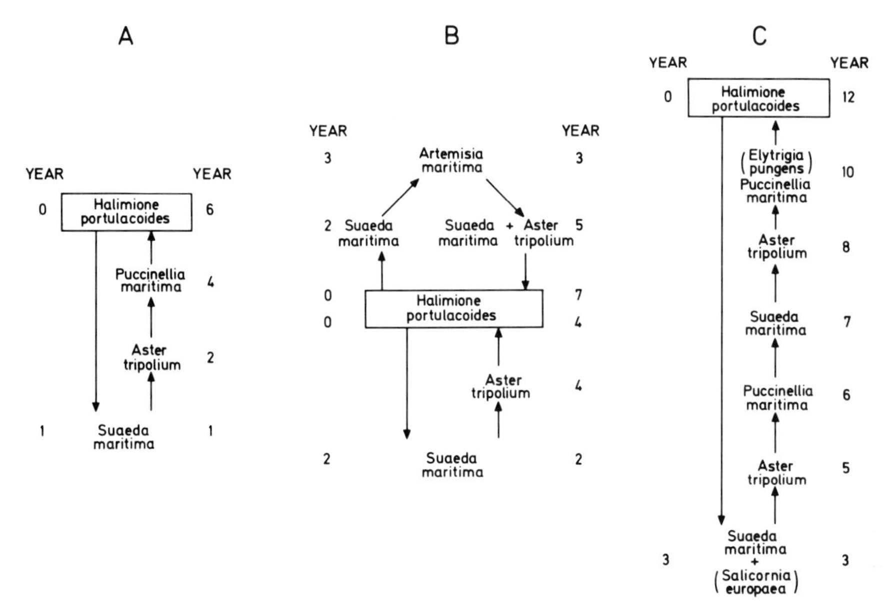

FIG. 5.3. Models of plant succession after three types of environmental disturbance in communities of *Halimione portulacoides*. A. After waterlogged conditions. B. After severe winter frost. C. After chemical destruction. The numbers indicate the years in which maximal population densities of species were observed after the year of disturbance (year zero) of *Halimione portulacoides*.

stages are represented. When the time required for recovery is long, the successional sequence outlined above shows tendencies to repeat itself, stressing the limited number of species involved (Figs. 5.2B and 5.3C).

In other salt marsh communities similar succession patterns occur. Sometimes however, changes in the species composition of the vegetation deviate from those found in the *Halimione* communities, and mostly involve shifts in the size of the population of *Salicornia europaea* or *Triglochin maritima* (Fig. 5.4). These variations in the succession pattern are caused by factors, such as:

(a) spatial and temporal differences in the pattern of seed production, seed dispersion and germination of seed of the participating species (cf. Ranwell 1972; Waisel 1972)
(b) other environmental changes in addition to those studied, both natural, for example the deposition of plant debris washed on to the research plots, and man-made, such as the effects of the barrage scheme of the Delta Plan
(c) differences in the effects of the herbicides on the vegetation as a consequence of local differences in environmental conditions and variation in the composition of the vegetation at sites.

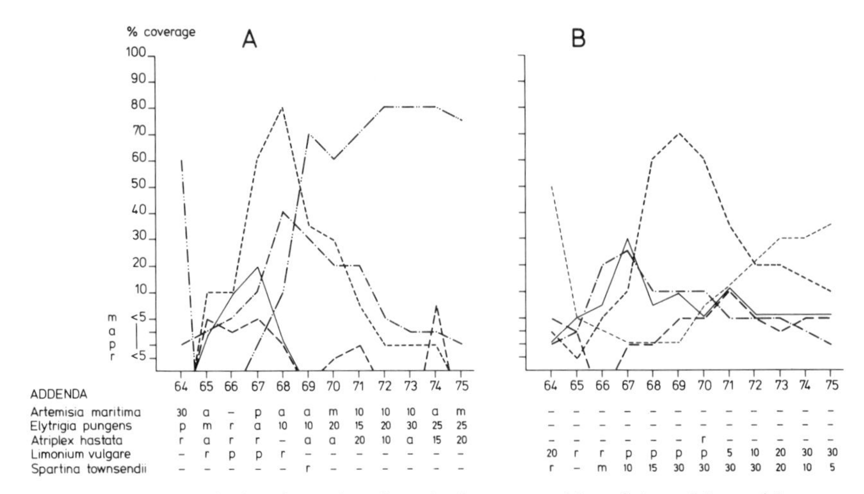

Fig. 5.4. Population dynamics of species in communities of *Artemisia maritima* (A) and *Puccinellia maritima* (B) after herbicide spraying in 1964 in the Stroodorpepolder salt marsh. Plot sizes were 2 × 6 m and 2·5 × 5 m respectively. Number of plants when cover is <5%: r = 1–3; p = 4–15; a = 16–40; and m = >40 individuals. (————), *Salicornia europaea*; (— — — — —), *Suaeda maritima*; (–·–·–·–·–), *Aster tripolium*; (– – – – –), *Puccinellia maritima*; (–· ·–· ·–· ·–· ·–), *Festuca rubra*; (-----), *Triglochin maritima*.

The effects of an increase in tidal flooding on vegetation

A sudden increase in mean high water level as a result of a changed tidal regime associated with the construction of a barrage, may induce permanent changes in the composition of the vegetation. Changes in the composition of vegetation following an increase in the duration of tidal submergence may result in the establishment of a community dominated by *Halimione portulacoides* (Beeftink *et al.* 1978). Maximum densities of populations of *Suaeda*, *Aster*, *Puccinellia* and *Halimione* occur successively. Minor increases in the mean high water level result in a longer time-lag before the populations change. Detectable changes in the sizes of the populations in response to the perturbation are absent when the increase of the mean high water level does not exceed 5 cm (Fig. 5.5). Under normal conditions most of the populations of these species probably only survive if micro-disturbances within the community allow iterative establishment.

The effect of increased flooding on the communities of the lower marsh may be compared with the effect on the communities of the higher marsh at the same locality. The increase in frequency of flooding of the creek-bank levées (higher marsh) is sevenfold, whereas in the depressions between creeks (lower marsh) it is about double. Consequently, the effect of this increase of

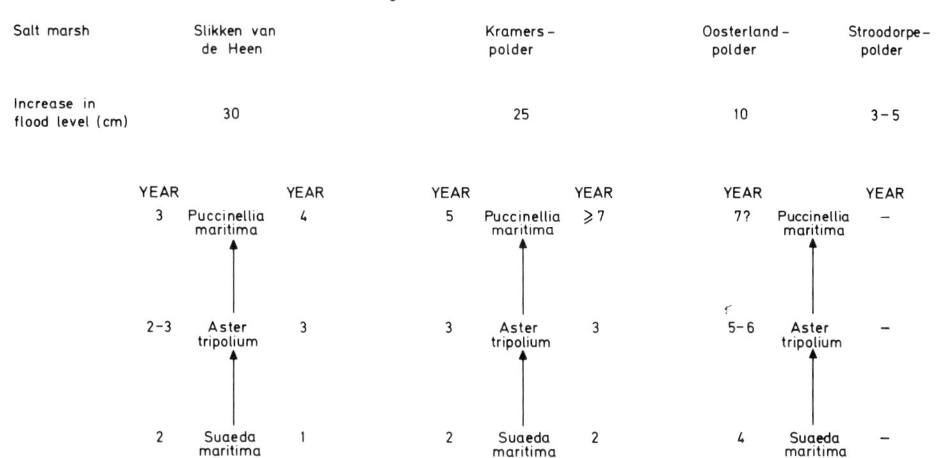

FIG. 5.5. Models of plant succession caused by different increases of the flood level (Mean High Water Mark) of tides. The numbers indicate the years in which maximal population densities of species occur after the year of disturbance (year zero) in different salt marshes.

about 30 cm in Mean High Water Mark of tides upon the vegetation is much more evident in the higher marsh than in the lower marsh.

Furthermore, it has been observed that when the disturbance persists, the extent to which the successive changes in the vegetation occur, as described above, is dependent on the frequency of flooding (Table 5.1). Very large increases in mean high water level, such as those exceeding 45 cm, result in the loss of the original community. Within one or two years the vegetation is replaced by *Salicornia stricta*, *Spartina townsendii*, *Triglochin maritima* and *Puccinellia maritima*. The final result is a regression of the original communities to communities characteristic of lower marsh (Table 5.1).

However, the original communities of the upper marsh are not equally susceptible to regression. Armerion communities, and especially those of the alliance Agropyro-Rumicion crispi which grow just beyond the permanent saline environments, are most susceptible, and are able to endure only a slight increase in the frequency of flooding induced by an increase of the mean high water level of only a few centimetres. The Puccinellion communities are able to tolerate an increase in the frequency of flooding associated with a tidal rise of about 10 cm. *Spartina* communities, and to a lesser extent *Aster* stands, are unaffected by tidal rises of up to 30–40 cm.

The time required for the development of a new community appears to differ according to the magnitude of the flood increase. After very low increases of less than 10 cm and after very high increases in excess of 35 cm in tidal height, the species composition of the vegetation generally stabilizes about 2–4 years later. Intermediate increases in the mean level of the high water

Table 5.1. *Final stages of regression of salt marsh plant communities as a result of an increase in the flood level (M.H.W. mark) of tides generated by barrage building in the Volkerak in 1969 (see arrow in Fig. 5.1). The marshes are arranged along a gradient in relation to the rise in tidal level associated with the disturbance and also in relation to the grazing regime. The table predicts whether a particular type of community will reappear after a given increase in the height of the tides has occurred.*

Salt marsh	Weipolderse gors	Drievrienden polder	Steenbergse Sas
Grazing regime	sheep	none	sheep
Increase in flood level (cm)	45	40	35
Original community			
Agropyro-Rumicion crispi (A-R)	Puccinellietum or Spartinetum	Absent	Juncetum
Juncetum gerardii with A-R elements (Armerion)	Puccinellietum with Salicornia	Puccinellietum	Puccinellietum with Armerion elements
Juncetum gerardii (Armerion)	Salicornietum strictae	Puccinellietum with Triglochin	Puccinellietum with Salicornia
Artemisietum maritimae (Armerion)	Absent	Puccinellietum with Triglochin	Absent
Atriplici-Elytrigietum pungentis	Absent	Puccinellietum with Halimione	Absent
Halimionetum portulacoidis (Puccinellion)	Absent	Absent	Absent
Puccinellietum maritimae with Armerion elements	Salicornietum strictae	Triglochin community	Absent
Plantagini-Limonietum (Puccinellion)	Absent	Absent	Absent
Puccinellietum maritimae (Puccinellion)	Absent	Spartinetum	Absent
Spartinetum communities	Absent	Spartinetum	Absent

* Estimated from data of Brouwer (1950) and from experiences in other salt marshes (Kwade Hoek).

mark of tides, however, generally result in the development of the sequence of population changes described earlier. At some sites they are still continuing at the present time, 8 years after the rise in tidal height.

The data obtained from the permanent plots can be used to produce a hypothetical model of the structure and function of salt marsh communities (Fig. 5.6). Each main community, arranged according to its general zonational position, and classified using the Braun-Blanquet system, may be restored via a series of developmental stages after disturbance. These successive stages are characterized by populations of individual species occurring at high

Table 5.1—*continued*

Slikken van de Heen none 30	Zijpe cattle 25	St. Annaland none 10	Stroodorpe polder none 3–5	Upper limit of flood increase for recovery (cm)
Absent	Armerion community	Absent	Absent	0–1 *
Absent	Juncetum with Puccinellia	Absent	Absent	1–3 *
Absent	Halimonetum with Armerion elements	Absent	Absent	3 *
Halimionetum or Puccinellietum	Puccinellietum	Halimionetum	Artemisietum with Puccinellia	3
Halimionetum with Elytrigia	Armerion community	Elytrigietum	Elytrigietum	10
Puccinellietum	Absent	Halimionetum	Halimionetum	10
Puccinellietum with Triglochin	Puccinellietum	Absent	Puccinellietum with Armerion elements	3–5
Triglochin community with Spartina	Absent	Plant.-Limonietum	Plant.-Limonietum with Puccinellia	10
Spartinetum	Puccinellietum or Spartinetum	Puccinellietum	Puccinellietum	15
Absent	Spartinetum	Spartinetum	Spartinetum	30–40

densities (Fig. 5.6, right). The species mentioned here may be considered as *r*-selected species. Halophytes which play only a minor role in post-disturbance successional events, establish only under more stable conditions and appear to be *K*-selected species (Fig. 5.6, left).

Effects of a decrease in the frequency of tidal flooding on vegetation

Another type of disturbance is that which takes place when a salt marsh is cut off from the sea by a barrage and the tidal influence abruptly disappears. Such disturbances lead to desiccation, desalinization, and increased aeration

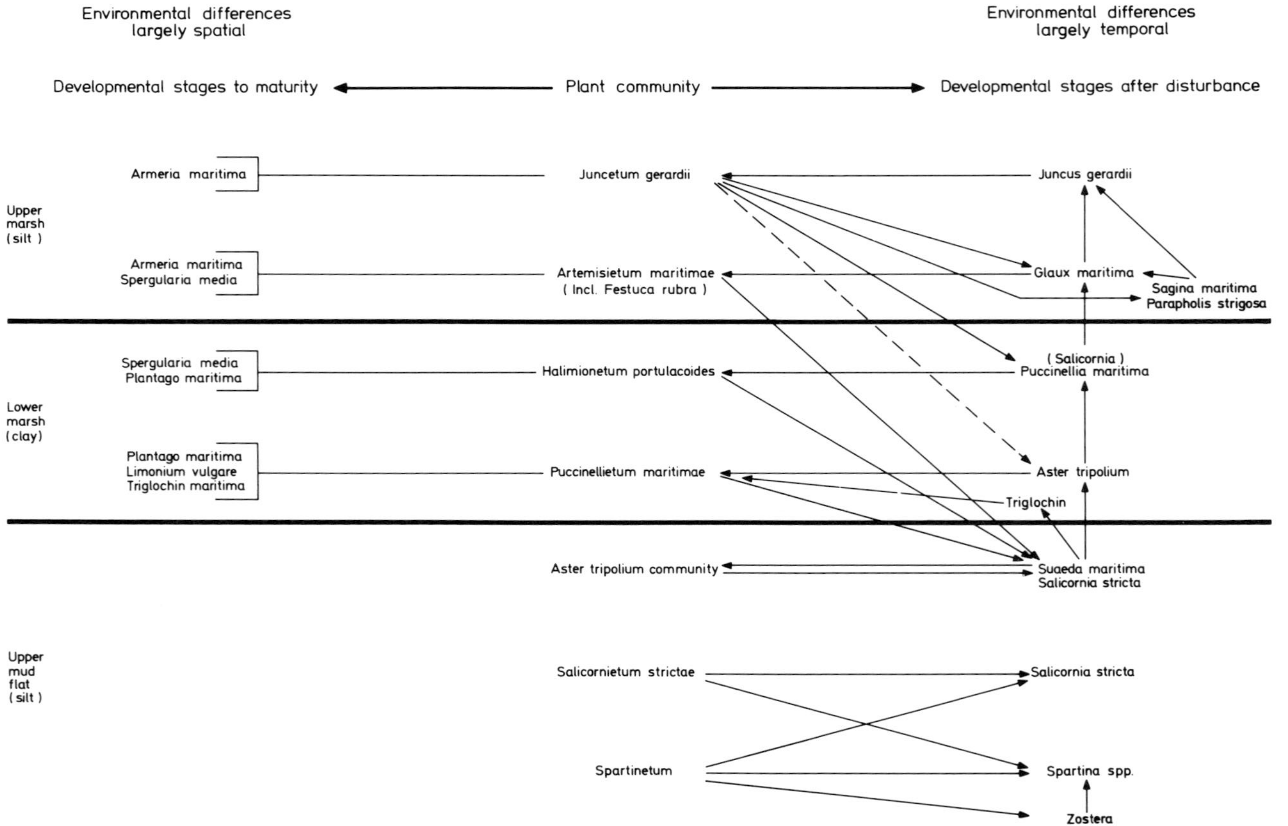

FIG. 5.6. Interrelationships between plant species and communities in the salt-marsh ecosystem, a hypothetical model.

of the soil, as well as an increased rate of mineralization of organic matter (Beeftink *et al.* 1971). The original halophytic plant community disappears in about one to seven years, and is replaced ultimately by glycophytic communities.

The first example is taken from studies in permanent plots established in the Springersgors salt marsh before (1962–69) and after (1971–76) the barrage had been built. From the permanent plots established in 1962 four plots were selected from each of the following community groups as representatives of the communities available: the *Puccinellia/Limonium*, the *Halimione/Limonium* and the *Festuca/Artemisia/Elytrigia* communities. In the successive relevés of each permanent plot, the association values of each species were compared with those obtained in other permanent plots. These analyses were based on a factorial design using a two-way analysis of variance (see Winer 1970, p. 228). Differences in affinity of nearly all species in these permanent plots indicated that most species were restricted to one or two of the three community groups. The only exceptions were *Suaeda* and *Glaux maritima*, where the values were not significant. All species did not show significant differences in affinity between communities during the period 1962–69, in spite of extensive frost damage to some species such as *Halimione portulacoides* in the winter of 1962–63. The data indicate a high affinity of the species to one or two communities (spatial constraints). In addition there was a marked continuity of the species in these communities over a number of years (temporal independence), indicating a high spatial diversity of species as reflected by these different communities, and a relatively high stability of the ecosystem.

Six representatives of the *Spartina/Aster*, the *Puccinellia/Limonium* and the *Halimione/Limonium* communities were selected from the plots during the period 1971–76 after the barrage had been built. Using the same kind of analysis of variance, the data indicate that the salt marsh species responded differently from the previous analysis. In Fig. 5.7 the relationships between the levels of significance of differences between species in space and time have been plotted, and the results indicate the degree to which the species show spatial and temporal patterns of abundance after the barrage had been built. A species group (*Suaeda maritima*, *Halimione*, *Spergularia*, *Puccinellia* and *Glaux*) can be recognized as appearing highly significant with respect to the three community types, and thus remaining spatially constrained. Another group appeared to be not significant (*Salicornia*, *Limonium*, *Aster*, *Parapholis*, *Elytrigia* and *Triglochin*). Similarly, Fig. 5.7 allows us to identify a group of species which is highly significant with respect to time (years) following barrage building (species from which $P_{\text{years}} < 0{\cdot}01$), and another group which is not significant. Populations of the first group generally experienced a sharp decline in numbers while those of the latter remained relatively stable. These two approaches allow the grouping of four spatial-temporal combinations of species, as the dashed lines in Fig. 5.7 indicate.

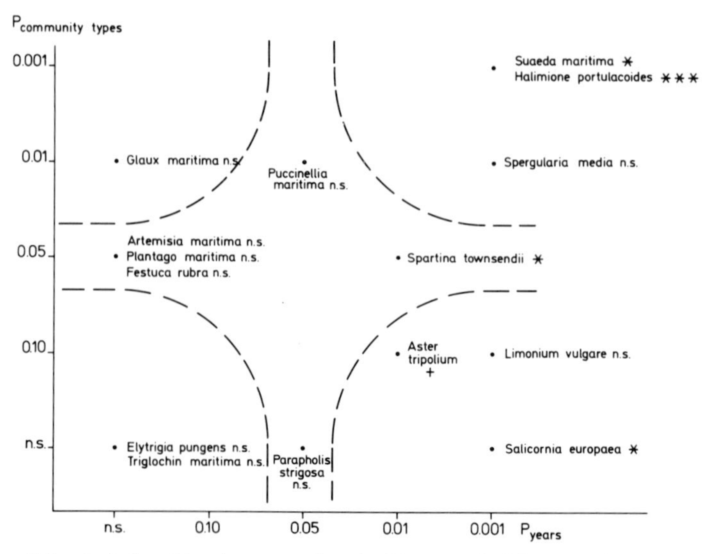

FIG. 5.7. Relationships between the significance of differences in space and time of halophytes after isolating the Springersgors salt marsh from the tides as a result of the construction of a barrage in the Grevelingen in 1971 (see arrow in Fig. 5.1). The significances of spatial differences and temporal changes were estimated using a two-way analysis of variance (Winer 1970). $P_{\text{community types}}$ is the level of significance at which the species show affinity to one or two of the three community groups mentioned in the text. P_{years} is the level of significance at which the species show pattern in time. Significance level of the interactions between community types and years has been included after the species names: ***, $P < 0{\cdot}001$; **, $P < 0{\cdot}01$; *, $P < 0{\cdot}05$; +, $P < 0{\cdot}10$; n.s., non significant.

Although *Suaeda* and *Glaux* did not show any affinity to one or two of the three community groups selected before the barrage has been built, these two species did do this after the construction of the barrage. These two species showed affinity with the *Spartina*/*Aster* and the *Puccinellia*/*Limonium* communities. The data demonstrated that *Glaux* temporarily increased in the *Puccinellia*/*Limonium* communities after the barrage construction. In that period *Suaeda* confined itself more to communities of the lower marsh, and increased in these communities for a time. Although the sets of data before and after the barrage had been built are not quite comparable, they show that the distribution of many other species changed as well. There was a sudden loss of species in a zone where previously they were dominant (e.g. *Salicornia*, *Spartina*, *Aster*, *Triglochin*, *Plantago*). Temporary increases in the population densities of some species in lower marsh occurred where previously these species were a minor component of the vegetation (e.g. *Glaux*, *Festuca*, *Elytrigia*). The increase of *Limonium* in the communities of the lower marsh (Fig. 5.7) compared with that in the pre-barrage period is a reflection of the fact that it is a member of the *Spartina*/*Aster* community rather than the *Festuca*/*Artemisia*/*Elytrigia* community.

Temporal changes in the rates of loss of populations of species varied considerably. There was a sharp decline of the population size of some species (e.g. *Salicornia*, *Spartina*, *Suaeda*, *Aster*), whereas in other species the rate of disappearance was slower (e.g. *Artemisia*, *Triglochin*, *Spergularia*, *Halimione*, *Plantago*). *Limonium* populations showed a gradual decrease. Species showing a temporary rise in their population densities after isolation of the marsh from tidal influence were *Puccinellia maritima*, *Glaux maritima* and *Festuca rubra*. *Elytrigia pungens* has increased continually since the establishment of the barrage. The discrepancy between the gradual decrease in *Limonium* and its high significance ($P_{years} < 0{\cdot}001$) in Fig. 5.7 is only apparent, as originally the species grew at very high densities in the marsh. Although its decline is highly significant, the species will be present for many years before it becomes extinct.

The same statistical procedures were used to examine the communities of glycophytes (Fig. 5.8). These glycophytic species entered the permanent plots after isolation of the salt marsh from the tides. In the absence of grazing the vegetation develops towards tall herb communities in which *Epilobium* spp., *Sonchus* spp., *Cirsium arvense*, *C. vulgare*, *Urtica dioica*, and *Solanum dulcamara* are present. It is thought that woodland will develop on these sites

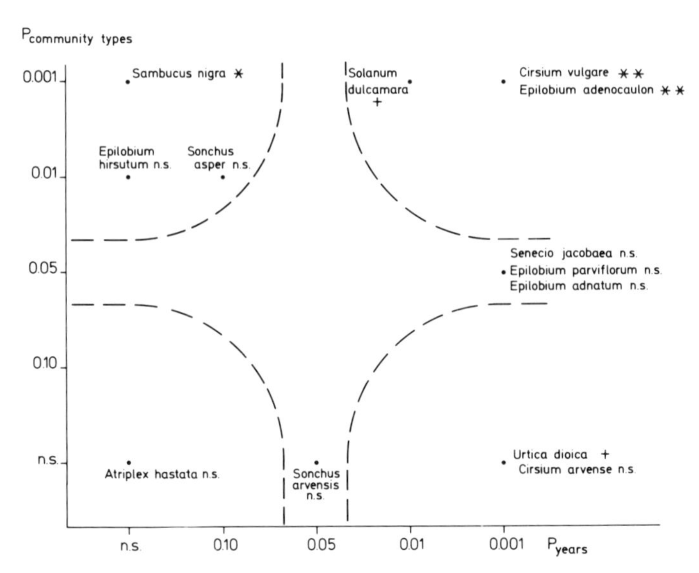

FIG. 5.8. Relationships between the significance of differences in space and time of glycophytes entering the Springersgors salt marsh after it was isolated from the tides by the building of a barrage in the Grevelingen in 1971 (see arrow in Fig. 5.1). For further explanation see Fig. 5.7.

Table 5.2. *Trends in plant succession in the Slikken van Flakkee salt marsh, expressed as sequence in peak densities of plant populations in the successive years after isolation from the tides as a result of barrage building in 1971.*

Year after barrage building	Original community Spartinetum townsendii	Puccinellietum maritimae	Halimionetum portulacoidis	Juncetum gerardii	Festuca rubra community with Elytrigia pungens
1		Suaeda maritima	Suaeda maritima	Plantago maritima	
		Salicornia europaea	Salicornia europaea	(Armeria maritima)	Glaux maritima
			(Spergularia media)		
2	Suaeda maritima	Puccinellia maritima	Puccinellia maritima	Spergularia media	Elytrigia pungens
			Suaeda maritima		
3	Atriplex hastata	Puccinellia maritima	Plantago maritima	Festuca rubra	Elytrigia pungens
			(Spergularia media)		Agrostis stolonifera
	Poa trivialis	Poa trivialis	Festuca rubra		
	Cerastium holosteoides	Plantago major	Plantago major		
4	Poa annua	Cerastium holosteoides	Cerastium holosteoides	Festuca rubra	Agrostis stolonifera
		Cirsium arvense	Agrostis stolonifera		Cerastium holosteoides
		Poa annua			
	Agrostis stolonifera	Poa trivialis	Poa trivialis	Trifolium repens	Poa trivialis
5		Trifolium repens	Trifolium repens	Cerastium holosteoides	Trifolium repens
		Trifolium fragiferum	Festuca rubra		
		Poa pratensis	Cerastium holosteoides		
	Lolium perenne	Lolium perenne	Lolium perenne	Lolium perenne	Lolium perenne
6	Agrostis stolonifera	Poa pratensis	Poa pratensis	Poa pratensis	Poa pratensis
	Trifolium repens	Stellaria media	Trifolium repens	Poa trivialis	Stellaria media
	Trifolium fragiferum		Trifolium fragiferum	Stellaria media	Cirsium arvense

eventually. Besides the differences in spatial and temporal patterns of the occurrence of species illustrated in Fig. 5.8, species show differences in the time at which maximum density of the population is observed. Some rise and fall 2–4 years after the barrage has been constructed (e.g. *Atriplex hastata, Solanum dulcamara, Sonchus asper*); others have a time-lag of 4–6 years before they increase (e.g. *Cirsium vulgare, Epilobium adenocaulon, E.parviflorum, E.adnatum, Sonchus arvensis*) and others are still increasing (*Urtica dioica, Epilobium hirsutum*). Other species show a combination of these characteristics. *Sambucus nigra* has a short germination period combined with local differences in the growth of the plants so that the growth is not uniform across the marsh. Temporal differences in the establishment of plants and an increase and decrease in population size are shown by different populations of some species (e.g. *Cirsium arvense, Senecio jacobaea*).

The second example is taken from permanent plot studies in the Slikken van Flakkee salt marsh (Table 5.2). Because this site was grazed by cattle even when the marsh was tidal, pasture plants were already present in the low embanked parts of the marsh and as well as the higher unembanked parts, and were locally dominant. Table 5.2 demonstrates that in the first years after the isolation of the site from the sea, some halophytes were abundant or even dominant, especially the following: *Suaeda maritima, Salicornia* spp., *Puccinellia maritima, Festuca rubra* and *Elytrigia pungens*. Later pasture plants took their place, especially *Lolium perenne, Poa trivialis, P.pratensis, Trifolium repens* and *Stellaria media*. The data suggest that after six years the newly established vegetation is much more uniform with respect to its species composition than the original salt marsh vegetation. This is of particular interest because the geomorphological features of the marsh have been maintained, and even reinforced by wind-borne accumulations of sand and silt blown over from the desiccated mud flats.

CONCLUSIONS

The present investigations have examined the effects of three types of disturbance on the vegetation of salt marshes in the south-west Netherlands. These are as follows:

(1) disturbances which do not result in a change of tidal conditions (e.g. severe winter frost);
(2) disturbances which lead to a sudden increase in tidal influence (e.g. barrage schemes);
(3) disturbances which result in a sudden decrease in the height of the tides (e.g. barrage schemes, embankments).

The vegetation generally recovers within 4 to 15 years after disturbances of the first type. During recovery the population densities of *Suaeda*, *Salicornia*, *Aster* and *Puccinellia* show sequential maxima. Dicks (1977) found similar results in the recovery of vegetation after the addition of oil to a Southampton salt marsh for twenty years.

Where the disturbances involve modifications to the soil, such as treading, grazing, cutting sods and digging the sediments, the vegetation generally does not recover its original composition and structure. The degree to which the vegetation is modified is highly dependent on the character of the disturbance itself. Soil conditions only recover gradually over a period of 10 to 50 years, and the original vegetation may return only after a number of developmental stages have occurred.

A sudden increase in the frequency of tidal immersion causes a regression towards those communities which are characteristic of earlier stages of the development of salt marshes. The extent of the regression is proportional to the increase in the tidal height. When the increase in the height of the tide fails to exceed about 45 cm intermediate changes in the composition of species on the marsh similar to those described earlier occur.

A sudden cessation of tidal cover causes the disappearance of the original vegetation within 1 to 7 years. The halophytic vegetation will be superseded by glycophytic vegetation of mainly tall herbs when the marsh is not grazed, and mainly pasture grasses when grazing takes place. In the early stages after this type of disturbance has occurred, some nitrophilous halophytes reach high densities for a short time. Differences in the way in which halophytic populations disappear reflect specific survival strategies of the populations.

It is concluded from this study that vegetational changes which take place when disturbances occur in salt marshes can be predicted with some accuracy, although ecological and physiological explanations of these changes are lacking.

ACKNOWLEDGMENTS

The author is much indebted to Dr A.G. Vlasblom and Mr H.J. Blok for processing data and providing statistical advice and to Dr R.L. Jefferies (Toronto) and Dr K.F. Vaas (Yerseke) for reviewing the English text.

REFERENCES

Beeftink W.G. (1975a) Vegetationskundliche Dauerquadratforschung auf periodisch überschwemmten und eingedeichten Salzböden im Südwesten der Niederlande. *Sukzessionsforschung* (Ed. by W. Schmidt), pp. 567–78. Cramer, Vaduz.

Beeftink W.G. (1975b) The ecological significance of embankment and drainage with respect to the vegetation of the South-West Netherlands. *J. Ecol.* **63**, 423–58.

Beeftink W.G. (1977a) The coastal salt marshes of Western and Northern Europe: An ecological and phytosociological approach. *Wet Coastal Ecosystems* (Ed. by V.J. Chapman), pp. 109–55. Elsevier, Amsterdam.

Beeftink W.G. (1977b) Salt marshes. A contribution to our understanding of its ecology and physiography in relation to land-use and management and the pressures to which it is subject. *The Coastline* (Ed. by R.S.K. Barnes), pp. 93–121. Wiley, London.

Beeftink W.G., Daane M.C. & Munck W. de (1971) Tien jaar botanisch-oecologische verkenningen langs het Veerse Meer (Ten years of phytoecological research on the borders of Lake Veere). *Natuur en Landschap* **25**, 50–63.

Beeftink W.G., Daane M.C., Munck W. de & Nieuwenhuize J. (1978) Aspects of population dynamics in *Halimione portulacoides* communities. *Vegetatio* **36**, 31–43.

Brouwer G.A. (Ed.) **(1950)** Griend. Het vogeleiland in de Waddenzee (*Griend, the bird-island in the Wadden Sea*). Nijhoff, The Hague.

Dicks B. (1977) Changes in the vegetation of an oiled Southampton Water salt marsh. *Recovery and Restoration of Damaged Ecosystems* (Ed. by J. Cairns, K.L. Dickson & E.E. Herricks), pp. 208–40. University of Virginia Press, Charlottesville.

Doing Kraft H. (1954) L'analyse des carrés permanents. *Acta Bot. Neerl.* **3**, 421–24.

Hydro Delft (1971) The Delta Project, halfway to completion. *Hydro Delft*, Nr. 23/24, 1–15.

Ranwell D.S. (1972) *Ecology of salt marshes and sand dunes.* Chapman and Hall, London.

Waisel Y. (1972) *Biology of halophytes.* Academic Press, New York.

Winer B.J. (1970) *Statistical principles in experimental design.* McGraw-Hill, London.

6. THE DEMOGRAPHY OF SAND DUNE SPECIES WITH CONTRASTING LIFE CYCLES

A.R. WATKINSON,[1] A.H.L. HUISKES[2] AND J.C. NOBLE[3]

School of Plant Biology,
University College of North Wales,
Bangor, Gwynedd, U.K.

SUMMARY

Sand dunes provide relatively simple and structured habitats in which the biological significance of different life cycles and growth forms can be analysed. In particular sand dune species can often be found in *r* and *K* phases of population development. Examples will be presented to illustrate the demographic analysis of annual and clonal perennial plants growing on the Anglesey dune system.

Five short-lived annuals have specialized life cycles and at least one of them has been shown to be subject to density-dependent regulation. The performance of individuals can be largely accounted for (69% of reproductive variance) as a function of distance from nearest neighbours. The absence of seed longevity is a feature of these species.

The demographic study of shoot systems in *Ammophila arenaria* and *Carex arenaria* reveals the high flux in the turnover of shoot systems that is disguised by conventional census methods. Application of fertilizers dramatically increases both birth and death rates in shoots of *C.arenaria*. The dynamics of the shoot population of *Ammophila arenaria* and of *Carex arenaria* appear to depend on the dynamics of the accumulated banks of meristems, the dormancy of which is controlled by the interaction of internal (correlative) inhibition and external (e.g. nutrient) factors.

RÉSUMÉ

Les dunes de sable présentent des habitats relativement simples et structurés dans lesquels nous analysons la signification biologique des différents cycles

[1] Present address: School of Biological Sciences, University of East Anglia, Norwich.

[2] Present address: Delta Institute for Hydrobiological Research, Vierstraat 28, Yerseke, The Netherlands.

[3] Present address: CSIRO Division of Land Resources Management, Deniliquin, N.S.W. 2710, Australia.

de vie et formes de croissance. Dans certaines dunes de sable on peut trouver des espèces dans les phases *r* et *K* de développement de population. Nous présentons des exemples pour illustrer les analyses démographiques de plantes annuelles et de plantes vivaces clonales du système de dune de Anglesey.

Cinq espèces à courte durée de vie ont spécialisé leur cycle de vie, et au moins l'une d'entre elles est sujette à une régulation dépendant de la densité. La performance des individus dépend largement (69% de la variance reproductible) de la distance des plus proches voisins. L'absence de longévité des graines est une propriété de ces espèces.

L'étude démographique des systèmes de pousses chez *Ammophila arenaria* et *Carex arenaria* révèle un flux important dans le «turnover» des systèmes de pousses, ce qui est masqué par les méthodes de recensement conventionnelles. L'utilisation des engrais augmente dramatiquement à la fois les taux de naissance et de mort des pousses de *C.arenaria*. Les dynamiques de population des pousses de *Ammophila arenaria* et *Carex arenaria* semblent dépendre des dynamiques d'accumulation des méristèmes dont la dormance est contrôlée par l'interaction de l'inhibition (corrélative) interne et les facteurs (e.g. nutriments) externes.

INTRODUCTION

Sand dune systems offer a specially suitable site for studying the ecological significance of the life cycles and growth forms of plants. They are appropriate because, unlike most other habitats, they provide sites which are permanently in a state of succession. They combine all the special interests of a successional sequence yet at the same time, because the process of dune formation is continuous, contain the earliest phases of the succession as a permanent feature of the area. These special features make them especially fascinating research areas. There are, however, dangers in extrapolating from sand dunes to general ecological theory because they are unusual. The consequence of the process of dune formation is that all stages of the succession (except sometimes the most advanced) are present all the time and, therefore, for any population of plants established at a site, new sites suitable for colonization are reliably nearby. This situation is fundamentally different from classical successional sequences in which colonization of new early stages depends on major episodes of dispersal. Neither dispersal of propagules in space, nor in time (buried long lived seed) have the same importance as in the life cycle of most successional species.

This paper is concerned with an analysis of the demography of plants which illustrate three contrasting life cycle patterns—the short lived annuals (e.g. *Vulpia fasciculata*), the perennial which is capable of extensive vegetative growth (*Carex arenaria*), and the tussock forming perennial (*Ammophila *arenaria*). Each of these life cycle patterns is considered as a strategy of meristem replacement and proliferation.

The procedure that has been adopted in these studies is both demographic and experimental. The demographic approach involves asking questions about numbers (numbers of shoots, genets, buds, seeds) and attempting to quantify the flux of numbers that occurs in natural populations. To this end it is essential to count units and to establish the parameters of population behaviour such as 'birth rates', 'death rates', 'expectation of life' and, so far as possible to discover the mechanisms that determine numbers and rates of change. The experimental approach involves deliberate perturbation of populations and study of their response. It is unlikely that causal relationships in ecology can be firmly established without altering the system. In the present studies the perturbation involved the application of nutrients, controlled alterations of plant density and the removal of species growing with the one under study.

The dunes on which the studies have been made are on the west coast of Anglesey at Aberffraw and Newborough Warren, and have been described in detail by Pemadasa (1973) and Ranwell (1958, 1959, 1960). There are three main dune ridges and the remains of an older ridge at Aberffraw and four main ridges at Newborough Warren which run approximately parallel to the coast with interdunal slacks. The dunes are relatively immobile and are formed from calcareous windblown sand—mainly silica with shell fragments (Pugh *et al.* 1974; Davis 1975). They are built and stabilized primarily by the growth of *Ammophila arenaria*, and the mobile dunes are dominated more or less entirely by this species. The inland dunes are more fixed, more stable and carry a more complete vegetational cover. The diversity of species is also greater, and *Festuca rubra* and *Carex arenaria* are particularly important on these more inland dunes.

THE LIFE CYCLES AND GROWTH STRATEGIES OF SAND DUNE ANNUALS

Annuals are generally restricted to the drier areas of the semi-fixed and fixed dune systems and are very rare or absent from wet stands in the slacks. Generally the abundance of annuals is negatively correlated with the vegetation cover of perennial species. These are the main conclusions from Pemadasa's careful ordination (Pemadasa 1973; Pemadasa *et al.* 1974) of the vegetation in the Aberffraw dunes. It would appear that annuals are present where drought prevents a full cover of perennials from developing: they are opportunists. Most of the annuals on the dunes at Aberffraw and Newborough Warren are winter annuals, and these include a number of grasses such as *Aira caryophyllea*, *A.praecox*, *Mibora minima*, *Phleum arenarium* and *Vulpia fasciculata* (= *V.membranacea*) and a number of forbs including *Cerastium atrovirens*, *Erophila verna* and *Saxifraga tridactylites*. All germinate in autumn, particularly October and November, and flower early in spring. The timing of flowering in *Mibora minima* is especially precocious. Pemadasa and Lovell (1974a) showed that there is a well defined flowering sequence as shown below.

Mibora minima	Late January–late March
Cerastium atrovirens	Late February–late April
Aira praecox	Mid April–early June
A.caryophyllea	Mid May–late June
Vulpia fasciculata	Early June–mid July

The species differ in quite subtle ways in their requirements for flowering. All require low temperatures ($<10°C$) for floral induction and high temperatures (10–20°C) and long days for rapid flower production. However, none of them has an absolute vernalization requirement, and *Vulpia fasciculata* is the only obligatory long day species. The *Aira* species flower more rapidly under long than short days, and *Cerastium atrovirens* and *Mibora minima* become day neutral once they have been vernalized. It would appear that the precise timing of flowering of the different species is determined by their differential response to an interaction of temperature regime and day length.

A very characteristic feature of the dune annuals in the field is their usually depauperate form compared with the vigour that they realize in culture. Many of the grasses tiller vigorously in fertile soil but bear only a single shoot in the field. Moreover some of the tillers may carry only one green leaf at a time (this is often also the case for the shoots of the perennials *Ammophila arenaria* and *Carex arenaria*). As a new leaf unfolds an older leaf dies and the process continues until the inflorescence is produced. At this stage the last leaf dies and the inflorescence is left as the only active photosynthetic organ. However, if the same plants are grown in a garden compost, leaves live longer and individual tillers commonly bear several green leaves even at the stage of seed ripening.

One possible reason for the short life of the leaves in nature is the mineral impoverished conditions of the dune system, particularly for an annual which from the starting nutrient capital in the seed has to accumulate in 2–6 months of slow growth the increased mineral capital necessary for distribution among the seeds that are produced as progeny. The plant behaves in the field as though one or more of the minerals available are insufficient to support more than one tiller and one live leaf and, as the plant develops, the limited nutrient capital is shunted from one leaf to another and finally to the inflorescence. In many ways the annual habit appears inappropriate for habitats which are nutrient deficient. In contrast to perennials in which a long pre-reproductive period may permit resources to be accumulated before a first episode of flowering, an annual must pre-empt from the environment in a single season the resources needed to convert a parent seed into progeny. Some plants of *Vulpia fasciculata* mature but fail to leave even a single seed and very many produce only one or two seeds. Similarly, many plants of *Cerastium atrovirens* may ripen only one capsule that contains only 3–4 seeds. Perennial plants, however, may devote a long juvenile period to a slow process of growth and accumulate nutrient resources over several years before any of these are distributed and hazarded in seed to subsequent generations.

The annuals in the sand dune complex are in a sense the feeders on the crumbs from what is a very poor man's table; their life cycle is fitted into the season of low light intensity and temperature, but when water is relatively abundant. They occupy sites bare of perennials. Nevertheless, they are highly responsive to increased fertility and grow with considerable speed and vigour in fertilized patches, for example near rabbit latrines. It is only in such sites that the plants attain anything near the size and vigour that they can show in culture.

Although the dune annuals are very small they interfere with each other's growth in a striking fashion. Mack and Harper (1977) sowed the seed of *Cerastium atrovirens*, *Mibora minima*, *Phleum arenarium*, *Saxifraga tridactylites* and *Vulpia fasciculata* in flats of dune sand. The seed was sown at random to give pure populations of each species at two densities, all the possible mixtures of two species and in addition mixtures of all five species. The plants were grown to maturity and the dry weight and seed production were measured. During the development of the seedlings, maps were prepared of the precise position of each individual. This made it possible to define the position of each plant in relation to its neighbour. Neighbours were defined in concentric circles around each plant and regressions were calculated between a summed measure of the weight of the neighbour population and the dry weight and reproductive output of the reference plant. Up to 69% of the dry weight and reproductive performance of plants could be accounted for simply in terms of the number, arrangement and species of neighbours within 2 cm. The remainder of the variation between plants was presumably due to variations in the time of seedling emergence, environmental gradients in the experiment and genetic variability. The method also gave a precise ordering of the aggressiveness of the species which was in the sequence *Vulpia* > *Phleum* > *Mibora* ⩾ *Cerastium* > *Saxifraga*. This order of aggressiveness is a rather direct function of the length of the growing season of the species. The tiny, fast maturing *Mibora minima* has a scarcely detectable influence on neighbouring plants. *Vulpia fasciculata*, although a very small plant, is larger relatively than the other annuals and more completely dominates them, presumably through competition for nutrients as at no stage in the experiment was there any indication of significant interference between the plants above ground. No comparable experiment has been made to include perennials in a comparison of neighbour influences, but from field observations and the experiments of Pemadasa and Lovell (1974b) it seems likely that species such as

Festuca rubra would occupy a yet higher position in the 'pecking order' of aggressiveness. The longer that an individual genet lives, the greater is its opportunity to pre-empt nutrient resources and to hold them against the demands of other individuals and the larger is the area from which it depletes the available nutrients.

THE DEMOGRAPHY OF *VULPIA FASCICULATA*

It is also possible to study the neighbour effects among annuals in natural populations though precise mapping is then much more difficult, and it is more realistic to look at neighbour effects generalized and expressed as effects which occur at different population densities. Watkinson and Harper (1978) manipulated natural field densities of *Vulpia fasciculata* on the fixed dunes at Aberffraw and Newborough Warren by deliberately adding excess seed to some plots and thinning the seedlings in others. In this way it was possible to obtain field densities varying from 100 to 10,000 plants per 0·25 m². The negatively density-dependent relationship between reproductive output and the density of flowering plants in the field is shown in Fig. 6.1. There was no

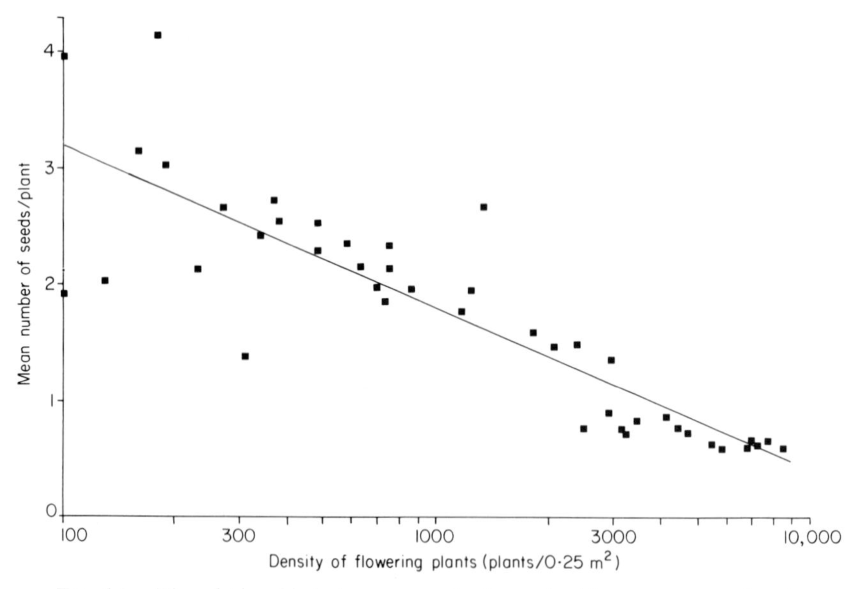

FIG. 6.1. The relationship between reproductive output (mean number of seeds per flowering plant) and the density of flowering plants of *Vulpia fasciculata* (plants per 0·25 m²) on several plots at Aberffraw, Anglesey (after Watkinson & Harper 1978).

evidence for any density-dependent mortality. Measurements of natural seed production were also made, and the fate of dispersed seed was followed by labelling the seeds with the radioactive tracer scandium-46 applied in nail varnish (Watkinson 1978a). A portable contamination monitor with a gamma probe was then used to track the seeds in the field. Permanent quadrats were mapped to obtain data on the fate of seedlings (Watkinson & Harper 1978), and these data together with those on the fate of seeds allowed the natural survivorship curve to be determined (Fig. 6.2). On average each plant of *V.fasciculata* produced 1·7 mature seeds of which 90% germinated, and 69% of the seedlings survived to flowering. The shape of this survivorship curve is markedly different from the concave or positively-skewed survivorship curves of species like *Spergula vernalis* (Symonides 1974a, b), which have a high seed output and a very high death risk. *Vulpia fasciculata* is somewhat like the small desert annuals (Went 1973) and *Cerastium atrovirens* (Mack 1976) that also have a very low seed output but a high probability of survival.

Combining the field estimates of density-dependent control of reproduction with the measured density-independent mortality was sufficient to account for the observed range of densities found in the natural field populations of *Vulpia fasciculata* on the fixed dunes (Watkinson & Harper 1978). This was an important finding because only for a very few species, e.g. the great tit, *Parus major* (Krebs 1970) has it been possible to extract from field data the necessary elements to permit the behaviour of populations to be predicted. This study appears to be the first example among plants. It is of some interest that the present densities of *V.fasciculata* on the Anglesey dunes are at levels that can be interpreted as self-regulating. In the past however the species was rare and it seems to have reached its present very common state on the Anglesey dunes after myxomatosis caused the rabbit population to crash. A.J. Willis (pers. comm.) has reported that *V.fasciculata* increased over 10,000 fold at Braunton Burrows in the years after myxomatosis.

Although the density of rabbits on the dunes at Aberffraw and Newborough Warren is now very low, compared with pre-myxomatosis levels, the intensity of predation by rabbits, which varies from year to year, is still an important element in the population dynamics of *V.fasciculata*. Rabbits graze the developing inflorescences of plants prior to ear emergence and also take complete, young inflorescences from flowering plants. The grazing of plants is, however, selective as the rabbits graze a larger proportion of the plants with several tillers and large spikes, which are commonly the individuals that germinate early (Watkinson 1975; Harper 1977). Thus high fecundity in *V.fasciculata* increases the risk of predation. The plants that are safest from predation are the depauperate single tillered forms. It appears that the fittest plants in the absence of the rabbit are those that germinate early, grow fast and produce several tillers and a large inflorescence. In the presence of the rabbit these same features are a disadvantage.

Populations of annuals depend, by definition, on seed to maintain a reserve of meristems from year to year; there is no other 'bank' of dormant buds. In most annual plants the seed has a phase of innate dormancy and in the dune annuals this in part prevents seed germination between seed shed and the late autumn (Pemadasa & Lovell 1975). Most annual plants (Harper 1977)

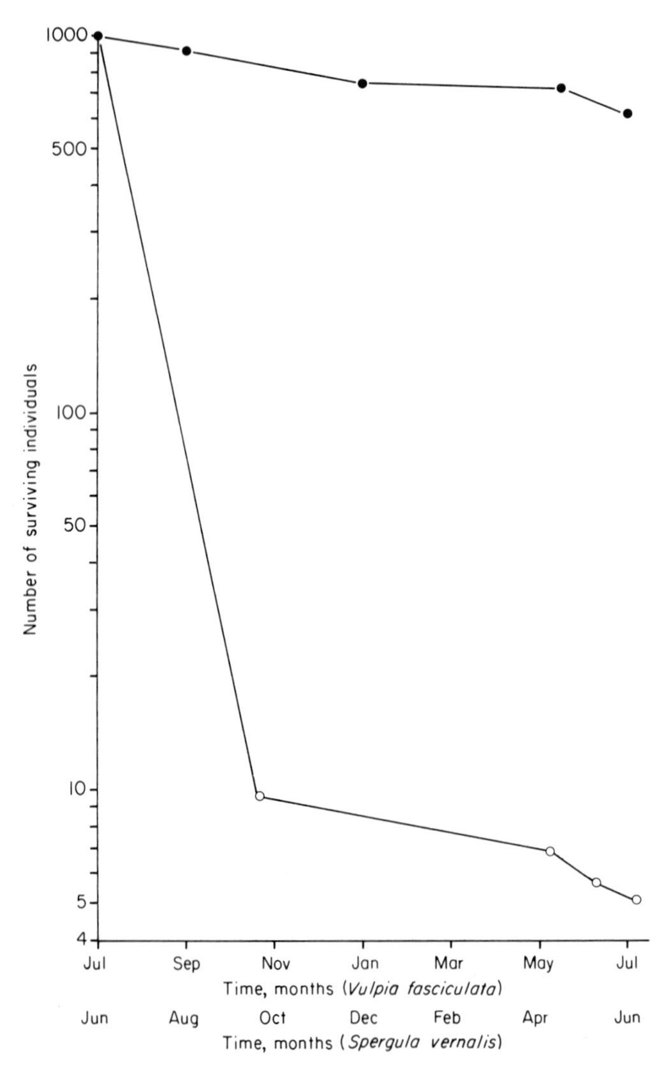

FIG. 6.2. Survivorship curves for natural populations of two species of dune annual from seed production to maturity. ● *Vulpia fasciculata* (from Watkinson & Harper (1978)) on fixed dunes in Anglesey; ○ *Spergula vernalis* (calculated from data of Symonides (1974a, b)) on dunes in the Toruń Basin, Poland.

also possess long-lived seeds and these accumulate to form an often dense population of seeds in the soil (the 'seed bank'). It is very surprising to find that the dune annuals lack this feature. Studies by Mack (1976) of *Cerastium atrovirens* showed that no buried seed populations were present after the main flush of germination in the field. Studies of scandium labelled seed of *Vulpia fasciculata* (Watkinson 1978a) showed that the fate of released seed could be accounted for almost wholly by germination or predation and that viable seed did not accumulate as a 'bank' in the soil. The extent of post-dispersal predation of seed varied quite markedly between the studied sites. At Aberffraw a maximum of 6% of the seeds were eaten before germination and in many plots there was no loss to predators at all. At Newborough Warren between 2 and 15% of the seeds were lost to predators. The presence of long-lived buried seed in most annual species is interpreted as a safeguard which has allowed individuals to leave descendants even if conditions during one year or season are wholly lethal to the growing population. The ability of arable weeds to persist through various rotational operations for their removal depends greatly on the buried long lived seed bank. The fact that there is no such seed bank in the dune annuals presumably implies that the dunes offer stable, reliable, safe habitats. Most of the dune annuals also have remarkably low reproductive capacities, and their potential intrinsic rate of natural increase r (e.g. after disasters) is therefore low. However, all viable seeds of dune annuals germinate in the autumn following their production and this means that the maximum rate of increase is gained from the limited fecundity. Seeds that remain dormant over several life cycles, such as those of the annuals of arable fields, represent a loss to the intrinsic rate of natural increase. This is compensated presumably by the stability that it gives to populations in environments that are temporally patchy (see discussion in Harper 1977).

THE DEMOGRAPHY OF *AMMOPHILA ARENARIA* AND *CAREX ARENARIA*

The main vegetation of the Anglesey dunes is perennial and overwhelmingly dominated by rhizomatous plants. The only significant woody species on the dune area is *Salix repens* and this is also 'rhizomatous'. The characteristic feature of the rhizomatous growth habit is that single genetic individuals (genets, each the product of a single zygote) may form expansive clones which both regenerate and proliferate from rhizome buds (Tomlinson 1974). Individual genets, growing clonally, may in theory form extensive stands and all derive from one individual zygote. It remains unknown how far the extensive communities formed by such species as *Ammophila arenaria* and *Carex arenaria* are single genets. There are few obvious genetic markers that can be used to distinguish genets, and the use of electrophoretic enzyme typing may be the most effective way at present to answer this question. There is however some evidence that stands of *Ammophila arenaria* may be genetically diverse. Huiskes (1977a, b) observed dense populations of seedlings at the foot of a dune slope and followed their fate for three years. At the end of that period a dense stand of *Ammophila* had developed to flowering and there were as many as 36 genets present per square metre (Fig. 6.3). It may be that, over the

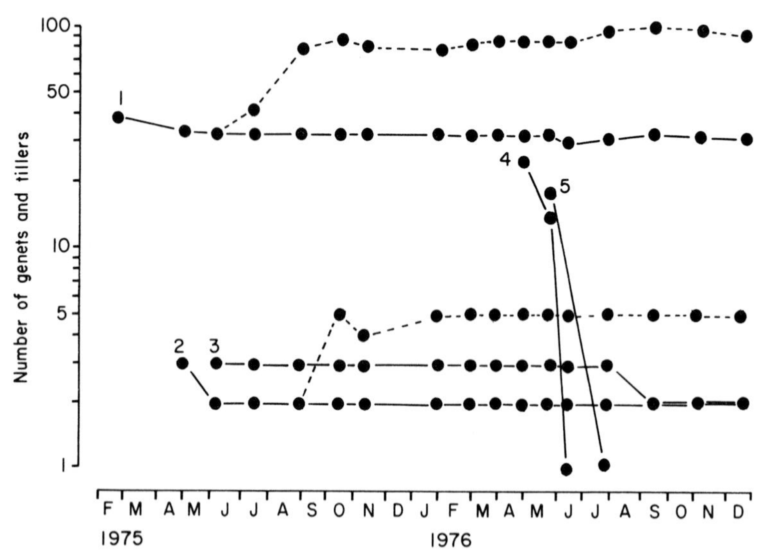

FIG. 6.3. Survivorship curves of successive cohorts of seedlings of *Ammophila arenaria* established naturally at the foot of a dune slope. Continuous lines (——) represent numbers of genets; broken lines (---) represent numbers of aerial shoots (from Huiskes 1977a).

years, one vigorous genet comes to dominate the population. It is important to know whether natural stands of marram grass are single clones or mixtures of genets, because this could determine optimal policy in the use of the species for reclamation. At present the species is generally planted as rooted tillers that have multiplied clonally, and it may be that deliberate genetic diversification involving rearing populations from seed, perhaps with some selection, would be a safer policy if natural populations are genetically diverse.

The growth habit of rhizomatous perennials involves the production of aerial shoots of relatively short life from a rhizome of longer life (see account in Harper 1977). The development of every new aerial shoot involves the loss of a bud from the rhizome system. At the same time the aerial shoot and the forward growing rhizome may add new buds to the subterranean bank of dormant meristems. In *Carex arenaria* for example, each aerial shoot contributes one basal bud that replaces it in the 'bud bank'. The sympodial rhizome grows forward to produce new shoots at intervals of four internodes (this pattern is rigid) and each of these shoots in turn contributes one basal bud to the 'bank' of meristems. In populations of *C.arenaria* at the base of retreating yellow dunes the number of viable dormant rhizome buds present may be as high as 1,400 per square metre and may exceed the number of aerial shoots by 11:1. Clearly the expectation of life of a dormant bud is much longer than that of aerial shoots for such a ratio to develop. There is an important sense in which the essence of such a plant is all below ground—the aerial shoots are evanescent, photosynthetic, reproductive appendages.

Any study of perennial vegetation that is based on repeated counts of shoots or estimates of cover or biomass must underestimate the flux that occurs in the birth and death of the shoots and their parts. Noble (1976) has studied the flux of shoot modules in *Carex arenaria* in a series of quadrats representative of the phasic development of this species at both Aberffraw and Newborough Warren. *C.arenaria*, like many other clonal plants, forms clearly defined growth phases. These are (a) a juvenile phase (the marginal belt or leading edge of a clone), (b) an adolescent phase with the density of shoots approaching the carrying capacity of the site, (c) a mature phase in which the population has attained the maximum carrying capacity and most shoots have reached a terminal (monocarpic) flowering state, (d) a senile phase with a sparser shoot population and (e) a slack phase which is equivalent to Watt's hinterland phase in bracken communities (Watt 1947). Ahead of the juvenile phase is a zone in which rhizomes extend beyond the 'front' of above ground shoots. In the juvenile phase the ratio of above ground shoot to rhizome rises steeply with increasing distance from the leading edge, reaches a peak in the early adolescent phase and then falls off as the proportion of biomass present as rhizomes increases steadily into the mature, senile and slack phases.

Shoots in the mature, senile and slack phases of development were mapped within replicate 0·5 m^2 quadrats and their fate was followed by repeated mapping at monthly or two-monthly intervals over two years. The population censuses showed a significant seasonality in the number of shoots present throughout the year with low numbers in winter and high numbers in summer. Over the period of study the sizes of the populations in the same month of succeeding years remained relatively stable but this apparent stability concealed a very rapid flux of births and deaths (Fig. 6.4). A remarkable feature of this flux is the close synchrony of the seasonal cycle of births and deaths. This suggests that the two processes are interlinked and are either the result of a common cause or one is the causal factor of the other. The probability of death for a tiller is greatest when new tillers are developing. It may be that the death of tillers creates the conditions suitable for the birth of new tillers.

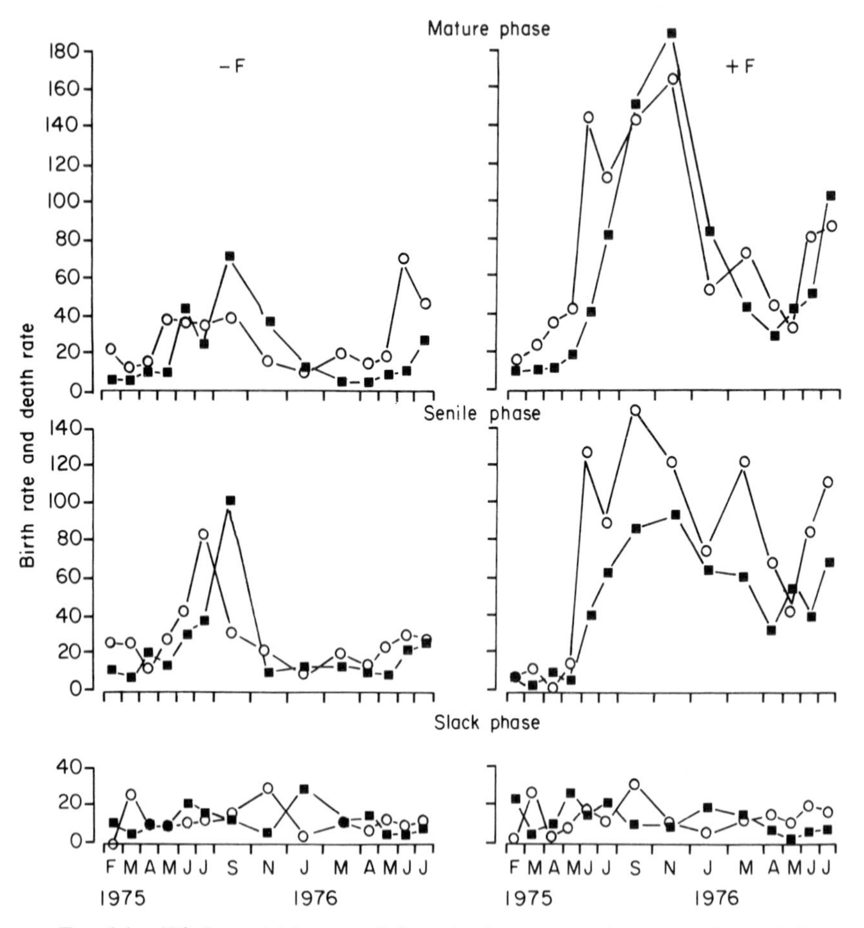

FIG. 6.4. 'Birth rate' (shoot modules gained every month or every 2 months) and death rate (shoots lost every month or every 2 months) in populations of *Carex arenaria* measured over time in the mature, senile and slack phases of development at Aberffraw (from Noble 1976). −F = control; +F = fertilized treatments; ○ = birth rate; ■ = death rate.

Reduced shading, reduced demand for nutrients or reduced correlative inhibition of the dormant buds may act as triggers that activate new tiller development. Alternatively the birth of new tillers may make demands on resources in old tillers, which die when these resources are withdrawn. The latter interpretation is closely analogous to the progressive transfer of nutrients from leaf to leaf in developing annuals discussed earlier. It is tempting to imagine tiller density approaching the carrying capacity, *K*, for a site, and at this density the growth of every new tiller demanding resources from and so the death of an old tiller.

When nitrogen, phosphorus and potassium in the form of a combined fertilizer were applied to a series of stands of *C.arenaria* in the various stages of the phase continuum the effect was to increase the number of shoots present per square metre, but only in those sites where *C.arenaria* was the monopolist inhabitant. Where other species were abundant (e.g. in the slacks) *C.arenaria* reacted sluggishly, if at all, to nutrient application. *C.arenaria* behaved as though it had received little or any of the nutrients whereas the vigorous growth of the associated species suggested that it was they that had absorbed the applied nutrients. Most dramatic was the very rapid increase in birth and death rates of shoots in all but the slack population (Fig. 6.4). This was a far larger response than was the corresponding change in shoot density. One manifestation of the change in flux was that the expectation of life of a tiller was much reduced. Again the effect was most strongly expressed in the sites where other species were sparse (e.g. mature and senile phases) and only weakly expressed in the species-diverse slack phase.

One obvious effect of the stimulation of tiller births and deaths was to change the age structure of the population, which became dominated by young tillers after the application of fertilizer. A second effect was that the bank of dormant rhizome buds was temporarily depleted. Each new shoot, in its time, contributes a new bud to the bank and restores the balance, but the immediate effect of fertilizer is to cause a striking reduction in the population of dormant buds. For example, the population in the plots of the senile phase of *C.arenaria* at Newborough Warren carried approximately 1,400 dormant buds per square metre, whereas in the corresponding fertilized plots the number fell to 1,080. The ratio of aerial shoots to dormant buds in the mature phase was 1:2, and in the fertilized plots was 1:1. Correspondingly in the senile phase the shoot: dormant bud ratio in the control plots was 1:11 and 1:1·5 in the fertilized plots. Apparently the application of fertilizer is responsible for breaking bud dormancy. This is probably a nitrogen effect and is a phenomenon that has been observed in other rhizomatous perennials, such as *Agropyron repens* (McIntyre 1965). Wareing (1964) described the 'late spring' dormancy of *Agropyron repens* buds in North America and attributed this to a depletion of labile nitrogen in rhizome tissue brought about by the spring flush of top growth.

It cannot be overemphasized in the context of this study, that the effects of the perturbation on the flux of shoots are far greater than on the number of shoots that form the standing crop. The application of fertilizer increases the death rate of shoots nearly as much as it increases the birth rate. However, the age structure and hence the expected physiological performance of the population are altered. The populations in the mature and senile phases behaved as though the numbers of shoots were not far from some limited carrying capacity. The application of heavy doses of fertilizer raised this carrying capacity somewhat but much more profoundly altered the pace of turnover of growth modules in the system. There were other effects of fertilizer application. Fertilized plots tended to suffer more than controls from grazing by rabbits and where other species were present, conspicuously in the slacks, it was they that made the major visible response to applied nutrients.

A demographic study of the leaf and shoot dynamics of *Ammophila arenaria* at Newborough Warren (Huiskes 1977a) involved a somewhat similar experimental design to that employed for *Carex arenaria*. Plots were laid out on a series of dune sites including mobile dunes, in which *Ammophila* was the only species present, semi-fixed dunes, grey dunes and intervening slacks. A series of plots aligned parallel to the controls, received either (a) nitrogen, phosphorus and potassium as a combined fertilizer application or (b) sand, repeatedly applied to simulate natural sand accretion. In a third treatment (c) the above ground parts of associated species were repeatedly removed in an attempt to decrease interspecific competition. The emphasis in this study was laid on the demography of leaves, rather than of tillers. This different procedure was followed because of the very different phenology and morphology of *Ammophila* compared with that of *Carex arenaria.* Only a small proportion of the tillers of *Ammophila* flower (normally no tillers flower in populations in the slacks) whereas a tiller of *C.arenaria* that survives into its second year normally flowers and dies. Tillers of *Ammophila* may apparently have a longer indeterminate life. Several essential features of the demography of *C.arenaria* are repeated in the behaviour of *Ammophila.* The birth and death rates of leaves are strongly synchronous through the season; the period of rapid leaf birth is the period of greatest death risk. Births and deaths are rare in winter. Fluxes in the populations of leaves are highest in the populations that (judged on the frequency of flowering) appear to be the most vigorous, and in dune slacks where associated species are present in abundance, the flux is low. The effect of applying fertilizer is to increase the rate of turnover of leaves but to a significant extent only in the areas where other vegetation is sparse.

Ammophila arenaria shows little response to applied nutrients in those slacks where it intermingles with other species such as *Festuca rubra*, although the associated species with more superficial root systems show a vigorous growth response. In experimentally established mixtures of *Ammophila* and *Festuca rubra* in 50 cm high pots of dune sand, *F.rubra* was aggressive and suppressed the growth of *Ammophila* when nutrients were applied to the sand surface. However, the species were much more evenly balanced in mutual aggressiveness when nutrients were applied to the bottom of the pots. The experiment was designed as a replacement series (de Wit 1960) and from the data it was possible to calculate the Relative Crowding Coefficients and the Relative Replacement Rates (de Wit & van den Burgh 1965) of the species in the mixtures. The R.C.C. of *F.rubra* towards *Ammophila* was 2·88 when nutrients were applied to the sand surface but 0·95 when nutrients were applied to the bottom of the pots. The corresponding values of the R.R.R. were 3·45 with surface application of nutrients and 2·06 for bottom application. The results from these experiments and the field data suggest that *Ammophila* suffers when other vegetation with surface roots can concentrate nutrients in the upper soil layers, thereby eventually depriving the deeper rooting *Ammophila* of nutrient resources.

The bud bank of *Ammophila* is relatively small and patchy in comparison with that of *Carex arenaria*, and most of the buds are found in the upper layers of the sand immediately underneath the clumps of aerial shoots. *Ammophila* has two distinct rhizome systems, and the aerial shoots are borne on the long vertical rhizomes which develop from the horizontal system. Rather few buds on the horizontal system form vertical rhizomes but there is extensive development of buds on the vertical system leading to the formation of the characteristic tussocks of this species (Greig-Smith *et al.* 1947; Gemmell *et al.* 1953). The bud bank on the horizontal rhizomes is sparse and the buds seldom develop. As a result the aerial clumps are distinct, and if they are damaged there is only a very small potential bud bank for the regeneration of new tussocks. The fact that few tillers of *Ammophila* flower each year (even in the most vigorous populations less than 10% of the tillers flower each year) gives individual shoots an expectation of life rather longer than those of *C.arenaria*. The longevity of shoot clusters and particularly the ability of the plant to elongate its vertical rhizomes quickly in response to sand accretion, give *Ammophila* its ability to stabilize sand. In contrast the horizontal growth of *C.arenaria* restricts it in the dune sequence to stable and particularly eroding dunes. However, *Ammophila* appears to bring about its own demise as a result of its ability to stabilize the sand because its habitat is then vulnerable to invasion by other species which cannot tolerate sand accretion. These successors, including *Festuca rubra*, concentrate recycled nutrients in the surface layers of the soil and *Ammophila* reacts by slowing its growth, slowing the flux of births and deaths, failing to flower and eventually declining in shoot density and height. On this interpretation, the vigour of *Ammophila* on the mobile, yellow dunes is accounted for largely by the failure of other species to maintain themselves under accreting sand.

These conclusions are based on perturbation experiments which involved the removal of associated species and the addition of nutrients and sand. Willis *et al.* (1959) reported that the water content of sand at depths between 60 and 90 cm, in dry seasons, was lower beneath dry dune pastures than beneath *Ammophila* on high dunes, and suggested that these differences might be due to the rapid absorption and exploitation of moisture by the shallow rooted species of the dune pasture. It would be interesting to carry out a series of perturbation experiments with water, similar to those reported here, to see if the decline in vigour of *Ammophila* in the older dunes is associated with a decline not only in the availability of nutrients but also with that of water, as a result of shallower rooting species concentrating both nutrients and water in the surface layers of the soil.

CONCLUDING REMARKS

The growth patterns and life histories of the annuals and of *Ammophila arenaria* and *Carex arenaria* represent only a few of the species of dune habitats but illustrate several critical features of both habitats and inhabitants. The dune succession, instead of spelling doom to early colonists, reliably and continuously provides colonizing habitats which are generally within easy reach of earlier colonists. None of the species studied here has significant seed dispersal over any distance. Establishment from seed is risky and rare on the mobile dunes, and in both accreting and eroding areas populations are maintained predominantly by perennial clonal growth of genets. Only in the stabilized but still open areas is the annual habit successful and here most of the annuals are short lived winter annuals. They occupy patches that have not yet become dominated by perennial vegetation. They are intensely sensitive to the growth of neighbours in the nutrient poor habitat and they survive in the spatial patches not yet occupied by perennials or else grow in the 'temporal patches' of winter and early spring when the perennials are making slow growth. It is interesting that the whole life cycle of the small annuals is completed during the period of the lowest flux in shoot modules in the perennials.

For each of the species studied the nutrient deficiency of the site appears to be overwhelmingly important, and competitive interactions for nutrient resources appear to be responsible for the decline of even the vigorous perennials *Ammophila arenaria* and *Carex arenaria*. The spatial interaction of the annuals in the experiments of Mack and Harper (1977) described early in this paper represent, in many respects, a microcosm of dune ecology which is dominated by nutrient resource shortage and by competition for these limiting resources; processes that are drawn out and extended in space and time by the perennial clonal growth strategy. Within this community there is of course migration and movement of the mosaic of species which change their position and sphere of influence and react to the successional changes in neighbours and physical conditions that follow primary dune formation. Surprisingly the rate of movement and migration of the perennial species does not seem to be very different from that of the annuals. The annual vegetative extension of genets of *Carex arenaria* at the advancing front is of the order of 2 m per annum and it is interesting to compare this with the dispersal distances achieved through seed by the annual *Vulpia fasciculata*. 79% of the dispersal units of *V. fasciculata* fell within 10 cm of the parent plants and the maximum distance of dispersal observed was 36 cm. Further movement occurred after the seeds landed on the soil surface but on the fixed dunes no seed moved further than 38 cm from its point of landing and on a very open site, where dispersal might have been expected to have been great, no seed moved further than 92 cm (Watkinson 1978b).

The annual *Vulpia fasciculata* is primarily an inbreeder (Cotton 1974; Watkinson 1975)—the perennials *Ammophila arenaria* and *Carex arenaria* are invading forms capable of rapid clonal growth. Both these life cycle patterns will tend to perpetuate a narrow range of genotypes in the face of a fast turnover of generations (in the annuals) or of growth modules (in the perennials). Taken together, the low seed output, the sluggish genetic systems, the poor dispersibility of seed and the dominance of species with persistent banks of vegetative meristems in the soil, suggest that the dune inhabitants have conservative evolutionary and ecological strategies. These are appropriate for a habitat that reliably contains all the repeated elements of a successional matrix.

ACKNOWLEDGMENT

We are indebted to Professor J.L. Harper for his considerable and stimulating assistance during the course of these studies and the preparation of this paper.

REFERENCES

Cotton R. (1974) *Cytotaxonomy of the genus* Vulpia. Ph.D. thesis. University of Manchester.

Davis A.M. (1975) *Temporal variations in the morphology and porosity of an Anglesey Beach.* M.Sc. thesis. University of Wales.

Gemmell A.R., Greig-Smith P. & Gimingham C.H. (1953) A note on the behaviour of *Ammophila arenaria* (L.) Link. in relation to sand-dune formation. *Trans. Proc. bot. Soc. Edinb.* **36**, 132–36.

Greig-Smith P., Gemmell A.R. & Gimingham C.H. (1947) Tussock formation in *Ammophila arenaria* (L.) Link. *New Phytol.* **46**, 262–68.

Harper J.L. (1977) *Population Biology of Plants.* Academic Press.

Huiskes A.H.L. (1977a) The population dynamics of *Ammophila arenaria* (L.) Link. Ph.D. thesis. University of Wales.

Huiskes A.H.L. (1977b) The natural establishment of *Ammophila arenaria* (L.) Link from seed. *Oikos* **29**, 133–36.

Krebs J.R. (1970) Regulation of numbers in great tit. *J. Zool.* **162**, 317–33.

Mack R.N. (1976) Survivorship of *Cerastium atrovirens* at Aberffraw, Anglesey. *J. Ecol.* **64**, 309–12.

Mack R.N. & Harper J.L. (1977) Interference in dune annuals: spatial pattern and neighbourhood effects. *J. Ecol.* **65**, 345–63.

McIntyre G.I. (1965) Some effects of the nitrogen supply on the growth and development of *Agropyron repens* (L.) Beauv. *Weed Res.* **5**, 1–12.

Noble J.C. (1976) The population biology of rhizomatous plants. Ph.D. thesis. University of Wales.

Pemadasa M.A. (1973) Ecology of some sand dune species with special reference to annuals. Ph.D. thesis. University of Wales.

Pemadasa M.A., Greig-Smith P. & Lovell P.H. (1974) A quantitative description of the distribution of annuals in the dune system at Aberffraw, Anglesey. *J. Ecol.* **62**, 379–402.

Pemadasa M.A. & Lovell P.H. (1974a) Factors controlling the flowering time of some dune annuals. *J. Ecol.* **62**, 869–80.

Pemadasa M.A. & Lovell P.H. (1974b) Interference in populations of some dune annuals. *J. Ecol.* **62**, 855–68.

Pemadasa M.A. & Lovell P.H. (1975) Factors controlling germination of some dune annuals. *J. Ecol.* **63**, 41–59.

Pugh K.B., Andrews A.R., Gibbs C.F., Davis S.J. & Floodgate G.D. (1974) Some physical, chemical and microbiological characteristics of two beaches of Anglesey. *J. exp. mar. Biol. Ecol.* **15**, 305–33.

Ranwell D.S. (1958) Movement of vegetated sand dunes at Newborough Warren, Anglesey. *J. Ecol.* **46**, 83–100.

Ranwell D.S. (1959) Newborough Warren, Anglesey. I. The dune system and dune slack habitat. *J. Ecol.* **47**, 571–601.

Ranwell D.S. (1960) Newborough Warren, Anglesey. II. Plant associates and succession cycles of the sand dune and dune slack vegetation. *J. Ecol.* **48**, 117–41.

Symonides E. (1974a) Populations of *Spergula vernalis* Willd. on dunes in the Torun Basin. *Ekol. pol.* **22**, 379–416.

Symonides E. (1974b) The phenology of *Spergula vernalis* Willd. in relation to microclimatic conditions. *Ekol. pol.* **22**, 441–56.

Tomlinson P.B. (1974) Vegetative morphology and meristem dependence—the foundation of productivity in sea grasses. *Aquaculture* **4**, 107–30.

Wareing P.F. (1964) The developmental physiology of rhizomatous and creeping plants. *Proc. Brit. Weed Contr. Conf.* **7**, 1020–30.

Watkinson A.R. (1975) The population biology of a sand dune annual, *Vulpia membranacea*. Ph.D. thesis. University of Wales.

Watkinson A.R. (1978a) The demography of a sand dune annual: *Vulpia fasciculata*. II. The dynamics of seed populations. *J. Ecol.* **66**, 35–44.

Watkinson A.R. (1978b) The demography of a sand dune annual: *Vulpia fasciculata*. III. The dispersal of seeds. *J. Ecol.* **66**, 483–98.

Watkinson A.R. & Harper J.L. (1978) The demography of a sand dune annual: *Vulpia fasciculata*. I. The natural regulation of populations. *J. Ecol.* **66**, 15–33.

Watt A.S. (1947) Pattern and process in the plant community. *J. Ecol.* **35**, 1–22.

Went F.W. (1973) Competition among plants. *Proc. natn. Acad. Sci. U.S.A.* **70**, 585–90.

Willis A.J., Folkes B.F., Hope-Simpson J.F. & Yemm E.W. (1959) Braunton Burrows: the dune system and its vegetation I, II. *J. Ecol.* **47**, 1–24 & 249–88.

Wit C.T. de (1960) On competition. *Versl. Landbouwk Onderz* **66**, 1–82.

Wit C.T. de & Bergh J.P. van den (1965) Competition between herbage plants. *Neth. J. Agric. Sci.* **13**, 212–21.

7. BRAUNTON BURROWS: DEVELOPING VEGETATION IN DUNE SLACKS, 1948–77

J.F. HOPE-SIMPSON AND E.W. YEMM

Department of Botany,
University of Bristol, U.K.

SUMMARY

1. Development of dune-slack vegetation from an early stage has been recorded on various sites on Braunton Burrows for periods of 20–29 years.
2. Transect data reveal a clear distinction between the role of pioneers preceding, and secondary species following, the attainment of *Salix repens* dominance within about ten years. Some increase is shown among secondary species after more than 20 years.
3. Twenty-five slack sites were recorded by a quick method three times in 20 years. Colonization is slowest on the driest sites.
4. Some shoreward slacks show a pronounced crust on macroscopically uncolonized ground. Its formation, and the establishment of varying vegetation, are discussed in relation to hydrology, which in shoreward slacks is such that it could well favour crust-formation. The crust is not easily colonized by means of seed. This fact, as well as sub-optimal hydrology for *Salix repens* establishment, may be a cause of major vegetational differences between shoreward and landward slacks.

RÉSUMÉ

1. Le développement, depuis les débuts, de la végétation des pannes (dépressions humides) des dunes a été enregistré dans divers sites de Braunton Burrows pour des périodes de 20–29 ans.
2. Les résultats provenants des transects révèlent une distinction nette entre le rôle des pionniers, qui ont precédé l'aquisition sur une dizaine d'années de la dominance par *Salix repens*, et les espèces secondaires qui l'ont suivi. On montre une certaine augmentation parmi les espèces secondaires après plus de vingt ans.
3. 25 sites des pannes ont été enregistrés par une méthode rapide, trois fois en vingt ans. La colonisation est la plus lente dans les sites les plus secs.
4. Quelques pannes près du rivage montrent une croûte prononcée sur le sol

macroscopiquement incolonisé. On discute de sa formation et de l'établissement de végétations différentes en relation avec l'hydrologie, qui est telle dans les pannes côtières qu'elle peut bien favoriser la formation de la croûte. Celle-ci n'est pas aisément colonisée au moyen de graine. Ce fait, ainsi que l'hydrologie suboptimale pour l'implantation de *Salix repens*, peut être une cause des différences majeures de végétation entre les pannes côtières et celles de l'intérieur.

INTRODUCTION

The vegetation of Braunton Burrows, a large calcareous dune system in south-west England, has been described at some length by Willis, Folkes, Hope-Simpson and Yemm (1959), with an outline of the succession in slacks. The investigations now reported began in 1948, only 3–4 years after at least 2 years of very severe use of the Burrows for military training. This had created widespread areas of bare sand, many of them near phreatic level and initially stabilized by capillary moisture. Consequently an exceptional opportunity existed to make a selection for long term study of juvenile slacks with plant colonization little advanced from a nearly simultaneous beginning in different slacks. On only one site considered here has investigation spanned the whole 29-year period indicated in the title, but on the others it has spanned 20 years or more.

This paper explains the methods used and gives a selection of results. The selection is of two kinds. Only the most plentiful species composing the vegetation are considered here, and only slacks in which *Salix repens* has become an important component. Thus the wettest slacks are omitted; but a wide variety of sites is included nevertheless. By the moisture criterion of Ranwell (1972) they would belong almost entirely within his category of 'wet slacks', in which the water table never—and we would add 'or very rarely'—falls more than 1 m below the soil surface.

METHODS

The data were obtained largely during field courses with students in mid- to late June every year. This fact had a controlling influence on procedure, since ideal research requirements could not be the sole consideration. Two kinds of investigation were made. One was by detailed recording along a few permanently marked transects, repeated at time intervals varying from one to five years. Results from three widely differing slacks are considered. The other method was by much more approximate recording of 25 sites in 1956, 1966 (or about then) and 1976. It is convenient to dispose of this second procedure first.

R-*sites*

The 25 sites, termed *R*-sites, were selected in the sense that they were intended to represent a good variety of situations. All are in the southern half of the Burrows, inland of the high dunes on the terrain which, though uneven, can be called the sand plain, following Ranwell (1972). Initially all were at quite early stages of succession; 19/25 sites have 1956 estimates of bare ground (% area) of 30% or more. The areas concerned vary from about 100 to 1,000 m^2 (rarely larger); this variation was inevitable because of the paramount need to choose units of ground recognizable again after many years. The whole site was recorded subjectively as a single unit, bare ground (including thin moss cover where present) by area and the species by relative bulk, on the principle explained in the next section but here expressed as %.

Transects

In slacks chosen for detailed recording by transect, 30 × 15 cm quadrats were placed accurately along a survey chain extended between permanent markers. For each quadrat the recording followed the method described fully by Willis *et al.* (1959). Ten marks are used. A share is given first to any bare ground present according to its area; within this component, 'bare sand' and 'crust' (explained later) were sometimes scored separately. The remaining marks are apportioned to the plant species according to their relative bulk in the quadrat.

With this scoring method it has to be remembered that the mark allocated to a plant species is an estimate of its proportion to other species in the one quadrat on the one occasion and not an absolute quantity. Often it is informative to have this deliberate expression of the balance between species in the composition of the vegetation. The deciding factor, however, in adopting a fixed total (10) was the discipline in scoring which it imposes when assessments are made by many different recorders. Tests of the variation between their assessments have been made in dune-slack vegetation. An example is shown in Fig. 7.1, where two features in particular may be noticed. Even among the species of smaller bulk the observation of presence as distinct from quantity shows few discrepancies. In estimated quantity, however, there are obvious variations. They form one reason for analysis of the field data in the manner shown under 'B3 transect' below.

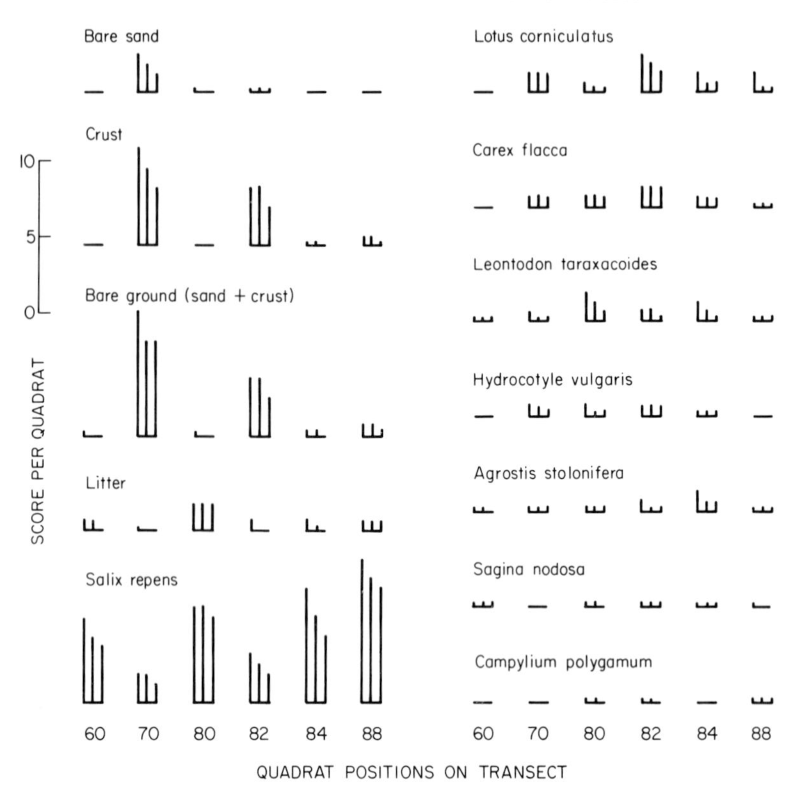

FIG. 7.1. Test of variation between recorders, at six quadrat positions in slack B3. Each quadrat was recorded by three recorder-pairs. The quantities shown are assessments out of a total of 10 marks per quadrat, by area for bare sand and crust, by bulk for litter and species. Some species are omitted. For clarity the quantities in each trio are arranged in order of magnitude, not by identity of recorders.

Site wetness

In comparing habitat differences between investigated sites a factor of outstanding importance is obviously ground wetness. This was determined for transects and *R*-sites alike. Ground levels were surveyed in detail and, with sufficient water levels, flooding indices (average months/year flooded; Willis *et al.* 1959) were calculated, nearly all by B.F. Folkes.

RESULTS

B3 *transect*

The small slack known as B3 is located on the sand plain in the southern part of the Burrows, close to slack site 10 of the Fig. 1 map of Willis *et al.* (1959).

The B3 transect was recorded in 13 of the 30 years 1948–77. At the outset the slack was almost a barren blow-out.

Attention is confined to the nearly level floor of the slack, with a transect length of 9 m containing 30 consecutive quadrats (Fig. 7.5, p. 124). Here the vegetation can, for purposes of a simplified analysis, be considered a single community. Accordingly it is permissible to combine the data from all 30 quadrats by obtaining, for each record-year, mean values for bare ground (area) and for the chief species (relative bulk) to give a representation of 'the average B3 quadrat' during 29 years (Fig. 7.2). The selection of chief species, confined to vascular plants, was decided by two criteria: attainment at some time of a mean bulk value of at least 0·2 and also, whether concurrently or not, a frequency of at least 50% (presence in 15/30 quadrats). The species thus included generally form more than 75% of the vegetation in the quadrats.

Fig. 7.3 shows the frequency changes for bare ground and the species. Frequency of course gives limited information in some respects, e.g. the apparently slight change in bare ground, or the similarity of value for the dominant *Salix repens* and the exiguous though widespread *Equisetum variegatum*. Nevertheless it is useful to be able to compare the two different parameters (Figs. 7.2 and 7.3). Their respective fluctuations have many coinciding rises and falls, but in obvious ways they are partly independent of each other. The relative bulk values, unlike frequency, are intended to, and do, reflect changes in species quantity within quadrats, but the necessary caution (p. 115) can be illustrated by reference to the falling estimates of mean relative bulk of the dominant *Salix repens* after 1966. Almost certainly these are attributable, at least in part, to the increases in other estimates during this period. Such an effect is inherent in the proportional nature of relative bulk estimates. As long as this limitation is borne in mind, the two types of graph together give much information about the temporal changes of the vegetation.

Largely they speak for themselves but a few points deserve comment. Most striking is the difference in role between the pioneers, reaching maxima before *Salix repens* predominates strongly, and the secondary invaders following that stage. *Agrostis stolonifera* is unique in versatility and persistence. Among the secondary species, *Hydrocotyle* and *Lotus* seem a little surprising in their simultaneous late increase, seen in both relative bulk and frequency, the former being a plant of wet slacks and the latter only of drier ones. But *Hydrocotyle* can thrive on drier ground where it has adequate shelter, and its invasion may reflect the sheltering effect of thickened *Salix repens*. *Lotus* has been plentiful from the start in the adjoining grassland but its seeds are relatively heavy.

Three slacks compared

Data from two further slacks have been analysed in the manner already presented for slack B3. The resulting information is summarized in Table 7.1.

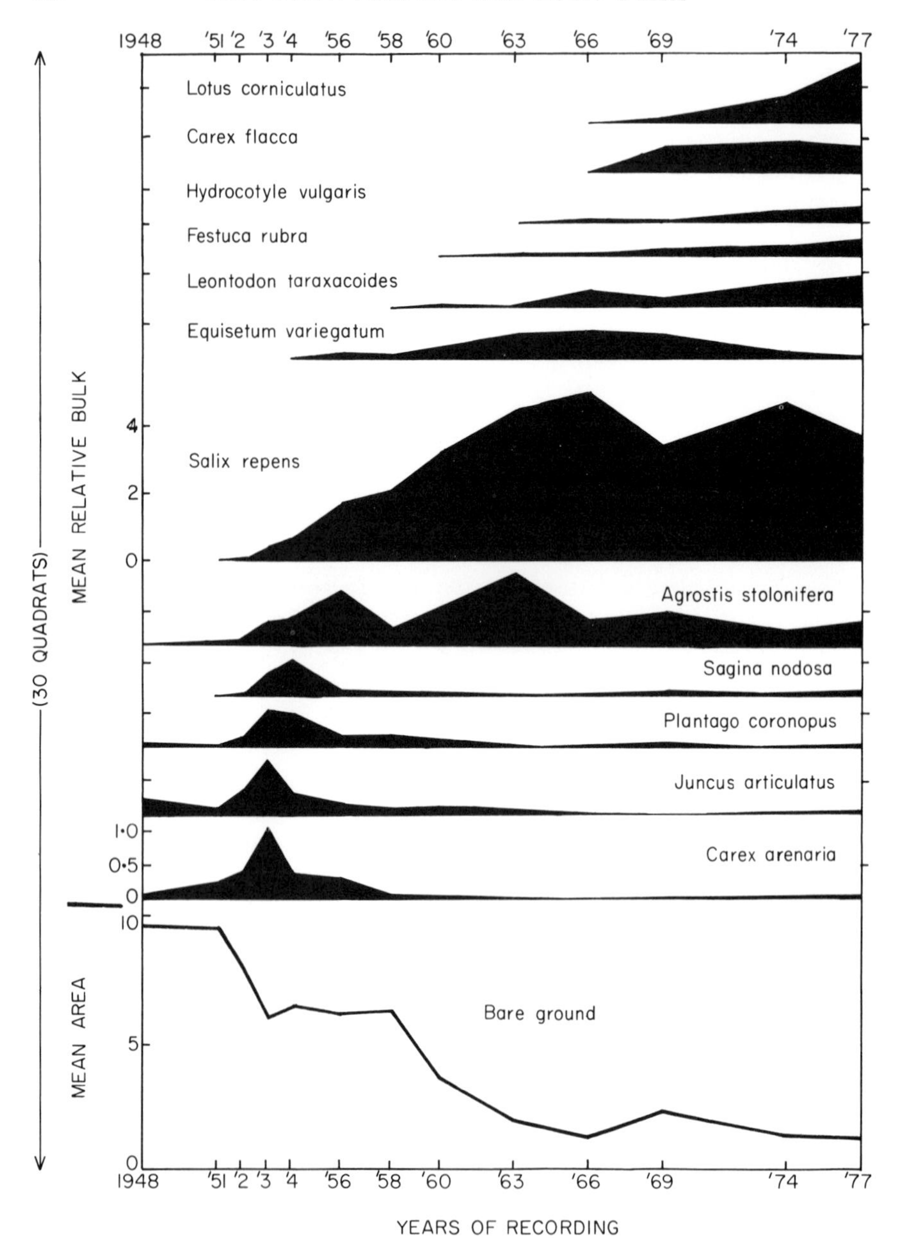

FIG. 7.2. Slack B3 transect: 29-year changes in mean quadrat scores of bare ground and of the chief species; explanation on p. 117. The vertical scale shown for *Carex arenaria* is the same for all other species except *Salix repens* (scale halved); the unmarked pointers on each side indicate mean relative bulk 0·5.

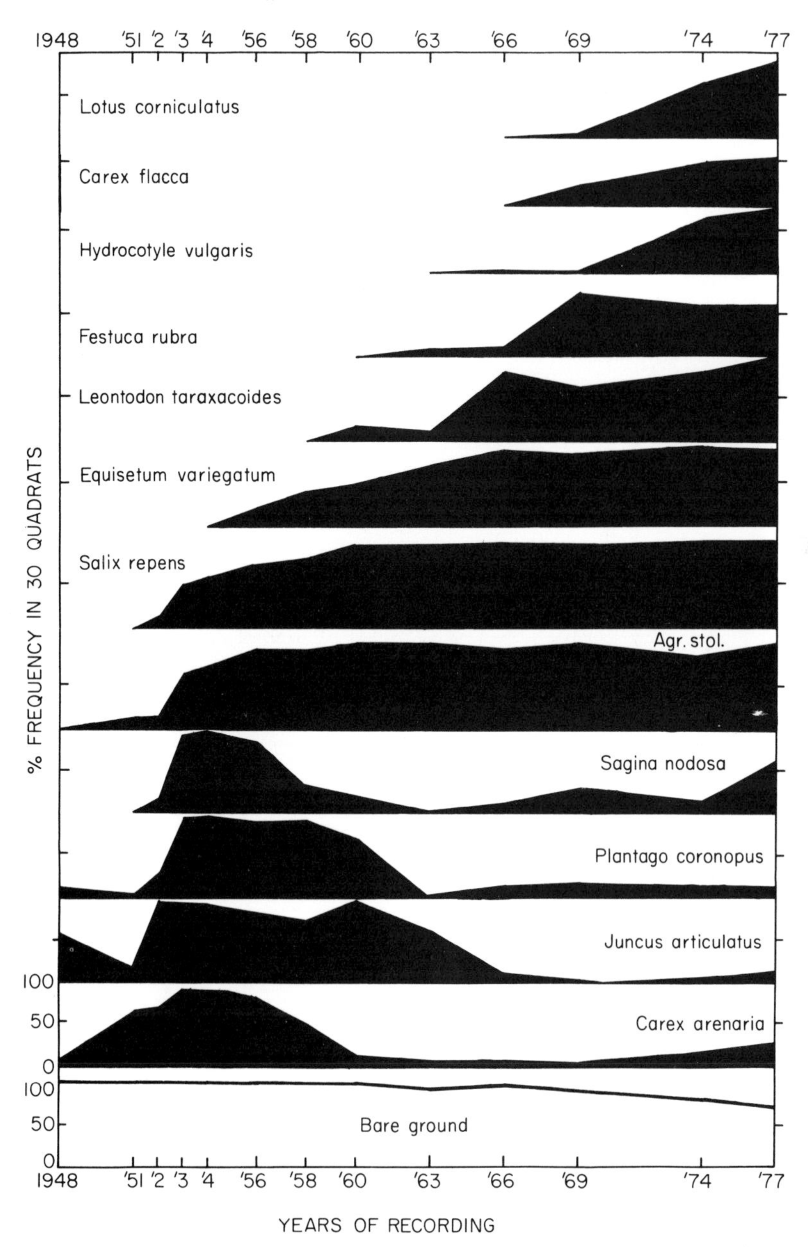

FIG. 7.3. Slack B3 transect: % frequency (presence) in 30 quadrats, derived from the same records as Fig. 7.2. The vertical scale is uniform for bare ground and for all the species; the unmarked pointers on each side indicate frequency 50%.

Table 7.1. *The development of Salicetum repentis in three slacks (B3, S23, E1/2). In 1,* Location and Topography, *the location notes refer to site numbers on maps (Figs. 1 and 8) in Willis* et al. *(1959). In 3, Vegetation, the facts shown for the chief species (selected as stated on p. 117) are derived from transect data analysed as in the production of Figs. 7.2 and 7.3.*

	Slack		
	B3 (recorded 1948–77)	**S23** (recorded 1949–72)	**E1/2** (recorded 1953, 1961–77)
1. *Location and Topography*	Close to slack site 10. The floor of a small blow-out adjoining 'dune pasture'	Slack site 23. Part of an extensive blow-out enclosed by low *Salix repens* dunes	Location shown by water level 21·7 in Willis *et al.* Fig. 8, c. 200 m from the shore line. Surrounded by mobile *Ammophila* dunes.
2. *Hydrology*			
(a) Mean and range of land heights (m O.D.)*	8·23 8·19–8·32	9·54 9·41–9·60	7·07 6·94–7·22
(b) Water level, June 1952 (m O.D.)	8·01	9·23	6·53
(c) Flooding index (months/year), mean and range on transect	4·7 5·0–2·7	4·4 5·3–3·2	Not flooded in years of average rainfall
3. *Vegetation*			
(a) Pioneer species (date of first record)	1948–52 { Carex arenaria, Agrostis stolonifera, Juncus articulatus, Plantago coronopus, Sagina nodosa }	1949 { Salix repens, Agrostis stolonifera, Juncus articulatus }	1961 { Carex arenaria, Agrostis stolonifera, Juncus articulatus, Glaux maritima }
(b) *Salix repens*: arrival to dominance	1952–58	1949–53	1961–67
(c) Secondary species (date of first record)	Equisetum variegatum (1956) Leontodon taraxacoides (1960) Festuca rubra (1963) Hydrocotyle vulgaris (1966) Carex flacca (1969) Lotus corniculatus (1969)	Hydrocotyle vulgaris (1953) Carex serotina (1953) Leontodon taraxacoides (1957) Epipactis palustris (1957) Lotus corniculatus (1967)	Trifolium fragiferum (1964) Leontodon taraxacoides (1964) Carex flacca (1967) Epipactis palustris (1967) Holcus lanatus (1969) Poa pratensis ssp. irrigata (1969)

* Metres above Ordnance Datum.

From the facts tabulated it is not difficult to extract some salient differences between the slacks in the (presumed) seral roles of the various chief species.

Slack S23, also on the sand plain, is especially distinctive in the pioneer role of *Salix repens* itself, plausibly attributable to the dense sources of seed surrounding the slack. *Lotus*, as in B3, is a chief species which was very late in arrival.

In the shoreward slack E1/2 (Fig. 7.6), the transect-length analysed may reasonably be compared as a *Salix repens* type, as is done in Table 7.1 This slack, however, presents several features of developmental interest which cannot be shown in the table. In 1953 it was virtually bare, but the surface was stabilized and colonization was just beginning. In 1961 *S.repens* was still very sparse; on the analysed (Salicetum) transect-length its dominance was not achieved until 1967 and, except in patches, has been less complete than in the other two compared slacks. Moreover E1/2, a fairly large slack, is varied in character and quite extensively devoid of *S.repens*. There is much herbaceous dominance, with *Carex flacca* widely prevalent. In addition a substantial proportion of the surface still (1977) lacks macroscopic vegetation cover and bears a firm crust. This feature, although not pronounced on the analysed transect-length, is persistent and conspicuous elsewhere (Fig. 7.6) in a manner distinctive of shoreward as compared with more inland slacks. The crust has two forms which occur both separately and mixed. Most extensively it consists of sometimes hardly recognizable moss debris with a large algal population. The other type is a thin deposit of calcium carbonate. Both kinds are clearly unfavourable surfaces for colonization by seed. This topic is considered further in the Discussion.

25 *approximately recorded sites* (R-*sites*)

Only generalized results can be given here, with *Salix repens* as the one species shown separately (Fig. 7.4). Consideration of the diagram requires first an explanation about the dates of recording. All were in June–July, the first in 1956 and the latest in 1976. The intermediate year was 1965 for 16 sites, 1966 for 7 and 1969 for 2. For simplicity the intermediate year can be regarded as 1966 and language approximated accordingly (e.g. 'ten-year' intervals for all sites), but in the individual site diagrams of Fig. 7.4 the actual year of recording is plotted.

One or two results contained in the figure as a whole are obvious, but otherwise extraction of convincing generalizations requires rather careful analysis of its contents, substantiated numerically. This has been done, and eight generalizations seem well founded. Five of them are stated dogmatically below; space is lacking for numerical substantiation. The first two are not new discoveries nor do they illustrate the primary objective of detecting changes with time.

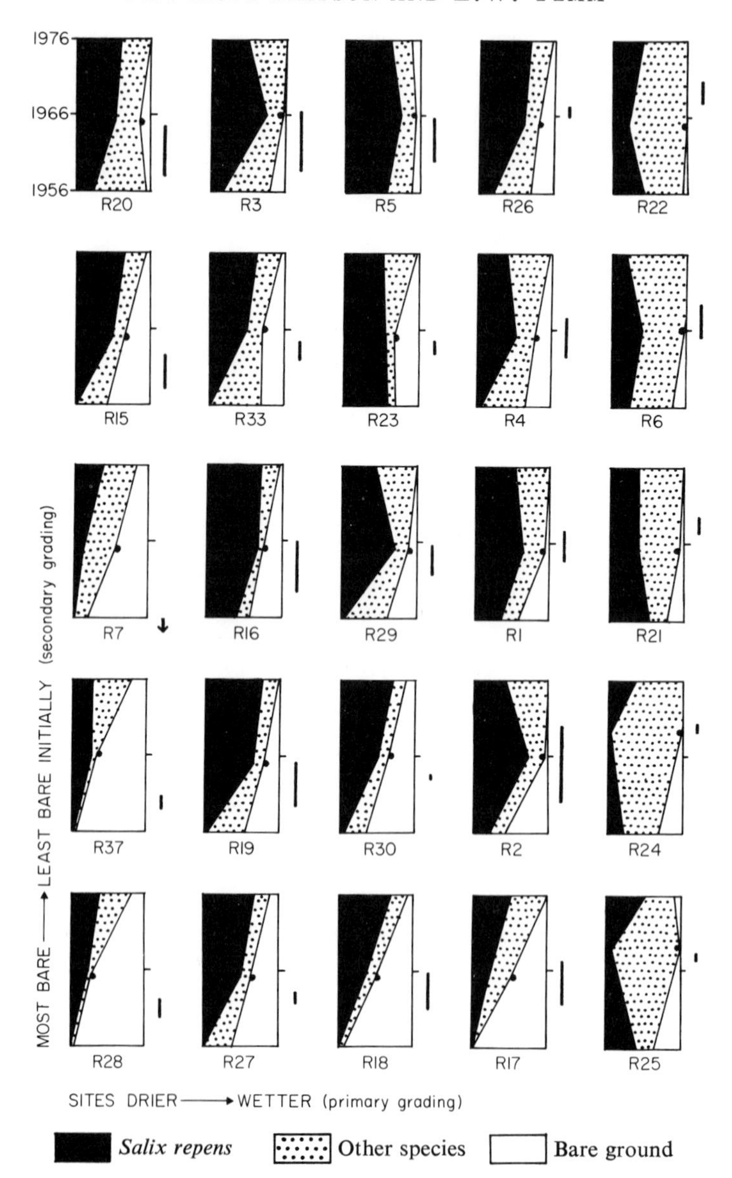

FIG. 7.4. 25 *R*-sites. Changes from 1956–76 in the three components: area of bare ground; relative bulk of *Salix repens*; bulk aggregate of all other species. The vertical time-scale is marked at top left. In each site diagram the width of each component at the recording years shows its % score (horizontal total = 100). The vertical black line beside each site diagram indicates the degree of wetness of the site expressed as flooding index. Index = 0 at the base of the rectangle and 5 half way up (pointer at right); the higher the index the wetter the site. The length of the line shows the index range, which is least on the flattest sites.

(1) The greatest dominance by *Salix repens* occurs in sites of medium wetness.
(2) The preponderance of 'other species' over *Salix* in the wettest sites, conspicuous in Fig. 7.4, is self-evident on the ground.
(3) The wettest sites were the earliest to become fully covered.
(4) In keeping with (3), the driest sites are the slowest in colonization of bare ground, of which some persists after 20 years; they show a continuing increase of *Salix*, usually accompanied by failure of other species to gain on *Salix*, in the second ten years.
(5) Excluding the wettest sites, those already substantially colonized in 1956 show a second-decade increase, not shown in the first decade, of other species against *Salix*. (These decades of recording of course run somewhat later than decades after first colonization.) The same trend is apparent in the B3 transect results (Fig. 7.2) and has occurred in S23 and E1/2 also.

DISCUSSION

A principal theme of this paper is variation in the development of *Salix repens* slacks. Hence habitat conditions affecting the establishment of *S.repens* itself are a prime concern. At Braunton all the evidence indicates that its colonization has been most complete in slacks where flooding fairly frequently lasts until well into spring, ensuring a wet ground surface at the season (about early June) of seed dispersal and germination. This conclusion was suggested by the earlier observations of Willis *et al.* (1959) and accords with those of Ranwell (1960) on *S.repens* seedlings. Confirmation comes especially from the *R*-sites. It must be remembered that an *R*-site flooding index, as shown in Fig. 7.4 to cover its variation on uneven ground, represents a mean year, so that the flooding indices of the best *Salix R*-sites, centred on values of about 4 (months), do not portray the longer flooding of divergent wet years. Such flooding (except in hollows too persistently wet for *S.repens*) is likely only where there is a wide seasonal range of water level, i.e. on the upper part of the domed water table of the Burrows, as Willis *et al.* have explained. This region embraces the *Salix*-dominated slacks, those on the sand plain and also the most landward interdunal ones. Less good slacks for *Salix* establishment are discussed later.

It would be surprising if the optimum establishment conditions stated above were unique to Braunton Burrows. A comparable long-period investigation is that of Londo (1971), whose records on the calcareous Kennemer dunes in Holland extended for 9–12 years (1956–68). The new slack habitat there was provided by the gently sloping margins of a shallow lake excavated in 1951–55. Communities dominated by *Salix repens* developed, as at Braunton, on sites of medium wetness; their flooding index range, derived from 10 years' observation of flooding, is about 2–6 according to position on the slope, corresponding quite closely to the values calculated for the Braunton sites of best Salicetum repentis establishment. *Juncus articulatus* and, to a lesser extent, *Agrostis stolonifera* were important pioneers, as at Braunton. Invasion by *Calamagrostis epigejos* and *Hippophaë rhamnoides* were later developments not relevant to the immediate comparison. The similarities mentioned testify to the controlling influence of ground wetness, particularly because there are marked differences between the Kennemer and Braunton slack habitats. Apart from climate and history, the habitats differ edaphically. At Kennemer Londo recorded, from 1960–68, the quick development of a well defined *A* horizon rich in humus, accompanied by a marked increase in nitrogen and phosphorus due, probably, to the large bird population. These nutrients are notably low in quantity at Braunton (Willis, 1963).

This paper gives no occasion to discuss Londo's thorough analysis of detailed records which he maintained annually. On the intensive plane the Braunton observations are much less complete. But with their long duration and wide spread over the dune system they offer an insight still to be considered. This concerns the shoreward slacks exemplified by E1/2 (pp. 120, 121), and involves some speculation. Their distinctive hydrology appears to have an instructive bearing on the subject of colonization in slacks, whether by *Salix repens* or other plants. It is true that other factors have operated, notably exposure to onshore wind and proximity of mobile *Ammophila* dunes yielding

Fig. 7.5. Braunton Burrows, slack B3, June 1977. The white tape lies along the transect length (9 m) from which data have been analysed.

FIG. 7.6. Braunton Burrows, slack E1/2, July 1977. The transect-length yielding analyzed data runs from side to side across the smooth ground in the middle distance. The crust surface referred to in the text is conspicuous (pale) on the low mounds but not confined to them.

wind-blown sand. These factors, however, need not enter into the present discussion of hydrology.

We have noted that the *Salix*-dominated slacks have been liable to fairly frequent occasions of late spring flooding or near-flooding, favouring *Salix* establishment when bare sand existed. Towards the shore, the lesser seasonal variation in water level has provided fewer such occasions but also means that the summer water table is rarely very low. Other vascular plants, therefore, have gained ground in the absence of *Salix*, so that in more shoreward slacks it is by no means a general dominant. When initially bare, the sand surface, kept moist for the relatively long periods when the water table remains within capillary range, has also been favourable to carpeting by mosses, notably *Barbula tophacea*; the rarity of inundation might perhaps be favourable too. The moss cover, however, does deteriorate, creating one kind of crust (p. 121; Fig. 7.6). The same moisture conditions could well be, by coincidence or not, the cause of the calcium carbonate crusting conspicuous in shoreward slacks. Both forms of crust, as stated earlier, severely hinder seedling establishment, but rhizomatous plants (notably *Carex flacca*) penetrate from below, leading to gradual alteration of crusted ground, while *Salix repens* cannot do so. The sporadic *Salix* patches do, however, extend prostrate branches which, by means of shading and litter, modify the crust and then become rooted. These vegetative methods of occupying ground are slow, compared with establishment by plentiful seed in the few initial seasons preceding moss-carpeting and crust-formation. If seed dispersal was sparse at that stage, fast colonization by means of seed has lost its chance. The foregoing interpretation, if broadly correct, would indicate how some of the differences in slack vegetation require knowledge of development as well as established condition if they are to be understood.

Finally, an obvious question arises from the research described here, suggested especially in the case of slack B3 by the appearance of Figs. 7.2 and 7.3: Is succession still proceeding, 30 years after its beginning? The rising curves of some secondary species during recent years certainly suggest that trends, rather than mere fluctuations, are continuing. It is impossible to consider the matter thoroughly now, because the present analysis omits numerous subsidiary species for which records exist. If we venture a forward glance nevertheless, we might expect that no sharp, definitive end-point to succession will be detectable; for instance additional species may continue to invade at long intervals; and weather-induced changes (ignored in this paper) may obscure attainment of any 'final' composition of the vegetation. Moreover, reverting to the recent trends seen in Figs. 7.2 and 7.3, if a retardation of these is imminent, it will not preclude the possibility of long-continued change at a much slower tempo, for example gradual change in the condition of *Salix repens*, which would in turn influence other species.

For such reasons, the present investigation of the first two or three decades of succession provides no vantage point for debating even more remote questions as to attainment of a climax status by *Salix repens* slacks and what the stabilizing factors might be (cf. Odum 1969). Any approach to that topic would first require, *inter alia*, a study of long-established slacks, not merely those chosen, as here, for their recent origin.

ACKNOWLEDGMENTS

The authors are grateful to many collaborators. Arthur Willis and Brian Folkes played leading parts for about 20 years. Other colleagues have participated on several occasions. About 300 students recorded data faithfully. Exceptionally plentiful help was given in 1976 by Juliet Brodie. The diagrams were drawn by Stella Sage. The University of Bristol provided generous financial support.

REFERENCES

Londo G. (1971) *Patroon en proces in duinvalleivegetaties langs een gegraven meer in de Kennemerduinen.* (Pattern and process in dune slack vegetations along an excavated lake in the Kennemer dunes.) Verhandeling No. 2, R.I.N., Leersum.

Odum E.P. (1969) The strategy of ecosystem development. *Science* **164**, 262–70.
Ranwell D.S. (1960) Newborough Warren, Anglesey. II. Plant associes and succession cycles of the sand dune and dune slack vegetation. *J. Ecol.* **48**, 117–41.
Ranwell D.S. (1972) *Ecology of Salt Marshes and Sand Dunes.* Chapman & Hall, London.
Willis A.J. (1963) Braunton Burrows: the effects on the vegetation of the addition of mineral nutrients to the dune soils. *J. Ecol.* **51**, 353–74.
Willis A.J., Folkes B.F., Hope-Simpson J.F. & Yemm E.W. (1959) Braunton Burrows: the dune system and its vegetation. *J. Ecol.* **47**, 1–24, 249–88.

8. SOME FACTORS AFFECTING THE GROWTH OF TWO POPULATIONS OF *FESTUCA RUBRA* VAR. *ARENARIA* ON THE DUNES OF BLAKENEY POINT, NORFOLK

CLARE ANDERSON AND K. TAYLOR

Department of Botany and Microbiology,
University College London,
Gower Street, London, U.K.

SUMMARY

Two populations of *Festuca rubra* var. *arenaria* have been studied at Blakeney Point, Norfolk, one in the mobile yellow dunes in association with *Ammophila arenaria* tussocks, and the other on the grey fixed dunes.

Whereas plants of *Festuca* from the yellow dunes appear to have a preferential advantage over plants from the grey dunes in being able to establish on the mobile dunes, plants of the fixed dunes show a larger growth response when grown in the grey sand in full sunlight.

RÉSUMÉ

Nous avons étudié deux populations de *Festuca rubra* var. *arenaria* à Blakeney Point, Norfolk, l'une située dans les dunes mobiles jaunes, en association avec les touffes d'*Ammophila arenaria*, et l'autre dans les dunes fixes grises.

Alors que les plantes de *Festuca* des dunes jaunes semblent avoir un avantage préférentiel sur les plantes des dunes grises dans la capacité à se fixer sur les dunes mobiles, les plantes des dunes fixes répondent par une plus grande croissance quand elles poussent sur les dunes grises et en plein soleil.

INTRODUCTION

At Blakeney Point on the north Norfolk coast, *Festuca rubra* var. *arenaria* occurs both in the mobile yellow phase where it is closely associated with the

Table 8.1. *Morphological description of two populations of* Festuca rubra *var.* arenaria *from the mobile yellow dunes and grey fixed dunes at Blakeney Point, Norfolk. 100 plants examined from collections made on 30 September 1974 and 25 June 1975.*

		Yellow sand population	Grey sand population
Vegetative characters			
Mean height (cm) ± S.E.M.		48·04 ± 1·74	13·02 ± 0·84
Mean leaf length (cm) ± S.E.M.		19·99 ± 1·16	5·01 ± 0.39
Width of leaf (mm) ± S.E.M.		1·77 ± 0·13	1·33 ± 0·09
No. abaxial sclerenchyma bundles		9	7
	(range)	(7–11)	(7–9)
No. vascular bundles		7	5
	(range)	(5–7)	(5–7)
Inflorescence			
Panicle length (cm) ± S.E.M.		13·44 ± 1.03	7·56 ± 0·61
Spikelet length (mm) ± S.E.M.		11·51 ± 0·99	7·22 ± 0·48
Lower glume (mm)		3·87	2·91
	(range)	(2·6–4·5)	(2·1–4·0)
Upper glume (mm)		5·31	4·43
	(range)	(3·8–7·3)	(3·0–5·7)
Lemma (mm)		6·02	5·52
	(range)	(5·0–7·5)	(4·5–6·9)
Lemma: upper glume ratio		1·13	1·20
	(range)	(1·08–1·19)	(1·16–1·24)

tussocks of *Ammophila arenaria*, and in the various facies of the grey fixed dunes, where it forms an open turf. There is a difference in the morphological appearance of plants at the two sites; on the yellow dunes the fescue grows to a height of 20–90 cm whereas on the grey dunes it attains a height of between 8–16 cm. There are also other measurable morphological differences between these populations (Table 8.1). Plants of both populations are hexaploids ($2n = 42$).

An attempt has been made to characterize the environmental factors which can be correlated with the distribution of these two divergent populations of *Festuca rubra* var. *arenaria*. In addition the relationships between the growth of plants from the two populations and these factors have been examined under natural conditions and in the laboratory. The factors include salinity, sand accretion, mineral nutrition, irradiance, competition, temperature, water relations and grazing.

DESCRIPTION OF THE SITES

Initially a transect, 70 m long was laid out perpendicular to the dune ridge (Fig. 8.1). The dune system is actively building at the west end of Blakeney

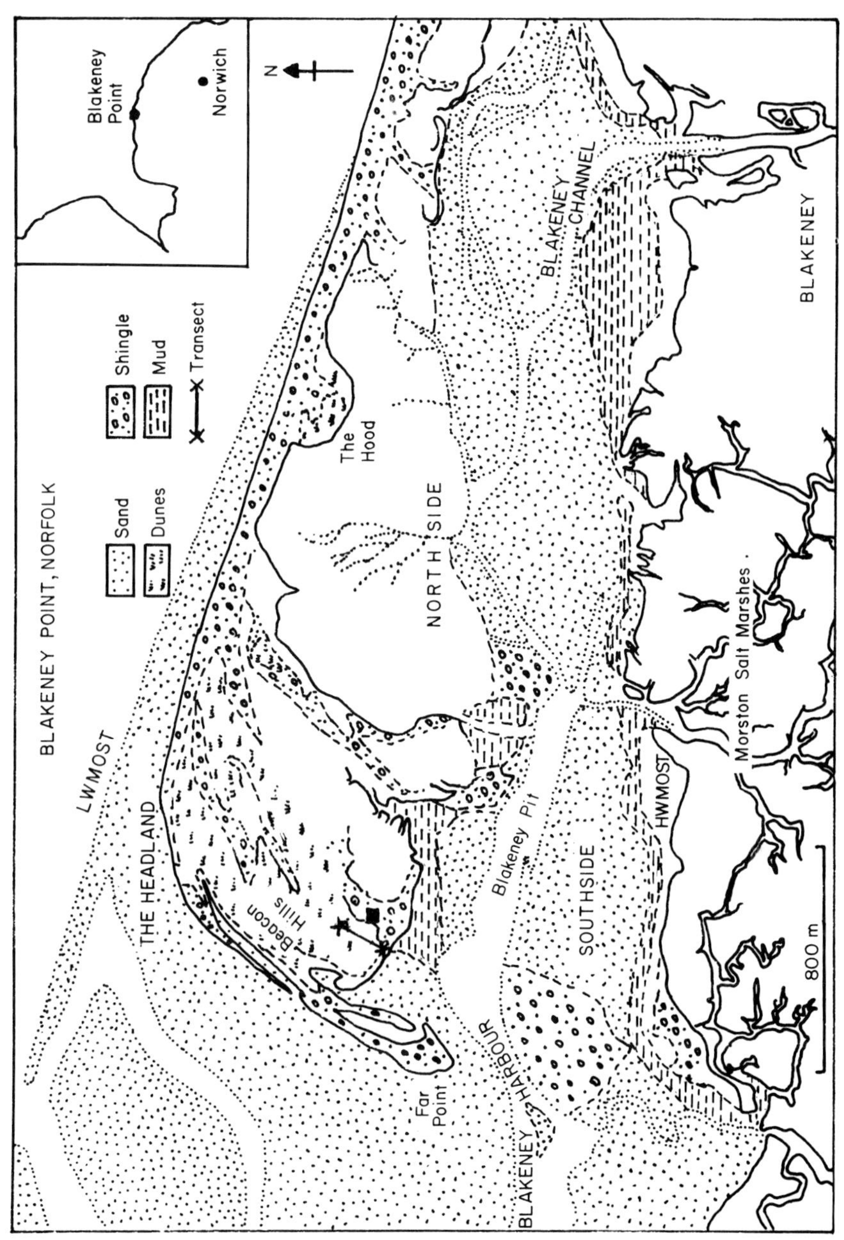

FIG. 8.1. Map of Blakeney Point, Norfolk, showing dunes, shingle banks and channels with the transect across Beacon Hills (modified from Wain and Bird 1963).

Point, overlooking Blakeney Harbour. The exposed windward slope of the dune ridge faces the estuary and the grey dunes are in the lee of the strong south-west winds but are open to the north and east winds from the North Sea. Both communities therefore are subjected to varying levels of salt deposition depending on the direction of the wind.

All the sands are acidic and low in nutrient content but there are significant differences in the soil nutrient status from the seaward to the landward end of the dunes. Marked differences in the concentrations of soluble ions occur between the yellow sand and the grey sand (Gorham 1958). The concentrations of calcium, magnesium, sodium and nitrate are higher in the mobile dune sand than in the grey dunes, whereas the opposite occurs with respect to concentrations of soluble inorganic phosphate and potassium in the two sands. Salisbury (1922) showed that the hydrogen ion concentration and the level of organic matter increases, but the total amount of calcium carbonate decreases, along a transect from the shore to the inland sites. He also stated that the water retention capacity of the sands is dependent upon the level of organic matter present and that this differs between mobile and fixed dunes.

On dunes where an onshore wind predominates, as at Blakeney Point, the Venturi effect causes maximum erosion close to the top of the windward face of the dune and a vortex develops on the lee slope producing a calm so that sand is deposited on this slope (Olson 1958). In the mobile dunes plants of *Ammophila arenaria* are able to keep pace with sand accretion in its building phase and plants of *Festuca rubra* var. *arenaria* at these sites show a similar response. They tiller vigorously, particularly in spring following fresh increments of sand.

The radiation levels in the two communities are different. In the mobile dunes there is considerable variation in the light regime from the dense shade of the *Ammophila* tussocks to the open areas in between tussocks. This contrasts with the very short turf of the grey fixed dunes where the vegetation is evenly exposed to incoming radiation at ground level.

Temperature fluctuations vary between different sand types, diurnal fluctuations being as great as 30°C at the surface of yellow sand but less so down the soil profile (Salisbury 1952). Mean temperature readings taken over one year indicate that grey sand is 1°C higher than yellow sand averaged over a depth of 45 cm.

The species compositions of the two communities differ greatly. On the mobile dunes *Ammophila arenaria* is the most frequent species of the yellow sand community and *Festuca rubra* var. *arenaria* appears to be competing for the same nutrients and light since the latter is found within and immediately outside the marram tussocks. There is greater species diversity of perennial plants in the grey sand community where one of the most frequent plants is *Carex arenaria*. Competitive ability of the grey dune fescues appears to depend to some extent on the degree of grazing by rabbits (White 1961).

At Blakeney Point, there is considerable rabbit grazing of fescue plants of the grey dunes but less grazing on the mobile dune population. A recent investigation into the diet of the rabbit at Holkham, Norfolk, has shown that 73% of the total intake throughout the year was *Festuca rubra* where the availability was only 9% of the shoot material which was produced (Bhadresa 1977).

SALINITY

Analyses of total sodium of washed and unwashed leaf laminae collected at monthly intervals showed differences in the amounts of this element present on the surface of the laminae of plants of the two fescue populations. Sub-samples of leaf material were washed in distilled water before all samples were oven-dried. Sodium was determined by emission spectrophotometry using a Unicam SP 90 spectrophotometer. Unwashed leaves gave total sodium values 8% and 2% higher than washed leaves from mobile and fixed dune populations respectively. These results are an average of estimates obtained from 16 collections between 30 April 1974 and 20 October 1975. There are probably corresponding differences in the salinity levels in the sands at the two sites but a wide variation in salinity according to season and amounts of salt leached out by rainfall is to be expected.

The effects of different salinities on the germination of seed of fescue were investigated under controlled conditions in a growth cabinet, where 25 seeds per treatment were sown on filter paper in Petri dishes under a 12 hour day-length light regime, at a temperature of 25°C during light hours and 5°C in the dark. The results of this experiment suggest that there is a greater inhibition of germination under conditions of increasing salinity of seed from the grey dune population of fescues compared with that of seed from the mobile dunes (Fig. 8.2). Seeds from the grey dunes showed a lower rate of germination than seed from the mobile dune population. When seeds of this population were immersed in seawater the germination was delayed longer than that of seed from the yellow dunes.

SAND ACCRETION

The mean annual net rate of sand accretion at sites along the dune transect from the mobile to the fixed dunes was estimated from records made at regular intervals of the height of the sand on fence posts and wooden blocks. Maximum sand deposition occurred immediately behind the crest of the mobile dunes where an average annual net increment of 14 cm was recorded, but this figure fails to indicate the dramatic changes in the height of the sand during this period.

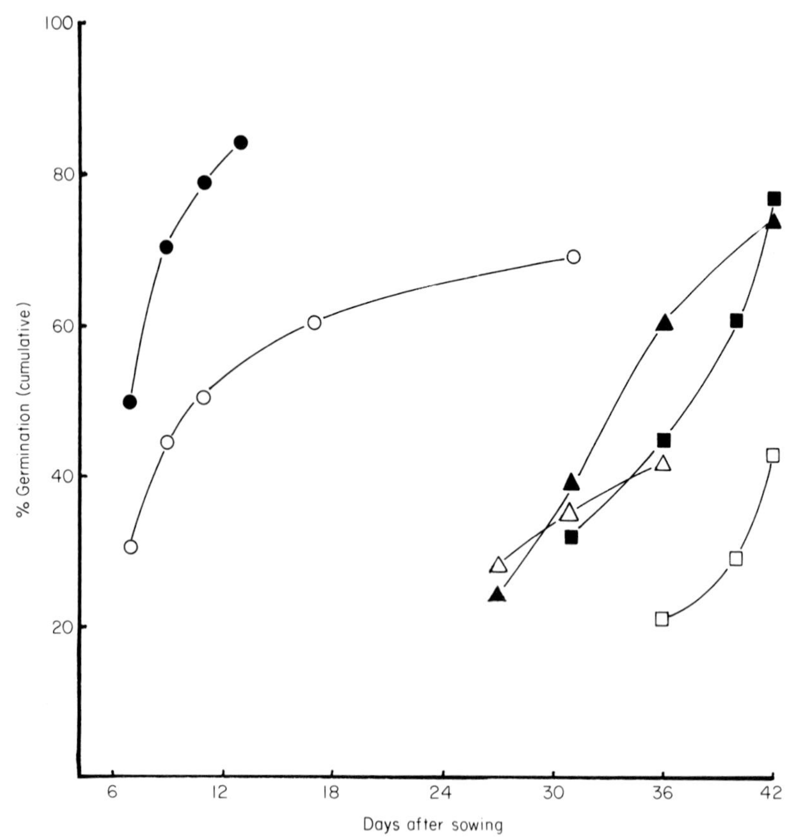

FIG. 8.2. Germination response of two populations of *Festuca rubra* var. *arenaria* to different salinity levels. (●), yellow dune population in dune rainwater; (○), grey dune population in dune rainwater; (▲), yellow dune population in dilute seawater (1:3) until 24th day after immersion when solution changed to distilled water; (△), grey dune population in dilute seawater (1:3) until 24th day when solution changed to distilled water; (■), yellow dune population in seawater until 24th day when solution changed to distilled water; (□), grey dune population in seawater until 24th day when solution changed to distilled water.

Glasshouse pot experiments were designed to determine the responses of *Festuca rubra* var. *arenaria* to different rates of accretion. These different rates of accretion were produced by fitting a series of plastic rings to the plastic pots and adding sand at the rate of one or two centimetres per fortnight, giving a total accretion of 10 or 20 cm during the period of the experiment. The sand used in all these experiments was collected from the surface 5 cm of either the yellow or grey sands at Blakeney. In one of these experiments, where 15 seeds

Table 8.2. *Mean values of the dry weight of seedlings, tiller numbers and leaves per plant of plants from two populations of* Festuca rubra *var.* arenaria *subject to different rates of sand accretion on two soils after a period of 20 weeks; each value is a mean for surviving seedlings (number surviving in parenthesis).*

	Mobile dune population			Fixed dune population		
	Dry weight (g) ± S.E.M.	Mean tiller number	Mean no. leaves per plant	Dry weight (g) ± S.E.M.	Mean tiller number	Mean no. leaves per plant
Yellow mobile sand						
Control	(8) 0·067 ± 0·012	1·0	5·5	(11) 0·091 ± 0·026	0·8	5·7
10 cm accretion	(10) 0·164 ± 0·029	1·9	6·9	(7) 0·311 ± 0·047	1·4	9·8
20 cm accretion	(5) 0·530 ± 0·149	2·3	11·6	(0) nil nil	nil	nil
Grey fixed sand						
Control	(6) 0·669 ± 0·160	2·8	19·0	(12) 0·415 ± 0·029	3·0	15·1
10 cm accretion	(6) 1·728 ± 0·498	4·5	21·3	(5) 0·897 ± 0·217	4·0	35·6
20 cm accretion	(3) 2·623 ± 0·301	5·7	38·3	(1) 0·347 —	4·0	11·0

were sown per treatment, pots were set up with saucers, to conserve water and nutrients, in a semi-random arrangement. Watering took place regularly using distilled water and the experiment continued for 20 weeks from May to September 1975.

Sand accretion appeared to stimulate growth (Table 8.2 and Fig. 8.3). The dry weight of plants from the yellow sand increased with increasing additions of sand from both the yellow mobile and grey fixed dunes. In contrast, plants of the fixed dune population showed a similar growth response as plants from the other population in the control treatment and in pots where the accretion was 10 cm. However at the highest level of sand accretion only one seedling remained in the grey sand at harvest and in the yellow sand all seedlings had been totally inundated by the eighth week. Differences in the growth pattern between the two populations are shown in Fig. 8.3. Where plants survive tiller numbers increase in both populations with additional increments of sand. There was particularly vigorous tillering and production of new roots at the nodes immediately below the sand surface, similar to the effects of rooting noticed by Marshall (1965) in his investigations into the response to accretion of *Corynephorus canescens.*

At the 10 cm level of accretion on both sand types, the mean number of leaves per plant for the fixed dune population at harvest are approximately one and a half times that of the mobile dune population. The greater lamina length of the mobile dune fescues compared with that of plants from the fixed dune population indicates that during periods of high rates of accretion they are able to present a larger area for photosynthesis than the more numerous but shorter leaves of the fixed dune population.

MINERAL NUTRITION

Ramets of the two populations were grown in both the yellow mobile and the grey fixed sands in pot culture in a glasshouse and the growth response of the plants examined. At the start of the experiment the fresh weight of a sub-sample of ramets was determined, the material was then oven-dried at 105°C to obtain the dry weight. The linear regression describing the relationship between fresh and dry weights of the ramets was used to predict the initial dry weight of material grown in the experiment.

A significant growth response was shown by plants of both populations to the higher nutritional status of the grey fixed sand (Table 8.3). The principal differences between the two sand types are in the amounts of organic matter (0·090% in yellow sand and 0·408% in grey sand) and in the exchangeable calcium levels (66·3 mg/100 g yellow sand and 10·7 mg/100 g grey sand).

Numbers of tillers produced by plants of each population also were higher in cultures containing grey fixed dune sand compared with the corresponding numbers of plants grown in the yellow mobile sand.

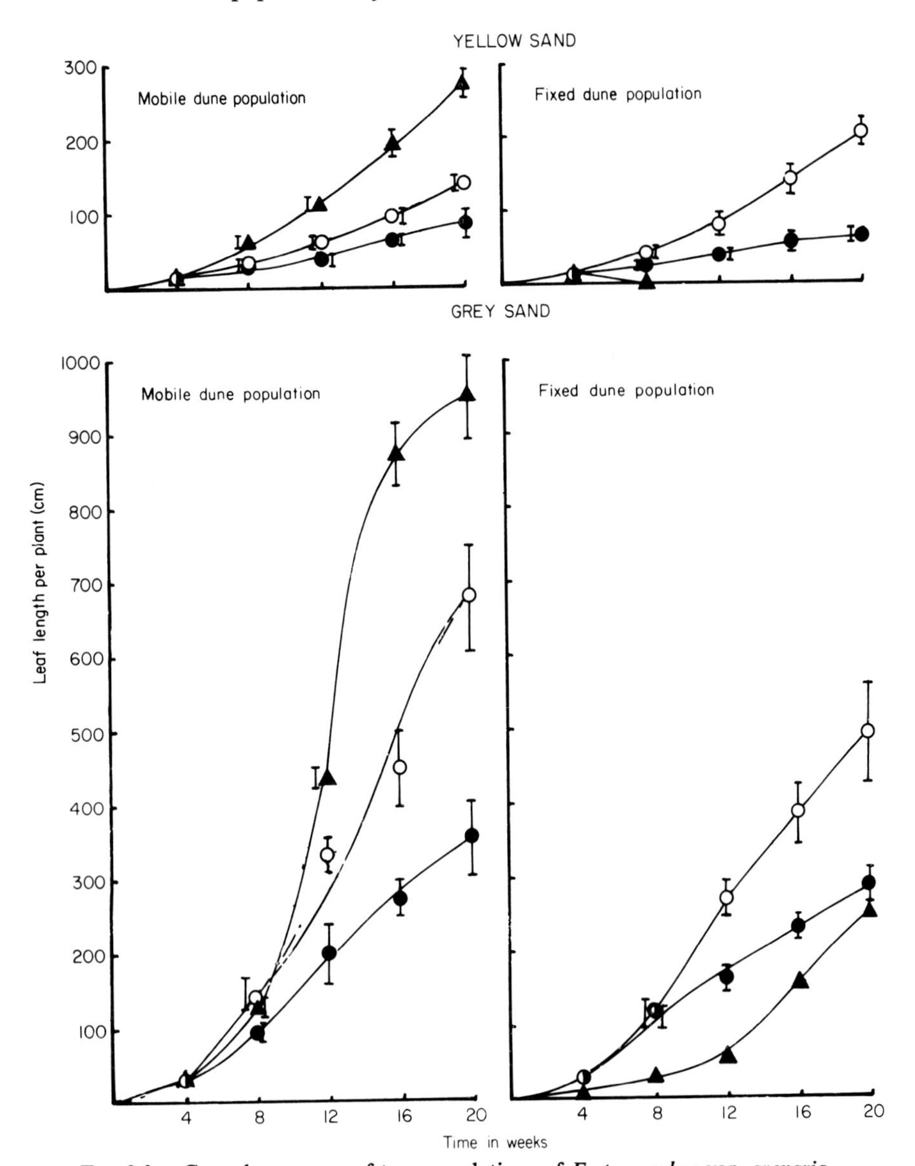

FIG. 8.3. Growth response of two populations of *Festuca rubra* var. *arenaria* to different rates of accretion; (●), control; (○), 1 cm of sand added every 14 days; (▲), 2 cm of sand added every 14 days. Vertical bars represent S.E.M.

Table 8.3. *Mean values for net dry weights, tiller numbers and root:shoot ratio of plants from two populations of* Festuca rubra *var.* arenaria *on two soils; each value is a mean of nine replicates.*

	Mobile dune population			Fixed dune population		
	Net dry weight (g) ± S.E.M.	Mean tiller number	root:shoot ratio	Net dry weight (g) ± S.E.M.	Mean tiller number	root:shoot ratio
Soil types						
Yellow sand	0·454 ± 0·056	2·44	0·647	0·419 ± 0·051	4·56	0·706
Grey sand	2·614 ± 0·281	14·44	0·891	2·586 ± 0·334	16·11	0·976

SOLAR RADIATION

Regular but discontinuous records of radiation collected at Blakeney Point were correlated with continuous records of radiation which were available from an automatic weather station at Holkham, Norfolk. A pair of tube solarimeters (Szeicz *et al.* 1964) were used to record total radiation at 2 m above the vegetation. Another pair were used to measure the mean radiation, over a distance of 1 metre, above the ground within tussocks of *Ammophila*, the space between the tussocks and in the open in the fixed dune community.

The level of the irradiance on the short turf of the fixed dune was approximately 99% of that at a height of 2 m above the sand. The reduction in the total radiation reaching *Festuca* plants within the marram tussock was approximately 88%.

In order to examine the growth response of the fescue plants to different levels of irradiance, plants from the two populations were grown in a garden under shade screens of Tygan netting which reduced the levels of irradiance to 79%, 55%, 33% and 12% respectively. One ramet was planted per pot and pots were arranged in random blocks under each screen. The two sands which were used, were collected from the top 5 cm layer of the mobile and fixed dunes respectively. Watering took place regularly with distilled water to augment natural precipitation. The experiment was carried out over 20 weeks between May and October 1975.

The growth response of the two populations to the different irradiance levels are shown in Fig. 8.4 and Table 8.4. There are significant differences in growth between plants grown in the two sands. The standard errors indicate a wide variation in the growth response of plants from the two populations. Whereas plants from the fixed dune population respond to increasing irradiance, as shown by the steady increase in growth, plants from the mobile dune population produce the largest growth response at an irradiance of about 55% daylight, appearing to behave as a shade tolerant plant.

COMPETITION

In a study of interspecific competition, ramets of each of the populations were planted in large plastic bins (35 cm diameter × 60 cm deep) in the dunes at Blakeney. Both pure stands and mixtures of two species were grown in separate bins. Mixed stands of ramets of *Ammophila arenaria* and of each of the *Festuca* populations were grown in the yellow sand. In another interspecific competition experiment ramets of *Carex arenaria* and *Festuca* from each of the two populations were planted in the grey sand. Sixteen plants of the two populations were planted in each of the monocultures and in the mixed

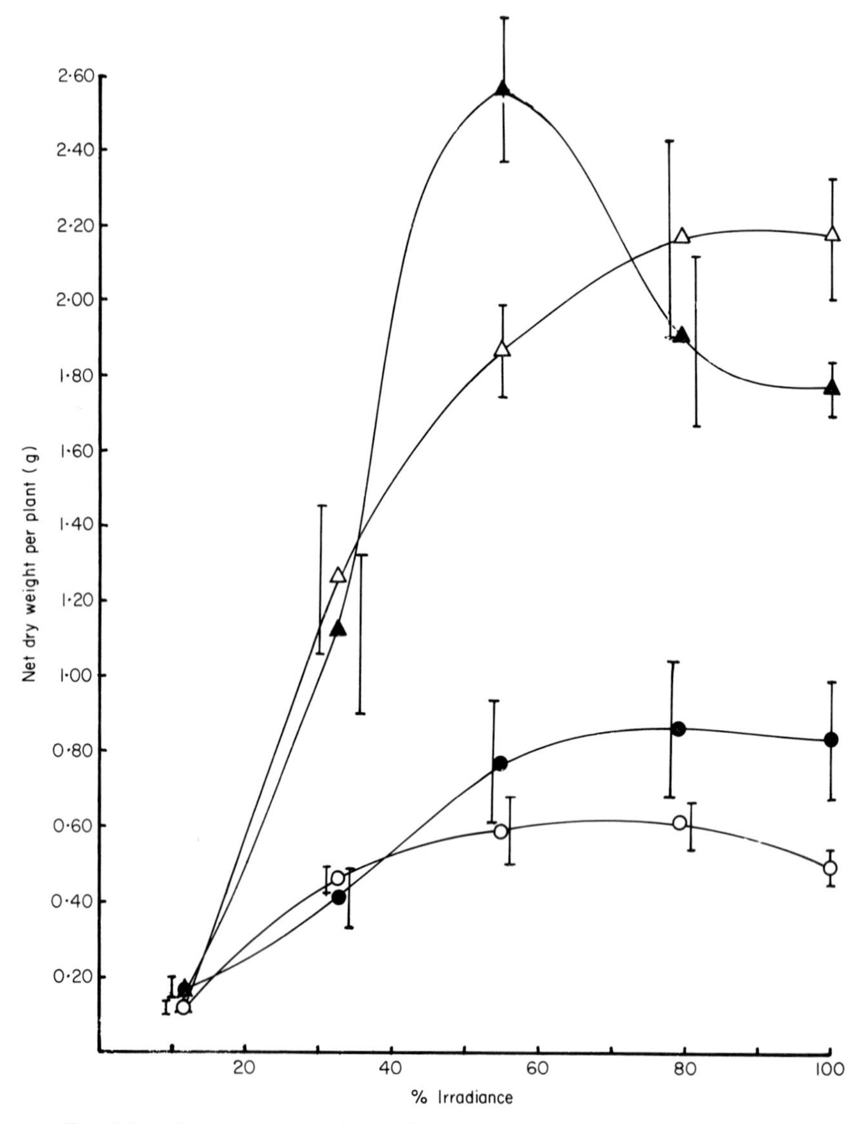

FIG. 8.4. Mean net dry weights of two populations of *Festuca rubra* var. *arenaria* at four different radiation levels on two soils; (●), mobile dune population in yellow sand; (○), fixed dune population on yellow sand; (▲), mobile dune population on grey sand; (△), fixed dune population on grey sand. Vertical bars represent S.E.M.

treatments eight plants of the two competing species were grown. Ramets were arranged in an alternating pattern so that they were equidistant from neighbours. Plants were watered with dune rainwater only at the time of planting on 23 August 1974. Bins were surrounded with 1-inch mesh netting to prevent grazing. Harvesting took place a year later on 20 August 1975.

Table 8.4. *Mean values for leaf area and tiller numbers of plants from two populations of* Festuca rubra *var.* arenaria *grown in two soils at five different radiation levels; each value is the mean of six replicates except for one treatment asterisked where five replicates only were available.*

	Mobile dune population		Fixed dune population	
	Leaf area (cm^2) ± S.E.M.	Mean tiller number	Leaf area (cm^2) ± S.E.M.	Mean tiller number
Yellow mobile sand				
100% irradiance	41·60 ± 6.62	5.83	26·96 ± 5·59	4·17
79% irradiance	42·28 ± 7·17	4·17	37·92 ± 7·53	5·00
55% irradiance	49·42 ± 4·10	4·67	34·64 ± 5·02	3·83
33% irradiance	43·53 ± 6·61	2·83	41·76 ± 2·60	4·88
12% irradiance	28·92 ± 4·07	1·17	30·81 ± 2·36	2·00
Grey fixed sand				
100% irradiance	86·21 ± 17·40	10·50	66·02 ± 5·94	9·33
79% irradiance	70·71 ± 6·16	7·50	74·37 ± 9·80	8·50
55% irradiance	*96·13 ± 18·51	9·50	83·77 ± 8·60	8·33
33% irradiance	76·23 ± 8·93	4·17	84·73 ± 9·51	6·00
12% irradiance	53·78 ± 6·47	1·67	50·25 ± 8·42	2·50

Although no statistical treatment was possible the results have been presented in a de Wit model graph (de Wit 1960), Fig. 8.5. Results from the mobile dunes in mixed stands with *Ammophila arenaria* indicate increased growth by both fescue populations over the corresponding monoculture treatments, particularly with the fixed dune plants. In monoculture, the native fescue of the mobile dunes appeared to produce three times the dry weight of that of the fixed dune population. The growth of the native fescues in the fixed dunes appear to be twice that of the mobile dune fescues in monoculture and there is little difference in growth of the two fescue populations in mixed or pure stands. Relative crowding coefficients show little variation between the associated species on the fixed dunes (k_{FC} = 3·42 with the mobile dune fescue and 3·75 with the fixed dune fescue). However, in the fixed dunes there are marked differences between the two associations (k_{FA} = 1·32 with the mobile dune fescue and 2·38 with the fixed dune population).

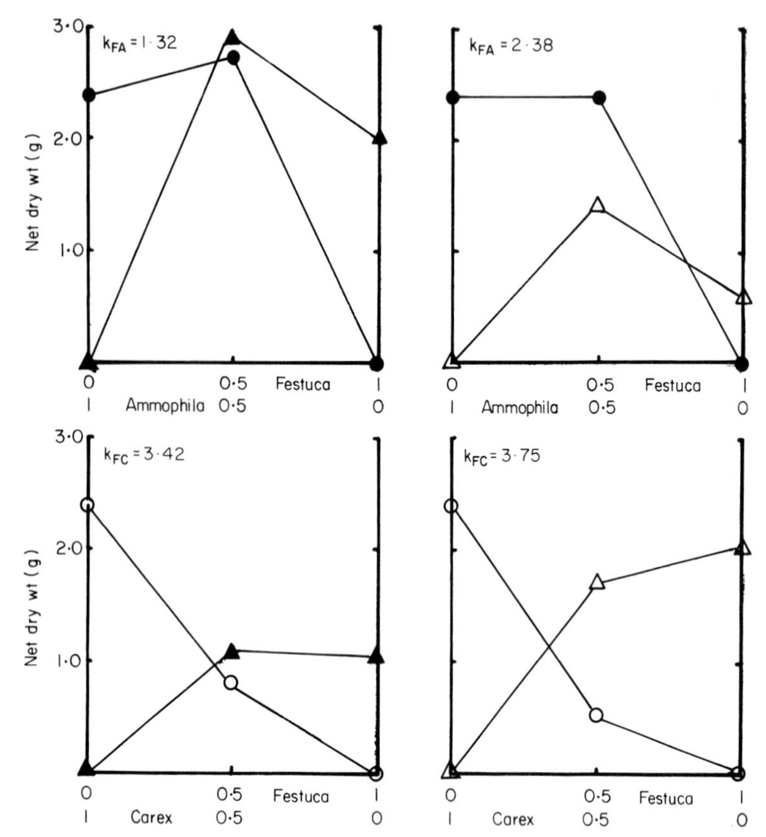

FIG. 8.5. The results of de Wit model competition experiments on the mobile dunes between *Festuca rubra* var. *arenaria* and *Ammophila arenaria* and on the fixed dunes between *Festuca sp.* and *Carex arenaria*; (●), *Ammophila arenaria*; (○), *Carex arenaria*; (▲), mobile dune fescue; (△), fixed dune fescue.

DISCUSSION

The only discrete morphological differences between the populations in the field are vegetative characters, but under cultivation these characters showed plasticity of form. However, the two populations do show different physiological responses to accretion, and to a lesser extent, salinity and radiation.

Apart from these, other possible differences separating the populations have not been examined. However, observations were made on the effects of rabbit grazing and also the time of anthesis in plants of each fescue population. The *Festuca* plants on the grey dunes are grazed heavily throughout the year. Although daily visits to Blakeney were not possible, it seemed that anthesis in the plants from the mobile dune occurs approximately ten days earlier than in those from the fixed dune. From records of wind velocity and direction, it is clear that during the period of anthesis and seed production there is considerable variation in force and direction (as much as 180°) of the wind which would allow gene flow via pollen and seed dispersal to occur between the two populations.

Without further investigation into possible isolating mechanisms it is difficult to assess exactly how far these two Blakeney populations are adapted to their particular niches. It is possible that the growth response to low levels of radiation is one of the factors implicated in favouring the establishment of the yellow dune fescue amongst the tussocks of the mobile dunes. The variation in response of seedlings to high rates of sand accretion may be the result of morphological differences in the early stages of growth between plants from the two populations favouring the mobile dune fescue and tending to prevent the establishment of the fixed dune fescue.

ACKNOWLEDGMENTS

Clare Anderson gratefully acknowledges Studentship Grants from University College London and a Mary Scharlieb Scholarship from the University of London. We wish to thank the Institute of Hydrology, Wallingford, Oxon, for access to data from the automatic weather station at Holkham, Norfolk.

REFERENCES

Bhadresa R. (1977) Food preferences of rabbits *Oryctolagus cuniculus* L. at Holkham sand dunes, Norfolk. *J. appl. Ecol.* **14**, 287–92.

Bird E.C.F. & Wain J. (1963) Changes at Blakeney Point since 1953. *Trans. Norf. and Nor. Nat. Soc.* **20**, 1.

de Wit C.T. (1960) On competition. *Versl. landbouwk, Onderz. Ned.* **66**, 1–82.

Gorham E. (1958) Soluble salts in dune sands from Blakeney Point in Norfolk. *J. Ecol.* **46**, 373–79.

Marshall J.K. (1965) *Corynephorus canescens* (L.) P. Beauv. as a model for the *Ammophila* problem. *J. Ecol.* **53**, 447–63.

Olson J.S. (1958) Lake Michigan dune development. 1. Wind-velocity profiles. *J. Geol.* **66**, 254–63.

Salisbury E.J. (1922) The soils of Blakeney Point; a study of soil reaction and succession in relation to the plant covering. *Ann. Bot.* **36**, 391–431.

Salisbury E.J. (1952) *Downs and Dunes.* Bell, London.

Szeicz G., Monteith J.L. & dos Santos J.M. (1964) Tube solarimeter to measure radiation among plants. *J. appl. Ecol.* **1**, 169–74.

White D.J.B. (1961) Some observations on the vegetation at Blakeney Point, Norfolk, following the disappearance of the rabbits in 1954. *J. Ecol.* **49**, 113–18.

9. RESPONSES OF ANIMALS TO SPATIAL AND TEMPORAL ENVIRONMENTAL HETEROGENEITY WITHIN SALT MARSHES

B. HEYDEMANN

Zoologisches Institut der Universität, Kiel, B.R.D.

SUMMARY

The distributions of different animal species within salt marshes of north-west Germany are described. Approximately 75% of the species of the lower salt marsh are not found in other habitats, whereas only 25% of the fauna of the upper salt marsh is restricted to such habitats. The responses of the fauna to changes in the frequency and duration of tidal inundations depend on whether the organisms are found on the aerial parts of vegetation (Phytobios), the soil surface (Epigaion) or in the soil (Endogaion). Among the factors which influence the ability of animals to colonize the low marsh are the prevailing salinity, the presence of host plants and animals and the availability of detritus which may act as a nutrient source for some groups of animals. In the upper marsh tolerance of organisms to salinity, submergence by sea water and passive dispersal by waves, appears to be important for successful colonization.

Besides spatial heterogeneity, temporal heterogeneity affects the distribution and activities of organisms within the marsh. Many terrestrial arthropods, like species of Araneae and Carabidae, are able to survive covered with sea water for 3 months during the winter. Although many animals of salt marshes exhibit circadian rhythms only a few species show a rhythm which is in phase with the 25 hour tidal cycle. A number of arthropods in salt marshes have extended periods of activity which are greater than six months, unlike those of inland populations of the same species. Reproduction in some spiders occurs during the period when the lowest number of tidal inundations takes place.

During primary succession, dispersal of animals to new habitats appears to be efficient. Many of the organisms are endophagous parasites of other animals and plants. A number of survival strategies in the different groups of organisms are evident. These include changes in morphology and reproductive rate, dispersal and migration, tolerance of salinity, osmotic and ionic regulation and behavioural changes. In addition there is evidence of population differentiation within coastal ecosystems.

Predacious species appear to show no ecological dependence on halobiont prey. Salt marsh aphids select plants with the lowest salt content and under experimental conditions select glycophytes in preference to halophytes. They also are able to regulate the osmotic pressure of their haemolymph when sucking sap which is hyperosmotic relative to that of the haemolymph. Spiders of salt marshes apparently drink rainwater or dew in order to regulate their internal osmotic pressure.

RÉSUMÉ

Nous décrivons la distribution de différentes espèces animales dans les marais salés du nord-ouest de l'allemagne. 75% des espèces des marais salés inférieurs ne sont pas trouvés dans d'autres habitats, contre 25% de la faune des marais salés supérieurs. Les réponses de la faune aux changements de fréquence et de durée des inondations par la marée varient selon que les organismes sont trouvés dans les parties aériennes de la végétation (phytobios), à la surface du sol (epigaion) ou dans le sol (endogaion). Parmi les facteurs qui influent la capacité des animaux à coloniser les marais salants inférieurs, on trouve en premier lieu la salinité, la présence de plantes hôtes, et d'animaux, et la présence de détritus comme nutriment pour quelques groupes d'animaux. Dans les marais supérieurs la tolérance des organismes à la salinité, la submersion par l'eau de mer, et la dispersion passive par les vagues, semble être importante pour le succès de la colonisation.

Avec l'hétérogénéité spatiale, l'hétérogénéité temporelle affecte la distribution et les activités des organismes dans le marais. Beaucoup d'arthropodes terrestres comme les espèces Araneae et Carabidae peuvent survivre en hiver jusqu'a trois mois dans l'eau de mer. Bien que beaucoup d'animaux des marais salants montrent des rythmes circadien, peu d'espèces montrent un rythme en phase avec le cycle de 25 heures de la marée. Nombre d'arthropodes dans les marais salés ont une durée d'activité supérieure à six mois, au contraire des populations de même espèce de l'intérieur. La reproduction de quelques araignées s'effectue durant la période d'inondations minima par la marée.

Durant la succession primaire, la dispersion des animaux vers de nouveaux habitats semble être efficace. Beaucoup d'organismes sont des parasites endophages d'autres animaux et de plantes. Nombre de stratégies de survie chez les différents groupes d'organismes sont visibles. Ils comprennent les changements de morphologie et de taux de reproduction, de dispersion et de migration, de tolérance à la salinité osmotique et la régulation ionique, et des changements de comportement. De plus il y a des évidences de différenciation de populations dans les écosystèmes côtiers.

Les prédateurs ne semblent pas dépendre écologiquement de leur proie hélobiont. Les aphides des marais salas sélectionnent les plantes de plus faible teneur en sel, et sous des conditions expérimentales préfèrent les glycophytes aux halophytes. Elles peuvent réguler également la pression osmotique de leur hémolymphe quand le suc est hyperosmotique par rapport à l'hémolymphe. Les araignées des marais salés ssemblent boire l'eau de pluie ou la rosée pour réguler leur pression osmotique interne.

INTRODUCTION

Two types of response of animals to environmental heterogeneity within salt marshes may be recognized:

(a) Response of individuals to short-term changes in environmental conditions. These responses include modifications in behavioral patterns, migration and the dispersion of individuals along ecological gradients.
(b) Alterations in the genetic structure of populations in response to long-term changes in environmental conditions. An adaptation in individuals of a population to particular environmental conditions within a habitat may enable these individuals to migrate into new habitats in which similar conditions prevail. Of particular interest are adaptations to high salinity, ability to tolerate periods of inundation and the capacity of organisms to float on the surface of water.

Animals which are tolerant of a wide range of environmental conditions are able to colonize different habitats when conditions within a habitat are unsuitable ('Euryökie'). In contrast some animals are highly specialized and are restricted to specific niches ('Stenökie').

Distribution of animals in salt marshes in relation to environmental heterogeneity

Within salt marshes of north-west Germany, where the vertical tidal range is 120 cm, the distribution of animals is related to the tidal amplitude. Approximately 75% of the fauna present in the lower salt marsh, between 0 and 40 cm above the mean high water mark of the tides (M.T.H.W.L.), is not found elsewhere. However at higher levels within the salt marsh this percentage falls. In the middle marsh (40–60 cm above M.T.H.W.L.) the corresponding values are between 40–45%, while in the high marsh (60–80 cm above M.T.H.W.L.) only 25% of the animal species are restricted to salt marshes. This value drops to 5–10% in the transition zone between salt marshes and fresh water marshes (80–120 cm above M.T.H.W.L.). The plant associations Puccinellietum maritimae, Festucetum rubrum littorale, Festucetum rubrum littorale-Artemisietum maritimae and *Leontodon autumnalis* dominate respectively the lower, middle, upper and brackish marshes of this region.

The responses of animals to the number of tidal inundations and to the duration of the submergences during each lunar cycle differ according to whether they live on the aerial parts of plants (phytobios), live on the surface of the soil (epigaion) or live within the soil (endogaion).

Animals which colonize the lower marsh are a highly specialized group of organisms, adapted to salinity of soil, water and food, and to an aquatic environment. In addition, tides disperse animals present in the Salicornietum and Puccinellietum zones of the lower marsh. Only a few species are able to live both under these conditions and those of a non saline habitat. They include the micryphantid spider, *Erigone atra* (Heydemann 1960); the oribatid mite, *Oppia minus*; the Collembola, *Tullbergia krausbaueri* and *Friesea mirabilis* (Weigmann 1973).

Species which colonize the upper levels of salt marshes often show a wide degree of tolerance to environmental factors such as salinity, submergence and passive dispersal by waves. In this habitat conditions fluctuate considerably unlike those of the lower levels of salt marshes. Hence those organisms present in marshes close to the sea only survive under a restricted range of environmental conditions but animals of the upper marsh show a wide ecological tolerance. These differences in tolerance reflect spatial and temporal differences in environmental conditions that exist between the upper and lower marshes of this region. Survival strategies of organisms in the different areas of a marsh involve both short-term responses of individuals and long-term changes in population structure (cf. Foster & Treherne 1976; Tischler 1976).

DIFFERENT SURVIVAL STRATEGIES OF ORGANISMS IN SALT MARSHES

Organisms which show little capacity for adjustment to changed conditions

Tolerance of metabolic reactions to the poikilosmotic values of tissues

The halobiont Lycosid spider *Pardosa purbeckensis* accumulates greater amounts of sodium in its tissues when the salinity is increased, compared with the amounts in the halotolerant spider *P.amentata.*

Tolerance of organisms to extreme and rapidly changing temperatures

Temperatures on the slopes of embankments or the shore may reach high values on hot summer days and some of the species which live in these habitats allow their internal temperature to rise.

Behavioural flexibility

Some spiders, such as *Erigone arctica* (Micryphantidae) and *Leptorrhoptrum robustum* (Linyphiidae), have the ability to spin nets under submerged conditions.

Presence of regulatory mechanisms in organisms

Internal physiological regulatory mechanisms

These mechanisms include osmotic and ionic regulation in hyperosmotic media, temperature regulation, regulation of oxygen consumption under submerged conditions and tolerance to still and turbulent water.

Morphological changes

Species of terrestrial invertebrates which live in saline habitats have a small body size compared with closely related species which are found in non-saline habitats. Small spiders (Micryphantidae, Linyphiidae) have a greater resistance to tidal coverage than the larger Lycosidae, although both have the same tolerance to salinity (e.g. *Erigone longipalpis*—Micryphantidae and *Pardosa purbeckensis*—Lycosidae).

It appears that as a result of the tidal coverage of a marsh the small Micryphantidae such as species of the genera *Erigone*, *Oedothorax*, *Enidia* and *Praestigia* are freed from aerial respiration which involves the use of lungbooks and trachea and instead the epidermis above the lungbooks acts as a tracheal gill. When these invertebrates are submerged under tidal water the epidermis appears as a 'white window' because of reflection of light from air trapped in the lungbook cavity.

Physical gills are widespread in salt marsh invertebrates. These physical gills take the form of bubbles or films of air which are held under water and act as an air store when organisms are submerged. Many spiders especially the Lycosidae, beetles and bugs show this type of adaptation. The presence of epidermal hairs is sufficient to ensure an air film is trapped between the hairs when the organisms are covered by tidal water. Many coastal invertebrates are adapted to live in both aquatic and terrestrial environments but this ability is not unique to this group of organisms. However it allows such organisms to colonize coastal habitats. Physical gills as films of air have an additional role in that they act as passive osmoregulators.

Plastron gills are present in many Arthropod groups which suggests the polyphyletic evolution of the plastron. Plastron respiration probably evolved in organisms living in the limnetic zone. Plastron-bearing gills are known to occur in inter-tidal beetles and Diptera (Hinton 1976) which enables the larval and pupal stages of these invertebrates to respire in both aquatic and terrestrial environments. The consequences of the evolution of these morphological structures is that the organisms have the ability to take up oxygen both when they are submerged and when they are exposed to the air. This evolution of a respiratory organ capable of functioning in both media is an alternative evolutionary response to other more complicated behavioural responses when changes in environmental conditions occur.

Morphological adaptations appear to be characteristic of organisms which move across the surface of the water. Many inhabitants of wet marshes in which there may be a considerable amount of standing water are adapted to move on the surface of water. Evolutionary trends may be recognized in some groups of organisms in which all stages of the life cycle are spent on the surface of the water. The spiders *Pardosa purbeckensis* and *Pirata piraticus* (Lycosidae) hunt prey on the surface of tidal water when calm conditions prevail. The end of the tarsal segments often have hydrofuge hairs which are of a different structure than normal hairs. Many species of the Lycosid spiders, especially species of the genus *Pirata*, are able to sail with the wind by lifting one or two legs into the air and the extremities act as a sail.

Members of the Diptera also show a series of adaptations which enable them to move across different substrates. For example many of the coastal Dolichopodid flies are able to move equally well across soil, leaves and the surface of water. They are not swamped by tidal water (Sommer, unpublished data). Only one dominant dolichopodid species of coastal marshes in Germany has specialized as a 'water strider' (Sommer, unpublished data). This species, *Hydrophorus oceanus* is mainly found in creeks and salt pans. Another species which shows a similar adaptation is *Ephydra riparia* (Diptera).

Indirect modes of independence from changes in environmental conditions

Endophagous habit

An interesting response to environmental changes is shown by many phytophagous insects. They avoid the deleterious effects of high salinity and inundation by living within the tissues of halophytic plants. A large proportion of the different species of coastal Lepidoptera (ca. 60%) spend the larval stages living inside roots, stems, shoots, leaves or flowers of salt marsh plants (Stüning, unpublished data). The selective advantage of the endophagous mode of life has favoured those groups of animals in salt marshes which have a small body such as the Microlepidoptera, the Curculionidae and the Cecidomyidae. Transition stages between the ectophagous and endophagous habits can be recognized in different species. There is an inverse relationship between the number of ectophagous species on marshes and the frequency of tidal inundations. In embanked higher marshes the number of ectophagous Noctuidae is three times greater than on an exposed shore.

Endogaeic habit

A parallel ecological response of animals to the changing environment is the endogaeic habit. Most species of beetles (c. 70% of all Coleoptera) in salt marshes live within 5 cm of the surface. Within that group of animals are types which burrow within the soil or survive in the channels of dead roots. Species within the genera *Bledius*, *Trogophloeus* (Staphylinidae) *Dichirotrichus* and *Dyschirius* (Carabidae) show this type of habit. The animals survive the adverse effects of inundation by utilizing air pockets within the soil as physical gills; and because they are below the surface of the sediment they are unlikely to be transported to unsuitable habitats.

Endoparasitic habit

Endoparasitism has evolved in a number of salt marsh species. This habit which occurs at the larval stages can be regarded as an adaptation whereby the organism avoids the uncertainties of the environment in coastal ecosystems. Hymenopterous parasites are found in over 100 different species in the salt marshes of northern Germany. These include 10 species in the Pteromalidae (Abraham 1970, 1973), 25 species in the Ichneumonidae (Horstmann 1970), 4 species in the Proctotrupid (Weidemann 1965), 70 species in the Braconidae (Konig 1969) and many Chalcidoidea, Proctotrupoidea and Mymaridae. Sometimes as many as 90% of the individuals of Micryphantid populations from the higher marshes contain parasites (Horstmann 1970).

RESPONSES OF ANIMALS TO TEMPORAL CHANGES

Unpredictable changes

Within salt marshes a large number of environmental changes are associated with unpredictable events such as the occurrence of exceptionally high tides. The latter may destroy islands of developing salt marsh (Halligen) so that these sites lose between 70–90% of their invertebrate fauna. Horstmann (1970) has found that nearly 75% of the species of Ichneumonidae present in these islands re-colonize the site each year. The females of the Ichneumonidae are fertilized immediately after hatching, hence only one fertilized female is necessary for the reestablishment of a population and the initial presence of both sexes on the island is not a prerequisite for colonization.

Most arthropods of the supralittoral marshes show no mortality following a period of tidal submergence of up to 8 hours. During the winter some terrestrial arthropods such as members of the Araneae and Carabidae are able to survive submergence for between one and three months provided the temperature is low.

Predictable changes: cyclical conditions

Circadian changes

Most arthropods of salt marshes show a circadian rhythm of activity. The rhythms are more pronounced in the animals which live in vegetation compared with soil inhabitants that are present below mats of grasses. Although most organisms show circadian rhythms in only a few species is the rhythm in phase with the 25 hour tidal cycle.

Seasonal changes

Seasonal changes in the structure of populations are pronounced in the majority of the salt marsh fauna compared with the marine fauna which lives below the mean high water mark of the tides; the changes are particularly marked in phytophagous species which live on dicotyledonous plants. The changes can be correlated with the seasonal growth patterns of individual plant species within the salt marshes. The predatory soil surface inhabiting Carabidae and Araneae which live under mats of grasses are active for much of the year as adults, unlike related species from inland areas.

The number of holeurychrone species (species which spend more than six months a year in an active state) is greater than the corresponding number of stenochrone species which spend less than 2 months each year in an active state (Fig. 9.1). In addition larval development in many Carabidae occurs in winter in these salt marshes, unlike the situation at inland sites. The long period of adult activity and the winter development of larvae appears to reflect the prevalence of oceanic climatic conditions in coastal habitats during this season. Another example of a seasonal pattern in the life cycle of a species is that the reproductive period of a number of different spiders in salt marshes coincides with the period of the year when minimal tidal inundations occur.

Biological rhythms associated with tidal cycles

In contrast to the animals which occur below the mean high water mark of the tides on the muddy foreshores of the Wadden sea and sandy and rocky foreshores, very few animals which live in salt marshes appear to show a rhythm of activity which is in phase with tidal cycles. Although the activities of most species follow a circadian rhythm, this rhythm is disrupted when the animals are covered with sea water. Unlike many sandy shores where the animals can move ahead of the advancing tide, salt marshes offer considerable obstacles to animal movement. In addition the distance across a marsh from the seaward edge to the upper levels is often large and upward migration of invertebrates as a means of escaping tidal inundation is limited because the vegetation is usually not more than 50 cm in height. For these reasons the activities of many salt marsh invertebrates appear to decrease when the animals are covered during the tidal cycle.

Successional changes in animal populations

Primary succession

On the north German coastline maximum rates of the net accumulation of sediment are 1 cm per year, although at many sites it is substantially less than

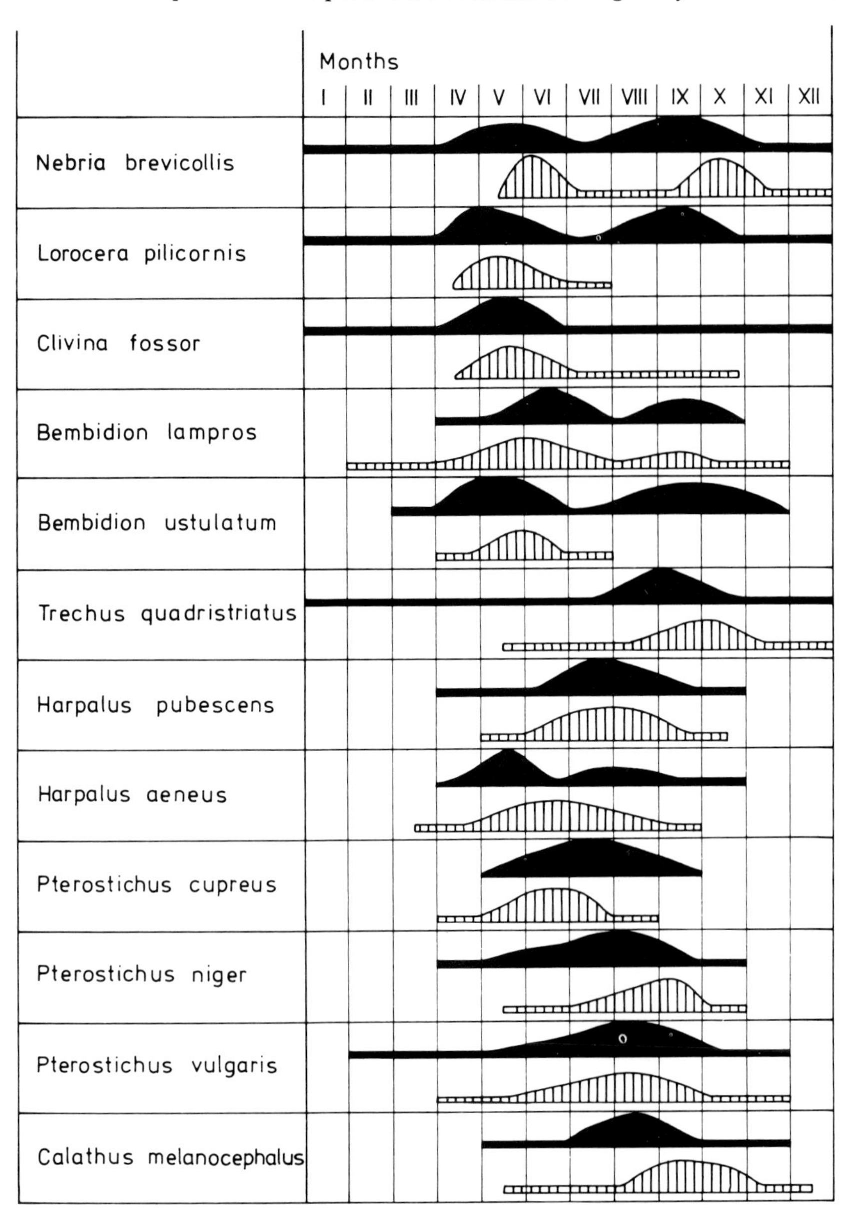

FIG. 9.1. Monthly changes of the relative activity of the dominant species of Carabidae, as measured by catches in pitfall traps, in salt marshes of north-west Germany (■) and in inland fields 80 km from the coast (◫). Populations from the coast have an extended period of activity throughout much of the year.

this. Under present conditions it takes between 25 and 40 years for the plant communities to change from the Salicornietum strictae and the Puccinellietum maritimae to the lower Festucetum rubrum littorale. A further 20–40 years is required for the upper Festucetum rubrum littorale to develop. This community is usually present at sites which are between 50–60 cm above the mean high water mark of the tides, whereas the lower Festucetum rubrum littorale occurs about 20 cm above the mean high water mark. After about 80 to 120 years freshwater plant communities such as the *Leontodon autumnalis* association develop in place of the existing salt marsh communities. Thus primary succession normally takes between 80 and 120 years on the north German coastline. However, recently as a result of the construction of dikes this sequence is only taking between 2 to 5 years to occur. Not only the distributions of species of phytophagous insects but also species of the Araneae and Ichneumonidae are closely correlated with the distributions of different plant communities associated with the successional stages (Figs. 9.2 and 9.3). One

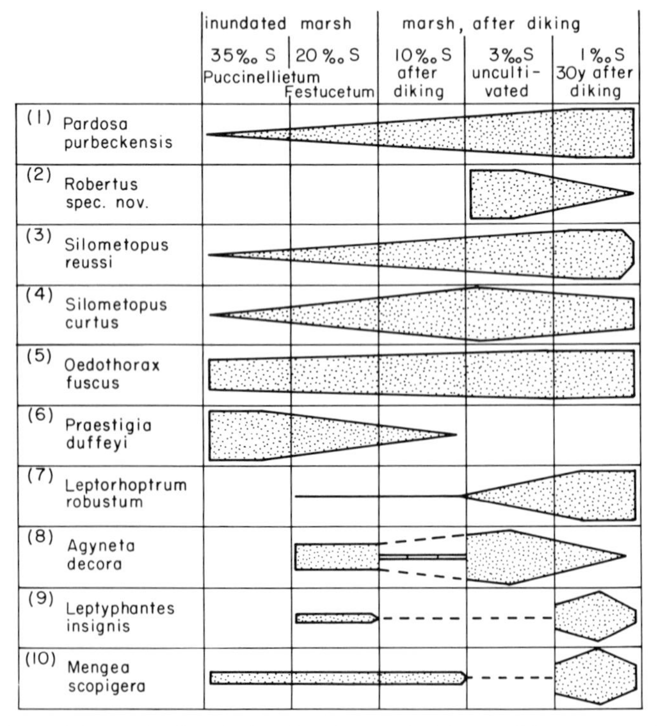

FIG. 9.2. Changes in the relative abundance of spiders at different stages in the development of salt marshes in north-west Germany. Soil salinities are shown for each stage of development. Species code number: (1) Family Lycosidae; (2) Family Theridiidae; (3–6) Family Micryphantidae; (7–10) Family Linyphiidae.

of the effects of diking the marshes is that the salt content of the sediment falls, and new species of invertebrates appear and populations of other species increase. Initially most species of the original salt marsh fauna are able to exist under these conditions, and the high mortality of individuals which occurred when the marsh was flooded is eliminated. These conditions result in

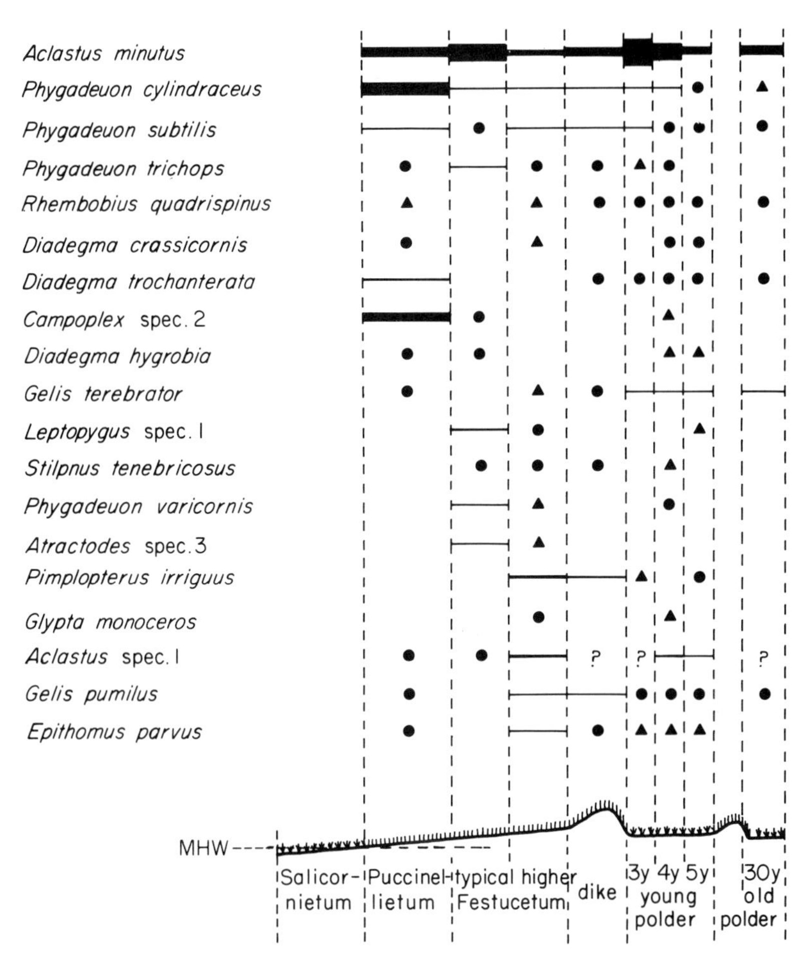

FIG. 9.3. Changes in abundance of Ichneumon flies (Ichneumonidae, Hymenoptera) at different stages in the development and reclamation of salt marshes in north-west Germany. 45% of all individuals and 4·3% of the species are parasites of spider egg cocoons. 36% of all individuals and 27% of the species are parasitic on flies and 12% of all individuals and 37% of the species are parasitic on moths (after Horstmann 1970). (▲) less than 1 specimen; (●) 1–5 specimens; (▬) 6–370 specimens shown by the width of the bar. Flies caught in a yellow light trap over a period of two months.

a species-rich ecosystem which may exist for five to six years after the diking of a marsh.

Secondary succession

Frequently an embanked area is cultivated after it is drained. These alterations to the environment result in a change in the species found on the marsh and an increase in the number of species, but an overall decrease in the abundance of arthropods after a few years. A comparison can be made with agricultural

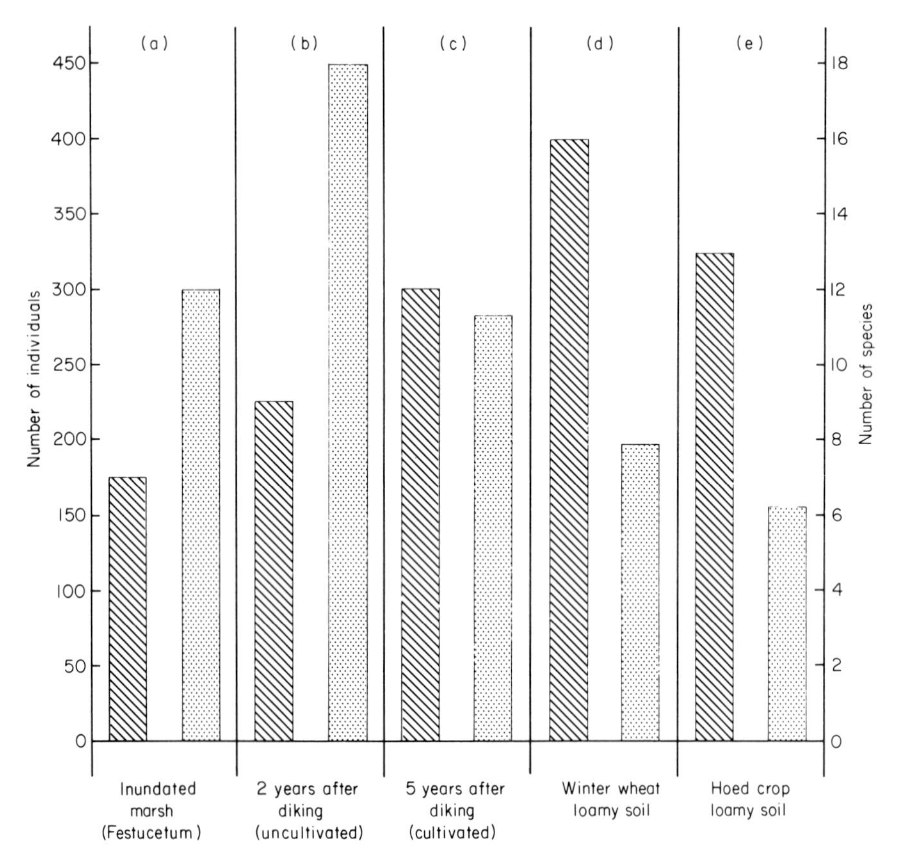

FIG. 9.4. Abundance (activity) of individuals and species of the Carabidae (imagines), based on catches in Barber traps, during a period of four weeks when maximum activity (mobility) of the invertebrates in different coastal and inland habitats in north-west Germany was evident. (a) inundated marsh, September; (b) reclaimed marsh, 1 to 2 years after diking, July; (c) reclaimed marsh, cultivated area 5 years after diking, August; (d) inland site in which winter wheat sown, June; (e) inland site where crop is hoed, mid-August to mid-September. (\\\\) number of species; (∴) number of individuals.

land in north-west Germany which has been cultivated for hundreds of years. In the family Carabidae although a greater number of species occurs in agricultural habitats compared with that of coastal environments, the densities of individuals are higher in coastal habitats.

Transition stages associated with reclamation

Changes in the abundance of different species after the construction of embankments are not only a response to an abrupt change in conditions but also to the large supplies of organic detritus which are available. These supplies which are derived from tidal inundations are limited in these reclaimed sites. Detritophagous species and the dependent carnivorous species are abundant from one to five years after the building of embankment. Thereafter as the supply of organic material decreases and is not replaced by fresh depositions of tidal litter, the numbers of these animals decline. The Carabidae in particular, show this trend (Figs. 9.4 and 9.5). The abundance of this group, based on captures from pitfall (Barber formalin) traps, reaches a maximum two years after embanking. Where reclaimed land is cultivated the abundance of these invertebrates five years after the change is lower than that in salt marsh habitats, but the living biomass of Carabidae is at a maximum five years after dike building. The values obtained are higher than comparable values for hoed crops in soils of similar nutrient status in inland areas of north-west Germany. However, the biomass of the Carabidae is lower than that recorded in winter wheat fields in Germany.

Responses of animal species during succession

During ecological succession from the Salicornietum of the lower salt marsh to the fresh water marshes, which develop after the building of embankments, two ecological groupings of species may be recognized. Some species appear to be relatively insensitive to the changing conditions whereas others show a marked increase or decrease in abundance as conditions alter. The carabid, *Bembidion minimum* (Heydemann 1964) and the thalassophilous littoral limoniid, *Symplecta stictica* (Wrage 1977) are present in the marshes before and after embanking and appear to be insensitive to the changed conditions. These euryoekous species are tolerant of a wide range of conditions.

Immediately after the construction of an embankment the nutritional status of the sediments undergoes considerable alteration. Under conditions of reduced salinity and improved drainage, supplies of inorganic nitrogen are released into the sediment as a result of mineralization (cf. Beeftink, this volume). This results in the growth of vegetation while supplies of nitrogen last. Associated with this increase in plant biomass is an increase in species of animals which have short life cycles and excellent dispersal mechanisms. These opportunists are especially prevalent among the inhabitants of the vegetation (Tischler 1976). It is evident that the dispersal mechanisms are efficient.

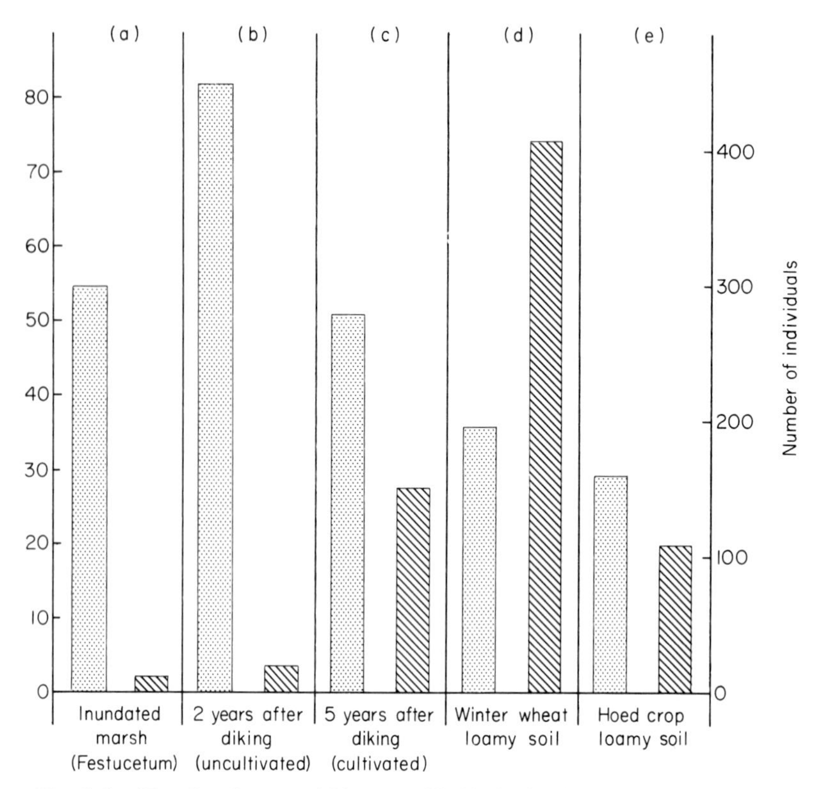

FIG. 9.5. The abundance and biomass of individuals of adult Carabidae based on catches in Barber traps during a period of four weeks when maximum activity (mobility) of the invertebrates in different coastal and inland habitats in north-west Germany was evident. (a) inundated marsh, September; (b) reclaimed marsh, 2 years after diking, July; (c) reclaimed marsh, cultivated area, 5 years after diking, August; (d) inland site in which winter wheat sown, June; (e) inland site where crop is hoed mid-August to mid-September. (\\\\) total biomass of individuals; (∴) number of individuals.

Embanked marshes which belonged to the marine ecosystems of the Wadden Sea contained only a marine fauna and very little vegetation before reclamation. Two years later a large number of animal species were recorded. The species included both carnivorous and parasitic species.

Responses of animals to changes in nutritional conditions

Dependence on salinity

For a large number of terrestrial predacious, halobiont species of arthropods it is uncertain why they are to be found only in coastal and inland saline areas, as they do not appear to be physiologically dependent on salinity. The

Micryphantid spider, *Erigone longipalpsis* shows little or no response to abrupt changes in soil salinity. Many predacious halobiont arthropods of the soil appear to be tolerant of different salinities, as indicated by their distribution under natural conditions. Whether all stages of the life cycle of these invertebrates are equally tolerant of different salinities is poorly known. In the case of plant-sucking insects there is some information which indicates that aphids such as *Macrosiphoniella asteris* select the stems of *Aster tripolium* which have a relatively low salt content. Apparently this aphid changes its sucking place on the plant in response to increases in the salt content of the phloem sap (Regge 1972).

Dependence of herbivorous insects on halophytic vegetation

A large number of species of herbivorous salt marsh insects appear to live as well on halophytes as on glycophytes but it has been suggested that the

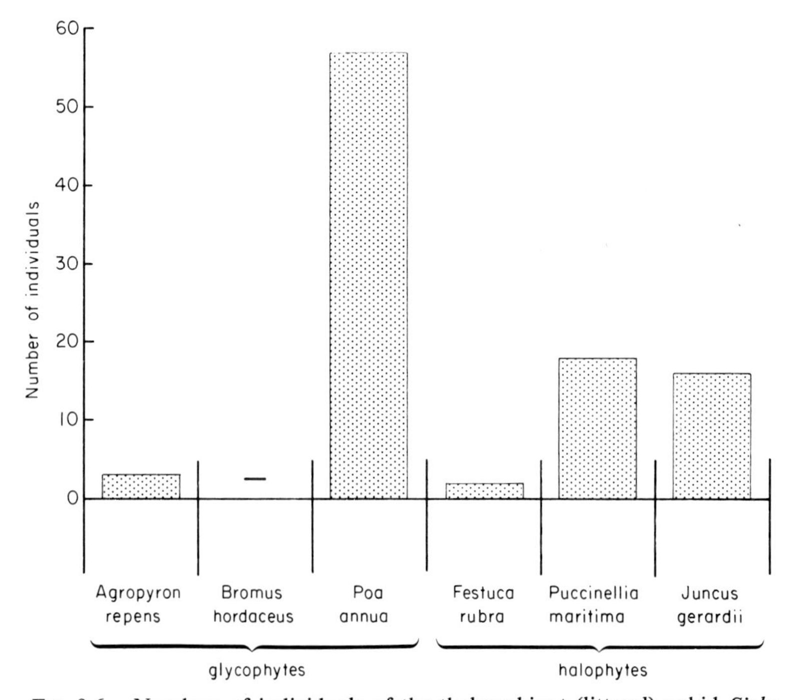

FIG. 9.6. Numbers of individuals of the thalassobiont (littoral) aphid *Sipha littoralis* which select different halophytes and glycophytes as a food source. The experimental period was five days. In salt marshes this aphid is found on *Puccinellia maritima*, *Spartina townsendii* and to a lesser extent *Festuca rubra*. It is not found on coastal populations of *Poa annua* adjacent to salt marshes, whereas a closely related species *S.glyceriae* does feed on this plant and also occurs on *F.rubra* in the upper salt marsh (after Regge 1972).

populations present in the saline and non-saline habitats have undergone population differentiation. However recent experimental work suggests that some species are not dependent on halophytic plants for at least part of their life cycle. The aphid *Sipha littoralis* which lives on *Spartina townsendii* and *Puccinellia maritima*, if given a choice of food materials, rejects these plants in favour of the glycophyte, *Poa annua* (Fig. 9.6), (Regge 1972). This dietary response is not observed under field conditions. The general response of the aphid species to variable environmental conditions is stenoekous but the individual behavioural response to food plants is euryoekous, where the organism appears tolerant of a wide dietary range. The persistent herbivorous invertebrates of salt marshes are more specialized in their feeding habits than

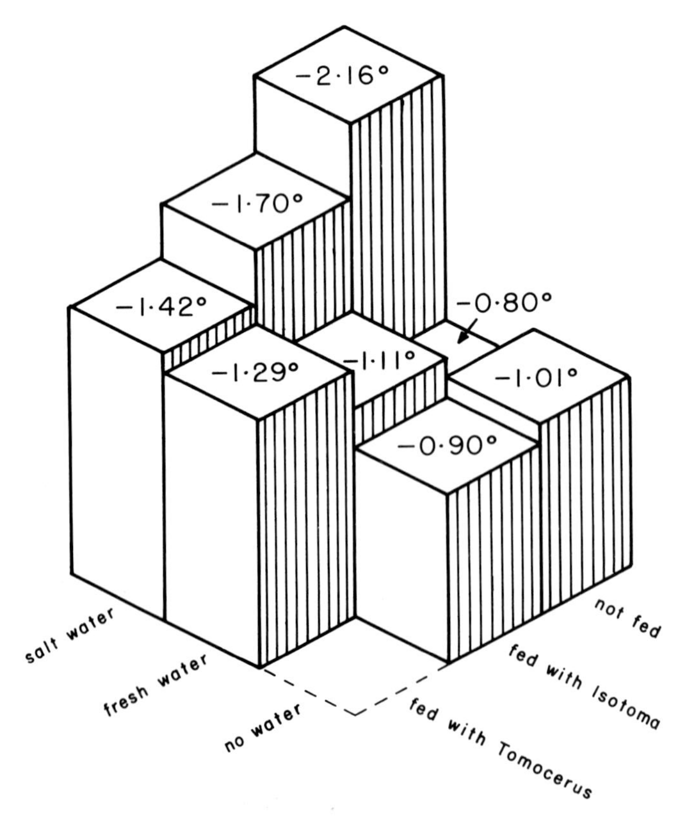

FIG. 9.7. Changes in the osmotic pressure of the haemolymph of the halobiont micryphantid spider, *Erigone longipalpsis*, based on the depression of freezing point (0°C), when the spiders are fed Collembola from salt marshes (*Isotoma* sp.) or inland forests (*Tomocerus* sp.) and allowed to drink either fresh water ($\Delta T = 0°C$) or salt water (30‰ S, $\Delta T = -1{\cdot}66°C$). The species is unable to survive for an indefinite period in the absence of fresh water (after Bethge 1973).

insects which are found in marshes only during certain seasons of the year. A number of halophytic species are known to contain high concentrations of organic solutes such as proline and sorbitol (cf. Stewart *et al.* & Jefferies *et al.*, this volume). It may be that herbivorous animals exploit this resource in spite of the high salinities which occur in plant tissues.

Dependence of carnivorous species on salinity

It is not known to what extent carnivorous species are influenced by changes in the salt content of their prey and whether this restricts them to salt marsh habitats. Experimental investigations with the halobiont species *Erigone longipalpsis* (Micryphantidae) show that this species will not only accept typical prey from a salt marsh habitat such as the Collembolan *Isotoma viridis* but also animals from inland forest habitats such as species of *Tomocerus* which lives in beech forests (Fig. 9.7). *Erigone longipalpis* obtains its water by drinking a drop of rain or dew in the salt marsh. Under these circumstances the haemolymph concentrations vary between 0·8° and 1·29°C based on the depression of freezing point (c. 9·7 to 15·6 atm). In the absence of fresh water

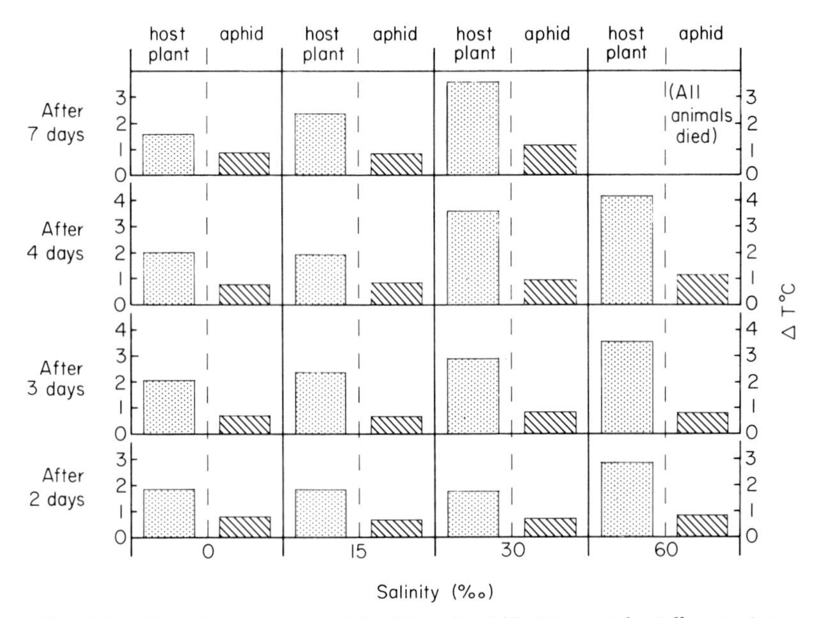

FIG. 9.8. Osmotic responses of the littoral aphid, *Macrosiphoniella asteris* to changes in the osmotic pressure of the host plant, *Aster tripolium*. The plants were placed in solutions of different salinity (0‰, 15‰, 30‰ and 60‰ S) for seven days and the aphids allowed to feed on the plants. Results of osmotic pressure of the aphids are shown as depression of freezing point (Δ°C) (after Regge 1972).

E.longipalpis drinks sea water. This can lead to a rise in the osmotic value of the haemolymph equivalent to a depression of freezing point of 2·16°C (c. 26·1 atm) which is lethal. The spider cannot tolerate seawater alone, some freshwater is required for survival.

Responses of phytophagous insects to changes in the salinity of the host plants

The ecological advantage of the phytophagous habit compared with the detritophagous habit is that the tissue of halophytes is subject to ionic and osmotic regulation so that the intake of food which has a near constant osmotic value does not require the animal to expend energy to maintain a constant osmotic pressure. However there are seasonal changes in the salt content of plant tissues. If plants of *Aster tripolium* are grown at different salinities under experimental conditions investigations of the aphid, *Macrosiphoniella asteris*, showed that the osmotic pressure of the haemolymph remained constant in spite of a rise in the salt concentration of the host plant. This resulted in a depression of the freezing point of the sap from 1·5° to 3·61°C (18·1 atm to 43·7 atm) (Fig. 9.8). If the osmotic pressure of the sap rose above 43·7 atm the animals died (Regge 1972).

REFERENCES

Abraham R. (1970) Ökologische Untersuchungen an Pteromaliden (Hym., Chalcidoidea) im Grenzraum Land-Meer an der Nordseeküste Schleswig-Holsteins. *Oecologia* **6,** 15–47.

Bethge W. (1973) Ökologisch-physiologische Untersuchungen über die Bindung von *Erigone longipalpis* (Araneae, Micryphantidae) an das Litoral. *Faun. Ökol. Mitt.* **4,** 223–40.

Foster W.A. & Treherne J.E. (1976) Insects of marine saltmarshes. Problems and adaptations. *Marine Insects* (Ed. by L. Cheng), pp. 5–41. North Holland Publishing Company, Amsterdam.

Heydemann B. (1960) Die biozönotische Entwicklung vom Vorland zum Koog. 1. Teil: Spinnen (Araneae) *Abh. math. naturw. Kl. Akad. Wiss. Mainz* **11,** 474–770.

Heydemann B. (1962) Die biozönotische Entwicklung vom Vorland zum Koog. 2. Teil: Käfer (Coleoptera). *Abh. math. naturw. Kl. Akad. Wiss. Mainz* **11,** 771–964.

Heydemann B. (1964) Die Carabiden der Kulturbiotope von Binnenland und Nordseeküste—ein ökologischer Vergleich (Coleopt./Carabidae) *Zool. Anz.* **172,** 49–86.

Heydemann B. (1967a) Der Überflug von Insekten über Nord-und Ostsee nach Untersuchungen auf Feuerschiffen. *Deutsche Ent. Z., N.F.* **14,** 185–215.

Heydemann B. (1967b) *Die biologische Grenze Land-Meer im Bereich der Salzwiesen.* Steiner-Verlag Wiesbaden.

Heydemann B. (1968) Das Freiland- und Laborexperiment zur Ökologie der Grenze Land-Meer. *Verh. dt. zool. Ges. Heidelberg* 1967, 256–309.

Heydemann B. (1973) Zum Aufbau semiterrestrischer Ökosysteme im Bereich der Salzwiesen der Nordseeküste. *Faun. Ökol. Mitt.* **4,** 155–68.

Hinton H.E. (1976) Respiratory adaptations of marine insects. *Marine Insects* (Ed. by L. Cheng), pp. 43–78. North Holland Publishing Company, Amsterdam.

Horstmann K. (1970) Ökologische Untersuchungen über die Ichneumoniden (Hymenoptera) der Nordseeküste Schleswig-Holsteins. *Oecologia* **4**, 29–73.

König R. (1969) *Zur Ökologie und Systematik der Braconidae von der Nordseeküste Schleswig-Holsteins (Hymenoptera, Braconidae)*. Diss. Kiel.

Meijer J. (1973) Die Besiedlung des neuen Lauwerszeepolders durch Laufkäfer (Carabidae) und Spinnen (Araneae). *Faun. Ökol. Mitt.* **4**, 169–84.

Regge H. (1972) *Zur Bionomie und Ökologie der Aphidoidea-Arten des Gezeitenbereiches.* Diss. Kiel.

Regge H. (1973) Die Blattlaus-Arten (Hexapoda, Aphidoidea) des Gezeitenbereichs der Nordseeküste Schleswig-Holsteins. *Faun. Ökol. Mitt.* **4**, 241–54.

Tischler W. (1976) *Einführung in die Ökologie.* Gustav Fischer Verlag, Stuttgart.

Weidemann G. (1965) Biologische und biometrische Untersuchungen an Proctotrupiden (Hymenoptera: Proctotrupidae, s. str.) der Nordseeküste und des Binnenlandes. *Z. Morph. Ökol.* **55**, 425–514.

Weigmann G. (1973) Zur Ökologie der Collembolen und Oribatiden im Grenzbereich Land-Meer (Collembola, Insecta-Oribatei, Acari). *Z. wiss. Zool.* **186**, 295–391.

Wingerden W.K.N.E. van (1973) Dynamik einer Population von *Erigone arctica* White (Araneae, Micryphantidae). Prozesse der Natalität. *Faun. Ökol. Mitt.* **4**, 207–22.

Wrage H.A. (1977) *Untersuchungen zur Ökologie der Limoniidae-Stelzenmücken (U.O. Nematocera) des Litorals und angrenzender Gebiete im Nordseeküsten-Bereich.* Diplom-Arbeit Kiel.

10. ADAPTIVE STRATEGIES OF AIR-BREATHING ARTHROPODS FROM MARINE SALT MARSHES

J.E. TREHERNE AND W.A. FOSTER
University of Cambridge, Department of Zoology, Downing Street, Cambridge, U.K.

SUMMARY

Air-breathing arthropods are the dominant terrestrial invaders of marine salt marshes. The problems associated with their dispersal and the synchrony of locomotory activity with tidal coverage are illustrated by recent work on three species. In response to the instability of its habitat the intertidal aphid (*Pemphigus trehernei* Foster) has developed an apparently novel mechanism of tidal dispersal. The well-defined nocturnal activity of the carabid beetle (*Dicheirotrichus gustavi* Crotch) is completely suppressed during periods of tidal coverage. The intertidal mite (*Bdella interrupta* Evans) adopts a different strategy in alternating between a non-entrained locomotory rhythm (during periods of tidal emergence) and a tidally entrained one.

RÉSUMÉ

Les arthropodes à respiration aérienne sont les principaux envahisseurs des marais salés. Un travail récent sur trois espèces illustre les problèmes de dispersion et de synchronisme de l'activité locomotrice avec la submersion par la marée. En réponse à l'instabilité de cet habitat, l'aphidien (*Pemphigus trehernei*) a développé un mécanisme de dispersion, apparemment nouveau, lors de la marée. L'activité nocturne bien définie d'une bitella carabidae (*Dicheirotrichus gustavi* Crotch) disparait complètement pendant les périodes de submersion par la marée. L'acarien (*Bdella interrupta* Evans) soumis aux balancements des marées, développe une stratégie différente, en alternant un rythme de locomotion de course libre (pendant les grandes marées) avec un entrainement par la marée.

INTRODUCTION

The intertidal zone is an extreme environment for air-breathing arthropods. The paucity of insect species in marine habitats is frequently assumed to result from the inability of terrestrial and freshwater species to adapt to the rigours of a hostile environment (Buxton 1926; MacKerras 1950; Hinton 1977). However, in one class of coastal environment, marine salt marshes, air-breathing arthropods can be a dominant component of the macrofauna (Foster & Treherne 1976a). Despite the ecological importance of marine salt marshes we possess only fragmentary knowledge of this dominant faunal component. Faced with this largely unexplored field we shall confine ourselves here to a description of recent field and experimental observations that illustrate some of the adaptive strategies which air-breathing arthropods have evolved in this extreme environment.

ENVIRONMENTAL EFFECTS OF TIDAL INUNDATION

Marine salt marshes are areas of angiosperm vegetation which are characteristically exposed to regular sequences of tidal inundation. The variable degrees of tidal submergence (Fig. 10.1) present a variety of potential difficulties for air-breathing arthropods. These include the following:

Exposure to high, and frequently fluctuating, salinities

It is, however, unlikely that this has been a limiting factor in long-term colonization, at least by insects, since species from a wide variety of families possess ionic and osmoregulatory mechanisms which enable them to withstand saline and hypersaline conditions (e.g. Shaw & Stobbart 1963; Stobbart & Shaw 1974; Foster & Treherne 1976a).

Respiratory problems

During periods of tidal inundation the normal supply of oxygen is cut off from air-breathing arthropods. During submergence, air tends to be trapped in soil cavities (Chapman 1960; Clarke & Hannon 1967), and this may be sufficient to support normal respiratory requirements. However, in poorly drained areas, which are widespread on salt marshes, conditions of soil aeration can deteriorate dramatically during prolonged submergence (Foster & Treherne 1976b).

Mechanical effects of tidal inundation

Air-breathing arthropods of both terrestrial and freshwater origin are potentially vulnerable to being washed away from the habitat during tidal coverage. In addition, many insects, especially those with delicate membranous wings such as aphids (Foster 1975) and parasitic Hymenoptera (Abraham 1970), are vulnerable to damage by wetting with seawater.

Environmental instability

Because of the poor drainage of most of the salt marsh surface, many soil-living air-breathing arthropods are restricted to the well-drained regions at

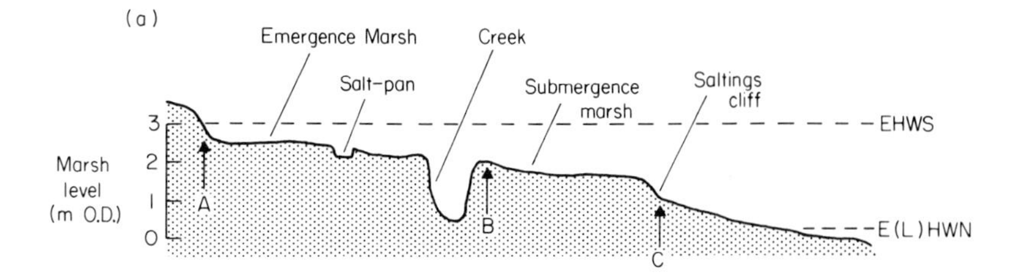

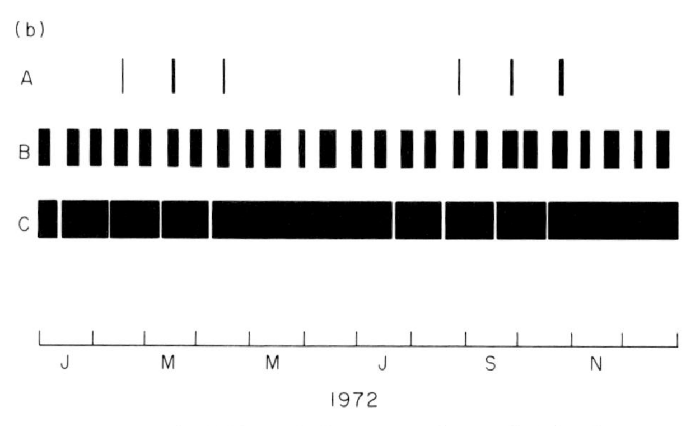

FIG. 10.1. Pattern of tidal inundation on a salt marsh. The data were obtained from Admiralty Tide Tables (1972) and from observations on Scolt Head, Norfolk, U.K. (a) Cross-section of a marsh, showing subhabitats and maximum tidal range. EHWS (Extreme high water spring tide), E(L) HWS (Extreme (lowest) high water neap tide): all high tides fall within this range. Heights of the three sites above Ordnance Datum (O.D.) in metres: A, 2·8 m; B, 2·0 m; C, 1·0 m. (b) Annual pattern of tidal inundation. Black bars indicate periods when the tide covered a particular site at least once daily for two or more consecutive days.

the edges of creeks and salt-pans and the saltings cliffs at the seaward edge of the marsh (see Fig. 10.1). Such areas are characteristically subject to considerable tidal erosion, and, at least in the case of the aster root aphid *Pemphigus trehernei*, entire local populations can be lost, at a stroke (Foster & Treherne 1978).

DISPERSAL

The extreme instability of much of the salt marsh environment in which air-breathing arthropods are abundant creates an obvious need for effective means of dispersal. The need for dispersal presents two major problems. First, dispersing individuals must leave their usual refuges and expose themselves to the mechanical dangers of tidal inundation. Second, the usual mode of insect dispersal, flight, is in many ways unsuitable for an intertidal insect, which might, by taking to the air, be rapidly transported to unsuitable regions of the sea or land. Partial or total winglessness is common in many marine insect taxa, and occurs for example in several intertidal Carabidae (Doyen 1976), many of the marine chironomids (Hashimoto 1976) and several of the shore bugs of the Saldidae and Omaniidae (Polhemus 1976). However, the loss of the ability to fly is not as marked in salt marsh insects as in those from rocky shores, probably because of the greater shelter and generally larger area of the salt marsh environment.

An unusual but effective means of dispersal is available to salt marsh insects: floating on the surface of the tides. It is known that males of some chironomid midges (*Clunio*, *Pontomyia*) glide on sea water, powered by their propeller-like wings (Hashimoto 1976). However, the only detailed investigation of tidal dispersal of an insect is that made on the root aphid *Pemphigus*

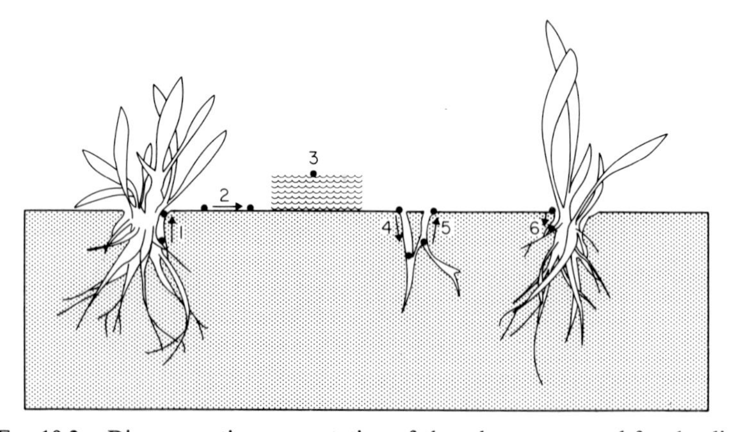

FIG. 10.2. Diagrammatic representation of the scheme proposed for the dispersal of 1st instars of the aster root aphid, *P.trehernei*. (1) Movement out of soil of young 1st instars. Positive response to light. Possibly encouraged by crowding, shortage of food and dryness of soil. (2) Movement on soil surface. (3) Floating on the tide. (4) Movement down into a soil crack by, (a) older 1sts (> 60 hr old) that have not necessarily dispersed on the tides, (b) younger 1sts after dispersal on the tides. Negative response to light over-ruling negative response to gravity. (5) Movement up to soil surface by young 1sts that have found no suitable aster roots. Positive response to light. (6) Movement of older 1sts, or younger 1sts after dispersal on tide, into cracks containing suitable aster roots, where the aphids may then find a colony. Negative response to light over-ruling negative response to gravity. (Modified from Foster & Treherne, 1976. Note that responses to gravity are stated incorrectly in the legend to Fig. 2.16.)

trehernei (Foster 1974; Foster & Treherne 1978). The 1st instar larvae crawl to the surface and are taken up by the flooding tide. Due to their waxy covering they float proud of the surface and are virtually unwettable. The floating aphids are sensitive to wind movements and are propelled rapidly across the surface of the sea. Experiments with model aphids (i.e. small polystyrene balls) indicate that the tide is a very effective means of dispersing floating objects from restricted sources over a wide area of marsh (Foster & Treherne 1978). The movements of these models was influenced by both water movements and by the wind; usually a period of tide-influenced movement was followed by a period of wind-influenced movement as the tide flooded the marsh. A significant proportion of the floating models was deposited in edge situations on the marsh. Field observations and experiments showed that 1st instar aphids that had floated on the tides were able to colonize aster plants in edge situations (Foster 1974).

Newly born 1st instar aphids differ from all subsequent stages in moving towards light (Foster 1974). This response wanes with age and can be rapidly reversed if the aphids are floated on sea water for a short period. These observations can be incorporated in a scheme for the dispersal behaviour of this salt marsh aphid (Fig. 10.2).

The dispersal mechanism of this aphid is neatly tailored to the demands of the salt marsh environment. Unlike most of its terrestrial relatives, for example the lettuce root aphid *Pemphigus bursarius* (L.), which disperses mainly by flight, this salt marsh species produces very few winged individuals (Foster 1975), but invests for dispersal in the 1st instar larvae.

SYNCHRONY WITH ENVIRONMENTAL CYCLES

The complex variable pattern of tidal inundation of marine salt marshes creates an obvious need for relatively accurate synchronization with tidal cycles to avoid the adverse effects of submergence. *A priori* it would seem reasonable to suppose that rhythms of activity associated with tidal cycles would be developed as has been well established in several rocky shore chironomids of the genus *Clunio* in which adult emergence is synchronized with extreme low water of spring tides (Neumann 1968, 1971, 1976; Pflüger & Neumann 1971). Laboratory observations indicate that the intertidal beetle *Thalassotrechus barbarae* Horn possesses a circatidal and circadian rhythm of locomotory activity (Evans 1976).

The surface locomotory activity of the carabid beetle *Dicheirotrichus gustavi* is related to the tidal cycle on a marine salt marsh (Treherne & Foster 1977). During periods of tidal emergence, when the beetle zone at creek edges is not submerged, this soil-living species exhibits well-defined nocturnal activity. Maximum numbers of individuals appear after dusk and decline, exponentially, until dawn when the remainder disappear abruptly into soil

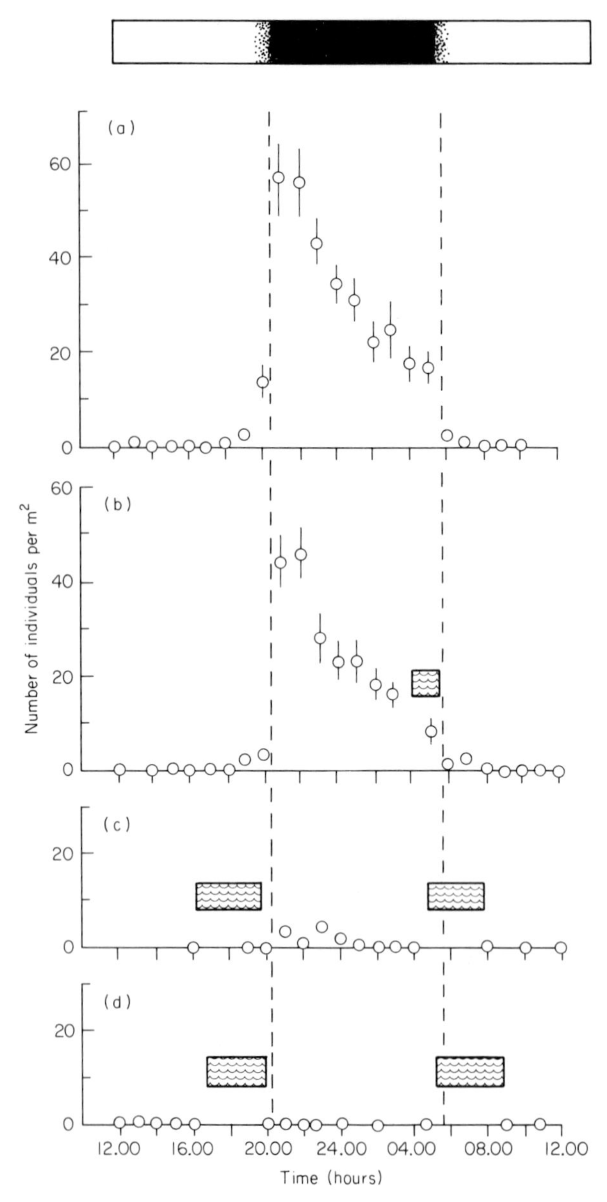

FIG. 10.3. Surface activity of *Dicheirotrichus gustavi* on four successive days, (a) before, and (b)–(d) during, a period of tidal submergences (horizontal bars) in a sequence of rising tides. The open circles indicate the mean number of individuals on the soil surface and the verticle lines the extent of twice the standard error of the mean. Dawn and dusk are indicated by the broken lines.

cavities (Fig. 10.3). During the period of observation illustrated in Fig. 10.2 the first tidal coverage in a sequence of spring tides occurred shortly before dawn, when minimal numbers of adult beetles were exposed to sea water, and the second before dusk, when the insects were subterranean. The timing of these initial tidal coverages resulted in exposure of only minimal numbers of beetles to direct exposure to seawater. As the first critical submersions in a sequence of rising tides are restricted to between 0300 and 0800 hours and 1400 and 1900 hours on the Scolt-Head salt marshes, Norfolk, most of the beetles will avoid direct contact with sea water (Treherne & Foster 1977).

Field and laboratory experiments showed that nocturnal surface locomotory activity is suppressed by two tidal submergences (Fig. 10.2) and is re-established after an absence of two consecutive tidal submergences (Treherne & Foster 1977). These changes in diel activity do not appear to result from changes in soil water content and are, therefore, presumed to result from unspecified changes in the conditions in the soil cavities.

There is thus no evidence that endogenous tidal or semilunar rhythms are involved in the control of locomotory activity of this salt marsh beetle. The diel activity of *D.gustavi* in periods of emergence is essentially similar to that of terrestrial carabid species (A.S. McClay, personal communication). The specific adaptation to the salt marsh environment is the suppression of nocturnal activity during periods of submerging tides. The beetles therefore remain subterranean throughout such periods.

Another air-breathing arthropod, the prostigmatid mite *Bdella interrupta* Evans, shows more specific adaptations to a tidal salt marsh environment. Like the rocky shore oribatid, *Ameronothrus marinus* Banks (Schulte 1976), the salt marsh mite shows two clearly defined peaks of locomotory activity per day on the soil surface (Treherne *et al.* 1977). The amplitude of the peaks appears to be related to soil temperature and thus tends to be reduced during the night in the natural environment, especially during periods of submerging tides.

The peaks of surface locomotory activity of *B.interrupta* do not show a constant relationship to the times of dawn and dusk. During periods of non-submerging tides the peaks of activity that occurred during daylight were 22·9 hr apart (Figs. 10.4 and 10.5). During periods of submerging tides the mites emerged approximately mid-way between successive tides, the daylight peaks of activity being 24·9 hr apart (Figs. 10.4 and 10.5). This activity rhythm parallels the bi-tidal frequency on the salt marsh (Fig. 10.4) and suggests that it is entrained by the tides. Since the activity peaks occur twice daily the locomotory activity can be defined as a circatidal rhythm governed by a clock with a period of approximately 12·5 hr. One explanation of this is that during non-submerging tides the rhythm is free-running, at a period of 11·5 hr, or alternatively that it is phase-shifting towards a second zeitgeber (e.g. light/dark cycle) (Foster *et al.* 1978).

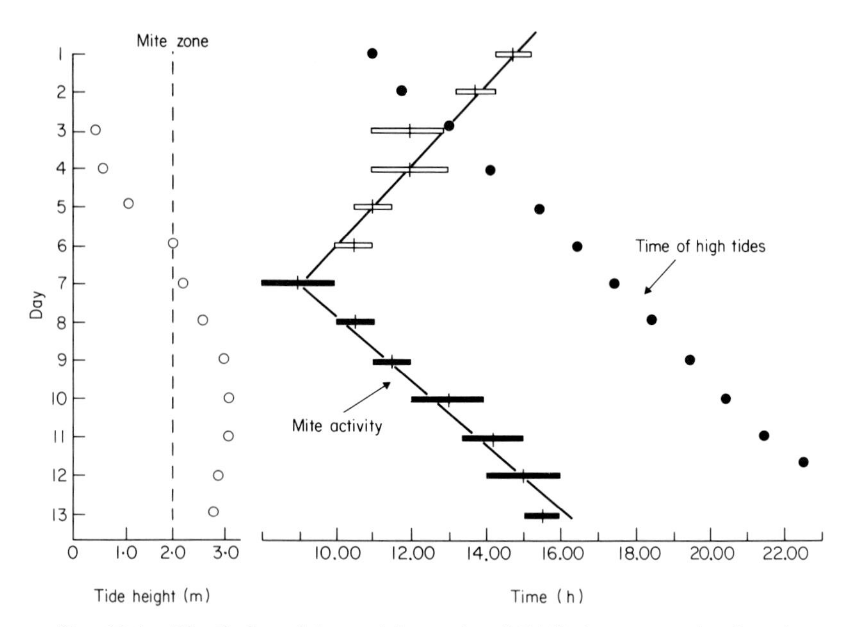

FIG. 10.4. The timing of day activity peaks of *Bdella interrupta* related to the time (closed circles) and the height (open circles) of high tides from 6 to 18 April 1976. The horizontal bars indicate the period during which mite activity was more than 50% of the maximum. The vertical lines are the mid-points of the bars and are taken as the phase reference point. The open bars represent activity during tidal emergence and the closed ones that during a subsequent period of submerging spring tides. The continuous lines are the calculated regression lines (for the period of tidal emergence: $r = 0{\cdot}9720$; $n = 6$; $P < 0{\cdot}005$; for the period of tidal submergence $r = 0{\cdot}9923$; $n = 7$; $P < 0{\cdot}001$). For these calculations the day sequence was used as the X-axis.

This is, as far as we are aware, the only example of an organism that shows regular short-term changes in its period of locomotor activity. Circadian activity rhythms have been shown to free-run temporarily in beavers under the ice-cover of their lakes in winter (Bovet & Oertli 1974) and in the continuous daylight of the arctic summer in, for example, fish (Müller 1969, 1973) and mice (Erkinaro 1969).

The alternation of a 12·5 hr periodicity (during periods of tidal submergence) with an 11·5 hr periodicity (during periods of tidal emergence) ensures that a peak of mite activity is continuously maintained in daylight hours. This would not occur if the circatidal periodicity (12·5 hr) continued during tidal emergence. It could be of selective advantage that the daytime peak never drifts into the hours of darkness but always occurs during daylight hours, when temperatures are generally higher and perhaps prey is more abundant, visible and easily caught. In addition a non-entrained period of 11·5 hr ensures that at the end of a period of tidal emergence mite activity is safely separated from the first, critical tidal inundation in the sequence of rising tides. Calculations (using data from the 1977 tide tables) show that with an intermediate non-entrained period, for example 12·0 hr, the activity peak is always within 2·5 hr of the first, critical tide, whereas with a non-

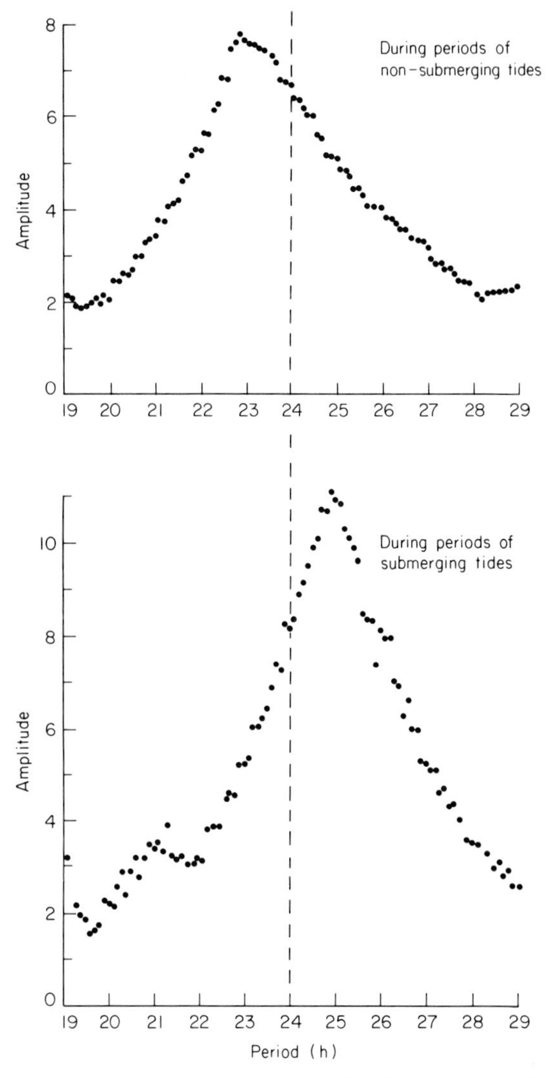

FIG. 10.5. Periodograms of data illustrated in Fig. 10.2, for the observed locomotory activity during periods of non-submerging and submerging tides.

entrained period of 11·5 hr the activity peak is within 2·5 hr of only 17% of the critical tides.

END-NOTE

The overwhelming importance of tidal inundation makes the salt marsh in certain respects a relatively simple environment. All potential salt marsh colonizers must modify their life-strategies in response to this single environmental factor. The range of adaptations employed by salt marsh angiosperms has been rather fully investigated (e.g. Waisel 1972; Ranwell 1972; Reimold & Queen 1974; Chapman 1977). Although we have discussed work on only three species, this is sufficient to indicate that the other important terrestrial colonizers of salt marshes, the animals, and in particular the dominant air-breathing arthropods, exhibit a range of adaptations and strategies at least as interesting and worthy of study as those shown by the flowering plants.

ACKNOWLEDGMENT

Figures 10.1 and 10.2 are reproduced by kind permission of North Holland and Fig. 10.3 by kind permission of the *Journal of Animal Ecology*.

REFERENCES

Abraham R. (1970) Ökologische Untersuchungen an Pteromaliden (Hym., Chalcidoidea) in Grenzraum Land-Meer an der Nordseeküste Schleswig-Holsteins. *Oecologia* **6**, 15–47.

Bovet J. & Oertli E.F. (1974) Free-running circadian activity rhythms in free-living beavers (*Castor canadensis*). *J. comp. Physiol.* **92**, 1–10.

Buxton P.A. (1926) On the colonization of the sea by insects: with an account of the habits of *Pontomyia*, the only known submarine insect. *Proc. zool. Soc. Lond.* 1926, 807–14.

Chapman V.J. (1960) The plant ecology of Scolt Head Island. *Scolt Head Island* (Ed. by J.A. Steers), pp. 85–163. Heffers, Cambridge.

Chapman V.J. (Ed.) (1977) *Wet Coastal Ecosystems.* Elsevier, Amsterdam.

Clarke L.D. & Hannon N.J. (1967) The mangrove swamp and salt marsh communities of the Sydney district. 1. Vegetation, soils and climate. *J. Ecol.* **55**, 753–71.

Doyen J.T. (1976) Marine beetles (Coleoptera excluding Staphylinidae). *Marine Insects* (Ed. by L. Cheng), Chapter 18. North-Holland, Amsterdam.

Erkinaro E. (1969) Der Verlauf desynchronisierter, circadianer Penodik einer Waldmaus (*Apodemus flavicollis*) in Nordfinnland. *Z. vergl. Physiol.* **54**, 407–10.

Evans W.G. (1976) Circadian and circatidal locomotory rhythms in the intertidal beetle *Thalassotrechus barbarae* (Horn): Carabidae. *J. exp. mar. Biol. Ecol.* **22**, 79–90.

Foster W.A. (1974) *The Biology of a Saltmarsh Aphid* (*Pemphigus* sp.) Ph.D. thesis (University of Cambridge, U.K.).

Foster W.A. (1975) The life history and population biology of an intertidal aphid, *Pemphigus trehernei* Foster. *Trans. R. ent. Soc. Lond.* **127**, 193–207.

Foster W.A. & Treherne J.E. (1976a) Insects of marine saltmarshes: problems and adaptations. *Marine Insects* (Ed. by L. Cheng), Chapter 2. North-Holland, Amsterdam.

Foster W.A. & Treherne J.E. (1976b) The effect of tidal submergence on an intertidal aphid, *Pemphigus trehernei* Foster. *J. Anim. Ecol.* **45**, 291–301.

Foster W.A. & Treherne J.E. (1978) Dispersal mechanisms in an intertidal aphid. *J. Anim. Ecol.* **47**, 205–17.

Foster W.A., Treherne J.E., Evans P.D. & Ruscoe C.N.E.R. (1978) Short-term changes in activity rhythms in an intertidal arthropod (Acarina: *Bdella interrupta* Evans). *Oecologia* (in press).

Hashimoto H. (1976) Non-biting midges of marine habitats (Diptera: Chironomidae). Chapter 14 in *Marine Insects* (Ed. by L. Cheng), North-Holland, Amsterdam.

Hinton H.E. (1977) Enabling mechanisms. *Proc. XV int. Congr. Ent. Washington*, pp. 71–83.

Mackerras I.M. (1950) Marine Insects. *Proc. R. Soc. Qld.* **61**, 19–29.

Müller K. (1969) Jahreszeitliche Wechsel der 24h-Periodik bei der Bachforelle (*Salmo trutta* L.) am Polarkreis. *Oikos* **20**, 166–70.

Müller K. (1973) Seasonal phase shift and the duration of activity time in the burbot, *Lota lota* (L.) (Pisces, Gadidae). *J. comp. Physiol.* **84**, 357–59.

Neumann D. (1968) Die Steuerung einer semilunaren Schlüpfperiodik mit Hilfe eines kunstlichen Gezeutenzyklus. *Z. vergl. Physiol.* **60**, 63–78.

Neumann D. (1971) The temporal programming of development in the intertidal chironomid *Clunio marinus* (Diptera: Chironomidae). *Can. Ent.* **103**, 315–18.

Neumann D. (1976) Adaptations of chironomids to intertidal environments. *A. Rev. Ent.* **21**, 387–414.

Pflüger W. & Neumann D. (1971) Die Steuerung einer gezeitenparallelen Schlupfrhythmik nach dem Sanduhrprinzip. *Oecologia* **7**, 262–66.

Polhemus J.T. (1976) Shore bugs (Hemiptera: Saldidae, etc.). *Marine Insects* (Ed. by L. Cheng), Chapter 9. North-Holland, Amsterdam.

Ranwell D.S. (1972) *Ecology of Salt Marshes and Sand Dunes*. Chapman & Hall, London.

Reimold R.J. & Queen W.H. (Eds.) (1974) *Ecology of Halophytes*. Academic Press, New York & London.

Schulte G. (1976) Gezeitenrhytmische Nahrungsaufnahme und Kotballenablage einer terrestrischen Milbe (Oribatei: Ameronothridae) im marinen Felslitoral. *Marine Biology*, **37**, 265–77.

Shaw J. & Stobbart R.H. (1963) Osmotic and ionic regulation in insects. *Adv. Insect Physiol.* **1**, 315–99.

Stobbart R.H. & Shaw J. (1974) Salt and water balance: excretion. In: *The Physiology of Insecta*. Vol. 5 (Ed. by M. Rockstein), pp. 361–446. Academic Press, New York & London.

Treherne J.E. & Foster W.A. (1977) Diel activity of an intertidal beetle, *Dicheirotrichus gustavi* Crotch. *J. Anim. Ecol.* **46**, 127–38.

Treherne J.E., Foster W.A., Evans, P.D. & Ruscoe, C.N.E. (1977) A free-running activity rhythm in the natural environment. *Nature, Lond.* **269**, 796–97.

Waisel Y. (1972) *Biology of Halophytes*. Academic Press, New York & London.

11. AN ECOLOGICAL STUDY OF THE SWANPOOL, FALMOUTH. IV: POPULATION FLUCTUATIONS OF SOME DOMINANT MACROFAUNA

R.S.K. BARNES,[1] ADÈLE WILLIAMS,[2]
COLIN LITTLE[2] AND A.E. DOREY[2]

[1] *Department of Zoology, University of Cambridge, Cambridge*
and
[2] *Department of Zoology, University of Bristol, Bristol, U.K.*

SUMMARY

Three species of Crustacea, *Gammarus chevreuxi*, *Palaemonetes varians* and *Neomysis vulgaris*, have been sampled over a total period of four and a half years by means of a hand operated grab sampler and a seine net. Populations of *Palaemonetes* and *Neomysis* showed regular seasonal cycles, few animals or none being present in winter and often very large numbers in summer. These variations are in part attributed to migration. There is probably no significant migration of *Gammarus* in and out of the pool, but the numbers of this species vary very widely without an obvious seasonal pattern. The variations are not fully accounted for, but some at least appear to be caused by variations of temperature and salinity. Specimens of *G.chevreuxi* from Swanpool are smaller than those found elsewhere: this may be due to the comparatively small particle-size of the available sediments. *Gammarus* is probably not severely affected in this environment by predation, competition, food shortage or drift.

RÉSUMÉ

Sur une période de quatre ans et demi nous avons échantillonné trois espèces de crustacés *Gammarus chevreuxi*, *Palaemonetes varians* et *Neomysis vulgaris*, au moyen d'un excavateur manuel et d'un réseau de filets. Les populations de *Palaemonetes* et *Neomysis* montrent des cycles saisonniers réguliers; en hiver il n'y a pas ou presque pas d'animaux, en été ils sont souvent en grand nombre. Ces variations sont dues pour une part à la migration. La migration de

Gammarus à l'intérieur et à l'extérieur du périmètre étudié n'est probablement pas significative, cependant le nombre de cette espèce varie très largement sans caractéristique saisonnière évidente. Ces variations ne sont pas complètement expliquées mais quelques unes semblent être causées par des variations de température et de salinité. Les spécimens de *G.chevreuxi* de Swanpool sont plus petits que ceux trouvés ailleurs : ceci peut être dû à la taille comparativement réduite des particules du sédiment disponible. *Gammarus* n'est probablement pas sévèrement affecté dans cet environnement par la prédation, ni par les variations ou l'insuffisance de l'alimentation.

INTRODUCTION

Swanpool is a coastal lagoon in Falmouth (Cornwall) separated from the sea by a shingle bar (Fig. 11.1). Descriptions of the pool, its fauna and flora, its hydrography and its origins are given in Barnes, Dorey and Little (1971), Dorey, Little and Barnes (1973) and Little, Barnes and Dorey (1973). In the present paper, we trace the fluctuations since January 1973 in the populations of three species of Crustacea: the amphipod *Gammarus chevreuxi* Sexton, the mysid *Neomysis vulgaris* (Thompson) (=*N.integer* (Leach)), and the decapod *Palaemonetes varians* (Leach). Some observations on other species of macrofauna are also presented, and an attempt to relate changes to environmental variation is made.

METHODS

Benthic samples were obtained by means of a hand-operated, van Veen style grab, sampling 0·05 m^2 and biting to a maximum depth of 12 cm. This was operated by means of two stout poles attached to the flat upper surfaces of the fibre-glass and metal quarter cylinders. The catch was washed through a 6·35 mm mesh sieve into a bucket, the contents of which were then washed in a 0·5 mm mesh sieve, preserved in alcohol and brought back to the laboratory for subsequent sorting. Preliminary samples were taken from five stations: storm drain, outlet, outlet tunnel, benches and stream entrance (Fig. 11.2). From January 1973, two samples were taken every three months from three stations only: storm drain, outlet I (in the main channel of sea water entry) and outlet II (shorewards of the channel). In May 1977 a series of samples was collected with the same grab, operated from a boat in somewhat deeper water, to a maximum of 1 m depth.

The deeper samples contained much organic matter and had to be sorted by hand. Sorting of the routine samples from shallower water was attempted

FIG. 11.1. A montage of two aerial photographs of Swanpool (1977), showing the nature of the surrounding land and the relationship of the lagoon to the adjacent beach (the intervening land, now 'developed', once bore sand-dunes). Cambridge University Collection: copyright reserved.

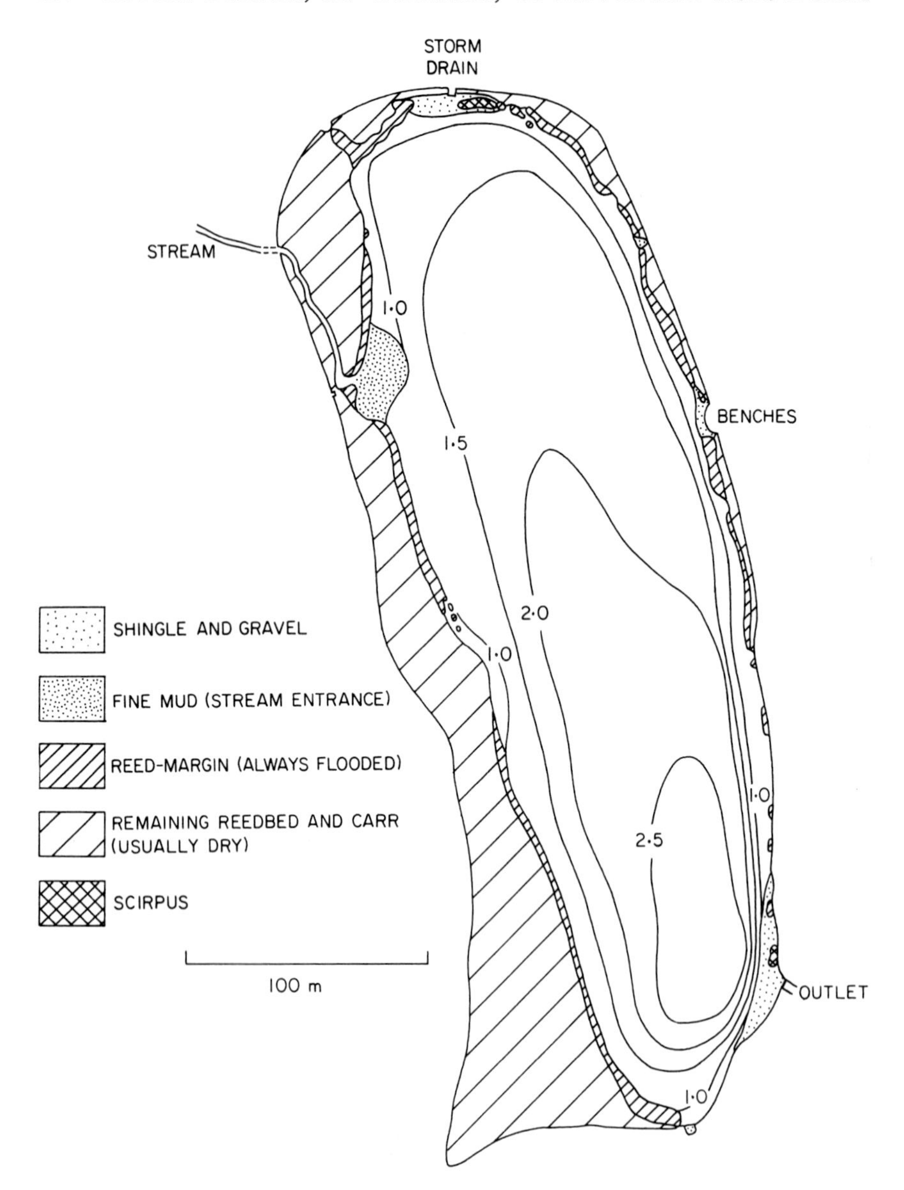

Fig. 11.2. Map showing sampling sites. Depth contours are in metres. The area near the stream entrance (stippled) has no *Gammarus* and was therefore excluded from the calculation of *Gammarus* numbers.

initially using an elutrifier (Lauff *et al.* 1961), but this failed to sort the samples completely. An air-lift pump designed to collect and sort samples simultaneously (Pearson *et al.* 1973) was equally unsuccessful, even after various modifications. Eventually, the sample retained by the 0·5 mm sieve was simply tipped into a 5 l polythene beaker, agitated by a controllable and manoeuvrable jet of water, and the water and suspended material decanted through the 0·5 mm mesh sieve. This process was repeated until two successive agitated washings and decantings yielded no further organisms. The material retained by the sieve was then transferred to a translucent dish (illuminated from below) from which all specimens of *Gammarus* were picked out, and counted. They were dried for 48 hours at 110°C and their total weight determined to 1 mg. Each result given below is the mean of the two replicate samples taken at each station, and is expressed only as no./m^2: the range of variation in these numbers is so great that the weight measurements follow similar patterns in spite of variations of mean weight.

Since the amount of sediment taken by the grab was very variable, it was necessary to determine whether this would cause serious error in the measurement of *Gammarus* numbers. The wet weight of the washed sediment was also recorded, therefore, but no correlation was found between this and the numbers of *Gammarus* (Spearman rank correlation coefficient ranged from $-0{\cdot}10$ to $-0{\cdot}34$, $P > 0{\cdot}05$, Siegel 1956).

To investigate particle-size composition of the sediments 5 core samples were taken at each station in April 1975. After drying, these were sorted with a standard series of ten sieves (from 62 μm to 32 mm mesh). Results are expressed as cumulative-percentage composition by weight in ϕ interval categories. (If the diameter of a particle is d mm, then its ϕ value is $-\log_2 d$ (Inman 1952).)

The population structure of *G.chevreuxi* was investigated by measuring the length of individuals to the nearest 0·5 mm, from the base of the antennae to the tip of the telson. Animals less than 4 mm long were regarded as juveniles, since Girisch *et al.* (1974) found no adults smaller than 3·9 mm, and in the present study no animals less than 4 mm long were found carrying eggs. All individuals longer than 4 mm were sexed by examination of the antennae (see Sexton 1913). The presence of ovigerous females in the grab samples was noted, but since a large and variable proportion of the eggs was discharged during preservation with alcohol, no quantitative information could be obtained. Proportions of ovigerous females and numbers of eggs per female were estimated on samples of live material in May 1977.

To sample *G.chevreuxi* swimming in open water, tows were made at the surface with a zooplankton net (0·25 mm mesh, mouth diameter 28 cm) from the benches to the southern end of the pool (220 m). Samples of animals leaving the pool were collected with a conical net, 3 m long and 0·5 mm mesh, fixed to a wooden frame in the outlet so as to filter the entire outflow. The net was left for varying periods of time, after which the catch was removed, subsampled where necessary and sorted by hand.

To estimate the numbers of *Gammarus* living in the wet areas of the reed-beds, a metal frame in the form of a cube with a 0·5 m side-length was used in place of a conventional quadrat. The vertical sides of this cube were covered with muslin. The reeds in a small area were cut down to water level and the cube was forced down into the bed. All *Gammarus* within the cube were then removed with a small sieve, sorted and counted alive.

Samples of *Palaemonetes*, *Neomysis*, *Sigara* and smaller fish were obtained by taking two replicate tows of a hand-pulled seine net (3·5 m long, 1·0 mm mesh) over a standard distance (c. 25 m) at the outlet and storm drain sites in water about 0·5 m deep. Material collected was preserved and transported to the laboratory, where animals were counted, dried and weighed as described above. The results are given below as mean numbers of the various species per standard tow, but data from the storm drain and from the outlet are not strictly comparable.

Measurements of salinity, oxygen and water temperature were made as described previously (Dorey *et al.* 1973). Measurements of air temperature were supplied by the Meteorological Office, Bracknell, for their Falmouth station. These are in the form of average monthly values derived from daily maximum and minimum measurements.

RESULTS

Nature of the environment

Descriptions of the sampling sites are given by Barnes *et al.* (1971). The storm drain, benches and outlet sites are regions of gravel and stones, whereas the stream entrance consists of fine mud with some large stones. Analyses of particle sizes are given in Fig. 11.3. The occasional large cobbles and boulders (>64 mm diameter) are excluded from these analyses. At the northern end of the pool the basic material is angular pebble-sized 'head' (see Little *et al.* 1973), and at the southern end most material is rounded beach shingle. Particles less than 1 mm diameter comprise less than 15% of the total sediment at all stations except the stream entrance, which has not been investigated in detail. The main channel at the outlet (outlet I) contains significantly fewer small particles than outlet II ($P < 0{\cdot}01$).

Variation in surface salinity of the pool is shown in Fig. 11.4a. It can be seen that in the period of this study, salinities were low in winter (<1‰), built up to a small peak in April–May, and then declined somewhat before reaching the annual maximum (up to 10‰) in August–September, after which they rapidly declined to the winter level. Inter-annual variation is also obvious: in 1976 the water was particularly saline, whilst in 1974 it was the most dilute.

Temperature variation (Fig. 11.4b) conformed to the expected seasonal pattern, although yearly differences are apparent. The summers since 1974 have been of increasing warmth, that of 1974 being the coolest of the period of study. The early part of the winters of 1972/73 and 1974/75 were comparatively mild, although monthly average minimum temperatures dropped below

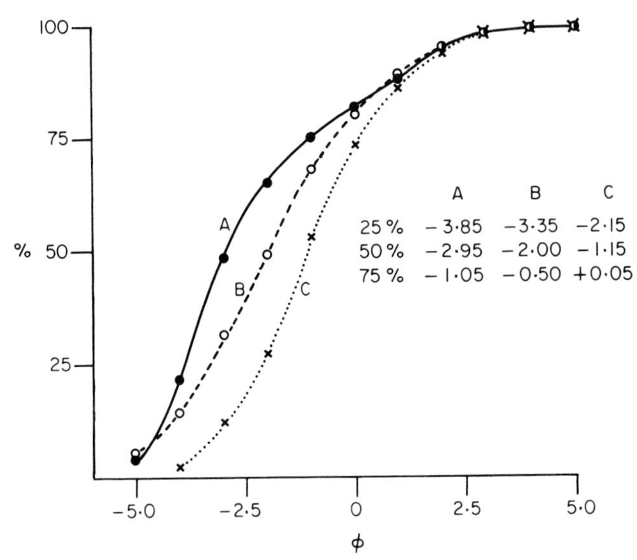

FIG. 11.3. Cumulative percentage frequencies of different size-fractions of the sediments at the Storm Drain (A), Outlet I (channel) (B), and Outlet II (C).

5°C every year and did not rise above this level until March/April. The winter cold spell was particularly short in 1974/75 and, to a lesser extent, in 1971/72, and it was, in 1973/74 and 1975/76, interrupted by a warmer period in January.

The chemical characteristics of the water in Swanpool are described by Crawford, Dorey, Little and Barnes (in press). For a considerable part of each year the deeper waters are devoid of oxygen, and even when oxygen is present at 2 m, the water near the bottom may be anoxic. In January 1975, for example, the concentration of oxygen at 2 m was 11·4 mg/l, but at 2·75 m it was only 0·08 mg/l. Only when a halocline is not present, which is an uncommon occurrence, is the bottom water oxygenated; this happened, for example, in the winters of 1972/73 and 1973/74.

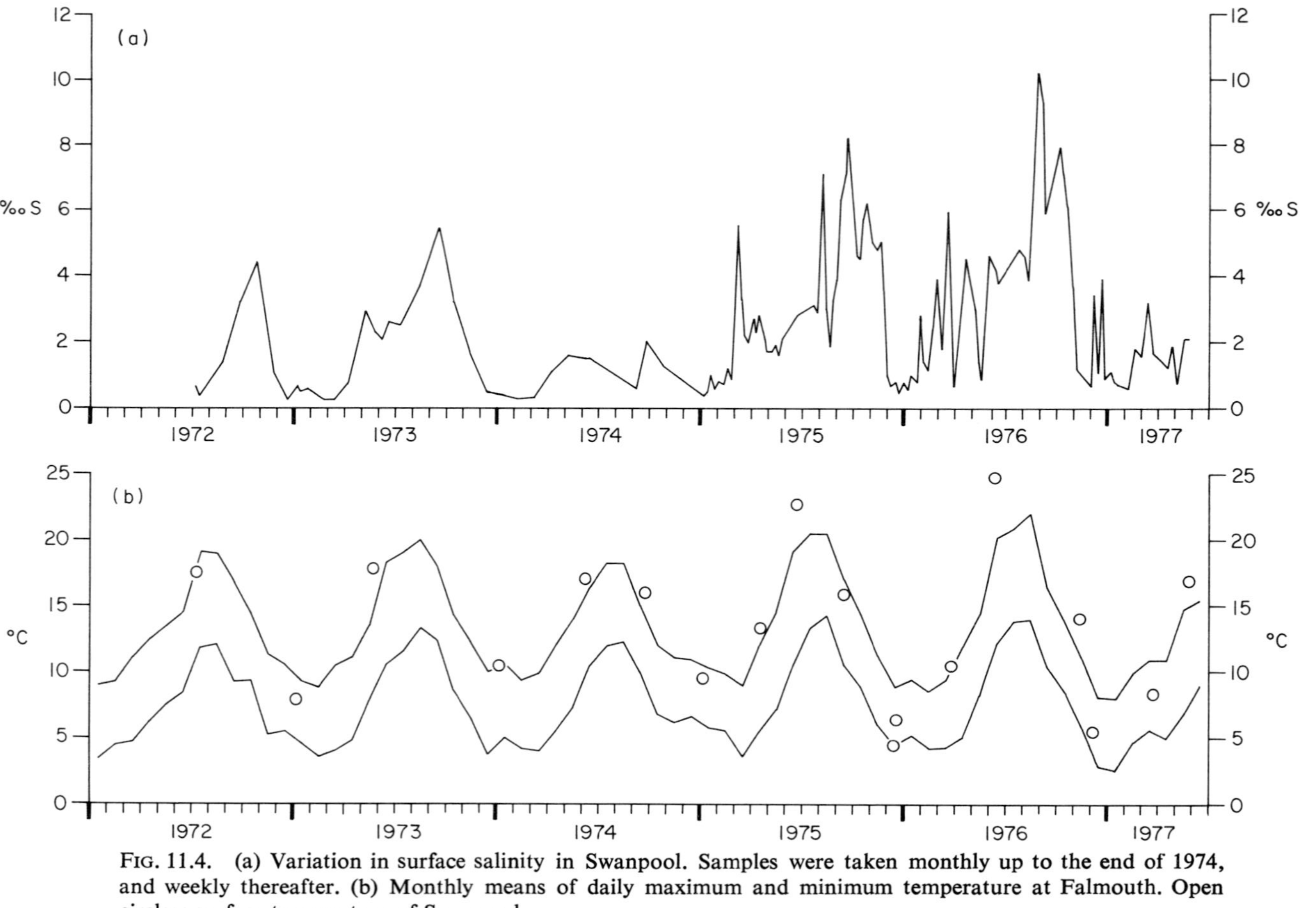

FIG. 11.4. (a) Variation in surface salinity in Swanpool. Samples were taken monthly up to the end of 1974, and weekly thereafter. (b) Monthly means of daily maximum and minimum temperature at Falmouth. Open circles: surface temperature of Swanpool.

Seine samples

The numbers of *Palaemonetes*, *Neomysis* and *Sigara* spp. are shown in Figs. 11.5, 11.6 and 11.7. In each case the trends for biomass were closely related to numbers, but the mean dry weights of *Palaemonetes* rose in the autumn of each year suggesting that the individuals were growing in the pool. In 1975 for example, at the outlet, a mean of 0·7 mg in June increased to 30 mg in September. In 1976, a mean of 1·0 mg in June increased to 10 mg in September.

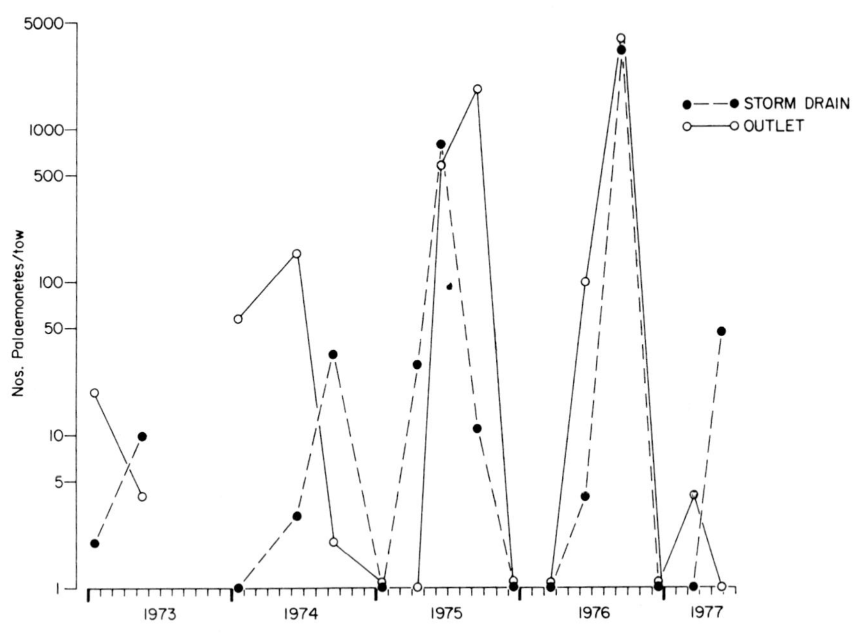

FIG. 11.5. Numbers of *Palaemonetes varians* per standard seine tow at the Outlet and Storm Drain.

The populations of all three species fluctuated markedly throughout the year, in general with peak numbers occurring in the period June–September. It was not possible to take seine samples in the summer of 1973, but sweep net samples in August showed *Palaemonetes*, *Neomysis* and *Sigara* all to be extremely abundant: 1973 was therefore comparable with later years. Besides the clear basic pattern of seasonal fluctuation, variations from year to year occurred. In particular these variations concerned the numbers still to be found in the winter months, the precise period of maximum abundance, and the value of the peak abundance attained. In the winters of 1971/72, 1972/73 and 1973/74, *Palaemonetes* and *Neomysis* were found at both sampling sites, and the winter of 1973/74 showed high numbers of both species. Surface salinity and temperature do not explain these high numbers, but the absence of a halocline in the winters of 1972/73 and 1973/74 may do so since the bottom waters were then oxygenated. A bottom trawl through the deeper layers of the pool in January 1973 captured both *Palaemonetes* and *Neomysis*, although numbers were low.

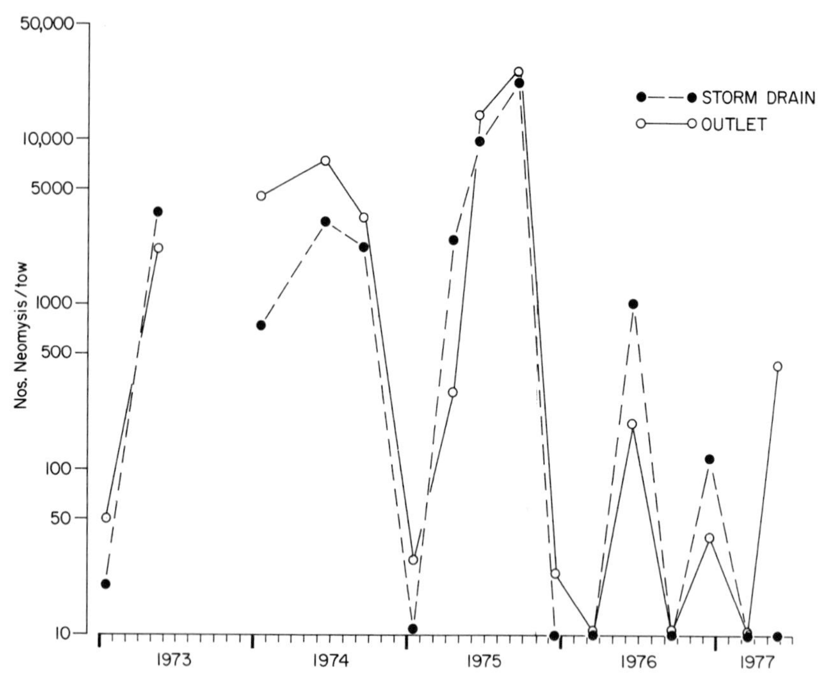

FIG. 11.6. Numbers of *Neomysis vulgaris* per standard seine tow at the Outlet and Storm Drain.

Since the numbers of fish caught by the seines were low, they are probably not representative and will not be reported in detail. Two species, *Pomatoschistus microps* and *Gasterosteus aculeatus*, were present throughout the year. Young *Crenimugil labrosus* were caught at the outlet in January of 1971, 1974 and 1975, and in May 1973, but were never taken later in the year. Young *Platichthys flesus* were uncommon, and were only caught in May 1973 and June of 1975 and 1976, and only at the outlet. *Anguilla anguilla* was only caught in seines from June to September. Only juveniles of these larger species were taken in the seine net but adult *Anguilla* and *Crenimugil* are also present.

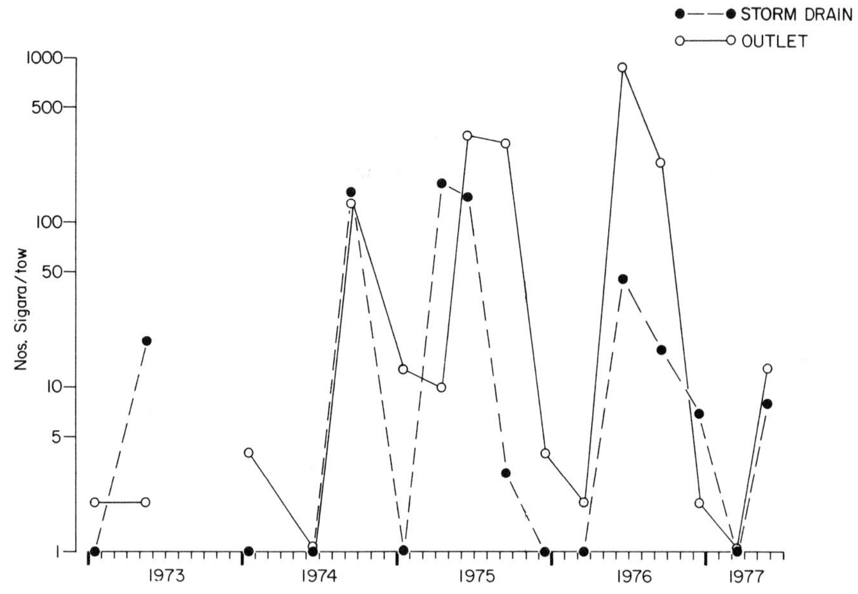

FIG. 11.7. Numbers of *Sigara* per standard seine tow at the Outlet and Storm Drain.

Grab samples

Total populations

The grab samples contained *Gammarus chevreuxi*, the gastropod *Potamopyrgus jenkinsi* (Smith), chironomid larvae, oligochaetes, and, locally, flatworms and polyzoans. Only the abundance of *Gammarus* will be considered here, and results are given in Fig. 11.8. No seasonal pattern is apparent. At all three stations populations were dense at the start of 1973 and from the end of 1974 to the summer of 1976, but the numbers were relatively low towards the end of 1973 and very low indeed in September 1976. Correlation between the numbers at the two outlet sites is fairly good throughout (Spearman rank correlation coefficient $+0{\cdot}64$, $P < 0{\cdot}01$), but during certain periods such as in early 1973 and from September 1974 to September 1975, the trends at the two outlet sites and at the storm drain seem strongly opposed.

For the pool as a whole, the major trends are a fall in numbers in late 1973 followed by growth in early 1974; a high level population from 1974 until September 1976 followed by a population crash. The early part of 1974 had a much lower salinity than in other years (see Fig. 11.3a), but was otherwise unexceptional. The period 1974 to 1976 has shown wide salinity fluctuations on a fairly regular basis, but with a gradually increasing mean salinity, and an increasing tendency for the halocline to persist in the winter months (Crawford,

Dorey, Little & Barnes in press). These observations suggest that there is no straightforward correlation between high or low salinity *per se* and growth rate of the *Gammarus* population.

The two periods of decline in numbers, late 1973 and late 1976, have in common, however, the fact that only at these times did the yearly temperature maximum coincide with the yearly salinity maximum (see Fig. 11.4). In these years average monthly maximum temperatures of 20°C or more coincided

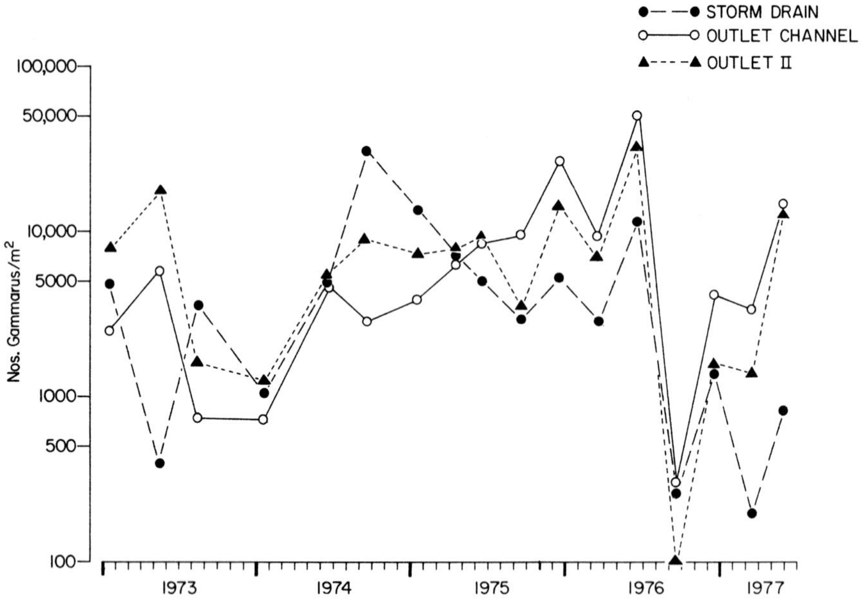

FIG. 11.8. Numbers per m² of *Gammarus chevreuxi* at the Storm Drain, Outlet channel and Outlet II.

with maximum salinities of 5 and 10‰ S. This combination of conditions may be unfavourable for the Swanpool race of *G.chevreuxi*, although for populations found in estuaries, mortality increases with increasing temperature particularly at *low* salinities (Girisch *et al.* 1974). The only other time at which high temperatures coincided with high salinity was in September 1975, but here, as in most years, the temperature maximum preceded the salinity maximum, and at a temperature of 20°C the salinity was never higher than 3‰ S. Even so, there is some evidence of decline in numbers at this time at the storm drain and at the outlet II station, though not at outlet I. High temperature therefore seems to be implicated in these sudden falls in population density, and the effect is enhanced by high salinity.

Population composition

G.chevreuxi is normally confined to the limits of the pool and the outlet tunnel. At the latter site it co-exists with *G.duebeni* which may account for 26% of the adult *Gammarus* population. In the stream entrance, where the water is entirely fresh, small numbers of *G.pulex* are occasionally present together with *G.chevreuxi.*

The composition of *G.chevreuxi* populations is shown in Fig. 11.9. The preponderance of juveniles at many sites is striking and periods where they constitute 80% or more of the population can be considered in three groups. In January 1973, this happened at the benches, when populations in the pool were high (Fig. 11.8). During the period of low numbers in 1973/74, there were few juveniles. Then again, when numbers were high from 1975 to June 1976, proportions of 80% juveniles were found at the outlet, with figures up to 76% at the storm drain. At the time of the population crash in September 1976, juveniles were few, but they had risen again to nearly 100% in the outlet channel as the population was recovering in December 1976. From this it can be concluded that most high population levels are due to high numbers of juveniles in the population, as might be expected. The only exception to this is in June 1974 when the populations had just recovered from the low figures at the beginning of the year. At this time there were very few juveniles, so that the high population levels must have been caused by growth of eggs and of juveniles too small to be taken early in the year. The reason for the low proportion of juveniles (i.e. the reduced breeding rate) at this time, is not known.

The drastic changes in population composition seen at some sites over very short periods may be due to the migration of some elements of the population from one site to another. More frequent samples would be necessary to analyse these changes. There is, at most sites, a tendency for adult males to be more common in summer than in winter, but this does not apply to the outlet tunnel, where adult males are more common even in January.

There is no clear cut change in the average size of animals at different times of year. At most stations, the average length of adult females is 4·5 to 5 mm, while males have higher means, from 5·5 to 6 mm. Mean lengths are slightly greater at the storm drain than at the outlet. In the outlet tunnel, average lengths are even greater: from 4·9 to 5·4 mm for females, and from 5·5 to 7·4 mm for males. This difference in the outlet tunnel is reinforced by the fact that no males longer than 8·5 mm have been found in the main pool, but in the outlet tunnel males sometimes reach 10 mm.

The adults in Swanpool are smaller than those in the populations examined by Sexton (1924) at Plymouth. 'Normal length' there was regarded as 7 mm for females and 11 mm for males. However, similar sizes to the Swanpool populations were found in the Dourduff Estuary in Brittany (Girisch *et al.*

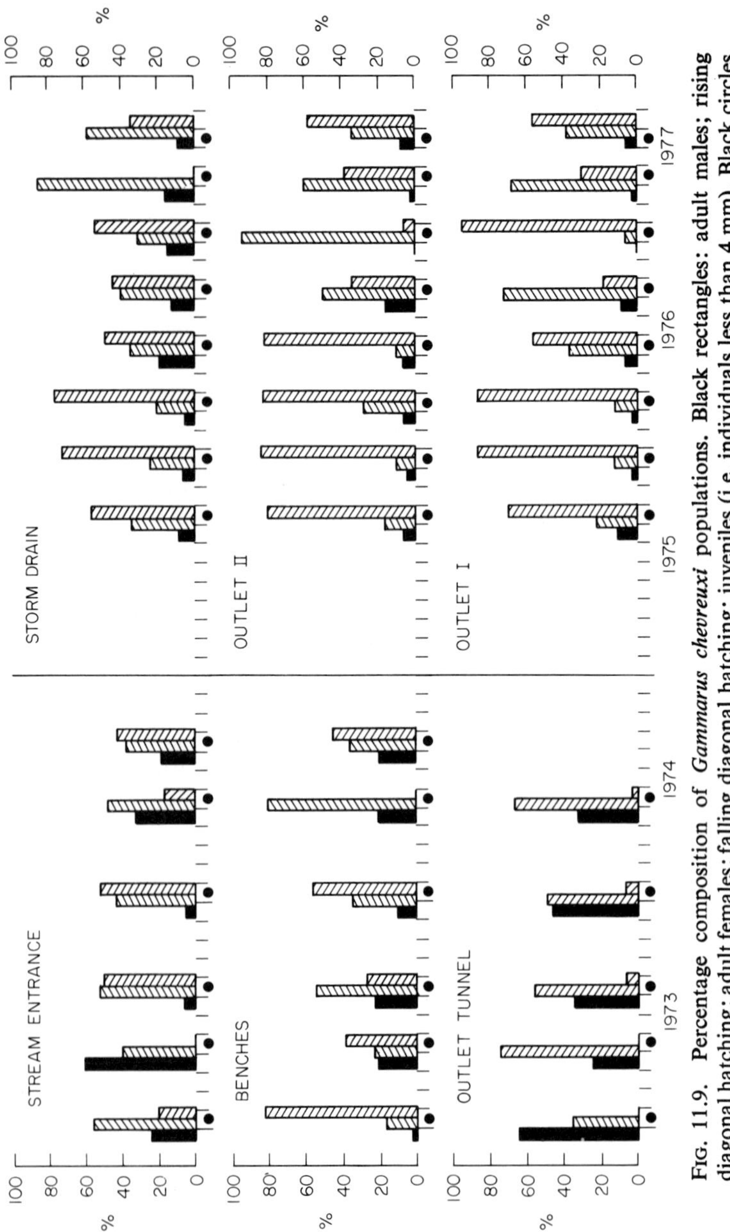

FIG. 11.9. Percentage composition of *Gammarus chevreuxi* populations. Black rectangles: adult males; rising diagonal hatching: adult females; falling diagonal hatching: juveniles (i.e. individuals less than 4 mm). Black circles show sampling dates.

1974). These animals grew to full size if kept in the laboratory (personal communication J.C. Dieleman), whereas the Swanpool animals did not do so, even though some of those from the outlet tunnel were almost full size. These observations suggest that periods of high salinity, but not continuous high salinity, may be necessary for growth in the Swanpool population.

Ovigerous females were found throughout the year in the outlet tunnel and at the storm drain, but only in the summer at the other sites. Breeding has been found to occur throughout the year at Plymouth (Sexton & Matthews 1913) and in Brittany (Girisch *et al.* 1974). Sexton and Matthews showed that size was related to egg number. However, they found that the normal number of eggs was 30 to 40 per female, whereas in Swanpool in May 1977, the average was 10 at outlet I, 11 at outlet II and 7 in the outlet tunnel.

Together with the small size of the Swanpool *Gammarus*, and their inability to grow in the laboratory, this suggests that the lagoonal population may be a different genetic race from the estuarine populations studied by Sexton (1913) and Girish *et al.* (1974).

Another possible explanation of the small size of adult *G.chevreuxi* could be the short flushing time of the epilimnion, which is only of the order of 10–14 days (Dorey *et al.* 1973). Since many *Gammarus* are found in open water at night, a large proportion of them may be expected to drift out of the outlet; if this loss were great enough, the average size of the *Gammarus* left in the pool would be reduced. To investigate this possibility, we estimated two parameters: the total *Gammarus* population, and the numbers leaving the pool in 24 hr. The area inhabited by *G.chevreuxi* is roughly defined by the 1 m depth contour, shown in Fig. 11.2. The area above 1 m depth is approximately 1,236 m^2. *Gammarus* is also found at the edge of the reedbeds, and a strip 2 m wide and the length of the reedbed has been taken as representing the area inhabited. This area is 1,210 m^2. The population was estimated in May 1977 to be 10 million (see Table 11.1). Numbers leaving the pool were estimated at various times of the day and night and are shown in Fig. 11.10. These varying

Table 11.1. *Estimates of* Gammarus *populations in the pool, May 1977*

Site	Area m^2	No. of *Gammarus* m^{-2}	Total no. *Gammarus*
Wet reedbeds	1,210	64	77,000
Area <1 m deep*	1,236	1,334	1,649,000
Storm drain	307	826	254,000
Outlet	610	13,305	8,116,000
		Total in pool	10,096,000

* The area adjacent to the stream entrance (Fig. 11.2) is excluded.

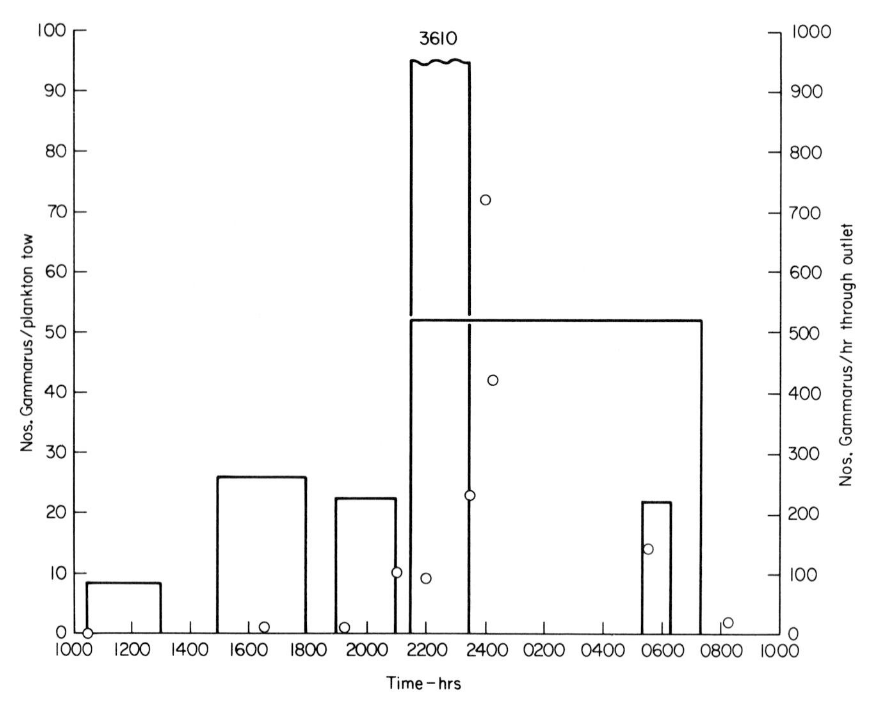

FIG. 11.10. *G.chevreuxi* in the plankton (open circles) and leaving the pool via the outlet (histograms) (June 1977). The width of the histograms indicates the times over which the samples were taken, and the area therefore represents the total number of animals in the sample. Each plankton tow probably sampled about 12 m^3 of water and the rate of flow at the outlet was about 120 m^3/hr. The circles would therefore coincide with the tops of the histograms if the planktonic populations were equally dense in the outflow and the main body of surface water.

numbers can be correlated with the numbers taken in surface plankton tows which are plotted in the same figure. The number leaving the pool in 24 hr was estimated to be 8,000 which is less than 0·1% of the total population.

DISCUSSION

Although capable of living permanently in shallow, enclosed pools and lagoons (Barnes & Jones 1975), both *Palaemonetes* and *Neomysis*, when able to do so, perform seasonal habitat changes or migrations—a phenomenon for which *Neomysis* is well known. These take the form of an autumnal/winter migration to deep water and a spring return to shallow regions in which to feed and breed (Vorstman 1951; Kinne 1955; Antheunisse *et al.* 1971; Rasmussen 1973). Such seasonal movements are commonly seen in shallow-water decapod crustaceans and are correlated with changes in maturity, salinity, temperature and food supply (Allen 1966); they are particularly characteristic of estuarine and other brackish habitats (Barnes 1974). Migrations of the two species under consideration here are likely to be closely linked because *Neomysis* is the main food source of *Palaemonetes* (Kemp 1910; Kaestner 1970).

Their seasonal cycles in Swanpool are best interpreted in terms of such seasonal movements, but where they winter is unknown. *Neomysis* is regarded as being sensitive to lack of oxygen (Tattersall & Tattersall 1951), and hence it is unlikely that it can regularly over-winter in the deeper parts of the pool. Although small numbers over-wintered in the pool in 1971/72–1973/74, most—and in some years all—presumably migrated out of Swanpool. It therefore seems likely that the size of the summer populations of both species is, at least in part, determined by conditions outside the pool.

Both species breed within the pool: ovigerous females and recently hatched young have been taken by plankton net and seine. The interval between samples is too long, however, to permit an accurate estimation of the number of generations produced. Young of *Neomysis* were taken by seines in April and, in some years, in September, whilst those of *Palaemonetes* were caught in June. Population increase in both species could be rapid (Figs. 11.5 & 11.6). Elsewhere, *Neomysis* can produce three generations per year (Vorstman 1951; Kinne 1955; Mauchline 1971) and *Palaemonetes* one or, more rarely, two (Gurney 1923; Thorson 1946).

Most of the fish species (*Anguilla anguilla* (L.), *Pomatoschistus microps* (Krøyer), *Gasterosteus aculeatus* L., *Platichthys flesus* (L.)) also appear to make seasonal migrations into the pool, resulting in marked summer maxima. *Gasterosteus* and *Pomatoschistus* are, however, present throughout the year and may be resident, and young (c. 9 months old) *Crenimugil labrosus* (Risso) occur in January but thereafter cease to be taken by the seine. We know little about the adult *Crenimugil*, but Hickling (1970) recorded this species in Swanpool as suffering mass mortality in 'severe weather'. We have also observed mass mortality of *Crenimugil*, but only in summer and in the presence of dramatic dinoflagellate blooms. All the fish species are predominantly benthic feeders (Wheeler 1969), taking oligochaetes, chironomid larvae, *Potamopyrgus* and *Gammarus*; some may also take *Neomysis*. All these have been recorded from the guts of *Crenimugil* (Hickling 1970), but the *Potamopyrgus* pass out undigested.

There has been no explosion of *Sigara* numbers comparable to that recorded earlier (Barnes *et al.* 1971). Although this hemipteran has a seasonal pattern of abundance comparable to that of the crustaceans and fish, it is most probable that its complete life cycle is spent within the pool.

Gammarus chevreuxi is also resident within the pool, often in large numbers, but the adults are smaller and females have fewer eggs than in the estuarine populations studied by Sexton and Matthews (1913). This might be due to heavy predation, to competition, to lack of food, or to physico-chemical parameters. In estuaries wading birds may take large numbers of crustaceans; in Swanpool the dominant bird species—gulls, duck and coot (Barnes *et al.* 1971)—probably take some *Gammarus* but would be unlikely to have a serious effect on the population. Fish such as *Pomatoschistus*, *Crenimugil*, *Anguilla* and *Gasterosteus* may depend heavily on *Gammarus*, but have never been found in large numbers. *Palaemonetes* may take *Gammarus*, but there is no apparent correlation in numbers of the two species. Competition may occur in the stream entrance with *G.pulex* but this would not affect the main population. In the outlet tunnel competition with *G.duebeni* may occur, but it seems unlikely that this is severe, since it is there that *G.chevreuxi* attains its greatest individual size. The food supply would appear to be more than adequate since Swanpool has a large amount of organic detritus, and the stones are usually covered by a good growth of epilithic flora (cf. Moore 1975).

The work of Morgan (1970) and Rees (1972) suggests that sediment particle-size is likely to be an important factor in relation to the individual size of *Gammarus*. Morgan showed that *Pectenogammarus planicrurus* (length c. 5 mm) preferred sediments with a median ϕ value of $-1{\cdot}7$ whereas Rees found *Gammarus pseudolimnaeus* (length c. 12 mm) mostly where ϕ was about -4. Rees also showed that in this latter species oxygen consumption increased as the particle size increased or decreased from $\phi = -4$. Comparable effects have also been shown for the isopods *Gnorisphaeroma oregonensis* and *Exosphaeroma applicauda* (Rees 1975). In Swanpool the median ϕ values range from -2 (outlet II) and $-2{\cdot}5$ (outlet I) to -3 (storm drain). In view of the findings of Rees and Morgan, these values may well explain why only comparatively small *G.chevreuxi* are found in Swanpool as a whole, and why somewhat larger specimens are obtained from the storm drain than from the outlet.

Population densities of *G.chevreuxi* have often been very high (up to 50,000/m^2), but they vary widely with no obvious relation to the seasons. We have found little evidence suggesting the importance of biological interactions, and suggest that physical and chemical factors may be the most important factors affecting numbers. It has been shown experimentally by Girisch *et al.* (1974) that an intermittent supply of saline water may increase survival. In Swanpool, low population densities were found during periods of prolonged high salinity with temperatures up to 20°C, and the most dramatic falls in population have been in those seasons (1973 and 1976) when the salinity maximum tended to coincide with the temperature maximum, rather than to follow it. The work of Girisch *et al.* showed an increased survival at all salinities when the temperature was raised from 5 to 9°C, but a slight decrease when it was raised as high as 14°C. No measurements were made at 20°C, but on the basis of our observations we should expect increased mortality. Our observations on the effects of low salinity and low temperature do not agree with the experimental work of Girisch *et al.* They found that survival at 5°C was poor even in water of approximately 14‰ S (as 8,000 mg/l Cl), whereas the Swanpool population increased dramatically in the low temperatures of the spring of 1974, and remained high even though in succeeding winters the temperature minimum fell below 5°C regularly. Here the evidence points to significant differences between the estuarine and lagoonal races of *G.chevreuxi*, although latitudinal variation may also be responsible, the more northern population having a greater tolerance of low temperature.

The losses of *Gammarus* from the pool via the outlet were clearly very small as compared with the potential increases through breeding, so that this mechanism cannot be a significant limiting factor. It may be worth noting, however, that during some periods (much of 1974 and 1975) the trends in population density were opposed to each other at opposite ends of the pool. Such a change could well be due to migration from one site to another, since as we have shown, quite large numbers of *G.chevreuxi* take to the plankton at night. Such a migration could be provoked by local shortage of food (although the pool as a whole contains plenty), since effects of this kind have been demonstrated by Hughes (1970) in *G.pulex* and by Girisch and Dennert (1975) in *G.zaddachi*.

There is much in the behaviour of *G.chevreuxi* populations that we cannot confidently account for, but we conclude that physical and chemical factors are probably predominant. Some of our evidence suggests that the race in Swanpool may be genetically different from those found in estuaries, where conditions are somewhat different, and where prolonged high temperature are not so common. Swanpool is probably not an atypical lagoon: it provides rich supplies of food for those few species which can withstand its rigours; but in an environment which fluctuates so markedly and so rapidly, species must either migrate in and out at appropriate times in order to utilize this bounty or suffer severe population crashes, even to the point of extinction, at times of environmental hostility.

ACKNOWLEDGMENTS

We wish to thank the following who have helped us in the course of this work: Mrs H.P. Barnes, Mrs P.E. Stirling, Dr D.J. Patterson and Dr R.M. Crawford who assisted us in the field at various times; the Bickford family and Dr C.R. Boyden who took numerous water samples for us at times when we were not able to visit the pool ourselves; Mr J. Fleming and, particularly, Mr M. Berry, who wrestled patiently for long hours with apparatus claimed to sort and/or obtain benthic samples, and Mr J.J. Clark, Mr A.W. Gentle, and Mr J.W. Rodford who gave other laboratory assistance.

REFERENCES

Allen J.A. (1966) The rhythms and population dynamics of decapod crustacea. *Oceanogr. mar. Biol., Ann. Rev.* **4**, 247–65.

Antheunisse L.J., Lammens J.J. & van den Hoven N.P. (1971) Diurnal activities and tidal migrations of the brackish water prawn *Palaemonetes varians* (Leach). *Crustaceana* **21**, 203–17.

Barnes R.S.K. (1974) *Estuarine biology.* Edward Arnold, London.

Barnes R.S.K., Dorey A.E. & Little C. (1971) An ecological study of a pool subject to varying salinity (Swanpool, Falmouth). An introductory account of the topography, fauna and flora. *J. anim. Ecol.* **40**, 709–34.

Barnes R.S.K. & Jones J.M. (1975) Observations on the fauna and flora of reclaimed land at Calshot, Hampshire. *Proc. Hants Fld Club* **29**, 81–91.

Crawford R.M., Dorey A.E., Little C. & Barnes R.S.K. (1978) Ecology of Swanpool, Falmouth. V. Phytoplankton and nutrients. *Est. coast. mar. sci.* (in press).

Dorey A.E., Little C. & Barnes R.S.K. (1973) An ecological study of the Swanpool, Falmouth. II. Hydrography and its relation to animal distributions. *Est. coast. mar. Sci.* **1**, 153–76.

Girisch H.B. & Dennert H.G. (1975) Simulation experiments on the migrations of *Gammarus zaddachi* and *G.chevreuxi. Bijdr. Dierk.* **45**, 20–38.

Girisch H.B., Dieleman J.C., Petersen G.W. & Pinkster, S. (1974) The migrations of two sympatric gammarid species in a French estuary. *Bijd. Dierk.* **44**, 239–73.

Gurney R. (1923) Some notes on *Leander longirostris* M. Edw., and other British prawns. *Proc. zool. Soc. Lond.* 1923, 97–123.

Hickling C.F. (1970) A contribution to the natural history of the English grey mullets (Pisces, Mugilidae). *J. mar. biol. Ass. U.K.* **50**, 609–33.

Hughes D.A. (1970) Some factors affecting drift and upstream movements of *Gammarus pulex. Ecology* **51**, 301–5.

Inman D.L. (1952) Measures for describing the size distribution of sediments. *J. sedim. Petrol.* **22**, 125–45.

Kaestner A. (1970) *Invertebrate zoology*, 3 (Transl. by H.W. Levi & L.R. Levi). Wiley Interscience, New York.

Kemp S. (1910) The Decapoda Natantia of the coasts of Ireland. *Scient. Invest. Fish. Brch Ire.* 1908, no. 1, 190 pp.

Kinne O. (1955) *Neomysis vulgaris* Thompson eine autökologisch-biologische Studie. *Biol. Zentral.* **74**, 160–202.

Lauff G.M., Cummins K.W., Eriksen C.H. & Parker M. (1961) A method for sorting bottom fauna samples by elutriation. *Limnol. Oceanogr.* **6**, 462–66.

Little C., Barnes R.S.K. & Dorey A.E. (1973) An ecological study of the Swanpool, Falmouth. III. Origin and history. *Cornish Stud.* **1**, 33–48.

Mauchline J. (1971) The biology of *Neomysis integer. J. mar. biol. Ass. U.K.* **51**, 347–54.

Moore J.W. (1975) The role of algae in the diet of *Asellus aquaticus* L. and *Gammarus pulex* L. *J. anim. Ecol.* **44**, 719–30.

Morgan E. (1970) The effect of environmental factors on the distribution of the amphipod *Pectenogammarus planicrurus*, with particular reference to grain size. *J. mar. biol. Ass. U.K.* **50**, 769–85.

Pearson R.G., Litterick M.R. & Jones N.V. (1973) An air-lift for quantitative sampling of the benthos. *Freshwat. Biol.* **3**, 309–15.

Rasmussen E. (1973) Systematics and ecology of the Isefjord marine fauna. *Ophelia*, **11**, 1–507.

Rees C.P. (1972) The distribution of the amphipod *Gammarus pseudolimnaeus* Bousfield as influenced by oxygen concentration, substratum and current velocity. *Trans. Am. micr. Soc.* **91**, 514–29.

Rees C.P. (1975) Competitive interactions and substratum preferences of two intertidal isopods. *Mar. Biol.* **30**, 21–25.

Sexton E.W. (1913) Description of a new species of brackish-water *Gammarus* (*G.chevreuxi* n. sp.). *J. mar. biol. Ass. U.K.* **9**, 542–45.

Sexton E.W. (1924) The moulting and growth-stages of *Gammarus*, with descriptions of the normals and intersexes of *G.chevreuxi*. *J. mar. biol. Ass. U.K.* **13**, 340–401.

Sexton E.W. & Matthews A. (1913) Notes on the life history of *Gammarus chevreuxi*. *J. mar. biol. Ass. U.K.* **9**, 546–56.

Siegel S. (1956) *Nonparametric statistics for the behavioural sciences.* McGraw-Hill, Kogakusha, Tokyo.

Tattersall W.M. & Tattersall O.S. (1951) *The British Mysidacea.* Ray Society, London.

Thorson G. (1946) Reproduction and larval development of Danish marine bottom invertebrates with special reference to the planktonic larvae in The Sound (Øresund). *Medd. Komm. Danm. fisk.-og Havunders.* (*Plankton*) **4** (1), 1–523.

Vorstman A.G. (1951) A year's investigations on the life cycle of *Neomysis vulgaris* Thompson. *Verh. int. ver. Limnol.* **11**, 437–45.

Wheeler A. (1969) *The fishes of the British Isles and north-west Europe.* Macmillan, London.

III

ECOPHYSIOLOGICAL RELATIONSHIPS OF ORGANISMS LIVING IN A SALINE ENVIRONMENT

12. THE IONIC AND WATER RELATIONS OF PLANTS WHICH ADJUST TO A FLUCTUATING SALINE ENVIRONMENT

J. DAINTY

Department of Botany, University of Toronto, Toronto, Ontario, Canada

SUMMARY

This paper contains a rather theoretical discussion of the problems faced by the cells of halophytes as compared with those faced by the cells of ordinary plants. Initially the water relations and later the ionic relations of plant cells are outlined in such a way as, hopefully, to lead to reasonable assessments of the basic situation faced by plant cells. Importance is placed on such parameters as water potential, osmotic pressure and, particularly, turgor pressure in discussing how plant cells 'osmoregulate'. Long-term 'osmoregulation' as faced by the cells of halophytes probably involves the production of appropriate organic osmotica. It also would seem to involve an extra expenditure of energy in maintaining a very asymmetric ionic distribution across the tonoplast and modifications of the tonoplast permeability would seem to be called for.

RÉSUMÉ

Cette publication contient une discussion plutôt théorique, de problèmes envisagés pour des cellules d'halophytes comparés à ceux des cellules de végétaux ordinaires. Initialement les relations de l'eau et plus tard les relations ioniques sont exposées d'une façon qui permet d'espérer une évaluation raisonnable de la situation de base affrontée par les cellules végétales. Nous insistons sur des paramètres tels que le potentiel de l'eau, la pression osmotique et en particulier la pression de turgescence, en discutant de la manière dont les cellules végétales s'osmorégulent. Ce long terme d'osmorégulation pour les cellules halophytes inclue probablement la production de molécules osmotiques organiques appropriées. Il semblerait inclure également une dépense d'énergie supplémentaire pour le maintien d'une distribution ionique très asymétrique, à travers le tonoplasme, et des modifications de perméabilité du tonoplasme sembleraient y contribuer aussi.

INTRODUCTION

I am trying, in this paper, to set out what seem to me to be the basic aspects of the water and ionic relations of plant cells and to discuss how it may be possible for halophytes to operate within these basic biophysical constraints. Much of what I have to say is general knowledge among plant cell 'membranologists'. It is perhaps not likely to be quite so familiar to ecologists, particularly to animal ecologists. Thus it is worth while referring to the following people who treat more fully and usually better some of the points I try to make; Slatyer (1967), Gutknecht and Dainty (1968), Dainty (1976), Hellebust (1976), Raven (1976), Flowers, Troke and Yeo (1977). And the same remarks apply to the speakers who followed me in the Symposium: Stewart *et al.*, Jefferies *et al.* and Winter.

WATER AND IONIC RELATIONS OF PLANT CELLS

A plant cell in general consists of a thin layer of living protoplasm surrounding a large central vacuole and this so-called protoplast (protoplasm + vacuole) is encased in an elastic box, the cell wall. There are thus *two* plasma membranes: one at the outer surface of the protoplasm, adjacent to the cell wall, the *plasmalemma* and one at the inner surface of the protoplasm, at its boundary with the central vacuole, the *tonoplast*. As will be seen later the properties, both relative and absolute, of these two membranes with respect to ionic movements probably play a key role in the operation of halophytes.

So far as the water relations of such a plant cell are concerned the determining parameter is the water potential, Ψ, which is rather simply related to the chemical potential of water (see Slatyer 1967; Dainty 1976). It can be easily shown that the water potential, Ψ^i, inside the cell, i.e. within the cell wall, is related to the hydrostatic pressure, P, and the internal osmotic pressure, π^i, by the equation

$$\Psi^i = P - \pi^i \tag{1}$$

Normally, in fact probably universally, because it is the driving force for growth, the hydrostatic pressure, P, which botanists call the turgor pressure, is positive and of the order of a few (up to 10 or 20) bars (1 bar approximately equals 1 atmosphere). Because the tonoplast cannot withstand any hydrostatic pressure differences across it, the turgor pressure *and therefore* the osmotic pressures will be the same in both the protoplasm and the central vacuole. If a plant cell is in water equilibrium with its external medium then its water potential, Ψ^i, will be the same as the external water potential, Ψ^o. If $\Psi^i \neq \Psi^o$ then water will move into or out of the cell; the force driving such water movement will be the water potential difference, $\Delta\Psi$, and the water will move in the direction of the more negative Ψ. (In practice all water potentials are negative.) For all plants, except those freshwater or marine algae which are never exposed, the water potential of the medium external to the cells is always changing and thus water is always entering or leaving the cells; i.e. the cells are swelling or shrinking to try and keep in approximate water equilibrium with their surroundings. Because they are encased in cell walls the extent of their swelling and shrinking is rather limited. In fact a very important parameter of the water relations of plant cells is the volume elastic modulus of the cell, ε, which is a quantitative expression of the extent to which a cell will expand or contract in response to a change of turgor pressure. It is given by the equation

$$\varepsilon = dP/(dV/V) \tag{2}$$

where dP is the change in turgor pressure associated with the fractional change in volume (dV/V). A very important consequence of the rather high values of ε for plant cell walls is that the volume changes involved in responding to a fluctuating external water potential, Ψ^o, are rather small. This means that the internal osmotic pressure, and therefore the internal ionic concentration, does not change much for the changes in Ψ^i necessary to respond to changes in Ψ^o are largely taken up by changes in the turgor pressure, P. In other words, at least over a certain range of external water potential, a plant cell does not need to osmoregulate because the change in internal water potential, Ψ^i, is accomplished largely by changes in P rather than π^i. This is an interesting difference from an animal cell which is not often stressed.

Water and ionic relations of glycophytes

With these rather simple, basic, ideas about water relations of plant cells let us see what their consequences are. For an ordinary plant, not a halophyte, the internal ionic concentration in the protoplasm is, for biochemical reasons, in the range 100–200 mM salt and mostly potassium salt; this implies a π^i for the protoplasm, and therefore for the central vacuole, of 5 to 10 bars. If such a cell were sitting in an external medium of water potential zero, corresponding to pure water at atmospheric pressure, then the turgor pressure would be +5 to +10 bars, a very reasonable value to sustain growth and also help keep the plant fairly rigid. Such a situation would apply, for instance, to the cells of the roots of an ordinary plant growing in fairly wet soil or to the cells of the whole of such a plant at night when there is no transpiration. For such plants the cells in the leaves, during the day, will be exposed to an external water potential less than zero, and fluctuating, because the plant will be transpiring, i.e. water will be moving from the soil, through the roots, up the xylem to the leaves and from thence evaporated into the air. Along such a transpiration stream the water potential must decrease (see Winter, this volume) for water can only flow *down* water potential gradients. Since for the leaf cells during the day $\Psi^o < 0$, and the fluctuations depend on the intensity of the transpiration stream, then the turgor pressure in the leaf cells will be reduced to adjust their Ψ^i according to Equation (1) and previous arguments. If the turgor pressure is reduced too much, growth will cease, at least temporarily, and the plant may wilt. These phenomena can occur over a long period if the soil dries out and certainly occur daily for the leaf cells in normal sunny weather. Providing that the range of external water potentials does not go too far beyond -5 to -10 bars for too long and too often, the ordinary plant—really because of its cell wall/turgor relation—can handle the water stresses and survive, without making any special ionic demands.

Water and ionic relations of halophytes

For the halophyte the water relations situation is rather different. The roots of the halophyte can be sitting in sea water, a NaCl solution of osmotic pressure about 30 bars, i.e. an external medium of Ψ^o of about -30 bars. Thus the internal water potentials of the root cells of such halophytes must be less than (more negative than) -30 bars. Since they must have a positive turgor pressure of several bars, for growth and other reasons, the internal osmotic pressures of these root cells must be of the order of 40 bars, corresponding to about 800 mM or more of salt. The figures would be a little greater, in the absolute sense, for the leaf cells. Thus one major difference, with all it could imply for the supply of the appropriate materials, between ordinary plants and halophytes is that the solute concentration in the cells of the halophytes is several times greater than that in the cells of the ordinary plants.

However the solute situation is more complex in the halophytes than a simple increase in concentration; the nature of the solutes and their partition between the vacuole and the protoplasm is crucial. For the normal biochemical functioning of all plant cells, including halophytes, it would appear that a protoplasmic concentration of potassium of between 100 and 200 mM is required. In the cells of halophytes the solute concentrations in *both* protoplasm and vacuole are of the order of 800 mM salt equivalent; where and what are the salts and/or other solutes? It seems, not surprisingly perhaps, that most of this 800 mM salt concentration is NaCl; after all NaCl is the major salt, by far, in the external medium. There is no biochemical difficulty in having 800 mM or more NaCl in the vacuole, for no metabolic processes take place there. (It should be kept in mind that the vacuole occupies about 90% of the volume of the tissue and thus that the protoplasm comprises only about 10% of the tissue volume.) There would be biochemical difficulties in having 800 mM or more of NaCl in the protoplasm. For there is no evidence that the enzymes of halophytes are different, in their sensitivity to NaCl, from the enzymes of other plants (see Flowers, Troke & Yeo 1977); thus low concentrations of NaCl would give optimal enzyme function and certainly the enzyme action would be strongly inhibited at concentrations of NaCl as high as 800 mM.

Since the protoplasm comprises only a few per cent of the cell volume it is very difficult to make any direct estimates of the Na^+ concentration in the protoplasm. In principle a so-called 'compartmentation analysis' of the uptake and efflux of radioactive sodium, both as a function of time and other parameters, into and out of appropriate tissues or tissue slices of halophytes could provide such an estimate as well as give estimates of other important quantities. (For a description of the theory and technique see Walker & Pitman 1976.) However compartmentation analysis is somewhat uncertain even when applied to the simplest possible situation of a single giant algal cell. It is notoriously difficult when applied to tissues or tissue slices of higher plants and the parameters estimated by this technique have to be treated with considerable reserve. Nevertheless the best assessment of the protoplasmic Na made by compartmentation analysis on *Triglochin maritima* by Jefferies (1973) indicates that the concentration of Na^+ in the protoplasm is about 100 mM, a relatively low figure.

A major, although again indirect, line of evidence has rather recently arisen suggesting that the NaCl is all in the vacuole and that something else is producing the appropriate osmotic balance in the cytoplasm. It has been observed under both natural and experimental conditions that various 'inocuous' organic solutes are synthesized in the cells of halophytes and other osmotically stressed plants. Such solutes are proline, betaines, polyhydric alcohols, etc. (see Hellebust 1976; Stewart & Lee 1974; Storey & Wyn Jones 1977; Wyn Jones *et al.* 1977, and the papers in this symposium by Stewart *et al.* and by Jefferies *et al.*) It has been demonstrated in some cases that high concentrations (up to 2 molar) of such substances have no effect on the function of enzymes tested, and this of course is a major requirement for their utility. And the observed concentrations of these special solutes are such that *if* they are confined to the protoplasm they would have just about the right concentration to provide the needed osmotic balance between the protoplasm and the vacuole. (But see Jefferies *et al.*, this volume.)

It would thus appear that the differences between a halophyte, at a given salinity level, and an ordinary plant reside both in the absolute concentration levels inside the cell and in the way in which the solutes are distributed between protoplasm and vacuole. And we must also consider how they handle fluctuating external water potentials in their possibly different ways. It may cost something to be a halophyte and the rest of this paper is concerned with what this cost might be.

As I have discussed earlier, the ordinary plant may well be able to handle its problems of a changing external water potential in a relatively passive way, by allowing the turgor pressure to adjust the internal water potential. But such a plant would still be spending energy on maintaining its typical ion balance. It is worth while to consider this and then see whether the halophyte differs markedly in the energy expended on pumping ions.

One of the main features of the ionic distribution in an ordinary plant cell is that for the cations, sodium and potassium, there are relatively small differences of concentration across the tonoplast. Also across this membrane there is only a small difference of electrical potential (less than 20 mV). Now the energy expended per second, W, in pumping an ion up a concentration ratio C_2/C_1 and against an electrical potential difference of E volts is given by the equation (see Penning de Vries 1975, for general discussion of energy expended by plant cells, and Lüttge & Pitman 1976, for specific discussion of energy involved in ion transport).

$$W = \text{Area} \times \text{flux} \times (RT \ln (C_2/C_1) + zFE) \tag{3}$$

or

$$W = \text{Area} \times \text{flux} \times n\Delta G_{\text{ATP}} \tag{4}$$

Equation (3) is the thermodynamic minimum energy used, whereas Equation (4) assumes that for each ion moved, n molecules of ATP are hydrolysed. In these equations 'Area' is membrane area, 'flux' is measured as moles/ $\text{cm}^{-2}\,\text{sec}^{-1}$, R is the gas constant, T the absolute temperature, z the number of electronic charges on the ion concerned, F is the Faraday, and ΔG_{ATP} is the free energy of hydrolysis of ATP. In order to be able to use either of these equations one must know whether the ion concerned is being actively transported and then what fraction of the observed flux is 'active', i.e. actually using metabolic energy; for the equations of course are only relevant to the actively transported ion fluxes. The small cation concentration differences and the small electrical potential difference across the tonoplast suggest that the cations are not far from being distributed according to electrochemical equilibium conditions and thus little work will be involved in cation movements across the tonoplast in the ordinary plant cell. More work may well be involved in anion movement from the cytoplasm to the vacuole; the anion (chloride or nitrate or malate) concentration in the vacuole is of the order of 150 mM, whereas it is probably ten or so times smaller in the protoplasm. Even so, anion movement from protoplasm to vacuole is likely to be assisted by the small electrical potential difference across the tonoplast. Thus the work expended on anion movement across the tonoplast, although it may be larger than that expended on the cation movements, it is likely to be relatively small. I therefore conclude that for ordinary plant cells relatively little metabolic energy is involved in maintaining the ionic distribution across the tonoplast, i.e. between protoplasm and vacuole. Most of the metabolic work involved in ion transport is spent at the plasmalemma where the major ion pumps, and of course the major electrochemical potential gradients, are situated. These are, for certain, a strong proton extrusion pump which may, or may not, be coupled with a sodium extrusion pump and inwardly directed potassium and anion pumps (see Raven 1976).

For the halophyte one should perhaps distinguish between two types of cells, or two types of situations in which halophytic cells find themselves. One type is exemplified by the leaf cell of a halophytic angiosperm. Here the external medium is probably not very different in its ionic contents from that of the ordinary cell I have previously been discussing. Furthermore the *ionic* contents of the cytoplasm are probably not very different from those of an ordinary cell, the osmotic balance being made up by one of the appropriate organic molecules. Thus the pumping activities at the plasmalemma of the leaf cell of a halophyte are likely to be rather similar to those at the plasmalemma of an ordinary cell; much the same amount of energy will be expended and thus halophyte leaf cells and ordinary cells may not differ substantially in this respect, although the halophyte leaf cell will have used more energy in producing a higher overall internal salt concentration.

The other type of halophyte cell or situation would be exemplified by a root cell or a marine algal cell where the external medium is much more concentrated in ions, chiefly NaCl, than in the previous case. It is known for marine algae (see Gutknecht & Dainty 1968; Raven 1976) that the plasmalemma ion fluxes are greater than in freshwater algae and therefore the plasmalemma ion pumps are probably using more metabolic energy in such halophytic cells. How much more is impossible to estimate for, in general, we know very little about the proton extrusion pumps in such cells and they may be the major users of metabolic energy at the plasmalemma pumping sites. So even for this kind of halophyte cell, not much more energy may be being expended at the plasmalemma than in an ordinary cell.

I am so far arguing that the situation at the plasmalemma, so far as energy expenditure is concerned, may well not be very different for the halophyte as compared with an ordinary cell. I have also argued that in the ordinary cell not much work is being done to maintain the, rather uniform except for the anions, ionic distribution across the tonoplast. But it is at the tonoplast where the big difference between a halophytic cell and an ordinary cell may reside.

Although the osmotic pressures in the protoplasm and the vacuole of an ordinary cell are the same, the ionic contents are, as previously discussed, probably very different. All the evidence, rather poor though it may be, suggests that the ionic concentrations of Na, K and Cl in the protoplasm of a halophyte are much the same as in an ordinary cell. The sodium concentration might be about 50 mM, the potassium concentration about 100–150 mM and the chloride concentration about 10 mM; the rest of the osmotically active molecules are made up of proline or a betaine or a polyhydric alcohol, depending on the plant. In the vacuole on the other hand the whole high osmotic pressure is provided by NaCl and rough concentrations could well be something like: Na, 1,000 mM; K, 10 mM; Cl, 1,000 mM. Thus, although my figures are only rough guesses, there will be large gradients of all ions across the tonoplast of halophyte cells and this is quite a different situation from that of ordinary cells where the ionic gradients across the tonoplast are relatively negligible. It is not known what the electrical potential difference is across the tonoplast of halophytes; its value would affect any calculations of the work done in maintaining this marked ionic asymmetry. But it remains certain that a halophytic cell must work considerably harder than an ordinary cell to keep the Na, K and anion content of the protoplasm low in the face of the large concentration of NaCl in the vacuole. There is another factor that could accentuate this problem for the halophytic cell; this is tonoplast permeability. In the very few cases where it has been measured (see Raven 1976), it seems that the tonoplast is much more permeable, perhaps up to 100 fold, than is the plasmalemma. If this is also true for halophyte cells then the ion pumps across the tonoplast must work much harder to maintain the ionic distribution across such a leaky membrane.

In the almost total absence of any reliable quantitative information about exact ionic concentrations and, particularly, ion fluxes across both plasmalemma and tonoplast for the appropriate cells it is a waste of time to try and make quantitative calculations about energy expended according to Equations (3) or (4). What I am suggesting from my semi-quantitative discussion and semi-educated guesses is that the tonoplast of halophyte cells is the place to look for at least some of their unique features. It is likely that they have evolved appropriate tonoplast ion pumps, the ability to direct some of their metabolic energy to powering these pumps, and have probably tightened up these membranes, i.e. made the tonoplast relatively less permeable than it is in non-halophytes.

When the saline environment of a halophyte changes, both the water potential and the external ionic concentrations change. Water potential changes can, and must in the initial stages, be adjusted to by changes in turgor pressure of the cells concerned; here it would be advantageous for the cell to have a high value of ε, the volume elastic modulus. However long-term fluctuations in the salinity can only be handled by appropriately changing the internal osmotic pressures. For although a cell does not seem to mind a temporary change in turgor, there seems to be a preferred turgor pressure for any particular cell which is almost certainly related to its capacity for growth and for many membrane functions which are turgor-sensitive. The internal osmotic pressure changes will involve both the movement of NaCl into or out of the vacuole and the synthesis or degradation of the appropriate organic molecule to adjust the cytoplasmic osmotic pressure. These internal osmotic pressure adjustments that a halophyte must make in the face of a fluctuating external saline environment will clearly impose an additional energy expenditure on the halophyte which, somehow, it has evolved to handle.

REFERENCES

Dainty J. (1976) Water relations of plant cells. *Encyclopedia of Plant Physiology*. New Series, Vol. 2, Part A (Ed. by U. Lüttge & M. Pitman), pp. 12–35. Springer-Verlag, Berlin, Heidelberg and New York.

Flowers T.J., Troke P.F. & Yeo A.R. (1977) The mechanism of salt tolerance in halophytes. *Ann. Rev. Plant Physiol.* **28**, 89–121.

Gutknecht J. & Dainty J. (1968) Ionic relations of marine algae. *Oceanogr. Mar. Biol. Ann. Rev.* **6**, 163–200.

Hellebust J.A. (1976) Osmoregulation. *Ann. Rev. Plant Physiol.* **27**, 485–505.

Jefferies R.L. (1973) The ionic relations of seedlings of the halophyte *Triglochin maritima* L. *Ion Transport in Plants* (Ed. by W.P. Anderson), pp. 297–321. Academic Press, London and New York.

Lüttge U. & Pitman M.G. (1976) Transport and Energy. *Encyclopedia of Plant Physiology*. New Series, Vol. 2, Part A (Ed. by U. Lüttge & M. Pitman), pp. 251–59. Springer-Verlag, Berlin, Heidelberg and New York.

Penning de Vries F.W.T. (1975) The cost of maintenance processes in plant cells. *Ann. Bot.* **39**, 77–92.

Raven J.A. (1976) Transport in algal cells. *Encyclopedia of Plant Physiology*. New Series, Vol. 2, Part A (Ed. by U. Lüttge & M. Pitman), pp. 129–88. Springer-Verlag, Berlin, Heidelberg and New York.

Slatyer R.O. (1967) *Plant-Water Relationships*. Academic Press, London and New York.

Stewart G.R. & Lee J.A. (1974) The role of proline accumulation in halophytes. *Planta (Berl.)* **120**, 279–89.

Storey R. & Wyn Jones R.G. (1977) Quaternary ammonium compounds in plants in relation to salt resistance. *Phytochemistry* **16**, 447–53.

Walker N.A. & Pitman M.G. (1976) Measurement of fluxes across membranes. *Encyclopedia of Plant Physiology*. New Series, Vol. 2, Part A (Ed. by U. Lüttge & M. Pitman), pp. 93–126. Springer-Verlag, Berlin, Heidelberg and New York.

Wyn Jones R.G., Storey R., Leigh R.A., Ahmad N. & Pollard A. (1977) A hypothesis in cytoplasmic osmoregulation. *Regulation of Cell Membrane Activities in Plants* (Ed. by E. Marré & O. Ciferri), pp. 121–36. North-Holland Publishing Company, Amsterdam, Oxford and New York.

13. NITROGEN METABOLISM AND SALT-TOLERANCE IN HIGHER PLANT HALOPHYTES

G.R. STEWART, F. LARHER,* I. AHMAD AND J.A. LEE

Department of Botany,
The University, Manchester, U.K.

SUMMARY

Many higher plant halophytes accumulate amino acids and methylated onium compounds, particularly when growing in saline media. The role of amino acids and methylated onium compounds in salt-tolerance is discussed. The patterns of nitrogen metabolism under saline and non-saline conditions have been examined in several species. These patterns are discussed in relation to the accumulation of NaCl and organic solutes. The species examined appear to exhibit three types of behaviour when grown in the presence of NaCl. Some species exhibit what can be regarded as a constitutive capacity for the accumulation of inorganic ions and organic solute; under non-saline conditions these species accumulate high levels of potassium and nitrate. The second group of species exhibit an adaptive accumulation of inorganic ions and osmotic solute. The third group is comprised of species which are able to regulate the accumulation of inorganic ions; the NaCl content in such species remains relatively constant over a wide range of external NaCl concentrations. A number of species are shown to have the capacity to accumulate more than one source of compatible osmotic solute and the possible ecological and physiological significance of this is discussed.

RÉSUMÉ

De nombreuses plantes supérieures halophytes accumulent les iminoacides et les composés méthyl-onium, particulièrement quand elles se développent dans un milieu salin. Nous discutons du rôle des iminoacides et des composés méthyl-onium dans la tolérance à la salinité. Nous avons examiné les modèles

* Permanent Address: Laboratoire de Biologie Végétale, Université de Rennes, 35000. Rennes, France.

du métabolisme de l'azote, dans des conditions de salinité ou non, pour plusieurs espèces. Nous discutons de ces modèles en relation avec l'accumulation de NaCl et des solutés organiques. Les espèces examinées semblent se comporter de trois manières, quand elles se sont développées en présence de sodium. Quelques espèces montrent ce que l'on pourrait considérer comme une capacité constitutive pour l'accumulation d'ions inorganiques et de solutés organiques. Sous des conditions de non salinité ces espèces accumulent à un haut niveau potassium et nitrate. Le second groupe montre une accumulation adaptative aux ions inorganiques et aux solutés osmotiques. Le troisième groupe comprend des espèces qui peuvent réguler l'accumulation des ions inorganiques. La teneur en NaCl de telles espèces reste relativement constante sur un large domaine de concentrations extérieures de NaCl. Nous montrons que certaines espèces ont la capacité d'accumuler plus d'une source de solutés osmotiques compatibles, et nous en discutons la signification physiologique et écologique possible.

INTRODUCTION

Nitrogen appears to be the major nutrient limiting the growth of many higher plant halophytes in coastal and estuarine marshes (Tyler 1967; Pigott 1969; Stewart *et al.* 1972 and 1973; Valiela & Teal 1974). The basic characteristics of nitrogen assimilation in halophytes appear qualitatively similar to those of other higher plants (see Stewart *et al.* 1973; Stewart & Rhodes 1978). There are however very marked differences in the intermediary nitrogen metabolism of some halophytes and glycophytes. In particular many halophytes exhibit very high levels of certain soluble nitrogen containing compounds, particularly when growing in saline conditions. The accumulation of imino acids, including proline (Goas 1965; Stewart & Lee 1974; Treichel 1975), pipecolic acid and 5-hydroxy pipecolic acid (Goas *et al.* 1970) and methylated quaternary ammonium compounds such as glycine betaine (Storey & Wyn Jones 1975; Storey *et al.* 1977; Larher 1977) and homobetaine (Larher & Hamelin 1975a) is then an unusual feature of the nitrogen metabolism of many halophytes. It has been suggested that some of these compounds play a role in the salt-tolerance mechanisms of halophytes (see e.g. Stewart & Lee 1974; Storey & Wyn Jones 1975) indicating the possibility of a close relationship between nitrogen metabolism and salt-tolerance.

The evidence regarding the mechanisms of salt-tolerance, while not entirely unequivocal, is consistent with a model in which sodium and chloride ions are compartmentalized in subcellular compartments which are isolated from the metabolic machinery of the cell (see review by Flowers *et al.* 1977). Part of the evidence for this model comes from studies of enzymes isolated from a range of halophytes; such studies indicate most enzymes are appreciably inhibited by concentrations of NaCl similar to those found in analyses of whole tissues (Greenway & Osmond 1972; Flowers 1972). *In vitro* protein synthesis of preparations from *Suaeda maritima* also has been shown to be inhibited by similar NaCl concentrations (Hall & Flowers 1973). Such observations suggest that normal metabolic activity would be severely disrupted if the NaCl concentrations at the sites of metabolism resembled those determined for 'whole tissue'. Studies of the subcellular localization of NaCl in halophytes have yielded somewhat conflicting results but Flowers *et al.* (1977), reviewing this area, suggest most of the evidence is consistent with a differential accumulation in the vacuole; they infer that cytoplasmic NaCl concentrations are half to one third those of the vacuolar levels; cytoplasmic NaCl concentrations in the range 150–300 mM are suggested.

Accepting that there is such an imbalance in the NaCl concentrations of cytoplasm and vacuole, then some mechanism must exist to equalize the water potentials of cytoplasm and vacuole. One mechanism which could achieve this equilibrium is an accumulation of 'compatible' solute (Brown & Simpson 1972) in the cytoplasm. The high levels of proline and methylated quaternary ammonium compounds present in the tissues of many halophytes may have such an osmoregulatory function (Stewart & Lee 1974; Storey & Wyn Jones 1975) and function as compatible cytoplasmic solutes which play an important role in maintaining the cytoplasmic and vacuolar water potentials in equilibrium. The observation that proline, even at concentrations as high as 750 mM, does not inhibit a range of enzymes (Stewart & Lee 1974) is consistent with this suggestion. Similarly glycine betaine has been found not to inhibit malate dehydrogenase (Wyn Jones *et al.* 1976). Studies of the effect of homobetaine and glycine betaine on a range of enzymes isolated from several halophytes indicate that, like proline, these compounds are not inhibitory at high concentrations (unpublished results).

There is at present little information on the subcellular localization of these compounds; however recent studies with red beet storage tissue indicate a cytoplasmic localization for glycine betaine (Wyn Jones *et al.* 1977). Histochemical examination of *Triglochin maritima* leaf tissue indicates the presence of high proline concentrations (Larher, unpublished results) in the chloroplasts, and chloroplasts isolated from *Armeria maritima* contain appreciable quantities of homobetaine (unpublished results).

These observations regarding the accumulation, characteristics and localization of proline and methylated quaternary ammonium compounds are consistent with their suggested role as compatible cytoplasmic solutes which function in the maintenance of osmotic balance between cytoplasm and vacuole in plants growing in saline conditions.

It is surprising that so many species of halophytes employ soluble nitrogenous compounds as compatible solutes when nitrogen seems to be a growth-limiting nutrient in salt marshes. The accumulation of high levels of proline and methylated quaternary ammonium compounds may impose, of course, higher nitrogen requirements on such species.

In this contribution we will consider the patterns of nitrogen metabolism and ion accumulation in a range of halophytes and we will discuss these observations in relation to strategies of salt-tolerance.

DISTRIBUTION AND LEVELS OF METHYLATED QUATERNARY AMMONIUM COMPOUNDS AND PROLINE IN COASTAL PLANTS

Many species of maritime plants are characterized by the presence of high levels of glycine betaine (see Storey *et al.* 1977; Larher 1977) or other methylated quaternary ammonium compounds including homobetaine (Larher 1977) and stachydrine (Larher & Stewart 1978; Wyn Jones, personal communication). In contrast, other species accumulate high levels of the imino acid, proline (see Stewart & Lee 1974). The results shown in Table 13.1 suggest that four groups can be recognized with respect to their capacity to accumulate these soluble nitrogenous compounds. In the first group are those which accumulate only glycine betaine, proline being present at relatively low concentrations. Most of the species in this group are members of the Chenopodiaceae or Gramineae. The second group contains species which accumulate only proline; in these species only low concentrations of methylated onium compounds are present. The third group comprises those species which appear to have the capacity to accumulate both methylated onium compounds and proline. The other group contains those species in which neither proline nor methylated onium compounds are accumulated. This group is of particular interest since, if the general model for salt-tolerance discussed earlier is applicable to such species, then some alternative source of compatible osmotic solute must be present in them. One of the species in this group, *Plantago maritima*, contains high levels of the polyol, sorbitol, and exhibits a salt-induced accumulation of sorbitol (see Ahmad *et al.*, in preparation, and Table 13.2). Other species in this group do not appear to accumulate sorbitol or dimethyl propiothetin. The latter is a methylated sulphonium compound, structurally related to homobetaine and is present in *Spartina anglica* (Larher *et al.* 1977).

Polyhydric alcohols and related compounds appear to be employed as compatible solutes in other plant groups (see Cram 1976; Hellebust 1976) and it may be that species not shown to accumulate imino acids or methylated quaternary ammonium compounds accumulate carbon compounds. The possibility of two groups, one accumulating soluble nitrogen, the other soluble carbon compounds, has important implications with respect to the nitrogen economy of different species.

Table 13.1. *Proline and methylated quaternary ammonium compounds in coastal plants.*

	μmoles g fwt.$^{-1}$	
	Proline	Methylated quaternary ammonium compounds
Group I. Methylated quaternary ammonium accumulators		
Agropyron junceiforme	<5	23*
Agropyron pungens	<5	80
Ammophila arenaria	<5	70
Atriplex hastata	<5	30
Atriplex patula	<5	25
Beta maritima	<5	40
Elymus arenaria	<5	77
Halimione portulacoides	<5	50
Salicornia europaea	<5	45
Salsola kali	<5	62
Suaeda maritima	<5	63
Group II. Proline accumulators		
Cochleria officinalis	35	<5
Glaux maritima	31	<5
Puccinellia distans	33	<5
Puccinellia maritima	60	<5
Spergularia marina	26	<5
Spergularia media	43	<5
Triglochin maritima	72	<5
Group III. Proline and methylated quaternary ammonium accumulators		
Agrostis stolonifera	40	15
Armeria maritima	38	32†
Aster tripolium	25	29
Festuca rubra	30	19
Limonium vulgare	60	40†
Spartina anglica	16	80‡
Group IV. Coastal plants not accumulating either proline or methylated quaternary ammonium compounds		
Carex arenaria	<5	<5
Eryngium maritimum	<5	<5
Juncus gerardii	<5	<5
Juncus maritimus	<5	<5
Plantago coronopus	<5	<5
Plantago maritima	<5	<5
Scirpus maritimus	<5	<5

Plants were collected from different coastal sites in north Wales and north-west Lancashire. Values are the average of several determinations made over two growing seasons. Analyses were made on shoot or leaf tissue.

* Unless otherwise indicated these are glycine betaine levels.

† Major compound in these species is homobetaine.

‡ Includes glycine betaine and dimethyl propiothetin.

See Stewart and Lee (1974) for details of proline determination and Larher and Stewart (1978) for details of methylated onium determinations.

Table 13.2. *Sorbitol accumulation in* Plantago maritima

NaCl External solution (mM)	Sorbitol (μmoles g fwt.$^{-1}$)		
	Leaves	Roots	Inflorescence
0	10	0·5	1
200	32	16	24
400	42	52	65

Plants were grown from seed collected from a salt marsh population. Analyses were performed on plants grown in solutions which contained different NaCl concentrations for a minimum of 14 days.

The presence of high levels of 'compatible osmotic solutes' in non-salt marsh plants (see Table 13.1 and Storey *et al.* 1977) is perhaps surprising since species such as *Ammophila arenaria* and *Elymus arenaria* are not characteristic of highly salinized soils. It may be that compounds such as glycine betaine play roles other than a simple osmoregulatory one; the possibility that they act as protectants of macromolecular structure has been suggested (Larher & Hamelin 1975b; Wyn Jones *et al.* 1977). It is interesting in relation to this possibility that compounds such as glycerol, mannitol and proline appear to protect membrane-bound enzymes against inactivation by high electrolyte concentrations which occur in frozen or dehydrated tissue (see Santarius 1969). Storey and Wyn Jones (1975, 1977) have suggested that glycine betaine levels are related to the genetic salt-sensitivity of particular species. There is therefore the possibility that some sand dune species have high salt-tolerance.

Differences between species as regards the nature of the compatible solute accumulated may not simply reflect differences in enzymic potential to synthesize specific compounds; certainly all species have the enzymic machinery to synthesize proline, but the capacity to accumulate it to high levels is exhibited by only some species. It may be that these species differences reflect the existence of different physio-types or are indicative of different ecological strategies. Storey *et al.* (1977) have suggested these differences may relate to quantitative differences in ion contents, species with a lower sodium to potassium ratio being proline accumulators.

PATTERNS OF NITROGEN METABOLISM IN DIFFERENT SPECIES

Species also appear to differ as regards the changes which occur in the levels of compatible osmotic solutes in response to increases in salt concentration. In *Suaeda monoica* for example high levels of glycine betaine are present in

Table 13.3. *Nitrogen assimilating enzymes in shoot tissue of halophytes.*

Species		Specific activity nmoles min^{-1} mg $protein^{-1}$			
		Nitrate reductase	Glutamine synthetase	Glutamate* synthase	Glutamate† dehydrogenase
Aster tripolium	(0 mM NaCl)	9	90	32	97
	(400 mM NaCl)	10	142	35	96
Plantago coronopus	(0 mM NaCl)	14	100	59	60
	(250 mM NaCl)	17	146	64	50
Puccinellia maritima	(0 mM NaCl)	28	225	84	121
	(200 mM NaCl)	30	245	73	123
Suaeda maritima	(0 mM NaCl)	15	248	150	121
	(400 mM NaCl)	17	253	143	123
Triglochin maritima	(0 mM NaCl)	8	122	162	192
	(200 mM NaCl)	9	185	173	182

Plants were grown from seed on 5 mM nitrate as sole nitrogen source.
Enzyme activity was determined as described by Stewart *et al.* (1973) and Stewart and Rhodes (1978).
* Ferredoxin-dependent activity.
† NAD-dependent activity.

plants grown at low NaCl concentrations and there is only a small increase in glycine betaine content of plants grown at high NaCl concentrations (Storey & Wyn Jones 1975). In contrast the proline content increases 15–30 fold in salt-stressed plants of *Triglochin maritima* (Stewart & Lee 1974), and a similar marked increase occurs with respect to sorbitol levels in *Plantago maritima* (Ahmad *et al.*, in preparation).

Species differences in the nature of the osmotic solute accumulated and in the behaviour of those solutes with respect to salinity levels pose a number of problems as regards the assimilation and metabolism of nitrogen in different species. In some species glycine betaine and/or proline can comprise over twenty per cent of plant nitrogen (see Storey *et al.* 1977). This suggests that a high capacity to assimilate nitrogen may be important and that in some species there may be an increased rate of assimilation under saline conditions.

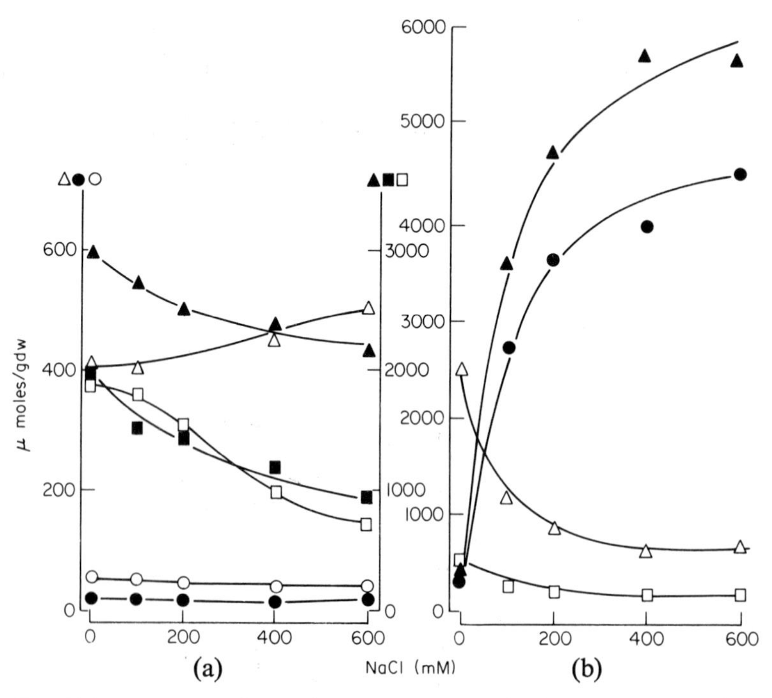

FIG. 13.1. Response of *Suaeda maritima* to NaCl. Plants were grown from seed on a modified Arnon and Hoagland medium. Salt concentrations were brought to the final concentrations shown above, by daily increments of 50 mM. Analyses were made on shoot tissue after a minimum of seven days on the final salt concentrations.

(a) Total organic nitrogen (▲), Soluble organic nitrogen (■), α-NH_2-N (○), Proline (●), NO_3 (□), Glycine betaine (△).

(b) K (△), Na (▲), Cl (●), Mg (□).

Studies of nitrogen assimilation in halophytes have shown that they can assimilate nitrate and that some species are characterized by having high levels of nitrate reductase (Stewart *et al.* 1972, 1973). The enzymes, glutamine synthetase, glutamate synthase (ferredoxin-dependent) and glutamate dehydrogenase are present in the shoot tissue of halophytes, and the relative levels of these enzymes suggest ammonia assimilation occurs via the combined action of glutamine synthetase and glutamate synthase (Stewart & Rhodes 1978).

There do not appear to be any marked differences in potential to assimilate nitrogen between species accumulating different osmotic solutes (Table 13.3). In addition none of the species studied show any very marked change in capacity to assimilate nitrogen when grown in saline medium (Table 13.2), although they exhibit some increase in glutamine synthetase level when grown under saline conditions.

The patterns of nitrogen metabolism and ion accumulation have been examined in a range of species and these results suggest several distinct patterns of response can be discerned. In *Suaeda maritima* (Fig. 13.1) glycine betaine levels show only a small increase with respect to external salinity;

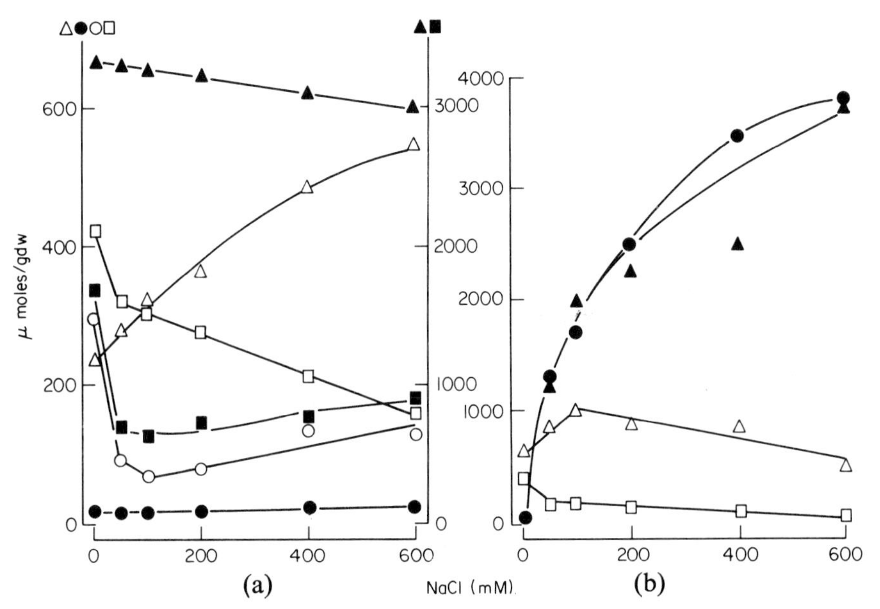

FIG. 13.2. Response of *Atriplex littoralis* to NaCl. See Fig. 13.1 for details. Analyses were made on leaf tissue.

(a) Total organic nitrogen (▲), Soluble organic nitrogen (■), α-NH_2-N (○), Proline (●), NO_3 (□), Glycine betaine (△).

(b) K (△), Na (▲), Cl (●), Mg (□).

there are however very marked changes in other nitrogenous compounds. In this species a major component of total plant nitrogen is the nitrate ion which in plants grown in the absence of NaCl accounts for 12% of the dry weight. Both nitrate and soluble organic nitrogen decrease markedly with an increase in external salt concentration. Neither proline nor α-amino acids show any appreciable variation with respect to external salinity. The high levels of soluble nitrogen present in plants grown in the absence of NaCl are not accounted for by glycine betaine and amino acids, while that of plants grown at 600 mM NaCl comprises largely glycine betaine, amino acids and amides. The nature of this 'unaccounted for' soluble organic nitrogen is unknown. An increase in salt-concentration results in marked decreases in potassium and magnesium levels, this being particularly evident in the case of potassium. Sodium and chloride accumulate to high levels and in plants grown at 600 mM NaCl they account for 30% of the dry weight.

A similar pattern of behaviour is exhibited by *Atriplex littoralis* (Fig. 13.2) but in this species there is an almost linear increase in glycine betaine content in response to an increase in external salt concentration. As with *Suaeda maritima* both soluble organic nitrogen, nitrate and potassium decrease with

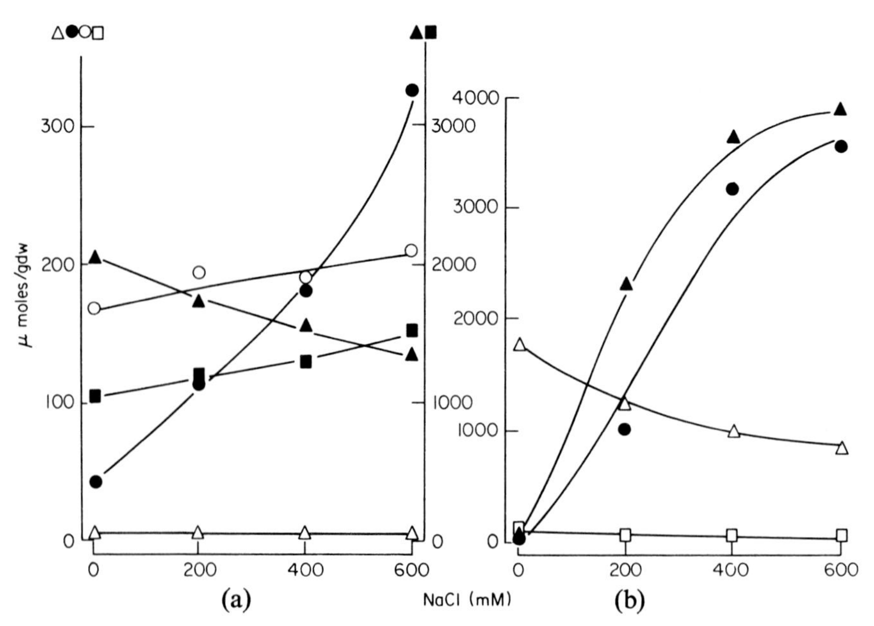

FIG. 13.3. Response of *Puccinellia maritima* to NaCl. See Fig. 13.1 for details. Analyses were made on shoot tissue.

(a) Total organic nitrogen (▲), Soluble organic nitrogen (■), α-NH_2-N (○), Proline (●), Total methylated onium compounds (△).

(b) K (△), Na (▲), Cl (●), Mg (□).

an increase in salinity. Plants of *Atriplex littoralis* grown under non-saline conditions appear to exhibit a lower capacity than plants of *S.maritima* to accumulate nitrate and potassium ions. *Atriplex littoralis* exhibits a progressive accumulation of sodium and chloride and at an external concentration of 600 mM NaCl these ions account for 21% of the dry weight.

The response of *Puccinellia maritima* to salt is similar in certain respects to that of *Atriplex littoralis*, but in this species there is an almost linear increase in proline content with respect to external salt concentration (Fig. 13.3). There is also an increase in α-amino nitrogen and this is accounted for by a 6–7 fold increase in glutamine content.

The patterns of response exhibited by *Limonium vulgare* and *Spartina anglica* are somewhat different. In both species the Na^+ and Cl^- contents remain relatively constant in plants grown between 50 and 600 mM NaCl (Figs. 13.4 and 13.5). In *Limonium* both proline and homobetaine increase with an increase in external salt concentration; the increase in proline content is particularly marked at salt concentrations above 200 mM, while the increase in homobetaine content occurs up to 200 mM. In *Spartina* (Fig. 13.5) there is a small increase in proline content at high external salt concentrations. Dimethyl propiothetin content is apparently unaffected by external salt concentration while glycine betaine increases up to 200 mM NaCl. The pattern of

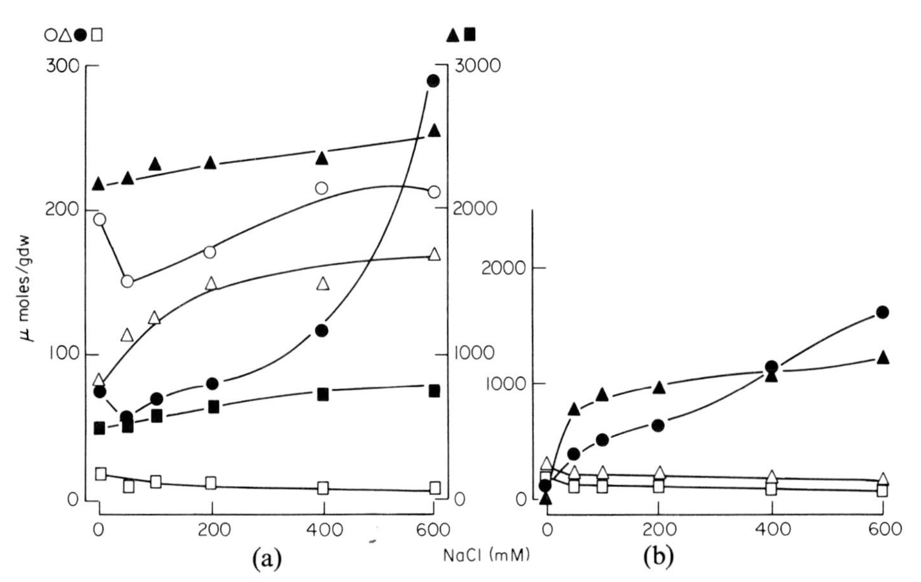

FIG. 13.4. Response of *Limonium vulgare* to NaCl. See Fig. 13.1 for details. Analyses were made on leaf tissue.

(a) Total organic nitrogen (▲), Soluble organic nitrogen (■), α-NH_2-N (○), Proline (●), Homobetaine (△), NO_3 (□).

(b) K (△), Na (▲), Cl (▲), Mg (□).

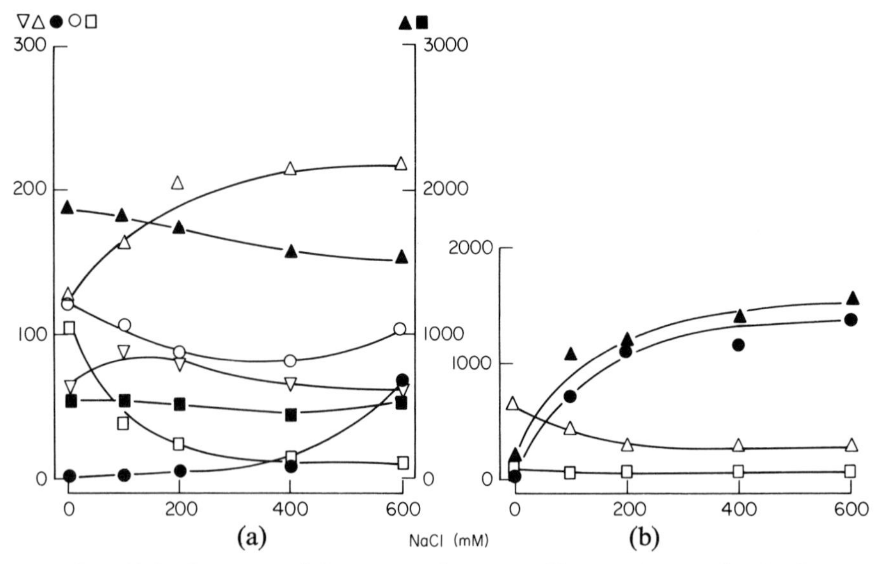

FIG. 13.5. Response of *Spartina anglica* to NaCl. See Fig. 13.1 for details. Plants were grown from tillers. Analyses were made on shoot tissue.

(a) Total organic nitrogen (▲), Soluble organic nitrogen (■), α-NH_2-N (○), Proline (●), Glycine betaine (△), Dimethyl propiothetin (▽), NO_3 (□).

(b) K (△), Na (▲), Cl (●), Mg (□).

sodium and chloride accumulation suggests both *Limonium* and *Spartina* regulate ion accumulation and this presumably is related to the functioning of the salt glands in both species.

With the exception of *Limonium vulgare* all the species examined show an apparent decrease in total organic nitrogen with an increase in external salt concentration. If, however, nitrogen content is calculated on the basis of organic matter rather than 'total dry weight' somewhat different results are obtained (see Table 13.4) in that only *Spartina anglica* and *Puccinellia maritima* show any decrease in nitrogen content. Even in these two species the reduction in nitrogen content is less than 15%. These results suggest that the accumulation of high internal concentrations of NaCl does not for the most part impair the plants capacity to assimilate nitrogen. All of the nitrogen assimilating enzymes are appreciably inhibited by high concentrations of NaCl, suggesting that salt is excluded from the sites of nitrogen assimilation. Proline and/or methylated quaternary ammonium compounds represent a greater proportion of plant nitrogen in salt-grown plants compared with those grown in the absence of salt and in the case of *Puccinellia maritima* and *Spartina anglica* these must be accumulated at the expense of other nitrogenous compounds. In general then most of the species examined appear able to maintain the rate of nitrogen assimilation under saline conditions but exhibit marked changes

Table 13.4. *Changes in plant nitrogen status in response to salinity.*

	Plant organic nitrogen as % organic matter		Proline as % organic nitrogen		Betaine as % organic nitrogen	
External salinity	0 mM NaCl	600 mM NaCl	0 mM	600 mM NaCl	0 mM NaCl	600 mM NaCl
Species						
Atriplex littoralis	4·9	5·3	1	1	6	17
Limonium vulgare	3·0	3·5	3	11	3	7
Puccinellia maritima	2·8	2·5	2	24	1	1
Spartina anglica	2·6	2·3	1	5	5	18
Suaeda maritima	5·2	5·4	1	1	13	23

Data calculated from results shown in Figs. 13.1–5.

in the patterns of nitrogen metabolism. These changes occur in the nature and levels of soluble nitrogenous compounds. At present the mechanisms underlying such changes are unknown and the physiological significance of some of them still has to be determined.

PHYSIOLOGICAL AND ECOLOGICAL CONSIDERATIONS

The results described above indicate species exhibit a spectrum of physiological responses when grown under saline conditions. At one end of this spectrum is a species such as *Suaeda maritima* which has a high capacity for inorganic ion accumulation and at the other end is a species such as *Limonium vulgare* which is able to restrict the accumulation of inorganic ions. In *Suaeda maritima* adaptation to saline growth conditions appears to occur largely through the replacement of ions such as nitrate and potassium by chloride and sodium. The high levels of glycine betaine present in plants of *Suaeda* grown in the absence of NaCl may be related to the high levels of potassium and nitrate accumulated in such plants. Studies of the effect of KNO_3 on isolated enzymes indicate that for some enzymes (e.g. glutamine synthetase) this salt is appreciably more inhibitory than NaCl and it seems likely that a large proportion of the nitrate accumulated could be located in the vacuole. Many species of the Chenopodiaceae are nitrate accumulators and a role for nitrate in maintaining low osmotic potentials in *Chenopodium album* has been proposed by Austenfeld (1972). It is likely that many nitrate accumulating species will also accumulate glycine betaine or some other compatible cytoplasmic solute. In many plant species the internal chloride levels are thought to influence the rate of nitrate influx (see Cram 1976) and this type of control may be important in the ionic relationships of nitrate accumulating halophytes.

The capacity of *Atriplex littoralis* and *Puccinellia maritima* to accumulate salt is somewhat intermediate between that of *Suaeda maritima* and *Limonium vulgare*. The nitrate and potassium contents of *A.littoralis* grown under non-saline conditions are lower than those in *Suaeda maritima*, and the lower glycine betaine content of *A.littoralis* may be related to this reduced capacity for ion accumulation. This species exhibits what can be regarded as an adaptive accumulation of ions and compatible solute whereas *Suaeda* exhibits a constitutive pattern of ion and compatible osmotic solute accumulation.

The ability of species such as *Limonium vulgare* and *Spartina anglica* to maintain relatively constant levels of sodium and chloride contrasts with the marked accumulation of these ions in *Suaeda* and *Atriplex*. This regulation of ion accumulation may be an energetically more expensive form of adaptation than that exhibited by constitutive or adaptive ion accumulators but its ecological significance may be that it affords the possibility of adjusting to rapid fluctuations in salinity.

Both *Limonium vulgare* and *Spartina anglica* appear to have the capacity to accumulate at least two sources of compatible solute; this may be of more general occurrence since under saline conditions *Puccinellia maritima* accumulates glutamine, *Agrostis stolonifera* asparagine, and *Beta maritima* soluble carbohydrates (Ahmad, unpublished results). These observations suggest the possibility that a range of compounds could be utilized as compatible osmotic solutes in a particular species. At present there are no clear indications as to why a range of compatible solutes might be utilized. One possibility is that the potential of a species to accumulate a range of compounds is related to the maintenance of osmotic potentials in different cell-types or sub-cellular compartments. The general model of salt-tolerance outlined earlier treats the cytoplasm as a simple, single component, ignoring the presence of highly specialized organelles. In halophytes, detailed studies of the osmotic relations of organelles are lacking, and it may be useful to examine these.

The accumulation of high concentrations of, for example, proline, would appear to require some spatial separation of the site of proline synthesis and the site of its accumulation, if the problem of product and end-product inhibition of the biosynthetic enzymes is to be resolved. It is possible that the accumulation of pipecolic and 5-hydroxypipecolic acid in proline accumulating species reflects a mechanism for resolving this problem, since these could be accumulated at the site of proline synthesis.

An alternative possibility is that the potential to accumulate several sources of compatible osmotic solute represents a flexible system of adjustment which permits the plant to accumulate necessary osmotic solutes with a minimum perturbation of essential metabolic processes. This could be of particular importance under conditions where the supply of nitrogen was limiting. Thus the accumulation of soluble carbohydrates, polyhydric alcohols or dimethyl propiothetin could substitute for proline or methylated quaternary ammonium compounds under conditions of nitrogen deficiency, conditions under which the levels of nitrogenous compatible solutes could not be maintained without major changes in protein levels.

Clearly much remains to be established regarding the possible ecological and physiological significance which may be attached to the type of osmotic solute accumulated in different species and to the apparent capacity of at least some species to accumulate several compounds which in theory, at least, could serve as compatible solutes. Hopefully the combination of laboratory and field studies will provide the answers to such problems.

ACKNOWLEDGMENTS

The technical assistance of Janice Coulson in much of this work is gratefully acknowledged. A large part of the work described here was supported by grants from the SRC and Royal Society (to G.R. Stewart).

REFERENCES

Austenfeld F.A. (1972) Untersuchungen zur Physiologie der Nitratspeicherung und Nitratassimilation von Chenopodium album, *Z. Planzen physiol.* **67**, 271–81.

Brown A.D. & Simpson J.R. (1972) Water relations of sugar tolerant yeasts: the role of intracellular polyols. *J. Gen. Microbiol.* **72**, 589–91.

Cram W.J. (1976) Negative feedback regulation of transport in cells. The maintenance of turgor, volume and nutrient supply. *Transport in Plants, II, Part A, Cells.* (Ed. by U. Luttge & M.G. Pitman), pp. 284–316. Springer-Verlag, Berlin and Heidelberg.

Flowers T.J. (1972) The effect of sodium chloride on enzyme activities from four halophyte species of Chenopodiaceae. *Phytochemistry* **11**, 1881–86.

Flowers T.J., Troke P.F. & Yeo A.R. (1977) The Mechanism of salt tolerance in halophytes. *Ann. Rev. Plant Physiol.* **28**, 89–121.

Greenway H. & Osmond C.B. (1972) Salt responses of enzymes from species differing in salt tolerance. *Plant Physiol.* **49**, 256–59.

Goas M. (1965) Sur le métabolisme azoté des halophytes: étude des acides aminés et amides libres. *Bull. Soc. Fr. Physiol. vég.* **11**, 309–16.

Goas M., Larher F. & Goas G. (1970) Mise en evidence des acides pipécolique et 5-hydroxy-pipécolique dans certaines halophytes. *C. R. Acad. Sci. Paris* **271**, 1368–71.

Hall J.L. & Flowers T.J. (1973) The effect of salt on protein synthesis in the halophyte *Suaeda maritima. Z. Planzenphysiol.* **71**, 200–6.

Hellebust J.A. (1976) Osmoregulation. *Ann. Rev. Plant Physiol.* **27**, 485–505.

Larher F. (1976) *Sur quelques particularites du métabolisme azoté d'une halophyte: Limonium vulgare* Mill. These Doct. Sci. Nat. Rennes.

Larher F. (1977) Les composés a groupement ammonium quaternaire trimethyle chez les plantes des vases salées. *Comm. Colloq. Soc. Bot. Fr.* (In press).

Larher F. & Hamelin J. (1975a) L'acide-trimethylamino-propionique des rameaux de *Limonium vulgare* Mill. *Phytochemistry* **14**, 205–7.

Larher F. & Hamelin J. (1975b) Mise en evidence de l'acide 2-trimethylamine 6-ceto-heptanoique dans les rameaux de *Limonium vulgare. Phytochemistry* **14**, 1798–1800.

Larher F., Hamelin J. & Stewart G.R. (1977) L'acides dimethylsulfonium-3 propanoique de *Spartina anglica. Phytochemistry* **16**, 2019–20.

Larher F. & Stewart G.R. (1978) The distribution of methylated quaternary ammonium compounds in coastal plants. *New Phytologist.* (In press).

Pigott C.D. (1969) Influence of mineral nutrition on the zonation of flowering plants in coastal salt marshes. In *Ecological Aspects of Mineral Nutrition of Plants* (Ed. by I. Rorison), pp. 25–35. Blackwell Scientific Publications, Oxford.

Santarius K.A. (1969) Der Einfluss von Elektrolyten auf Chloroplasten beim Gefrieren und Trocknen. *Planta* **89**, 23–46.

Stewart G.R., Lee J.A. & Orebamjo T.O. (1972) Nitrogen metabolism of halophytes. I. Nitrate reductase activity in *Suaeda maritima. New Phytol.* **71**, 263–67.

Stewart G.R., Lee J.A. & Orebamjo T.O. (1973) Nitrogen metabolism of halophytes. II. Nitrate availability and utilization. *New Phytol.* **72**, 539–46.

Stewart G.R. & Lee J.A. (1974) The role of proline accumulation in halophytes. *Planta* **120**, 279–89.

Stewart G.R. & Rhodes D. (1978) Nitrogen metabolism of halophytes. III. Enzymes of ammonia assimilation. *New Phytol.* **80**, 307–16.

Storey R. & Wyn Jones R.G. (1975) Betaine and choline levels in plants and their relationship to NaCl stress. *Plant. Sci. Lett.* **4**, 161–68.

Storey R. & Wyn Jones R.G. (1977) Quaternary ammonium compounds in plants in relation to salt-resistance. *Phytochemistry* **16**, 447–53.

Storey R., Ahmad N. & Wyn Jones R.G. (1977) Taxonomic and ecological aspects of the distribution of glycine betaine and related compounds in plants. *Oecologia* **27**, 319–32.

Treichel S. (1975) Der Einfluss von NaCl auf die Prolinkonzentration verschiedener Halophyten. *Z. Planzenphysiol.* **76**, 56–68.

Tyler G. (1967) On the effect of phosphorus and nitrogen supplied to Baltic Shore meadow vegetation. *Bot. Notiser* **120**, 433–48.

Valiela I. & Teal J.M. (1974) Nutrient limitation in salt marsh vegetation. *Ecology of Halophytes* (Ed. by R.J. Reimold & W.H. Queen), pp. 547–63. Academic Press, New York and London.

Wyn Jones R.G., Storey R. & Pollard A. (1976) Ionic and osmotic regulation in plants particularly halophytes. *Transmembrane ionic exchanges in plants* (Ed. by J. Dainty & M. Thellier), Colloques internationaux. Paris C.N.R.S.

Wyn Jones R.G., Storey R., Leigh R.A., Ahmad N. & Pollard A. (1977) *Regulation of Cell Membrane Activities in Plants* (Ed. by E. Marre & O. Ciferri), pp. 121–36. Elsevier/North Holland Biomedical Press, Amsterdam.

14. POPULATION DYNAMICS AND ECOPHYSIOLOGICAL ADAPTATIONS OF SOME COASTAL MEMBERS OF THE JUNCACEAE AND GRAMINEAE

JELTE ROZEMA

Department of Ecology, Free University,
De Boelelaan 1087, *Amsterdam-Buitenveldert,*
The Netherlands

SUMMARY

From measurements of the annual increment of growth of tussocks of different *Juncus* species of known age, a relationship between size of tussock and age was established and from this relationship the age of all tussocks was estimated and the tussocks placed in different age classes. As no mortality of established individuals occurred during a period of 20 years, population growth curves were constructed. The choice of either number of individuals or plant cover as a parameter for population size may lead to markedly different conclusions. The species of *Juncus* which were studied did not colonize the area intensively, instead there was a steady increment in the numbers of individuals per annum to the site. In contrast plant cover at the site increases exponentially. The different strategies of growth of the species are discussed.

Proline accumulation in leaf tissues associated with increasing salinity only occurred in salt and drought tolerant members of the *Juncaceae*. This appears to be a response to osmotic stress, as mannitol induced the same effect. Soluble sugars, especially saccharose, also increase at higher salinity. Unlike the effect of high proline concentrations and mixtures of salts and proline on enzyme activity, high saccharose levels strongly depressed enzymic activity, suggesting that proline is a protective, non-toxic osmoticum maintaining osmotic equilibria within the cell.

RÉSUMÉ

A partir de l'augmentation annuelle de la croissance des touffes d'herbe de différentes espèces de *Juncus* d'un âge connu, nous avons établi une relation entre la taille et l'âge des touffes, et de cette relation, nous avons estimé l'âge de toutes les touffes et les avons réparties dans différentes classes d'âge. Nous avons construit la courbe de croissance de la population en négligeant la mortalité individuelle sur une période de vingt ans. Le choix du nombre d'individus ou celui de la couverture végétale comme paramètre pour la taille de la population peut conduire à des conclusions très différentes. L'espèces de *Juncus* étudiéent ne présentes pas une extension importante, par contre à l'état stationnaire, elles augmentent le nombre de ses individus par unité de surface. Au contraire, la couverture végétale de ce site croit exponentiellement. Nous discutons des différentes stratégies de croissance de cette espèces.

Seules les *Juncaceae* tolérant la sécheresse et la salinité, accumulent la proline dans leurs tissus foliaires à la suite d'une augmentation de salinité. Le manitol provoquant le même effet, ceci apparaît donc comme une réponse au choc osmotique. Les sucres solubles, en particulier le saccharose, augmentent également aux fortes salinités. Une forte concentration de saccharose, au contraire de la proline ou d'un mélange proline et de sel, inhibe l'activité enzymatique, suggérant que la proline soit une substance osmotique, non toxique, protectrice, maintenant l'équilibre osmotique dans la cellule.

INTRODUCTION

The recognition of an individual and determinations of the age of perennial herbs present difficulties in studies of plant demography. Information on demographic studies of salt marsh plants is lacking and at present only speculations are possible (Jefferies 1972). Detailed studies of the demography of different *Juncus* species have been made in a sandy salt marsh at the Island of Schiermonnikoog in the Netherlands for the following reasons:

1. Easy recognition of individuals because of their tussock structure.
2. The time of the birth of the populations is known (1959–60) when an artificial sand ridge was constructed across a very extensive sandy beach. Seed which originated from salt marsh vegetation southwards of the study area was transported by Waddensea tidal water via creeks during the winter months.
3. Relatively homogeneous habitat in both space and time, where abiotic factors have not changed appreciably since 1959.
4. As the vegetation is open at the site, intra- and interspecific competition appear to play a relatively small role in influencing the distribution of different species of *Juncus*.
5. Reliable age classes could be obtained from the results of measurements of clearly recognizable annual growth.
6. The presence of at least five species of *Juncus* in the same area enabled comparison of 'strategies of growth' of species within one genus.

7. Absence of mortality of individuals (apart from the seedling stage) facilitated the construction of the growth curves of populations over a twenty-year period.

MATERIAL AND METHODS

A survey of environmental characteristics of the study area is given elsewhere (Rozema 1975; Rozema 1976b; Rozema *et al.* 1977). Methods of chemical analyses are described by Rozema (1976) and Rozema and Blom (1977). Field work was performed from 1974 onwards and is still continuing. All individuals of all species were counted and measured annually in permanent quadrats of different size (ranging from 160–875 m^2). In absence of recognizable annual growth 50 plants of *Juncus gerardii*, were marked and their position and size recorded annually.

RESULTS AND CONCLUSIONS

Population dynamics

General remarks

In tussocks of *Juncus maritimus*, *J.alpino-articulatus* ssp. *atricapillus* and *J.balticus* slowly decaying shoots remain visible for at least one year. They may be clearly distinguished from the newly developed shoots by their greyish colour. It is possible to measure tussock diameter and obtain values for the annual increment in the diameter of tussocks of these species (Fig. 14.1a). Using tussocks of known age determined from early vegetation maps of the area, a relationship between age and diameter was obtained (Fig. 14.1b) and from this relationship each individual tussock was placed in a particular age class of the population. Fig. 14.1 illustrates the different steps which were used to determine the age structure of a population. Data from a population of *J.maritimus* are used to illustrate the different calculations and reconstructions of the size, age and growth of tussocks of *J.maritimus* that were made. A non-linear relationship between tussock size and the annual increment (Fig. 14.2) is evident. Because of a low overall plant cover at the sites intraspecific competition is unlikely to be important. In addition to selective rabbit grazing of the older tussocks, the rapidly growing grasses *Agrostis stolonifera* and to a lesser extent *Festuca rubra* occupy the centre of the tussocks and may depress the rate at which the tussocks spread as a result of competition for light and minerals. In an old population of *J.maritimus*, 6-year-old tussocks predominated in the age distribution (Fig. 14.1d), but if cover per age class is used as a criterion for population size, the relatively few older tussocks represent the

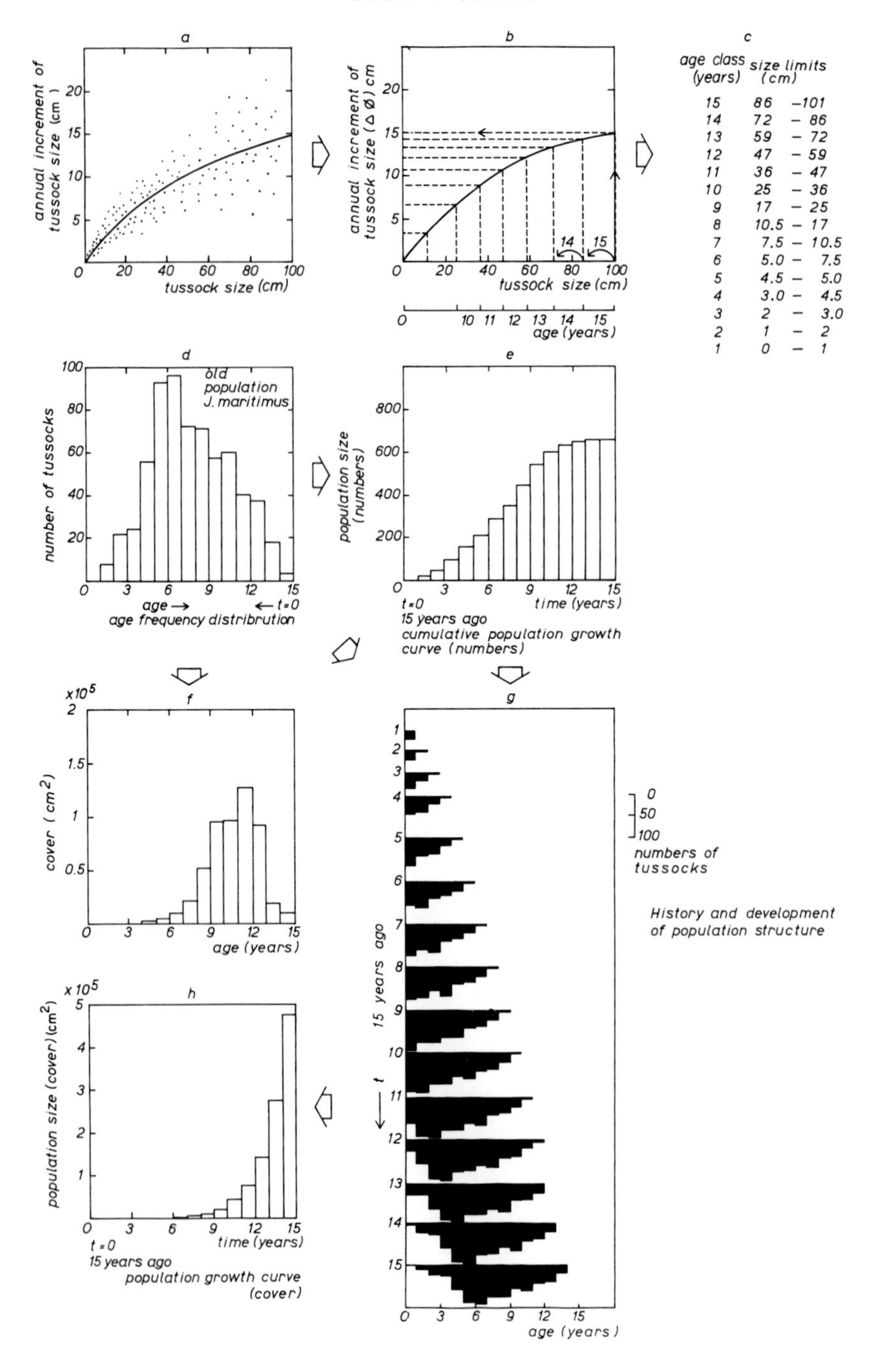
a
annual increment of tussock size (cm)
tussock size (cm)
b
annual increment of tussock size (Δ Ø) cm
tussock size (cm)
age (years)
c
age class (years)
size limits (cm)
15 86 –101
14 72 – 86
13 59 – 72
12 47 – 59
11 36 – 47
10 25 – 36
9 17 – 25
8 10.5 – 17
7 7.5 – 10.5
6 5.0 – 7.5
5 4.5 – 5.0
4 3.0 – 4.5
3 2 – 3.0
2 1 – 2
1 0 – 1
d
old population J. maritimus
number of tussocks
age →
← t=0
age frequency distribrution
e
population size (numbers)
t=0
15 years ago
time (years)
cumulative population growth curve (numbers)
f
cover (cm²)
age (years)
g
15 years ago
numbers of tussocks
History and development of population structure
age (years)
h
population size (cover) (cm²)
t=0
15 years ago
time (years)
population growth curve (cover)

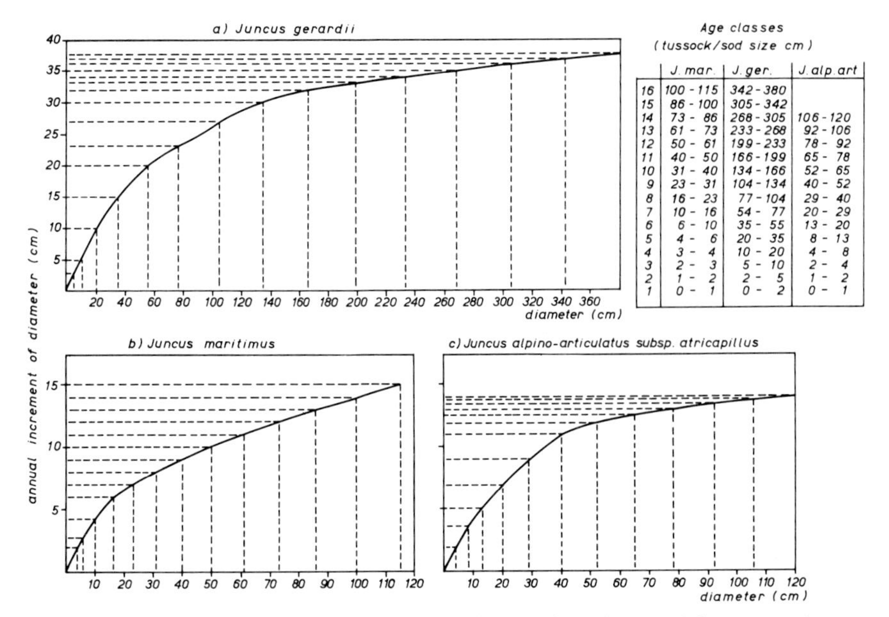

	J. mar.	J. ger.	J. alp. art
16	100 - 115	342 - 380	
15	86 - 100	305 - 342	
14	73 - 86	268 - 305	106 - 120
13	61 - 73	233 - 268	92 - 106
12	50 - 61	199 - 233	78 - 92
11	40 - 50	166 - 199	65 - 78
10	31 - 40	134 - 166	52 - 65
9	23 - 31	104 - 134	40 - 52
8	16 - 23	77 - 104	29 - 40
7	10 - 16	54 - 77	20 - 29
6	6 - 10	35 - 55	13 - 20
5	4 - 6	20 - 35	8 - 13
4	3 - 4	10 - 20	4 - 8
3	2 - 3	5 - 10	2 - 4
2	1 - 2	2 - 5	1 - 2
1	0 - 1	0 - 2	0 - 1

FIG. 14.2. Relationships between size of tussock and annual increment in diameter of tussock for (a) *Juncus gerardii*, (b) *Juncus maritimus* and (c) *Juncus alpino-articulatus* ssp. *atricapillus*, together with a table relating tussock size (cm) to age class (1–16 years).

bulk of the photosynthetic and reproductive apparatus of the population (Fig. 14.1f). As no mortality of these older tussocks has been observed, the addition of tussocks results in a cumulative population growth curve, showing a regular increase of numbers (Fig. 14.1e). Construction of a population growth curve using plant cover as an index reveals an exponential or logistic growth curve (Fig. 14.1h). Comparison of the two curves shows a decline in the annual increment of numbers in recent years after a long period of continuous linear growth. The plant cover of the population increases exponentially as a result of the continuous growth of long lived established individuals (Figs. 14.1e and 14.1h). Although the annual increment of number of individuals to the population has decreased or stopped, those individuals

---

FIG. 14.1. A survey of the different calculations and conversions of field measurements relating the growth and age of plants to the number and size of tussocks, using a population of *J.maritimus* as an example. (a) Relation between tussock size and annual increment in diameter of tussock, (b) construction of age classes, (c) relation between tussock size (diameter) and age, (d) age frequency distribution of a *J.maritimus* population, (e) cumulative growth curve of numbers of tussocks, (f) cover represented by each age class, (g) reconstruction of the history of a particular *J.maritimus* population, (h) cumulative growth curve of the total plant cover of the population.

present continue to grow, which results in an increase in vegetational cover at the site.

Comparisons of the growth of different species

In Fig. 14.2 correlation diagrams of tussock size and the annual increment of growth in diameter of three *Juncus* species are presented. From a knowledge of the relationship between the age of some tussocks based on early maps of the vegetation, it is possible to place each tussock in an age class. These age classes for each species are given in Fig. 14.2; the data together with those shown in Table 14.1 provide evidence of three different 'strategies of growth'

Table 14.1 *A comparison of different populations of* Juncus *species studied at a sandy beach plain on the Island of Schiermonnikoog, the Netherlands.*

Species		Number of tussocks	Quadrat area (m^2)	Number of tussocks (m^{-2})	Total plant cover (m^2)	Plant cover as % of total area
J.maritimus	young	462	300	1·54	5·34	1·8
	old	657	300	2·19	57·35	19·1
J.gerardii	young	340	625	0·54	78·47	12·6
	old	182	875	0·21	289·33	33·1
J.alpino-articulatus		89	160	0·56	7·08	4·4

in a brackish habitat, in which frequent fluctuations in soil salinity occur. Although *J.gerardii* is a medium salt-tolerant species (Rozema 1976), which shows maximum growth in culture solutions devoid of sodium chloride, 16-year-old tussocks reach a diameter of 4 m and the total cover of the population is as high as 33·1% of the quadrat area. In a population of this species relatively few individuals grow rapidly during the less saline periods which occur frequently. In contrast, old tussocks of the salt-tolerant *J.maritimus*, exhibit the same slow growth rate both in saline and weakly saline culture solutions, and on the Island tussocks seldom exceed 1 m in diameter. Populations of *J.maritimus*, produce relatively little cover. Thirdly, the salt-sensitive *J.alpinoarticulatus* (Rozema 1976) has a high growth rate in hydroculture where sodium chloride is absent but growth is strongly depressed at high salinites. The tussocks apparently suffer from salinity and summer drought and this results in relatively small tussock sizes and scanty plant cover on the Island.

The two ways, outlined above, of describing the population are compared for the three species of *Juncus* in Figs 14.3a and 3b, and the graphs indicate, that cover as a parameter of the size of the population discriminates between the three species. Results from Figs 14.3a and 3b and Fig. 14.4 (cumulative

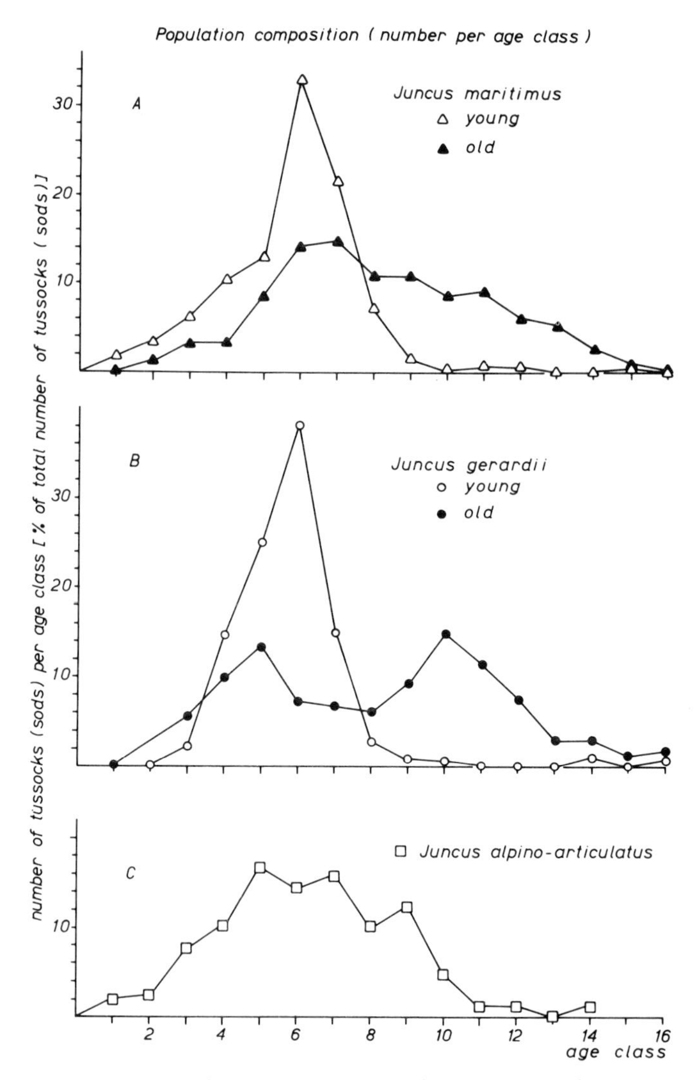

FIG. 14.3a. Number of tussocks in each age class are expressed as a percentage of the total number of tussocks in a population.

population growth curves) indicate, that the number of individuals which establish each year is decreasing at present. Also it is seen, that the type of statistical distribution describing the process of establishment (the number of establishing plants per annum) appears to be a normal distribution, although an adequate ecological explanation of the phenomenon remains difficult. The

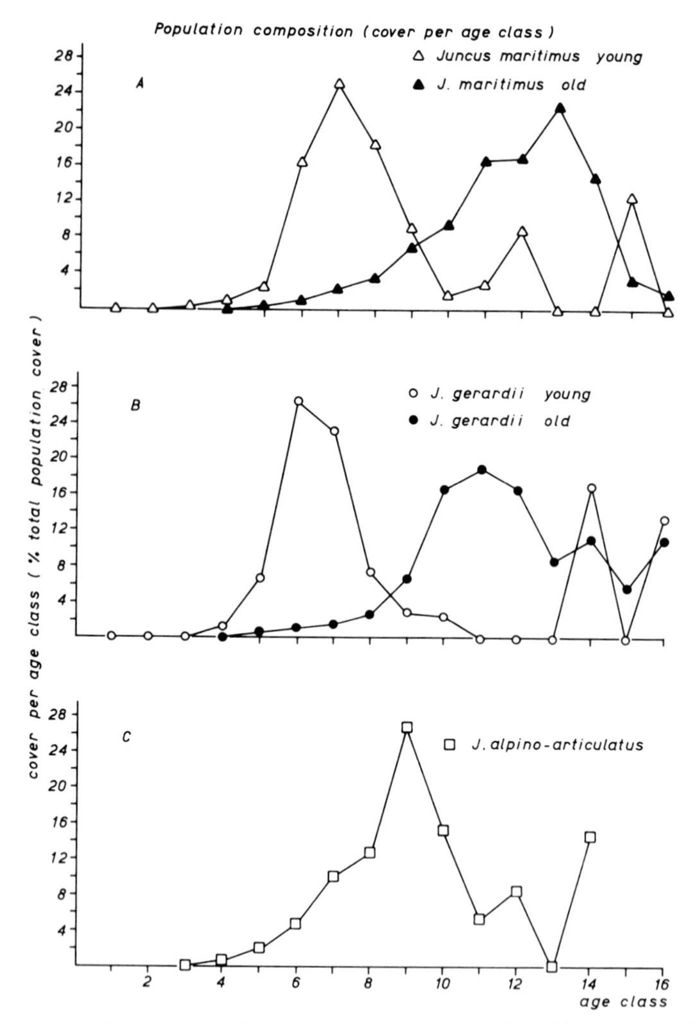

FIG. 14.3b. The amount of cover occupied by tussocks within an age class expressed as a percentage of the total cover of a population.

decrease in the number of recruits to the population at present does not reflect the inability to detect small established seedlings, as these were found elsewhere on the Island. Conditions necessary for germination may be unsatisfactory but this is unlikely as the development of an extensive moisture retaining mat of the green alga *Rhizoclonium riparium* in the area provides ideal germination conditions, and total vegetation cover of higher plants seldom reaches 25%. Cumulative growth curves of the increase in cover of the population plotted on a logarithmic scale (Fig. 14.5) are of the same type for all species studied, the slope of the curve for *J.gerardii* curves being the steepest. The shape of the curves suggests logistic growth, the decrease of growth possibly reflecting the relatively slow increase in diameter of old tussocks (Fig. 14.2). Estimates of additional time needed to reach a 100 m^2 cover yield the following results, *J.gerardii* (old) 4 years, *J.maritimus* (old) 5–6 years, and *J.alpino-articulatus* 10–11 years. These figures should be regarded as underestimates as interspecific competition as a result of the rapid growth of

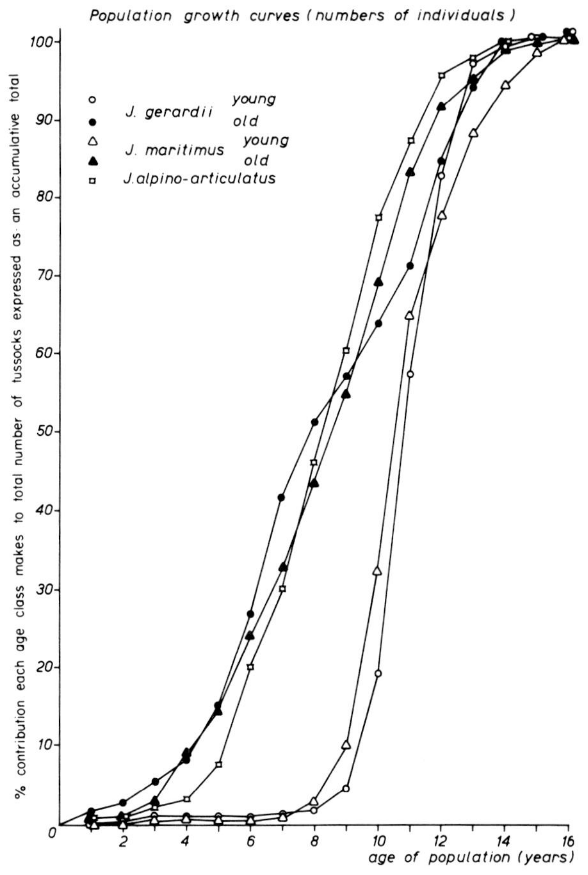

FIG. 14.4. Cumulative growth curves of three Juncus species. Accumulative total of the number of tussocks in relation to the age structure of populations of *Juncus* expressed as a % of the total number of tussocks in the population at present (see Table 14.1). ○ (young), ● (old) *J.gerardii*; △ (young), ○ (old) *J.maritimus*; (□) *J.alpino-articulatus* ssp. *atricapillus*.

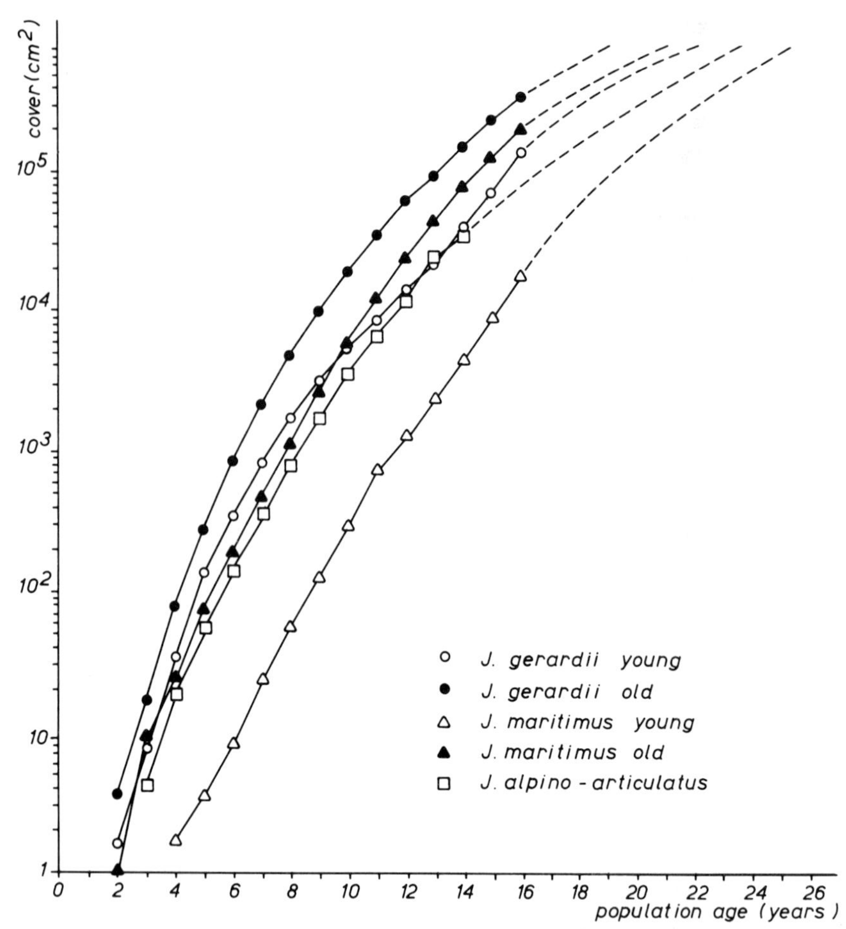

FIG. 14.5. Cumulative increase in total cover (expressed as cm^2) of populations of *J.maritimus* (△ young, ▲ old), *J.gerardii* (○ young, ● old), *J.alpino-articulatus* ssp. *atricapillus* (□). The curves have been extrapolated to produce an estimate of the age of the oldest individuals in a population when the value for plant cover reaches 10^6 cm^2 (100 m^2).

Agrostis stolonifera and *Festuca rubra* in tussocks of species of *Juncus* is likely to be more severe as the plants age and therefore this will modify growth of species.

Ecophysiological adaptations to salinity

Many possible adaptations of members of the *Juncaceae* and *Gramineae* to salinity have been investigated (Rozema 1976a; Rozema & Blom 1977). Abscission of salt-saturated shoots, increase in the degree of succulence, organic acid metabolism and sensitivity of enzymes do not appear to be important components of salt tolerance mechanisms in species studied.

Functional adaptations in salt-tolerant species include the exclusion of sodium and chloride from the root system of individuals (Rozema 1976a), differences in salt-tolerance of populations such as between populations of *Festuca rubra* (Rozema *et al.*, in preparation), preferential selective uptake of potassium (Rozema 1976a; Rozema & Blom 1977), low growth and transpiration rates and the use of organic compounds such as proline in osmotic adjustment. Only salt tolerant species of *Juncus* show an increase in

Table 14.2. *The influence of increasing salinity (mM, NaCl) or increasing mannitol (mM) on the concentration of proline (mM) in leaves of different* Juncus *species. Values in brackets refer to mannitol treatments. Proline content of plant material collected at Schiermonnikoog is also given.*

	Concentration of osmoticum (mM)				
NaCl	0	60	150	300	Proline content field material
mannitol	0	120	300	600	(μg g^{-1} dry wt)
Species					
J.maritimus	1·2 (1·7)	1·7 (2·0)	4·0 (7·1)	11·7 (12·8)	3760
J.gerardii	0·4 (0·6)	0·9 (1·9)	1·3 (3·5)	2·8 (6·3)	460
J.alpino-articulatus	0·2 (0·2)	0·3 (0·3)	0·2 (0·3)	0·3 (0·3)	138
J.articulatus	0·2 (0·3)	0·2 (0·2)	0·3 (0·3)	0·3 (0·3)	117
J.bufonius	0·1 (0·1)	0·1 (0·3)	0·2 (0·1)	0·2 (0·2)	96

proline content in response to an increase in salinity. If mannitol is used instead of sodium chloride to decrease the osmotic potential of the external medium, proline accumulates in the tissues (Table 14.2) which suggests that this accumulation is a response to the osmotic effects of salinity (Kluge 1976). This is supported by the finding of marked differences in salt tolerance between populations of *Festuca rubra*. The increase of proline content was pronounced in the drought tolerant dune population (Rozema *et al.*, in preparation). The ability of *J.maritimus* and to a lesser extent *J.gerardii* to increase the levels of proline in the tissues in response to salinity and drought and the almost complete inability of the salt-sensitive species *J.alpino-articulatus*, *J.articulatus* and *J.bufonius* to raise proline levels supports the conclusion that the former two species are relatively drought-resistant whereas the latter species are drought-sensitive (Rozema, unpublished). The ecological significance of proline accumulation is supported by proline measurements of plant material collected in the field (Table 14.2), which is in accordance with the results found in nutrient culture experiments.

Mono and di-saccharides have been proposed as non-toxic osmotica, that are involved in salt tolerance (e.g. Waisel 1972). Measurements of the soluble sugar content, calculated on a glucose basis, are summarized in Table 14.3. The concentrations of sugars are relatively high as compared with the concentrations of proline. In the latter case the assumption of localization within the cytoplasm or other cell compartments is required in order that this substrate reaches a sufficiently high concentration to act as an osmoticum.

Table 14.3. *The influence of increasing salinity on the sugar content of plant sap of different* Juncus *species (mM) calculated on a glucose basis.*

Salinity (mM NaCl)	0	60	150	300
J.maritimus	152·6	176·8	165·6	251·1
J.alpino-articulatus	65·8	76·9	119·5	286·7
J.articulatus	67·4	55·8	100·7	193·4
J.gerardii	188·0	206·7	290·0	396·2

Analyses based on the use of gas-liquid and paper chromatography demonstrated that saccharose forms a major component of the soluble sugars measured. In Table 14.3 it is seen, that the largest increase of sugar content occurs in salt sensitive and drought sensitive species, which raises doubts regarding the function of sugars in salt tolerance. It may be that the carbohydrate metabolism of the salt-tolerant species is relatively unaffected by increasing salinity. *In vitro* assays of the effects of different osmotica on the activities of malate dehydrogenase and pyruvate kinase were investigated. The enzymes were extracted both from salt-sensitive species and salt-tolerant species cultured in hydroculture in which the concentrations of sodium chloride in the nutrient solution were either 0 or 300 mM. As the results from all experiments showed the same trend only one example is presented in Table 14.4 (*J.articulatus*, 0 mM NaCl). Surprisingly, a 2 M proline concentration and mixtures of NaCl (0·5 M) and KCl (0·5 M) with proline (1 M) do not depress enzyme activity, in contrast with the strong inhibition associated with solutions of NaCl and KCl alone. Iso-osmotic concentrations of saccharose also inhibited enzymic activity. This leads to the conclusion, that proline represents a protective, non-toxic osmoticum maintaining osmotic equilibria within the cell. The well-known salt sensitivity of cytoplasmic enzymes of halophytes is consistent with the localization of proline within the cytoplasm (Greenway & Osmond 1972). However, the existence of other functional protective osmotica (e.g. glycine betaine and related compounds (Storey & Wyn Jones 1977) must not be excluded.

Table 14.4. *The influence of different osmotica on the activity of soluble malate dehydrogenase extracted from leaf tissue of* J.articulatus *(expressed as percentage of the control in the absence of added solutes in the cuvette).*

Control (buffer)	1 M NaCl	1 M KCl	2 M proline	2 M glucose	2 M saccharose
100	31	25	119	47	14

Control (buffer)	0·5 M NaCl + 1 M proline	0·5 M KCl + 1 M proline	0·5 M NaCl + 1 M glucose	0·5 M NaCl + 1 M saccharose
100	130	125	70	48

ACKNOWLEDGMENTS

Thanks are due to Miss W.B. Cramer for careful and expert analyses, to Dr R.L. Jefferies and Dr A.J. Davy for revision of the English text and valuable comments, to Mr D.J. Tolsma, who made thorough and critical studies of the demography, and to Mr F. Stroeve, who mapped and studied the growth of *Juncus gerardii*. The author thanks Mr G.W.H. van den Berg for the drawings, Mr J.H. Huysing for photography and Miss T. Moens for typing the manuscript.

REFERENCES

Greenway H. & Osmond C.B. (1972) Salt responses of enzymes from species differing in salt tolerance. *Plant Physiol.* **49**, 256–59.

Jefferies R.L. (1972) Aspects of salt marsh ecology with particular reference to inorganic plant nutrition. *The Estuarine Environment* (Ed. by R.S.K. Barnes & J. Green), pp. 61–85. Applied Sci. Publ., London.

Kluge M. (1976) Carbon and nitrogen metabolism under water stress. *Water and Plant Life* (Ed. by O.L. Lange, L. Kappen & E.D. Schulze), pp. 243–52. Springer-Verlag, Berlin.

Rozema J. (1975) The influence of salinity, inundation and temperature on the germination of some halophytes and non-halophytes. *Oecol. Plant* **10**, 341–53.

Rozema J. (1976a) An ecophysiological study on the response to salt of four halophytic and glycophytic *Juncus* species. *Flora* **165**, 197–209.

Rozema J. (1976b) Vegetation zonation at the Beach Plain of Schiermonnikoog. *Waddenbulletin* **11**, 144–48 (In Dutch).

Rozema J. & Blom B. (1977) Effects of salinity and inundation on the growth of *Agrostis stolonifera* and *Juncus gerardii*. *J. Ecol.* **65**, 213–22.

Rozema J., Nelissen, H.J.M., van der Kroft M. & Ernst W.H.O. (1977) Nitrogen mineralization in sandy salt marsh soils in the Netherlands. *Z. Pflanzenern. Bodenk* **140**, 707–17.

Storey R. & Wyn Jones R.G. (1977) Quaternary ammonium compounds in plants in relation to salt resistance. *Phytochemistry* **16**, 447–53.

Waisel Y. (1972) *Biology of Halophytes*. Academic Press, New York and London.

15. THE GROWTH STRATEGIES OF COASTAL HALOPHYTES

R. L. JEFFERIES,[1] A. J. DAVY[2] AND T. RUDMIK[1]

[1] *Department of Botany, University of Toronto, Toronto, Canada*
and
[2] *School of Biological Sciences, University of East Anglia, Norwich, U.K.*

SUMMARY

Population differentiation, the phenology of plants and the allocation of resources to maintenance, growth and reproduction have been examined in relation to environmental heterogeneity within a salt marsh at Stiffkey, Norfolk. Although there is seasonal variation in the concentrations of soluble inorganic nitrogen in the sediments of the upper and lower levels of the salt marsh, the values obtained are similar for both levels but the upper level, unlike the lower level, becomes hypersaline in summer.

Both intra- and interspecific variation in plants from this marsh have been observed. Populations of perennial and annual species from the upper marsh show a poor growth response to nitrogen both in the greenhouse and in the field, compared with populations from the lower marsh.

Unlike the results from the lower marsh, the growth and development of young plants of *Salicornia europaea* agg. in the upper marsh is delayed until late summer. Frequent additions of water and nitrogen to the upper marsh throughout the summer in order to reduce hypersalinity and eliminate nitrogen deficiency have no effect on growth until late in the season when tides cross the upper marsh again. The population from the lower marsh responds to the influence of the perturbations early in the summer.

Perennial plants of the upper marsh show a burst of vegetative and reproductive growth early in the season before hypersaline conditions develop. Much of the nitrogen required for this growth appears to be recycled from the underground storage organs. Levels of highly soluble organic compounds such as proline, quaternary ammonium compounds, sorbitol and reducing sugars in the leaves of these perennials fall during the summer. It is suggested that these compounds are involved in osmoregulation and serve as a nitrogen or carbon source for flower and fruit formation and root growth. The perennial plants flower infrequently and it is likely that there is competition for resources between maintenance of the individual, growth and sexual reproduction. Unlike vegetative growth there is some evidence of developmental plasticity in relation to flowering and fruit formation in the face of interseasonal unpredictability in these perennial plants.

RÉSUMÉ

Dans un marais salé, et en relation avec l'hétérogénéité environnementale, nous avons examiné la différenciation de la population, la phénologie des végétaux, et l'allocation de ressources pour l'entretien, la croissance et la reproduction. Malgré des variations saisonnières de la concentration en azote soluble inorganique dans les sédiments des marais supérieurs et inférieurs de Stiffkey, Norfolk, nous avons obtenu des valeurs similaires pour les deux marais, avec cependant une hypersalinité du marais supérieur en été.

Nous avons observé à la fois les variations intra et interspécifiques dans les plantes de ce marais.

Les populations d'espèces annuelles ou vivaces du marais supérieur comparées à celles du marais inférieur, montrent une faible réponse de croissance à l'azote, aussi bien dans la serre que dans les champs. Au contraire, d'après les résultats du marais inférieur, la croissance et le développement des jeunes plantes de *Salicornia europaea* agg. s'étalent jusque tard dans l'été. Des apports fréquents d'eau et d'azote au marais supérieur, durant l'été, afin de réduire l'hypersalinité et éliminer l'insuffisance d'azote, n'ont aucun effet sur la croissance, sauf tard dans l'été quand la marée traverse de nouveau le marais. La population de cette espèce dans le marais inférieur répond aux perturbations, tôt dans l'été.

Les plantes vivaces du marais supérieur montrent un saut de croissance végétative et reproductive tôt dans la saison, avant que ne se développent les conditions d'hypersalinité. La plupart de l'azote nécessaire à cette croissance semble être recyclée dans les organes de réserve souterrains. Les niveaux des composés organiques solubles tels que la proline, les composés ammonium quaternaires, le sorbitol et les sucres réducteurs dans les feuilles de ces plantes vivaces, chutent en été. Nous suggérons que ces composés sont inclus dans l'osmorégulation et servent au même titre que l'azote ou la source de carbone, à la formation des fleurs et des fruits et à la croissance des racines. La floraison de ces plantes est peu fréquente, comme s'il y avait compétition entre les ressources pour l'entretien de l'invidu, la croissance et la reproduction sexuelle. A la différence de la croissance végétative, on trouve une certaine plasticité dans le développement de la floraison et de la formation des fruits, en raison de l'imprédictabilité intersaisonnière dans ces plantes vivaces.

INTRODUCTION

Within salt marshes there is considerable spatial and temporal environmental heterogeneity, much of which is related to the frequency and duration of tidal cover. Odum (1969) has recognized that the tidal regime sets limits on the extent to which the vegetation at a site within a marsh may change. Accordingly he has described salt marshes as pulse-stabilized systems, although the effects of vegetation on the rate of accretion of sediment over a period may result in changes to the tidal regime. Because tidal cover at any particular marsh throughout the year is predictable, seasonal trends in the salinity, water potential and osmolarity of the sediments in different parts of that marsh also may be predicted. The magnitude of these seasonal changes also depends on the prevailing weather conditions, resulting in variations from year to year in edaphic conditions. However salt marshes display cyclical stability, as defined by Orians (1975), in as much as the salinity, water potential and osmolarity of the sediments all fluctuate around the corresponding values for sea water, which change little during a year. To this extent salt marshes are highly predictable, stable environments. This contrasts with small scale spatial variation associated with changes in the rates of erosion or accretion which accompany physical instability, such as occurs in the vicinity of drainage channels or at the seaward edge of the marsh.

Since the saline environment is an important independent variable influencing the evolution of plants within salt marshes it is a determinant of the growth of plants and their response to perturbations. Levins (1968) has indicated that a general theory of the responses of organisms to a fluctuating environment must include the predictability and regularity of occurrence of a particular set of environmental conditions, the length of time that these conditions exist relative to the life cycle of an organism and their effects on its fitness and the total investment required for reproduction. This paper is concerned with an investigation of some of these factors in relation to the growth strategies of annual and perennial plants present within a salt marsh.

As pointed out by Pomeroy (1970) salt marshes are nutrient sinks where normally there is a large excess of all essential elements, with the exception of nitrogen (Tyler 1967; Pigott 1969; Stewart *et al.* 1972, 1973; Valiela & Teal 1974; Patrick & Delaune 1976). The response of many halophytes to the availability of a limited resource, nitrogen, is of particular interest because of the known role of soluble organic nitrogen compounds in osmoregulation in algae and higher plants (Goas 1965; Stewart & Lee 1974; Larher & Hamelin 1975; Storey & Wyn Jones 1975; Hellebust 1976; Storey *et al.* 1977).

In this study the following questions have been examined.

(a) What are the magnitudes of the seasonal changes in salinity, osmolarity and water potential in different parts of the marsh?
(b) Can these changes be related to the different stages of development of areas within the marsh?
(c) Can the genetic differentiation that exists between populations of species

of salt marsh plants (cf. Gray *et al.*, this volume) be related to the environmental heterogeneity which occurs in salt marshes?

(d) If a considerable amount of a limited resource is allocated to the maintenance of plants growing in a saline environment, is there competition for this resource among the processes of maintenance, productivity and reproduction? Is there a redistribution of nitrogen to different parts of both annual and perennial plants during the growing season in relation to these different requirements?

(e) If additions of inorganic nitrogen are made to sites where nitrogen is a limited resource in a salt marsh, does this perturbation result in increased growth and enhanced reproductive output of the different species?

ENVIRONMENTAL HETEROGENEITY

The salt marsh which is used for this long term study is at Stiffkey on the north Norfolk coast of England (Fig. 15.1). The ecology of the salt marshes in this area has been described previously by Chapman (1938, 1939, 1960) and Jefferies (1976) has given a general account of the north Norfolk coastline. The upper and lower levels of this tidal marsh, which are separated by an old dune, reflect a major discontinuity in time in the evolution of the marsh. The marsh on the seaward side of the dunes is probably not more than two hundred years old, on the basis of known rates of changes in the vegetation in young marshes elsewhere on the Norfolk coast (Steers 1960, 1971); whereas the upper level marsh is at least 1,000 years old (Jefferies 1976).

In Fig. 15.2 both annual and seasonal changes in the sodium concentration of interstitial water of the upper layer of the sediments for different sites within the marsh are shown as an example of spatial and temporal environmental heterogeneity. Details of the sampling and analytical procedures are described elsewhere (Jefferies 1977). The sites include one in the lower marsh seaward of the old dune, and two in the upper marsh; one of the latter was adjacent to a drainage channel and the other was in a flat area away from the creeks. An important characteristic of the marsh is that in most years the tides fail to flood the upper marsh during periods of at least six weeks immediately prior to the summer and winter solstices. The upper marsh is usually flooded within two to three weeks after these dates. The absence of tides during these periods, together with high rates of evapotranspiration in summer and the fact that precipitation exceeds evapotranspiration in winter, result in a wide annual amplitude in salinity in the sediments at sites in the upper marsh. In some years, such as in 1974, the peak sodium concentration exceeded 0·9 M, although the amplitude achieved in any one year strongly reflects the prevailing weather conditions, particularly those of the summer months when evapotranspiration exceeds precipitation. The data also indicate

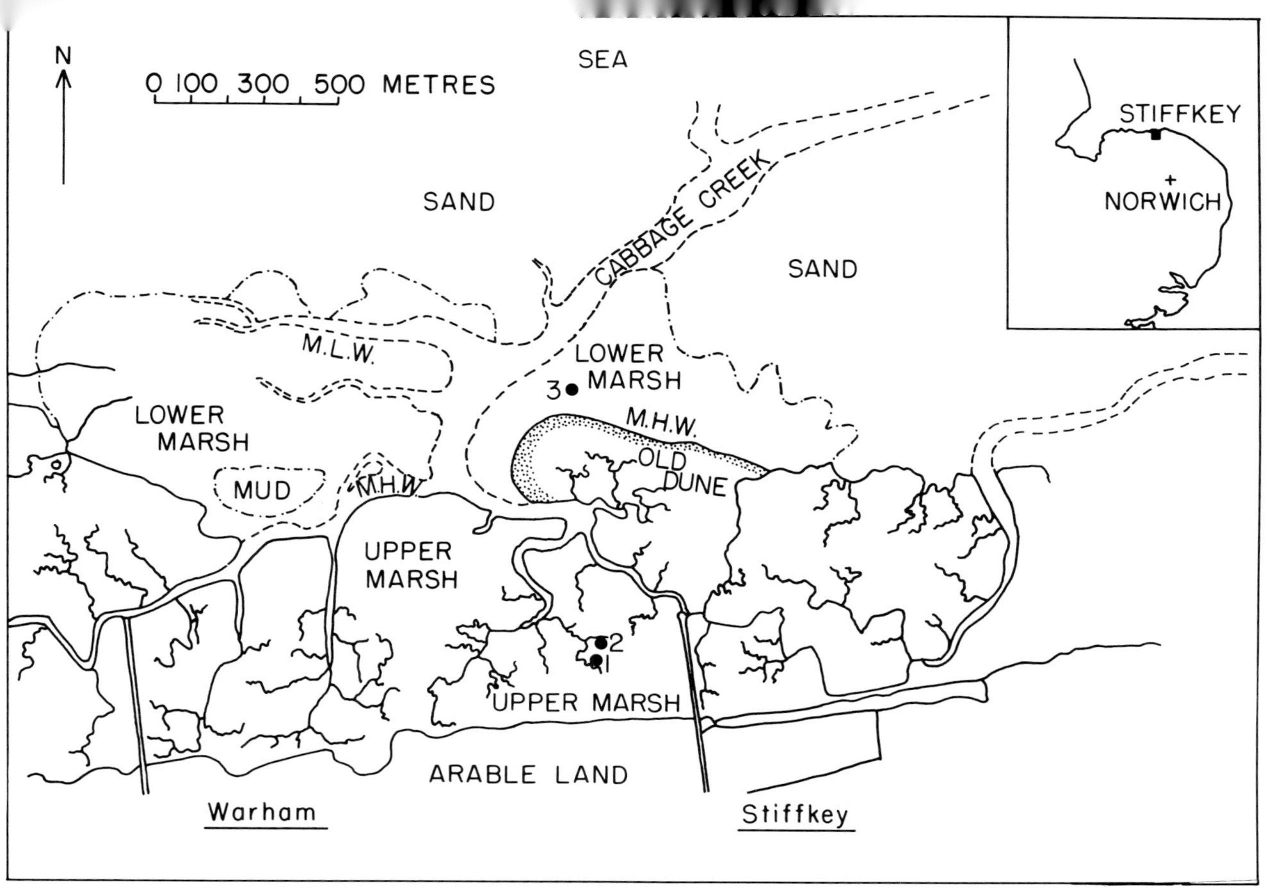

FIG. 15.1. Stiffkey salt marsh, Norfolk, England, showing old dune, tidal creeks, and location of sampling sites in upper and lower marshes. (1) upper marsh site away from drainage channels; (2) upper marsh, bank of drainage channel; (3) lower marsh.

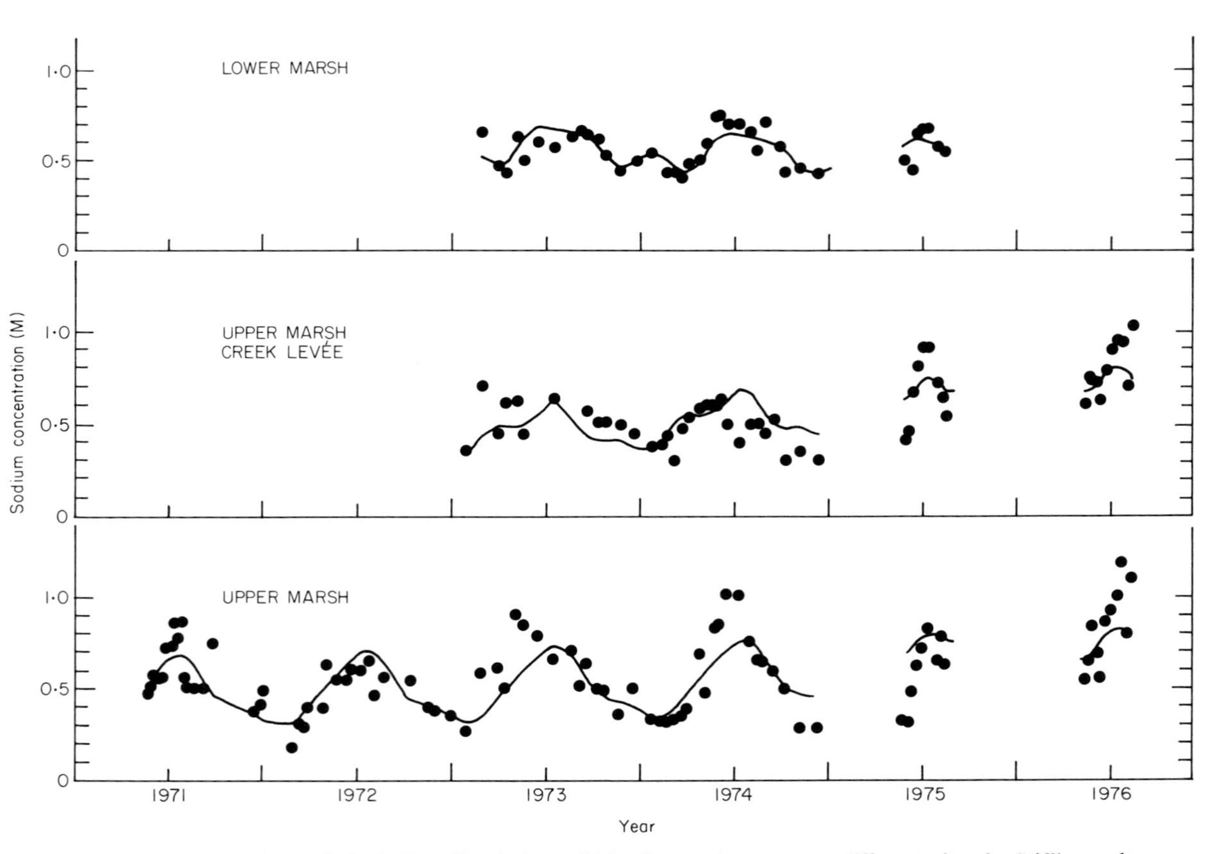

FIG. 15.2. Soil salinity of the bulk soil solution within the rooting zone at different sites in Stiffkey salt marsh. A modified Fourier analysis has been used to fit the curves.

that the highest values for salinity did not occur at exactly the same time each year but were dispersed about the time of the summer solstice, again reflecting differences between years in the weather conditions and the frequency of flooding of the marsh. However during the period 1971 to 1976 hypersaline conditions ($>0{\cdot}5$ M Na^+) prevailed from June to September each year at the site in the upper marsh away from drainage channels.

In contrast the tides failed to flood the site in the lower marsh on a total of not more than ten days during each lunar cycle and consequently at this site the amplitude of salinity is small and the values lie relatively close to that of sea water (c. 0·5 M Na^+). Although the upper marsh is not flooded during much of the summer and winter, tidal water penetrates the creeks and these sites can be regarded as an extension of the lower marsh in that the larger drainage channels are flooded regularly with tidal water. The annual amplitude of salinity in the interstitial water of sediments at a site in a creek bank is also relatively small compared with corresponding data for sites in the upper marsh away from the drainage channels.

A further example of spatial and temporal environmental heterogeneity is shown in Fig. 15.3. *In situ* measurements of water potentials in the upper layers of the sediments at the site in the upper marsh have been made during the summers of 1976 and 1977 with Wescor soil psychrometers in conjunction with a Wescor HR-33 dew point microvoltmeter. The results reflect the contrasting weather during the summers of 1976 and 1977 in eastern England; the first summer was hot and dry while the more recent was cool with considerable rainfall. During the early part of the summer of 1976 the water potential in the upper layers of the sediment fell steadily to below −5·0 mega Pascals (−50 bars). The rise in July reflects the ameliorating effect of the tides flooding the upper marsh again and reducing the extreme hypersalinity. The fall in the middle of August was during a short period of fine settled weather when the tides failed to flood the marsh. In 1977 the water potentials were barely below that of sea water (c. −2·3 MPa). The results also indicate spatial heterogeneity within the sediments. Usually water potential increased down the sediment profile; on some occasions values close to the surface were as much as 1·0 to 1·5 MPa lower than corresponding values 15 to 40 cm below the surface. In late winter and early spring each year this gradient was reversed as a result of low rates of evapotranspiration and high amounts of rainfall but was re-established in late spring. Clearly plants with shallow rooting systems, such as the annual species of *Salicornia* which have a root system that often extends only a few centimetres into the sediment, are subject to very different edaphic conditions from perennial plants with deeper root systems.

The development of hypersaline conditions in the upper levels of salt marshes has been reported by Chapman (1960), Ranwell *et al.* (1964), Beeftink (1965, 1966) and Tyler (1971). However the ecological significance of these

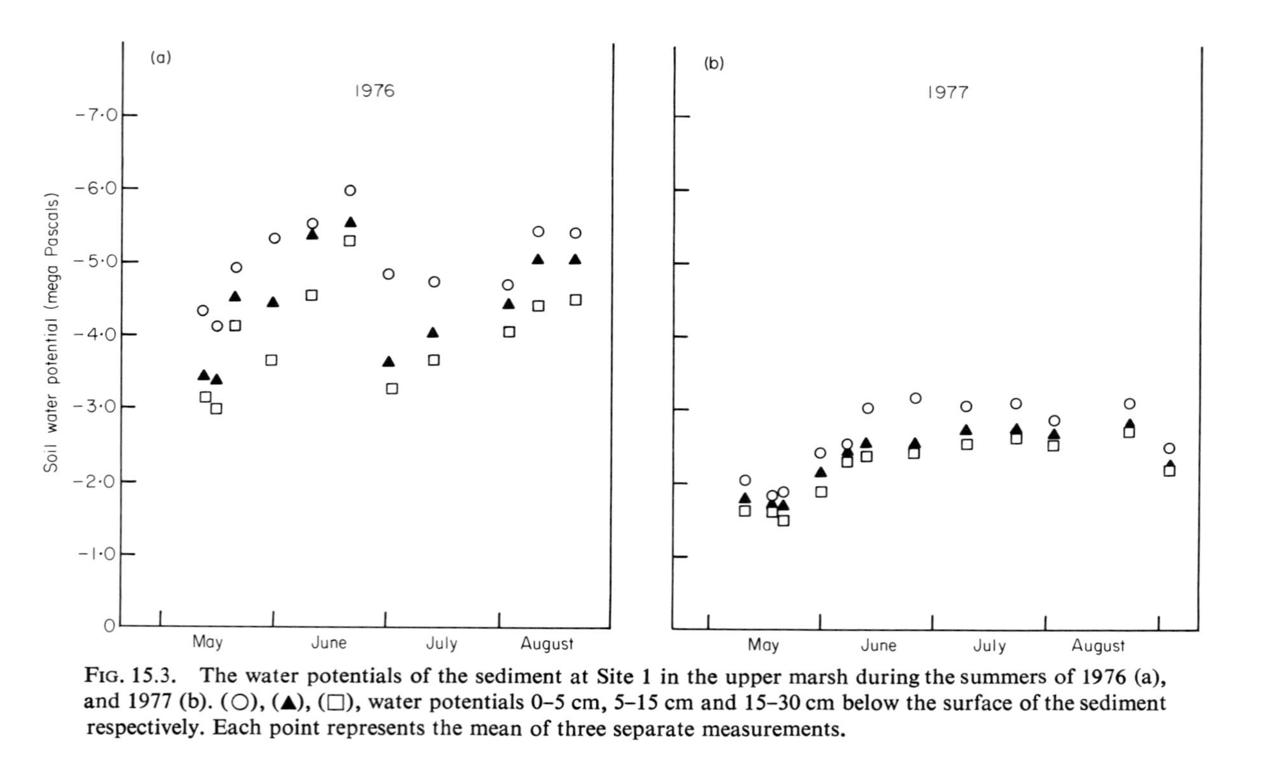

FIG. 15.3. The water potentials of the sediment at Site 1 in the upper marsh during the summers of 1976 (a), and 1977 (b). (○), (▲), (□), water potentials 0–5 cm, 5–15 cm and 15–30 cm below the surface of the sediment respectively. Each point represents the mean of three separate measurements.

differences between the upper and lower levels of salt marshes does not appear to have been fully explored.

POPULATION DIFFERENTIATION IN RELATION TO ENVIRONMENTAL HETEROGENEITY

Population differentiation

Population differentiation within a species in relation to environmental heterogeneity is most likely to occur where environmental conditions vary greatly in space but are relatively stable in time (Levins 1962, 1963; Maynard Smith 1966; Snaydon & Davies 1972). As indicated above conditions differ considerably between the upper and lower marshes but seasonal trends in salinity and associated factors show cyclical stability, thereby fulfilling the conditions under which population differentiation is likely to occur. In addition estate records at Holkham Hall indicate that in 1780 the upper limit of spring tides was similar to the level reached at present and that the upper marsh was flooded on only 144 out of a possible 730 occasions, a frequency similar to present estimates (120). These records show that the tidal regime of the upper marsh has changed little during the last three centuries and the findings suggest a high degree of constancy in environmental conditions of the upper marsh. Adaptive traits of populations of plants in the marsh are likely to reflect this long history of the constancy of environmental heterogeneity.

Significant differences in growth rate, in response to additions of inorganic nitrogen (NH_4^+ or NO_3^-), were observed between populations of *Salicornia europaea* agg., *Triglochin maritima*, *Aster tripolium* and *Plantago maritima* from the upper and lower marsh (Jefferies 1977). The plants were grown in sand culture in an unheated greenhouse. Populations of *Aster tripolium* and *Plantago maritima* from the upper marsh away from the drainage channels also grew more slowly in culture than populations which occurred close to the creeks of the upper marsh. It is likely that there is a significant genetic component to these intraspecific differences. Besides intraspecific variation in growth response to nitrogen, species characteristic of the low marsh at Stiffkey, such as *Halimione portulacoides*, *Aster tripolium* and *Suaeda maritima*, have fast growth rates compared with the rates of species abundant in the upper marsh such as *Armeria maritima* and *Limonium vulgare*. The interesting finding (Jefferies 1977) is that although there is seasonal variation in the concentrations of soluble inorganic nitrogen in the sediments of the two marshes, the values obtained are similar for both marshes. However, the upper marsh, unlike the lower marsh, becomes very hypersaline in summer and so it is suggested that the presence of lower soil water potentials and hypersaline conditions during the growing season has resulted in selection for plants of low growth potential which are able to tolerate these extreme conditions.

If this hypothesis is correct, additions of inorganic nitrogen to sites in the upper marsh should result in little change in the growth rates of the plants. In many experiments of this type, species which have high relative growth rates have exploited the additional resource and become dominant at the expense of species with slow relative growth rates. The latter may disappear from the site as a result of interspecific competition. Regular additions of inorganic nitrogen have been made over a four- or five-year period to sites in the upper marsh at Stiffkey and the plant communities exhibited a strong degree of constancy towards the nutritional perturbations (Jefferies & Perkins 1977). There was little or no change in the frequency of the dominant species as a result of the perturbations and no species disappeared from the site. These results are in striking contrast to those of Willis (1963), who added mineral nutrients to dune soils and reported that at the end of a period of three years considerable changes in the composition of the vegetation had occurred. Species such as *Agrostis stolonifera* and *Festuca rubra* which have high relative growth rates dominated the plots and there was a large fall in species diversity on those plots treated with nitrogen and phosphorus. In our investigations vegetative growth of many of the species was poor; a number of the abundant species failed to show significant differences in yield of above-ground dry matter between treated and control plots. *Aster tripolium*, *Halimione portulacoides* and *Suaeda maritima* were the only species which showed increases in shoot frequencies in those plots treated with inorganic nitrogen compared with the control plots; these species also exhibited a marked growth response to nitrogen in sand culture (Jefferies 1977). It is significant that these species, which are characteristic of the lower marsh where water and nitrogen often are not limiting for plant growth, have high relative growth rates. In the upper marsh they are present at low frequencies and because of the effects of hypersalinity and low water potentials, they show only modest increases in shoot frequency and biomass in those plots treated with nitrogen. The results provide further evidence to support the suggestion that in the upper marsh there has been selection for plants with relatively slow growth rates which have much of their biomass below ground and which are capable of surviving the hypersaline conditions and low water potentials of the sediments. In this respect there is a similarity to desert ecosystems where 'the price of survival for desert plants and plant communities is the utter restriction of growth and production' (Evenari *et al.* 1975).

Phenology of coastal halophytes

Although these inter- and intraspecific responses can be correlated with environmental heterogeneity, perhaps of more interest is the evolution of various life history traits in halophytes in relation to this heterogeneity. In particular the phenology of growth and reproduction in relation to seasonal changes in environmental conditions and the allocation of resources to the processes of maintenance, productivity and reproduction in both annual and perennial plants provide a basis for the analysis of life histories. Lewontin (1974) and Stearns (1977) have discussed in detail the difficulties associated with a mechanistic interpretation of life history traits. Although in our own studies we are some way from meeting the criteria laid down by these two authors for resolving the ecological basis of the different life histories of plants in salt marshes, some of the results are worth examining.

Annual habit

Initially the life cycle of annual species such as *Salicornia europaea* agg. will be discussed. Populations of this species from the upper and lower marshes appear to show genetic differentiation (Jefferies 1977). Measurements of standing crop indicate that growth of *Salicornia* plants in the lower marsh occurred throughout the summer months whereas in the upper marsh it took place in early and late summer (Fig. 15.4). The lack of growth in individuals in the latter populations during much of the summer might have been either a phenotypic response of individuals or else genetic adaptation of the population to the predictable extreme conditions of the summer.

In an attempt to resolve this question, additions of sea water and nitrogen as nitrate or ammonium salts were made to experimental plots in the upper marsh, and ammonium or nitrate salts were added to corresponding plots in the lower marsh. If the lack of growth of individuals is a phenotypic response to either the deleterious effects of hypersalinity or a deficiency of nitrogen, addition of water and nitrogen should alleviate these stresses and provide conditions suitable for growth. In 1976 two experimental plots of similar design to those described by Jefferies and Perkins (1977) were set up in the upper marsh. Applications of ammonium or nitrate salts, as described in the above paper, were made to the appropriate sub-plots every ten days; other sub-plots were not treated and served as controls. One of the experimental plots was watered three times a week with 25 litres of sea water per square metre on each occasion. From each of five sub-plots which received a similar treatment five cores were taken every ten days throughout the summer and plants of *Salicornia* harvested, dried and weighed. The yield of this species for the control and treated plots is shown in Fig. 15.4. In the summer of 1977, which was much wetter than the previous summer, the study was extended to include the lower marsh. In both the upper and lower marsh inorganic salts of nitrogen were applied to experimental plots once every fourteen days, cores were taken and the dry weight of *Salicornia* plants measured (Fig. 15.4). As shown in Fig. 15.3 the salinity (water potential) of the sediments was barely above that of sea water during the summer of 1977.

The growth pattern of *Salicornia* in the upper marsh appears to be strongly determinate (Harper & White 1974) since plants undergo a relatively

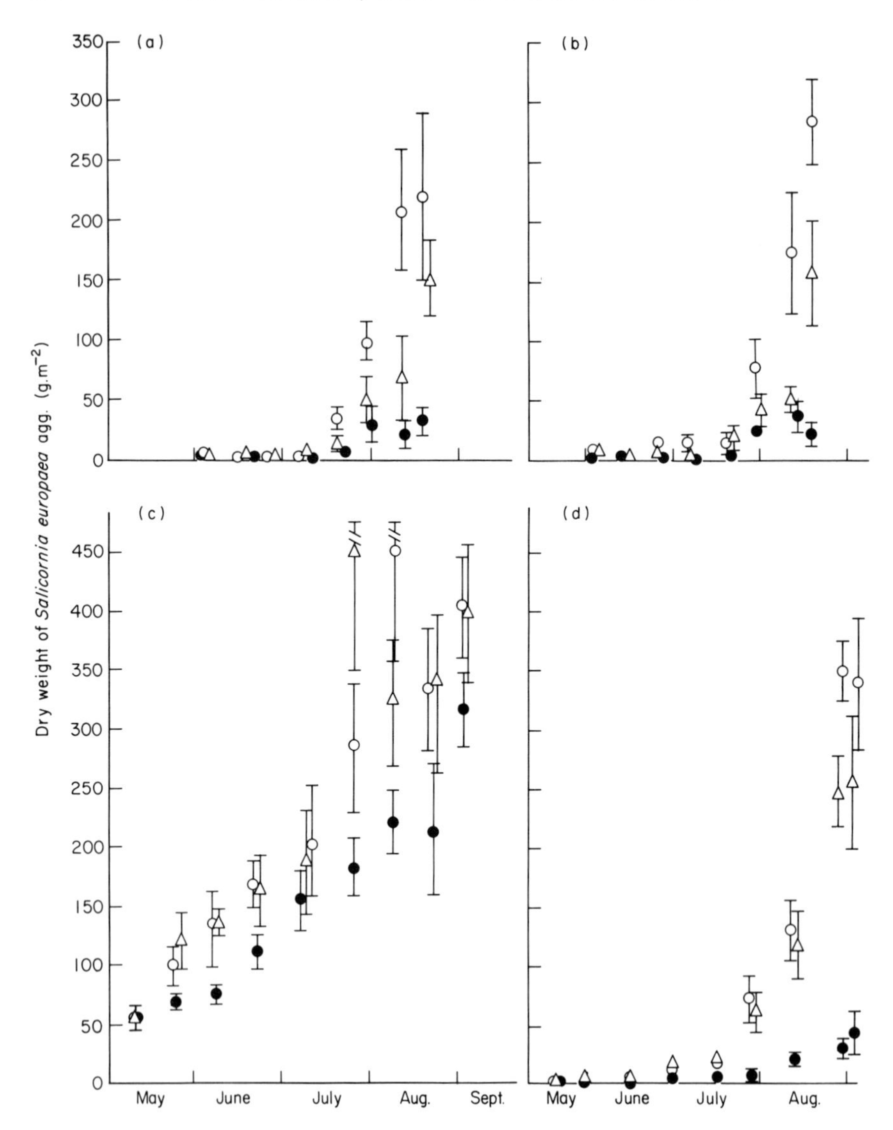

FIG. 15.4. Dry weight yield of living material of plants of *Salicornia europaea* agg. growing in control plots (●) or plots treated with either ammonium (△) or nitrate (○) salts. (a) plot in upper marsh watered with sea water; (b) plot in upper marsh not watered; (c) plot in lower marsh; (d) plot in upper marsh. Vertical bars denote S.E.M. Data shown in (a) and (b) were obtained in 1976, those shown in (c) and (d) were obtained in 1977.

long period of vegetative growth before a sudden transition to the flowering and fruiting stage in late summer and early autumn when the tides flood the marsh. The vegetative stage coincides with the period of hypersalinity and low water potentials of the sediments, and the perturbations completely fail to modify the growth rates of these annual plants during this period. Instead there is a delayed response to the perturbations of approximately six weeks, before a sudden burst of growth occurs associated with sexual reproduction. This lag phase is characteristic of organisms living in an environment exhibiting cyclical stability (Orians 1975) where there is selection for organisms which can time their growth and reproduction to coincide with favourable environmental conditions. Harper and White (1974) conclude that this determinate growth is the optimal behaviour in an environment of high predictability. At present we do not know the environmental triggers that initiate the onset of this phase of active growth in individuals but there is some circumstantial evidence that daylength might be important. In any event the effect of the supplementary nitrogen on the final yield is evident. Analytical data to be published elsewhere indicate that during the vegetative stage, plants from plots treated with nitrogen contain large quantities of soluble nitrogen so that nitrogen is readily available for growth.

The response of plants of *Salicornia* from the lower marsh to additions of nitrogen to the sediments is very different (Fig. 15.4). In this population the growth of individuals is continuous throughout the period from May to early September and there is no delayed response to the addition of nutrients as described above. The final yields realized are comparable between the two populations from the lower and upper marshes. In the lower marsh where the tide frequently covers the site, there appears to have been selection for plants capable of maintaining growth throughout the season. Although the effects of nitrogen addition on yield were small, seed production of plants subject to the various treatments was very different (Table 15.1). Plants in those sub-plots which received inorganic nitrogen produced significantly more seed than those

Table 15.1. *Numbers of seed produced by plants of* Salicornia europaea *agg. growing in the lower marsh in control plots or plots treated with either ammonium or nitrate salts.*

Treatment	Mean number of seed per plant	SEM	Number of plants examined
Control	31·4	6·6	25
Ammonium addition	148·8	20·4	43
Nitrate addition	104·7	14·2	44

Details of the experimental methods are given in the text and by Jefferies and Perkins (1977).

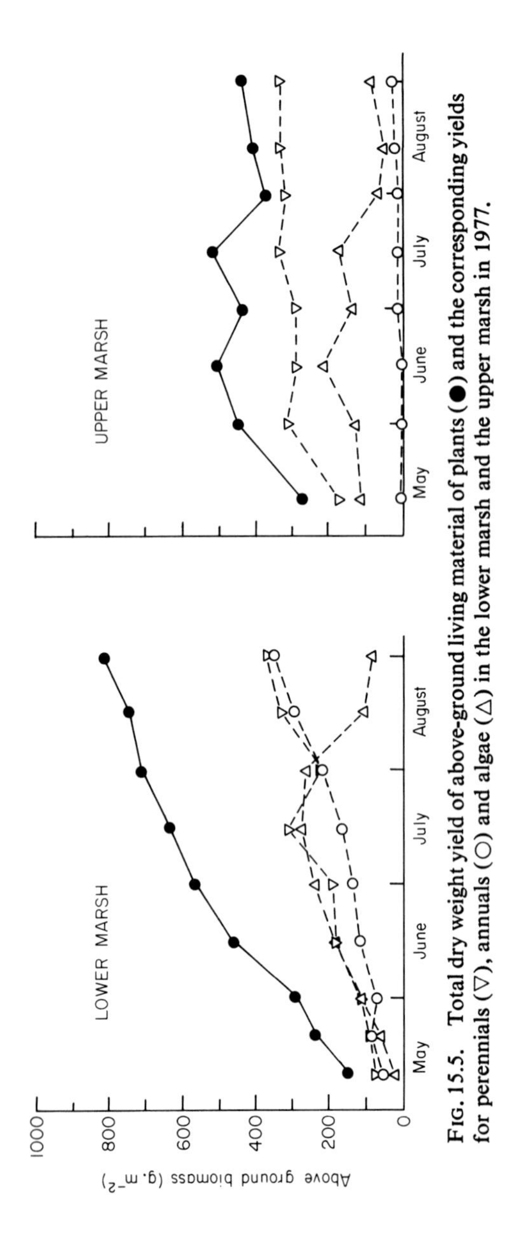

FIG. 15.5. Total dry weight yield of above-ground living material of plants (●) and the corresponding yields for perennials (▽), annuals (○) and algae (△) in the lower marsh and the upper marsh in 1977.

plants from the control plots. The weight of seeds is a very small percentage of the total weight of plants but the seeds are a nitrogen sink, hence under conditions of apparent nitrogen limitation the number of seeds could be reduced. The weights of seeds from plants subject to the different treatments were similar. This type of plastic response has been known for a long time (Harper 1960).

The effects of nitrogen limitation on plants of *Salicornia* from the upper marsh are more severe. At this site the yield is appreciably lower than that on the treated plots. A consequence of delayed growth is that there is a lack of available nitrogen in the sediment when hypersalinity ameliorates at the end of summer (Jefferies 1977). The annual habit appears poorly adapted to this type of environment. This, in part, provides a possible explanation why the perennial habit predominates among coastal halophytes.

Perennial habit

The seasonal growth patterns of perennial plants are also different in the upper and lower marshes (Fig. 15.5). The characteristic feature of perennial plants of the upper marsh such as *Plantago maritima*, *Limonium vulgare* and *Triglochin maritima*, which contribute over 60% to the annual net primary production at this site (R.L. Jefferies, unpublished), is that they show an appreciable increase in above-ground biomass early in the season before the development of hypersalinity in the sediment. In most years the above-ground biomass of these perennial plants is highest in late May or very early June; thereafter the amount of vegetative material declines or else fails to increase (Fig. 15.5). In Fig. 15.6 seasonal changes in biomass of one species, *Limonium vulgare*, from the upper marsh are given as an example and the rapid increase in above-ground biomass early in the season is evident. In this species, unlike *Plantago maritima*, *Triglochin maritima* and *Armeria maritima*, the onset of flowering is delayed until later in the season: the development of flowering shoots, which first appear in May, takes place in July and August. In all of these perennial plants with woody rootstocks and appreciable quantities of biomass below ground (Fig. 15.6), much of the annual increment in the biomass below the surface of the sediments occurs from July onwards. Hence in the perennial plants in the upper marsh the different stages of growth during the period April to September are partitioned in time, so that they occur either early or late in the season.

The perennial plants of the lower marsh, which include *Aster tripolium*, *Puccinellia maritima* and *Halimione portulacoides*, show a steady increase in above-ground biomass throughout the season (Fig. 15.5) to values that are comparable with those of perennial plants from the upper marsh. No rapid increase in the amount of standing crop early in the season is evident and with the exception of *Puccinellia maritima*, flowering of all species (*Aster tripolium*, *Halimione portulacoides*, *Spartina anglica*, *Salicornia perennis*, *Salicornia*

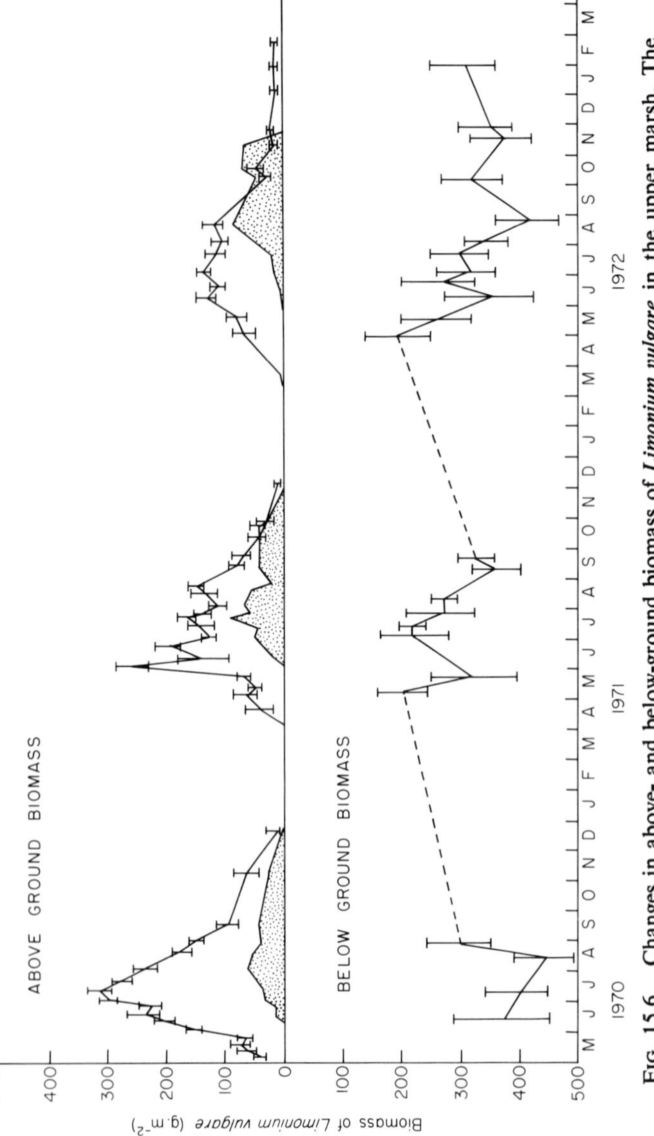

FIG. 15.6. Changes in above- and below-ground biomass of *Limonium vulgare* in the upper marsh. The graph of above-ground biomass is divided into vegetative growth (non-stippled area) and growth associated with sexual reproduction (stippled area). Vertical bars denote S.E.M.

europaea agg.) from areas at the seaward end of the lower marsh occurs in late summer. The possible ecological significance of these different growth strategies in perennial plants will be discussed later in relation to the allocation of resources in these plants.

ALLOCATION OF RESOURCES AND OSMOREGULATION

The annual plants depend totally on the ability of a small root system restricted to the upper layers of the sediment to take up nitrogen. However in the perennial plants, particularly those species with well-developed rhizomes which are characteristic of the upper marsh, the burst of vegetative and reproductive growth in spring creates a high demand for nitrogen which can be met by a combination of uptake and transfer from older tissue. Evidence of internal recycling of nitrogen between above- and below-ground organs of plants of *Limonium vulgare* from the upper marsh is shown in Fig. 15.7. During the winter months the rhizomes contain relatively high levels of nitrogen but in summer this sink is depleted temporarily and the leaves become sinks, particularly early in the season. These source-sink interactions

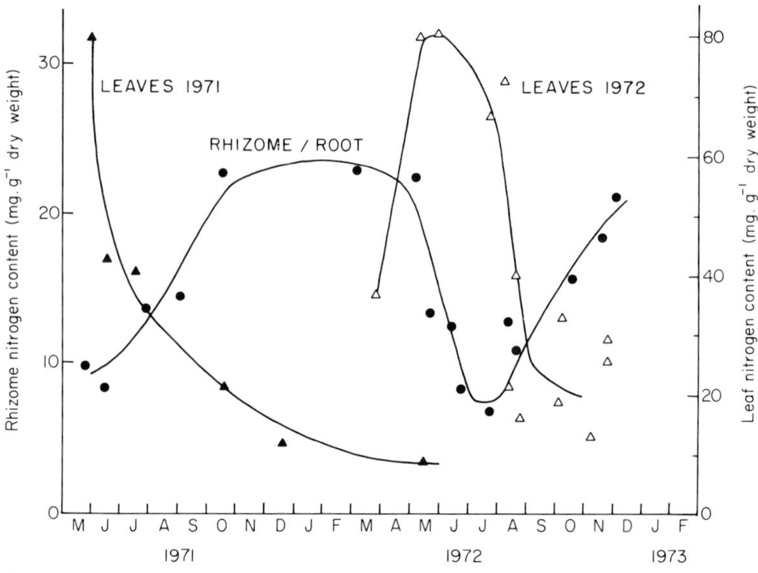

FIG. 15.7. Seasonal changes in the total nitrogen content of the leaf and rhizome/root tissue of plants of *Limonium vulgare* from the upper marsh. Note that different scales are used for each tissue.

are well known in agricultural crops (Carr & Pate 1967; Pate 1968; Milthorpe & Moorby 1974) especially in relation to sudden changes in growth which occur when new demands are made, as with a changed nutritional or photoperiodic regime or at flowering and fruiting (Wareing & Seth 1967). Pate (1968) has shown that the leaves act as reservoirs for nutrients: leaves when not fully expanded accumulate nitrogenous compounds, such as amino acids, but thereafter there is a gradual loss of nitrogen until senescence. The results of changes in the nitrogen content of leaves of *Limonium vulgare* are consistent with the earlier findings of Pate. These losses probably reflect the development of competing metabolic sinks such as flowering, fruiting and the growth of new roots, as there is little vegetative growth in most perennial plants of the upper marsh after the initial burst of growth in spring. Seasonal changes in the activity of nitrate reductase of the leaves of these plants indicate that there are two phases when the activity is high and the internal level of nitrate in the tissues low; one early in the season associated with leaf development and the other in late June and July at the onset of fruiting and root growth (E. Dillon & R.L. Jefferies, unpublished). The high level of activity in mid-summer is consistent with the mobilization of nitrogen for competing sinks.

These requirements of nitrogen and carbon for growth and reproduction are of particular interest in halophytes because of the known or presumed role of soluble organic nitrogenous compounds and carbohydrates in osmoregulation (cf. Hellebust 1976; Flowers, Troke & Yeo 1977; Dainty & Stewart *et al.*, both in this volume, for reviews on the subject). Evidence from several different approaches indicates that cytoplasmic ionic concentrations are low compared with those in the vacuole and that for the osmotic potential to be in equilibrium between these compartments, other substances in the cytoplasm must contribute to the lowering of the osmotic potential in order to achieve equilibrium. The reviews mentioned above stress the diversity of substances which fulfil this apparent role in osmoregulation: they include amino acids, polyhydric alcohols, quaternary ammonium compounds and reducing sugars. Furthermore each species may contain high concentrations of one or more of these compounds. These substances are of low molecular weight, highly soluble and unlike ions, do not inhibit enzymic reactions when present at high concentrations (Sims & Greenway 1973). As mentioned earlier, if a considerable amount of a limited resource such as nitrogen is allocated to osmoregulation within the plant the availability of this element for growth and reproduction may be restricted. In culture the imposition of elevated levels of salinity results in an increase in the concentrations of these organic substrates in plants (Stewart & Lee 1974; Storey *et al.* 1977; Stewart *et al.*, this volume). However information on seasonal trends in the concentrations of these substances in halophytes growing in salt marshes in relation to salinity, growth and reproduction is lacking. During the period 1975 to 1977 we have measured the concentrations of some of these substances in different halophytes from Stiffkey salt marsh. Results from 1977 from the leaf tissues of three perennial species, *Plantago maritima*, *Limonium vulgare* and *Triglochin maritima* of the upper marsh, are given in Figs. 15.8–10, together with concentrations of the dominant inorganic cations in the tissues and the water potentials of the leaves. Where applicable, corrections have been made for dry weight and internal salt content and the results are based on the water content of the tissues.

In all three species there is a fall in the concentrations of these organic solutes during late June or July and a recovery in some cases in August. Changes in the overall mean concentrations of these substances in the leaf tissues do not show the same trends as changes in the water potential of the leaves. The mid-summer fall in the concentration of proline in leaf tissues of

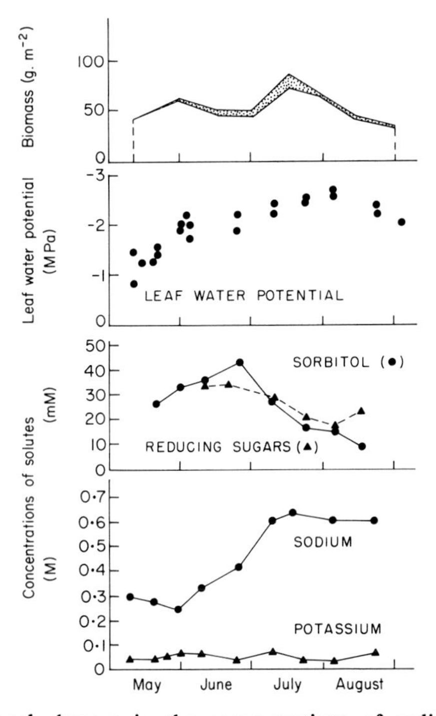

FIG. 15.8. Seasonal changes in the concentrations of sodium, potassium, sorbitol and reducing sugars in leaves of *Plantago maritima* from the upper marsh, together with data of leaf water potential and above-ground biomass. The stippled area of the graph of the above-ground biomass represents the biomass associated with sexual reproduction. Molarities are expressed on the basis of the water content of the leaves. Flame spectrophotometry was used to estimate sodium and potassium; sorbitol was estimated with a gas chromatograph (after acetylation) and the dinitrosalycilic acid method was used to measure reducing sugars. Glucose was used as a standard.

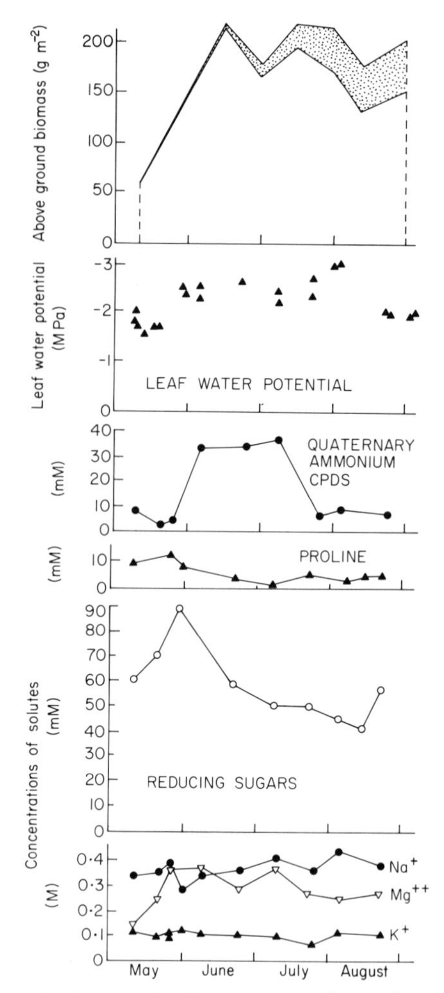

FIG. 15.9. Seasonal changes in the concentrations of sodium, potassium, magnesium, proline, quaternary ammonium compounds and reducing sugars in leaves of *Limonium vulgare* from the upper marsh together with data of leaf water potential and above-ground biomass. The stippled area of the graph of above-ground biomass represents that component associated with sexual reproduction. Molarities are expressed on the basis of the water content of the leaves. Proline estimated after Chinard (1952) and quaternary ammonium compounds estimated by method of Storey and Wyn Jones (1977). Other substances estimated by use of methods given in legend of Fig. 15.8.

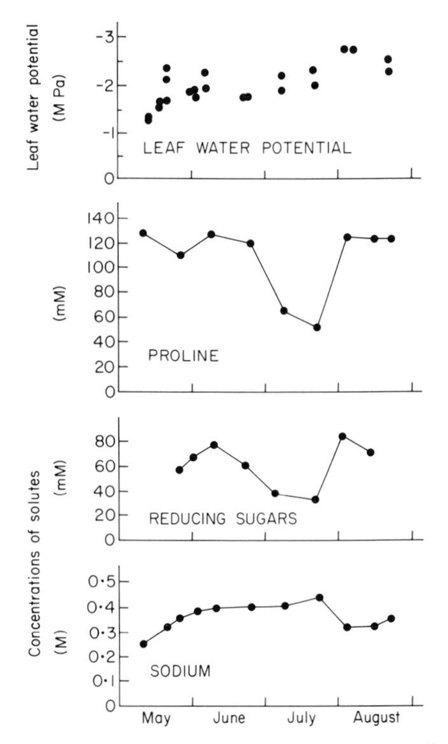

FIG. 15.10. Seasonal changes in the concentrations of sodium, proline and reducing sugars in leaves of *Triglochin maritima* from the upper marsh, together with data of leaf water potential. Molarities are expressed on the basis of the water content of the leaves. Methods used to estimate substances are given in the legends of Figs. 15.8 and 15.9.

Limonium vulgare and *Triglochin maritima* has been recorded for three successive summers, between which weather conditions differed considerably (E. Dillon, T. Rudmik & R.L. Jefferies, unpublished). Corresponding falls are seen in the levels of soluble nitrogen in the leaf tissues of *Halimione portulacoides*, *Limonium vulgare* and *Triglochin maritima* (Table 15.2). In leaves of these three species soluble nitrogenous compounds may contain over 20% of the total nitrogen content of the leaf, a finding similar to that reported by Stewart *et al.* in this volume in their studies on the response of halophytes in culture to salinity. The leaves of *Plantago maritima*, which contain relatively high concentrations of sorbitol and reducing sugars (Fig. 15.8), have a small percentage of the total nitrogen as soluble nitrogen. The soluble nitrogen content of annuals appears to fall in late summer at a time when flowers and seeds are being formed. Much of the soluble nitrogen is sequestered in proline or quaternary ammonium compounds in these perennial and annual plants.

Table 15.2. *Percentage of the total nitrogen present as soluble nitrogen in the leaves of coastal halophytes from the upper marsh, Stiffkey.*

	Species				
1977	Halimione portulacoides	Limonium vulgare	Triglochin maritima	Salicornia europaea	Plantago maritima
5 May	35·2	16·6	15·8	14·8	4·5
24 May	33·3	20·4	16·3	13·5	5·8
2 June	—	—	23·4	—	—
8 June	10·9	12·3	15·2	18·0	2·8
24 June	15·8	8·0	12·2	12·9	2·8
8 July	15·0	6·5	8·5	17·6	3·1
22 July	9·7	11·3	7·5	15·0	5·3
4 August	4·8	10·2	18·1	12·0	2·2
22 August	22·0	8·4	5·5	10·5	3·7

For example, the percentage of soluble nitrogen present as proline in the leaves of *Triglochin maritima* is high throughout the season but particularly during June and early July (Table 15.3). The data strongly suggest that these compounds are mobilized as nitrogen and carbon sources for competing metabolic sinks associated with root growth and flower and fruit formation.

How can these results be reconciled with the possible roles of these solutes in osmoregulation and growth and reproduction? The values of concentrations shown in Figs. 15.8–10 and Tables 15.2 and 15.3 are mean concentrations and no attempt has been made to determine in which compartments of the cells these substances are located. One assumption often made is that these solutes are restricted to the cytoplasm, however it can be shown that the concentrations of some of the organic solutes in the cytoplasm would be very high indeed if this is the situation. A more tenable explanation is to assume that a proportion of these solutes are located in the vacuole and that as demands for growth and reproduction are made, these solutes can be mobilized and replaced by inorganic ions. In this way the integrity of the cytoplasm is

Table 15.3. *Percentage of soluble nitrogen present as proline in the leaves of* Triglochin maritima *from the upper marsh.*

1977				
Date	10 May	24 May	8 June	24 June
%	76·8	70·0	97·5	96·3
Date	8 July	22 July	4 August	22 August
%	93·4	81·5	89·8	85

maintained as the accumulation of inorganic ions in the vacuole does not adversely affect enzymic reactions. It is significant that in all three species there is an increase in the mean sodium concentrations as the season progresses (Figs. 15.8–10). Alternatively other substances, as yet unknown, may fulfil this role. Clearly this discussion is speculative but it does offer an explanation of the possible dual role of these highly soluble organic substances in osmo-regulation and nutrition. It has been suggested also that some of these solutes, such as the quaternary ammonium compounds, may serve a third role as protective agents against insect predation (T.R.E. Southwood, personal communication). Phytophagous insects feeding on these plants deplete the pool of soluble organic substances (cf. Levin 1976; Heydemann, this volume).

Under conditions of extreme hypersalinity, as occurred during the summer of 1976, mean concentrations of proline in tissues of *Triglochin maritima* and *Limonium vulgare* were very much higher than corresponding data for 1977. Values in excess of 400 mM were recorded in *Triglochin*. Of particular interest are the control mechanisms in these long-lived perennial plants which govern the allocation of resources to sexual reproduction and to the maintenance of the individual. Under environmental conditions which require large allocations of nitrogen and carbon to maintenance, sexual reproduction may fail to occur or the development of seeds may be aborted. These perennial plants do not flower each year (Jefferies & Perkins 1977) and during the hot summer of 1976 there was a very low incidence of flowering of all species. If additional nitrogen is added to plots in the salt marsh, species such as *Limonium vulgare* show enhanced sexual reproduction but the amount of vegetative growth is not significantly different from that of the control plots (Jefferies & Perkins 1977). These results suggest that in these perennial species, the process of sexual reproduction with its high resource requirements for both the inflorescence and seed development may be restricted to years when conditions are favourable. Thus there appears to have been selection for individuals capable of developmental plasticity with respect to flowering and fruit formation in the face of interseasonal unpredictability. The effect of rabbit grazing on *Triglochin* populations is further evidence of developmental plasticity in this species. As discussed by Jefferies and Perkins (1977) the inflorescences of this species are grazed in April and early May of most years. New inflorescences may develop in late May or June but the seeds fail to swell and indeed in many years some seeds on inflorescences produced early in the season abort. This type of post-reproductive plasticity in relation to allocation of resources has been discussed by Janzen (1975) in his studies of tropical trees. In summary therefore there appears to be relatively little developmental plasticity with respect to above-ground vegetative growth in these long-lived herbaceous plants. Sexual reproduction, however, is infrequent and its occurrence appears to be linked to the availability of nitrogen and an absence of hypersalinity and drought.

CONCLUSIONS

Many of the ideas advanced in this paper are speculative. Although there is circumstantial evidence for most of them, detailed studies of the population biology of salt marsh species over a number of years in different geographical localities are needed to substantiate the tentative conclusions of studies at Stiffkey. However it is unlikely that results of studies of the responses of plants to the physical constraints of a saline environment alone will provide a basis for the interpretation of the growth strategies of coastal halophytes. The effects of plant–plant interactions and plant–animal interactions on the growth and reproduction of these species also need to be included in such studies.

Questions concerning the allocation of resources and the fate and localization of substrates within plants in relation to growth demand a quantitative approach: a requirement which incidentally is not unique to salt marsh ecology. The findings reported here draw attention to the need to estimate the cost of maintenance processes in plant cells, possibly using the approach of Penning de Vries (1972, 1975) or Thornley (1977). Since such a large part of assimilated carbon apparently is released by respiratory processes (Zelitch 1971), it is necessary to establish how much is coupled to synthetic processes and what proportion is involved, for example, in such processes as ionic regulation in plants. Two important characteristics of halophytes appear to be the ability to divert large quantities of photosynthate towards maintenance and the capacity of these plants to maintain large pools of soluble organic compounds.

ACKNOWLEDGMENTS

We thank Dr D. Andrews, Mrs E. Dillon, Mr L. Brown, Ms N. Lem, Dr W.O. Hassell and Mr N. Perkins for advice and help in these investigations and Dr S.A. Barratt and Dr A. Watkinson for their valuable criticisms of the manuscript. One of us (R.L.J.) acknowledges a grant from the National Research Council of Canada.

REFERENCES

Beeftink W.G. (1965) De zoutvegetatie van ZW-Nederland beschouwd in Europees verband. *Meded. Landbouwhogeschool Wageningen* **65**, 1–167.

Beeftink W.G. (1966) Vegetation and habitat of the salt marshes and beach plains of the south-western part of the Netherlands. *Wentia* **15**, 83–108.

Carr D.J. & Pate J.S. (1967) Ageing in the Whole Plant. *Aspects of the Biology of Ageing* (Ed. by H.W. Woolhouse), pp. 559–99. Society for Experimental Biology, Cambridge.

Chapman V.J. (1938) Studies in salt marsh ecology I–III. *J. Ecol.* **26**, 144–79.

Chapman V.J. (1939) Studies in salt marsh ecology IV–V. *J. Ecol.* **27**, 160–201.

Chapman V.J. (1960) *Salt Marshes and Salt Deserts of the World.* Leonard Hill, London.

Chinard F.P. (1952) Photometric estimation of proline and ornithine. *J. Biol. Chem.* **199**, 91–95.

Durrant A. (1972) Studies on reversion of induced plant weight changes in flax by outcrossing. *Heredity* **29**, 71–81.

Evenari M., Bamberg S., Schulze E.-D., Kappen L., Lange O.L. & Buschbom U. (1975) The biomass production of some higher plants in Near-Eastern and American deserts. *Photosynthesis and productivity in different environments* (Ed. by J.P. Cooper), pp. 121–27. Cambridge University Press, London.

Flowers T.J., Troke P.F. & Yeo A.R. (1977) The mechanism of salt tolerance in halophytes. *Ann. Rev. Plant Physiol.* **28**, 89–121.

Goas M. (1965) Sur le métabolisme azoté des halophytes: étude des acides aminés et amides libres. *Bull. Soc. Fr. Physiol. vég.* **11**, 309.

Greenway H. & Sims A.P. (1974) Effects of High Concentrations of KCl and NaCl on Responses of Malate Dehydrogenase (Decarboxylating) to Malate and various Inhibitors. *Aust. J. Plant Physiol.* **1**, 15–29.

Harper J.L. (1960) Approaches to the Study of Plant Competition. *Mechanisms in Biological Competition* (Ed. by F.L. Milthorpe), pp. 1–39. Society for Experimental Biology, Cambridge.

Harper J.L. & White J. (1974) The Demography of Plants. *Ann. Rev. Ecol. Syst.* **5**, 419–63.

Hellebust J.A. (1976) Osmoregulation. *Ann. Rev. Plant Physiol.* **27**, 485–505.

Janzen D.H. (1975) *Ecology of Plants in the Tropics.* Edward Arnold, London.

Jefferies R.L. (1976) The North Norfolk coast. *Nature in Norfolk a Heritage in Trust.* The Norfolk Naturalists Trust, pp. 130–38. Jarrold & Sons, Norwich.

Jefferies R.L. (1977) Growth responses of coastal halophytes to inorganic nitrogen. *J. Ecol.* **65**, 847–65.

Jefferies R.L. & Perkins N. (1977) The effects on the vegetation of the additions of inorganic nutrients to salt marsh soils at Stiffkey, Norfolk. *J. Ecol.* **65**, 867–82.

Larher F. & Hamelin J. (1975a) L'acide β-trimethylaminopropionique des rameau de *Limonium vulgare* Mill. *Phytochemistry* **14**, 205.

Levin D.A. (1976) The chemical defenses of plants to pathogens and herbivores. *Ann. Rev. Ecol. Syst.* **7**, 121–59.

Levins R. (1962) Theory of fitness in a heterogeneous environment. I. The fitness set and its adaptive function. *Am. Nat.* **96**, 361–73.

Levins R. (1963) Theory of fitness in a heterogeneous environment. II. Developmental flexibility and niche selection. *Am. Nat.* **97**, 75–90.

Levins R. (1969) *Evolution in Changing Environments.* Princeton University Press, New Jersey.

Lewontin R.C. (1974) *The Genetic Basis of Evolutionary Change.* Columbia University Press, New York.

MacRobbie E.A.C. (1971) Fluxes and compartmentation in plant cells. *Ann. Rev. Pl. Physiol.* **22**, 75–96.

Maynard-Smith J. (1966) Sympatric speciation. *Am. Nat.* **100**, 637–50.

Milthorpe F.L. & Moorby J. (1974) *An Introduction to Crop Physiology.* Cambridge University Press, London.

Odum E.P. (1969) The strategy of ecosystem development. *Science* **164**, 262–70.

Orians G.H. (1975) Diversity, stability and maturity in natural ecosystems. *Unifying concepts in ecology* (Ed. by W.H. van Dobben and R.H. Lowe-McConnell), pp. 139–50. Junk B.V. Publishers, The Hague.

Pate J.S. (1968) Physiological Aspects of Inorganic and Intermediate Nitrogen Metabolism (with special reference to the legume, *Pisum arvense* L.). *Recent Aspects of Nitrogen Metabolism in Plants* (Ed. by E.J. Hewitt and C.V. Cutting), pp. 219–40. Academic Press, New York.

Patrick W.H. Jr. & Delaune R.D. (1976) Nitrogen and phosphorus utilization by *Spartina alterniflora* in a salt marsh in Barataria Bay, Louisiana. *Estuar. Coast. Mar. Sci.* **4**, 59–64.

Penning de Vries F.W.T. (1972) Respiration and growth. *Crop Processes in Controlled Environments* (Ed. by A.R. Rees, K.E. Cockshull, D.W. Hand and R.J. Hurd), pp. 327–47. Academic Press, London and New York.

Pigott C.D. (1969) Influence of mineral nutrition on the zonation of flowering plants in coastal salt marshes. *Ecological Aspects of the Mineral Nutrition of Plants* (Ed. by I.H. Rorison), pp. 25–35. Blackwell Scientific Publications, Oxford.

Pomeroy L.R. (1970) The strategy of mineral cycling. *Ann. Rev. Ecol. Syst.* **1**, 171–90.

Ranwell D.S., Bird E.C.F., Hubbard J.C.E. & Stebbings R.E. (1964) *Spartina* salt marshes in southern England. V. Tidal submergence and chlorinity in Poole Harbour. *J. Ecol.* **52**, 637–41.

Snaydon R.W. & Davies M.S. (1972) Rapid population differentiation in a mosaic environment. II. Morphological variation in *Anthoxanthemum odoratum. Evolution Lancaster Pa.* **26**, 390–405.

Stearns S.C. (1977) The Evolution of Life History Traits: a Critique of the Theory and a Review of the Data. *Ann. Rev. Ecol. Syst.* **8**, 145–71.

Steers J.A. (1960) Physiography and Evolution. *Scolt Head Island* (Ed. by J.A. Steers), pp. 12–66. Heffer and Sons, Cambridge.

Steers J.A. (1971) The Physical Features of Scolt Head Island and Blakeney Point. *Blakeney Point and Scolt Head Island* (Ed. by J.A. Steers), pp. 13–25. The National Trust, Norfolk.

Stewart G.R., Lee J.A. & Orebamjo T.O. (1972) Nitrogen metabolism of halophytes. I. Nitrate reductase activity in *Suaeda maritima. New Phytol.* **71**, 263–67.

Stewart G.R., Lee J.A. & Orebamjo T.O. (1973) Nitrogen metabolism of halophytes. II. Nitrate availability and utilization. *New Phytol.* **72**, 539–46.

Stewart G.R. & Lee J.A. (1974) The role of proline accumulation in halophytes. *Planta* **120**, 279–89.

Storey R. & Wyn Jones R.G. (1975) Betaine and choline levels in plants and their relationship to NaCl stress. *Plant Sci. Lett.* **4**, 161–68.

Storey R. & Wyn Jones R.G. (1977) Quaternary ammonium compounds in plants in relation to salt resistance. *Phytochemistry* **16**, 447–54.

Storey R., Ahmad A. & Wyn Jones R.G. (1977) Taxonomic and ecological aspects of the distribution of glycine betaine and related compounds in plants. *Oecologia* **27**, 319–32.

Thornley J.H.M. (1977) Growth, maintenance and respiration: a re-interpretation. *Ann. Bot.* **41**, 1191–1203.

Tyler G. (1967) On the effect of phosphorus and nitrogen, supplied to Baltic shore-meadow vegetation. *Bot. Notiser* **120**, 433–47.

Valiela I. & Teal J.M. (1974) Nutrient limitation in salt marsh vegetation. *Ecology of Halophytes* (Ed. by R.J. Reimold and W.H. Queen), pp. 547–63. Academic Press, New York and London.

Wareing P.F. & Seth A.K. (1967) Ageing and Senescence in the Whole Plant. *Aspects of the Biology of Ageing* (Ed. by H.W. Woolhouse), pp. 534–58. Society for Experimental Biology, Cambridge.

Willis A.J. (1963) Braunton Burrows: the effects on the vegetation of the addition of mineral nutrients to the dune soils. *J. Ecol.* **51**, 353–74.

Zelitch I. (1971) *Photosynthesis, photorespiration and plant productivity.* Academic Press, New York.

16. ECOPHYSIOLOGICAL RESPONSES OF *GAMMARUS DUEBENI* TO SALINITY FLUCTUATIONS

A.P.M. LOCKWOOD AND C.B.E. INMAN

Department of Oceanography,
University of Southampton,
Southampton, U.K.

SUMMARY

(1) Studies on the blood osmotic and sodium concentration, urine osmotic concentration, influx of sodium and apparent permeability to water in the amphipod *Gammarus duebeni* are reviewed.
(2) It is shown that the physiological responses of the animal differ when it is experiencing cyclical changes of salinity or sudden shifts of concentration compared with those when it is acclimatized to a given salinity.
(3) The rate of uptake of sodium is greater, at any given blood sodium concentration, in animals suffering sudden dilution of the medium than it is in acclimatized animals. Similarly, urine hyptonic to the blood is formed at higher blood osmotic concentrations by animals experiencing sudden dilution of the medium. It is postulated that control of the rate of active uptake of sodium is not based solely on blood osmotic or ionic concentration.
(4) Changes in apparent permeability to water occur in response to changes in external salinity.
(5) The versatility of physiological response of the species is discussed in relation to the environmental conditions to which it may be exposed in its habitat.

RÉSUMÉ

(1) Nous avons revu les études sur la concentration osmotique du sang et sa concentration en sodium, sur la concentration osmotique de l'urine, sur l'influx de sodium et l'apparente perméabilité de l'eau, chez l'amphipode *Gammarus duebeni.*
(2) La réponse physiologique de l'animal diffère s'il est soumis à des changements cycliques de salinité ou à des variations brusques de concentration, ou s'il est accoutumé à une certaine salinité.

(3) Quelque soit la concentration en sodium du sang, le taux d'absorption du sodium d'un animal soumis à de brusques dilutions du milieu est supérieur à celui d'un animal accoutumé. De même, l'hypotonicité de l'urine par rapport au sang a lieu pour des concentrations osmotiques du sang supérieures quand l'animal est soumis à une dilution soudaine du milieu. Il est postulé que le contrôle de l'absorption active de sodium ne soit pas basée uniquement sur la concentration osmotique ou ionique du sang.
(4) Des changements de la perméabilite apparente de l'eau répondent à des changements de la salinité extérieure.
(5) La variabilité des réponses physiologiques de l'espèce est discutée en relation avec les conditions d'environnement auxquelles elle peut être exposée dans son habitat.

INTRODUCTION

The majority of previous studies of the osmoregulatory capabilities of euryhaline organisms have examined animals acclimatized to specific salinities. Such an approach, whilst defining the responses of species to a particular situation, may be misleading as, in an estuary, environmental fluctuations can be sufficiently extensive and rapid to ensure that full acclimatization to a particular salinity does not occur.

Variety and unpredictability of conditions within the estuarine habitat is exemplified at an extreme level by environmental fluctuations which occur in the gullies and pans of salt marshes and tidal water meadows. Recorded salinity change in a marsh gully off the estuary of the River Test at Redbridge, Southampton ranged between 1‰ and 22‰ on a single tidal cycle (Ralph 1965). The concentration range experienced in gullies is, however, influenced not only by the tidal cycle itself but also by non-cyclical factors such as river flow, wind speed and direction and barometric pressure, which together affect the tidal range and the relative balance between fresh and saline waters within an estuary.

On the surface of a marsh the salinity of pools and pans cannot be easily predicted. Such pools generally are flushed at high water on spring tides when the marsh is submerged but the salinity of the inundating water can vary widely in the case of estuarine marshes. For example, at Redbridge it ranges from 23‰ (7 October 1975) to virtually fresh water (4 March 1977). Between spring tidal excursions the marsh pools are exposed to evaporation, precipitation or both according to prevailing meteorological conditions.

Animals inhabiting such variable bodies of water must not only be able to tolerate a range of salinities but also be able to respond rapidly to change in salt concentration if the body fluids are not to vary to a degree deleterious to cell function or the maintenance of blood volume (Lockwood 1976).

It is the purpose of this review to summarize the extent and versatility of the physiological responses of one of the typical inhabitants of this ecosystem, the amphipod crustacean, *Gammarus duebeni*. Changes in sodium fluxes and in the sodium concentrations in blood and urine of this organism in response to sudden changes in salinity, as well as acclimatization to a single salinity, have been studied. The results obtained on the two groups of animals differ and thereby serve to indicate that measurements made on animals maintained at given salinities within a range do not adequately represent the situation in individuals exposed to fluctuations of salinity within the same range.

METHODS

Osmotic pressures of blood, urine and medium were determined by the cryoscopic method of Ramsay and Brown (1955); sodium by flame emission spectrophotometry utilizing a Unicam SP900; water fluxes by liquid scintillation counting of tritiated water. Flux measurements were supported by ^{51}Cr EDTA studies of urine filtration rate and U/B ratios for EDTA. Levels of ^{51}Cr were determined by well-type solid state scintillation methodology (Panax Reigate Series AAUCO, Tim 10, SON 10, RTM 10, PSU 13 systems). Sodium tracer, ^{22}Na, was also determined by Panax equipment.

Salinity cycles, created in a simulator, were monitored by sodium electrode (Orion) and recorded by a Servoscribe IS.

BLOOD CONCENTRATION

Acclimatized animals

The salinity range of *Gammarus duebeni* in natural waters extends from fresh water (*G.duebeni celticus*, Reid 1939; Hynes 1954) to 45‰ when the organism remains active or to 73‰ when it is in a sluggish condition (Forsman 1951). For maintenance of individuals within the laboratory the optimum salinity is within the range 5 to 15‰ (Kinne 1953, 1959).

At salinities in excess of about 50–60% sea water the blood tends towards isotonicity with respect to the medium although if animals are acclimatized to sea water (33‰) for several days the blood remains hypertonic by about 8 m osmoles (Lockwood & Inman 1973). At salinities below about 16‰ the body fluids are maintained hypertonic to the medium (Fig. 16.1) (Beadle & Cragg 1940; Lockwood 1961; Haywood 1970).

Animals exposed to cyclical changes in salinity

Cyclical changes in salinity in which the cycle is completed within the time scale of a typical tidal period (c. 12 hours 50 min) result in only a relatively small variation in the sodium concentration of the blood, provided that the

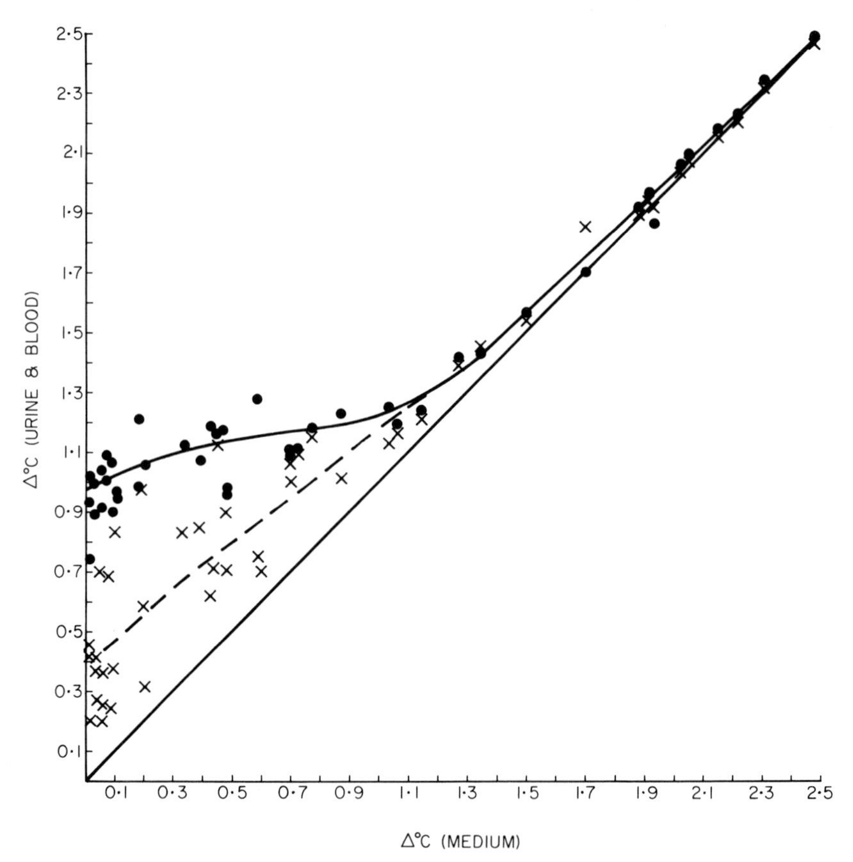

FIG. 16.1. Blood and urine osmotic concentrations in *G.duebeni* acclimatized to a range of salinities. (Based on data from Lockwood (1961) and Heywood (1970).) ●, Blood osmotic pressure; ×, urine osmotic pressure. Δ°C, depression of freezing point.

external variation in salinity is within the range most commonly experienced by the animal. Individuals exposed to a 12 hr cycle created in the laboratory by continuously mixing different percentages of fresh water and 50% sea water (salinity 17‰) to produce salinities between 0 and 50% sea water (0/50 cycle) display, on average, a blood concentration range of only some 18 mM/l Na, whilst the limits of the external medium are 10 and 240 mM/l Na (Fig. 16.2).

Exposure to a cycle between 100% sea water and fresh water produces a variation in sodium concentration in the blood which, though greater than before, is still restricted to the range 280–340 mM/l Na (Fig. 16.2).

Where the upper limit of salinity reached is 50% sea water or less the animals remain hypertonic to the medium throughout, and the peaks and troughs of sodium concentrations of the blood and medium approximately

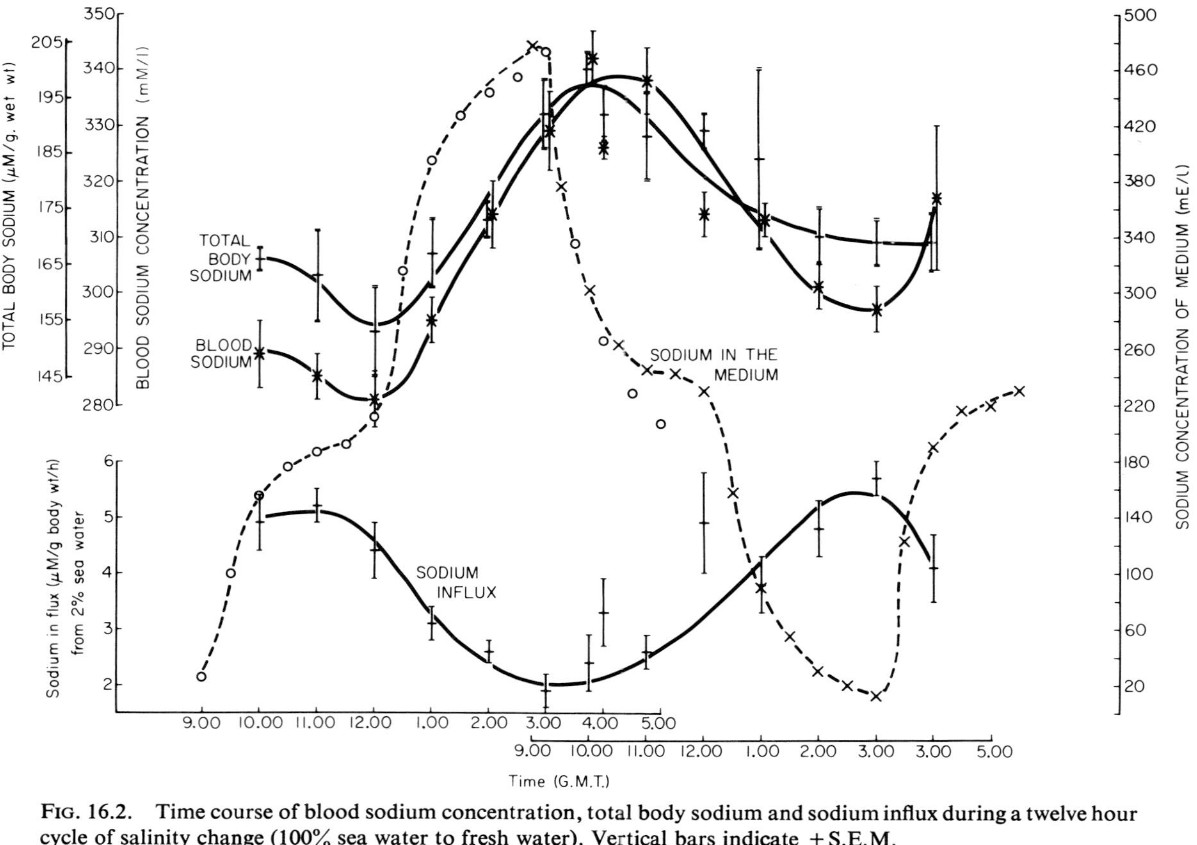

FIG. 16.2. Time course of blood sodium concentration, total body sodium and sodium influx during a twelve hour cycle of salinity change (100% sea water to fresh water). Vertical bars indicate ±S.E.M.

coincide. If the upper limit of the external concentration of sodium is above that of 50% sea water however, the body fluids become hypotonic to the medium for part of the cycle (Fig. 16.3) and the peaks of both blood sodium concentration and total body sodium are no longer coincident in time with the upper limit of the sodium concentration in the medium (Fig. 16.2). The time lag between the peaks can be attributed to the influence of the hypotonicity of the blood during the latter part of the rising phase of the external concentrations since the blood concentration continues to increase as long as the external medium is hypertonic to the body fluids. Hence blood sodium concentration and total body sodium continue to rise in the hour or so that it takes for the falling external concentration to drop from its peak level to equiconcentration with the body fluids.

Both active and passive mechanisms are involved in limiting the change in

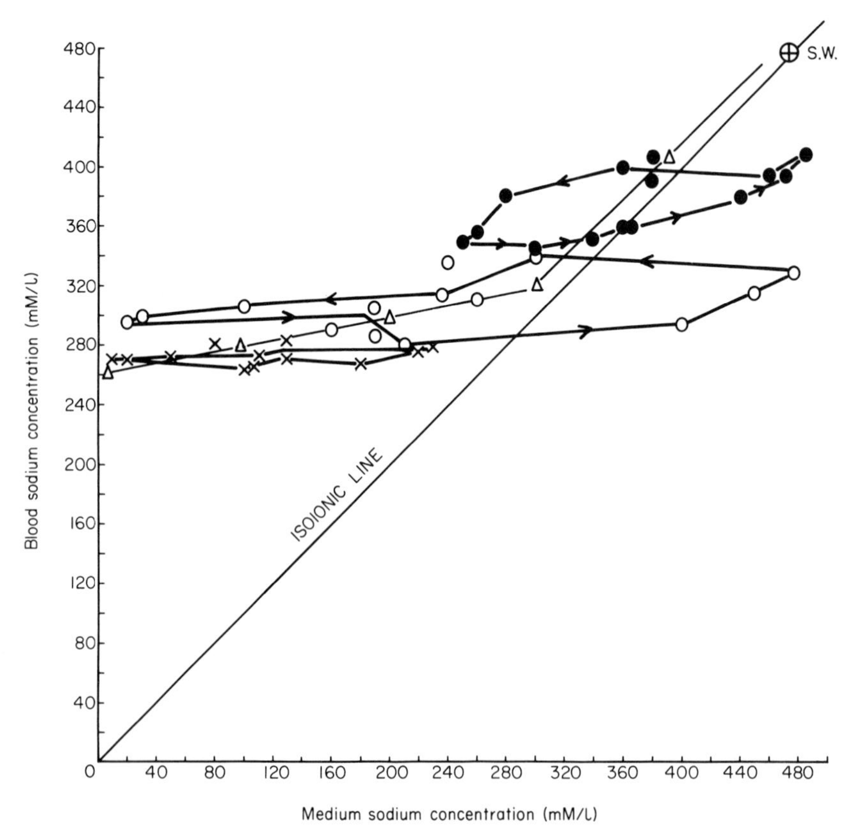

FIG. 16.3. Blood sodium concentration of *G.duebeni* experiencing twelve hour cycles of salinity change; (△), acclimatized animals; (×), cycle of 50% sea water to fresh water; (○), cycle of 100% sea water to fresh water; (●), cycle of 100% sea water to 50% sea water; ⊕, sea water.

sodium concentration in the blood. The passive mechanism involves general restriction of water permeability at the body surface, *G.duebeni* being less permeable to water than its fresh water co-genor, *G.pulex* (Lockwood 1961). Active components include (a) regulation of the rate of active ion uptake, (b) control over urine volume and concentration, (c) variation in water permeability.

SODIUM INFLUX

Acclimatized animals

In fresh water crustaceans such as the crayfish, *Austropotamobius* (Shaw 1959) and *Asellus* (Lockwood 1960) a small decrease in the sodium concentration of the blood is followed by an increase in the active uptake of sodium into the body. For *Gammarus duebeni* acclimatized to salinities in the range 2% to 100% sea water there also appears to be a direct correlation between sodium influx and blood sodium concentration. Thus, animals capable of actively taking up sodium when transferred from the acclimatization medium to 2% sea water labelled with ^{22}Na for a short standard period, show a relationship between the sodium concentration of the blood and sodium influx which reflects the gradient maintained between blood and medium. In media more concentrated than 50% sea water influx of the ion remains small, as is appropriate for the small osmotic gradient maintained when the animals are in high salinities (Fig. 16.4).

Such ion transport as there is in high salinities has been attributed to the need to bring fluid into the body for subsequent excretion as urine, in conditions where blood and medium are essentially isotonic (Lockwood & Inman 1973; Lockwood 1970). Hypertonicity of the blood is maintained by the uptake of ions when the medium is dilute. The rate of sodium uptake varies with the concentration gradient (Fig. 16.4).

Non-acclimatized animals

Abrupt unidirectional change of salinity

A different response emerges if animals, previously acclimatized for eight days to 100% sea water are suddenly transferred to 2% sea water. In these circumstances the sodium influx starts to rise almost immediately after transfer (Lockwood 1964) and rates of uptake of sodium are considerably in excess of those shown by acclimatized animals with identical blood sodium concentrations (Fig. 16.4).

Cyclical variation in the salinity of the medium

Individuals exposed to a 100/0 cycle of salinities for eight cycles (4 days) subsequently show a level of sodium influx intermediate between that of animals acclimatized to steady-state conditions and those responding to sudden dilution (Fig. 16.4). A novel feature of the response of the animals

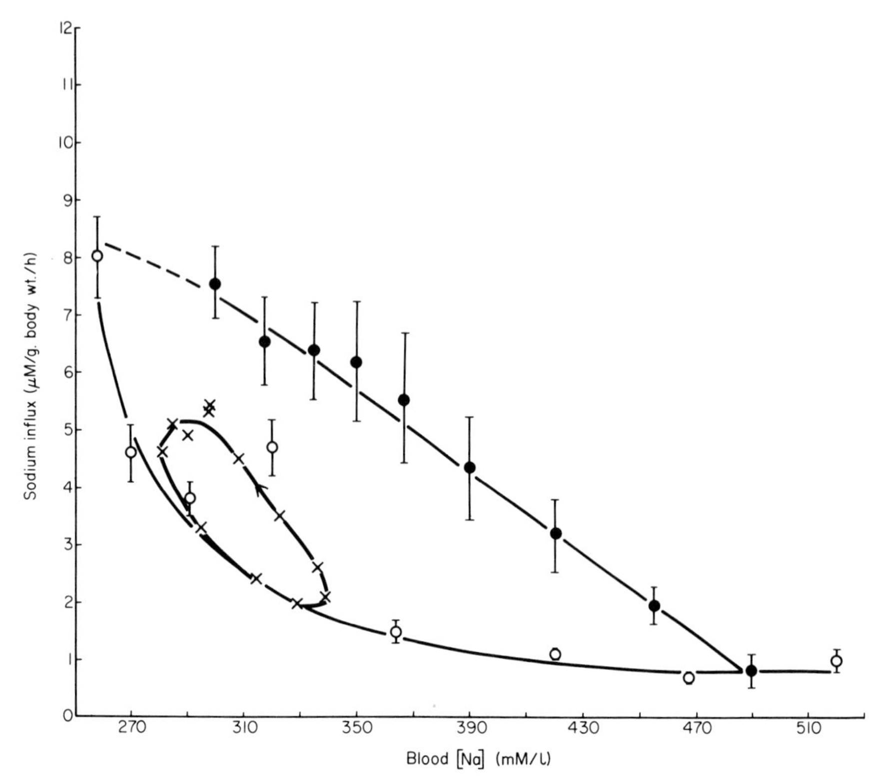

FIG. 16.4. Sodium influx (μM/g body wet wt/hr) from ^{22}Na-labelled 2% sea water by animals acclimatized to a range of salinities from 100% sea water to 2% sea water, (●); animals transferred from 100% sea water to 2% sea water and measured prior to completion of acclimatization to 2% sea water, (○); animals from a 12 hour cycle fluctuating between 100% sea water and fresh water, (×). Vertical bars indicate ±S.E.M.

subjected to the cycle is the difference in sodium influx in relation to blood concentration when the medium is declining in concentration and when it is rising. For any given blood concentration there is a faster influx if the external concentration is falling. A plot of influx against blood sodium concentration thus appears as an elipse rather than as a line (Fig. 16.4).

OSMOTIC CONCENTRATION OF THE URINE

Acclimatized animals

As in the case of sodium influx, the relationship between the osmotic concentration of the urine and blood varies according to the environmental conditions pertaining.

When *G.duebeni* are acclimatized to media in the range fresh water to 160% sea water, the urine is isotonic with the blood at external salinities exceeding about 40–50% sea water, but becomes progressively more hypotonic as the medium is diluted below this level (Fig. 16.1) (Lockwood 1961; Haywood 1970).

Non-acclimatized animals

The situation differs from that outlined above when individuals previously acclimatized to salinities in the range 110–160% sea water are suddenly transferred to fresh water. The urine is isotonic with blood initially but usually becomes hypotonic after some 1½ to 2 hours (Fig. 16.5). At this time the osmotic concentration of the blood is still one and a half to two times the lowest level at which the production of hypotonic urine would be anticipated in acclimatized animals (Lockwood 1961).

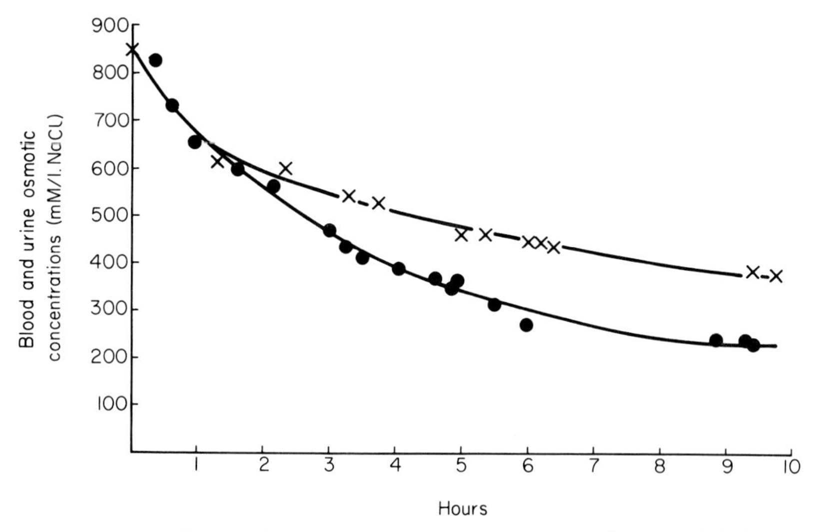

FIG. 16.5. Decline in blood and urine osmotic concentrations of *G.duebeni* transferred from 160% sea water to fresh water at zero time (●), urine; (×), blood. (After Lockwood 1961.)

DECREASE IN CONCENTRATION OF SODIUM IN BLOOD ON SUDDEN DILUTION OF THE MEDIUM

The combined interaction of an increment in the rate of ion uptake and the production of hypotonic urine would be expected to result in reduction in the rate of decline of blood sodium concentration in animals experiencing sudden

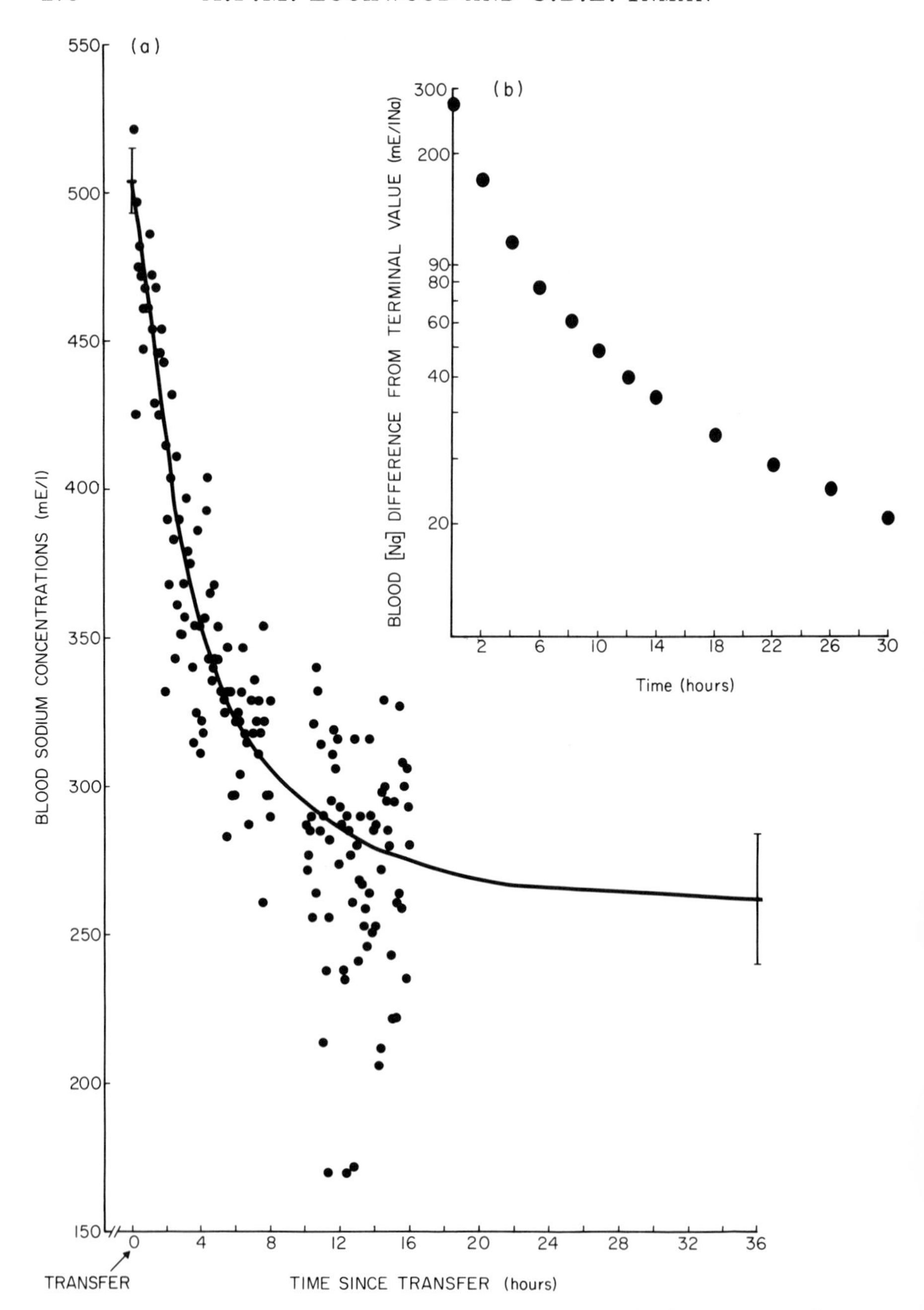

FIG. 16.6. Decline in blood sodium concentration following transfer of *G.duebeni* from 100% sea water to 2% sea water after 8 days in 100% sea water. (a) linear plot, (b) semi-logarithmic plot. The vertical bar represents ± the standard deviation of the blood sodium concentration 36 hr after such transfer.

dilution of their medium. This indeed seems to be the case as a semi-logarithmic plot of blood concentration against time following transfer of animals from 100% to 2% sea water is a curve rather than a straight line (Fig. 16.6a, b). However, the production of dilute urine and the increment in ion uptake, although effective in reducing the net rate of ion loss on dilution, do not appear quantitatively adequate to account for the observed pattern of decrease in blood sodium concentration.

In 100% sea water *G.duebeni* produces a volume of urine which is equivalent to 3·4% of the body weight per day, and about half of this urine flow can be attributed to osmotic uptake of water (Lockwood & Inman 1973). On transfer to 2% sea water from 100% sea water the osmotic gradient increases

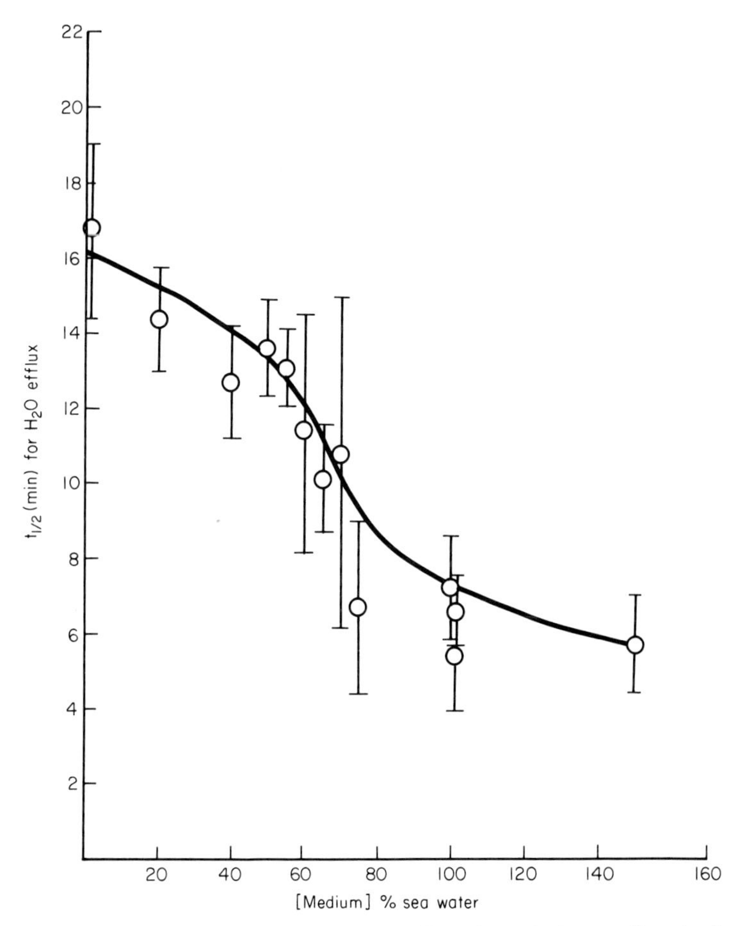

FIG. 16.7. Half time for ^3H_2O exchange (efflux) in *G.duebeni* acclimatized to salinities in the range 150% sea water to fresh water. Vertical bars delimit ± the standard deviation. (After Lockwood, Inman & Courtenay 1973.)

more than 100-fold and it is estimated that this should result in an ion loss via the urine sufficient to cause an initial decline in the concentration of some 90 mM/l blood/hr. Such a presumed rate of loss takes no account of losses via the body surface which actually account for some 20% of the total loss of sodium in such circumstances (Lockwood 1965). The expected initial decline in the sodium concentration in the blood ought therefore to be more than 90 mM/l/hr. In fact the observed rate of fall in blood sodium concentration is only about half the expected level (Fig. 16.6a). Conclusions based on such calculations are necessarily speculative but, if the discrepancy is genuine, the implication must be that a reduction in excretory rate has occurred because of a reduction in the osmotic entry of water. Some evidence has been found suggesting that reduction in water permeability does indeed occur on dilution of the medium.

APPARENT PERMEABILITY TO WATER

Acclimatized animals

Efflux of ^3H_2O from *Gammarus* acclimatized to a range of salinities from fresh water to 150% sea water alters in relation to the external salinity. The rate is

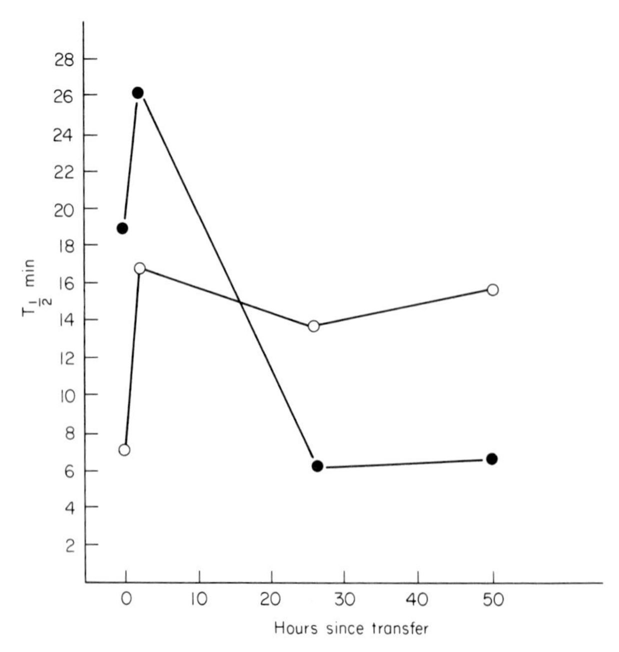

FIG. 16.8. Half time for ^3H_2O exchange (influx) in animals initially acclimatized to 100% sea water (○) and 2% sea water (●) and then transferred respectively to 2% sea water (○) and 100% sea water (●) after the first point.

some 2·5 times slower at the lower end of the salinity range than it is in 150% sea water (Fig. 16.7). Influx values in 100% and 2% sea water are fairly similar to the corresponding efflux values suggesting that the observed flux changes reflect differences in permeability of the organisms to water (Lockwood, Inman & Courtenay 1973).

Non-acclimatized animals

Data from measurements of ^3H_2O flux made on animals acclimatized to 100% sea water and then two hours after transfer to 2% sea water suggest that a large part of the change in apparent permeability occurs at or soon after dilution of the medium (Fig. 16.8) (Lockwood, Inman & Courtenay 1973). The magnitude of the apparent change is sufficient to account for the discrepancy in urinary loss of ions mentioned above. Data from confirmatory experiments in which efflux is measured are required however before such apparent permeability changes can be accepted as the actual cause of reduction in ion loss in animals transferred from a high to a low salinity.

The transfer of individuals from 2% sea water to 100% sea water results in temporary decrease in apparent permeability to ^3H_2O prior to a subsequent increase (Fig. 16.8) (Lockwood, Inman & Courtenay 1973).

DISCUSSION

In its natural environment *Gammarus duebeni* requires the ability to tolerate three types of salinity conditions: (a) the relatively stable salinities in salt marsh pans, which can occur for several days at a time; (b) sudden changes in salinity following precipitation or inundation of the marsh by tidal water and (c) cyclical changes of salinity of water in tidal creeks. The present review of the osmotic and ionic responses of the species suggest that its physiological capabilities are well suited to such a demanding habitat.

When exposed to twelve hour cycles of changes in salinity with the limits of 50% sea water and fresh water, the blood concentration is maintained within narrow limits, the fluctuation for sodium being only 10 mM/l either side of the mean value. Relative stability is also observed if the cycle is from 100% sea water to fresh water (Fig. 16.2). Much wider ranges in blood concentration can however be tolerated for long periods should the medium concentration remain relatively static (Fig. 16.1); animals acclimatized to salinities in the range 50–150% sea water become virtually isotonic with the medium. Even higher salinities may be tolerated for a time though with reduced levels of locomotory activity.

Sudden dilution of the medium from 100% sea water or other high salinities results in a rapid decline in blood osmotic concentration. The rate of fall is however reduced by activation of an increased intake of inorganic ions and production of urine hypotonic to the blood. These factors, acting in concert with a reduction in surface permeability of the organism to water, are sufficiently effective to slow the rate of decline of blood concentration to a level such that osmotic adjustments in the cells can compensate. In consequence, individuals transferred from 100% to 2% sea water show no significant changes in blood volume as a result of internal water shifts between blood and cells one hour after the change in salinity.

Following dilution of the medium rapid changes are effected by the animals in respect of urine volume, urine concentration, active uptake of sodium and surface permeability to water.

An increase in the rate of urine production occurs within five minutes of transfer of the animals from a high to a low salinity and, irrespective of the blood osmotic concentration, the urine becomes hypotonic to the blood about one and a half hours after such transfer, Lockwood (1961). Increased active uptake of sodium from the medium usually occurs within twenty minutes and alteration of apparent permeability to water is also rapid following sudden dilution of the medium from 100% sea water to 2% sea water. The capacity to take up sodium from 2% sea water increases steadily over the few hours after transfer from 100% sea water to 2% sea water (Fig. 16.4). The linear nature of this increment suggests that there is a finite rate at which the transport system can increase its capacity.

As a result of the increase in uptake, a much higher influx of sodium at any given blood sodium concentration occurs in animals whose blood concentration is falling compared with the rate in animals whose blood concentration is stable. Similarly urine hypotonic to the blood is found at higher blood concentrations in non-acclimatized individuals whose blood concentration is falling than in animals fully acclimatized to their medium. This fact, coupled with the observation that in cycling salinity regimes the rate of ion transport at any given ionic concentration of the blood is lower when the external salinity is rising than when it is falling, leads to the conclusion that the ionic concentration of the blood *per se* is not the only factor determining transport rates.

The system functions to buffer the blood concentration against extreme changes in external concentration. Its effectiveness in the salinity range most commonly experienced by salt marsh *G.duebeni* of c. 1% to 50% sea water, is demonstrated by the very narrow variation recorded in blood sodium concentration during cyclical changes in the salinity of the medium (Fig. 16.1). A system in which a fixed transport rate was associated with a given level of blood ionic concentration would be expected to be less effective in regulating blood concentration in fluctuating conditions.

The role of changes in water permeability in fluctuating salinities remains somewhat enigmatic because of the difficulty of constructing experiments, the results of which are unequivocal. Flux measurements in both directions in animals in a steady state suggest that the water permeability of individuals acclimatized to 2% and 100% sea water differs by a factor of about 2·5, the animals in the lower concentration being the less permeable (Fig. 16.8). Flux measurements of animals in cycling salinity regimes have been made only unidirectionally so far and hence the interpretation of the results may be open to question due to potential interference from factors such as bulk flow or drinking. Some changes must occur however in water permeability after sudden dilution of the external medium if the observed rates of ion loss are to be explained. Further study of water relations of estuarine animals under cyclical changes in salinity would seem a fruitful field for the future.

From the experiments described it appears that *Gammarus duebeni* is physiologically capable of responding both to extended exposure to particular salinities in the range 1% to 150% sea water and to sudden or cyclical changes of salinity within this range. The need to respond to some of the more extreme experimental conditions described may be required relatively rarely in even such unpredictable an environment as a salt marsh but may on occasions have survival value. Certainly the animal must normally be ready to adapt to a change in its medium rather than avoid extreme situations by directional behaviour since, as shown by Bettison and Davenport (1976), it displays a less well-developed sense of selection for a particular concentration, than other amphipod species, though tending to avoid 100% sea water if lower salinities are available.

G.duebeni is often found on land amongst the roots of grass tussocks or under bricks or pieces of wood near the water's edge and Lagerspetz (1963) has shown that it can not only survive out of water for extended periods of time (50% survival after 172 hours in water-saturated air at 19°C) but also, if given a choice, it can avoid low humidities. Presumably, therefore, individuals experiencing unacceptable salinity regimes in salt marsh pools can, if necessary, leave the water until the conditions ameliorate. Night migration of individuals from pool to pool, over rocks, has been observed (ref. in Lagerspetz 1963).

The behavioural patterns and the physiological responses of this species reflect its versatility and its ability to adjust to the demands of an exceptional environment—the sight of a dead *G.duebeni* on the marsh at Redbridge is a rarity, virtually restricted to occasions when the salt pans have completely dried out.

REFERENCES

Beadle L.C. & Cragg J. (1940) Studies on adaptation to salinity in *Gammarus* species 1. Regulation of blood and tissues and the problem of adaptation to fresh water. *J. Exp. Biol.* **17**, 153–63.

Bettison J.C. & Davenport J. (1976) Salinity preference in gammarid amphipods with special reference to *Marinogammarus marinus* (Leach). *J. mar. biol. Assoc. U.K.* **56**, 135–42.

Forsman B. (1961) Studies on *Gammarus duebeni*. Lillj. with note on some rock pool organisms in Sweden. *Zool. Bidr. Uppsala* **29**, 215–37.

Haywood G.P. (1970) *A study of the osmotic changes in the blood and urine of three species of gammarid exposed to varying external salinities.* M.Sc. dissertation, The University, Southampton.

Hynes H.B.N. (1954) The ecology of *Gammarus duebeni* Lilljeborg and its occurrence in fresh water in western Britain. *J. Anim. Ecol.* **23**, 38–84.

Kinne O. (1953) Zur Biologie und Physiologie von *Gammarus duebeni* Lillj. *I. Z. wiss Zool.* **157**, 427–91.

Kinne O. (1959) Ecological data on the amphipod *Gammarus duebeni*. A monograph. *Ver Inst. Meeres. Bremenhaven* **6**, 177–202.

Lagerspetz K. (1963) Humidity reactions of three aquatic amphipods, *Gammarus duebeni*. *J. Exp. Biol.* **40**, 105–10.

Lockwood A.P.M. (1960) Some effects of temperature and concentration of the medium on the ionic regulation of the isopod *Asellus aquaticus* (L.) *J. Exp. Biol.* **37**, 614–30.

Lockwood A.P.M. (1961) The urine of *Gammarus duebeni* and *G.pulex*. *J. Exp. Biol.* **38**, 647–58.

Lockwood A.P.M. (1964) Activation of the sodium uptake system at high blood concentrations in the amphipod *Gammarus duebeni*. *J. Exp. Biol.* **41**, 447–58.

Lockwood A.P.M. (1965) The relative losses of sodium in the urine and across the body surface in the amphipod, *Gammarus duebeni*. *J. Exp. Biol.* **42**, 59–69.

Lockwood A.P.M. (1970) The involvement of sodium transport in the volume regulation of the amphipod crustacean *Gammarus duebeni*. *J. Exp. Biol.* **53**, 737–51.

Lockwood A.P.M. (1976) Physiological adaptation to life in estuaries. *Adaptation to Environment* (Ed. by R.C. Newell), 539 pp. Butterworths, London, Boston.

Lockwood A.P.M. & Inman C.B.E. (1973) Water uptake and loss in relation to the salinity of the medium in the amphipod crustacean *Gammarus duebeni*. *J. Exp. Biol.* **58**, 149–63.

Lockwood A.P.M., Croghan P.C. & Sutcliffe D.W. (1976) Sodium regulation and adaptation to dilute media in crustacea as exemplified by the isopod *Mesidotea entomon* and the amphipod *Gammarus duebeni*. *Perspectives in Experimental Biology. I. Zoology* (Ed. by P. Spencer Davies). 525 pp. Pergamon, Oxford and New York.

Lockwood A.P.M., Inman C.B.E. & Courtenay T.H. (1973) The influence of environmental salinity on the water fluxes of the amphipod crustacean *Gammarus duebeni*. *J. Exp. Biol.* **58**, 137–48.

Ralph R. (1965) *Some aspects of the ecology and osmotic regulation of Neomysis integer* (Leach). Ph.D. thesis. Southampton University.

Reid D.M. (1939) On the occurrence of *Gammarus duebeni* (Lillj.) (Crustacea, Amphipoda) in Ireland. *Proc. R. Irish Acad.* (B) **45**, 207–14.

Shaw J. (1959) The absorption of sodium ions by the crayfish *Astacus pallipes* Lereboullet. I. The effect of external and internal sodium concentrations. *J. Exp. Biol.* **36**, 126–44.

17. LES RÉPONSES DES COPÉPODES (CRUSTACÉS) AUX CHANGEMENTS DES CONDITIONS PHYSICO-CHIMIQUES DANS LES EAUX TEMPORAIRES CAMARGUAISES PROVENÇALES ET CORSES

A. CHAMPEAU

Laboratoire de Biologie Générale Ecologie,
Université de Provence, Marseille, France

SUMMARY

In the French mediterranean coastal marshes many lagoons dry up in summer when the rate of evaporation is high.

These marshes contain an abundant and interesting aquatic fauna. In particular, certain species of copepods are able to survive the drought because they burrow in the sediment (e.g. harpacticids and cyclopids) or else, lie on the marsh surface (e.g. calanids). The rapid growth of copepods during favourable periods and their ability to survive different environmental conditions (e.g. a range of salinities, variation in the dryness of sediments) explain their success in a habitat which appears to be inhospitable to aquatic life.

The resting stage is characteristic of the following:

(1) Quiescence in the fertilized females of the multivoltine harpacticid *Cletocamptus retrogressus*,
(2) Facultative diapause in the copepodid IV stage of the bivoltine cyclopid *Diacyclops bicuspidatus odessanus*, and in the egg stage of the calanids, *Mixodiaptomus kupelwieseri*, *Arctodiaptomus wierzejskii* and *Eurytemora velox*,
(3) Obligatory diapause in the egg stage of univoltine calanids, *Hemidiaptomus roubaui lauterborni* and *Diaptomus cyaneus*.

The ability of copepods, particularly the calanids, to adapt to the desiccation of their habitat suggests that they are extremely sensitive to changes in environmental conditions. These invertebrates were the first group of animals to disappear from the eastern coast of Corsica and the Camargue following the development of land for agricultural and industrial purposes as well as for tourism.

RÉSUMÉ

Les marais temporaires de la région méditerranéenne française abritent une faune aquatique abondante et originale dominée par des espèces de Copépodes capables de survivre à l'asséchement en vie latente enfouis dans le sédiment (Harpacticides et Cyclopides) ou posés sur le fond (Calanides). D'une espèce à l'autre la vie latente correspond à une quiescence, à une diapause facultative ou à une diapause obligatoire. La précision des adaptations à l'assèchement chez les Calanides font qu'ils sont particulièrement sensibles aux changements des conditions du milieu résultant du développement agricole touristique et industriel.

INTRODUCTION

Dans la région méditerranéenne française, en dehors des grands étangs largement ouverts sur la mer, il existe de nombreuses lagunes peu profondes qui occupent des surfaces importantes, surtout en Camargue et dans la plaine orientale corse. Dans ces lagunes, l'évaporation estivale entraîne un asséchement partiel ou total qui permet de séparer 2 catégories de marais :

1. *Les marais semi-temporaires*, plus ou moins salés en fonction de l'importance des apports d'eau douce. La salinité, le plus souvent inférieure à celle de la mer, atteint son maximum en été quand le niveau est le plus bas. Les écarts saisonniers sont de l'ordre de 20‰ S. Dans ces milieux où les Copépodes sont peu abondants, les populations de crustacés benthiques sont les plus nombreuses (Gammares, Sphaeromes). L'euryhalinité de ces espèces leur permet de coloniser des plans d'eau différents par leur salinité moyenne, où ils peuvent se reproduire toute l'année.
2. *Les marais temporaires*, où l'alternance des périodes d'inondation durant la saison froide et pluvieuse et d'asséchement durant la saison chaude et sèche, crée des conditions de vie à première vue très difficiles pour des animaux aquatiques. Cependant, une faune très abondante de crustacés planctoniques de petite taille au développement rapide colonise les eaux temporaires. On y récolte des Ostracodes (19 sp.) des Cladocères (29 sp.) mais surtout des Copépodes (75 espèces de Calanides, Cyclopides et Harpacticides).

LE MILIEU

Trois caractéristiques des marais temporaires déterminent essentiellement l'importance des Copépodes : la gamme étendue de la salinité des eaux, l'abondance du matériel nutritif au niveau primaire durant l'inondation, le degré de dessiccation du sédiment pendant l'asséchement.

La salinité

L'éventail de la salinité va de l'eau douce dans les marais les plus éloignés du littoral—en haute Camargue, en Provence et dans le sud-est de la Corse—à des eaux 2 fois plus salées que la mer en basse Camargue. Dans les marais d'eau douce ou peu salés les variations annuelles de la salinité sont peu importantes ou comparables à celles des marais semi-temporaires. Ces variations augmentent dans les stations salées. Elles peuvent dépasser 100‰ S dans les stations sursalées de basse Camargue.

Dans la région méridionale la quantité annuelle des précipitations varie assez pour que la salinité moyenne double ou triple dans un même marais, d'une année humide à une année sèche.

La nourriture

L'abondance du matériel nutritif au niveau primaire résulte de l'alternance de périodes d'inondation et d'assèchement. La part du phytoplancton produite pendant la mise en eau et directement consommable par le zooplancton ne constitue pas l'essentiel de la nourriture disponible. C'est l'importance des déchets végétaux accumulés durant l'été et consommés l'hiver suivant qui explique la plus forte densité du zooplancton en milieu temporaire. Les tapis de macrophytes et d'algues filamenteuses qui, au printemps, constituent une part importante de la production primaire n'est pas utilisable par les populations les plus denses appartenant à des espèces «filtreuses». Ces végétaux sont desséchés et finement divisés durant l'été. Il s'y ajoutent les débris des plantes hydrophiles. Dans les marais peu profonds ces particules sont constamment remises en suspension par le vent et sont alors disponibles pour le zooplancton. D'après les premiers examens de contenus stomacaux on peut noter que les espèces zooplanctoniques d'eau temporaire auraient des régimes alimentaires beaucoup moins spécialisés que ceux des espèces lacustres et marines.

La dessiccation du sédiment

La durée et la rigueur de la dessiccation estivale varient d'une station à l'autre. Le sédiment reste humide si la proximité de la nappe phréatique et la texture permet à l'eau capillaire de remonter en surface, ou si l'importance de la couverture végétale limite les effets de l'évaporation. Au contraire, la teneur en eau des sédiments et l'humidité relative dans les interstices du sol peut être très faible lorsque la frange capillaire n'atteint pas le fond dénudé des stations.

Souvent, en Camargue, la dessiccation osmotique diminue encore la teneur en eau disponible quand du sel de la nappe phréatique se dépose près de la surface, après évaporation de l'eau capillaire.

L'enfoncement estival de la nappe phréatique dépendant de la hauteur des pluies, la dessiccation du sédiment change d'une année à l'autre dans le même marais temporaire.

LES RÉPONSES DES COPÉPODES DANS LES MILIEUX NATURELS

La faune aquatique des eaux temporaires, dominée par les Copépodes, vit dans un milieu particulierement instable. Les nombreuses espèces de Cyclopides, Harpacticides et Calanides peuvent exploiter au maximum les possibilités trophiques du milieu. Elles couvrent l'éventail le plus large de salinités et ont acquis les modes de survie qui permettent de répondre à toutes les conditions d'asséchement.

L'exploitation des possibilités trophiques

Dans les marais temporaires de Camargue et du sud de la Corse on rencontre, ensemble, jusqu'à 5 espèces de Copépodes dominantes. Par exemple, en moyenne Camargue :

1. *Cletocamptus retrogressus*, harpacticide benthique de petite taille (0·8 mm) broute les diatomées qui tapissent le fond,
2. *Diacyclops bicuspidatus odessanus*, Cyclopide omnivore de taille moyenne (1·5 mm) mange des algues microscopiques qui recouvrent les tiges des plantes aquatiques et chasse de petites proies (Rotifères, jeunes Daphnies),
3. *Hemidiaptomus roubaui lauterborni*, *Diaptomus cyaneus* et *Arctodiaptomus wierzejskii*, Calanides filtreurs de tailles différentes—respectivement 6·4 et 2 mm—utilisent le phytoplancton et les débris végétaux en suspension dans l'eau. Les filaments des maxilles qui forment le filtre collecteur de particules alimentaires sont les plus espacés chez l'espèce la plus grande. Ils retiennent les grosses particules et laissent passer les particules plus fines dont se nourrissent successivement les deux espèces suivantes.

Les réponses aux variations de la salinité

L'éventail de salinité supporté à l'état actif par les différentes espèces de Copépodes d'eau temporaire va de 0 à 120‰ S. Certaines espèces sont parmi les animaux les plus euryhalins, comme l'harpacticide *Cletocamptus retrogressus* qui résiste à des salinités encore plus élevées en vie latente. D'autres espèces relativement stenohalines mais adaptées à des salinités différentes se remplacent quand la salinité augmente d'une station à l'autre. C'est le cas en

Camargue pour les 3 Calanides *Mixodiaptomus kupelwieseri* dominant dans les stations d'eau douce ou peu salée, *Arctodiaptomus wierzejskii* dominante dans les stations moyennement salées et *Eurytemora velox* dominante dans les stations très salées.

Les modes de survie à l'assèchement en vie latente

La quiescence des femelles fécondées de Cletocamptus retrogressus

Le cycle de cet harpacticide se développe pendant toute la phase de mise en eau. La période de ponte dure plusieurs mois. Le développement rapide des larves permet à plusieurs générations de se succéder de l'automne au printemps.

A la fin du printemps, les copépodites V, les adultes—mâles et femelles—s'enfouissent dans le sédiment pour échapper aux fortes températures. L'enfouissement précède l'assèchement des stations. Les individus se déplacent dans le substrat tant que la teneur en sel n'atteint pas des valeurs limites qui provoquent leur inactivité. Ces valeurs, précisées en expérience, varient d'une population à l'autre mais ne correspondent pas à des différences génotypiques (Champeau 1967). Peu après l'assèchement, les copépodites V, puis les mâles meurent. Ces stades supportent une courte inactivité sans grande valeur adaptative si l'on considère la durée de la période défavorable. Au contraire l'état inactif de la femelle peut se prolonger pendant plusieurs années quand la station reste à sec. On peut réactiver les femelles du commencement à la fin de la période de latence, qui correspond donc à une quiescence. Après l'inondation, si la salinité se situe dans la zone favorable au retour à l'état actif et à l'ovogenèse, les femelles émergent du sédiment et pondent des œufs fécondés par les spermatozoïdes restés vivants tout l'été dans leurs réceptacles séminaux.

La diapause des Copépodites IV de Diacyclops bicuspidatus odessanus

Deux générations, ou exceptionnellement trois se succèdent durant la période froide. Au printemps, tous les stades évolutifs—nauplii, copépodites et adultes—nagent dans l'eau des mares. Les copépodites IV mâles et femelles, possèdent déjà les caractères distinctifs de ceux récoltés en été dans le sédiment. Leur enfouissement peut commencer bien avant la baisse du niveau. Après l'assèchement, les copépodites V et quelques adultes supportent une quiescence de courte durée, puis meurent. Au contraire, comme les femelles de *C.retrogressus*, les copépodites IV survivent plusieurs mois, voire plusieurs années dans le sédiment où ils sont protégés de la chaleur et de la dessiccation excessives qui règnent en surface. Leur inactivité correspond à une diapause : leur réactivation d'abord possible dans les prélèvements de sédiment effectués à la fin du printemps, devient impossible en été. Les températures froides accélèrent l'élimination de la diapause. Quand l'inondation des stations se produit bien après le refroidissement d'octobre, les copépodites IV attendent en quiescence le retour d'un environnement favorable.

On reconnaît ces trois phases dans l'évolution de la diapause des copépodites IV des 7 espèces des autres Cyclopides abondants, en hiver, dans les eaux temporaires méridionales :

1. Une phase d'installation de la diapause pendant laquelle l'état physiologique change et le comportement fouisseur s'accentue; l'apparition de gouttelettes lipidiques rouges est la conséquence visible de ce changement,
2. Une phase de diapause vraie durant laquelle il est impossible de réactiver les individus latents,
3. Une phase de quiescence durant laquelle la réactivation est possible.

La diapause des œufs de résistance des Calanides

Pendant l'inondation, en période froide, les femelles des espèces bivoltines (*M.kupelwieseri, A.wierzejskii, E.velox*) pondent en majorité des œufs donnant naissance à des nauplii au bout d'une semaine. Au printemps elles pondent en majorité des œufs qui éclosent après l'asséchement estival. On distingue ainsi, des œufs immédiats et des œufs de résistance. Les œufs de résistance ont un chorion dont l'épaisseur dépasse 1·5 μ, les œufs immédiats un chorion dont l'épaisseur reste proche de 1 μ.

Dans les conditions naturelles, l'apparition des œufs de résistance suit l'élévation de la température,

1. d'une part, dans les prélèvements successifs effectués dans une seule mare,
2. d'autre part, dans les prélèvements simultanés effectués dans les mares ombragées et dans les mares ensoleillées. Dans les mares saumâtres l'apparition des œufs de résistance peut correspondre à la seule augmentation de la salinité.

Après la ponte, les œufs de résistance tombent sur la vase. L'activité des animaux fouisseurs fait qu'une partie des œufs pondus restent posés sur le fond et que les autres sont enterrés sous plusieurs centimètres de sédiment. Ces derniers rejoignent une réserve d'œufs constitués au cours des années précédentes. Inversement chaque année de nombreux œufs du stock se trouvent ramenés en surface.

Pendant la période d'asséchement, les œufs demeurés ou remis en surface sont soumis à une dessiccation intense qui altère la résistance du chorion au niveau d'une ligne équatoriale de dehiscence. Le développement embryonnaire est bloqué par une diapause. La diapause est levée par l'exposition des œufs à des températures élevées pendant tout l'été.

En automne les températures baissent. Les premières pluies gorgent le sédiment d'eau. Le développement reprend. A la fin du développement embryonnaire la membrane d'éclosion se forme. Le nauplius attend en quiescence l'inondation des stations, nécessaire à son éclosion : l'eau entre par osmose dans l'œuf, la membrane de l'éclosion se gonfle et sépare le chorion en deux demi-sphères, le nauplius s'en échappe.

Les femelles des espèces univoltines (*H.roubaui lauterborni*, *D.cyaneus*) ne pondent que des œufs de résistance. Aussi, dans la nature deux générations sont toujours séparées par une période d'assèchement. Le chorion de leurs œufs est plus épais que celui des œufs de résistance des espèces bivoltines. Ils ont besoin, pour éclore, d'une dessiccation intense : on les rencontre dans les stations les plus arides en été, ou seulement pendant les années sèches dans les autres marais temporaires.

En définitive, malgré l'extrême instabilité des milieux aquatiques temporaires, les Copépodes peuvent répondre à tous les changements des conditions physico-chimiques rencontrées dans les conditions naturelles, même lorsqu'il s'agit de situations exceptionnelles comme :

1. une inondation ou un assèchement qui se prolonge puisque la vie latente peut se maintenir durant plusieurs années,
2. des assèchements précoces avant la reproduction de la première génération puisque tous les œufs de résistance des Calanides n'éclosent pas en même temps, puisque tous les copépodites IV des Cyclopides ne sont pas soumis à des diapauses de même durée,
3. des changements très importants d'une année à l'autre dans le même marais en ce qui concerne la salinité de l'eau et la dessiccation du substrat puisque le sédiment contient les stades de résistance de nombreuses espèces qui peuvent émerger si l'environnement est favorable à leur vie active ou au contraire attendre plusieurs années le retour à de meilleures conditions.

Au contraire, la précision des adaptations des Copépodes à ces changements dans les conditions naturelles font qu'ils sont particulièrement sensibles aux transformations des marais temporaires résultant de l'aménagement intensif des zones humides.

LES CONSEQUENCES DU DEVELOPPEMENT INDUSTRIEL AGRICOLE ET TOURISTIQUE

Dans le sud-est de la France l'impact des activités humaines sur les marais temporaires intéresse des surfaces considérables. De plus, il s'agit le plus souvent d'une détérioration définitive, fatale en particulier aux Calanides qui font l'originalité du peuplement en Copépodes de ces milieux.

En Camargue

L'extension des marais salants et des rizières ont provoqué une évolution rapide des milieux aquatiques et de leurs biocénoses.

L'impact des marais salants

Les bassins de préconcentration et les tables saunantes occupent une superficie d'environ 15,000 hectares. Ils ont été implantés sur l'emplacement d'étangs et de marais localisés au sud-est de la Camargue et, à l'ouest, en Petite-Camargue. La faune de ces marais a été détruite lors de l'aménagement des bassins de préconcentration où l'eau de mer introduite par pompage est concentrée, par évaporation, jusqu'à une salinité de 80‰ S. Ces bassins, vidés lorsque l'eau préconcentrée est transférée dans les marais salants proprement dits, constituent encore des milieux aquatiques temporaires. Mais on y rencontre uniquement quelques Harpacticides très euryhalins dont *Cletocamptus retrogressus* qui côtoient les populations surabondantes de l'Anostracé *Artemia salina.*

L'impact des rizières

20,000 hectares de marais ont été transformés en rizières qui occupent actuellement 10,000 hectares. Ces rizières forment des mares temporaires alimentées par les eaux déjà polluées du Rhône. Elles sont inondées d'avril en septembre quand les marais naturels sont à sec et constituent un «milieu relais». Une étude a été entreprise au Centre d'Ecologie de Camargue pour connaître l'intérêt de leur faune aquatique où les Copépodes sont nombreux (Pont 1977).

On y récolte seulement des espèces communes de Cyclopides. Les causes de l'élimination des Harpacticides et des Calanides ne sont pas encore bien établies mais il semble, qu'elles pourraient résulter, pour les Calanides, d'un asséchement insuffisant du substrat éxondé en hiver et d'une sensibilité plus grande aux herbicides et insecticides répandus dans l'eau. Les eaux d'irrigation dont le volume annuel est évalué à près de 40 millions de mètres cubes sont transférées, en grande partie, des rizières vers les étangs centraux de la réserve par un réseau de canaux de drainage. Cet apport considérable d'eau douce polluée est tel que les milieux temporaires caractéristiques de la moyenne Camargue se rarefient au profit de milieux aquatiques permanents. Le niveau du Vaccarès, autrefois semi-temporaire demeure très élevé en été. La salinité de ses eaux a fortement diminué. *Eurytemora velox* a été éliminé au profit d'un autre Calanide, *Calanipeda aquae dulcis.* Des espèces banales, franchement limnétiques sont introduites en remplacement des grands Calanides univoltins dont le périmètre d'extension de plus en plus restreint fait craindre leur disparition prochaine du territoire de la Camargue. La concentration des produits toxiques dans les milieux naturels n'atteint pas encore des valeurs léthales pour les premières espèces de Copépodes testées, mais elle peut être très élevée en bout de chaîne alimentaire (facteur de concentration : $\times$ 10,000 pour le Lindane, Heurteaux *et al.* 1973).

Remarque

De vastes opérations de démoustication ont été entreprises sur le littoral languedocien, jusque dans les marais du Plan du Bourg à l'est de la Camargue.

Les quelques Copépodes testés sont moins sensibles au fénitrothion que les larves de moustiques (Raibaut 1967; Champeau *et al.* 1977). Cependant leur sensibilité varie notablement d'une espèce à l'autre. Aussi les tests seront étendus à d'autres espèces et à d'autres produits toxiques rencontrés dans les milieux naturels. D'ores et déjà la banalisation du peuplement en Copépodes par l'élimination de nombreux Calanides est évidente dans les zones démoustiquées.

En Corse

Ces dernières années, l'aménagement agricole de la plaine orientale a détruit de nombreuses «Padule» qui ont été asséchées. Leur peuplement en Calanides était remarquable par ses affinités avec celui des «dayas» de l'Afrique du Nord (Schachter et Champeau 1969). Il s'y ajoute un projet près d'aboutir pour transformer en port de plaisance le marais Del Sale, le plus grand et le plus beau marais temporaire de l'île.

En Provence

Les marais de Grimaud au fond du Golfe de Saint-Tropez ont été transformés en ports de plaisance. D'autres «marinas» sont prévues à l'emplacement des marais d'Hyères et de Saliers. Par contre la situation est stationnaire dans les marais temporaires de l'arrière pays varois.

REFERENCES

Champeau A. (1967) Etats de quiescence déterminés chez le Copépode Harpacticoïde *Cletocamptus retrogressus* Schmankevitsch par des variations de chlorinité, de température et par dessiccation. *C. R. Acad. Sci. Paris* **265**, 248–51.

Champeau et al. (1977) Les invertébrés indicateurs de la qualité des eaux en Camargue. *L'Eau et l'Industrie* **17**, 87–91.

Heurteaux P. et al. (1973) Contamination des milieux aquatiques camarguais par les résidus de produits phytosanitaires. *La Terre et la Vie* **1**, 33–61.

Pont D. (1977) Structure et évolution saisonnières des populations de Copépodes, Cladocères et Ostracodes des rizières de Camargue. *Amnls Limmnol.* **13**, 15–28.

Raibaut A. (1967) Recherches écologiques sur les Copépodes Harpacticoïdes des étangs côtiers et des eaux saumâtres temporaires du Languedoc et de Camargue. *Thèse Fac. Sci. Montpellier.*

Schachter D. et Champeau A. (1969) Contribution à l'étude écologique de la Corse I—Les Copépodes des eaux stagnantes. *Vie et Milieu* **20** (1-C), 41–56.

IV

PHOTOSYNTHESIS AND PRIMARY PRODUCTION

18. PHOTOSYNTHETIC AND WATER RELATIONSHIPS OF HIGHER PLANTS IN A SALINE ENVIRONMENT

KLAUS WINTER*

Institut für Botanik der Technischen Hochschule, Schnittspahnstr. 3–5, *D*-6100 *Darmstadt, B.R.D.*

SUMMARY

Halophytes are a group of higher plants naturally adapted to the high concentrations of electrolytes prevailing in saline environments.

Halophytes respond to increasing soil salinity with a net solute increase in their cells in order to avoid a drop in turgor pressure.

Increasing NaCl concentrations reduce transpiration. This has been attributed to increased stomatal and mesophyll resistances to water vapour loss in the leaves. Reduction of water supply to the leaves may be indirectly involved in the reduced transpiration. The combined effects of reduced transpiration and salt-stimulated growth result in a higher water use efficiency.

In most species a more or less pronounced decrease in net CO_2 fixation (per unit of fresh weight, dry weight or leaf area) with rising salinity has been reported. However, on a chlorophyll basis, net CO_2 fixation may increase considerably, particularly at salt levels which stimulate growth.

Most halophytes so far examined, especially those occupying coastal habitats, fix CO_2 via the C_3 pathway of photosynthesis. Nevertheless, numerous salt-tolerant grasses and many halophytic Chenopodiaceae (particularly those from inland salines) are characterized by the C_4 pathway. Crassulacean Acid Metabolism (CAM) is of importance in the adaptation of members of the Aizoaceae to salinity and water stress. *Mesembryanthemum crystallinum*, for example, shifts from C_3 photosynthesis to CAM when grown under highly saline conditions in the laboratory, and in its natural coastal environment. In general, however, although most halophytes are rather succulent, a property which is associated with CAM, CAM does not play a major role in the adaptation of halophytes to a saline environment.

* Present address: Department of Environmental Biology, Research School of Biological Sciences, The Australian National University, P.O. Box 475, Canberra City, A.C.T., 2601, Australia.

RÉSUMÉ

Les halophytes sont un groupe de plantes supérieures adapté aux fortes concentrations d'électrolytes, régnant dans les environnements salins.

Les halophytes répondent à un accroissement de la salinité du sol par une augmentation nette de soluté dans leurs cellules, afin d'éviter une chute de la pression de turgescence.

Une augmentation des concentrations de NaCl réduit la transpiration. Ceci a été attribué à l'augmentation des résistances des stomates et du mésophylle contre la perte de vapeur d'eau dans les feuilles. La réduction de l'apport d'eau dans les feuilles peut jouer indirectement sur la réduction de la transpiration. Les effets combinés de la transpiration réduite et de la croissance stimulée par la salinité résultent en une plus grande efficacité de l'utilisation de l'eau.

Dans la plupart des espèces, une diminution plus ou moins prononcée dans la fixation de CO_2 (par unité de poids frais, poids sec, ou surface de feuille) avec l'augmentation de la salinité, a été rapportée. Cependant par rapport au taux de chlorophylle, la fixation nette de CO_2 peut augmenter considérablement, particulièrement à des niveaux de salinité qui stimulent la croissance.

La plupart des halophytes examinées, en particulier celles des habitats côtiers fixent le CO_2 par la voie C_3 de la photosynthèse. Néanmoins nombre d'herbes tolérant la salinité et beaucoup de Chenopodiaceae (en particulier celles des environnements salins intérieurs) sont caractérisées par la voie C_4. Le métabolisme acide des Crassulacean (CAM) est important pour l'adaptation des Aizoaceae à la salinité et au choc osmotique. Par exemple, *Mesembryanthemum crystallinum* dévie sa photosynthèse en C_3 vers le CAM quand elle est cultivée au laboratoire dans des conditions de forte salinité; il en est de même dans son environnement naturel côtier. En général, malgré la succulence de la plupart des halophytes, propriété associée au CAM, le CAM ne joue cependant pas un rôle majeur dans l'adaptation des halophytes à l'environnement salin.

INTRODUCTION

Among higher plants only a limited number of species are capable of growing in saline habitats. Sodium chloride is often present at high concentrations in these environments. Halophytes usually tolerate concentrations of sodium chloride higher than 400 mM when grown in culture solutions. For example, *Salicornia europaea* survived a salinity of 6·3% in culture (Montfort & Brandrup 1927a) and *Diplanthera wrightii*, a marine angiosperm (seagrass) persisted even at a salinity of 9·5% which is about 2½ times as saline as sea water (McMillan & Mosely 1967; McMillan 1974). Most halophytes merely do not tolerate high salinity, they appear to need considerable amounts of electrolytes for maximum growth (Flowers *et al.* 1977).

The plant–water relations of halophytes have interested scientists since the end of the last century. Yet only recently, as appropriate methods became available, has it been possible to investigate the influence of salinity on water relations and photosynthesis of halophytes (Gale 1975). It is to be regretted that there is only a small number of field studies on this topic.

WATER RELATIONS

The soil–plant–atmosphere continuum

The water balance of a plant depends on the flow of water through the soil–plant–atmosphere continuum. The flow of water (J) can be expressed analogously to an Ohm's system as

$$J = \frac{\Psi_{soil} - \Psi_{root}}{R_1} = \frac{\Psi_{root} - \Psi_{leaf}}{R_2} = \frac{C_{leaf} - C_{atmosphere}}{r_3} \quad (1)$$

The differences in water potential (Ψ) and the difference in water vapour concentration (C), respectively, are the driving forces, and R_1, R_2 and r_3 are resistances to water movement through different parts of the system.

Osmotic adjustment

Salinity decreases the water potential of the soil and hence of the plants. A decrease in water potential of plant cells may be composed of changes in the turgor pressure P and the osmotic pressure π:

$$\Psi = P - \pi \quad (2)$$

In order to avoid a drop in turgor pressure when Ψ decreases halophytes absorb ions from the root medium which are stored in suitable cell compartments. As a consequence π increases (= osmotic adjustment) accommodating the decrease in Ψ and maintaining turgor. The osmotic pressures of expressed leaf saps of maritime halophytes range between 30 and 40 bars (Flowers 1975) whereas π of sea water is approximately 23 bars. However, it should be noted, that osmotic adjustment apparently does not occur in the xylem. Mangroves, for example, show low values of π for the xylem sap, but highly negative values of P were found in the xylem vessels (Scholander *et al.* 1965).

The phenomenon of osmotic adjustment to soil salinity was neglected in early studies of the water relations of halophytes. Schimper (1898) regarded a saline medium, no matter if wet or dry, as 'physiologically dry' and attributed xerophytic characteristics of plants growing in a saline environment to such a condition. Although the original ideas of Schimper have been disputed, there are indications that the 'physiological drought hypothesis' remains valid.

Physiological drought hypothesis—modern aspects

In exudation experiments using a Scholander pressure bomb, a reduced flow of water was measured through the roots of beans (O'Leary 1969) and the halophyte *Atriplex halimus* (Kaplan & Gale 1972), when the plants were grown on NaCl. The reduction in water flow has been ascribed to an increased resistance to water in the roots. It has been argued that, at least in glycophytes, a reduced water supply to the leaves may result in a water deficit in the leaf cells when evaporative demand is high.

At present the nature of this increased resistance to the flow of water in the roots is not understood. Oertli (1976) doubted that resistance changes due to salinity are true resistance changes but rather gradients in water potential which are inefficient for water flow. It has also been questioned whether the root resistance determined from exudation experiments is the same as the root resistance of an intact transpiring plant (Shalhevet *et al.* 1976).

With respect to the osmotic adjustment of leaf cells Oertli (1968) pointed out that the estimation of osmotic pressure of leaf sap is complicated by solutes which may accumulate in the extracellular space. Under certain circumstances, the osmotic pressure of the extracellular solution may be sufficiently high to cause an intracellular water deficit although the osmotic pressure of the expressed leaf sap is higher than that of the root medium. On the other hand, Slatyer (1966) has calculated that the back diffusion of solutes carried forward in the transpiration stream should prevent significant solute accumulation in the cell walls. The ion concentrations in the xylem sap of some halophytes are low but data on extreme salt-accumulating species are unavailable (Flowers *et al.* 1977).

Turgor pressure

A better understanding of the water relations of halophytes surely would result from reliable, direct determinations of P, particularly because of the crucial role of turgor pressure in cell growth (Hsiao 1973). However, the small size of higher plant cells limits the use of the methods which are at present available for direct measurement of P. Only one attempt has been made so far in higher plants. With the pressure probe technique developed by Steudle and Zimmermann (1971) it was possible to perform direct measurements of P on the extremely large epidermal bladder cells of the halophyte *Mesembryanthemum crystallinum* (Steudle *et al.* 1975). Although the authors were primarily interested in the water relations between the bladder cells and the sub-

Table 18.1. *Values of the turgor pressure P_b (obtained by direct measurement) and of the osmotic pressure π_b (determined cryoscopically) of bladder cells of* Mesembryanthemum crystallinum *grown under different saline conditions (after Steudle* et al. *1975, modified). Based on these data, the water potential of the bladder cells ψ_b and the gradient in water potential between the bladder cells and the culture solution $\psi_b - \psi_c$ was calculated. For conversion of units a temperature of 25°C was assumed. All data are given in bars. Subscript b refers to bladder cell, subscript c to culture solution.*

Treatment	P_b	π_b	ψ_b	$\psi_b - \psi_c$
(1) soil culture without addition of NaCl	+3·5 ± 0·8 (24)	+16·4 ± 4·0 (7)	−12·9	
(2) 5 weeks in culture solution containing 100 mM NaCl $\psi_c = -4{\cdot}9$	+2·2 ± 0·9 (5)	+22·0 ± 0·3 (2)	−19·8	−14·9
(3) 1–5 weeks in culture solution containing 400 mM NaCl $\psi_c = -18{\cdot}5$	+1·5 ± 0·7 (15)	+33·9 ± 6·8 (5)	−32·4	−13·9

epidermal cell layers, some conclusions about the overall plant water balance can be drawn (Table 18.1). The water potential of the bladder cells depends mainly on π, especially when the plants are grown under saline conditions, i.e. at 100 and 400 mM NaCl. In the latter case, the water potential of the bladder cells is clearly more negative than that of the root medium. Interestingly, the data suggest a decreased turgor pressure with increasing salinity. However, in interpreting this result, the special role of the bladder cells should be considered. They form reservoirs for salts and water and probably function as a buffering system, protecting the smaller mesophyll cells during periods of water stress. Therefore, a relation between salinity level and turgor pressure as indicated above cannot simply be extrapolated to the carbon assimilating mesophyll cells or even to cells of halophytes in general.

Transpiration

Under field conditions

Schimper's hypothesis was the stimulus for numerous studies on the transpiration of halophytes. A major point of controversy was the idea that halophytes have xerophytic features, such as succulence, which diminish water loss. However, a rather high water loss per day, in some cases equivalent to almost twice the plants' water content, was reported by Stocker (1925) for succulent halophytes of the North Sea shore (e.g. *Suaeda maritima, Salsola kali, Cakile maritima*) irrigated with sea water. The transpiration of these species per unit leaf area was not reduced in comparison with corresponding data of non-salt treated glycophytes. In his review on the water relations of halophytes Adriani (1956) summarized that, in their natural environments, most halophytes are distinguished by a relatively low transpiration rate per unit of fresh weight compared to non-halophytes. The differences become blurred when transpiration is calculated on an area basis. The interpretation of the absolute transpirational rates is complicated however by the lack of information on the gradients in water vapour concentration between leaf and air during measurements (see next chapter).

Under controlled conditions

Halophytes reduce transpiration when grown in culture solutions of increasing NaCl-content (Fig. 18.1). This result is well established but the degree of reduction varies according to the way in which the rate of transpiration is expressed. A reduced rate of transpiration can be brought about by one or more of the resistances to water flow in the soil–plant–atmosphere continuum (see Equation 1). For example, transpiration may be reduced directly by diminishing the water loss from the leaves to the air or indirectly by diminishing the water supply to the shoots.

A decreased permeability of the roots to water may be one factor which reduces the flow of water to the leaves. This point has already been discussed. A similar effect could result from an increased shoot:root ratio which is frequently observed in halophytes as a response to increasing salinity (Schratz 1935; Baumeister & Schmidt 1962; Webb 1966).

The transpirational loss of water vapour (W) from the leaf can be described as

$$W = \frac{C_w - C_a}{r_m + r_s + r_a} \tag{3}$$

The quantities C_w and C_a are the water vapour concentrations (=absolute humidities) at the evaporating surfaces within the leaf (i.e. the mesophyll cell walls) and the bulk air, respectively. The terms r_m, r_s and r_a represent resistances to the diffusion of water vapour from the mesophyll cell walls, through the stomata and through the boundary layer of unstirred air adhering to the leaf surface.

The influence of salinity on diffusion resistances to water vapour loss has been studied in detail for *Atriplex halimus* (Kaplan & Gale 1972). Increases of both r_s and r_m were reported to be involved in the reduction of transpiration.

There are several explanations for an increase of r_s. Salinity stimulates succulence in many halophytes. This is often accompanied by a decrease in the number of stomata per unit of leaf area (Bickenbach 1932; Waisel 1972). At high salt concentrations, a reduced turgor may cause a decrease in stomatal aperture. Bearce (1968) (cited by Gale & Poljakoff-Mayber 1970) raised the possibility that there is a lack of osmotic adjustment of the guard cells relative to the adjacent epidermal cells. Plant hormones, particularly kinetin and

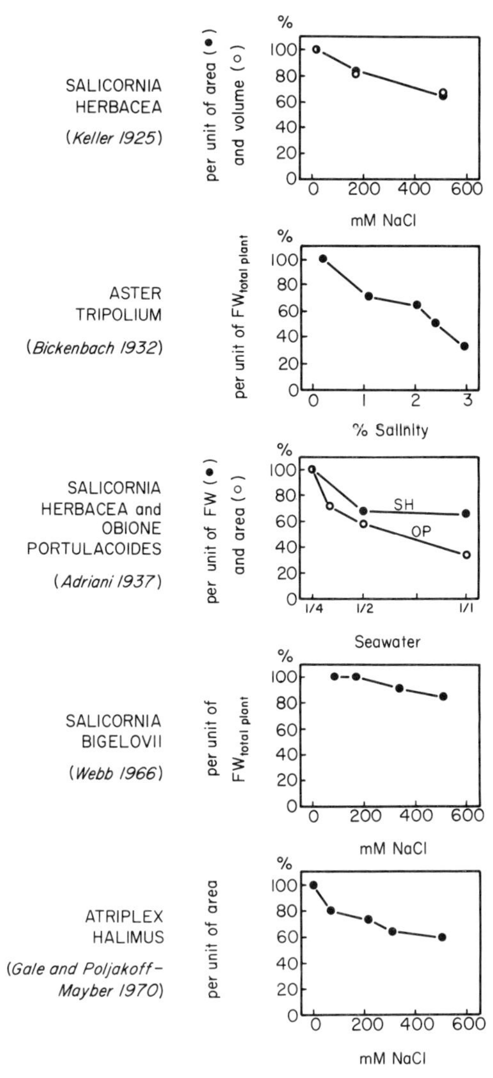

FIG. 18.1. Percentage changes in transpiration of some halophytes exposed to increasing levels of salinity in the root medium. Transpiration at the lowest salinity level is taken as 100%. FW = fresh weight.

abscisic acid (ABA) play an important role in controlling the stomatal function (Itai & Benzioni 1976). Kinetin favours and ABA reduces opening of stomata. Hitherto, effects of NaCl on the balance between kinetin and ABA have been studied only in glycophytes, and there is no information available for halophytes. One could suppose that, in halophytes, the influence of ABA dominates over that of kinetin when the plants are exposed to salt concentrations which reduce growth.

The determination of r_m is not without problems (Gale *et al.* 1967; Farquhar & Raschke 1978) and high r_m values as a consequence of salinity are difficult to explain. There is evidence, at least in some plants, for the existence of a waxy layer on the mesophyll cell walls facing the substomatal cavities (Esau 1967). Such a hydrophobic layer may form a barrier against diffusion of water vapour (Nobel 1974) or against flow of liquid water (Jarvis & Slatyer 1970).

Using Equation 3, one generally assumes C_w to be the saturation water vapour concentration of pure water. In halophytes this is not necessarily so, because with increasing salinity, solutes may accumulate in the cell walls as has been discussed by Oertli (1968). However, large changes in the solute concentration in water, i.e. in π, lead to only minor changes in the activity of water a_w. This can be seen from Equation 4,

$$RT \ln a_w = -\bar{V}_m \pi \qquad (4)$$

where R, T and $\bar{V}_m$ refer respectively to the ideal gas constant, the absolute temperature and the partial molal volume of water. Since a_w corresponds to $C_w/C_{w,\text{sat}}$, where $C_{w,\text{sat}}$ represents the saturation water vapour concentration, values of C_w and $C_w - C_a$ were calculated for osmotic pressures of 0 and 50 bars of the liquid in the mesophyll cell walls. (Assuming a temperature of 25°C, $RT/\bar{V}_m$ is 1,373 bars and $C_{w,\text{sat}}$ is 23 μg/cm^3.) The result is shown in Table 18.2. The data in parentheses indicate the percentage reduction of $C_w - C_a$ at 50 bars compared to $C_w - C_a$ at 0 bars.

Table 18.2. *The effect of a high (hypothetically assumed) osmotic pressure π of the liquid at the evaporating surfaces in the leaf on the gradient in water vapour concentration $C_w - C_a$ between the leaf interior and the ambient air (Temperature = 25°C). Explanations see text.*

		$C_w - C_a$ (μg cm^{-3})			
		100%	95%	50%	20%
(Bars)	C_w (μg cm^{-3})	Relative humidity of the bulk air			
0	23·0	0	1·1	11·5	18·4
50	22·2	−0·8	0·3	10·7	17·6
			(72·7)	(7·0)	(4·4)

Obviously 50 bars represents an extremely high value of π corresponding to a solution of about 1,080 mM NaCl. Nevertheless, $C_w - C_a$, the driving force of the transpirational water loss, decreases by less than 7% at relative humidities of the bulk air lower than 50%. At high relative humidities, the relative decrease of $C_w - C_a$ is not of great importance, since the absolute values of $C_w - C_a$ and therefore the transpirational rates are very small under these conditions, and this is true whether π is high or low.

Water requirement

The water requirement (WR) of a plant during growth may be expressed as the ratio of water loss to the weight of dry matter produced. A reduction of WR with increasing salinity has been found for numerous halophytes (Eaton 1927; Ashby & Beadle 1957; Önal 1971). This suggests an improved water economy of halophytes, when grown in the presence of NaCl, which is of importance in dry, saline habitats. (A more efficient use of water is also indicated by an increase in the ratio of CO_2 uptake to water loss which has been found in *Atriplex halimus* following exposure of plants to NaCl (Kaplan & Gale 1972); see also '*Balance between C_3 pathway and CAM*' in the present paper.)

The main drop in WR takes place between plants grown without addition of NaCl to the root medium and plants cultivated at salt concentrations which strongly stimulate growth. This decrease in WR probably reflects the requirement of halophytes for a certain amount of NaCl for optimum growth and is only of limited ecological significance, since growth in media with only traces of NaCl represents a rather artificial situation for halophytes.

Low values of WR at high NaCl levels in the root medium can be explained by a more pronounced reduction of transpiration relative to the decrease in dry matter production. However, it should be noted that large amounts of Na^+ and Cl^- in the shoots may substantially contribute to the dry weight of the plants, leading to much lower values of WR at high salinity levels than if only the organic matter component of the dry weight is considered. Recent studies on the water requirements of the halophytes *Arthrocnemum macrostachyum* and *Suaeda monoica*, performed in the Negev desert of Israel, show no significant changes in WR between plants irrigated with solutions containing 10, 100 and 400 mM NaCl (Winter, unpublished results).

RATES OF PHOTOSYNTHESIS

General aspects

The flow of CO_2 from the ambient air into the leaf and then to the site of CO_2 fixation is schematically described in Fig. 18.2. Various resistances to the flow

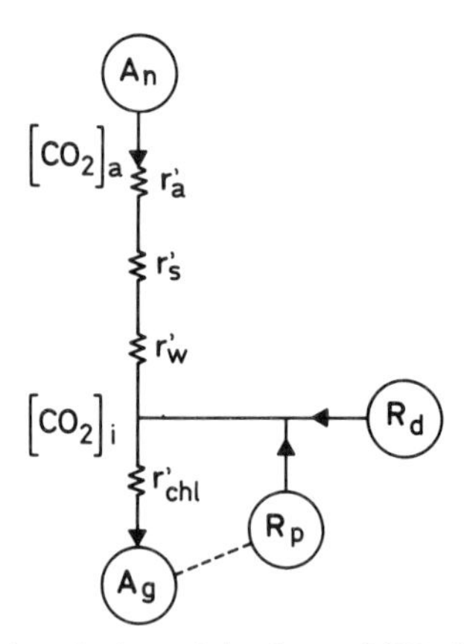

FIG. 18.2. Schematical description of the flow of CO_2 from the bulk air into the leaf (after Hall & Björkman 1975, slightly modified). Explanations see text.

of CO_2 are considered, as well as the CO_2 evolution by mitochondrial respiration (R_d) and photorespiration (R_p). A_n is the rate of net photosynthesis and A_g the rate of gross photosynthesis. $[CO_2]_a$ and $[CO_2]_i$ are the concentrations of CO_2 in the ambient air and in the protoplasts of the mesophyll cells. The resistances r'_a and r'_s are used in the same way as the corresponding terms r_a and r_s in Equation 3. The term r'_w represents the resistance to the flow of CO_2 in the liquid phase from the mesophyll cell walls into the protoplasts of the mesophyll cells and r'_{chl} refers predominantly to biochemical and photochemical limitations to CO_2 fixation.

Field investigations

Early studies by Neuwohner (1938) and Beiler (1939) show no striking peculiarities in the carbon assimilation of halophytes from the North and Baltic Sea shores. Corresponding to transpirational rates under natural conditions, photosynthetic rates per unit of fresh weight in succulent halophytes are usually lower than in glycophytes whereas the photosynthetic rates of halophytes may equal those of glycophytes when calculated on an area basis.

Under controlled conditions

More recent data on carbon assimilation of halophytes refer mainly to experiments under *more or less* controlled conditions. Net photosynthesis of halophytes is not drastically affected over a wide range of NaCl levels in the root medium (Fig. 18.3). In most species, net photosynthesis per unit of fresh weight, dry weight and/or leaf area is gradually reduced with increasing NaCl salinity. In *Atriplex halimus*, there is a concomitant increase in the stomatal resistance r'_s (Gale & Poljakoff-Mayber 1970) which may be due to similar reasons as discussed above for the increase in r_s to water vapour.

Large increases in r'_w due to the presence of solutes in the liquid phase, e.g. the mesophyll cell walls, are questionable, since the diffusivity of CO_2 is not greatly reduced in NaCl solutions ($1{\cdot}72\ cm^2\ sec^{-1}\ 10^{-5}$ at 1,041 mM NaCl, 25°C and 1 atm) compared with pure water ($1{\cdot}92\ cm^2\ sec^{-1}\ 10^{-5}$) (Ratcliff & Holdcroft 1963). However, the actual concentration of CO_2 in the aqueous phase is unknown because it may also diffuse as H_2CO_3 or HCO_3^-.

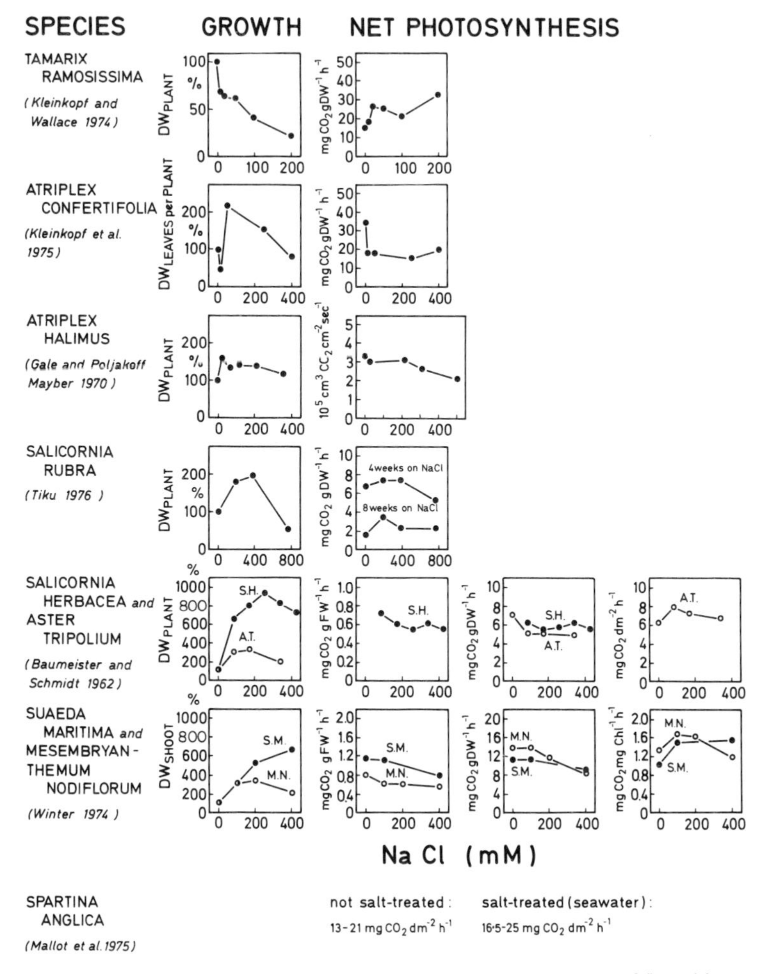

FIG. 18.3. Changes in growth (as a percent of the control = 100%) and in net photosynthesis of some halophytes (including the salt tolerant species *Tamarix ramosissima*) exposed to increasing levels of salinity in the root medium. DW = dry weight, FW = fresh weight.

Especially at high external NaCl concentrations there may be a significant increase in r'_{chl}. Swelling of the chloroplasts has been observed in *Atriplex halimus* when grown in culture solutions containing 264 and 456 mM NaCl (Blumenthal-Goldschmidt & Poljakoff-Mayber 1968). Cl^- inhibition of ribulose-1,5-diphosphate carboxylase and phosphoenolpyruvate carboxylase should also be considered (Osmond & Greenway 1972). An increased r'_{chl} may indirectly increase r'_s by raising the intercellular CO_2 concentration which would favour stomatal closure.

CO_2 evolution due to mitochondrial respiration (R_d) and photorespiration (R_p) may considerably diminish net photosynthesis. In glycophytes, increased R_d is assumed to be a major factor in the reduced carbon gain under conditions of salinity (Gale 1975). However, in the case of the leaves of halophytes, there is hardly evidence for an increased R_p in response to salinity (Flowers 1972) and there appears to be no report on the influence of NaCl on photorespiration.

The salt-tolerant species *Tamarix ramosissima* shows somewhat exceptional behaviour (Fig. 18.3). Growth is depressed with increasing salinity whereas the rate of photosynthetic carbon fixation remains relatively stable and even tends to increase under these conditions.

Growth and photosynthesis

Many effects of salinity on the CO_2 (and also water vapour) exchange of halophytes may be simply a consequence of changes in size and anatomy of the leaves. Thus, the growth stimulation of halophytes by NaCl is thought to be due to larger leaf areas as a consequence of increased cell expansion, and is thought not to be due to an increased photosynthetic efficiency (Greenway 1968; Kaplan & Gale 1972). The greater leaf area would lead to increased photosynthesis and growth.

One major problem in the interpretation of the correlation between growth and photosynthesis in halophytes is the selection of appropriate units. In particular, fresh and dry weights do not seem to be an appropriate basis since they partly reflect the water and salt content of the carbon assimilating organs; water and salt content vary considerably with the salt content of the growth medium. Interestingly, net photosynthesis in *Mesembryanthemum nodiflorum* and *Suaeda maritima* has been found to increase per unit of chlorophyll at NaCl concentrations which stimulate growth, whereas net photosynthesis decreases when calculated on a fresh weight or dry weight basis (Fig. 18.3). There is possibly a relationship to the findings of Gale and Poljakoff-Mayber (1970) with *Atriplex halimus*: the ability of the mesophyll to take up CO_2 appears to increase when the plants are grown in media containing NaCl up to 216 mM, since there is a decrease in the mesophyll resistance to CO_2 (corresponding mainly to r'_w and r'_{chl} in Fig. 18.2).

When the chlorophyll concentrations and rates of net photosynthesis are compared, similar results to those demonstrated for *Mesembryanthemum nodiflorum* and *Suaeda maritima* are indicated in the study of Tiku (1976) on *Salicornia rubra*. This suggests an increased photosynthetic efficiency per unit of chlorophyll and may point to a modification in the photochemical properties of the leaves (i.e. in r'_{chl}) in response to salinity, although it is unlikely that 'chlorophyll' limits the rates of photosynthesis at high light intensities. The fluorescence behaviour of halophytes grown at different salinity levels deserves study in relation to this problem.

Seagrasses

Photosynthesis of submerged marine angiosperms such as the seagrasses *Zostera marina*, *Zostera nana* and *Posidonia oceanica* is diminished when plants are transferred from their natural environment (full-strength sea water) to diluted sea water (Montfort & Brandrup 1927b; Gessner & Hammer 1960; Ogata & Matsui 1965). Photosynthesis may rapidly return to almost the initial value after re-transfer of plants from fresh water to full-strength sea water. Changes in carbon supply and pH as well as osmotic effects are considered as major factors in determining photosynthetic rates in these experiments (Hammer 1968).

PHOTOSYNTHETIC PATHWAYS

General aspects

Among higher plants, 3 types of photosynthetic carbon assimilation have been characterized: the C_3 pathway, the C_4 pathway and Crassulacean Acid Metabolism (CAM). The C_3 pathway is typical for the majority of species. Atmospheric CO_2 is fixed via ribulose-1,5-diphosphate carboxylase into 3-phosphoglycerate. Species with C_4 photosynthesis fix carbon via phosphoenolpyruvate carboxylase into C_4 dicarboxylic acids. The C_4 pathway is usually correlated with the presence of the Kranz-type leaf anatomy (Brown 1975) and occurs mainly in members of the Poaceae (Smith & Brown 1973) and Centrospermae (Chenopodiaceae, Amaranthaceae etc.) (Downton 1975). CAM is distributed among many succulent species, belonging, for instance, to the Cactaceae, Crassulaceae, Euphorbiaceae and Asclepiadaceae families (Black & Williams 1976). It is characterized by net CO_2 dark-fixation via phosphoenolpyruvate carboxylase in chloroplast containing tissues which results in the accumulation of significant quantities of free malic acid. The malic acid is metabolized to carbohydrate next day (Osmond 1976).

$\delta^{13}C$ values are frequently used to classify species into C_3, C_4 and CAM types. $\delta^{13}C$ values reflect the degree of discrimination against $^{13}CO_2$ during carbon assimilation and the respective ranges for C_3, C_4 and CAM plants are -22 to -38, -11 to -19 and -13 to -34‰ (Black & Bender 1976).

Several physiological characteristics associated with C_4 photosynthesis and CAM are considered to be advantageous under hot and dry conditions.

C_3 and C_4 pathway

The number of species in which C_4 photosynthesis has been detected continues to increase. The current status of the distribution of the C_4 pathway is indicated by the check-lists of Welkie and Caldwell (1970), Smith and Brown (1973), Osmond (1974), Carolin *et al.* (1975), Downton (1975) and Winter and Troughton (1978). In order to give a rough idea of the occurrence of the C_4 pathway among plants occupying saline habitats, some salt-tolerant C_4 species are listed in Table 18.3.

Apparently, most coastal halophytes are characterized by the C_3 pathway whereas many inland halophytes, particularly those of dry, saline habitats, are characterized by the C_4 pathway. This was shown, for instance, by a survey of internal salt concentrations and photosynthetic pathways of plants from Israel and the Sinai (Winter *et al.* 1976; Winter & Troughton 1978). However, more species from different regions need to be investigated to confirm this picture.

Many C_4 halophytes belong to the genera *Atriplex*, *Suaeda* and *Salsola* of the Chenopodiaceae family. Nearly all of the 56 Australian *Atriplex* species have been characterized as C_4 plants (Osmond 1974) and many of the Chenopodiaceae from the Irano-Turanian salt deserts of Asia (see Zohary 1973) seem to have C_4 photosynthesis. All members of the genus *Salicornia* so far examined are C_3 plants. Among the Poaceae, *Spartina angustifolia* and *Spartina townsendii* represent important C_4 grasses, which dominate many salt marshes. Interestingly, these grasses, as well as some Chenopodiaceae like *Salsola kali* and *Atriplex laciniata*, are capable of occupying rather cool temperate climatic regions (Northern America, Northern Europe) and therefore differ from the majority of C_4 plants which prefer subtropical and tropical regions (Long *et al.* 1975).

Balance between C_3 and C_4 pathway?

There are reports on a shift from the C_3 to the C_4 pathway and vice versa in certain species, based mainly on experiments indicating relative changes in the activities of ribulose-1,5-diphosphate carboxylase and phosphoenolpyruvate carboxylase as well as relative changes in the incorporation of ^{14}C into C_3 and C_4 compounds during ^{14}C fixation (review: Huber & Sankhla 1976). Shomer-Ilan and Waisel (1973) applied these criteria in their study on the highly salt-tolerant grass *Aeluropus littoralis* and stated that there was a change in

Table 18.3. *List of some species exhibiting characteristics of C_4 photosynthesis and occupying saline environments.*

Family	Species	References
Chenopodiaceae	Anabasis	
	aphylla L.	2
	setifera Moq.	2, 18
	Atriplex	
	confertifolia (Torr. Frem.) Wats.	17
	halimus L.	14, 15, 18
	nummularia Lindl.	6, 9, 14, 15, 16
	paludosa R. Br.	10
	polycarpa (Torr.) Wats.	11, 14, 17
	semibacatta Moq.	5, 6
	spongiosa Mueller	15, 16
	vesicaria (Benth.) Heward	2, 9, 14
	Girgensohnia	
	oppositiflora (Pall.) Fenzl	2
	Halogeton	
	glomeratus (Bieb.) Meyer	17
	Petrosimonia	
	triandra (Pall.) Simonkai	2
	Salsola	
	kali L.	2, 6, 7, 9, 17, 18
	soda L.	2, 18
	tetrandra Forssk.	18
	vermiculata L.	2
	Seidlitzia	
	rosmarinus Ehrenb.	2, 18
	Suaeda	
	altissima (L.) Pall.	2
	californica Wats.	17
	monoica Forssk.	2, 13, 18
	splendens Gen. et G.	2, 18
	torreyana Wats.	16, 17
Poaceae	Distichlis	
	spicata (L.) Greene	1, 14
	Monanthochloe	
	littoralis Engelm.	14
	Panicum	
	virgatum L.	1, 6
	Spartina	
	alternifolia Lois.	14
	anglica Hubbard	8
	townsendii	7
Portulacaceae	Portulaca	
	oleracea L.	1, 4, 6, 12, 16, 17
Zygophyllaceae	Zygophyllum	
	simplex L.	3, 6

References: (1) Bender 1971, (2) Carolin *et al.* 1975, (3) Crookston & Moss 1972, (4) Hofstra *et al.* 1972, (5) Johnson & Hatch 1968, (6) Krenzer *et al.* 1975, (7) Long *et al.* 1975, (8) Mallott *et al.* 1975, (9) Osmond 1970, (10) Osmond 1974, (11) Philpott & Troughton 1974, (12) Sankhla *et al.* 1975, (13) Shomer-Ilan *et al.* 1975, (14) Smith & Epstein 1971, (15) Tregunna *et al.* 1970, (16) Troughton *et al.* 1974, (17) Welkie & Caldwell 1970, (18) Winter & Troughton 1978.

photosynthesis from the C_3 to the C_4 pathway following irrigation of the plants with 100 mM NaCl. Completely contradictory data were obtained by Downton and Törökfalvy (1975). Up till now, there is no convincing report showing that a given species of higher plant can attain net carbon gain alternatively by the C_3 and the C_4 pathway.

Photosynthetic carbon metabolism of seagrasses

None of the seagrasses so far investigated possesses the Kranz-type leaf anatomy known from C_4 land plants (Jagels 1973; Doohan & Newcomb 1976). Some physiological characteristics indicate the presence of a photorespiratory mechanism similar to that of terrestrial C_3 plants, such as the effect of oxygen on respiration in the light (Hough 1976; Downton *et al.* 1976), glycine and serine production (Burris *et al.* 1976), and phosphoglycollate phosphatase activity (Randall 1976). However, seagrasses have $\delta^{13}C$ values of about -7 to $-12‰$ which appear to place them among C_4 plants (Smith & Epstein 1971; Doohan & Newcomb 1976; Black & Bender 1976; Benedict & Scott 1976). The possible carbon source for photosynthesis in aquatic plants is HCO_3^- which has a $\delta^{13}C$ value 7 to 8‰ less negative than that of atmospheric CO_2 (Deuser & Degens 1967). Even after this correction is applied, the $\delta^{13}C$ value of seagrasses resembles that of C_4 land plants more closely than C_3 plants. Indeed, Benedict and Scott (1976) reported an apparent C_4 type carbon assimilation in the marine grass *Thalassia testudinum*, using ^{14}C-fixation experiments in addition to the $\delta^{13}C$ technique. On the other hand, in *Halophila spinulosa* and *Thalassia hemprichii* the pattern of early ^{14}C-labelled intermediates was typical of the C_3 pathway (Andrews & Abel 1977).

CAM

Many halophytes are succulent and this property may represent a means of diluting high salt concentrations within the cells. Furthermore, succulence of the carbon assimilating organs is a property of most CAM plants and is thought to be related to the capacity of the vacuoles to accumulate malic acid during the dark. Among halophytes however, CAM seems to be mainly restricted to halophytic members of the Aizoaceae as shown by examination of the malic acid content (measured at dawn and dusk) and $\delta^{13}C$ value of a large number of coastal and inland halophytes of Israel and the Sinai (Winter *et al.* 1976; Winter & Troughton 1978). More recently, von Willert (personal communication) surveyed numerous species of the Aizoaceae in South Africa. All the species which he studied showed CAM-like diurnal fluctuations in malic acid content under natural conditions and most of them were characterized by high Na^+ and Cl^- concentrations in the leaf sap.

Balance between C_3 pathway and CAM

Under controlled conditions

In the Aizoaceae, the degree of CAM (i.e. the degree of net CO_2 fixation and malate accumulation during the dark) is very dependent on plant water relations. Many Aizoaceae exhibit features of CAM only when exposed to water stress (including high salinity) and display C_3 photosynthesis under well-irrigated conditions (Winter & von Willert 1972; Winter 1973; Treichel & Bauer 1974; Treichel 1975; Winter, unpublished results). The shift between C_3 and CAM has been investigated most extensively in the halophyte *Mesembryanthemum crystallinum* (Winter 1975; review: Winter & Lüttge 1976).

Fig. 18.4 shows an experiment in which a 6-week-old plant of *Mesembryanthemum crystallinum* grown in culture solution without added NaCl (day 1) was gradually transferred to solutions with a final concentration of 400 mM NaCl (day 9). Net CO_2 exchange and transpirational water loss were measured over the whole experimental period. Net CO_2 exchange at day 1 was characterized by a constant net CO_2 uptake during the light and a constant net CO_2 production during the dark period. With increasing NaCl concentration of the culture solution (day 2 to day 9) the daily net CO_2 exchange pattern slowly changed to that of CAM plants. After 11 and 33 days treatment with 400 mM NaCl (day 19 and day 41) a pronounced depression of net CO_2 uptake during the light and net CO_2 uptake during the dark was observed. The daily course of transpirational water loss responded to increased salinity in the same way as CO_2 uptake, suggesting the development of an inverted stomatal rhythm characteristic of CAM. The assimilation/transpiration quotient (mg CO_2 day^{-1}/g H_2O day^{-1}) was 6·9 on day 1 and 16·4 on day 41, indicating a considerable increase in the efficiency of water use by the plant.

In *Mesembryanthemum crystallinum*, CAM develops not only in response to NaCl salinity, but also following growth in nutrient solutions of low temperature or low oxygen content which is known to reduce the water uptake of many plants. This result implies that salinity diminishes water supply to *Mesembryanthemum crystallinum* and thus represents a situation of 'physiological drought'. A verification of this suggestion requires a more detailed study of the water relations in this species.

In the field

A change from C_3 to CAM has also been observed in *Mesembryanthemum crystallinum* growing in its natural environment. *Mesembryanthemum crystallinum* is an annual species and found predominantly in regions with a mediterranean climate characterized by winter rain and warm, dry summers. We have studied the mode of carbon assimilation in specimens growing on a coastal cliff of the Mediterranean Sea near Caesarea, Israel, during the first six months of 1977. Germination took place during the rainy season in

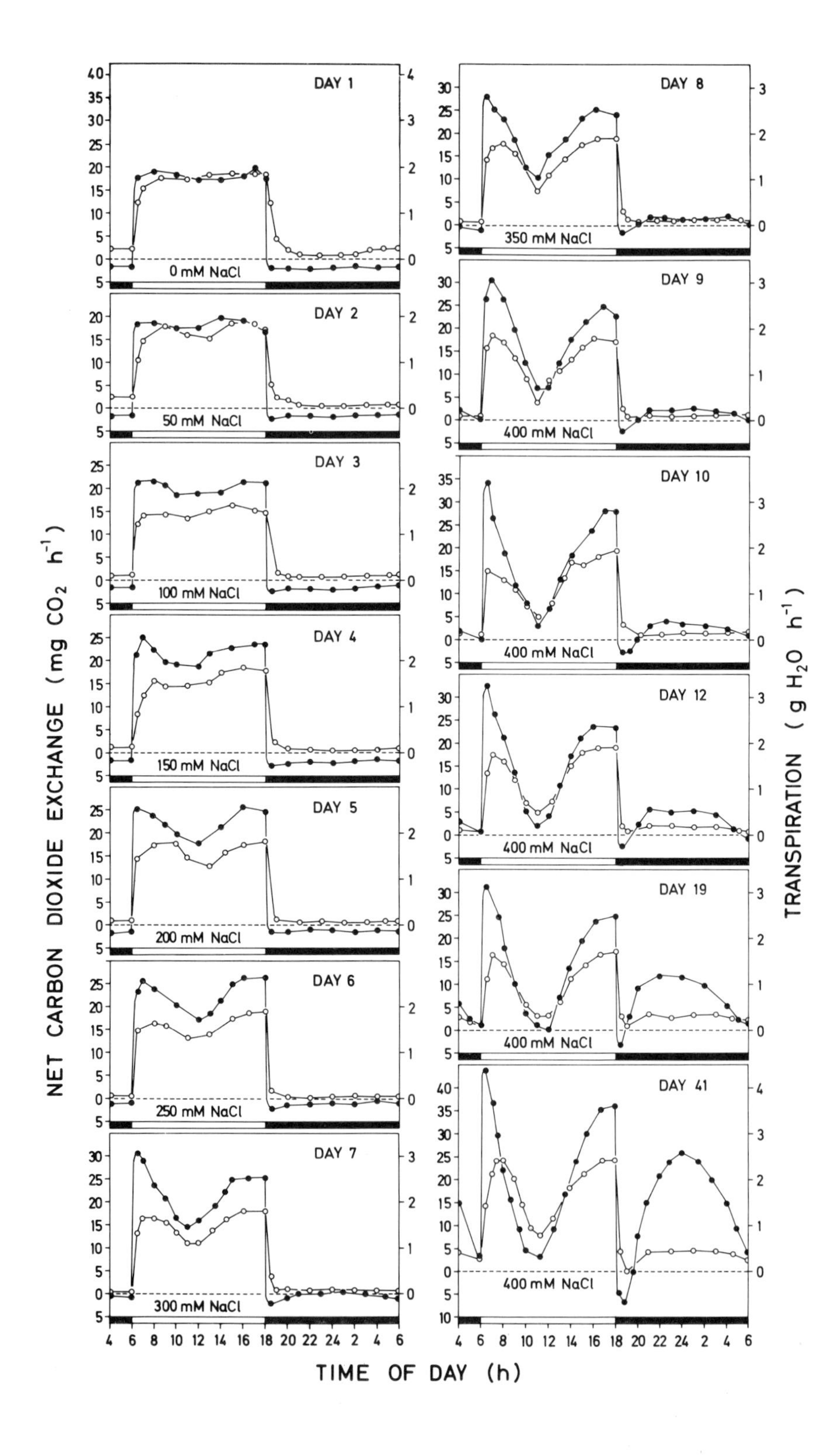

DAY 1
0 mM NaCl
DAY 2
50 mM NaCl
DAY 3
100 mM NaCl
DAY 4
150 mM NaCl
DAY 5
200 mM NaCl
DAY 6
250 mM NaCl
DAY 7
300 mM NaCl
DAY 8
350 mM NaCl
DAY 9
400 mM NaCl
DAY 10
400 mM NaCl
DAY 12
400 mM NaCl
DAY 19
400 mM NaCl
DAY 41
400 mM NaCl
NET CARBON DIOXIDE EXCHANGE (mg CO2 h-1)
TRANSPIRATION (g H2O h-1)
TIME OF DAY (h)

December but diurnal malate fluctuations in the leaves were not detected in January, February and March (Table 18.4). At the start of the dry season in April and May, pronounced diurnal changes in malate were observed. In the field, salinity *per se* does not seem to be the main factor controlling the balance between C_3 and CAM in *Mesembryanthemum crystallinum*. Rather, factors such as aridity and plant age are probably more important.

Table 18.4. *Changes in malate levels during the night (Δ malate) at various dates during 1977 in fully developed leaves of* Mesembryanthemum crystallinum *growing at the Mediterranean Sea shore near Caesarea, Israel.*

Date	Δ malate (μeq g fresh weight^{-1})
26/27 January	−3
24/25 February	−4
17/18 March	−1
26/27 April	+113
25/26 May	+113

CONCLUSIONS: FUTURE RESEARCH

There are many aspects of future research which appear to be necessary or at least interesting for a further understanding of the photosynthetic and water relationships of halophytes. These include:

(1) Simultaneous measurements of CO_2 exchange and transpiration under controlled experimental conditions, both *in situ* as well as in the laboratory, after long-term exposure of plants to different levels of NaCl in the root medium. Emphasis on a comparison between C_3, C_4 and CAM halophytes may be fruitful.
(2) Evaluation of possible changes in resistance to the flow of water in the roots and shoots, to the flow of water vapour from the leaf to the atmosphere and the flow of CO_2 from the atmosphere into the leaf following salinity

FIG. 18.4. Course of net CO_2 exchange (●) and transpiration (○) of a *Mesembryanthemum crystallinum* plant (shoot) in response to a gradual increase of the NaCl content in the culture solution up to 400 mM. ▭ light, ▬ dark. Concerning net CO_2 exchange, positive values refer to net CO_2 uptake, negative values represent net CO_2 loss from the shoot (from Winter 1975).

treatments. In particular, the effect of salinity on leaf anatomy and its consequences on plant gas exchange should be considered.
(3) The effect of salinity on photorespiration.
(4) The effect of salinity on the photochemical reactions (fluorescence responses, for example).
(5) Experimental analysis of the balance between kinetin and abscisic acid following growth of plants in saline media.
(6) Direct measurements of cell turgor in response to salinity.
(7) Studies on osmoregulation and photosynthesis in seagrasses.

ACKNOWLEDGMENTS

I am grateful to Prof. H. Greenway, Dr G.D. Farquhar, Prof. C.B. Osmond and Prof. U. Lüttge for critical comments on the manuscript. My own investigations reported in this paper were supported by grants from the Deutsche Forschungsgemeinschaft.

REFERENCES

Adriani M.J. (1937) Sur la transpiration de quelques halophytes cultivées dans des milieux différents en comparaison avec celle de quelques non-halophytes. *Proc. Kon. Ned. Akad. V. Wetensch.* **40**, 524–29.

Adriani M.J. (1956) Der Wasserhaushalt der Halophyten. *Handbuch der Pflanzenphysiologie III* (Ed. by W. Ruhland), pp. 902–14. Springer-Verlag, Berlin, Göttingen and Heidelberg.

Andrews T.J. & Abel K.M. (1977) Photosynthetic carbon metabolism in tropical seagrasses. *Abstracts, 4th International Congress on Photosynthesis*, 4–9 September 1977, Reading, U.K., 8.

Ashby W.C. & Beadle N.C.W. (1957) Studies in halophytes. III. Salinity factors in the growth of Australian saltbushes. *Ecology* **38**, 344–52.

Baumeister W. & Schmidt L. (1962) Über die Rolle des Natriums im pflanzlichen Stoffwechsel. *Flora* **152**, 24–56.

Bearce B.C. (1968) Adjustment of stomates to salinity. Ph.D. thesis. University of California, Davis.

Beiler A. (1939) Untersuchungen über die Kohlensäureassimilation der Strand- und Dünenpflanzen. *Jahrb. wiss. Bot.* **87**, 356–407.

Bender M.M. (1971) Variations in the $^{13}C/^{12}C$ ratios of plants in relation to the pathway of photosynthetic carbon dioxide fixation. *Phytochemistry* **10**, 1239–44.

Benedict C.R. & Scott J.R. (1976) Photosynthetic carbon metabolism of a marine grass. *Plant Physiol.* **57**, 876–80.

Bickenbach K. (1932) Zur Anatomie und Physiologie einiger Strand- und Dünenpflanzen. *Beitr. Biol. Pflanz.* **19**, 334–70.

Black C.C. & Bender M.M. (1976) $\delta^{13}C$ values in marine organisms from the Great Barrier Reef. *Aust. J. Plant Physiol.* **3**, 25–32.

Black C.C. & Williams S. (1976) Plants exhibiting characteristics common to Crassulacean Acid Metabolism. *CO_2 Metabolism and Plant Productivity* (Ed. by R.H. Burris & C.C. Black), pp. 407–24. University Park Press, Baltimore, London and Tokyo.

Blumenthal-Goldschmidt S. & Poljakoff-Mayber A. (1968) Effect of salinity on growth and submicroscopic structure of leaf cells of *Atriplex halimus* L. *Aust. J. Bot.* **16**, 469–78.

Brown W.V. (1975) Variations in anatomy, associations and origins of Kranz tissue. *Amer. J. Bot.* **62**, 395–402.

Burris J.E., Holm-Hansen O. & Black C.C. (1976) Glycine and serine production in marine plants as a measure of photorespiration. *Aust. J. Plant Physiol.* **3**, 87–92.

Carolin R.C., Jacobs S.W.L. & Vesk M. (1975) Leaf structure in Chenopodiaceae. *Bot. Jahrb. Syst.* **95**, 226–55.

Crookston R.K. & Moss D.N. (1972) C-4 and C-3 carboxylation characteristics in the genus *Zygophyllum* (Zygophyllaceae). *Ann. Missouri Bot. Garden* **59**, 465–70.

Deuser W.G. & Degens E.T. (1967) Carbon isotope fractionation in the system $CO_{2(gas)}$–$CO_{2(aqueous)}$–$HCO^-_{3(aqueous)}$. *Nature* **215**, 1033–35.

Doohan M.E. & Newcomb E.H. (1976) Leaf ultrastructure and $\delta^{13}C$ values of three seagrasses from the Great Barrier Reef. *Aust. J. Plant Physiol.* **3**, 9–23.

Downton W.J.S. (1975) The occurrence of C_4 photosynthesis among plants. *Photosynthetica* **9**, 96–105.

Downton W.J.S. & Törökfalvy E. (1975) Effect of sodium chloride on the photosynthesis of *Aeluropus litoralis*, a halophytic grass. *Z. Pflanzenphysiol.* **75**, 143–50.

Downton W.J.S., Bishop D.G., Larkum A.W.D. & Osmond C.B. (1976) Oxygen inhibition of photosynthetic oxygen evolution in marine plants. *Aust. J. Plant Physiol.* **3**, 73–79.

Eaton F.M. (1927) The water requirement and cell-sap concentration of Australian saltbush and wheat as related to the salinity of the soil. *Amer. J. Bot.* **14**, 212–26.

Esau K. (1967) Plant Anatomy. John Wiley & Sons, New York, London and Sydney.

Farquhar G.D. & Raschke K. (1978) On the resistance to transpiration of the sites of evaporation within the leaf. *Plant Physiol.* (in press.)

Flowers T.J. (1972) Salt tolerance in *Suaeda maritima* (L.) Dum. *J. exp. Bot.* **23**, 310–21.

Flowers T.J. (1975) Halophytes. *Ion Transport in Plant Cells and Tissues* (Ed. by D.A. Baker & J.L. Hall), pp. 309–34. North-Holland Publishing Company, Amsterdam and Oxford.

Flowers T.J., Troke P.F. & Yeo A.R. (1977) The mechanism of salt tolerance in halophytes. *Ann. Rev. Plant Physiol.* **28**, 89–121.

Gale J. (1975) Water balance and gas exchange of plants under saline conditions. *Plants in Saline Environments* (Ed. by A. Poljakoff-Mayber & J. Gale), pp. 168–85. Springer-Verlag, Berlin, Heidelberg and New York.

Gale J., Poljakoff-Mayber A. & Kahane I. (1967) The gas diffusion porometer technique and its application to the measurement of leaf mesophyll resistance. *Isr. J. Bot.* **16**, 187–204.

Gale J. & Poljakoff-Mayber A. (1970) Interrelations between growth and photosynthesis of salt bush (*Atriplex halimus* L.) grown in saline media. *Aust. J. biol. Sci.* **23**, 937–45.

Gessner F. & Hammer L. (1960) Die Photosynthese von Meerespflanzen in ihrer Beziehung zum Salzgehalt. *Planta* **55**, 306–12.

Greenway H. (1968) Growth stimulation by high chloride concentrations in halophytes. *Isr. J. Bot.* **17**, 169–77.

Hall A.E. & Björkman O. (1975) Model of leaf photosynthesis and respiration. *Perspectives of Biophysical Ecology* (Ed. by D.M. Gates & R.B. Schmerl), pp. 55–72. Springer-Verlag, Berlin, Heidelberg and New York.

Hammer L. (1968) Salzgehalt und Photosynthese bei marinen Pflanzen. *Marine Biol.* **1**, 185–90.

Hofstra J.J., Aksornkoae S., Atmowidjojo S., Banaag J.F. Santosa, Sastrohoetomo R.A. & Thu L.T.N. (1972) A study on the occurrence of plants with a low CO_2 compensation point in different habitats in the tropics. *Ann. Bogor.* **5**, 143–57.

Hough R.A. (1976) Light and dark respiration and release of organic carbon in marine macrophytes of the Great Barrier Reef region. *Aust. J. Plant Physiol.* **3**, 63–68.

Huber W. & Sankhla N. (1976) C_4 pathway and regulation of the balance between C_4 and C_3 metabolism. *Water and Plant Life—Problems and Modern Approaches* (Ed. by O.L. Lange, L. Kappen & E.-D. Schulze), pp. 335–63. Springer-Verlag, Berlin Heidelberg and New York.

Hsiao T.C. (1973) Plant responses to water stress. *Ann. Rev. Plant Physiol.* **24**, 519–70.

Itai C. & Benzioni A. (1976) Water stress and hormonal response. *Water and Plant Life—Problems and Modern Approaches* (Ed. by O.L. Lange, L. Kappen & E.-D. Schulze), pp. 225–42. Springer-Verlag, Berlin, Heidelberg and New York.

Jagels R. (1973) Studies of a marine grass, *Thalassia testudinum.* I. Ultrastructure of the osmoregulatory leaf cells. *Amer. J. Bot.* **60**, 1003–9.

Jarvis P.G. & Slatyer R.O. (1970) The role of the mesophyll cell wall in leaf transpiration. *Planta (Berl.)* **90**, 303–22.

Johnson H.S. & Hatch M.D. (1968) Distribution of the C_4-dicarboxylic acid pathway of photosynthesis and its occurrence in dicotyledonous plants. *Phytochemistry* **7**, 375–80.

Kaplan A. & Gale J. (1972) Effect of sodium chloride salinity on the water balance of *Atriplex halimus. Aust. J. biol. Sci.* **25**, 895–903.

Keller B. (1925) Halophyten- und Xerophyten-Studien. *J. Ecol.* **13**, 224–61.

Kleinkopf G.E. & Wallace A. (1974) Physiological basis for salt tolerance in *Tamarix ramosissima. Plant Sci. Lett.* **3**, 157–63.

Kleinkopf G.E., Wallace A. & Cha J.W. (1975) Sodium relations in desert plants: 4. Some physiological responses of *Atriplex confertifolia* to different levels of sodium chloride. *Soil Sci.* **120**, 45–48.

Krenzer E.G., Moss D.N. & Crookston R.K. (1975) Carbon dioxide compensation points of flowering plants. *Plant Physiol.* **56**, 194–206.

Long S.P., Incoll L.D. & Woolhouse H.W. (1975) C_4 photosynthesis in plants from cool temperate regions with particular reference to *Spartina townsendii. Nature* **257**, 622–24.

Mallott P.G., Davy A.J., Jefferies R.L. & Hutton M.J. (1975) Carbon dioxide exchange in leaves of *Spartina anglica* Hubbard. *Oecologia (Berl.)* **20**, 351–58.

McMillan C. (1974) Salt tolerance of mangroves and submerged aquatic plants. *Ecology of Halophytes* (Ed. by R.J. Reimold & W.H. Queen), pp. 379–90. Academic Press, New York and London.

McMillan C. & Moseley F.N. (1967) Salinity tolerances of five marine spermatophytes of Redfish Bay, Texas. *Ecology* **48**, 503–6.

Montfort C. & Brandrup W. (1927a) Physiologische und pflanzengeographische Seesalzwirkungen. II. Ökologische Studien über Keimung und erste Entwicklung bei Halophyten. *Jahrb. Wiss. Bot.* **56**, 902–46.

Montfort C. & Brandrup W. (1927b) Physiologische und Seesalzwirkungen. III. Vergleichende Untersuchungen der Salzwachstumsreaktionen von Wurzeln. *Jahrb. wiss. Bot.* **57**, 105–73.

Neuwohner W. (1938) Der tägliche Verlauf von Assimilation und Atmung bei einigen Halophyten. *Planta (Berl.)* **28**, 644–79.

Nobel P.S. (1974) *Introduction to biophysical plant physiology.* W.H. Freeman and Company, San Francisco.

Oertli J.J. (1968) Extracellular salt accumulation, a possible mechanism of salt injury in plants. *Agrochimica* **12**, 461–69.

Oertli J.J. (1976) The soil-plant-atmosphere continuum. *Water and Plant Life—Problems and Modern Approaches* (Ed. by O.L. Lange, L. Kappen & E.-D. Schulze), pp. 32–41. Springer-Verlag, Berlin, Heidelberg and New York.

Önal M. (1971) Der Einfluss steigender Natriumchlorid-Konzentrationen auf den Transpirationskoeffizient einiger Halophyten. *Rev. Fac. Sci. Univ. Ist.* Serie B36, 1–8.

Ogata E. & Matsui T. (1965) Photosynthesis in several marine plants of Japan as affected by salinity, drying and pH, with attention to their growth habitats. *Botanica mar.* **8**, 199–217.

O'Leary J.W. (1969) The effect of salinity on permeability of roots to water. *Isr. J. Bot.* **18**, 1–19.

Osmond C.B. (1970) C_4 photosynthesis in the Chenopodiaceae. *Z. Pflanzenphysiol.* **62**, 129–32.

Osmond C.B. (1974) Leaf anatomy of Australian saltbushes in relation to photosynthetic pathways. *Aust. J. Bot.* **22**, 39–44.

Osmond C.B. (1976) CO_2 assimilation and dissimilation in the light and dark in CAM plants. *CO_2 Metabolism and Plant Productivity* (Ed. by R.H. Burris & C.C. Black), pp. 217–33. University Park Press, Baltimore, London and Tokyo.

Osmond C.B. & Greenway H. (1972) Salt responses of carboxylation enzymes from species differing in salt tolerance. *Plant Physiol.* **49**, 260–63.

Philpott J. & Troughton J.H. (1974) Photosynthetic mechanisms and leaf anatomy of hot desert plants. *Carnegie Inst. Wash. Yearb.* **73**, 790–93.

Randall D.D. (1976) Phosphoglycollate phosphatase in marine algae: isolation and characterization from *Halimeda cylindracea. Aust. J. Plant Physiol.* **3**, 105–11.

Ratcliff G.A. & Holdcroft J.G. (1963) Diffusivities of gases in aqueous electrolyte solutions. *Trans. Instn Chem. Engrs.* **41**, 315–19.

Sankhla N., Ziegler H., Vyas O.P., Stichler W. & Trimborn P. (1975) Eco-physiological studies on Indian aride zone plants. V. A screening of some species for the C_4-pathway of photosynthetic CO_2-fixation. *Oecologia (Berl.)* **21**, 123–29.

Schimper A.F.W. (1898) Pflanzengeographie auf physiologischer Grundlage. Jena.

Scholander P.F., Hammel H.T., Bradstreet E.D. & Hemmingsen E.A. (1965) Sap pressure in vascular plants. *Science* **148**, 339–46.

Schratz E. (1935) Beiträge zur Biologie der Halophyten. II. Untersuchungen über den Wasserhaushalt. *Jahrb. wiss. Bot.* **81**, 59–93.

Shalhevet J., Maas E.V., Hoffman G.J. & Ogata G. (1976) Salinity and hydraulic conductance of roots. *Physiol. Plant.* **38**, 224–32.

Shomer-Ilan A. & Waisel Y. (1973) The effect of sodium chloride on the balance between the C_3- and C_4-carbon fixation pathways. *Physiol. Plant.* **29**, 190–93.

Shomer-Ilan A., Beer S. & Waisel Y. (1975) *Suaeda monoica*, a C_4 plant without typical bundle sheaths. *Plant Physiol.* **56**, 676–79.

Slatyer R.O. (1966) Some physical aspects of internal control of leaf transpiration. *Agric. Meteorol.* **3**, 281–92.

Smith B.N. & Epstein S. (1971) Two categories of $^{13}C/^{12}C$ ratios for higher plants. *Plant Physiol.* **47**, 380–84.

Smith B.N. & Brown W.V. (1973) The Kranz syndrome in the Gramineae as indicated by carbon isotopic ratios. *Amer. J. Bot.* **60**, 505–13.

Steudle E. & Zimmermann U. (1971) Hydraulische Leitfähigkeit von *Valonia utricularis. Z. Naturforsch.* **26b**, 1302–11.

Steudle E., Lüttge U. & Zimmermann U. (1975) Water relations of the epidermal bladder cells of the halophytic species *Mesembryanthemum crystallinum*: direct measurements of hydrostatic pressure and hydraulic conductivity. *Planta (Berl.)* **126**, 229–46.

Stocker O. (1925) Beiträge zum Halophytenproblem. II. Standort und Transpiration der Nordsee-Halophyten. *Z. Bot.* **17**, 1–24.

Tiku B.L. (1976) Effect of salinity on the photosynthesis of the halophyte *Salicornia rubra* and *Distichlis stricta. Physiol. Plant.* **37**, 23–28.

Tregunna E.B., Smith B.N., Berry J.A. & Downton W.J.S. (1970) Some methods for studying the photosynthetic taxonomy of the angiosperms. *Can. J. Bot.* **48**, 1209–14.

Treichel S. (1975) Crassulaceen-Säurestoffwechsel bei einem salztoleranten Vertreter der Aizoaceae: *Aptenia cordifolia. Plant Sci. Lett.* **4**, 141–44.

Treichel S. & Bauer P. (1974) Unterschiedliche NaCl-Abhängigkeit des tagesperiodischen CO_2-Gaswechsels bei einigen halisch wachsenden Küstenpflanzen. *Oecologia* (*Berl.*) **17**, 87–95.

Troughton J.H., Card K.A. & Hendy C.H. (1974) Photosynthetic pathways and carbon isotope discrimination by plants. *Carnegie Inst. Wash. Yearb.* **73**, 768–80.

Waisel Y. (1972) Biology of Halophytes. Academic Press, New York and London.

Webb K.L. (1966) NaCl effects on growth and transpiration in *Salicornia bigelovii* a salt-marsh halophyte. *Plant and Soil* **24**, 261–68.

Welkie G.W. & Caldwell M. (1970) Leaf anatomy of species in some dicotyledon families as related to the C_3 and C_4 pathways of carbon fixation. *Can. J. Bot.* **48**, 2135–46.

Winter K. (1973) NaCl-induzierter Crassulaceensäurestoffwechsel bei einer weiteren Aizoacee: *Carpobrotus edulis. Planta* (*Berl.*) **115**, 187–88.

Winter K. (1974) Wachstum und Photosyntheseleistung der Halophyten *Mesembryanthemum nodiflorum* L. und *Suaeda maritima* (L.) Dum. bei variierter NaCl-Salinität des Anzuchtmediums. *Oecologia* (*Berl.*) **17**, 317–24.

Winter K. (1975) Die Rolle des Crassulaceen-Säurestoffwechsels als biochemische Grundlage zur Anpassung von Halophyten an Standorte hoher Salinität. Dissertation, Technische Hochschule Darmstadt.

Winter K. & von Willert D.J. (1972) NaCl-induzierter Crassulaceensäurestoffwechsel bei *Mesembryanthemum crystallinum. Z. Pflanzenphysiol.* **67**, 166–70.

Winter K. & Lüttge U. (1976) Balance between C_3 and CAM pathway of photosynthesis. *Water and Plant Life—Problems and Modern Approaches* (Ed. by O.L. Lange, L. Kappen & E.-D. Schulze), pp. 322–34. Springer-Verlag, Berlin, Heidelberg and New York.

Winter K., Troughton J.H., Evenari M., Läuchli A. & Lüttge U. (1976) Mineral ion composition and occurrence of CAM-like diurnal malate fluctuations in plants of coastal and desert habitats of Israel and the Sinai. *Oecologia* (*Berl.*) **25**, 125–43.

Winter K. & Troughton J.H. (1978) Photosynthetic pathways in plants of coastal and inland habitats of Israel and the Sinai. *Flora*, **167**, 1–34.

Zohary M. (1973) Geobotanical foundations of the Middle East. Vol. I. Gustav Fischer Verlag, Stuttgart.

19. ECOPHYSIOLOGICAL INVESTIGATIONS OF PLANTS IN THE COASTAL DESERT OF SOUTHERN AFRICA. ION CONTENT AND CRASSULACEAN ACID METABOLISM

D.J. v. WILLERT, E. BRINCKMANN AND E.-D. SCHULZE

Lehrstuhl Pflanzenökologie,
Universität Bayreuth, Am Birkengut,
*D-*8580 *Bayreuth, B.R.D.*

SUMMARY

In their natural habitat plants of the Mesembryanthemaceae show a considerable accumulation of inorganic ions especially Na^+ and Cl^-. The accumulation seems to be independent of the concentration of these ions in the soil. All investigated plants of Mesembryanthemaceae can accumulate malate overnight (i.e. show a CAM under certain atmospheric conditions). Naturally occurring increases in temperature and decreases in humidity can greatly reduce CAM. From this it is considered that CAM in the Mesembryanthemaceae may be an adapted but not an adaptable metabolism.

RÉSUMÉ

Dans leur habitat naturel, les plantes de Mesembryanthemaceae montrent une accumulation considérable d'ions inorganiques, en particulier Na^+ et Cl^-. L'accumulation semble indépendante de la concentration de ces ions dans le sol. Toutes les plantes de Mesembryanthemaceae étudiées peuvent accumuler le malate, la nuit (i.e. montrent un CAM sous certaines conditions atmosphériques). Une augmentation naturelle de la température et une diminution de l'humidité peuvent réduire significativement le CAM. De ce fait on considère que le CAM chez Mesembryanthemaceae peut être un métabolisme adapté, mais non adaptable.

INTRODUCTION

Walter (1936) was one of the first who provided data on the ion contents of various plants of the Mesembryanthemaceae in their natural habitat of the

Abbreviations: CAM: Crassulacean acid metabolism. PEP: Phosphoenolpyruvate.

Namib desert. He assumed that the Mesembryanthemaceae might be more or less halophilic plants that accumulate NaCl to a considerable extent even when they are not exposed to high salinity. A more recent investigation done on greenhouse grown plants substantiates Walter's assumptions (v. Willert *et al.* 1977a). All of the 27 investigated members of the Mesembryanthemaceae, although grown under low salt conditions, exhibited high internal ion concentrations, especially the concentrations of sodium and chloride which also varied from species to species. The enormous enrichment of NaCl in plant cells seems to be characteristic for members of the Mesembryanthemaceae. As all plants have been grown under uniform conditions, the results imply that there might be a marked genetic component associated with the ability of the plants to accumulate sodium and chloride ions.

It is unknown why plants of the Mesembryanthemaceae accumulate such high amounts of NaCl. One possible explanation could be that NaCl might be involved in the establishment of a CAM, as has been reported for four members of the Mesembryanthemaceae (Winter & v. Willert 1972; Treichel & Bauer 1974; Treichel 1975). In these cases a prolonged treatment with NaCl was necessary to establish a CAM. As many species of the Mesembryanthemaceae show a CAM even when grown under low salt conditions (Cockburn 1974; v. Willert *et al.* 1977a) it seems doubtful whether NaCl really plays a key role in triggering a CAM. Although all plants investigated possess rather high NaCl concentrations, no clear cut correlation between any one ion and the extent of malate accumulation could be detected (v. Willert *et al.* 1977a).

It was the aim of this study to investigate the influence of climatic and soil conditions on the physiological behaviour, especially CAM and ion content, of the Mesembryanthemaceae in their natural habitat of the Southern Namib desert.

ENVIRONMENT AND METHODS

The investigations were done in the Knersvlakte and the northern part of the Richtersveld near Numees, South Africa. The Knersvlakte is a more or less flat area while the northern Richtersveld is a mountainous region with very steep slopes. Both regions belong to the 'succulent desert' where rainfall only occurs in winter. The experiments were done during March 1977. At this time the vegetation at both places was represented by members of the Mesembryanthemaceae, mostly perennial short shrubs, but also a few annual plants such as *Mesembryanthemum crystallinum* were still alive. The plant coverage was patchy as it depended on soil structure characteristics. It showed rapid changes within short distances.

Measurements were made of air and soil temperatures, relative humidity and evaporation as indicated in the legends of Figs. 19.1 and 19.2. Leaf

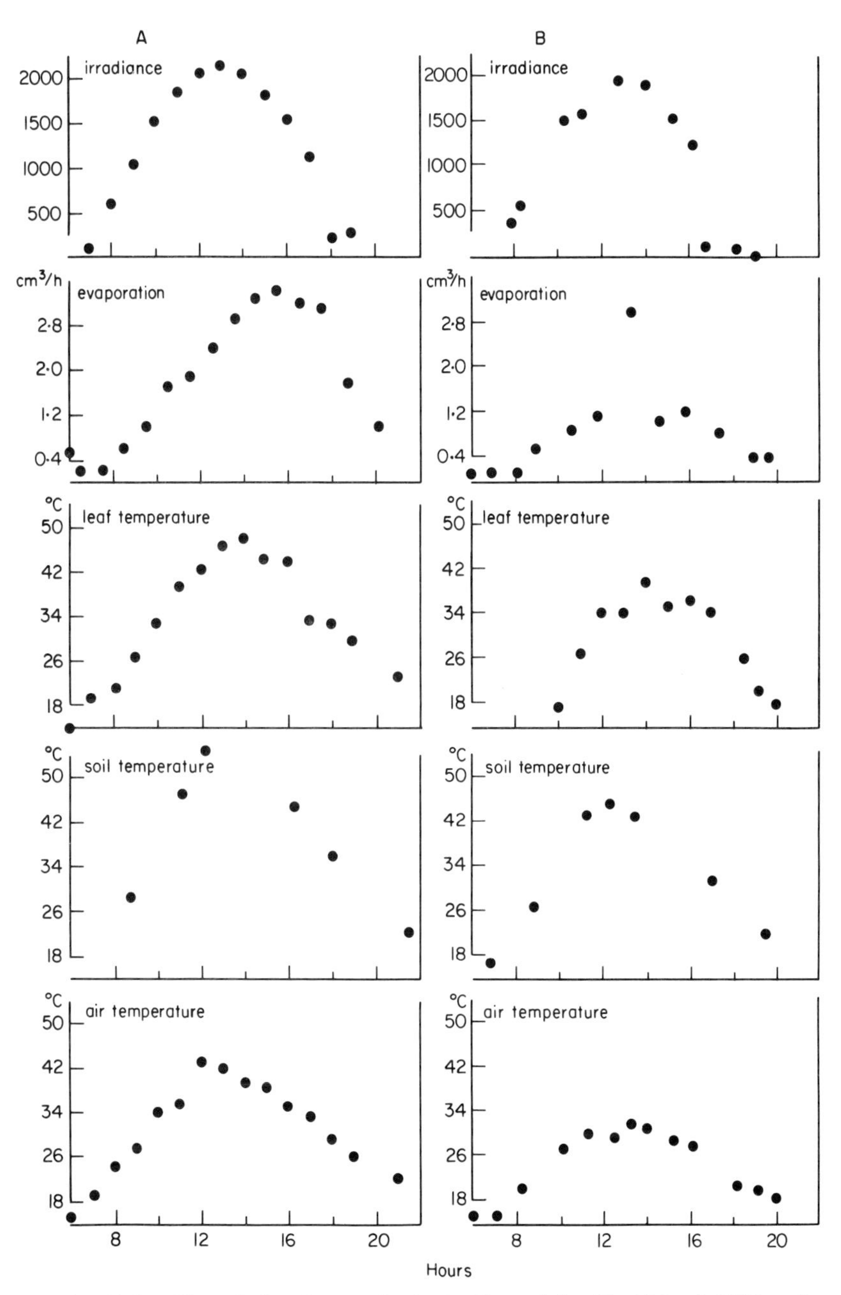

FIG. 19.1. Climatic data for one day in the Knersvlakte (A, 4 March 1977) and in the Richtersveld near Numees (B, 9 March 1977). Irradiance is given in μEinstein m^{-2} sec^{-1}. Evaporation of a disk of green filter-paper (3 cm ϕ) was measured 6 cm (A) or 20 cm (B) above ground. Air temperature was measured at the same positions as evaporation with the use of shaded thermometers. Soil temperature was measured 1–2 cm below the surface with thermocouples and a Wescor HR 33-T Microvoltmeter.

temperature was measured in the upper cell layers of the leaf with copper/constantan thermocouples (wire size 0·25 mm ϕ) which were connected to a Wescor HR 33-T Microvoltmeter. Photosynthetically active radiation was measured with a quantum sensor and a Quantum/Radiometer/Photometer (Lambda Inst. Corp., Model LI-185). Leaf samples were taken at sunrise and sunset each day. In all samples malate was determined enzymatically in the field according to the method described by Bergmeyer (1970). In the same samples Na^+, K^+, Ca^{++} and Cl^- were determined later in the laboratory, as has been described earlier (v. Willert *et al.* 1977a). Parallel sampling was done for determination of water content of leaves.

RESULTS

Meteorological data

Air, soil and leaf temperatures as well as irradiance, and evaporation on sample days in the Knersvlakte and in the Richtersveld are given in Fig. 19.1. The automatically recorded air temperature and the relative humidity during the day/night cycle in the Knersvlakte is shown in Fig. 19.2A. The increase in relative humidity during the night is the result of fog moving in from the coast. The same is true for the increase in relative humidity in the Richtersveld (Fig. 19.2B). During the third night in the Richtersveld a heat storm blowing from inland towards the sea prevented coastal fog, and caused a drastic decrease in relative humidity and an increase in air temperature. This heat storm lasted throughout the rest of our stay in the Richtersveld.

Ion content of the plants and the soil

Figs. 19.3 and 19.4 give the concentration of Na^+, K^+ and Cl^- in the cell sap of the plants. Since ion content depends markedly on leaf age, Na^+ increases while K^+ and Cl^- decrease with increasing age. The leaves with the highest Na^+ concentration have been used as a basis for comparison in Figs. 19.3 and 19.4.

The ion contents of the soils in the Knersvlakte and the Richtersveld vary considerably. While in the Knersvlakte all measured ions make up about 0·4% of the dry weight of the soil the percentage in the Richtersveld reaches only 0·04%. Nevertheless, no corresponding differences in the ion content of the plants of both areas were found (Table 19.1).

A further indication that leaf ion composition is independent of ion supply in the soil is the fact that plants of the genus *Conophytum* behave similarly in both habitats.

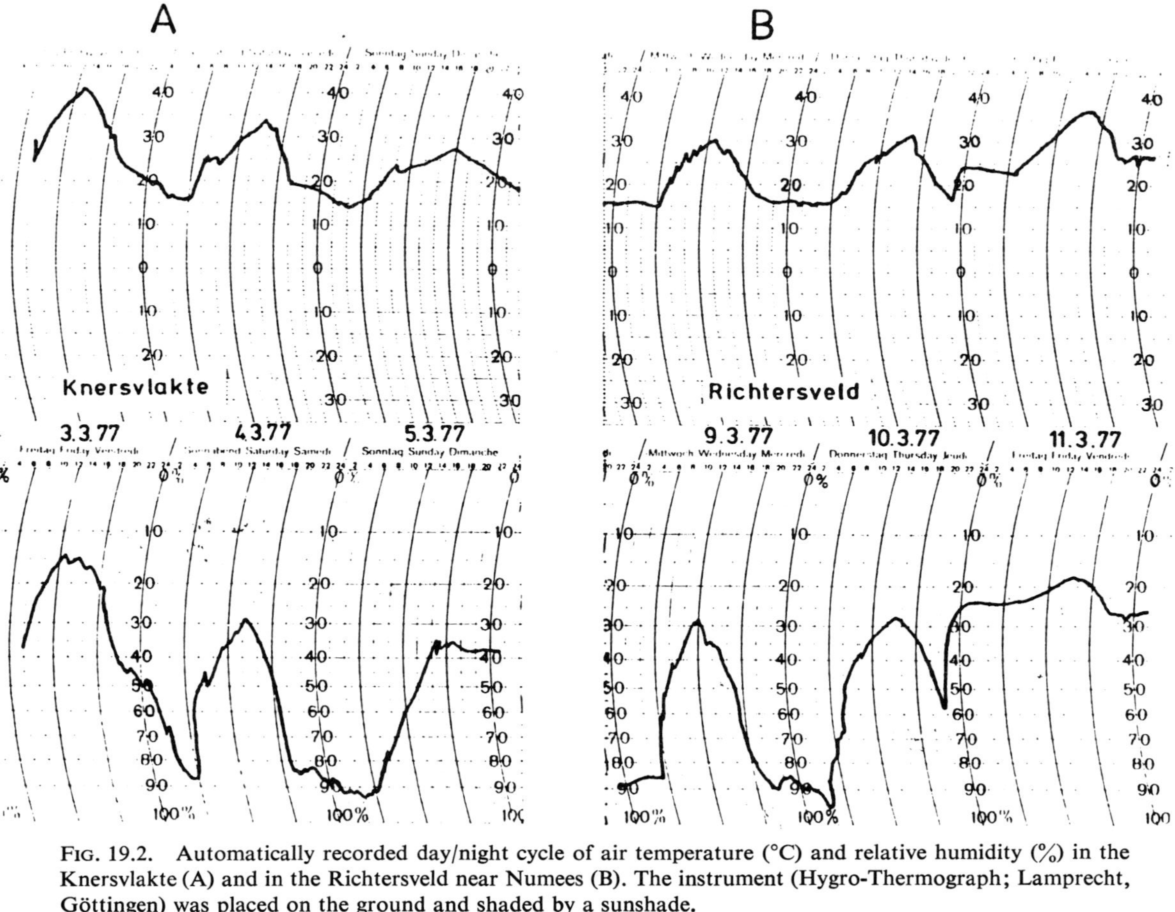

FIG. 19.2. Automatically recorded day/night cycle of air temperature (°C) and relative humidity (%) in the Knersvlakte (A) and in the Richtersveld near Numees (B). The instrument (Hygro-Thermograph; Lamprecht, Göttingen) was placed on the ground and shaded by a sunshade.

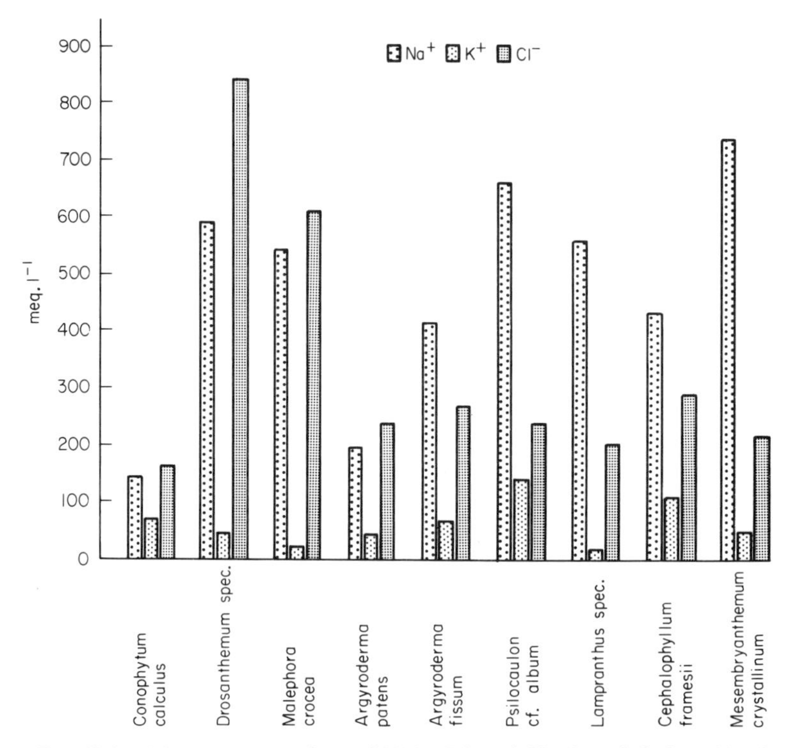

FIG. 19.3. Mean concentrations of Na^+, K^+ and Cl^- (meq l^{-1}) found in the cell sap expressed from leaves of nine members of the Mesembryanthemaceae growing in the Knersvlakte. Samples were collected during 3 to 6 March 1977.

Evidence of genetic control of ion accumulation, as was suggested earlier (v. Willert *et al.* 1977a), can be seen from Table 19.2. In this table data of the ionic composition of plants of *Argyroderma* grown in a greenhouse (data taken from v. Willert *et al.* 1977a) are compared with corresponding results of the ion composition of the same or of a closely related species in their natural habitat. Although the ion composition in the soils is very different the ionic concentrations and the Na:Cl ratios in the leaves of the different plants are nearly the same.

Malate accumulation

A main feature of members of the Mesembryanthemaceae is their capacity to fix CO_2 in the night under certain conditions. It was the aim of this study to look for the occurrence of an overnight malate accumulation in the plants in their natural habitat. At the two sites we did not find a member of the Mesembryanthemaceae which was unable to accumulate malate at night,

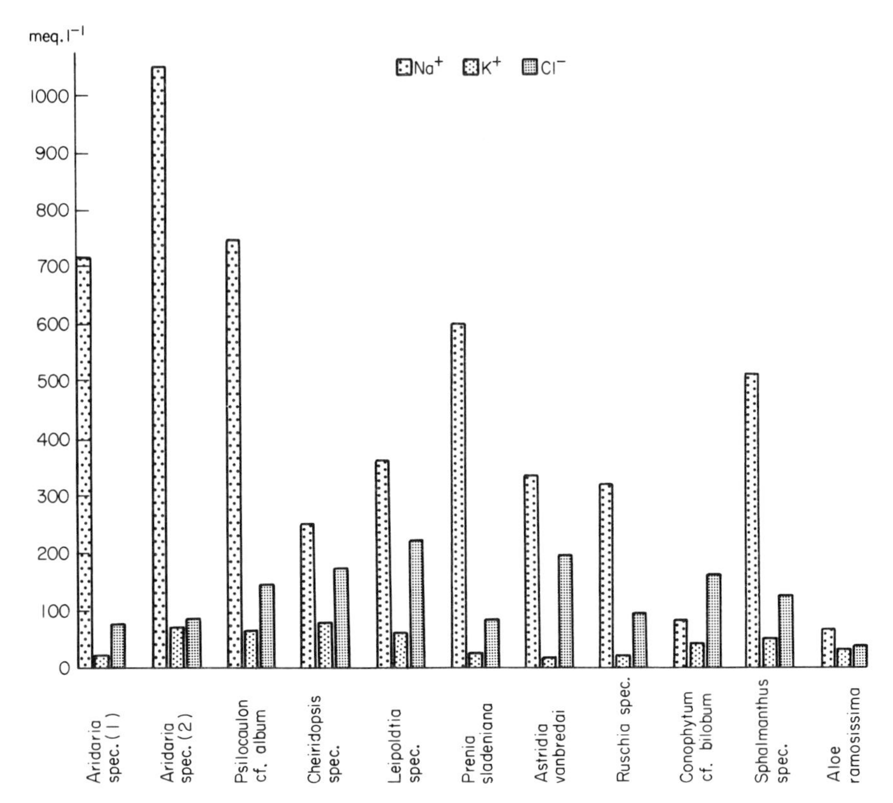

FIG. 19.4. Mean concentrations of Na^+, K^+ and Cl^- (meq l^{-1}) found in the cell sap expressed from leaves of 10 members of the Mesembryanthemaceae and *Aloe ramosissima* in the Richtersveld near Numees. Samples were collected during 8 to 12 March 1977.

although the total amount accumulated varied considerably from species to species. The highest concentration of malate in the morning was found in leaves of members of the genus *Aridaria* (188 mM), and the lowest in leaves of *Malephora crocea* in which the concentration was only 8·7 mM. The reasons for the observed variability have been discussed in detail in a previous paper (Willert *et al.* 1977a). Here we focus on a surprising fact; with the appearance of the heat storm which started at 22.30 hr of 10 March, malate formation in nearly all investigated plants fell and was completely arrested during the subsequent hot night. Only two species (*Prenia sladeniana* and *Aridaria sp.*), were able to accumulate malate under these changed conditions. However, even in these plants the accumulation was greatly reduced in comparison to that shown on a cooler night. Fig. 19.5 gives an example of what happened between 9 and 12 March in three species differing in their amount of overnight

Table 19.1. *Soil and leaf tissue ion content in the Knersvlakte and Richtersveld regions of South Africa. The data are given as mean values of all plants investigated (9 in the Knersvlakte, 11 in the Richtersveld, see Figs. 19.3 and 19.4) and the soil at their rooting area. The given errors are standard deviations (S.D.).*

Ion content of the soil		
Ionic species	Site	
	Knersvlakte mg g dwt.$^{-1}$	Richtersveld mg g dwt.$^{-1}$
Na^+	0·900 ± 0·704	0·155 ± 0·139
K^+	0·229 ± 0·612	0·035 ± 0·025
Ca^{++}	0·228 ± 0·023	0·058 ± 0·034
Cl^-	2·090 ± 1·890	0·096 ± 0·085
Ion content of the plants		
Ionic species	Site	
	Knersvlakte mg g dwt.$^{-1}$	Richtersveld mg g dwt.$^{-1}$
Na^+	79·9	98·7
K^+	19·1	14·9
Cl^-	86·0	41·9

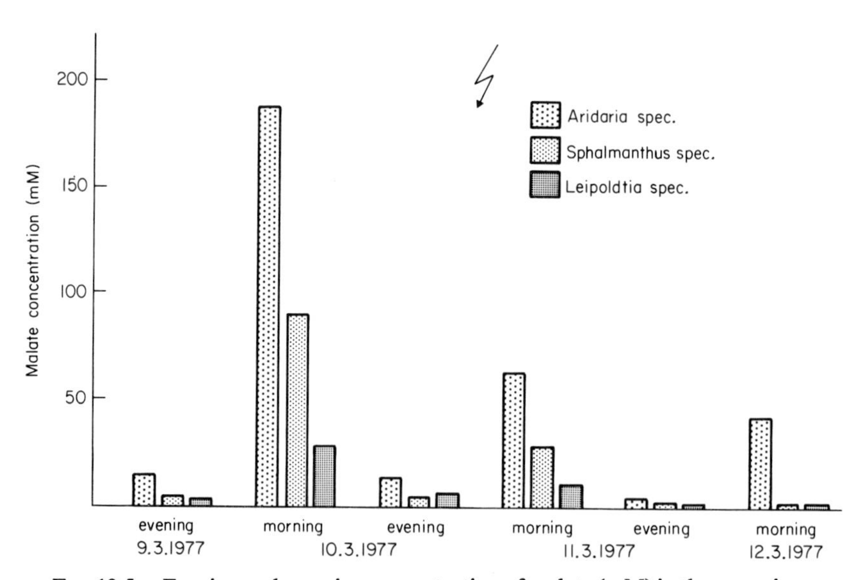

FIG. 19.5. Evening and morning concentration of malate (mM) in three species of Mesembryanthemaceae before and during a heat storm in the Richtersveld. The arrow indicates when the storm arose. For details of climatic conditions see Fig. 19.2.

malate accumulation. From the 25 species investigated in the Richtersveld 25 exhibited a CAM but three days later only two species showed features characteristic of CAM.

DISCUSSION

Firstly it seems that in all members of the Mesembryanthemaceae studied there is a marked accumulation of Na^+ and Cl^- in comparison to K^+. Yet, a calculation of the enrichment factors of Na^+ and K^+ shows that some plants discriminate against K^+ while others do not. It is surprising that regardless of soil ion content the same species accumulate ions approximately to the same concentrations (See *Conophytum* and *Psilocaulon*, Figs. 19.3 and 19.4). If ion uptake is under genetic rather than environmental control, and the results obtained with *Argyroderma* species (Table 19.2) tend to confirm this, the level of ion accumulation as well as the ionic composition might be a useful tool for taxonomic studies in this family.

One might argue that the high salt concentration and associated lowered water potential in the plants would facilitate water uptake from the soil. However, it remains obscure why the associated flora does not show as high a salt concentration as the Mesembryanthemaceae. One example is given in Fig. 19.4 for *Aloe ramosissima.* No connection between endogenous salt levels and malate accumulation exists, as *Aloe* accumulates malate during the night to a similar extent as that found in members of the Mesembry-anthemaceae. At the present stage of our knowledge it remains unknown why the Mesembryanthemaceae accumulate Na^+ and Cl^- to an extent which is toxic in glycophytes and some halophytes.

Table 19.2. *Comparison of the contents of* Na^+, K^+, Cl^- *and the ratio Na: Cl in the cell sap of the leaves for two* Argyroderma *species, either grown in the greenhouse in Hamburg or the Knersvlakte (South Africa), their natural habitat. The amounts of* Na^+, K^+, *and* Cl^- *in the greenhouse soil were about 0·03 mg g dry soil*$^{-1}$ *for each element.*

Plant species	Na^+ meq l^{-1}	K^+ meq l^{-1}	Cl^- meq l^{-1}	Na:Cl
Argyroderma delaetii (greenhouse cultivated)	207	22·4	355	0·58
Argyroderma patens (Knersvlakte)	214	30·8	288	0·76
Argyroderma fissum (greenhouse cultivated)	416	65·0	196	1·89
Argyroderma fissum (Knersvlakte)	414	68·5	265	1·56

It has been shown that watering may convert a CAM plant to daytime CO_2 uptake (v. Willert *et al.* 1976; Hartsock & Nobel 1976). In salt stressed *Mesembryanthemum crystallinum* this conversion is accompanied by a decrease in the activity of the PEP-carboxylase (v. Willert *et al.* 1976). An increase in PEP-carboxylase activity has been reported to occur during NaCl promoted establishment of a CAM in various Mesembryanthemaceae (Treichel *et al.* 1974; Treichel 1975; v. Willert *et al.* 1977b). We do not know what happened to the PEP-carboxylase of the plants in the Richtersveld when the heat storm arose. But it appears that here a severe increase of water stress switched off an existing CAM, which seems to be in contradiction to what has been reported. A possible explanation which might lead out of the dilemma and which is under further investigation is that CAM may be an adapted but not an adaptable metabolism.

ACKNOWLEDGMENTS

We gratefully acknowledge the support of Prof. Dr H. Walter, who made this research possible by dedicating the Schimper Stipendium to this project. The work was also supported by the DFG and the generous help of South African Airways. We thank also B. Scheitler, P. Ernst, M. Wartinger, M. Küppers and N. Lange for their help in taking samples and measurements, Dr H. Hartmann's guidance in the desert, and Dr D.A. Thomas for critical advice in preparing the manuscript.

REFERENCES

Bergmeyer H.U. (Ed.) (1970) *Methoden der enzymatischen Analyse.* Verlag Chemie, Weinheim.

Cockburn W. (1974) Crassulacean acid metabolism in *Lithops insularis*, a non-halophytic member of the Mesembryanthemaceae. *Planta* (Berl.) **118**, 89–90.

Hartsock T.L. & Nobel P.S. (1976) Watering converts a CAM plant to daytime CO_2 uptake. *Nature* **262**, 574–76.

Treichel S. (1975) Crassulaceen Säurestoffwechsel bei einem salztoleranten Vertreter der Aizoaceae: *Aptenia cordifolia. Plant Sci. Lett.* **4**, 141–44.

Treichel S. & Bauer P. (1974) Unterschiedliche NaCl-Abhängigkeit des tagesperiodischen CO_2-Gaswechsels bei einigen halisch wachsenden Küstenpflanzen. *Oecologia (Berl.)* **17**, 87–95.

Treichel S.P. et al. (1974) Veränderung der Aktivität der Phosphoenolpyruvat-Carboxylase durch NaCl bei Halophyten verschiedener Biotope. *Z. Pflanzenphysiol.* **71**, 437–49.

Walter H. (1936) Die ökologischen Verhältnisse in der Namib-Nebelwüste (Südwestafrika) unter Auswertung der Aufzeichnungen des Dr G. Boss (Swakopmund). *Jb. wiss. Bot.* **84**, 58–221.

von Willert D.J. et al. (1976) Environmentally controlled changes of phosphoenolpyruvate carboxylases in *Mesembryanthemum. Phytochem.* **15**, 1435–36.

von Willert D.J. et al. (1977a) Ecophysiologic investigations in the family of the Mesembryanthemaceae. Occurrence of a CAM and ion content. *Oecologia* (*Berl.*) **29**, 67–76.

von Willert D.J. et al. (1977b) Veränderungen der PEP-Carboxylase während einer durch NaCl geförderten Ausbildung eines CAM bei *Mesembryanthemum crystallinum. Biochem. Physiol. Pflanzen* **171**, 101–7.

Winter K. & von Willert D.J. (1972) NaCl-induzierter Crassulaceen-Säurestoffwechsel bei *Mesembryanthemum crystallinum. Z. Pflanzenphysiol.* **67**, 166–70.

20. PRIMARY PRODUCTION IN *SPARTINA* MARSHES

S.P. LONG[1] AND H.W. WOOLHOUSE[2]

[1] *Department of Biology, University of Essex, Colchester, Essex, U.K.*
and
[2] *Department of Plant Sciences, University of Leeds, Leeds, U.K.*

SUMMARY

Salt marshes dominated by species of the genus *Spartina* are a feature of many hundreds of miles of coastline in the temperate zones of the Western Hemisphere, and thus represent major coastal ecosystems. The world, and European, distributions of *Spartina* are considered. By comparison to many other natural ecosystems, reported values of both net and gross primary production in *Spartina* marshes are high. Net primary production in these marshes may exceed 40 tonnes of dry matter per hectare per year.

Coastal *Spartina* marshes have been the subject of many productivity studies. However, many of these studies fail to take account of below-ground biomass and biomass turnover. Thus, few conclusions can be reached on the relationships of production processes in these marshes to environmental variables. A characteristic of these marshes is that the *Spartina* species present usually occur only in monotypic stands. Thus, these marshes represent relatively simple systems for which the modelling of primary production processes is a realistic aim. The complex of factors influencing production processes in *Spartina* is considered in terms of a compartmental electrical analogue model. This model is used to critically examine the present state of knowledge of the relationship of production to environmental variables in these marshes.

The high productivity of these marshes may be associated with the occurrence of C_4 photosynthesis in the genus *Spartina*. Unlike the majority of C_4 grasses, *Spartina* is not confined to tropical and sub-tropical climatic regions, but has a distribution extending into cool temperate regions. Above leaf temperatures of 10° leaf photosynthetic rates in the European *S.townsendii* greatly exceed those of many temperate C_3 grasses. Unlike C_4 species which have previously been examined, leaf photosynthetic rates are not inhibited by leaf temperatures of 10°C and below, but are equivalent to those of temperate C_3 grasses. Laboratory and field studies of the responses of leaf photosynthetic rate in *S.townsendii* to climatic variables are discussed.

RÉSUMÉ

Les marais salés dominés par l'espèce du genre *Spartina* caractérisent des centaines de miles de côtes des zones temperées de l'hémisphère ouest, et ainsi représentent le principal écosystème côtier. Nous considérons les distributions mondiale et européenne de *Spartina*. Les valeurs rapportées de la production primaire brute et nette dans les marais de *Spartina* sont élevées par rapport à de nombreux autres écosystèmes naturels. La production nette de ces marais peut dépasser 40 tonnes de matière sèche par hectare et par année. Les marais côtiers de *Spartina* ont fait le sujet de nombreuses études productives. Cependant beaucoup de ces études ont négligé la masse biologique du sous-sol et son «turnover».

Ainsi, de ces relations des processus de production dans les marais avec les variables de l'environnement, on ne peut tirer que peu de conclusions. Une caractéristique de ces marais est la présence de l'espèce *Spartina* uniquement dans les endroits monotypiques. Ainsi ces marais représentent des systèmes relativement simples pour lesquels il est vraisemblable d'utiliser le modèle du processus de production primaire. Nous considérons l'ensemble des facteurs influençant les processus de production de *Spartina* comme un modèle analogique électrique compartimental. Nous utilisons ce modèle pour examiner d'une manière critique l'état actuel des connaissances sur la relation entre la production et les variables environnementales de ces marais.

La haute productivité de ces marais peut être associée à la photosynthèse du genre *Spartina* par la voie des C_4. A la différence de la majorité des herbes en C_4, *Spartina* ne se limite pas aux régions tropicales et subtropicales, mais s'étend jusque dans les régions tempérées. En europe, au dessus d'une température de feuille de 10°, les taux de photosynthèse des feuilles de *S.townsendii* dépassent fortement ceux de beaucoup d'herbes en C_3 tempérées. Contrairement à la majorité des espèces en C_4 précédemment examinées, les taux de photosynthèse des feuilles ne sont pas inhibés par des températures de feuilles de 10° ou moins, mais sont équivalents à ceux des herbes en C_3 tempérées. Nous discutons des études en laboratoire ou dans le milieu sur les réponses des taux de photosynthèse des feuilles chez *S.townsendii* en relation avec les variations climatiques.

INTRODUCTION

World distribution of Spartina

Marshes dominated by species of the genus *Spartina* Schreb. are a feature of hundreds of miles of the coastline of the Western Hemisphere, and thus represent a major coastal ecosystem. Undoubtedly the largest areas of *Spartina* marsh occur on the Atlantic coast of North America between the mouth of the St Lawrence river and the Gulf of Mexico. However, large *Spartina* marshes are also to be found on the Atlantic coast of South America, the coast of California, and the coasts of north-west Europe.

In his monograph on *Spartina*, Mobberley (1956) considered that the genus consisted of sixteen species (Table 20.1). All of these species are rhizomatous perennials and all are tolerant of saline soil conditions. In common with other species of the tribe *Chlorideae*, species of *Spartina* possess numerous two-celled salt glands which actively secrete a concentrated solution of salts onto the leaf surface when the plants are grown in saline conditions (Skelding & Winterbotham 1939; Levering & Thomson 1971; Lipschitz *et al.* 1974). Although it is predominantly a genus of plants of salt marshes and other saline habitats the natural distribution of the genus *Spartina* is not confined to saline soils. For example, the typical habitats of *S.pectinata* include dry prairie, railway and road embankments, and fresh-water marsh (Mobberley 1956).

The extensive marshes of the east coast of North America may contain up to four or five species of *Spartina*, each occupying a distinct part of the marsh often as dense monotypic stands. *S.alterniflora* commonly occupies inter-tidal mudflats and salt-pans whilst *S.patens* occupies drier parts of the marsh above the mean high-water mark. Where there is a fresh-water inlet, *S.pectinata* may also be present in the brackish or fresh-water areas behind the marsh and *S.cynosuroides* in the brackish marshland of estuaries (Mobberley 1956). If there is disturbed land, such as a rubbish tip or railway embankment, close to the salt marsh, the *S.patens* × *S.pectinata* hybrid, *S.caespitosa*, may be found (Marchant 1970).

Spartina *in Europe*

Although habitats suitable for the genus undoubtedly exist outside the Americas, *Spartina* is not found in such diversity and abundance in the Old World, indeed, only three species *S.maritima*, *S.neyrautii* and *S.townsendii* (*s.l.*) appear to have originated outside the American continents. Until the 19th century introduction of *S.alterniflora* from North America, only *S.maritima* was known in north-west Europe. This latter species is confined to the eastern side of the Atlantic, its range extends from South Africa to north-west Europe. South-east Britain appears to be the northern limit of its range. Specimens of *S.maritima* from Britain and other sites in north-west Europe rarely exceed 0·2 m in height and would appear to lack vigour by comparison to African specimens which are larger and can exceed 0·5 m in height (Marchant 1967). Chevalier (1923) suggests that this may be the result of an African origin of this species, which was later distributed by shipping to less favourable habitats in north-west Europe. In north-west Europe today *S.maritima* has been

Table 20.1. *Summary of the world distribution of* Spartina. (*Data from Mobberley, 1956.*)

Species	Distribution and habitat
Complex I	
S.arundinacea Carmich.	Rare species, only known from two widely separated groups of islands: Tristan de Cunha (37·0°S and 12·0°W) and the islands of St Paul and Amsterdam (38·5°S and 77·5°E).
S.ciliata Brong.	Sandy beaches and dunes above the inter-tidal zone, along the east coast of South America, between c. 30°S and 45°S.
S.spartinae Hitchc.	On sandy beaches and dunes by the coast and on a variety of inland habitats around the Gulf Coast in North America and inland in northern Argentina and Paraguay.
Complex II	
S.alterniflora Lois.	Inter-tidal salt marsh along the east coast of North America (25°N–50°N) and along the east coast of South America (10°N–40°S).
S.foliosa Trin.	Inter-tidal salt marsh on the west coast of North America (20°N–40°N).
S.longispica Haum.	Inter-tidal salt marsh on the east coast of South America (30°S–40°S).
S.maritima Fern.	Inter-tidal salt marsh and lagoons on the west coast of Europe (36°N–52°N) and on the east and south coasts of Africa (c. 15°N and c. 34°S).
S.townsendii (*sensu lato*)	Inter-tidal salt marsh on the west coast of Europe (48°N–57°N).
S.neyrautii Fouc.	Very rare species of inter-tidal salt marsh around San Sebastian in Northern Spain (44°N).
Complex III	
S.bakeri Merr.	Sandy beaches on the shores of freshwater and saline lakes in Florida and Southern Georgia (26°N–32°N).
S.caespitosa Fern.	Disturbed land near the east coast of North America (40°N–45°N).
S.cynosuroides Roth.	Borders of salt marsh on brackish water estuaries along the east coast of North America (30°N–40°N).
S.densiflora Brong.	Dry parts of coastal salt marsh on the east and west coast of South America (30°S–50°S).
S.gracilis Trin.	Around the margins of inland alkali lakes in North America (25°N–60°N).
S.patens Muhl.	Salt marshes and sandy beaches above the inter-tidal zone on the east coast of North America (20°N–50°N).
S.pectinata Link.	Brackish water marshes by the coast, freshwater marshes and dry prairie inland, in North America (30°N–50°N).

replaced at many of its former positions by *S.townsendii* (Goodman 1969). Where it remains in Britain it is usually found at the edges of salt-pans or in other poorly drained parts on consolidated sediments in a well-established salt marsh community, only rarely is it found as a primary colonizer. Even, where the species is locally abundant, as on the marshes around the Colne estuary in Essex, it constitutes less than 1% of the vegetative cover (Broadhurst, personal communication), so that its contribution to the primary production of the marsh is negligible.

Spartina has only become a major component of salt marsh vegetation in north-west Europe since the appearance of the *S.maritima* × *S.alterniflora* hybrid *S.* × *townsendii* H. & J. Groves, and the fertile amphidiploid derived from it, *S.anglica* Hubbard. For the purposes of this article we will refer to both the hybrid and the morphologically very similar, and sometimes indistinguishable, amphidiploid collectively by a single specific name, *S.townsendii* (*sensu lato*). Since the first recorded occurrence of *S.townsendii* in 1870 (Marchant 1967), this species has spread both as a result of deliberate plantings and natural dispersal to occupy an estimated total area of about 25,000 hectares in north-west Europe (Ranwell 1967). Its distribution in Europe extends from Udale Bay in Scotland (57·5°N) to the Pontrieux estuary in France (48°N) (Ranwell 1967). Extrapolating from data for Bridgewater Bay (Ranwell 1961), and for the Colne estuary in Essex (Masca 1977), the estimated total net aerial primary productivity of *Spartina* marshes in north-west Europe is of the order of $2{\cdot}5 \times 10^5$ tonnes of dry matter per year. Whether expansion of the inter-tidal area occupied by *S.townsendii* in north-west Europe will continue is open to question. On many of the older *Spartina* marshes in south-east England expansion seems to have been checked and in some areas the total area is receding (Hubbard 1965). However, in the north-west of England around the Dee and Ribble estuaries the area occupied by *S.townsendii* is rapidly increasing, and means of checking this expansion are being sought (Taylor & Burrows 1968).

NET PRIMARY PRODUCTION

Review

The major inter-tidal *Spartina* marshes of the world are the *S.alterniflora* marshes of the east coasts of South and North America, the *S.foliosa* marshes of California and the *S.townsendii* marshes of north-west Europe. It is the productivity of these inter-tidal marshes, and in particular the European *S.townsendii* marshes, which we will consider in this article, although many of our considerations will be applicable to other communities dominated by *Spartina* species.

The primary production of inter-tidal *Spartina* marshes has been the subject of a number of reviews, the most recent and comprehensive being that of Turner (1976). Thus, we will not attempt to add to these comprehensive reviews, but merely try to assess the current state of our knowledge of primary production in these marshes. At a first glance inter-tidal *Spartina* marshes appear to represent one of the most thoroughly studied ecosystems from the standpoint of primary production. Turner (1976) catalogued estimates of net primary production for more than 140 *Spartina* marshes, whilst a *S.alterniflora* marsh in Georgia was one of the first ecosystems on which a detailed study of energy-flow was made (Teal 1962). In addition, by comparison with other natural ecosystems inter-tidal *Spartina* marshes are relatively simple, especially at the primary producer level. Many of these marshes consist of extensive monotypic stands of no greater complexity as ecosystems than a field of an arable crop. It might therefore be expected that in view of both the relative simplicity and the mass of data on these ecosystems that we should have a very good understanding of primary production processes and their interaction with the environment; this however proves to be far from the truth.

There is very little information on the net primary production of *S.townsendii* in Europe, the only published data being estimates of above-ground standing crop at Bridgewater Bay, Somerset (Ranwell 1967). However, in view of the similarities in morphology, size, habitat and growth patterns, together with the close evolutionary relationship of *S.townsendii* and *S.alterniflora*, it would seem reasonable to expect that many of the conclusions reached from studies of net primary production in *S.alterniflora* marshes also should apply to *S.townsendii* marshes. Most of our current knowledge of net primary production in *Spartina* marshes consists of net aerial primary production estimates made on North American *S.alterniflora* marshes. Estimates of net annual primary production for *S.alterniflora* range from 279 g (dry wt.) m^{-2} (Stroud & Cooper 1968) to 6,000 g m^{-2} (extrapolating from Wiegert, this volume); a figure close to the highest dry matter yields reported for intensively managed arable crops (Boardman 1977). Net primary production of these marshes does show some correlation with latitude, production appears to increase from an average of about 500 g m^{-2} yr^{-1} at 38°N to about 700 g m^{-2} yr^{-1} at 28°N (Turner 1976). However, variation of net primary production within a *S.alterniflora* marsh can greatly exceed latitudinal variation. Stroud and Cooper (1968) reported for one marsh values of net primary production ranging from 279 g m^{-2} yr^{-1} for the 'short form' of *S.alterniflora* on the upper marsh to 1,563 g m^{-2} yr^{-1} for the 'tall form' of *S.alterniflora* on the lower marsh. Transplantation and isoenzyme studies have suggested that these different forms are not genetically distinct (Shea *et al.* 1975). Broome *et al.* (1975a) correlated the differences in net primary production within the marsh with differences in soil properties, in particular available nitrogen and phosphorus. The two or threefold increases in net primary production of

S.alterniflora marshes obtained by the addition of nitrogen and phosphate fertilizers demonstrate the limitations due to availability of nitrogen and phosphorus (Broome *et al.* 1975b; Valiela *et al.* 1975).

Net primary production in *S.alterniflora* marshes has been determined by a number of techniques, although the majority are based on peak biomass measurements or the method outlined by Smalley (1958). Both of these methods may grossly underestimate net primary production, the extent of underestimation varying between sites (Turner 1976). The comparability and usefulness of much of the data on the net primary production of *S.alterniflora* marshes is therefore doubtful. Only a few workers have made measurements of the turnover of biomass, and only four studies present any data on the below-ground biomass. This latter point could represent a very serious omission when we consider the observation that the below-ground biomass can constitute up to 75% of the total biomass of *S.alterniflora* (Cammen *et al.* 1974). Even if we ignore the many deficiencies in much of the data on net primary production of *Spartina* marshes, such measurements alone can tell us little as to why variation in net primary production exists. Estimates of net primary production alone will not enable us to answer such questions as why British material of *S.townsendii* cannot be established in sub-tropical regions (Ranwell 1967) or how increase in available nitrogen increases net primary productivity (Valiela *et al.* 1975)? In this context, many estimates of net primary production must be seen to have very limited value. Indeed, one cannot help asking whether at least part of the tremendous effort which has been put into these numerous studies of net primary production might have been more profitably spent in establishing the comparability and most suitable methods of measuring net primary production and in examining the major processes underlying production. To understand the basis of variation in net primary production of *Spartina* in different environments it is necessary to consider the processes underlying net primary production. These processes are considered here in the framework of a compartmental model.

Model of primary production

The complex of processes which in summation we term production, and the interaction of these processes with the environment, might be summarized most simply, in a compartmental model. Inevitably such a model will be a very crude and incomplete approximation of the actual interaction of biological and environmental processes. However, such a model has a number of important functions. First, it supplies us with an ordered framework for the examination of the completeness of our knowledge of production processes. Secondly, it should enable us to pinpoint the most important processes and interactions underlying variation in productivity. Ultimately, such a model should allow us to predict the responses of production to a new environment or changes in the existing environment.

The compartmental model which we will describe here is an analogue computer simulation model. In this model, developed from the ideas of Olson (1964), voltages are used continuously to simulate both the flow of carbon between compartments and the accumulation of carbon within compartments.

Construction of model

Olson (1964) has described in detail the construction and application of analogue computer simulations of ecosystems, so only a brief account will be given here. In such a model the system which is being studied is represented on an analogue computer circuit board, these are constructed using four types of component.

1. Amplifiers can be used for the summation of two, or more, input voltages. Where these are of opposite electrical sign, algebraic summation results in subtraction. An important feature is that the output is of opposite electrical sign to the input. Thus, if the inputs to an amplifier consist of a negative voltage representing respiratory rate and a positive voltage representing photosynthetic rate, the output will be a negative voltage representing the assimilation rate ($-A'$). If this then forms the input to a second amplifier, the output from that amplifier will be A' (Fig. 20.1a).
2. Integrators are amplifiers to which a capacitor has been added. These provide a continuously running summation or integration of inputs. Thus, if in our first example (Fig. 20.1a) we replace the second amplifier by an integrator its output will not be A', but its integral net primary production (NPP), where NPP $= \int A' \cdot dt$ (Fig. 20.1b).
3. Multipliers allow the multiplication of two input voltages. Thus, if we have two input voltages, one representing leaf photosynthetic rate (F_c) and one representing leaf area (L_a), the output from the multiplier will be the product of the inputs, stand photosynthetic rate or the rate of gross primary productivity (Fig. 20.1c).
4. Potentiometers are used to specify the fraction of voltage from one compartment that is to represent input to another compartment of the model. Thus, the rate of leaf biomass death can be simulated as a fraction of the voltage representing leaf biomass (Fig. 20.1d). These fractions (f) are not necessarily constants, but can vary as a function of environmental parameters. For example, f shown in Fig. 20.1d could be varied to account for seasonal differences in the rate of loss of leaves.

The analogue computer model of primary production processes in *Spartina* is presented in Fig. 20.2. Although this model is designed for use on an analogue computer the model can also be used with a digital computer where an analogue simulation language such as Block-CSMP is available. The model

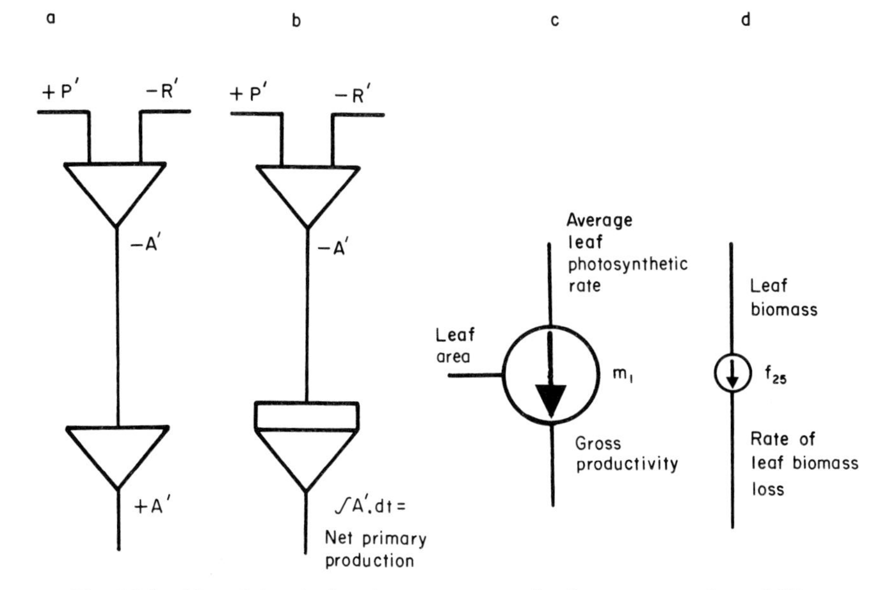

FIG. 20.1. Use of electrical analogue computer circuit components in modelling 'carbon flow'. a. Two amplifiers are used to calculate the assimilation rate (A'), i.e. the net primary productivity, by summing the respiratory losses of carbon ($-R'$), represented by a negative voltage, with photosynthetic gains (P'), represented by a positive voltage. b. In this model the second amplifier of Fig. 20.1a has been replaced with an integrator, to provide the integral of A', i.e. the net primary production. c. A multiplier (m_1) allows the computation of gross productivity, i.e. the rate of stand or community photosynthetic carbon accumulation, by multiplying a voltage representing the average leaf photosynthetic rate with a voltage representing the stand, or community, total leaf area. d. A potentiometer (f_{25}) allows the computation of rate of leaf biomass loss, through death, by specifying this rate as a function of the input voltage which represents the total leaf biomass.

(Fig. 20.2) consists of five interconnected compartments, which are described below.

Compartment 1: The input or gross primary productivity compartment. The input of carbon from the atmosphere is considered at the individual leaf level. Leaf photosynthetic rate is represented as an input voltage which is varied according to environmental conditions. The rate of gross productivity or stand photosynthesis is then obtained by multiplying leaf photosynthetic rate by the leaf area of the stand. This multiplication by leaf area represents a positive feedback from the leaf compartment. A negative feedback is also necessary as increase in the leaf area will increase the amount of shading within the canopy and thus depress the mean leaf photosynthetic rate.

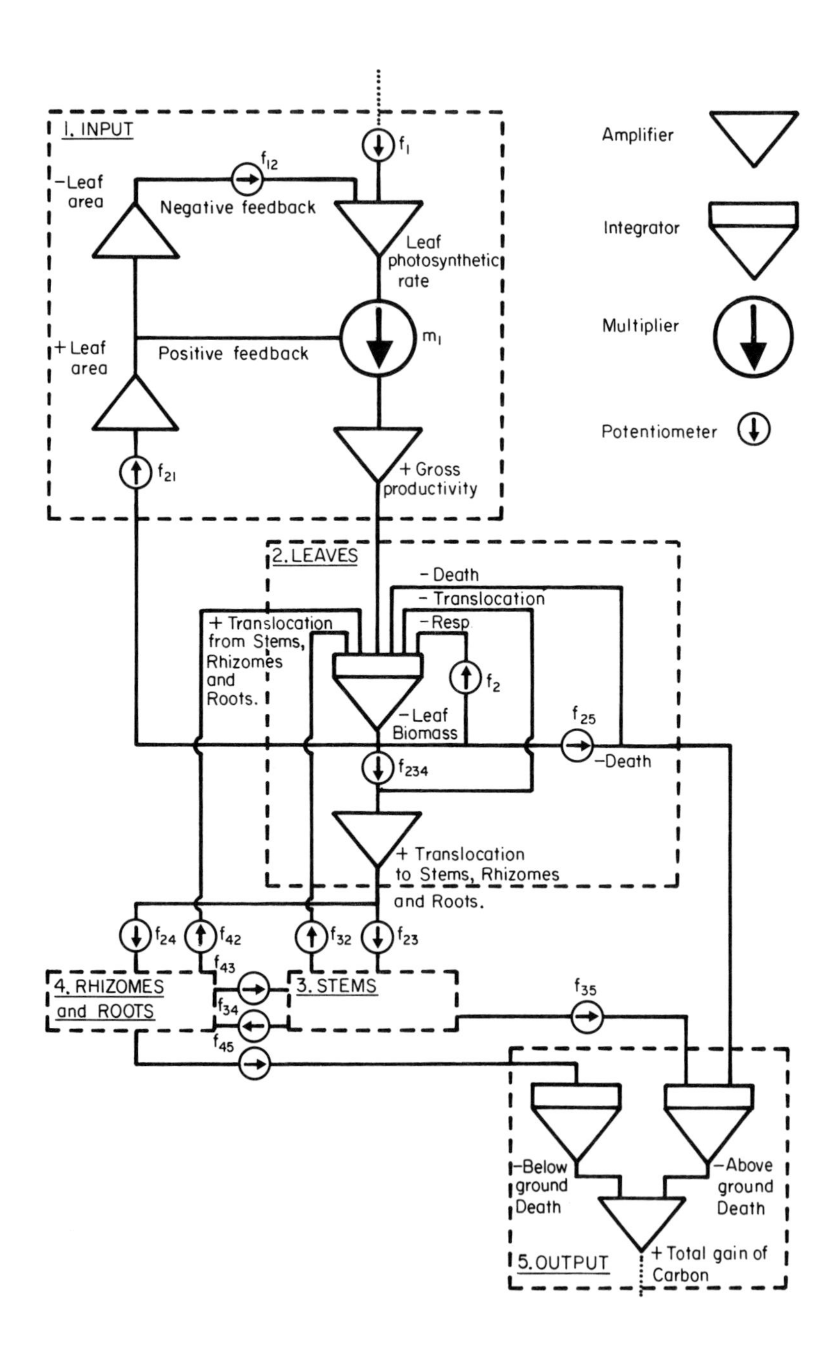

1. INPUT
f_1
f_{12}
−Leaf area
Negative feedback
Leaf photosynthetic rate
m_1
+Leaf area
Positive feedback
f_{21}
+Gross productivity
2. LEAVES
−Death
−Translocation
−Resp.
+Translocation from Stems, Rhizomes and Roots.
f_2
−Leaf Biomass
f_{25}
−Death
f_{234}
+Translocation to Stems, Rhizomes and Roots.
f_{24}
f_{42}
f_{32}
f_{23}
f_{43}
f_{34}
f_{45}
f_{35}
4. RHIZOMES and ROOTS
3. STEMS
−Below ground Death
−Above ground Death
5. OUTPUT
+Total gain of Carbon
Amplifier
Integrator
Multiplier
Potentiometer

Compartments 2, 3 and 4: The leaf, the stem and the rhizome and root biomass compartments respectively. The construction of each of these compartments is basically identical. Each consists of an integrator for the computation of biomass of that compartment. The integrator has positive inputs representing translocation from other organs, and negative inputs simulating losses, due to respiration, death, and translocation to other parts of the plant. Rates of loss from a compartment are computed as a function of the biomass of that compartment. Summation of the respiratory losses from each of these compartments represents the amount of carbon lost by *Spartina.*

Compartment 5: The output or net primary production compartment. In this compartment rates of loss of carbon, due to death of plant parts are integrated to give the cumulative losses for both above- and below-ground parts. These cumulative losses are then summed to give the total gain of carbon by other parts of the ecosystem from the primary producer level.

Application of the model

Influence of the environment on rates of transfer of carbon between compartments may be specified at the points on the model marked f_{11} (Fig. 20.2). Thus, rates of translocation, death and respiration can be considered as functions of environmental variables, such as soil temperature, day-length and available nitrogen. However, there is virtually no quantitative information available for inter-tidal *Spartina* marshes on the rates of these processes and their dependence on environmental variables. If information was available on the turnover of above- and below-ground biomass and its relationship to environmental variables, then hypotheses on the effects of environmental variables on rates of translocation, respiration and death could be tested by the use of this model. However, much of the published work on net primary production of *Spartina* supplies us only with estimates of above-ground biomass, as we have mentioned there is little information on turnover of biomass and no information on the relationships of turnover with environmental variables (Turner 1976). Measurements of gross primary production and the influence of environmental variables upon gross primary production are even fewer. There is neither detailed information on leaf canopy structure and

---

FIG. 20.2. A compartmental model of carbon fluxes within a *Spartina* stand. The model is presented as an electrical analogue computer circuit diagram. For clarity, the circuits within compartments 3 and 4 have been omitted from the diagram since these are identical in construction to the circuit of compartment 2. Inputs and outputs to amplifiers and integrators, together with the signs of the voltages by which they are represented, are indicated in the diagram. Subscripts to the letter *f*, which denotes potentiometers in the diagram, indicate which compartments each potentiometer interconnects. The construction of the model is explained in the text.

development, nor is there any data on leaf area indices, for inter-tidal *Spartina* marshes. In 1972 when we first considered this problem we could find no data on leaf photosynthetic rates for *Spartina.* Thus, even in the context of the simple model presented here (Fig. 20.2) it is evident that knowledge of primary production processes in inter-tidal *Spartina* marshes is far from complete. Thus, of the three functions which we ascribed to the model proposed here, only the first, that of providing a framework for the examination of the completeness of our knowledge of primary production processes, can be fulfilled. Application of the model in the prediction of net primary production of *Spartina* in response to a new environment or changes in the existing environment requires a far more detailed knowledge of the fluxes of carbon between the compartments of the model and their dependence on environmental variables, than is available at present.

Our objective has been to obtain a more complete picture of factors determining net primary production in *Spartina* marshes. We will report here results of our investigations into gross primary productivity in *S.townsendii*, in particular the responses of leaf photosynthetic rate to climatic variables. Integration of studies of gross productivity with a study of above- and below-ground biomass turnover forms part of a project currently being conducted on a *S.townsendii* marsh on the Stour Estuary in Essex.

GROSS PRIMARY PRODUCTION IN *S.TOWNSENDII*

In the analogue computer model (Fig. 20.2) the input of carbon is considered at the level of photosynthetic assimilation of CO_2 by individual leaves. The rate of photosynthetic CO_2 assimilation will be a function of a number of environmental variables. Obviously the amount of light (photon flux density) and the leaf temperature will be among the most important variables determining photosynthetic rate. Detailed laboratory studies of photosynthesis in higher plants suggest a wide range of other factors which also influence leaf photosynthetic rate including: leaf-air vapour pressure deficit (VPD), leaf age, leaf water potential (ψ_l), leaf temperature pre-history, mineral nutrition and the presence of pathogens (Sestak *et al.* 1971). In the inter-tidal salt marsh we must add to this list, salinity and period of submergence. To assess the importance of some of these factors on leaf photosynthetic rate in *S.townsendii*, laboratory and field studies of photosynthesis were made on plants from an inter-tidal salt marsh on the Humber Estuary.

Laboratory studies

Plants of *S.townsendii* were grown in a controlled environment cabinet in light, temperature, water vapour pressure deficit and daylength conditions roughly equivalent to the average conditions for June at the site on the Humber Estuary from which the plants were collected. The effects of individual climatic variables on leaf photosynthetic rate were determined in an open gas-exchange system (Long & Woolhouse 1978).

C_4 photosynthesis

One of the first observations made in these studies was that the CO_2 compensation point of photosynthesis was close to zero, c. 10 mg m^{-3} of CO_2 in air (Long *et al.* 1975). This result implied that *S.townsendii* was a species utilizing the C_4 pathway of photosynthetic carbon metabolism, i.e. the first product of CO_2 assimilation is a C_4-dicarboxylic acid. Further studies have confirmed that C_4-dicarboxylic acids are the first compounds which become radioactive when leaves of *S.townsendii* are allowed to photosynthesize in air containing $^{14}CO_2$ (Thomas 1977). Besides a low CO_2 compensation point, 'Kranz' leaf anatomy and a higher ratio, in all tissues, of $^{13}C/^{12}C$ are also considered to be features distinguishing C_4 species from plants in which the first product of photosynthetic carbon metabolism is the C_3 compound, 3-phospho-glycerate (Downton 1971). On these bases it can be suggested that at least ten of the sixteen species of the genus *Spartina* are C_4 species, these include the parents of *S.townsendii*, *S.alterniflora* and *S.maritima* (Table 20.2). There is no evidence to suggest that there are any C_3 species in the genus *Spartina*.

Table 20.2. *Summary of evidence for C_4 photosynthesis in the genus Spartina.*

	'Kranz' leaf anatomy	High C^{13}/C^{12} ratio	Low CO_2 compensation point
S.alterniflora	3	4	—
S.cynosuroides	2	—	7
S.foliosa	5	—	—
S.gracilis	6	—	7
S.maritima	6	—	7
S.patens	7	—	—
S.pectinata	—	1	7
S.spartinae	—	4	—
S.townsendii	6	—	6

1. Bender (1971).
2. Haberlandt (1904).
3. Harschberger (1909).
4. Johnson and Brown (1973).
5. Laetsch (1971).
6. Long *et al.* (1975).
7. Long, unpublished data.

C_4 photosynthesis is also associated with high potential rates of CO_2 assimilation, some two or three times the maximum rates reported for many C_3 species. However, studies of other C_4 species have suggested that this high potential would not be realized in cool temperate regions, such as north-west Europe (Black 1971). The C_4 species previously studied have all originated from tropical or sub-tropical regions. Their poor performance in cool temperate conditions might therefore be a result of their origin rather than any inherent feature of C_4 photosynthesis. *S.townsendii* appears unusual, and perhaps unique, as a C_4 species which has both originated and is largely confined to cool temperate regions (Long *et al.* 1975).

Temperature, light and VPD responses of leaf photosynthesis

In contrast to other C_4 grasses, *S.townsendii* shows rates of leaf photosynthesis at 5°C and 10°C equal, and not below, those of temperate C_3 grasses, whilst above 10°C the leaf photosynthetic rate in *S.townsendii* is considerably higher

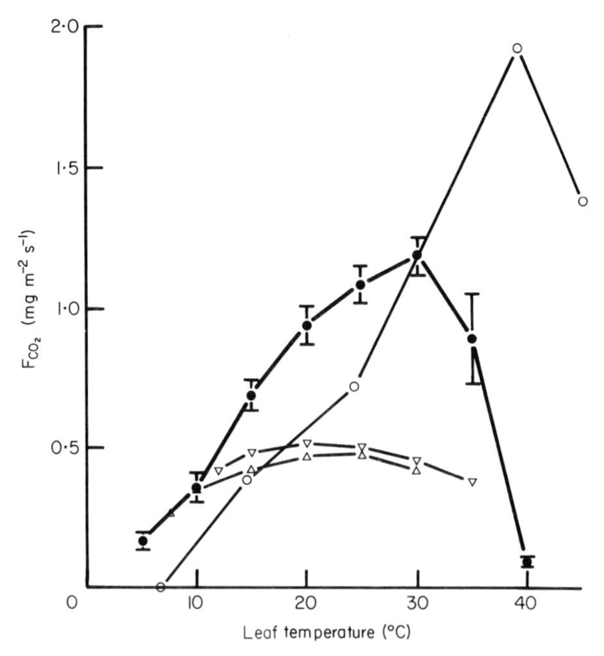

FIG. 20.3. Mean ± s.e. response of the leaf photosynthetic rate of CO_2 assimilation (F_{CO_2}) for *S.townsendii* (●) to leaf temperature compared to responses for the tropical C_4 grass *Pennisetum purpureum* (○), the alpine C_3 grass *Sesleria albicans* (△) and the temperate C_3 grass *Festuca arundinacea* (▽). All measurements were made at a photon flux density of about 2,000 μmol m^{-2} s^{-1} on plants which were all grown under similar controlled environment conditions. Data from Long *et al.* (1975).

(Fig. 20.3). The mean maximum rate of CO_2 assimilation per unit of leaf area recorded for *S.townsendii* in these studies was 1·2 mg (CO_2) m^{-2} s^{-1}, approximately twice the maxima reported for temperate C_3 herbage grasses (Cooper 1969). Although leaf photosynthetic rate is not saturated at photon flux densities equivalent to full-sunlight, c. 2,500 μmol m^{-2} s^{-1}, there is only a small increase in rate between half and full-sunlight (Fig. 20.4). Leaf photosynthetic rate was found to be independent of water vapour pressure deficit (VPD) below 1·0 kPa, above this value leaf photosynthetic rate decreased with increase in VPD. However, water vapour pressure deficits in excess of 1·0 kPa were rarely recorded at the field site.

Field studies

The laboratory studies provide information on the relationship of photosynthetic rate for randomly selected mature leaves to both the range of photon flux densities and leaf temperatures which would be experienced by the plants at the field site (Long 1976). From these studies, leaf photosynthetic rates in the field may be predicted, given the incident photon flux density and the leaf

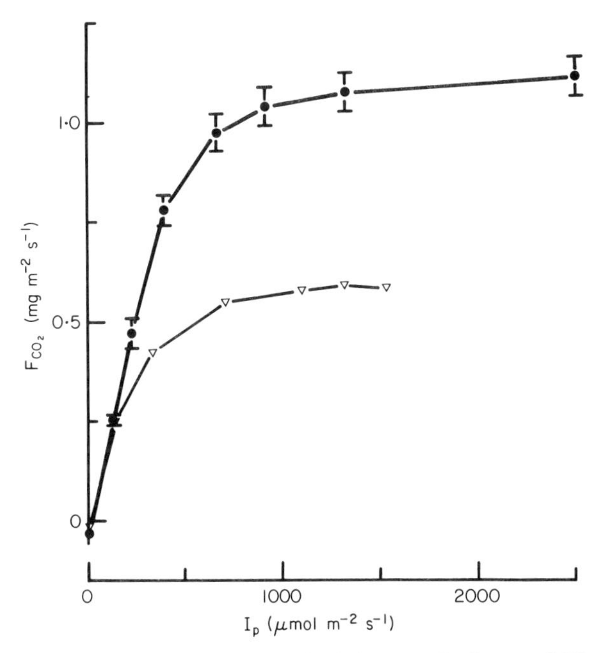

FIG. 20.4. Mean ± s.e. response of the leaf photosynthetic rate of CO_2 assimilation (F_{CO_2}) to photon flux density (I_p) for *S.townsendii* (●) at a leaf temperature of 25°C. Data on the response of F_{CO_2} to I_p for leaves of the temperate C_3 grass, *Lolium perenne* (▽), (Van Laar & De Vries 1973) is included for comparison.

temperature. The validity of these predictions will depend on the relative importance of other factors which might influence leaf photosynthetic rates and which were not examined in the laboratory studies. There were many features of the field site which were not simulated in the laboratory controlled environment. Change in daylength, exposure to low air temperatures and saline soil conditions might all markedly modify the responses of leaf photosynthetic rate to light and temperature observed in the laboratory studies.

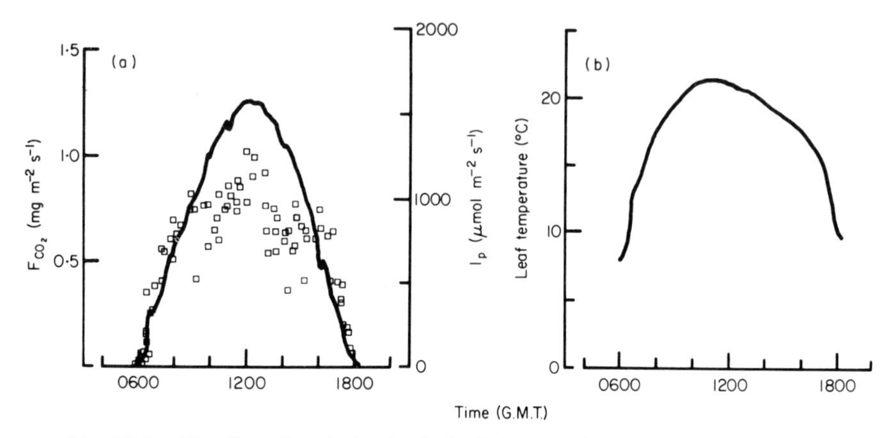

FIG. 20.5. The diurnal variation in the leaf photosynthetic rate of CO_2 assimilation (F_{CO_2}) of *S.townsendii* at a field site on the Humber Estuary, Yorkshire, on 18 September 1974. Each point (□) represents a single measurement of F_{CO_2}. The solid line in Fig. 20.5a represents the diurnal course of the photon flux density (I_p) incident in the horizontal plane above the *S.townsendii* leaf canopy and the solid line in Fig. 20.5b represents the diurnal course of mean leaf temperature through the same day.

To test the validity of laboratory predictions and to examine whether the high potential rates of leaf photosynthesis observed in the laboratory were actually realized in the field, field measurements of leaf photosynthesis were made on an inter-tidal *S.townsendii* marsh on the Humber Estuary. At three-weekly intervals between March and December 1974 field measurements of leaf photosynthetic rate were made using a ^{14}C technique (Incoll 1977). Measurements were made on randomly selected mature leaves. Fig. 20.5 illustrates individual measurements of leaf photosynthesis through the course of a day in September, co-plotted with photon flux density and mean leaf temperature. Even during this day in September, some individual measurements show a rate of leaf photosynthesis around mid-day of 0·90 mg (CO_2) m^{-2} s^{-1}, 150% of the value considered to be a maximum for temperate C_3 herbage grasses (Cooper 1969). Table 20.3 compares rates of leaf photosynthesis measured in the field with those predicted for the same leaves using

Table 20.3. *Comparison of the mean values of leaf photosynthetic rate predicted from laboratory studies (F_c^1) with the actual values recorded in the field (F_c).*

Date	F_c	F_c^1
12/4/74	0·09	0·12*
21/4/74	0·39	0·58*
12/5/74	0·29	0·36*
1/6/74	0·59	0·67
15/6/74	0·50	0·53
11/7/74	0·49	0·53
25/7/74	0·76	0·85
23/8/74	0·78	0·80
18/9/74	0·90	0·86
29/9/74	0·66	0·73
23/10/74	0·51	0·54
17/11/74	0·39	0·41

F_c is the mean of all measurements of leaf photosynthetic rate, in mg (CO_2) m (leaf area)$^{-2}$ s^{-1}, made between 11.00 and 13.00 hr GMT on each date. F_c^1 is the mean leaf photosynthetic rate predicted from the laboratory data using the photon flux densities and leaf temperatures recorded at the time that the field measurements were made. An asterisk indicates that the difference between F_c and F_c^1 is significant at the 95% confidence interval of Student's *t*-test.

the laboratory data. Although the predictions are based only on the light and temperature responses of leaf photosynthetic rate in the laboratory grown plants, the predicted and actual rates agree closely for much of the year. Only during April and May of 1974 were the measured rates significantly lower than the predicted rates. Thus, for much of the year leaf photosynthetic rate can be predicted simply from a knowledge of light and temperature conditions. The field studies also confirm the observation that the relatively high potential rates of leaf photosynthesis associated with C_4 photosynthesis can be realized in *S.townsendii* even in the temperature conditions of a cool temperate climate.

Estimate of gross primary production

The leaf area per unit area of ground at the field site increased from 0 in February to a maximum of 3·0 ± 1·0 s.e.m. in September. Since the field measurements were made on randomly selected leaves from within the canopy, stand photosynthetic rate can be computed using the estimates of total leaf area. Extrapolating from these estimates of stand photosynthetic rate and assuming the photosynthetic rate is zero when the canopy is submerged at high tide, the estimated gross primary production for the field site during 1974 was 2,500 gC m^{-2}. No estimates of net primary production were made at this field site for 1974. However, a preliminary investigation on an *S.townsendii* marsh on the Colne Estuary with a similar stem density and mean stem height to the site on the Humber Estuary, yielded an estimated net primary production for 1976 of 900 gC m^{-2}. The exact reasons for this large difference between gross and net primary production are far from clear. Even if the large errors which must be attached to these estimates are taken into account, net primary production represents between 45 and 20% of gross primary production. One major, and at present poorly understood, source of this loss may be through the respiratory costs of maintaining a large non-photosynthetic root and rhizome system, as two-thirds of the estimated net primary production was of below-ground biomass.

REFERENCES

Bender M.M. (1971) Variations in the $^{13}C/^{12}C$ ratios of plants in relation to the pathway of photosynthetic carbon dioxide fixation. *Phytochemistry* **10**, 1239–44.

Black C.C. (1971) Ecological implications of dividing plants into groups with distinct photosynthetic production capacities. *Adv. ecol. Res.* **7**, 87–114.

Boardman N.K. (1977) Solar energy conversion in photosynthesis and its potential contribution to world demand for liquid and gaseous fuels. *Proc. 4th Int. Congr. Photosyn.* (Ed. by D.O. Hall, J. Coombs & T.W. Goodwin), pp. 635–44. Biochem. Soc., London.

Broome S.W., Woodhouse W.W. Jnr. & Seneca E.D. (1975a) The relationship of mineral nutrients to growth of *Spartina alterniflora* in North Carolina: I. Nutrient status of plants and soils in natural stands. *Soil Sci. Soc. Amer. Proc.* **39**, 295–301.

Broome S.W., Woodhouse W.W. Jnr. & Seneca E.D. (1975b) The relationship of mineral nutrients to growth of *Spartina alterniflora* in North Carolina: II. The effects of N, P and Fe fertilizers. *Soil Sci. Soc. Amer. Proc.* **39**, 301–7.

Cammen L.M., Seneca E.D. & Copeland B.J. (1974) Animal colonization of salt marshes artificially established on dredge spoil. *Univ. of N. Carolina Sea Grant Publ.* UNC-SG 74-15.

Chevalier A. (1923) Note sure les *Spartina* de la flore française. *Bull. Soc. bot. Fr.* **70**, 54–63.

Cooper J.P. (1969) Potential production and energy conversion in temperate and tropical grasses. *Herb. Abstr.* **40**, 1–15.

Downton W.J.S. (1971) Adaptive and evolutionary aspects of C_4 photosynthesis. *Photosynthesis and Photorespiration* (Ed. by M.D. Hatch, C.B. Osmond & R.O. Slatyer), pp. 3–17. Wiley-Interscience, New York.

Goodman P.J. (1969) Biological Flora of the British Isles, *Spartina* Schreb. *J. Ecol.* **57**, 285–313.

Haberlandt G. (1904) *Physiologische Pflanzenanatomie*. 4th Edn., Engelmann, Leipzig.

Harschberger J.W. (1909) The comparative leaf structure of the strand plants of New Jersey. *Trans. Am. Phil. Soc.* **48**, 72–89.

Hubbard J.C.E. (1965) *Spartina* marshes in Southern England. VI. Pattern of invasion in Poole Harbour. *J. Ecol.* **53**, 799–813.

Incoll L.D. (1977) Field studies of photosynthesis: monitoring with $^{14}CO_2$. *Environmental Effects on Crop Physiology.* (Ed. by J.J. Landsberg & C.V. Cutting), pp. 137–55. Academic Press, London.

Johnson C. & Brown W.V. (1973) Grass leaf ultrastructural variations. *Am. J. Bot.* **60**, 727–35.

Laetsch W.M. (1971) Chloroplast structural relationships in leaves of C_4 plants. *Photosynthesis and Photorespiration* (Ed. by M.D. Hatch, C.B. Osmond & R.O. Slatyer), pp. 323–49. Wiley-Interscience, New York.

Levering C.A. & Thomson W. (1971) The ultrastructure of the salt gland of *Spartina foliosa. Planta* **97**, 183–96.

Lipschitz N., Shomer-Ilan A., Eshel A. & Waisel Y. (1974) Salt glands on leaves of Rhodes Grass (*Chloris gayana* Kth.) *Ann. Bot.* **38**, 459–62.

Long S.P. (1976) *C_4-photosynthesis in cool temperature climates, with reference to* Spartina townsendii (s.l.) *in Britain.* Ph.D. thesis. University of Leeds.

Long S.P., Incoll L.D. & Woolhouse H.W. (1975) C_4 photosynthesis in plants from cool temperate regions with particular reference to *Spartina townsendii. Nature, Lond.* **257**, 622–24.

Long S.P. & Woolhouse H.W. (1978) The responses of net photosynthesis to vapour pressure deficit and CO_2 concentration in *Spartina townsendii* (*sensu lato*). *J. exp. Bot.* **29**, 567–77.

Marchant C.J. (1967) Evolution of *Spartina* (Gramineae): I. The history and morphology of the genus in Britain. *J. Linn. Soc. Lond. Bot.* **60**, 1–24.

Marchant C.J. (1970) Chromosome pairing and fertility in *Spartina* × *caespitosa. Can. J. Bot.* **48**, 183–88.

Masca C. (1977) *Nitrogen levels and the growth performance of* Spartina townsendii. B.Sc. project report, Univ. of Essex. 69 pp.

Mobberley D.G. (1956) Taxonomy and distribution of the genus *Spartina. Iowa State Coll. J. Sci.* **30**, 471–574.

Olson J.S. (1964) Gross and net production of terrestrial vegetation. *J. Ecol. Jub. Symp. Sup.*, pp. 99–118.

Ranwell D.S. (1961) *Spartina* salt marshes in southern England. I. The effects of sheep grazing at the upper limits of *Spartina* marsh in Bridgewater Bay. *J. Ecol.* **48**, 385–95.

Ranwell D.S. (1967) World resources of *Spartina townsendii* (*sensu lato*) and economic use of *Spartina* marshland. *J. appl. Ecol.* **4**, 239–56.

Sestak Z., Jarvis P.G. & Catsky J. (1971) Criteria for selection of suitable methods. *Plant Photosynthetic Production. Manual of Methods* (Ed. by Z. Sestak, J. Catsky & P.G. Jarvis), pp. 1–48. Dr W. Junk, The Hague.

Shea M.L., Warren R.S. & Niering W.A. (1975) Biochemical and transplantation studies on the growth form of *Spartina alterniflora* on Connecticut salt marshes. *Ecology* **56**, 461–66.

Skelding A.D. & Winterbotham J. (1939) The structure and development of hydathodes of *Spartina townsendii* Groves. *New Phytol.* **38**, 69–79.

Smalley A.E. (1958) The growth cycle of *Spartina* and its relations to the insect populations in marshes. *Proc. Salt Marsh Conf., Univ. Ga., Mar. Inst.*, pp. 25–28.

Stroud L.M. & Cooper A.W. (1968) Color-infrared aerial photographic interpretation and net primary productivity of a regularly flooded North Carolina salt marsh. *Water Resources Res. Inst. Univ. of N.C. Rept.* No. 14.

Taylor M.C. & Burrows E.M. (1968) Studies on the biology of *Spartina* in the Dee estuary Cheshire. *J. Ecol.* **56**, 795–809.

Teal J.M. (1962) Energy flow in the salt marsh ecosystem of Georgia. *Ecology* **43**, 614–24.

Thomas S.M. (1977) C_4 photosynthesis in a cool temperate plant. *Rothamsted Experimental Station Report for 1976*, p. 40. Lawes Agricultural Trust, Harpenden.

Turner R.E. (1976) Geographic variations in salt marsh macrophyte production: a review. *Contrib. Mar. Sci.* **20**, 47–68.

Valiela I., Teal J.M. & Sass W.J. (1975) Production and dynamics of salt marsh vegetation and the effects of experimental treatment with sewage sludge. *J. appl. Ecol.* **12**, 973–92.

Van Laar H.H. & De Vries F.W.T.P. (1973) CO_2-assimilation light response curves of leaves; some experimental data. *Inst. Biol. Scheikg. Onderz. Landb., Versl.* **62**, 1–55.

21. FIXATION, ACCUMULATION AND RELEASE OF ENERGY BY *AMMOPHILA ARENARIA* IN A SAND DUNE SUCCESSION

IAN K. DESHMUKH

Department of Biological Sciences and
Tay Estuary Research Centre,
*University of Dundee, U.K.**

SUMMARY

The production ecology of *Ammophila arenaria* was studied in three zones, 45 m, 70 m and 95 m from the high-tide line. Net above-ground primary production was estimated and secondary production was accounted for almost entirely by microbial respiration in the detritus food web. With estimates of dune age, the three zones were linked by a model of energy fixation and accumulation in relation to succession. The predicted energy accumulation agreed closely with that obtained from harvested samples of plants.

RÉSUMÉ

Nous avons étudié l'écologie de production de *Ammophila arenaria* à trois distances, 45 m, 70 m et 95 m de la ligne de haute marée. Nous avons estimé la production primaire nette du sol et comptabilisé la production secondaire dans le réseau alimentaire des détritus, principalement par la mesure de la respiration microbienne. D'après l'estimation de l'âge des dunes, un modèle de fixation de l'énergie et d'accumulation relierait la succession dans les trois zones. L'accumulation de l'énergie théorique correspond très étroitement à celle mesurée chez les échantillons récoltés.

INTRODUCTION

Changes in the vegetation of sand dunes represent one of the classic examples of terrestrial primary succession. Since Cowles (1889), numerous studies have

* Present address: Department of Zoology, University of Nairobi, Nairobi, Kenya.

concentrated on the structural dynamics of dune systems, but as Elton (1966) and Ranwell (1972) have emphasized, information about energy flow in this type of ecosystem is lacking.

The early stages of sand dune development in Britain are dominated by marram grass (*Ammophila arenaria**) and in this paper the production ecology of this species is considered in relation to succession. Emphasis is placed on relating results from different zones in the succession rather than on details of methodology which are given in full by Deshmukh (1974).

The study site

Tentsmuir Point National Nature Reserve is a dune system of more than 30 km^2 situated between the Tay and Eden estuaries in eastern Scotland. At the study site a series of dune ridges, parallel to the sea, is orientated along a north–south axis.

Three 10 m wide belts were chosen in the fore-dunes and were designated mobile, semi-fixed and fixed zones which were centred 45 m, 70 m and 95 m, respectively, from the marine strand line (Fig. 21.1). All zones were dominated by *A.arenaria.* In the mobile zone, this grass formed tussocks interspersed with bare sand and occasional plants of *Elymus arenarius.* In the semi-fixed zone the tussocks were less pronounced, but marram grass covered more of the sand surface and some areas of loose sand remained. *Festuca rubra* and mosses were sparsely represented within the shelter of *A.arenaria.* The fixed zone had less *A.arenaria* cover than the semi-fixed zone, but this species still predominated in terms of above-ground biomass. *F.rubra* and mosses were wide-

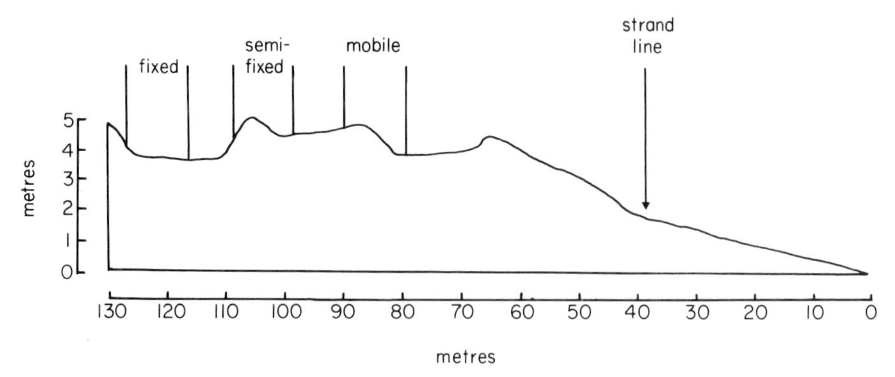

FIG. 21.1. Cross-section to show dune morphology at the study site and the positions of the study zones.

* Plant nomenclature follows Clapham, Tutin and Warburg (1962).

spread and these 'grey dunes' also contained scattered herbs such as *Lotus corniculatus*, *Hieracium pilosella* and *Senecio jacobaea.*

ENERGY FLOW AND SUCCESSION

The objective of the study was to construct a model of the changes in fixation and accumulation of energy by *Ammophila arenaria* through time, as the sand dune development proceeds from mobile to fixed dunes. For this, it was necessary to estimate net primary production, rate of litter decay and the age of the dunes in each study zone. Above-ground production only was studied, because adequate sampling of roots and rhizomes could have caused serious erosion. Energy content of live leaf, dead leaf and litter in each zone was estimated by bomb calorimetry. Ash content was also determined by ignition in a muffle furnace so that energy estimates could be expressed on an ash-free basis.

The age of the dunes

Lateral extension seaward of the dunes over 160 years was estimated from changes in the high-water line (Fig. 21.2). From 1946 to 1972 there was no substantial alteration in the low-water line. The mean annual rate of lateral net accretion was estimated from transects a to e in Fig. 21.2, as $9{\cdot}3 \pm 3{\cdot}2$ (SE) m yr^{-1}. The study site, centred on transect e appeared to have the smallest, but most consistent, rate of lateral accretion of $5{\cdot}5 \pm 1{\cdot}6$ (SE) m yr^{-1}. Estimates of dune age from these data indicate that the centres of the mobile, semi-fixed and fixed zones were 13·5, 18·9 and 24·3 years old, respectively.

However, dune building is not a continuous process, but occurs rapidly for short periods which may be separated by several years when little or no change occurs. Thus use of the 160 year mean annual rate of net accretion is a simplification which may only be strictly applicable if considering changes over several years when these discontinuities will average out.

Net above-ground primary production

As defined by Milner and Hughes (1968),

$$P_n = \Delta B + L + G,$$

where P_n = net production,
ΔB = change in biomass,
L = losses by death
and G = losses to herbivores; all data expressed over a defined time period.

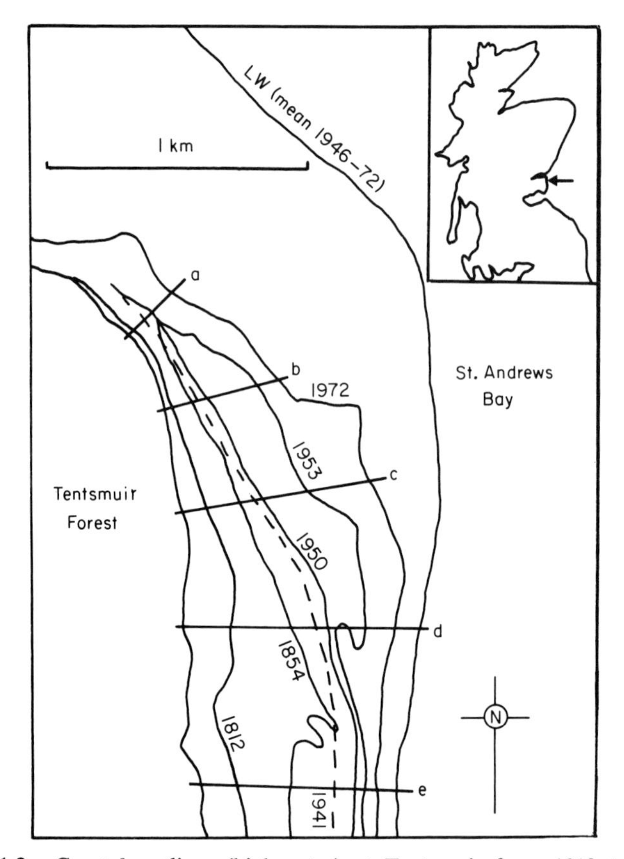

FIG. 21.2. Coastal outlines (high-water) at Tentsmuir from 1812 to 1972. (a) to (e) denote lines along which rates of accretion were estimated. Mean low-water mark for the period 1946 to 1972 is shown. The study site was adjacent to transect (e). Sources: 1812, 1854 Ordnance Survey; 1941 anti-tank defences placed at high-water; 1950, 1953 from Grove (1953); 1972 compass traverse. Inset shows the position of the study area on the mainland of Scotland.

Green biomass change (ΔB), as dry weight, was estimated from harvest data of the standing crop of green aerial plant parts throughout 1970 and 1971. Dry weight of the total dead material in harvests did not change significantly with time and so in 1971 leaf mortality (L) was estimated from the results of a study of necrosis in individual leaves. Throughout the growing season changes in the lengths of green and necrosed sections of marked leaves were measured. This was correlated with leaf dry weight and subsequently the dry weight per unit area of leaves which were dying. Correction for leaf mortality is often ignored in estimates of primary production (cf. Wiegert & Evans 1964;

Williamson 1976). If the correction had not been made, net primary production would be underestimated by approximately 30% in this study.

Net above-ground primary production is given in Table 21.1 in terms of energy on an ash-free basis. This was considered to be equivalent to above-ground litter production since losses to herbivores (G) were found to be negligible. Clearly production of *A.arenaria* drops markedly at the inland sites. If it is assumed that this decline in primary production proceeds linearly with distance from the mobile zone, then for each metre further inland, net above-ground primary production of *A.arenaria* is reduced by 72 kJ m^{-2} yr^{-1}. It was not possible to verify this assumption accurately with only three data points.

Table 21.1. *Estimates of annual above-ground net primary production* (P_n) *of* Ammophila arenaria *for 1971 as the sum of biomass increment* (ΔB) *and loss by death* (L) *in each zone. Mean* ± *SE, kJ (ash-free)* m^{-2}.

	Zones		
	Mobile	Semi-fixed	Fixed
ΔB	3,367·2 ± 412·8	2,342·2 ± 183·8	939·6 ± 145·3
L	1,360·8 ± 115·6	969·7 ± 125·2	498·3 ± 52·3
P_n	4,727·3 ± 415·8	3,311·9 ± 400·7	1,437·8 ± 150·3

Litter breakdown

This is usually described as a complex phenomenon involving micro-organisms, soil animals and physical processes, but with most of the energy being dissipated by micro-organisms. The overall loss of energy from the leaf litter of *A.arenaria* was followed over a period of 18 months using the litter bag technique (Bocock & Gilbert 1957). Three replicate bags were collected at intervals of three months from each zone, both at the sand surface and buried at 10 cm depth, and the dry weight, energy and ash content of the litter determined. From these determinations it was estimated that 39·8 ± 3·5 (SE)% of energy (ash-free basis) was lost in the first year of litter decay, with no clear differences between bags at different zones and depths.

Microbial energy flow was estimated as the respiration rate of litter from the field, per unit weight of litter, at various temperatures. Every three months litter from each of the bags collected was placed in a Gilson respirometer and oxygen uptake measured during the following 24 hours at a range of temperatures encompassing those experienced in the field. A total of 54 samples was taken from bags at different zones and depths of burial. In each case there was a significant relationship between temperature (°C) and log respiration rate ($P < 0{\cdot}01$). Howard (1971) has suggested that the use of logarithmic temperature metabolism curves for mixed decomposer populations may not be the best model in all cases, but the use of this method was considered statistically valid for present purposes. Temperature in the field was measured at hourly intervals, just below the sand surface in each zone. Since these records were incomplete, covering a total of about half of the period required, they were supplemented by data from Shanwell Meteorological Station (4 km north-west of the study site, in a dune area) for periods with no sand temperature estimates. These air temperature data were converted to the equivalent of sand temperature on the basis of the difference between sand and air temperature for periods when both records were complete. An estimate of mean field temperature over periods centred on the dates of respiration measurements was thus obtained and used to obtain total metabolism in the field using a standard conversion to energy values (Petrusewicz & Macfadyen 1970).

From data of the energy content of litter when originally placed in the bags, it was estimated that microbial respiration released 42·8 ± 2·3 (SE)% of this energy in one year, combining data from all zones. This is strikingly similar to the result obtained by the use of the litter bag method. Additional evidence comes from a study revealing the very small energy flow to soil animals. The millipede, *Cylindroiulus latestriatus* (Curtis) was the only animal of those found in the soil that fed upon *A.arenaria* litter in feeding trials. At its highest annual mean density (100 individuals m^{-2}, in the semi-fixed zone) this animal consumed less than 0·25% of litter production annually. Thus it was concluded that release of energy from litter was approximately 40% in the first year and almost solely due to the respiration of micro-organisms.

Litter breakdown is often described in terms of an exponential function (e.g. Olson 1963) although this has been questioned by Minderman (1968). More complex models have been suggested (Flanagan & Bunnell 1976), but with the limited data available in this study, energy loss from litter was fitted to an exponential function from which the half-life of the litter is estimated as approximately 1·3 years.

The model

In the preceding sections a number of simple assumptions were made concerning the production, mortality and decay of leaves of *Ammophila arenaria* and of the age of the study zones (Table 21.2). A graphical model of energy fixation and accumulation by *A.arenaria* was developed from these assumptions. The model considers changes in production and accumulation of energy in dead grass through time as though production and decomposition were estimated initially in the mobile sand zone, but then this zone progressively changed to become the semi-fixed zone and eventually the fixed sand zone, as the shoreline moved seawards.

Table 21.2. *Assumptions used to produce a graphical model (Fig. 21.4) of energy fixation and accumulation by* A.arenaria *at different stages of sand dune development.*

1. Lateral accretion of the dunes seaward occurs uniformly at a rate of 5·5 m yr^{-1}.
2. Net above-ground primary production of *A.arenaria* declines linearly with dune age at a rate of 306 kJ m^{-2} yr^{-1}.
3. Energy release from leaf litter occurs exponentially at a rate of 40% per year, irrespective of the successional stage.
4. There is an accumulation of 5,685 kJ m^{-2} of litter produced in previous years in the mobile zone as estimated from harvest data.

The computations begin at the centre of the mobile zone where there was an accumulation of dead *A.arenaria* of x_1 kJ m^{-2} as estimated from harvest data (Table 21.2). This zone had a net primary production of y_1 kJ m^{-2} and, assuming that an equivalent amount dies during the year, then $(x_1 + y_1)$ kJ m^{-2} is the total energy available to decomposers in that year. During the following year p% of this total is released by microbial respiration, leaving $[1 - (p/100)](x_1 + y_1) = x_2$ kJ m^{-2} remaining as litter accumulation after one year of decomposition. In this second year net primary production (y_2 kJ m^{-2}) will have decreased (Table 21.2) and this is added to the litter accumulation already present to give total energy available to decomposers at the end of this second year, $(x_2 + y_2)$ kJ m^{-2}. These computations (given diagrammatically in Fig. 21.3) were carried through for the number of years necessary to reach the estimated age of the fixed zone. Fig. 21.4 shows predictions for energy accumulation as dead *A.arenaria*, with curves showing energy available to decomposers at the beginning of any year $(x + y)$ and energy remaining as accumulation (x) at the end of any year before net primary production (y) for that year is added.

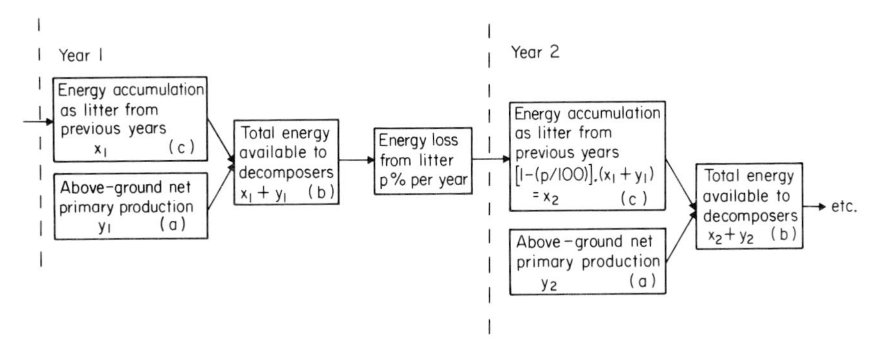

FIG. 21.3. Flow diagram of computations used in construction of the graphical model (Fig. 21.4) of the above-ground energy fixation and accumulation by *A.arenaria*, through time. (a), (b) and (c) refer to the appropriate curves in Fig. 21.4.

The model follows changes through time and predicts energy accumulation in litter as dune accretion proceeds. Equivalent estimates of energy accumulation in space are available as the annual mean of dead *A.arenaria* harvested in the semi-fixed and fixed zones. These are also shown in Fig. 21.4 where they lie, as expected from the predictions of the model, between the curves representing the amount available to decomposers $(x + y)$ and the accumulation after decomposition (x) respectively.

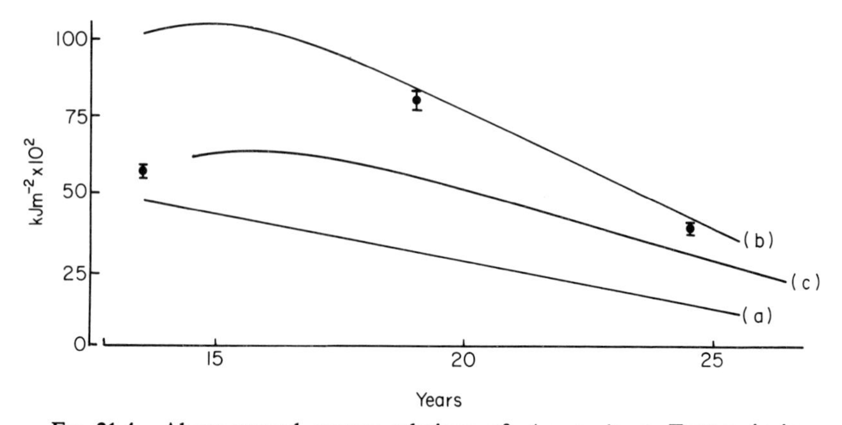

FIG. 21.4. Above-ground energy relations of *A.arenaria* at Tentsmuir in relation to dune age, based on the assumptions of Table 21.2. Net primary production (a); amount of litter available to decomposers at the beginning of any year (i.e. net primary production plus litter from previous years) (b); and the amount of litter after decomposition at the end of any year (c). Vertical bars are mean ± SE of energy accumulation of dead material based on the mean of one year of harvests of standing crop.

The sensitivity of the model may be examined by altering the assumptions in Table 21.2 and considering the effect on the predictions of energy accumulation. A number of such tests were made, but in all cases the predictions fitted the independently assessed harvest estimates of litter accumulation less well. Major alterations such as regarding litter breakdown as a linear rather than an exponential function or assuming that net primary production did not decline with dune age, but maintained a medium value, produced very poor predictions. Smaller changes, for example, an exponential litter breakdown rate of 35% or 45%, also produced less accurate predictions.

Clearly, the model could be refined to take account of annual variations in accretion and of primary production, which are here regarded as simple linear functions. Also, addition of primary production to the litter could be studied in more detail so that the model would reflect this as a continuing process throughout the year, rather than as a discrete, once yearly, event. However, adequate data on these phenomena were not available.

DISCUSSION

The model presented uses simple mathematical abstractions of complex biological and physical processes. Nevertheless, the predictions of litter accumulation are close to the independent harvest estimates. However, extrapolation beyond the study zones would suggest that *A.arenaria* production ceases after 29 years and that dead material is absent after 33 years, whereas, in reality, moribund *A.arenaria* tussocks occur in dunes more than 100 years old, although their contribution to community energy flow is probably very small. Clearly other factors must take effect beyond the study zones as the community becomes more complex.

Odum (1960) considered secondary succession in a fallow field and later listed a number of ecosystem attributes which he believed would show certain trends if one compared developmental with mature ecosystems (Odum 1969). For example, he suggested that the ratio of production to respiration of the community would initially be less or greater than one, but a balance would be approached during succession. At Tentsmuir, production initially exceeds respiration, but by the semi-fixed zone respiration exceeds production and continues to do so in the fixed zone (Fig. 21.3). It must be remembered that only *A.arenaria* is being considered, but it is proposed that other species are insufficiently abundant to counteract this trend. Landward of the study zones further organic accumulation must occur (for example in the dune scrub), but there seems to be a gap between reduction in the production of *A.arenaria* and the increase of that of other species. This probably represents a transitional zone of semi-stabilized sand which is not an optimal substrate either for *A.arenaria*, or for other species. Thus it appears that the tendency for community production and respiration to balance may have several phases of accumulation and reduction of organic matter.

Odum (1969) also suggests that 'grazing food chains' predominate in developing ecosystems and 'detritus food chains' in mature ecosystems. However, grazing of *A.arenaria* was negligible during this study. Rabbits (*Oryctolagus cuniculus*) are known to feed on marram grass (Rowan 1913, Gimingham, personal communication), but faecal analyses revealed that they were not doing so at Tentsmuir. Indeed, the early stages of most terrestrial primary successions seem unsuited to abundant grazers, although they are often important in secondary and aquatic successions.

ACKNOWLEDGMENTS

I would like to thank the following: my supervisors, Dr J.J.D. Greenwood and Dr M.J. Cotton, the University of Dundee for provision of finance and facilities and the Nature Conservancy for permission to work on their Reserve. The study of coastal changes was carried out in collaboration with Dr C.D. Green.

REFERENCES

Bocock K.L. & Gilbert O.J.W. (1957) The disappearance of leaf litter under different woodland conditions. *Plant and Soil* **9**, 178–85.

Clapham A.R., Tutin T.G. & Warburg E.F. (1962) *Flora of the British Isles.* (2nd edn.) Cambridge University Press, London.

Cowles H.C. (1889) The ecological relations of the vegetation on the sand dunes of Lake Michigan. *Bot. Gaz.* **27**, 361–69.

Deshmukh I.K. (1974) *Primary production and leaf litter breakdown of* Ammophila arenaria *in a sand dune succession.* Unpublished Ph.D. thesis, University of Dundee.

Elton C.S. (1966) *The Pattern of Animal Communities.* Methuen, London.

Flanagan P.W. & Bunnell F.L. (1976) Decomposition models based on climatic variables, substrate variables, microbial respiration and production. *The Role of Terrestrial and Aquatic Organisms in Decomposition Processes* (Ed. by J.M. Anderson & A. Macfadyen), pp. 437–57. Blackwell Scientific Publications, Oxford.

Grove A.T. (1953) Tentsmuir, Fife. Soil Blowing and Coastal Changes. Unpublished report. Nature Conservancy, Edinburgh.

Howard P.J.A. (1971) Relationship between the activity of organisms and temperature and the computation of the annual respiration of micro-organisms decomposing leaf litter. In: *Proc. 14th Colloq. of Zool. Committee of Intern. Soc. of Soil Sci.* Dijon, pp. 197–205.

Milner C. & Hughes R.E. (1968) *Methods for the Measurement of the Primary Production of Grassland.* Blackwell Scientific Publications, Oxford.

Minderman G. (1968) Addition, decomposition and accumulation of organic matter in forests. *J. Ecol.* **56**, 355–62.

Odum E.P. (1960) Organic production and turnover in an old field succession. *Ecology* **41**, 34–49.

Odum E.P. (1969) The strategy of ecosystem development. *Science* **164**, 262–70.

Olson J.R. (1963) Energy storage and the balance of producers and decomposers in ecological systems. *Ecology* **44**, 322–31.

Petrusewicz K. & Macfadyen A. (1970) *Productivity of Terrestrial Animals.* Blackwell Scientific Publications, Oxford.

Ranwell D.S. (1972) *Ecology of Salt Marshes and Sand Dunes.* Chapman & Hall, London.

Rowan W. (1913) Note on the food plants of rabbits on Blakeney Point, Norfolk. *J. Ecol.* **1**, 273–74.

Wiegert R.G. & Evans F.C. (1964) Primary production and disappearance of dead vegetation on an old field in S.E. Michigan. *Ecology* **45**, 49–63.

Williamson P. (1976) Above-ground primary production of chalk grassland allowing for leaf death. *J. Ecol.* **64**, 1059–75.

22. NITROGEN LIMITATIONS ON THE PRODUCTIVITY OF *SPARTINA* MARSHES, *LAMINARIA* KELP BEDS AND HIGHER TROPHIC LEVELS

K.H. MANN

Department of Biology, Dalhousie University, Halifax, Nova Scotia, Canada

SUMMARY

Evidence is presented for nitrogen limitation affecting the productivity of *Spartina* marshes, kelp beds (*Laminaria*) and sea urchins. *Spartina* communities show greatly increased growth in response to nitrogen fertilization, and plants have nitrogen-fixing bacteria in and around their roots. The rate of nitrogen fixation is inversely proportional to the ammonium content of the soil, and is higher in the centre of marshes than at the edges of creeks. *Laminaria* grows at an accelerating rate throughout winter, and at this time, when the nitrate content of the water is highest, builds up internal reserves of nitrogen. When, at the time of spring bloom of phytoplankton, the nitrate content of the water falls dramatically, the plants continue to grow using internal reserves.

Sea urchins which feed on the *Laminaria* have nitrogen-fixing bacteria in their guts, and the rate of fixation is inversely proportional to the nitrogen content of the food. Detritus derived from *Spartina* and *Laminaria* becomes coated with bacteria which obtain much of their nitrogen from the surrounding water. Dissolved nitrogen appears to control the rates of processes at both the primary and secondary trophic levels.

RÉSUMÉ

Nous montrons que la limitation de l'azote affecte la productivité des marais de *Spartina*, des lits de varech (*Laminaria*) et des oursins. La croissance des communautés de *Spartina* est due à la fertilisation par l'azote et ces plantes ont des bactéries qui fixent l'azote autour de leurs racines. Le taux de fixation de l'azote est inversement proportionnel à la teneur du sol en ammonium et il est plus important dans le centre du marais que sur les bords de la crique.

La croissance des *Laminaria* s'accélère en hiver, quand la teneur de l'eau en nitrate est la plus élevée, elles constituent alors des réserves d'azote. A la floraison du phytoplancton, quand le taux de nitrate de l'eau décroît dramatiquement, les plantes continuent leur croissance—à une vitesse cependant plus réduite qu'en hiver—en utilisant leur réserve interne.

Les oursins qui se nourrissent de *Laminaria* ont dans leur intestin des bactéries fixatrices d'azote et leur taux de fixation est inversement proportionnel à la teneur en azote de leurs aliments. Les détritus provenant de *Spartina* et *Laminaria* sont recouverts de bactéries qui obtiennent la plupart de leur azote de l'eau environnante. L'azote dissous semble contrôler les vitesses du processus à deux niveaux trophiques, primaire et secondaire.

INTRODUCTION

Work at Dalhousie in the last 5 years has demonstrated that in nearshore marine environments the productivity of macrophytes is very high, and that it is limited in a very fundamental way by seasonal variation in the availability of dissolved nitrogen compounds. The plants involved, both salt marsh plants and macrophytic algae, have evolved elaborate strategies for optimizing the use of available nitrogen and for coping with periods of deficiency. Furthermore, this nitrogen limitation extends to the animal food chains which depend on the macrophytes, so that the availability of nitrogen in surface waters of the coastal zone now appears to be a key factor influencing productivity at all trophic levels. Since one of the strategies that has evolved is nitrogen fixation, this process can be seen as one of crucial importance. Even if it adds only 10–20% to the overall nitrogen budget of an area, it may serve an indispensable role if it permits organisms to survive periods of nitrogen deficiency.

NITROGEN RELATIONS IN *SPARTINA* MARSHES

The evidence for nitrogen being a limiting nutrient in the growth of salt marsh vegetation has been well documented by Valiela and Teal (1974), and is referred to elsewhere in this volume. In North American salt marshes the area below the mean high tide level is normally dominated by *Spartina alterniflora*, and there is a clear distinction between the plants growing close to the edges of creeks, which are tall, thick and dark green, and those more remote from the creeks, which are shorter, thinner and have less chlorophyll per g fresh weight. This suggests that the inflowing tides bring some limiting nutrient which particularly benefits those plants along the creek bank. When plants in the interior of the marsh were fertilized with urea, they too became tall, heavy and dark green. Fertilization with phosphate did not produce this effect. Hence it seems probable that the limiting nutrient is nitrogen. Areas of marsh above mean tide level are usually dominated by *Spartina patens* and *Distichlis spicata*. These again are smaller and less productive than the *Spartina alterniflora* plants in the low marsh, and again responded strongly to fertilization with nitrogen.

In temperate coastal waters there is a seasonal cycle in the availability of dissolved nitrate, with high levels in winter but a decline to very low levels at the time of the spring bloom of phytoplankton. This condition persists until the end of summer stratification. Regeneration of ammonium compounds from sediments may be an important source of nitrogen, especially in summer, but it appears to be taken up very rapidly by phytoplankton and concentrations in the water column in summer are often very low (Nixon *et al.* 1976). Hence, if salt marsh plants were to rely entirely on nitrogen supplied by tidal flushing, they would suffer shortages on both a seasonal basis and according to their distance from a creek bank.

It is now clear that nitrogen fixation in the rhizosphere of salt marshes is an important method of combating nitrogen shortages. When Patriquin and Denike (1978) tested for nitrogenase activity by the acetylene reduction technique, using cylinders placed on the marsh surface so as to enclose the plants and soil, they obtained rates of 25–55 μ moles C_2H_2 m^{-2} h^{-1} on the creek edges, and up to 267 μ moles C_2H_2 m^{-2} h^{-1} in the interior of the marsh. A separate study of the algae on the marsh surface showed that they were responsible for no more than 20% of the acetylene reduction, the remainder was occurring primarily in the rhizosphere of the *Spartina*. Bacteria have been isolated from the surfaces of the roots and from within their tissues which appear to be responsible for the nitrogenase activity. When roots were removed from the plants, washed and assayed for nitrogenase activity in the laboratory there was a lag period of 12–24 hr during which no activity occurred. This was thought to be due to destruction of anaerobic micro-environments around the bacteria, which were restored under experimental conditions by the respiration of the plant roots. Optimal conditions for nitrogenase activity appear to be provision of carbon substrates by the roots, and a low-oxygen environment for the bacteria (Patriquin 1978). The nitrogenase activity is also inversely proportional to the ammonium content of the soil water. All of this is consistent with the ideas given above. Salt marsh soils subject to periodic inundation are likely to have partially anaerobic soils, and ammonium content of the soil water is likely to be lower in the centre of a marsh than at the creek banks.

Conversion of acetylene reduction rates to potential nitrogen fixation is a controversial topic. However, Patriquin's (unpublished) best estimate is that potential nitrogen fixation is of the same order of magnitude as the nitrogen accumulated by the plants in a growing season. A similar conclusion was reached in a Georgia marsh by Haines *et al.* (1977). Since much of the plant production enters detritus food chains, this represents a nitrogen subsidy to a nitrogen-deficient system. However, there are also indications that denitrifying bacteria are active in salt marshes (Kaplan *et al.* 1977; Haines *et al.* 1977), and good measurements of the net balance of nitrogen fixation have yet to be obtained. Consideration of the strategies adopted by the algae (see below) suggests that diversion of energy to nitrogen fixation is probably important for survival at certain times of year, but of no great importance to the gross nitrogen budget of a marsh.

NITROGEN RELATIONS OF A *LAMINARIA* KELP BED

The important differences between *Laminaria* and *Spartina* for the present discussion are that whereas *Spartina* is exposed to the air part of the time, and in temperate climates dies down in winter, *Laminaria* is permanently submerged and is capable of photosynthesis and growth throughout winter. An additional difference is that *Spartina* is a flowering plant which takes up its nutrients from the soil through roots, whereas *Laminaria* is an alga with a holdfast but no roots, and takes up its nutrients from the water which bathes the whole blade.

Investigation of seasonal growth patterns of *Laminaria* (Mann 1972) showed that increase in length, which takes place at the meristem at the junction of stipe and blade, is least rapid in summer and autumn, accelerates through winter and reaches a maximum in very early spring, when water temperatures are at the seasonal minimum (about 1°C) or a little above. These observations raised a whole series of questions about whether these plants can carry out photosynthesis effectively at 1°C, and what could possibly be the advantage of making rapid growth in winter and slow growth in summer.

Hatcher *et al.* (in press) provided the answer to the first of these questions, by incubating the plants on the sea floor in an apparatus designed to provide large light and dark chambers filled with prefiltered, constantly moving water. Respiration and photosynthesis were measured by the oxygen method over 24 hr periods, once every 2 weeks for a year. Continuous light measurements were used to extrapolate the data to one week before and one week after each field measurement, thus providing estimates of photosynthesis and respiration for the whole year. It was found that there was a net photosynthetic surplus on every day of measurement except for one in November. The photosynthetic surplus was large enough to account for the observed growth, except for a period during autumn, between October and December. During this time it was shown that carbon reserves were depleted and used to support growth. However, during January, February and March there was a clear photosynthetic surplus and the plants grew at an accelerating rate, although water temperatures were never above 3°C.

The phenomenon of winter growth turned out to be an adaptation to seasonal fluctuations in the availability of nitrate (Chapman & Craigie 1977). Growth in summer is clearly nitrogen limited, and fertilization of a kelp bed with sodium nitrate led to a doubling of growth rates. When nitrate becomes relatively abundant in surface waters the plants not only accelerate their

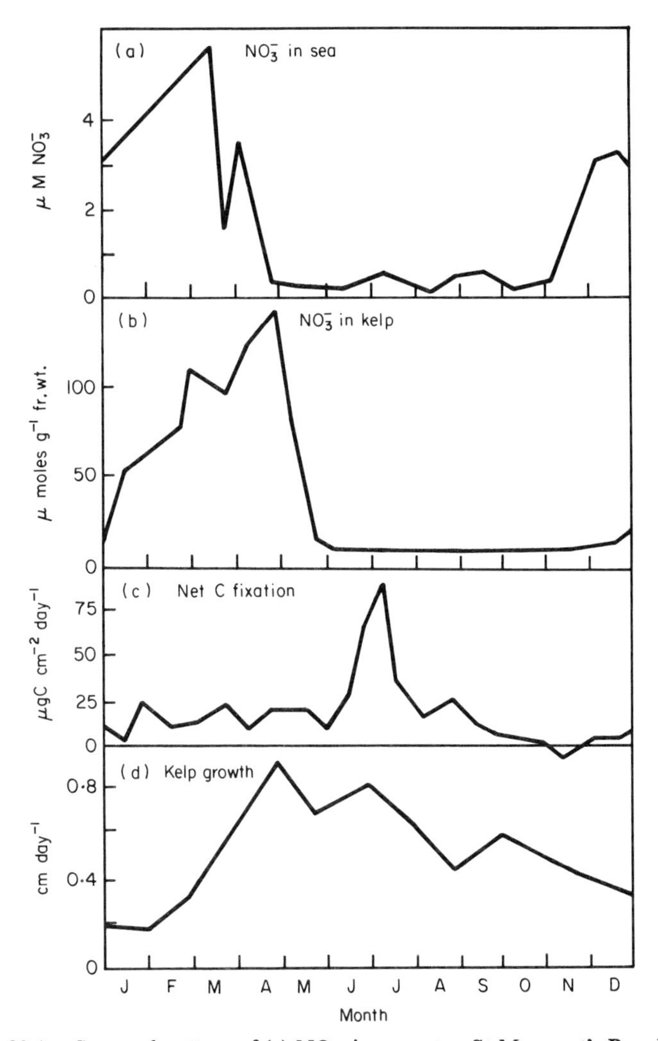

FIG. 22.1. Seasonal pattern of (a) NO_3^- in seawater, St Margaret's Bay, Nova Scotia, (b) NO_3^- in tissues of *Laminaria longicruris*, (c) net carbon fixation (diel surplus of photosynthesis over respiration) in *L.longicruris* and (d) growth at the meristem of *L.longicruris* growing at a depth of 18 m. (a), (b) and (d) from Chapman and Craigie (1977). (c) from Hatcher *et al.* (1978).

growth but accumulated internal stores of nitrate up to 28,000 times the concentration in sea water. This is used to sustain growth during spring and early summer (Fig. 22.1).

HIGHER TROPHIC LEVELS

On the Atlantic coast of Nova Scotia the dominant invertebrate in the kelp beds is the green sea urchin *Strongylocentrotus droebachiensis*. It is primarily a herbivore, and can exist for long periods, perhaps indefinitely on a diet of *Laminaria*. The C/N ratio of this food material is in the range 13:1 to 27:1 (Mann 1972a), while it is commonly held that to sustain animal life and growth a ratio of at least 17:1 is needed (Russell-Hunter 1970). It therefore seems likely that this abundant herbivore is also nitrogen limited. Guerinot *et al.* (1977) have recently shown that the sea urchins have in their guts organisms capable of carrying out nitrogen fixation, and that the rate of fixation is highest when the nitrogen content of the food is lowest. It has yet to be demonstrated that the nitrogen thus fixed is incorporated in the tissues of the urchins, but the probability of it happening appears to be very high.

It is well known that a wide variety of marine invertebrates depend for their food on detritus derived from marine macrophytes. (See Mann 1972 for review.) The process of 'maturing' of detritus in the sea involves colonization by micro-organisms and an increase in the nitrogen content by uptake from the water (Gosselink & Kirby 1974). Although rates have not been measured accurately, it is fairly clear that in this process detritus food chains are also nitrogen limited.

PRODUCTION OF DISSOLVED ORGANIC MATTER

The data of Hatcher *et al.* (1978) showed that in summer, when growth was nitrogen limited, the rate of carbon fixation of *Laminaria* vastly exceeded the amount incorporated in the tissues, and the annual budget suggested that about one third of the carbon fixed did not appear in new tissue or storage products. Johnston *et al.* (1977) found that *Laminaria saccharina* in Scotland had a 13% surplus of fixation over storage and growth. It seems probable that the excess carbon is lost as dissolved organic matter, as described by Sieburth (1969) and Khailov and Burlakova (1969). Such dissolved organic carbon is readily taken up by bacteria and enters the detritus food chains (Mann 1976). The proportion released in the soluble form may well be determined by the degree of nitrogen limitation experienced by the algae. An analogous situation, of production of dissolved organic matter by *Spartina* and its uptake by bacteria, was documented by Gallagher *et al.* (1976). The amount involved was at least 61 kg ha^{-1} yr^{-1}. The rate of loss recorded was at a maximum in March and June, but the relationship to nitrogen limitation, if any, was not recorded.

CONCLUSIONS

Of the evidence reviewed here, the clearest case of nitrogen limitation is that of the *Laminaria*. To adapt to seasonal variation in the availability of nitrate the plants have evolved a strategy of more rapid growth in winter and slower growth in summer. Photosynthesis in summer leads to production of excess organic carbon which is probably leaked to the environment. Internal reserves of nitrogen are accumulated in winter and used to sustain declining growth rates in spring and summer.

Evidence of nitrogen limitation on *Spartina* growth is clear, but the seasonal pattern is not. Nitrogen fixation occurs in the rhizosphere of the plants, carried out by bacteria living in an associative symbiosis, but spatial and temporal variations in the amount of fixation, and the extent to which it is offset by denitrification have yet to be worked out. In north temperate latitudes the shoots die down completely in winter and the plants may not be able to benefit from the seasonal abundance of nitrate in the tidal waters.

The idea that nitrogen limitation may extend to higher trophic levels in coastal ecosystems has not to my knowledge been explicitly stated previously. The evidence is still fragmentary, but scattered evidence from other ecosystems leads me to suppose that a nitrogen limitation on secondary productivity may well be widespread and prove to be a useful predictive tool.

ACKNOWLEDGMENTS

The author is indebted to Drs A.R.O. Chapman, J.C. Craigie and D.G. Patriquin for helpful discussions and permission to quote from unpublished work. Much of the research quoted here was supported by a Negotiated Development Grant to Dalhousie University from the National Research Council of Canada.

REFERENCES

Chapman A.R.O. & Craigie J.S. (1977) Seasonal growth in *Laminaria longicruris*: Relations with dissolved inorganic nutrients and internal reserves of nitrogen. *Mar. Biol.* **40**, 197–205.

Gallagher J.L., Pfeiffer W.J. & Pomeroy L.R. (1976) Leaching and microbial utilization of dissolved organic carbon from leaves of *Spartina alterniflora*. *Estuar. and Coast. Mar. Sci.* **4**, 467–71.

Gosselink J.G. & Kirby C.J. (1974) Decomposition of salt marsh grass, *Spartina alterniflora* Loisel. *Limnol. Oceanogr.* **19**, 825–32.

Guerinot M.L., Fong W. & Patriquin D.G. (1977) Nitrogen fixation (acetylene reduction) associated with sea urchins (*Strongylocentrotus droebachiensis*) feeding on seaweeds and eelgrass. *J. Fish. Res. Board Can.* **34**, 416–20.

Haines E., Chalmers A., Hanson R. & Sherr B. (1977) Nitrogen pools and fluxes in a Georgia salt marsh. *Estuarine Processes* Vol. II (Ed. by M. Wiley), pp. 241–54. Academic Press, New York.

Hatcher B.G., Chapman A.R.O. & Mann K.H. (1978) An annual carbon budget for the kelp *Laminaria longicruris*. *Mar. Biol.* **44**, 85–96.

Johnston C.S., Jones R.G. & Hunt R.D. (1977) A seasonal carbon budget for a laminarian population in a Scottish sea-loch. *Helgoländer wiss. Meeresunters.* **30**, 527–45.

Kaplan W.A., Teal J.M. & Valiela I. (1977) Denitrification in salt marsh sediments: Evidence for seasonal temperature selection among populations of denitrifiers. *Microbial Ecol.* **3**, 193–204.

Khailov K.M. & Burlakova Z.P. (1969) Release of dissolved organic matter by marine seaweeds and distribution of their total organic production to inshore communities. *Limnol. Oceanogr.* **14**, 521–27.

Mann K.H. (1972a) Ecological energetics of the seaweed zone in a marine bay on the Atlantic coast of Canada. I. Zonation and biomass of seaweeds. *Mar. Biol.* **12**, 1–10.

Mann K.H. (1972b) Ecological energetics of the seaweed zone in a marine bay on the Atlantic coast of Canada. II. Productivity of the seaweeds. *Mar. Biol.* **14**, 199–209.

Mann K.H. (1976) Decomposition of marine macrophytes. *The Role of Terrestrial and Aquatic Organisms in Decomposition Processes* (Ed. by J.M. Anderson & A. Macfadyen), pp. 247–67. Blackwell Scientific Publications, Oxford.

Nixon S.W., Oviatt C.A. & Hale S.S. (1976) Nitrogen regeneration and the metabolism of coastal marine bottom communities. *The Role of Terrestrial and Aquatic Organisms in Decomposition Processes* (Ed. by J.M. Anderson & A. Macfadyen), pp. 269–83. Blackwell Scientific Publications, Oxford.

Patriquin D.G. (1978) Factors affecting nitrogenase activity (acetylene reducing activity) associated with excised roots of the emergent halophyte *Spartina alterniflora* Loisel. *Aquat. Bot.* **4**, 193–210.

Patriquin D.G. & Denike D. (1978) *In situ* acetylene reduction assays of nitrogenase activity associated with the emergent halophyte *Spartina alterniflora* Loisel: methodological problems. *Aquat. Bot.* **4**, 211–26.

Russell-Hunter W.D. (1970) *Aquatic Productivity: A Introduction to Some Basic Aspects of Biological Oceanography and Limnology*, 306 pp. MacMillan, New York.

Sieburth J.McN. (1969) Studies on algal substances in the sea. III. The production of extracellular organic matter by littoral marine algae. *J. exp. mar. Biol. Ecol.* **3**, 290–309.

Valiela I. & Teal J.M. (1974) Nutrient limitation in salt marsh vegetation. *Ecology of Halophytes* (Ed. by R.J. Reimold & W.H. Queen), pp. 547–63. Academic Press, New York.

V

NUTRIENT AND ENERGY FLOW

23. NITROGEN MINERALIZATION IN A SALT MARSH ECOSYSTEM DOMINATED BY *HALIMIONE PORTULACOIDES*

KAJ HENRIKSEN[1] AND ARNE JENSEN[2]

[1] *Institute of Ecology and Genetics, University of Aarhus, Ny Munkegade, DK-8000 Aarhus C, Denmark*

and

[2] *Botanical Institute, University of Aarhus, Nordlandsvej 68, DK-8240 Risskov, Denmark*

SUMMARY

In the aerobic sediment of the outer part of the Skalling marsh nitrate concentrations were low throughout the year except for a spring peak in April/May. Ammonia concentrations usually were high (0·5–1·5 μmol cm^{-3}) for much of the year, but depletion of ammonia was recorded in late summer of years with high summer temperatures. Net nitrogen mineralization rates of 43, 374 and 591 nmol cm^{-3} week^{-1} and net nitrification rates of 19, 318 and 629 nmol cm^{-3} week^{-1} at 5°C, 15°C and 25°C respectively were measured in long-term incubations of sediment cores in the laboratory. From these results and those of the sediments in the marsh, net rates per month of mineralization and nitrification were calculated. Net mineralization per year was 11 g N m^{-2}. More than 80% of the mineralized nitrogen probably was converted to nitrate during the growing season from May to October.

RÉSUMÉ

Dans les sédiments aérobies de la partie extérieure des marais Skalling, les concentrations en nitrate ont été faibles durant l'année sauf un pic au printemps, en avril mai. Les concentrations d'ammonium ont été normalement élevées (0·5–1·5 μmol cm^{-1}) la plupart de l'année, mais une diminution de l'ammonium a été enregistrée à la fin de l'été, pour les années avec de fortes températures estivales. Les taux de minéralisation nette d'azote par semaine de 43, 374 et 591 nmol cm^{-3} respectivement à 5°C, 15°C et 25°C ont été mesurés dans le laboratoire pour des incubations à long terme, d'échantillons de sédiments. A partir de ces résultats et de ceux des sédiments dans le marais, nous avons calculé le taux net mensuel de minéralisation et de nitrification. La minéralisation nette annuelle a été de 11 g N m^{-2}. Plus de 80% de l'azote minéralisé ont été convertis en nitrate durant la saison de croissance, de mai à octobre.

INTRODUCTION

Deficiency of nitrogen appears to limit the primary production of salt marshes, particularly at sites far from the sea (Pigott 1969; Stewart *et al.* 1972, 1973; Valiela & Teal 1974). Several authors have suggested that these gradients in the availability of nitrogen in marshes reflect differences in the input of inorganic nitrogen associated with the frequency and duration of tidal inundation at these sites. Concentrations of inorganic nitrogen in coastal waters, however, are usually low during much of the growing season (Ryther & Dunstan 1971).

Most investigations of whether primary production of halophytes is limited by the availability of nitrogen and phosphorus have examined the response of plants to additions of fertilizer and little attempt has been made to measure rates of nitrogen mineralization in marsh sediments themselves. The aims of the present paper are to present data for seasonal variations in concentrations and net mineralization rates of inorganic nitrogen (NH_4^+, NO_3^-) in the sediments of a marsh and to relate the results to variations in temperature of the sediments and the frequency and duration of tidal inundation at the different sites.

DESCRIPTION OF STUDY SITE

The Skallingen marsh is a young marsh of the off-shore bare type situated on the west coast of southern Jutland. The marsh has developed since the start of the century. A thorough description of developmental stages is given by Jacobsen (1952). In the outer ungrazed parts of the marsh the sedimentation layer varies between 5 and 20 cm in depth and the level of the underlying sandflat is between 70 and 90 cm above mean water level.

The marsh is drained by a well developed creek system and the mean tidal amplitude is 130 cm. The salinity of the incoming tide is about 30‰. Precipitation is 750 mm per year and mean annual temperature is 8·4°C (Jan. 0·6°C, July 16·7°C).

The study site is dominated by *Halimione portulacoides*. According to the point-frame method (Levy & Madden 1933) the percentage cover for the various species is:

Halimione portulacoides	85%	*Suaeda maritima*	1%
Puccinellia maritima	25%	*Aster tripolium*	<1%
Limonium vulgare	7%	*Salicornia europaea* agg.	<1%
Plantago maritima	6%	*Spergularia marginata*	<1%
Triglochin maritima	1%		

The sedimentation layer is fine textured with a high content of silt (60%) and clay (20%) by weight. Nevertheless the sediment is well aerated throughout the year with a remarkably constant air-filled pore column of 14% ± 3%. The upper layer is nearly impenetrable to water. During flood tide the soil air in this layer is trapped between the rising soil water table and the tide. This mechanism creates a constantly aerated upper sediment layer as described by Chapman (1974).

Two sampling sites were used in the investigation, both with a yearly tidal submergence of approximately 1,000 hours. The depths of the sedimentation layers were 18 cm and 12 cm for sites I and II respectively. The organic N content of the upper 10 cm was 330 g m^{-2} and the C/N ratio was 8–10 W/W with the highest values at the surface for both sites. The pH varied between 6 and 7.

MATERIALS AND METHODS

Field studies

Seasonal changes in ammonia and nitrite plus nitrate available for plant growth (i.e. soluble + exchangeable) were followed from May 1973 to September 1974 at sampling site I and from September 1974 to October 1975 at sampling site II.

On each sampling occasion 10 cores of the sedimentation layer were taken randomly with use of random number tables. The cores were sectioned as follows:

Site I 0–9 cm and 9–18 cm below the surface of the sediment.
Site II 0–6 cm and 6–12 cm below the surface of the sediment.

There was no evidence of stratification of the sediment, but more than 90% of the root biomass was restricted to the upper 10 cm of the sediment.

The seasonal variations in the periods of tidal inundation at the sampling sites were calculated from tidal data obtained in the harbour of Esbjerg. Regression curves of water levels at Esbjerg and Skallingen were constructed (Jensen 1973) and the theoretical submergence time at the sampling sites calculated and expressed as hours of inundation per month. This figure is an underestimate of the actual time of submergence (Jensen 1975), but it does show the seasonal variation of inundation.

During the spring and summer of 1976 measurements of the temperature in the sediment at site II were carried out with data-logging equipment at intervals of one hour. Measurements were made at depths of 2·5 cm, 5 cm, 10 cm and 20 cm and the results were correlated with the mean air temperature from the nearest meterological station at Nordby, Fanø.

Laboratory experiments

In an attempt to investigate the potential net nitrogen mineralization rate of the sediment, intact cores (9 cm long, 7 cm diameter) were incubated in the dark at 27°C in plastic beakers covered with parafilm, in order to maintain the water content of the cores as well as aerobic conditions. The redox potential (Eh) never fell below 400 mV. Thirty cores were incubated and on each sampling occasion 10 cores were analysed separately for ammonia and nitrate ions. Long-term incubations of intact cores from both sites were made over a period of five months. Again the Eh never fell below 400 mV, whereas in pilot experiments with homogenized samples the interior of the samples became anaerobic after one week.

The influence of temperature on the net rates of ammonification and nitrification was studied in cores from site II. The cores were incubated for 3 months at 5°, 15° and 25°C and on each sampling occasion five cores were analysed separately for soluble plus exchangeable ammonium, nitrite and nitrate ions.

Analytical procedures

All analyses were made on fresh sediment within 24 hours of sampling. Analyses were made in duplicate on all samples and analytical errors were less than 4%. All concentrations are expressed as μmol cm^{-3} of fresh sediment. Ammonia and nitrite plus nitrate were measured by the method of Bremner and Keeney (1965) after extraction of the samples with 0·2 N KCl. The indophenol-blue method was used for determination of ammonium ions after steam distillation of the extracts (Solorzano 1969).

Nitrite was measured separately in the extracts (KCl) using the method described by Strickland and Parsons (1968) and pH was measured directly in the KCl extract with a combined glass-calomel electrode.

RESULTS

Field observations

Seasonal variation in the quantities of exchangeable ammonium and nitrite plus nitrate ions at sites I and II are shown in Fig. 23.1. Nitrite concentrations were always low (<10 nmol N cm^{-3}). Nitrate concentrations were high in spring, both in 1973 and 1974. With the onset of the growing season for *Halimione portulacoides* in the second half of May the concentrations fell to near zero in July 1973 and in July 1974, although in 1974 there was still a big pool of exchangeable ammonium ions (1·6 μmol cm^{-3}) present at this time.

Throughout the autumn of 1974 the exchangeable levels of ammonium ions were high while the corresponding levels of nitrate ions were close to zero. Measurements stopped in the spring of 1975 and recommenced in July the same year. From August to September the pool of inorganic exchangeable nitrogen changed from being predominantly ammonium to predominantly nitrate ions.

The seasonal variations in air temperatures and estimates of the hours of

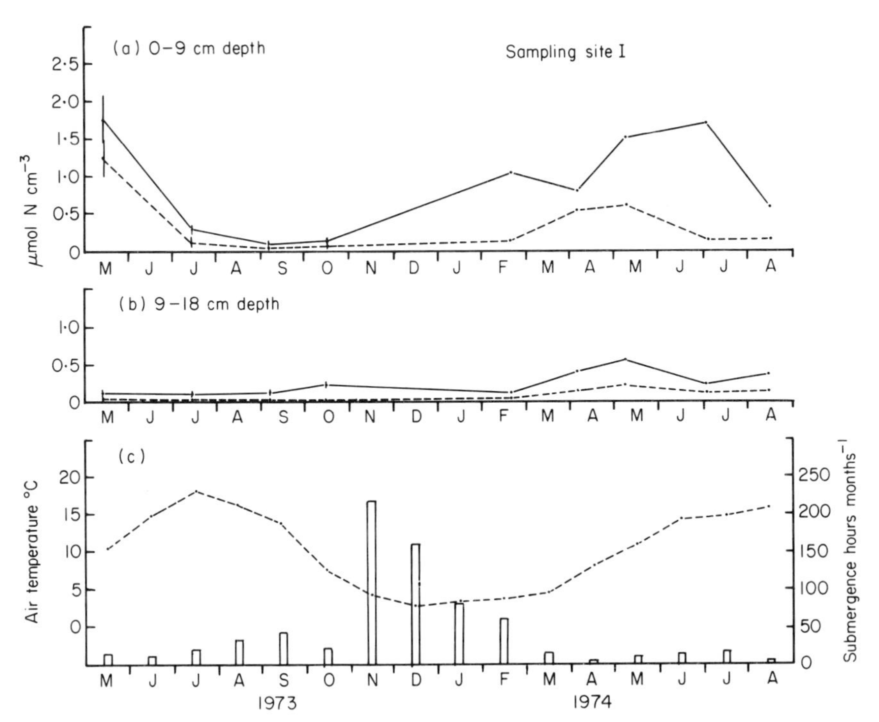

FIG. 23.1. Seasonal variations in amounts of soluble plus exchangeable ammonium ions (·——·) and nitrite plus nitrate ions (·- - -·) (μmol N cm^{-3}) at different depths (a, b) together with mean monthly air temperature and hours of submergence (c) at site I (May 1973 to August 1974) and site II (August 1974 to September 1975). Each point is the mean of 10 samples chosen by random number table. 95% confidence intervals are indicated for 1973. In 1974 and 1975 the samples were pooled, homogenized and three subsamples taken out for analyses.

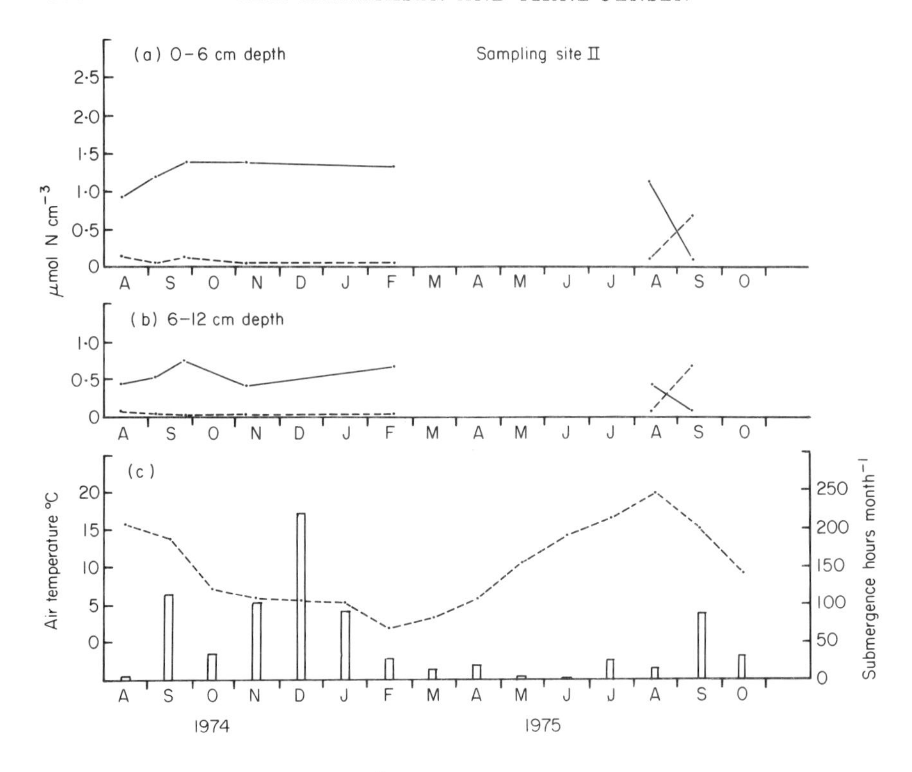

Fig. 23.1.—*continued*

submergence per month are shown also in Fig. 23.1. The duration of submergence was very similar for the different years with low values recorded in spring and summer. The mean air temperature was more than 2°C higher in July 1973 than in July 1974. August 1975 was abnormally warm (20°C).

Linear regressions based on mean weekly temperatures of the sediments and air are shown in Fig. 23.2. The lines which related the temperatures of the sediments at depths of 5 and 10 cm with air temperatures were not significantly different from one another and the data were pooled. The close relationship between air and sediment temperatures ($r = 0{\cdot}99$ and $r = 0{\cdot}98$) makes it possible to use data from the local meteorological station to estimate the sediment temperature.

Results of laboratory experiments

Results of the net nitrogen mineralization in intact cores from site I are shown in Fig. 23.3. The maximum rate for the upper 0–9 cm was 1,228 nmol cm^{-3} $week^{-1}$ and mean rate for the whole period was 918 nmol cm^{-3} $week^{-1}$. The rate of net mineralization in the cores taken at a depth of 9–18 cm was only

30% of that in the upper layer. The net mineralization rates for site II were slightly lower, but here too all excess ammonia apparently was converted to nitrate.

There appeared to be no exhaustion of substrate during the long incubation period, indicating a large supply of easily mineralizable organic nitrogen in the cores. The magnitude of the rate of net mineralization is comparable to

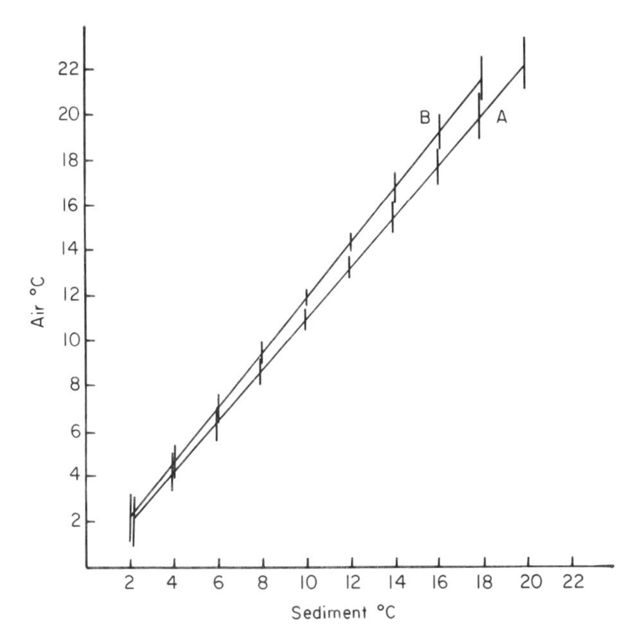

FIG. 23.2. Regression of air temperature on sediment temperature (weekly mean) for 2·5 cm (A, $y = 0{\cdot}39 + X \times 1{\cdot}22$, $r = 0{\cdot}98$) and 5 cm and 10 cm (B, $y = 0{\cdot}29 + X \times 1{\cdot}13$, $r = 0{\cdot}99$) depth of sediment. Regressions for depths of 5 and 10 cm were not significantly different and the figures were pooled. 95% confidence intervals are indicated.

that of good agricultural soils, indicating a high flux of nutrients through the primary producer-decomposer chain in this ecosystem.

The influence of temperature on the net rates of ammonification and nitrification is shown in Fig. 23.4. Data of the first set of samples are omitted from the calculations of net rates as they are considered to represent the lag period. This is especially evident in samples incubated at lower temperatures. The optimal pH conditions for nitrifying bacteria are between pH 7 and 9 with growth considerably inhibited at pH 6 compared to pH 7 (Morrill & Dawson 1962). The mean of the pH values between sets of samples was constant throughout the incubation period ($6{\cdot}25 \pm 0{\cdot}25$). There was, however,

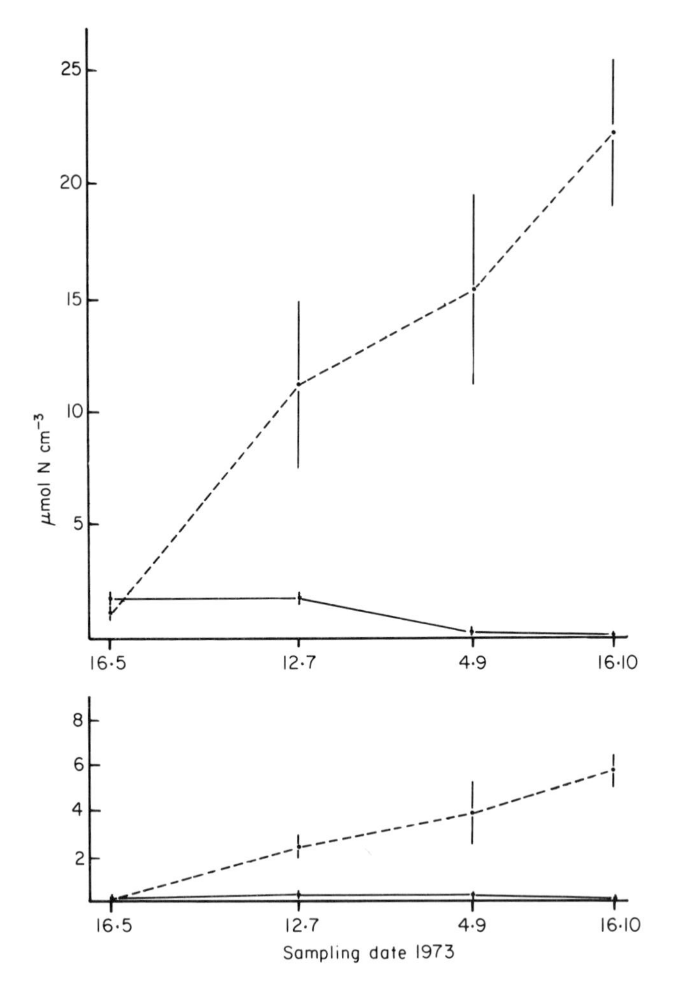

FIG. 23.3. Soluble plus exchangeable ammonium ions (·——·) and nitrite plus nitrate ions (·– – – –·) (μmol N cm^{-3}) in intact sediment cores (9 cm × 7 cm) taken from two depths (a, 0–9 cm, b, 9–18 cm) at site I and incubated at 27°C for periods of time up to 5 months. 95% confidence intervals are indicated ($n = 10$).

considerable deviation in pH values between individual samples (maximum range 0·8 pH units). An inhibition of nitrification occurred in samples where the pH was low after 4 weeks of incubation, but these differences were not observed after 8 and 12 weeks of incubation. Low sub-optimal pH conditions in the marsh sediment are therefore considered to be responsible for the long lag preceding nitrification even at 15°C and 25°C.

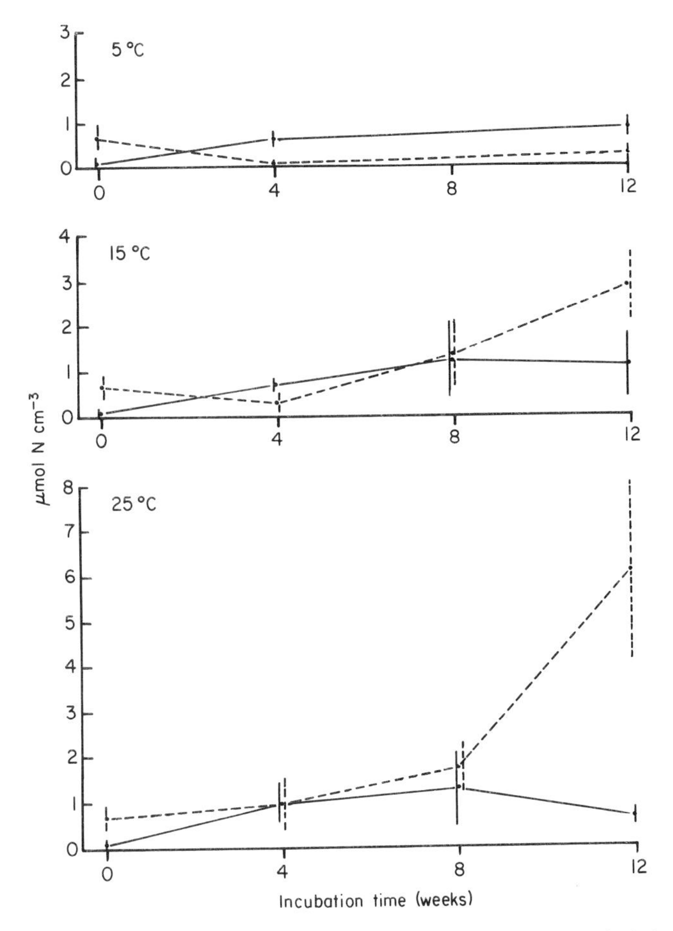

FIG. 23.4. Soluble plus exchangeable ammonium ions (·——·) and nitrite plus nitrate ions (·- - - -·) (μmol N cm^{-3}) in intact sediment cores (9 cm × 7 cm) incubated at 5°C, 15°C and 25°C. Cores were taken from site II in the Skalling marsh at a depth between 0 and 10 cm and cores were harvested at intervals of four weeks after incubation commenced. S.E.M. are indicated ($n = 5$).

Nitrite concentrations were low in all samples (1–10 nmol cm^{-3}). The calculated net mineralization and nitrification rates are given in Table 23.1. There was an almost linear relationship between the measured net rate and temperature. From this relationship the rates of net mineralization and net nitrification for each month were estimated from the actual mean temperature at depths of 5 and 10 cm in the sediment. The results are shown in Fig. 23.5. In the calculations it has been assumed that no nitrification occurs below 5°C (November to April).

Table 23.1. *Net nitrogen mineralization ($NH_4^+ + NO_2^- + NO_3^-$) and nitrification rates ($NO_2^- + NO_3^-$) at different temperatures (5°C, 15°C and 25°C) in intact sediment cores from site II, incubated at a constant water content in the laboratory.*

	nmol cm^{-3} $week^{-1}$		
	5°C	15°C	25°C
$NH_4^+ + NO_2^- + NO_3^-$	43	374	591
$NO_2^- + NO_3^-$	19	318	629

It should be emphasized that the calculations reflect only the influence of temperature and do not take account of possible seasonal variations in net rates of ammonification and nitrification. Higher rates do often occur in spring due to increases in easily mineralizable organic nitrogen and decreases in bacterial population during the winter months (Davy & Taylor 1974).

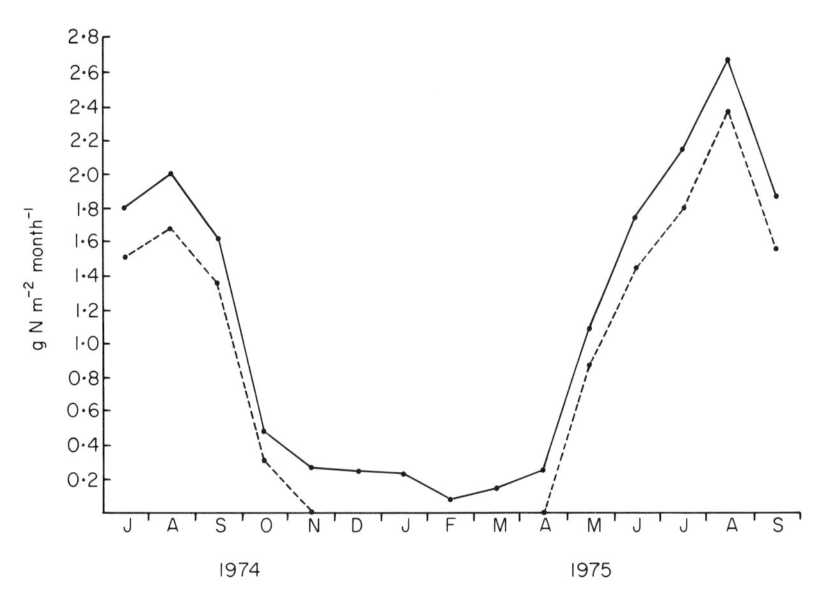

FIG. 23.5. Estimated rates of net nitrogen mineralization (·——·) and net nitrification (·– – –·) (g N m^{-2} $month^{-1}$) in sediment (0–12 cm below surface) at site II, Skalling marsh during 1974/1975. The rates were calculated from data of rates of net mineralization and net nitrification in sediments incubated at different temperatures in the laboratory and from data of sediment temperatures at site II.

Net nitrogen mineralization per year (September 1974 to September 1975) was 11 g N m^{-2} and net nitrification 8·3 g N m^{-2}.

DISCUSSION

The aerobic sediments of the Skallingen marsh appear to justify the use of simple incubation techniques to measure net mineralization rates. The water content and air-filled pore volume show little annual variation, the supply of organic nitrogen which can be easily mineralized is large and the sediments have a low C/N ratio; these characteristics of the sediments allow the results of mineralization to be examined with some confidence.

The nitrate which is accumulated during nitrification does not participate normally in the internal bacterial nitrogen cycle (Harmsen & Kolenbrander 1965) because of a microbial preference for ammonia. All of the nitrate produced during nitrification can therefore be considered as available for plant growth provided there is no denitrification.

Temperature appears to be an important factor affecting the quantity of inorganic nitrogen available for plant growth. At low temperatures ammonia accumulates together with nitrate in the sediment, but at higher temperatures the amount of ammonia is between 0·5 and 1·0 μmol cm^{-3}. One reason for this could be an inhibition of the activities of nitrifying bacteria as a result of the sub-optimal pH conditions of the sediment.

Calculated net production of inorganic nitrogen from September 1974 to September 1975 is 11 g N m^{-2}, which corresponds well with the measured N-assimilation of 12 g N m^{-2} by the vegetation in this period (Jensen, unpublished data). Of the total amount of nitrogen which was mineralized 80% or 8·3 g N m^{-2} probably was converted to nitrate in the growing season from May to October. Whether this amount is produced depends, in part, on the competition between plants and nitrifying bacteria for the excess ammonia produced.

Nitrate concentration in the sediment was low throughout the year except for April and May, while ammonia concentrations (0·5–1·5 μmol cm^{-3}) usually were high for most of the year. Depletion of ammonia in the sediment occurred in the late summer of 1973 and 1975. This may reflect the high rates of nitrification at the high temperatures followed by assimilation or denitrification.

It is not likely that inorganic nitrogen supplied by the tide could play a major role in supplying nitrogen to the study sites. Concentrations of ammonia and nitrate in coastal waters are low usually during the growing season (Ryther & Dunstan 1971), and this also applies to the Wadden Sea (Postma 1966). Even at the lowest concentrations of nitrate in the sediment in the period from May to October, the concentrations in the interstitial water are between 50 and 200 μmol NO_3^- l^{-1}. It is more likely that ammonia is exported with the tide during most of the year.

ACKNOWLEDGMENTS

This work was supported, in part, by a grant from the National Research Council of Denmark. We are grateful to Professor T.H. Blackburn for useful discussions and critical reading of the manuscript.

REFERENCES

Chapman V.J. (1974) *Salt Marshes and Salt Deserts of the World*, 2nd edn., Verlag von J. Cramer, Lehre.

Bremner J.M. & Keeney D.R. (1965) Steam distillation methods for determination of ammonium, nitrate and nitrite. *Anal. Chim. Acta* **32**, 485–95.

Davy A.J. & Taylor K. (1974) Seasonal patterns of nitrogen availability in contrasting soils in the Chiltern Hills. *J. Ecol.* **62**, 793–807.

Harmsen G.W. & Kolenbrander G.J. (1965) Soil inorganic nitrogen. *Agronomy No. 10. Soil Nitrogen* (Ed. by W.V. Bartholomew & F.E. Clark), pp. 43–92. Amer. Soc. of Agron., Madison, Wisconsin.

Jacobsen B. (1952) Landskabsudviklingen i Skallingmarsken. *Geografisk Tidsskrift* **52**, 147–59.

Jensen A. (1973) Kormofytvegetationen i den ydre marsk på Skallingen—succession i relation til edafiske, hydrologiske og klimatiske faktorer. Cand. scient. thesis, Institute of Ecological Botany, Copenhagen University.

Jensen A. (1975) A method of measuring salt marsh inundation. *Oikos* **25**, 252–54.

Levy E.B. & Madden E.A. (1933) The point method of pasture analyses. *N.Z. J. Agric.* **46**, 267–79.

Morrill L.G. & Dawson J.E. (1962) Growth rates of nitrifying chemoautotrophs in soil. *J. Bacteriol.* **83**, 205–6.

Pigott C.D. (1969) Influence of mineral nutrition on the zonation of flowering plants in coastal salt marshes. *Ecological Aspects of Mineral Nutrition of Plants* (Ed. by I.H. Rorison), pp. 25–37. Blackwell Scientific Publications, Oxford.

Postma H. (1966) The cycle of nitrogen in the Wadden Sea and adjacent areas. *Neth. J. Sea Res.* **3**, 186–221.

Ryther J.H. & Dunstan W.M. (1971) Nitrogen, phosphorus and eutrophication in the coastal marine environment. *Science* **171**, 1008–13.

Solorzano L. (1969) Determination of ammonia in natural waters by the phenol-hypochlorite method. *Limnol. Oceanogr.* **14**, 799–810.

Stewart G.R., Lee J.A. & Orebamjo T.O. (1972) Nitrogen metabolism of halophytes. I. Nitrate reductase activity in *Suaeda maritima*. *New Phytol.* **71**, 263–67.

Stewart G.R., Lee J.A. & Orebamjo T.O. (1973) Nitrogen metabolism of halophytes. II. Nitrate availability and utilization. *New Phytol.* **72**, 539–46.

Strickland J.D.H. & Parsons T.R. (1968) *A Practical Handbook of Seawater Analysis.* Fish. Res. Bd. Can. No. 167.

Valiela I. & Teal J.M. (1974) Nutrient limitation in salt marsh vegetation. *Ecology of Halophytes* (Ed. by R.J. Reimold & W.H. Queen), pp. 547–63. Academic Press, New York.

24. MICROBIAL NITROGEN TRANSFORMATIONS IN THE SALT MARSH ENVIRONMENT

S.A. ABD. AZIZ AND D.B. NEDWELL

Department of Biology,
University of Essex, Colchester, U.K.

SUMMARY

Organic nitrogen was the most abundant form of nitrogen in salt marsh sediments at all sites investigated. Only a small part of this organic nitrogen in the upper layer of the sediment was transformed annually to inorganic nitrogen as a result of microbial activity, the remainder being refractory and unavailable for recycling. Mineralization of organic nitrogen results in production of ammonium ions but no nitrate is produced.

The rates of mineralization of organic nitrogen in intact sediment cores are described by Arrhenius functions and these data together with results of sediment temperatures are used to calculate the annual rates of ammonia production at each site. Estimates of the nitrogen assimilated by higher plants on the salt marsh gave values of similar magnitude to those obtained for the mineralization of organic nitrogen. Annual rates of nitrogen fixation were an order of magnitude lower than corresponding rates for mineralization and assimilation of nitrogen.

Nitrification could not be detected readily in the salt marsh sediments. The microbial community of the sediments had a high capacity to denitrify, but the actual *in situ* rates of denitrification were negligible because of nitrate limitation.

It is suggested that in this salt marsh the majority of nitrogen cycling is internal and only a comparatively small proportion of the annual nitrogen turnover is exported.

RÉSUMÉ

L'azote organique est la forme la plus abondante de l'azote dans les marais salés de tous les sites étudiés. Chaque année, seule une petite part de cet azote organique de la couche supérieure du sédiment est transformée en azote inorganique par l'activité microbienne, le reste étant réfractaire et inutile pour le cycle. La minéralisation de l'azote organique produit des ions ammonium, mais pas de nitrate.

Le taux de minéralisation de l'azote organique dans le sédiment intact suit la fonction d'Arrhenius, et nous calculons les taux annuels de production d'ammonium dans chaque site grâce à ces expériences auxquelles s'ajoutent les résultats des températures du sédiment. Des estimations de l'azote assimilé par les plantes supérieures du marais salé ont donné des valeurs similaires à celles obtenues pour la minéralisation de l'azote organique. Les taux annuels de fixation de l'azote sont d'un ordre de grandeur inférieur à ceux correspondant à la minéralisation et l'assimilation de l'azote.

INTRODUCTION

Pigott (1969) and Valiela and Teal (1974) have demonstrated that nitrogen is commonly the major growth-limiting nutrient for plants present on a salt marsh. Enhancement of the nitrogen concentration in the sediment by artificial fertilization greatly stimulates primary production, as well as producing profound changes in the algal species present (van Raalte *et al.* 1976).

The classic work of Teal (1962) showed that in the *Spartina*-dominated salt marshes of the North American coast there is a net export of energy from the marsh community to the adjacent estuarine or coastal water in the form of detrital biomass. Organically bound nitrogen is lost with the detritus, and inorganic forms of nitrogen may also be exported as a result of tidal movement (Odum & de la Cruz 1967; Gardner 1975). Nitrogen fixation occurs within marshes (Jones 1974), and pollution may sometimes provide an additional source of inorganic nitrogen in coastal waters. If this latter source provides adequate supplies of nitrogen to the marsh, nitrogen fixation by organisms may be depressed (van Raalte *et al.* 1974).

Notwithstanding the obvious importance of the nitrogen cycle in the functioning of the salt marsh system, little data are available on the relative importance of the different forms of nitrogen, and on their interconversions. Particular stages in the nitrogen cycle have previously been individually investigated, but in the work described here all nitrogen transformations have been investigated to determine their relative importance in the function of the salt marsh ecosystem.

The study site is the Colne Point Salt Marsh owned by the Essex Naturalist Trust, at the mouth of the River Colne, Essex, U.K. The area, which is a salting, is a relatively stable area of high marsh, some 4·8 m above tidal datum, dominated by a mixed community of *Halimione portulacoides*, *Puccinellia maritima*, *Limonium vulgare* and some *Spartina townsendii*. The marsh is drained by a creek system and the levées of these creeks are being eroded by tidal water. There is little of what might be termed a primary colonization zone.

CHEMICAL ANALYSES

As a first step in the investigation the major pools of nitrogen in the sediments were analysed periodically. Ten sites were selected which represented the different habitats of the marsh. Two creek sites, six sites on the salt marsh flat, and two salt marsh pans were sampled. The sites on the marsh flat were selected areas with distinct plant associations (S. Osman, personal communication). Two distinct associations were present in the plant community in a 50 × 20 m study area. On one end of the area the plant community was composed of *Halimione portulacoides* and *Puccinellia maritima* which strongly shaded the underlying sediment. The sediment contained about 50% water by weight. The other portion of the area was covered by a mixed growth of *Limonium vulgare*, *Armeria maritima*, *Triglochin maritima*, *Spartina townsendii* and *Puccinellia maritima*. This plant association did not shade the underlying sediment as much as the *Halimione* and *Puccinellia* association. The sediment contained 70–75% (w/w) water. Shorter *Puccinellia* plants were found in this wetter region of the study area. Of the six study sites on the salt marsh flat, two were in each of these major plant associations, while two were in the transitional region between the two associations.

Duplicate vertical cores of sediment (5 cm deep, diameter 1·1 cm) were taken from within each site using cut off 5 ml plastic hypodermic syringes. Complete randomization of sampling within each site was not possible as clumps of plant stems prevented coring and cores were, therefore, only taken from areas of base sediment between plant stems. However, the cores removed would have included plant roots cut off by the coring process. The monthly samples were removed as evenly as possible from over the entire sampling site to ensure that no particular part of the site was changed by locally concentrated removal of cores. The hypodermic syringes were capped with rubber teats to exclude air and returned to the laboratory for immediate analysis. The surface layer of sediment which was thus sampled had been previously shown to encompass the zone of major microbial activity (Nedwell & Abram 1978).

Exchangeable ammonium was steam distilled directly from a sediment core suspended in 2N potassium chloride solution (Bremner 1965). The distillate was trapped in dilute sulphuric acid and ammonium ion assayed by a colorimetric technique (Harwood & Kuhn 1970). Recovery of added ammonium was shown to be at least 98% efficient by this method.

A saturated calcium sulphate solution was used to extract nitrate from the sediments, nitrate in the extract being assayed colorimetrically (Strickland & Parsons 1968). The efficiency of recovery of nitrate added to sediment samples was only 83%. Therefore, in all analyses nitrate standards were added to replicate sediment samples to check the efficiency of extraction, and the values for nitrate extracted from unsupplemented replicates were corrected accordingly.

Total nitrogen was determined by Kjeldahl digestion of nitrogenous compounds to ammonia (Bremner 1965). The ammonia was then recovered from the digest by steam distillation into dilute sulphuric acid and assay of the ammonium ion concentration as described previously. Organic nitrogen was calculated as the difference between these results and the estimates of nitrate and ammonium ions.

At the end of six months analyses of variance were performed on the accumulated data. Plots were grouped in pairs and if there was no significant difference between a pair of plots in the same habitat, one of the plots was abandoned. Five sites remained (Table 24.1). One difference between the salt

Table 24.1. *Selected sampling sites used on Colne Point salt marsh, Essex.*

Site	Description
Site 1.	On the marsh flat, dominated by mixed *Spartina townsendii* and *Puccinellia maritima.*
2.	On the marsh flat, dominated by *Halimione portulacoides.*
3.	Bottom of drainage creek.
4.	Salt marsh pan with brown benthic layer.
5.	Salt marsh pan with black benthic mat dominated by *Oscillatoria limosa.*

marsh pans was the presence of a well developed black benthic mat dominated by *Oscillatoria limosa* at site 5 which was absent at site 4. The two remaining sites on the marsh flat were represented by plot 1 from the region consisting of *Limonium vulgare*, *Armeria maritima*, *Triglochin maritima*, *Spartina townsendii* and *Puccinellia maritima*, and plot 2 from the region composed of *Halimione portulacoides* and *Puccinellia maritima.*

Fig. 24.1 shows the analytical data of the concentrations of different species of nitrogen in the sediments at the sites. Organic nitrogen was by far the greatest source of nitrogen at all sites, although in general the amounts present in the salt marsh flat were twice as high as the amounts in the pans, and corresponding values for the sediments at the creek site were even lower. For the marsh flat the amount of organic nitrogen present in the sediments at site 2 was higher than at site 1. Ammonium was the dominant inorganic ion with the highest level found in the sediments at site 5 and lowest in the sediments at site 3, while nitrate ions were present only in very small amounts in all the sediments. No distinct seasonal trends were present except that ammonium ions in the sediments were found to be lower in late autumn and in winter.

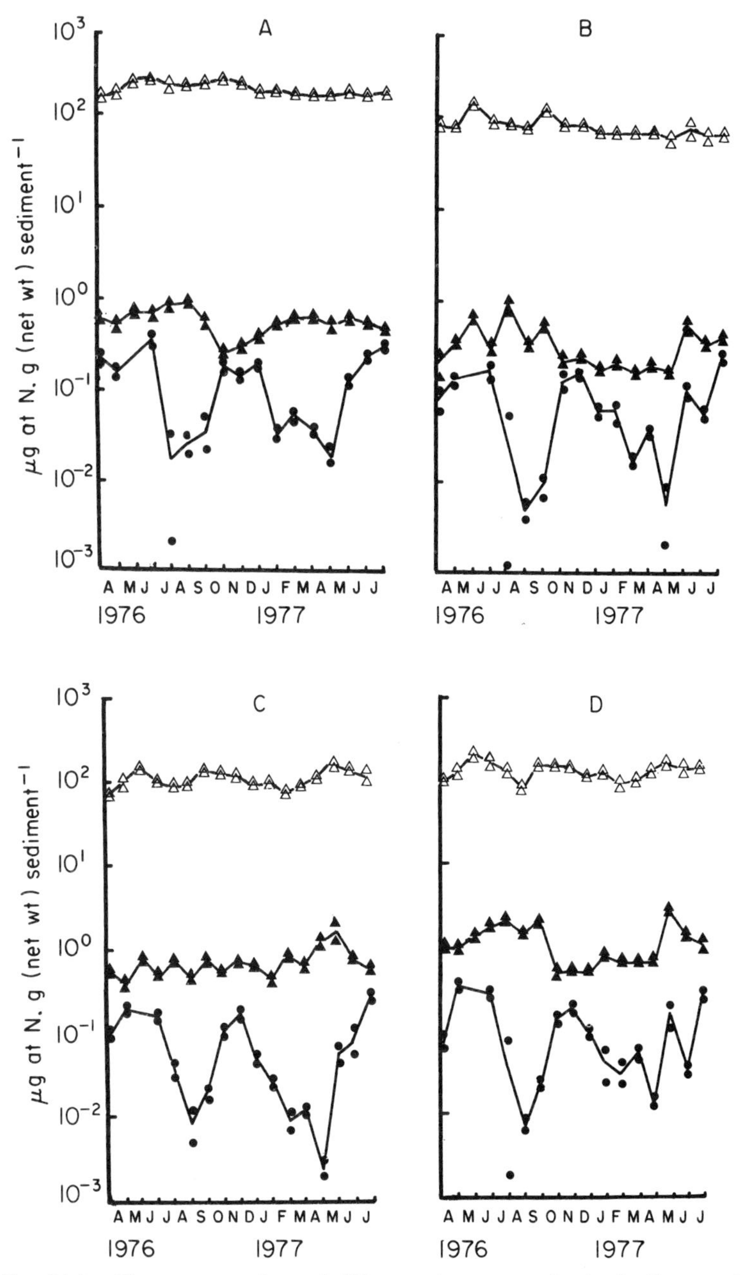

FIG. 24.1. The concentrations of different nitrogen species in duplicate sediment cores from the surface 5 cm layer: A, site 1 (marsh flat); B, site 3 (creek bottom); C, site 4 ('brown' salt marsh pan); D, site 5 ('black' salt marsh pan). △ = organic nitrogen, ▲ = ammonium nitrogen, ● = nitrate nitrogen. Site 2, the second site on the marsh flat, is not illustrated as it was not significantly different from site 1.

MINERALIZATION (AMMONIFICATION)

The ability of microbes to mineralize this abundant organic nitrogen was examined by incubating replicate cores of sediments collected in capped hypodermic syringes as described earlier and periodically analysing cores for the amounts of exchangeable ammonium and nitrate ions present. In preliminary experiments amounts of exchangeable ammonium ions increased during incubation, but the amounts of nitrate remained negligible and, therefore, this latter species of nitrogen was discounted as a product of mineralization. Evolution of ammonium ions was, therefore, used as a measure of mineralization of organic nitrogen in the sediment samples.

Two types of experiments were performed. In the first type replicate samples of cores were taken from three depths below the sediment surface (0–5, 10–15 and 20–25 cm) at three sites (1, 3 and 5) in order to examine the rates of mineralization of organic nitrogen at different sites and at different depths. Sediment cores at different sites were obtained using tubes with an internal diameter of 7·5 cm and subsampling from this large vertical core at the required depths with 5·0 ml plastic hypodermic syringes. The samples were incubated at 20°C in the dark and duplicate samples taken at monthly intervals for analysis of exchangeable ammonium ions.

Secondly, in order to investigate the influence of temperature upon mineralization, replicate samples from the surface sediments were taken from

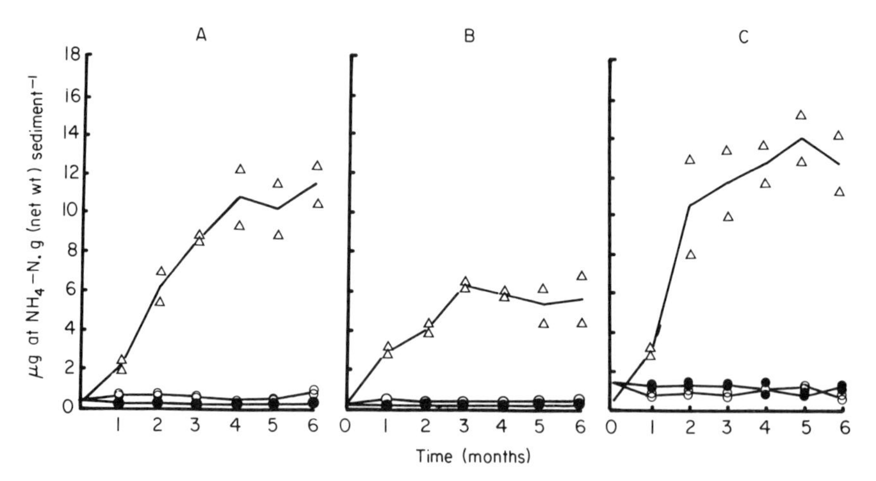

FIG. 24.2. Mineralization of organic nitrogen to ammonia in samples of sediment taken from three sites in hypodermic syringes and incubated at 20°C: A, sediment from site 1 (marsh flat); B, sediment from site 3 (creek bottom); C, sediment from site 5 ('black' salt marsh pan). Sediment samples were taken from three depths at each site. △ = sediment from 0–5 cm depth, ○ = sediment from 10–15 cm depth, ● = sediment from 20–25 cm depth.

all five sites with 5·0 ml cut-off plastic hypodermic syringes which were incubated at a variety of temperatures (5, 10, 15, 20 and 25°C). Duplicate cores from each site incubated at each different temperature were removed at monthly intervals and analysed for exchangeable ammonium ions.

The results of the first experiment (Fig. 24.2) show that in the cores of sediment from the creek, salt pan and the marsh flat the amounts of liberated ammonium ions decreased sharply with depth. Only in cores taken from the surface layer (0–5 cm) was there any measurable mineralization. Presumably in the deeper, older, sedimentary layers the labile nitrogenous organic molecules already had been recycled and only refractory organic nitrogen compounds remained. This suggests that microbial transformations in the surface sediments to a depth of 5 cm are sufficient to describe the major part of the microbial nitrogen cycling which occurs in these sediments. Although abun-

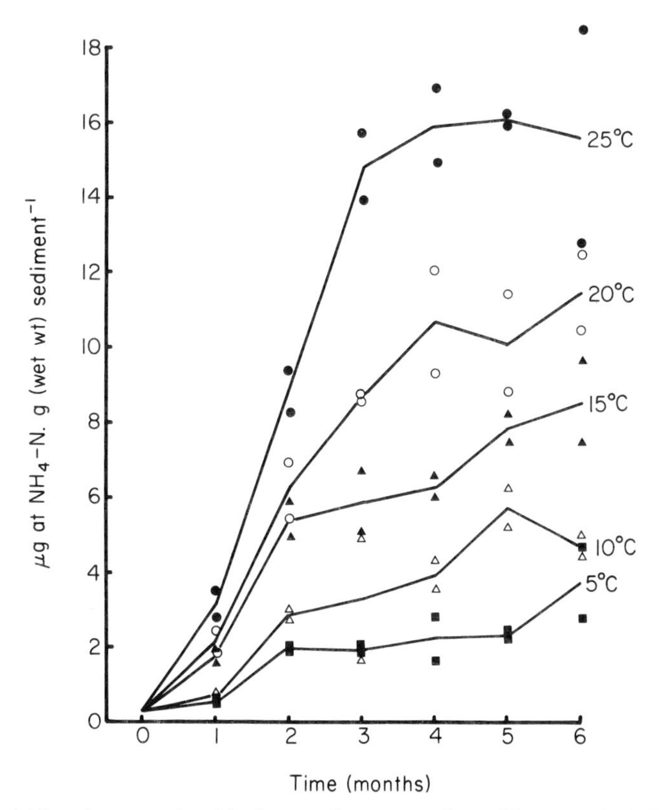

FIG. 24.3. An example with site 1 sediment samples to illustrate the influence of temperature upon the release of ammonia in sediment cores. Duplicate hypodermics of sediment from each temperature were analysed at monthly intervals. Temperatures of incubation are indicated upon the figure.

dant nitrogen is present in the salt marsh sediments in the form of organic nitrogen, it is obvious that this is converted to inorganic nitrogen at only a relatively slow rate.

On the average there were lower amounts of organic nitrogen present in the sediments of the pans than in the cores from the salt marsh flat, although the amounts of ammonium ions released from cores collected from site 1 and site 5 were similar. This suggests that less refractory material is produced by the algae of the pans than by the higher plants growing on the marsh flat.

Fig. 24.3 shows the results of the effect of temperature upon ammonification in cores from site 1. Typically, the rate at which ammonium ions were released increased with an increase in temperature, and the rates at different temperatures were linear over at least the first three months of incubation. The first three months' data for each treatment were subjected to a linear regression analysis (Fig. 24.4). The rates of ammonification of sediment from the different sites for the five incubation temperatures were plotted as an Arrhenius plot, and in all cases the effect of an increase of temperature upon

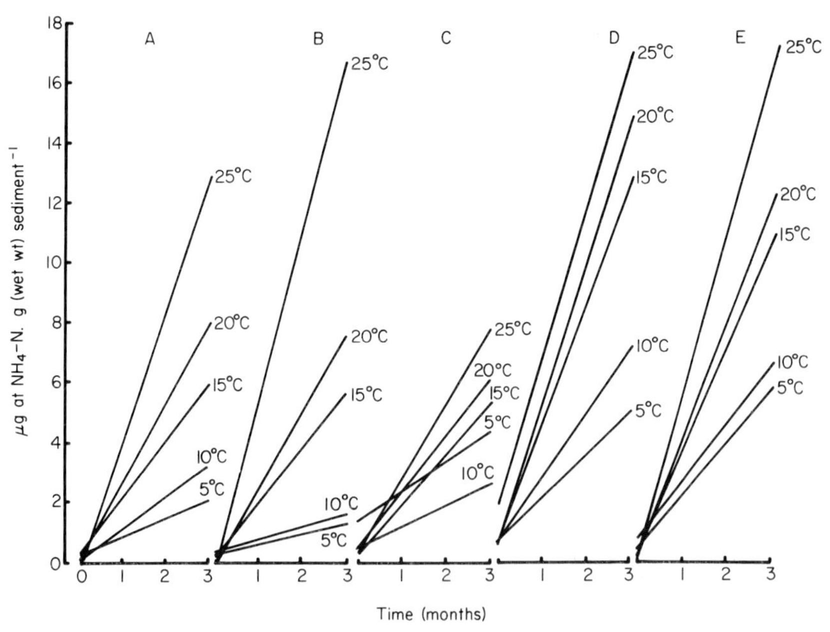

FIG. 24.4 The best-fit rates of ammonia release (ammonification) in samples of sediment from each site incubated at a variety of temperatures. The rates were determined by linear regression analysis of the change in ammonia concentrations during the first three months of incubation. Incubation temperatures are indicated upon the figure: A, site 1 (marsh flat); B, site 2 (marsh flat); C, site 3 (creek bottom); D, site 4 ('brown' salt marsh pan); E, site 5 ('black' salt marsh pan).

Table 24.2. *Arrhenius relationships describing mineralization of organic nitrogen (ammonification) in cores of surface sediment (0–5 cm depth) from each sampling site over a temperature range of 0–25°C.*

Site	Log rate of mineralization	Significance
1 (marsh flat)	$-3{,}080\ (1/T°A) + 10{\cdot}99$	$P < 0{\cdot}05$
2 (marsh flat)	$-5{,}358\ (1/T°A) + 18{\cdot}73$	$P < 0{\cdot}05$
3 (creek bottom)	$-2{,}006\ (1/T°A) + 7{\cdot}11$	$P < 0{\cdot}01$
4 ('brown' pan)	$-2{,}378\ (1/T°A) + 8{\cdot}78$	$P < 0{\cdot}01$
5 ('black' pan)	$-2{,}209\ (1/T°A) + 8{\cdot}17$	$P < 0{\cdot}05$

the rate of ammonification conformed closely to an Arrhenius relationship (Table 24.2).

The Arrhenius relationship of the rate of ammonification in relation to temperature for each site was used in conjunction with a record of the temperatures of the sediments (obtained by monthly measurements of *in situ* temperatures at each of the five sites) to calculate the total annual amount of ammonia released in the top 5 cm of sediment at each site. The results are expressed as an estimate of the annual amount of ammonium ions released from a 1 $cm^2 \times 5$ cm deep core of sediment (Table 24.3). The calculations assume that the Arrhenius relationships are constant throughout the year, but it is possible that variations in the bacterial community, or seasonal changes in the amounts of labile organic nitrogen present in the sediments may modify the Arrhenius functions. The experiments are being repeated periodically to resolve this uncertainty, but for the moment this approach permits us to obtain an estimate of the amount of organic nitrogen which is mineralized annually.

Table 24.3. *Estimated annual rates of mineralization (ammonification) in the surface 0–5 cm of sediment at each sampling site.*

Site	Rate (μg atoms NH_4–N $cm^{-2}\ y^{-1}$)
1 (marsh flat)	127
2 (marsh flat)	83
3 (creek bottom)	87
4 ('brown' pan)	209
5 ('black' pan)	199

ASSIMILATION

The amount of nitrogen assimilated during primary production must be equal to or greater than the amount liberated during mineralization. Initial estimates of primary production at Colne Point indicate an annual value of 500–1,000 g dry weight of plant biomass per square metre (A. Hussey, personal communication). The average nitrogen:dry weight ratio for this material is 1,834 μg atoms N g dry weight^{-1}. Therefore, a rough estimate of nitrogen assimilated, assuming a primary production value of 750 g dry weight m^{-2} yr^{-1} is 137·5 μg atoms N cm^{-2} yr^{-1} for the sites on the salt marsh flat. We have no information on primary production at the other sites.

Although any conclusions about the proportionate remineralization of primary biomass are probably premature in view of the errors of the above estimate, the calculation at least shows that the amount of organic nitrogen mineralized appears to be of similar magnitude to the amount assimilated by the plants.

DENITRIFICATION AND NITRIFICATION

These processes have been investigated using ^{15}N techniques. In order to measure denitrification rates sodium ^{15}N nitrate solution (5 μg atoms ^{15}N per sediment core) was injected as uniformly as possible throughout the length of 5 cm deep sediment cores from each site, held in 10 ml plastic hypodermic syringes with 2 ml of air present above the sediment surface. The syringes were capped and incubated at 20°C for 4 days. After this time the air above each sediment core was transferred into a second hypodermic syringe and subsequently measured for enrichment of the ^{15}N/^{14}N ratio in a mass spectrometer. The rate of denitrification was calculated according to the method of Hauck *et al.* (1957).

Measurements showed that denitrification was rapid. Nitrification was, therefore, detected indirectly by adding an amount of $^{15}NH_4$ equivalent to the natural *in situ* ammonium concentration. If nitrification of $^{15}NH_4$ occurred the rapid subsequent denitrification of nitrate would result in enrichment of the gas atmosphere with $^{15}N_2$. In addition an attempt was made to culture nitrifying bacteria from the sediment of each site by inoculating petri plates of agar medium with suspension of sediment in sterile sea water. The medium used was entirely inorganic (Rodina 1972) containing ammonium salts and able to support the growth of *Nitrosomonas* sp., the bacteria responsible for the conversion of ammonium to nitrite. The petri dishes were incubated aerobically at 20°C for three weeks and inspected for formation of bacterial colonies.

Denitrification was rapid in cores from all sites (Table 24.4), except apparently in cores from site 5, where there was little ^{15}N enrichment of the atmosphere. It would appear likely that at this site, which is the salt pan covered with *Oscillatoria*, either nitrate is reduced directly to ammonium or that the evolved nitrogen is being refixed by the blue-green algae at this site.

Table 24.4. *Denitrification in sediment cores (0–5 cm depth) taken from each site, injected with $^{15}NO_3$ and incubated for 4 days at 25°C.*

Site	Total NO_3 in 5 cm deep surface core (μg atoms NO_3–N cm^{-2})	Proportion of ^{15}N released as N_2	*In situ* rate of denitrification (μg atoms N_2–N cm^{-2} day^{-1})
1 (marsh flat)	1·070	0·181	0·0482
2 (marsh flat)	0·8377	0·473	0·1037
3 (creek bottom)	0·737	0·190	0·0351
4 ('brown' pan)	0·780	0·345	0·0672
5 ('black' pan)	0·681	0·005	0·0008

The anomalous nature of this result was evident from the results of other experiments in which nitrate was added to the sediment cores. After an incubation lasting four days nitrate concentrations in cores from all sites had fallen to the normal *in situ* values as the nitrate was reduced. Thus it is certain that the bacterial communities at all sites on the marsh are capable of rapid reduction of nitrate.

The addition of $^{15}NH_4$, however, resulted in no enrichment of the atmosphere by $^{15}N_2$, and in addition attempts made to culture nitrifying bacteria were unsuccessful. No colonies of bacteria developed upon the medium, indicating an absence of *Nitrosomonas* sp. from the sediments of all five sites. It appears, therefore, that nitrification does not occur in the sediments investigated, at least to any measurable extent. Vanderboght and Billen (1975) also failed to detect nitrification in reduced estuarine sediments. They also mentioned that the rapid decrease of nitrate in the interstitial water could be due to the presence of denitrification.

In the absence of further data, therefore, we conclude that nitrification is absent from these salt marsh sediments, probably because of the reduced nature of the environment. While the bacterial populations of the sediments have large capacities for reduction of nitrate, the *in situ* levels of nitrate represent a background level below which the bacteria are unable to further deplete the nitrate concentration. The addition of nitrate to the system is followed by its rapid depletion back to background levels as a result of stimulation of bacterial anaerobic respiration of nitrate. Such a capacity for nitrate removal results in salt marsh being potentially capable of acting as a nitrogen sink in the presence of nitrate pollution.

NITROGEN FIXATION

Finally nitrogen fixation must be considered as the major source of replacement for that nitrogen lost from the salt marsh by either denitrification, or export as organic or inorganic nitrogen. Fixation was estimated with the use of the acetylene reduction method (Hardy *et al.* 1971). Acetylene (0·2 atmospheres) was injected into the gas phase above each short vertical core of surface sediment in clear plastic hypodermic syringes. The samples of sediment were incubated *in situ* for 24 hours and the atmosphere above the core assayed for ethylene by the use of gas chromatography. Nitrogen fixation rates were calculated per unit area of sediment surface by assuming a conversion factor of 3 moles of acetylene reduced to ethylene for every 1 mole of nitrogen fixed.

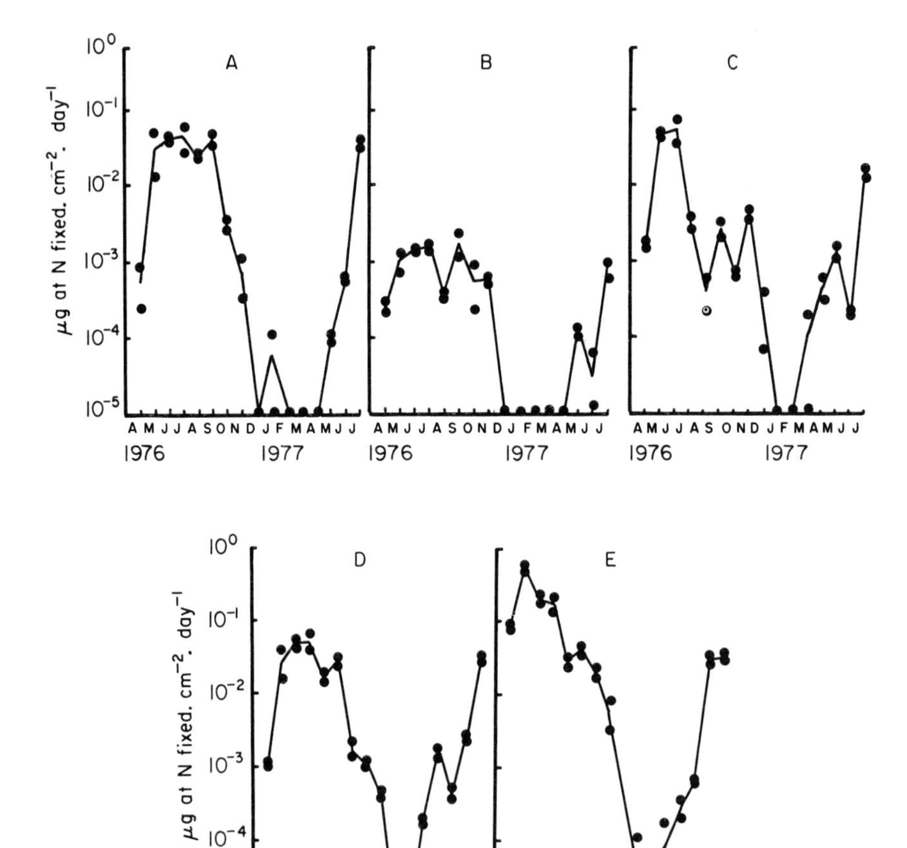

FIG. 24.5. Nitrogen fixation (acetylene reduction) measured at monthly intervals with duplicate cores of sediment from each site, incubated at field temperatures: A, site 1 (marsh flat); B, site 2 (marsh flat); C, site 3 (creek bottom); D, site 4 ('brown' salt marsh pan); E, site 5 ('black' salt marsh pan).

The estimated rates of nitrogen fixation for each site typically were low during winter, but the rates increased during spring and reached a maximum in summer (Fig. 24.5). On the two sites on the salt marsh flat fixation was principally associated with the presence of *Oscillatoria* and *Nostoc* sp. but at site 2 the shrubby plant *Halimione portulacoides* shaded the blue-green algae on the sediment surface and the rate of fixation in summer was reduced compared to the other sites.

In order to obtain the annual nitrogen fixation, the area under the curve for each site was calculated. The area under the curve between two successive sampling dates approximated to a trapezium and the annual value for the nitrogen fixation was calculated by summation of all the areas of trapezia between successive months over the period of one year. The results are summarized in Table 24.5.

Table 24.5. *Estimate of annual rates of nitrogen fixation (acetylene reduction) at each site obtained by integration of the area under the curve of monthly rates of fixation.*

Site	Rate (μg atoms N_2 fixed cm^{-2} y^{-1})
1 (marsh flat)	5·78
2 (marsh flat)	0·23
3 (creek bottom)	3·42
4 ('brown' pan)	5·45
5 ('black' pan)	32·27

Quite obviously cores from the 'black' pans which contained the *Oscillatoria* benthic mat showed the greatest capacity of all the sediments to fix nitrogen. The overall importance of these pans in the nitrogen economy of the salt marsh will depend upon what proportion of the marsh area they occupy. R. Butler (personal communication) has estimated from a study of 7 sites at Colne Point that on the average the salt marsh pans occupy only 3% of the marsh surface, although this varies 2–8% locally. We must conclude, therefore, that the presence of the *Oscillatoria* benthic mat is not of major importance in the nitrogen economy of the marsh. The dense shrubby *Halimione* with some *Puccinellia* at site 2 results in a very low annual rate of fixation at this site. Results from the other sites are intermediate in value.

Probably the most obvious conclusion to be made is that the nitrogen fixation rates in general are between 7 and 100 times lower than the annual rates of nitrogen assimilation and mineralization at each site. Assuming that the salt marsh approaches a steady-state condition then net export of nitrogen cannot exceed the replacement rate by nitrogen fixation. It follows, therefore, that net export of nitrogen can only represent a small proportion of the annual turnover of nitrogen in the system, the majority of the nitrogen being cycled within the salt marsh between the organic nitrogen and ammonium fractions.

REFERENCES

Bremner J.M. (1965) Total nitrogen. *Agronomy No. 9. Methods of Soil Analysis. Part 2* (Ed. by C.A. Black *et al.*), pp. 1149–77. Amer. Soc. of Agron., Madison, Wisconsin.

Bremner J.M. (1965) Inorganic forms of nitrogen. *Agronomy No. 9. Methods of Soil Analysis. Part 2* (Ed. by C.A. Black *et al.*), pp. 1179–237. Amer. Soc. of Agron., Madison, Wisconsin.

Gardner L.R. (1975) Runoff from intertidal marsh during tidal exposure. *Limnol. Oceanogr.* **20**, 81–89.

Hardy R.W.F., Burns R.C., Herbert R.R., Holsten R.D. & Jackson E.K. (1971) Biological nitrogen fixation: A key to world protein. *Plant Soil Special Volume*, 561–90.

Harwood J.E. & Kuhn A.L. (1970) A colorimetric method for ammonia in natural water. *Water Res.* **4**, 805–11.

Hauck R.D., Melsted S.W. & Yankwich P.E. (1957) Use of N-isotope distribution in nitrogen gas in the study of denitrification. *Soil Sci.* **86**, 287–391.

Jones K. (1974) Nitrogen fixation in a salt marsh. *J. Ecol.* **62**, 553–65.

Nedwell D.B. & Abram J.W. (1978) Bacterial sulphate reduction in relation to sulphur geochemistry in two contrasting areas of salt marsh sediment. *Estuarine Coastal Mar. Sci.* **6**, 341–52.

Odum E.P. & de la Cruz A.A. (1967) Particulate organic detritus in a Georgia salt marsh–estuarine ecosystem. *Estuaries* (Ed. by G. Lauff), pp. 383–88. A.A.A.S., Washington.

Pigott C.D. (1969) Influence of mineral nutrition on the zonation of flowering plants in coastal salt marshes. *Ecological Aspects of the Mineral Nutrition of Plants* (Ed. by I.H. Rorison), pp. 25–35. Blackwell Scientific Publications, Oxford.

Rodina A.G. (1972) *Methods in Aquatic Microbiology* (Ed. by R.R. Colwell and M.S. Zambruski). Butterworths, London.

Strickland J.D.H. & Parsons T.R. (1968) *A Practical Handbook of Seawater Analysis*, 2nd edn. Fisheries Res. Bd., Canada, Ottawa.

Teal J.M. (1962) Energy flow in a salt marsh ecosystem of Georgia. *Ecology* **4**, 614–24.

Valiela T. & Teal J.M. (1974) Nutrient limitation in salt marsh vegetation. *Ecology of Halophytes* (Ed. by R.J. Reimold and W.H. Queen), pp. 547–63. Academic Press, London and New York.

Vanderboght J.P. & Billen J. (1975) Vertical distribution of nitrate concentration in interstitial water of marine sediments with nitrification and denitrification. *Limnol. Oceanogr.* **20**, 953–61.

van Raalte C.D., Valiela T. & Teal J.M. (1976) The effect of fertilization on the species composition of salt marsh diatoms. *Water Res.* **10**, 1–4.

van Raalte C.D., Valiela T., Carpenter E.J. & Teal J.M. (1974) Inhibition of nitrogen fixation in salt marshes measured by acetylene reduction. *Estuarine Coastal Mar. Sci.* **2**, 301–25.

25. INPUTS, OUTPUTS AND INTERCONVERSIONS OF NITROGEN IN A SALT MARSH ECOSYSTEM*

IVAN VALIELA[1] AND JOHN M. TEAL[2]

[1] *Boston University Marine Program, Marine Biological Laboratory, Woods Hole, Massachusetts, U.S.A.*

and

[2] *Woods Hole Oceanographic Institution, Woods Hole, Massachusetts, U.S.A.*

SUMMARY

Ground and rain water provided large amounts of nutrients, particularly nitrate, to Great Sippewissett marsh. Nitrogen fixation, primarily carried out by bacteria, added a smaller additional input of nitrogen. Tidal exchanges resulted in net annual losses of both dissolved and particulate nitrogen. The seasonal uptake and release of tidally transported nutrients reflected the activity of the vegetation and decomposers. The total particulate losses were equivalent to 40% of the net annual above-ground production of *Spartina alterniflora*, the dominant marsh plant. Denitrification was the other major loss of nitrogen, substantially exceeding fixation of nitrogen. Most of the nitrogen was in the sediments, with slow turnover rates. The nitrogen content of the vegetation is a balance between uptake and leaching rates, and the balance changes as the growing season progresses. Other components of the salt marsh are also involved in very active interconversions and exchanges of nitrogen. The net effect of the processing of nitrogen by the marsh ecosystem is to convert nitrate in ground water to ammonium and particulate nitrogen and to export both to coastal water. Additional nitrogen losses occur when nitrogen gas is transferred from the marsh to the atmosphere.

RÉSUMÉ

L'eau de pluie et du sol produit de larges quantités de nutriments, en particulier les nitrates, au marais de Great Sippewissett. La fixation d'azote effectuée en

* Research supported by NSF grants GA 43008, GA 43009 and GA 41506. Contribution No. 4050 from the Woods Hole Oceanographic Institution.

premier lieu par les bactéries est une plus petite source qui s'ajoute au pool de l'azote. Les échanges dus aux marées se traduisent par des pertes nettes annuelles d'azote, à la fois sous forme dissoute et solide. L'accumulation saisonnière et la libération des nutriments transportés par les marées reflètent l'activité de la végétation et des décomposeurs. Les pertes totales des particules équivalent à 40% de la production nette annuelle au dessus du sol, de *Spartina alterniflora*, la plante dominante du marais. La dénitrification est l'autre perte d'azote, pouvant dépasser substantiellement sa fixation. La plupart de l'azote se trouve dans les sédiments avec un faible taux de transformation. La quantité d'azote dans les végétaux représente un équilibre entre le taux d'absorption et celui de filtration, et la balance varie à mesure que la saison de la croissance progresse. D'autres composants du marais sont aussi en jeu dans les très actives interconversions et échanges d'azote. Le résultat de la transformation de l'azote par l'écosystème du marais est la conversion du nitrate de l'eau du sol en ammonium et l'azote solide est exporté vers l'eau des côtes, avec des pertes additionnelles d'azote gazeux vers l'atmosphère.

INTRODUCTION

The amount of available nitrogen is critical for many important components of salt marsh ecosystems. Several studies have shown that primary producers in salt marshes are nitrogen-limited (Tyler 1967; Pigott 1969; Valiela & Teal 1974). The morphology of *Spartina alterniflora* (Valiela *et al.* 1978a), as well as its biomass and production (Valiela *et al.* 1975; Valiela *et al.* 1976) are strongly influenced by the supply of nitrogen. We also have data showing that the nitrogen content of marsh grasses provided with added nitrogen is high, and consequently the quality of this fertilized grass as a herbivore food is improved (Caswell *et al.* 1973). As a result, the biomass of herbivores is also influenced by nitrogen supply (Vince *et al.*, in preparation).

The decomposer food web is also altered by changes in the nitrogen supply. Measurements of oxygen uptake in fertilized plots of marsh grasses show that decomposition rates are enhanced where nitrogen is added, probably as a response to the lowered C/N ratios of the detritus of the dead fertilized grasses. The increased microbial fauna and flora associated with the enriched detrital particles leads to increased biomass of detritus-feeding invertebrates in the fertilized plots. We also have evidence that fish and other predators that feed on the detritus feeders have increased growth rates. There are many other parts of a salt marsh that are affected by the nitrogen supply. The above examples suffice to show that, at least in Great Sippewissett marsh, Massachusetts, where we are carrying out our studies, nitrogen seems to be a keystone nutrient, responsible for establishing rates of activity and standing crops of most of the ecosystem. Before we can delve into the intricacies of

FIG. 25.1. Air view of Great Sippewissett marsh, Falmouth, Massachusetts, U.S.A.

just how nitrogen (probably in combination with other factors) determines rates, we need to have some idea of the supplies and losses, as well as the major pools, of nitrogen in the system. In this paper we present an account of work on the inputs and outputs of nitrogen in a salt marsh ecosystem and discuss some of the transfers of nitrogen among components of the salt marsh.

Great Sippewissett marsh has a single entrance (Sippewissett creek, Fig. 25.1) through which the entire tidal volume flows. Water from Buzzards Bay floods the marsh twice daily with a maximum tidal excursion of about 1·6 m. The marsh is a mosaic of habitats, the major ones being:

(1) Creek bottoms usually bare of vegetation. In Fig. 25.1 the light areas in the creeks are sandy bottoms (42,500 m^2) while the dark areas are covered by muddy, organic sediments (47,100 m^2).
(2) Salt pannes, small waterlogged depressions with soft flocculent surface sediments and dwarfed plants (2,190 m^2). Pannes are not clearly visible in Fig. 25.1.
(3) Algal mats (6,200 m^2) occurring on mud flats (top right, Fig. 25.1). The sand is very unstable except where consolidated by mats of blue-green algae and purple sulphur bacteria.
(4) Low marsh, by far the most widespread vegetation zone. This habitat has two distinct subareas, the major one dominated by short *Spartina alterniflora* (90,900 m^2) and another visible as a very narrow dark ribbon on the creek banks (Fig. 25.1) covered by tall *Spartina alterniflora* (7,530 m^2) (Valiela *et al.* 1978a).
(5) High marsh (28,000 m^2), visible as areas of lighter vegetation in Fig. 25.1, consisting of the grasses *S.patens* and *Distichlis spicata* and found above the elevation of low marsh.

The marsh is surrounded by glacial moraine on the south (top right, Fig. 25.1), east and north and by sand dunes to the west, both covered by upland vegetation.

INPUTS AND OUTPUTS

The major sources and sinks of nitrogen in Great Sippewissett marsh are shown in Fig. 25.2. The inputs of nitrogen include ions brought in by rain and snow, probably associated with particulate matter in the air. The fresh water table follows the topography of the upland, resulting in a slope towards the sea, which in turn leads to flow of fresh water into Great Sippewissett marsh. Ground water flow is usually through many underground springs, and the flow of ground water brings in dissolved nitrogen species. This marsh does not have any major surface streams feeding into it. Another source of nitrogen is the fixation of atmospheric N_2 by micro-organisms in the various habitats of the marsh, converting N_2 into combined nitrogen compounds.

Tidal exchanges result in both input and losses of nitrogen compounds. Large volumes of sea water enter and leave the marsh during the twice-daily tides, and the chemical composition of ebbing and flooding water differs substantially.

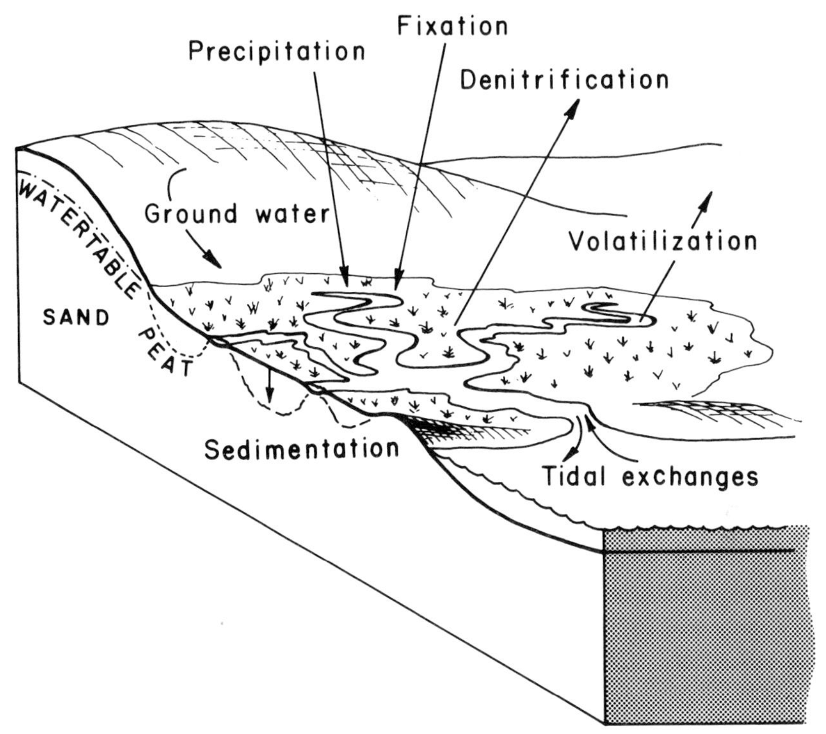

FIG. 25.2. Diagram of major inputs and outputs of nitrogen into a salt marsh ecosystem.

An additional mechanism whereby nitrogen leaves the marsh is denitrification, the microbial process by which nitrate is converted to nitrogen gas and is removed from the marsh to the atmosphere. Other mechanisms of nitrogen loss, probably minor, are volatilization of NH_3 from marsh water and movement of nitrogen into sediments below the root zone of the marsh vegetation.

We are in the process of publishing our measurements of each of these mechanisms elsewhere, and we only provide a brief account here.

Precipitation

The amount of precipitation and the concentration of ammonium, nitrite, nitrate and dissolved organic nitrogen (DON) were measured on water samples collected in rain gauges (Valiela *et al.* 1978b).

There was no seasonal pattern in the amount of precipitation or the concentration of nutrients in rainfall brought into the marsh by rain. Summing over the entire year, we obtained values for inputs of each nitrogen species, multiplied by the total area of Great Sippewissett marsh (226,000 m^2), as listed in Table 25.1. The major form of nitrogen brought to the marsh by precipitation was dissolved organic nitrogen, with substantial amounts of nitrate, and to a lesser extent, ammonium nitrogen.

The filters through which we filtered the rain water were often surprisingly darkened by the filtered particulates. The C/N ratios of the particles varied

Table 25.1. *Nitrogen budget for Great Sippewissett marsh (kg of nitrogen yr^{-1}).*

Processes	Input		Output		Net exchange
Precipitation		179			179
NO_3–N	52				
NO_2–N	0·2				
NH_4–N	31				
DON*	89				
Particulate N	7				
Ground water flow		5,471			5,471
NO_3–N	2,495				
NO_2–N	29				
NH_4–N	492				
DON*	2,455				
N_2 fixation		1,592			1,592
Algal	145				
Rhizosphere bacteria	1,273				
Non-rhizosphere bacteria	174				
Tidal water exchange		26,252		31,604	−5,352
NO_3–N	386		1,215		
NO_2–N	154		166		
NH_4–N	2,623		3,539		
DON*	16,346		18,479		
Particulate N	6,743		8,205		
Denitrification				1,558 (+1,273)	−2,831
Sedimentation				25	−25
Volatilization of NH_3				8	−8
Deposition of bird feces		9			9
Shellfish harvest				9	−9
		33,503		34,574	−974

* Dissolved organic nitrogen.

from 10–30, with occasional higher values. Since the ratios of particulates in sea water average between 7 and 12, particulates in rain did not contribute much to the nitrogen supply of the marsh.

Ground water

The concentrations of ammonium, nitrite and nitrate ions as well as dissolved organic nitrogen were measured in samples of ground water collected throughout the year from underground springs emptying into Great Sippewissett marsh (Valiela *et al.* 1978b). The volumes of fresh water were obtained by calculating the volume of fresh water necessary to dilute the salinity of flooding tidal water to the lower salinity of ebbing tidal water. The salinity of ebbing water was 2–9‰ S lower than that of flooding sea water of each tide during the year.

We obtained values for volumes of fresh water lost from the marsh during tides through the year and multiplied these values by the nutrient concentration to obtain nutrient fluxes.

Nitrate was the predominant nitrogen species entering the marsh via ground water flow, and the concentration of ions in the water averaged about 50 μg-at l^{-1}. Considering all the forms of N, ground water brought into the marsh about 27 times as much nitrogen as rain water (Table 25.1).

Nitrogen fixation

Both blue-green algae and bacteria carry out the fixation of atmospheric nitrogen in Great Sippewissett marsh. The rates of fixation in each recognizable habitat of the marsh showed different seasonal patterns so that fixation was estimated in each habitat separately.

The measurements of N_2 fixation were carried out by the use of an adaptation of the acetylene-reduction technique. The methods used to estimate nitrogen fixation by blue-green algae are detailed in Carpenter *et al.* (1978). In order to measure the rate of fixation by bacteria, we incubated cores of sediment for 24 hours under a gas mixture of 86% nitrogen, 10% acetylene and 4% oxygen. We then reflushed the headspace above the core with gas of a similar composition and measured ethylene production 24 hours later with the use of gas chromatography. The procedures and justification of the methods will be detailed elsewhere.

Total fixation by blue-green algae integrated for each habitat over the entire area of the marsh and over the year was 145 kg (Table 25.1). The rate of bacterial fixation per unit area, especially at sites where bacteria were associated with plant roots, was up to nine times as large as the rate of algal fixation. Both blue-greens and bacteria showed low rates of fixation during the cold months compared with the high rates in midsummer.

Tidal exchanges

We measured the total flow of sea water and nutrients directly at the single tidal entrance to Great Sippewissett marsh. We used mechanical flow meters to obtain flow rates through an entire tidal cycle. We sampled water at frequent intervals throughout the cycle and analysed it for dissolved and particulate nutrients. This procedure was repeated monthly through the year and included instances of neap, spring and intermediate tides.

By integrating over a tidal cycle, we calculated the net exchange of nutrients between the flood and ebb tides. Annually the amounts of all nutrients carried by ebb tides exceeded the amounts introduced into the marsh by flood tides. However, there was seasonal variation such that Great Sippewissett marsh actually imported dissolved inorganic nitrogen during mid-summer. In order to show the direction of the overall net nutrient exchanges we have plotted (Fig. 25.3) the difference in total nutrients carried by flood and ebb tidal water (solid dots). Also plotted is the amount of dissolved inorganic nitrogen (principally nitrate) brought in by ground water (open circles). In January and February dissolved inorganic nitrogen (primarily nitrate) was exported from Great Sippewissett marsh. This is probably nitrate from the ground water which is transported unused through the marsh and

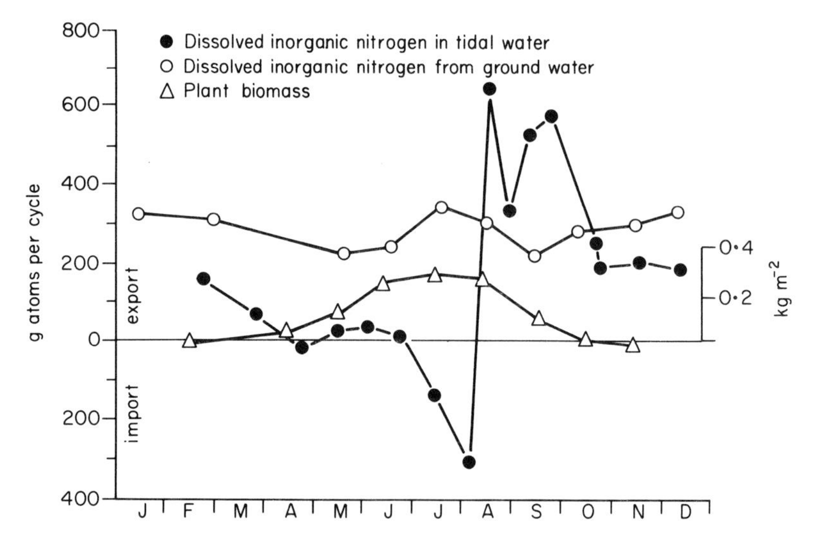

FIG. 25.3. Net exchanges of dissolved inorganic nitrogen (DIN) as g-atoms of nitrogen from Great Sippewissett marsh. The terms import and export refer to whether there was a net import or export from the marsh at the end of the tidal cycle. The left-hand vertical scale should be used for both the tidal water and the ground water values. Above-ground plant biomass as dry wt. m^{-2} (right-hand vertical scale) is given for comparison of tidal exchanges with plant growth within the marsh.

enters the sea. During spring, rates of denitrification increased and little if any nitrate escaped from the marsh. Throughout this period there was no significant exchange of ammonium, a condition that lasted until summer, when the marsh imported this ion and which resulted in the trough seen in Fig. 25.3. This import coincided in time with the period of maximum plant biomass (triangles) and it took place while the grasses were flowering and setting seed, a period of great nitrogen demand. In mid-August *Spartina* plants flowered. It is well known from the horticultural literature (Tukey 1970) that at that stage of their life cycle the tissues of plants are susceptible to leaching. We hypothesize that the sharp increase in ammonia lost by the Great Sippewissett marsh reflects the leaching of soluble nitrogen from marsh plants. Currently we are carrying out measurements of the rates at which ammonium is released from *S.alterniflora* and *S.patens*, the two most common plants in the marsh. Preliminary calculations from the data show that leaching from plants may indeed make a substantial contribution to the amount of ammonium-nitrogen lost from Great Sippewissett marsh in August. Exports of dissolved inorganic nitrogen continued to October (Fig. 25.3), which probably reflects ammonium released by the decay of senescent and dead plant matter.

The amounts of dissolved organic nitrogen were very large in the tidal water (Table 25.1). The ecological significance of these compounds is poorly understood, since biologically useful compounds such as amino acids and urea are likely to be only a very minor fraction of the total.

Table 25.2. *Annual exchanges of major forms of nitrogen for Great Sippewissett marsh (all values in kg yr^{-1}).*

Form of N	Input	Output	Net	% of Input
NO_3–N	2,933	1,215	1,718	59
NH_4–N	3,046	3,539	−493	−16
DON*	18,880	18,479	401	2
Particulate N	6,750	8,205	−1,455	−22
N_2	1,592	2,831	−1,239	−78
Totals	33,201	34,259	−1,058	−3·2

* Dissolved organic nitrogen.

The export of particulate matter (Table 25.2) is noteworthy because there are few direct measurements of movement of marsh-produced organic matter to the sea, yet evidence of the export of such materials is the prime argument used for marsh conservation in the United States. The amount of particulate organic material exported from Great Sippewissett (Table 25.2) was about 40% of the net annual above-ground production of *S.alterniflora* (Valiela *et al.* 1976), a very substantial export considering the high productivity of *S.alterniflora*.

Denitrification

The rates at which oxidized inorganic nitrogen compounds were converted to nitrogen gas were measured *in situ* monthly in each of the habitats of Great Sippewissett marsh. The primary technique used involved placing a darkened bell jar over marsh sediments. The headspace was filled with helium and a small amount of argon. We measured the evolution of N_2 by calculating the N_2/Ar ratio based on gas chromatographic measurements (Kaplan 1977). This technique measures net denitrification, since nitrogen fixation by bacteria in the sediment took place during the measurement of denitrification.

The creek bottoms were the habitat where the greatest rates of net denitrification were observed, but in all habitats net denitrification rates increased in May and peaked in mid-summer (Kaplan 1977). These increases in the rates of denitrification coincided in time with the low amount of nitrate exported from Great Sippewissett marsh during summer. Some nitrate was lost from the marsh in winter, at a time of the year when denitrification rates were lowest.

Other sources and losses of nitrogen

A number of minor mechanisms are also included in our nitrogen budget (Table 25.1). From analysis of the nitrogen in profiles of marsh sediment and knowledge of the accumulation rate, losses to the sediments below the root zone (20 cm in depth) were calculated to be low. Similarly, as the pH of the marsh mud was about 5–8·5, we estimated that the amount of ammonia volatilized was not large. Flocks of gulls feed primarily outside the marsh but roost within it. We calculated the amount of nitrogen contributed by the birds from data on the nitrogen content of faecal deposits, the number of birds visiting the marsh, and the frequency of defecation. Great Sippewissett marsh is frequented by shell fishermen, and we estimated the removal of nitrogen via this pathway from the shellfish warden's records of take. All of these exchanges were small compared to the inputs and outputs discussed above.

INTERCONVERSIONS WITHIN THE SALT MARSH

The nitrogen budget of Great Sippewissett marsh (Table 25.1) provides an account of the sources and losses of nitrogen for an entire year for the marsh. We are now in the process of examining the transformations of nitrogen within the marsh itself. Any description of an ecosystem is necessarily complex. To simplify our presentation, while including some idea of the relations among the principal parts of a marsh, we have chosen in this paper to depict the standing crops and exchanges of nitrogen taking place during a day in early

August, the time of peak biomass of all the grasses (Fig. 25.3). The choice of a day as a time unit is arbitrary. There are some components such as tidal water (Fig. 25.4) that have a smaller time scale than a day. The nitrogen content of tidal water can change more during a single tidal cycle than from one month to another. Nitrogen movements in other components of the ecosystem, such as the sediments, may be examined more instructively if the unit of time is longer. The box and arrow model of a marsh makes it hard to deal, therefore, with the evaluation of the nitrogen in the tidal water box. The different time scales used for the frequency of measurements are extremely important in obtaining adequate estimates of pools and rates, and these different time scales should be retained in the construction of any descriptive or analytical model of the system under study.

The rates of exchanges were calculated from data obtained by our research group, some of which are published (Valiela *et al.* 1975; Valiela *et al.* 1976). Other values were obtained from work in progress.

During early August, gains of nitrogen from fixation and rain were similar to losses as a result of denitrification. Ground water contributed about twice as much nitrogen as was removed by tidal action. At this time of the year, the marsh intercepted more nitrogen than it released to the sea. In addition,

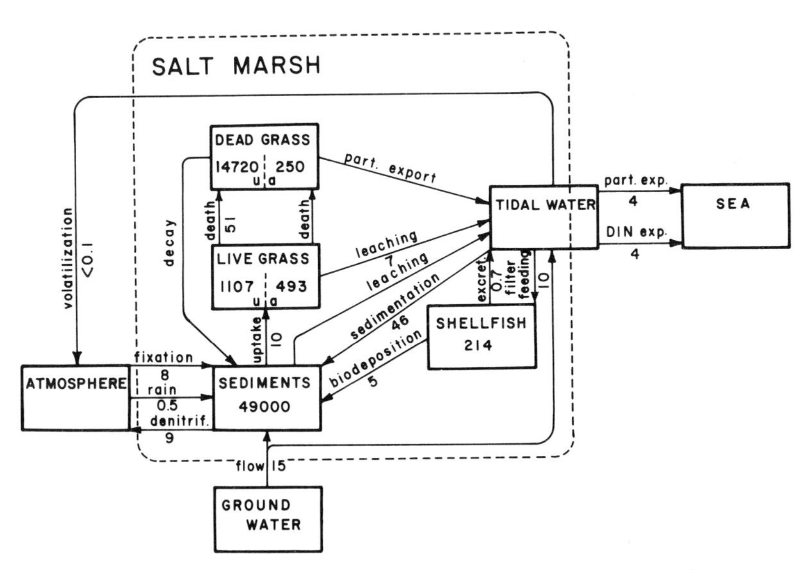

FIG. 25.4. Standing stocks and fluxes of nitrogen among some major components of the Great Sippewissett ecosystem. The values are in kg N for the entire marsh (226,000 m^2), while exchanges are in kg day^{-1}. The values are for one day in early August. Letters 'u' and 'a' refer to under- and above-ground respectively and 'part. export' means particulate matter exported. Sedimentary N is also resuspended but this is not shown.

the marsh converted a part of the nitrate brought in by flow of ground water into particulate nitrogen and ammonium. This is important because ultimately the export of this particulate nitrogen affects the detritus feeders of the neighbouring coastal ecosystem. If the marsh were not there, a very different part of the coastal food web, the phytoplankton, would be affected by the export of this nitrate in the ground water to the sea.

The sediments contained by far the largest amount of nitrogen. Judging from the rates of input and output, the turnover of the pool of nitrogen in sediment must be slow. Included in our designation as sediments are the mineral fractions as well as degraded organic matter not recognizable as plant parts. There were some losses of nitrogen to sediments below 20 cm, but roots present below this depth may recover some of this nitrogen. These losses were low (Table 25.2) and are not shown in Fig. 25.4.

The marsh grasses absorbed nitrogen from the sediments and incorporated it into organic nitrogen. The estimated uptake rate was high in early summer and tapered off towards the end of August, when the maximum grass biomass is achieved.

In the live grasses, the underground parts contained more than twice the amount of nitrogen than the above-ground parts. The standing stock of nitrogen in live roots and rhizomes did not change very much seasonally (Valiela *et al.* 1976). The underground biomass was primarily rhizomes, the part of the plant that survives from one growing season to another. In contrast, the above-ground parts contained high levels of nitrogen in August, but throughout the winter levels were close to zero since nearly all above-ground biomass dies in winter (Valiela *et al.* 1975). In one day in August, the amount of nitrogen leached from leaves as ammonia was nearly equal to the amount of nitrogen absorbed from the sediments. Perhaps in part this loss of ammonium is involved in the senescence and death of the sward. From numbers given in Fig. 25.4, we can see that the plants can lose their entire nitrogen content over a period of about 20 days in September. This calculated period agrees with field observations of the period between onset of senescence to death. We should note here that in salt marsh plots where we experimentally added nitrogen, the grasses remained green in the autumn longer than unfertilized plots. Although leaching was higher in fertilized than in unfertilized plots, the fertilization may have provided enough available nitrogen to forestall senescence, if indeed nitrogen balance and senescence are related.

Dead grass below-ground contained much more nitrogen than the above-ground dead matter. In fact, the nitrogen accumulated in dead underground parts was larger than the yearly turnover of nitrogen in live roots and rhizomes. It may be that the nitrogen in dead underground parts is fixed in the cells of micro-organisms associated with the dead tissues. We have observed very large decay rates for underground biomass (Valiela *et al.* 1976), and have measured also substantial activity of sulphate reducers as well as other decomposers. It would seem reasonable that large numbers of active micro-organisms (leading to large amounts of cellular nitrogen) would be required to carry out all this degradation of organic matter.

The large pool of nitrogen in sediments raises an apparent paradox. We have already noted that marsh vegetation is nitrogen limited. Further, the dissolved nitrogen in tidal marsh water is very high (Valiela *et al.* 1974) compared to that in sea water. We and others have shown that nitrogen fixation is reduced in the presence of available nitrogen, the micro-organisms using the already-reduced nitrogen compounds. Yet in salt marshes microbial fixation of nitrogen is among the highest of any habitat in the world (Carpenter *et al.* 1978). We are clearly missing pieces of the puzzle here; it may be that the abundant sediment and tidal water nitrogen is unavailable for plant growth through some as yet undetermined mechanism.

The dead above-ground grass found in early August was mostly left from the previous year's growth. About one half of the amount of nitrogen accumulated during the growing season remained from one year to the next. The fragments of dead grass become particulate nitrogen, and these were exported by tidal water. Once in the water, the particulate nitrogen was subject to sedimentation within the marsh or consumption by the marsh bivalves. To balance the amount of nitrogen lost by sedimentation and feeding by bivalves, the supply of particles of dead grass into tidal water must be at least 10 kg N day^{-1}.

The shellfish, while intercepting particulates, excreted some ammonia and deposited faecal particulates. Their role, therefore, is to act as scrubbers of particulates which would otherwise be flushed out to sea, incorporating particles into large heavy fragments which are not as readily resuspended. In the course of doing this, however, they excreted ammonia which accounted for about a fourth of the losses of this species of nitrogen from the entire marsh.

Each box shown in Fig. 25.4 has a complex internal structure. As an example, we can look at the tidal water. The nitrogen content of tidal water depends on the stage of the tidal cycle. We can compare the amounts of diverse nitrogen compounds in tidal water as it leaves the vegetated upper reaches of the marsh and as it exits into the sea. Fig. 25.5 shows a set of data, including the particulate, ammonium and nitrate forms of nitrogen in the water through a whole tidal cycle taken from these two sites.

The tidal height at different stages of the tidal cycle on 5 August 1975 at the upper reaches and at Sippewissett creek, the main channel which empties into the sea, is shown in Fig. 25.5. The changes in salinity during the cycle (middle left, Fig. 25.5) reflect the marked effect of ground water in the upper reaches of the marsh. When this water was finally mixed with the sea water at Sippewissett creek, the depression of salinity was still present but much smaller. There are pronounced differences in the amount of particulate

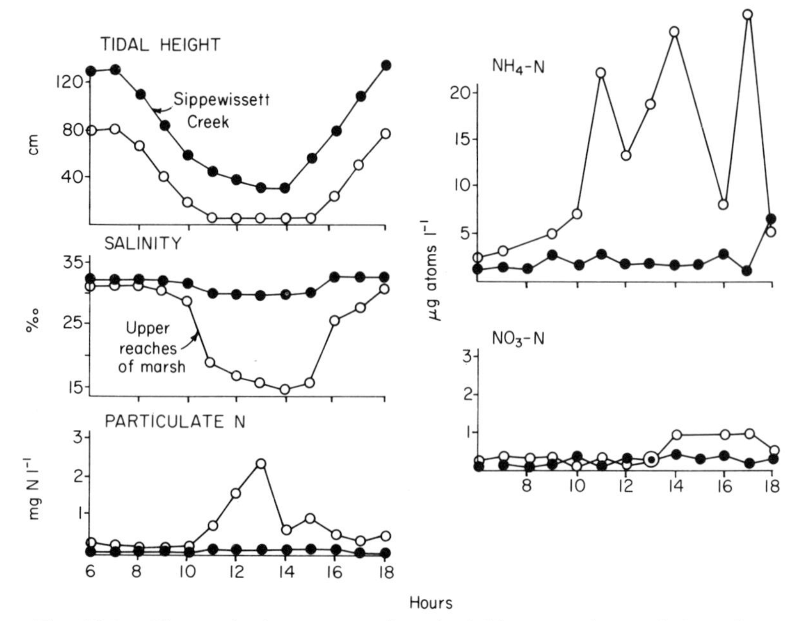

FIG. 25.5. Changes in the concentration of soluble ammonium and nitrate ions and particulate nitrogen and salinity during a tidal cycle as shown by the tidal height in the upper reaches of the marsh (○) and in Sippewissett creek (●) during 5 August 1975.

nitrogen and ammonium-nitrogen between water at the upper reaches and water leaving the marsh. There are several reasons which may account for this difference. Certainly dilution of the water from the upper reaches into the larger volume of less nutrient-rich water in part accounts for these differences. A graph of ammonium-nitrogen versus salinity (a passive tracer) gives a more or less linear relation with a negative slope, suggesting that dilution is the primary cause of the difference. For particulate nitrogen the same graph shows a concavity, implying consumption as the water moved from the upper reaches to the sea. Consumption by shellfish and sedimentation of particles are likely mechanisms for such a depletion (Fig. 25.4). A graph of nitrate-nitrogen content versus salinity shows no trend at all. We need to look more closely at the processes involved in nitrogen exchanges.

DISCUSSION

The major sources of nitrogen for Great Sippewissett marsh were flooding tidal water, flow of ground water and nitrogen fixation. The losses, principally through ebbing tidal water and denitrification, exceed the gains by 974 kg N yr^{-1} (Table 25.1). This amounts to only 2·9% of the inputs primarily because of the overwhelming magnitude of the dissolved organic nitrogen component.

The nature of the nitrogen compounds involved in the exchanges and the seasonality of the fluxes are two aspects that need emphasizing, because these properties reflect how the salt marsh functions.

The total amounts of each form of nitrogen entering and leaving Great Sippewissett marsh can be compiled (Table 25.2). Nitrate, primarily provided by ground water, was consumed by the marsh and converted to ammonium and particulate nitrogen and exported in these two forms. Although the fluxes of dissolved organic nitrogen were very large, similar amounts entered and left the marsh (Table 25.2). This, as discussed above, may suggest that this material may be biologically unaffected as it moves through the marsh. Abd. Aziz and Nedwell (in this volume) report evidence supporting the refractory nature of dissolved organic nitrogen in salt marsh sediments. The salt marsh also exported much more nitrogen than it imported. Again the totals of Table 25.2 show only a small discrepancy between gains and losses.

While the provision of nitrate by ground water was fairly uniform through the year, there was a pronounced seasonality in the export of ammonium and particulate nitrogen (Fig. 25.3). The exports of ammonium were related probably to seasonal peaks in leaching of plants and excretion by animals, while the patterns of particulate export were probably related to the comminution and mineralization of plants by decomposers.

The full significance of the export of nitrogen by salt marshes to coastal ecosystems remains to be explored, but this export must be of importance since the production of coastal phytoplankton is nitrogen-limited and particulates serve as the nitrogen source of detrital food webs.

Woodwell *et al.* (this volume) and other workers in the eastern coast of North America report examples of marshes that show a net import of particulates. These examples seem related to topographical features that result in either sediment traps (as in Woodwell's study) or reduced flow of tidal water. Whether marshes export organic matter is an important issue, since this export to coastal waters poor in organic matter is a principal reason advanced for conservation of salt marshes in the eastern U.S. Our view is that salt marsh ecosystems grow abundant grass that becomes detritus and the detritus then becomes available for export. If the physiographic setting enables sufficient water to flow and if no deep water areas act to trap suspended particles, then export of organic matter to coastal water will take place.

ACKNOWLEDGMENTS

We thank the following members of our research group for supplying us with results from their work: Tom Jordan, Jim Morris, Terri Goldberg, Bob

Howarth and Mike Connor. Wendy Wiltse criticized an earlier version of the text. We also thank Norma Butler, Suzanne Volkmann, Charlotte Coggswell and Debbie Shafter for technical assistance. We are indebted to the Arnold Gifford family for the use of their land in Great Sippewissett Marsh.

REFERENCES

Carpenter E.J., Van Raalte C.D. & Valiela I. (1978) Nitrogen fixation by algae in a Massachusetts salt marsh. *Limnol. Oceanogr.* **23**, 318–27.

Caswell H., Reed F., Stephenson S.N. & Werner P.A. (1973) Photosynthetic pathways and selective herbivory: a hypothesis. *Am. Nat.* **107**, 465–80.

Kaplan W. (1977) Denitrification in a Massachusetts salt marsh. Ph.D. dissertation. Boston University.

Pigott C.D. (1969) Influence of mineral nutrition on the zonation of flowering plants in coastal salt marshes. *Ecological Aspects of Mineral Nutrition in Plants* (Ed. by I.H. Rorison), pp. 25–35. Blackwell Scientific Publications, Oxford.

Tukey H.B. Jr. (1970) The leaching of substances from plants. *Ann. Rev. Plant Physiol.* **21**, 305–24.

Tyler G. (1967) On the effect of phosphorus and nitrogen, supplied to Baltic shore meadow vegetation. *Bot. Notiser* **120**, 433–47.

Valiela I. & Teal J.M. (1974) Nutrient limitation in salt marsh vegetation. *Ecology of Halophytes* (Ed. by R.J. Reimold & W.H. Queen), pp. 547–63. Academic Press, New York and London.

Valiela I., Teal J.M. & Sass W.J. (1975) Production and dynamics of salt marsh vegetation and effect of sewage contamination. Biomass, production and species composition. *J. appl. Ecol.* **12**, 973–82.

Valiela I., Teal J.M. & Persson N.Y. (1976) Production and dynamics of experimentally enriched salt marsh vegetation: belowground biomass. *Limnol. Oceanogr.* **21**, 245–52.

Valiela I., Teal J.M. & Deuser W.G. (1978a) The nature of growth forms in the salt marsh grass *Spartina alterniflora*. *Am. Nat.* (in press).

Valiela I., Teal J.M., Volkmann S., Shafer D. & Carpenter E.J. (1978b) Nutrient and particulate fluxes in a salt marsh ecosystem: Tidal exchanges and inputs by precipitation and ground water. *Limnol. Oceanogr.* **23**, 798–812.

26. THE DISTRIBUTION OF FUNGI IN COASTAL REGIONS

G.J.F. PUGH

Department of Biological Sciences, University of Aston, Birmingham, U.K.

SUMMARY

Studies of the occurrence of pathogenic fungi on coastal plants show that the rusts, powdery mildews and ergot diseases have been most frequently noted. Mycorrhizal associations in sand dune plants occur most abundantly in the fixed-dune open plant communities. These associations have been observed in salt marsh plants but have not been intensively studied.

The patterns of distribution of fungi in the primitive soils in coastal regions show that there is a rise in fungal abundance with increasing higher plant cover. In salt marsh muds 'transient' fungi are distinguished from 'inhabitants' by their relative distribution.

Studies on phylloplane fungi show that the mycoflora of coastal plants is essentially similar to that of inland plants but there is some evidence of physiological adaptation of *Sporobolomyces roseus* to saline conditions.

Some of the difficulties encountered in estimating the activities of fungi in the soil are discussed. They play an important, but at present a not fully quantifiable role in the decomposition processes.

RÉSUMÉ

L'étude des cas des champignons pathogènes des végétaux côtiers montre que l'on a noté le plus souvent les rouilles, les mildious poudreux, et les maladies d'ergot. Les associations mycorrhyzales chez les végétaux des dunes de sable ont lieu le plus abondamment dans les communautés végétales clairsemées des dunes fixes. Ces associations ont été observées dans les marais salés, mais n'ont pas été intensivement étudiées.

Les propriétés de distribution des champignons dans les sols primitifs des régions côtières montrent que l'abondance des champignons augmente avec une plus grande couverture végétale. Dans les boues des marais salés on distingue les champignons «transitoires» des «habitants», par leur distribution relative.

Les études des champignons phyllophanes montrent que la mycoflore des végétaux côtiers est essentiellement similaire à celle des plantes de l'intérieur, avec cependant quelques évidences d'adaptation physiologiques de *Sporobolomyces roseus* aux conditions salines.

Nous discutons de quelques unes des difficultés rencontrées dans les estimations de l'activité des champignons dans le sol. Ils jouent un rôle important dans les processus de décomposition, mais celui-ci n'est pas à présent entièrement quantifiable.

INTRODUCTION

Fungi are heterotrophic organisms which may obtain their nutrients from a living host, or from dead organic matter. A fungus may be associated with a living host essentially in one of two ways: it may be parasitic, obtaining its food at the expense of the host, or it may live in close association with the green plant in such a way that both partners derive benefit. This type of association is usually referred to as symbiosis, but in the original usage of symbiosis, de Bary (1887) referred to the relationship between a parasite and its host. It might be appropriate to regard lichens and mycorrhizal associations which are commonly referred to as symbiotic, as green plants living in a state of controlled parasitism, in which the host derives some benefit, especially in the form of increased uptake of minerals. Saprophytic fungi obtain their food from dead organic matter, which may be of plant or of animal origin.

All three nutritional groups of fungi occur in coastal areas. Parasites and some saprophytes which form obvious lesions or spots on plants have been recorded for more than a hundred years. The pioneering studies on soil fungi in salt marshes were carried out by Bayliss Elliott (1930), while the distribution of mycorrhizal associations on salt marsh plants was described by Mason (1928). More recently there have been many accounts of the fungi in salt marsh muds and in sand dunes from different parts of the world. Thus mycorrhizal associations have been studied in British sand dunes (Nicolson 1959, 1960), and in France by Boullard (1958) who worked on estuarine plants in Normandy. The occurrence of mycorrhizas on inland halophytes in Bohemia, Czechoslovakia was investigated by Klecka and Vukolov (1937). The plants studied included several which are usually considered in north-west Europe to be coastal in distribution such as *Salicornia herbacea, Suaeda maritima* and *Triglochin maritima.* Fries (1944) also studied mycorrhizas in halophytes in Sweden and found vesicular-arbuscular associations to be common.

Sand-dune fungi have been examined by Webley, Eastwood and Gimingham (1952) and by Brown (1958a, b), while Pugh and Williams (1968) studied the fungi associated with *Salsola kali*. Specialized nutritional groups such as cellulose-decomposers were studied by Pugh, Blakeman, Morgan-Jones and Eggins (1963) and keratinophilic fungi by Pugh and Mathison (1962).

Salt marsh fungi have received considerable attention: in Japan by Saito (1955), in Britain by Pugh (1960, 1974), Pugh and Dickinson (1965), in German estuaries by Siepman (1959) and in United States estuaries and marshes by, amongst many others, Borut and Johnson (1962), Meyers (see 1974) and Wagner (1969).

FUNGAL PATHOGENS ON PLANTS IN COASTAL AREAS

Ergots, mildews and rust fungi are probably the most easily noticed diseases on plants. Ergot disease, caused by *Claviceps purpurea* is common on *Agropyron* spp. in the Wash area, and is locally abundant on *Spartina* in Southampton Water. The disease can probably be found wherever it is looked for on grasses in coastal areas. The seed is replaced by a mass of fungal tissue, forming a sclerotium, or ergot. This germinates, normally in the spring of the following year when a stroma containing perithecia is formed (Fig. 26.1). The ascospores infect the ovary which becomes converted into a sclerotium.

Powdery mildew occurs locally abundantly on *Limonium vulgare* at Gibraltar Point, Lincolnshire. The leaves become grey in colour because of the superficial growth of mycelium, on which the chains of conidia are formed, and in which, later in the year, the cleistocarps develop.

A new species of parasite has been reported on *Ruppia maritima* by Feldman (1959) who described *Melanotaerium ruppiae*. She also mentioned *Tetramyxa parasitica* as occurring on this host, and *Plasmodiophora bicaudata* as a parasite on *Zostera nana*.

Rust diseases have been recorded on many coastal plants, and have been listed in Table 26.1, where it can be seen that the genera *Puccinia* and *Uromyces* occur frequently. *Phragmidium violaceum* with its predominantly four-celled teleutospores occurs more commonly in coastal areas than *P.rubrum* which has teleutospores normally six-celled, and which is more inland in its distribution.

MYCORRHIZAL FUNGI

On salt marsh plants

The study of mycorrhizas on salt marsh plants has not advanced beyond the collection of data on the distribution of vesicular-arbuscular (VA) associations on different plants. Mason (1928) examined plants collected from the salt

FIG. 26.1. Ergots of *Claviceps purpurea* on inflorescences of *Lolium perenne*. The ergots germinate the following spring to form stromata in which the perithecia develop.

Table 26.1. *Rust fungi which occur on plants in coastal areas (from Grove 1913)*

Host	Parasite
Limonium spp.	*Uromyces limonii* Leveille
Armeria maritima	*Uromyces armeriae* Lev.
Silene maritima	*Uromyces behenis* Unger
Suaeda maritima	*Uromyces chenopodii* Schrot.
S.fruticosa	*Uromyces chenopodii* Schrot.
Salicornia europaea	*Uromyces salicorniae* de Bary
Beta maritima	*Uromyces betae* Lev.
Artemisia maritima	*Puccinia absinthii* DC.
Smyrnium dusatrum	*Puccinia smyrnii* Corda
Carex arenaria	*Puccinia schoeleriana* Plowr. et Magn.
Carex arenaria	*Puccinia arenariicola* Plowr.
Elymus arenarius	*Puccinia glumarum* Er. et Henn.
Rubus fruticosus	*Phragmidium violaceum* Wint.

marshes at Borth and Talsanau, on the Welsh coast. She recorded the following species as possessing mycorrhizas: *Agrostis stolonifera, Armeria maritima, Aster tripolium, Cochlearia officinalis, Glaux maritima, Puccinellia maritima, Plantago coronopus* and *P.maritima.* She examined, but did not find mycorrhizas in *Juncus gerardii, J.maritimus, Salicornia europaea, Spergularia marginata* or *Triglochin maritima.* She observed that the VAs were found in young rather than old roots, and that the mycelium was a septate. The vesicles collected in July–August showed vacuolated cytoplasm and many nuclei, but none had been found sporulating. However, in *Agrostis stolonifera, Armeria maritima* and *Puccinellia maritima* there was, in addition a septate mycelium.

Klecka and Vukolov (1937) found VAs in *Salicornia europaea, Juncus gerardii* and *Triglochin maritima,* in contrast to the findings of Mason (1928), but confirmed their presence in *Aster tripolium* and *Plantago maritima.* In addition they recorded the presence of VAs in *Carex secalina, Samolus valerandi, Suaeda maritima* and *Taraxacum leptocephalum.* Their plants were obtained from salty ground at Neusiedler Lake and Auschitz and Loumy in Bohemia.

Fries (1944) found VAs to be common in plants on the Swedish coast, and recollected (1974) that VAs were 'almost omnipresent even in the most exposed places'.

It would be advantageous to have an up-to-date survey, preferably one based upon regular monthly collections throughout a growing season, of the occurrence, distribution and abundance of VAs in salt marsh plants. Are they mainly to be found in areas of mature saltings, or are they ubiquitous? Can they be found on seedlings early in the year, or are they best found in the summer months? Can the spores and other propagules be found and recovered from salt marsh muds—indeed, how well can they withstand saline conditions? Firm, quantitative data on the occurrence of VAs are needed before their full importance on salt marsh plants can be estimated.

On sand-dune plants

The most important work on VAs in sand-dune plants has been that by Nicolson. In 1959 he reported on the external mycelial phase of the VAs and drew attention to the genus *Endogone*. In fixed sand dunes he found profuse external mycelium associated with living roots, but could also obtain extensive clumps directly from the sand after removal of the roots. These clumps grew around fragments of organic matter, which were often old decaying roots, but included other types of organic debris.

Penetration of living roots can occur via root hairs or through a cell of the piliferous layer, or an exodermal cell in older roots. In *Festuca rubra* var. *arenaria*, he found that 92% of the root hairs had either viable hyphae or hyphal remains. Invasion through epidermal cells showed on average that one penetrating hypha gave rise to about a 1 mm length of invasion of the root, but he added that 'a small number may be responsible for extensive lengths of root infection'.

Nicolson (1960) discussed the development of VAs and paid particular attention to the sand-dune habitat. His work was mainly carried out at Gibraltar Point, Lincolnshire, but he also examined samples collected in Lancashire and in Scotland. He found VAs abundantly on *Agropyron junceiforme*, *Ammophila arenaria*, *Festuca rubra* var. *arenaria* and *Poa pratensis*. The most abundant external mycelium, and the greatest number of infection penetrations were found in the fixed yellow dunes. The open communities there appeared to provide the optimum conditions for endophyte development, with little sand movement and a relatively stable sand surface. This would reduce the number of new roots just below the surface, and would enable the endophyte to accumulate around the roots. At the same time the stability of the sand allows some accumulation of organic debris from earlier colonizing plants, which would be utilized by the endophyte mycelium. At the same time the quantity would be insufficient to stimulate the development of a large microbial flora which could be antagonistic to the endophyte.

In the fixed dune but open plant communities, where *Ammophila*, *Festuca* and *Poa* all occurred together, the extensive external mycelium and penetration phases of the fungus may directly benefit the higher plants which are to some degree competing against each other. By comparison, in the mobile dune *Ammophila* has no competition from other plants and can therefore utilize all the available nutrients. In the fixed, closed dunes, a soil is being formed, enriched with organic debris, and consequently, with a more generous supply of nutrients. Here, *Ammophila* will be at a competitive disadvantage: the improved edaphic conditions allow more plants to grow, while the reduced amount of blown sand does not give an advantage to *Ammophila*. It therefore grows less strongly, and is likely to have fewer root exudates. This in turn could adversely affect the abundance of VA associations.

SOIL FUNGI

Salt marshes

Fungi have been isolated from muds uncolonized by higher plants, from rhizospheres of pioneer plants in developing salt marshes, and from the continuous rhizospheres which are present in mature saltings. An overall trend was found in which the mycoflora increased in abundance and in numbers of species in an upshore direction (Pugh 1960, 1962). The species with this type of distribution pattern were regarded as 'salt marsh inhabitants'. They included *Cephalosporium acremonium*, *Dendryphiella salina*, *Mortierella alpina* and probably *Fusarium* spp. The heavily sporing fungi of the genera *Aspergillus*, *Mucor*, *Penicillium* and *Trichoderma* were all more frequently isolated towards the bottom of the mud flats and from sea water, than from higher up the marsh. They were regarded as 'salt marsh transients' which could be deposited on the mud flats by every tide, being most abundant where tidal coverage was most frequent. This view is supported by the results of Saito (1952) who showed that the genera *Penicillium* and *Trichoderma* were much more abundant in the surface muds than in the anaerobic deeper layers. *Fusarium*, *Mucor* and *Aspergillus* were infrequent in occurrence, but with more isolates from the surface muds.

The increase in abundance of the salt marsh inhabitants up the shore could be a result of better aeration and greater amounts of soil organic matter in the higher regions. However at least one of the species, *Dendryphiella salina*, needs regular coverage by the sea to maintain its presence. Pugh (1974) has shown that in the mature saltings this species becomes replaced by *Gliocladium roseum*, which is the more successful competitive saprophyte in the absence of sea water. With increasing salinity, however, its rate of growth is rapidly diminished, and then *D.salina* becomes dominant.

Thus, many of the inhabitants may well be terrestrial fungi which are invading salt marsh muds as conditions become tolerable for them. The mycoflora of the British salt marshes which have been studied is thus similar to that of mangrove swamp muds studied by Swart (1963) who regarded the species he isolated as being terrestrial in origin. Others, and particularly *D.salina* may use the saline conditions to escape the competition which exists in terrestrial soils.

This type of up-shore increase has been found in marine muds at Gibraltar Point, Lincolnshire (Pugh 1960, 1962) in a sandy littoral area near Madras (Pugh 1966), and with the nutritionally specialized group of keratinophilic fungi at Gibraltar Point (Pugh & Mathison 1962) and near Vancouver, B.C. (Pugh & Hughes 1975). A similar picture was found in fresh water muds during the reclamation and subsequent cultivation of polder soils in Holland (Pugh & van Emden 1969).

Sand dunes

Studies of the mycoflora in transects from open beach to mature dune pasture show essentially a similar increase in abundance to that seen in salt marsh muds. Webley, Eastwood and Gimingham (1952) found that 'numbers rose steadily with the development of the vegetation and increasing complexity of the plant community'. Representatives of the genera *Cephalosporium*, *Cladosporium*, *Fusarium*, *Mucor*, *Penicillium*, *Stemphylium* and *Trichoderma* were isolated. *Cephalosporium* was the most frequently isolated genus from the rhizosphere and root surfaces of *Agropyron juncieforme* and *Ammophila arenaria*. *Fusarium* was also abundant on *Agropyron*, but relatively less common on *Ammophila*.

Brown (1958a) compared the mycoflora of two dune systems, one alkaline and the other acid, and found that they were distinct from each other. She found that a succession of fungal species occurred across the dune systems. From her lists of fungi isolated, it is obvious that they are terrestrial in origin with populations dominated by various Phycomycetes and *Penicillium* species, with *Trichoderma viride* becoming very common in acid mature dune soils. Her lists, and those of Dickinson and Kent (1972) also show that hyaline fungi rather than pigmented forms tend to predominate in sandy soils.

Working with the fungi which occur on *Salsola kali*, Pugh and Williams (1968) showed that the seeds were mainly colonized by the hyaline genera *Cephalosporium* and *Fusarium*, and the pigmented *Alternaria* and *Cladosporium*. Following germination, the roots became rapidly colonized by the hyaline forms, with only a few isolations of the pigmented fungi being obtained one month after germination. On the aerial parts of the plant, however, while hyaline fungi were common, the pigmented forms increased dramatically as the plants matured. Thus while *Cephalosporium* and *Fusarium* increased by 16% and 21% respectively, *Alternaria* increased by 49% and *Botrytis*, *Camarosporium*, *Epicoccum* and *Stemphylium* more than trebled their abundance. This increase of the pigmented forms was attributed to their ability to withstand exposure to ultraviolet radiations, mainly by the screening effects of melanin deposited in the hyphal walls. *Cephalosporium*, although hyaline, possesses carotenoid pigments which may help the fungus to derive benefit from light. Codnor and Platt (1959) have discussed the stimulation of carotenoid production in response to light, while Leach (1965) showed that sporulation in several fungi is induced by ultra-violet absorbing substances.

Heavily pigmented fungi have been reported as representing a majority of the species isolated from sands in the Sahara region by Nicot (1960). She listed, as protective mechanisms against desiccation and strong light, the possession of brown pigmentation, generally thick membranes, a high rate of production of generally multicellular spores and a tendency for chlamydospores to be present in groups. These features can all be seen in fungi which occur on the aerial parts of plants in general: sand-dune plants are no exception.

FUNGI ON AERIAL PARTS OF PLANTS

The presence of fungi on the aerial parts of living plants in coastal areas has been mainly studied by Dickinson (1965) who worked on *Halimione portulacoides* and by Lindsey and Pugh (1976a, b) who examined leaves of *Hippophäe rhamnoides*. Apinis and Chesters (1964) gave an account of ascomycetes which they recorded on senescent and dead parts of several grasses, including *Agropyron junceiforme*, *A.pungens*, *Ammophila arenaria*, *Puccinellia maritima* and *Spartina townsendii*.

The leaf surface, or phylloplane, fungi are ideally situated to utilize any nutrients which may be exuded from the leaf. Tukey (1971) has discussed the qualitative and quantitative aspects of leaching, indicating that in addition to all essential minerals, leachates include free sugars, pectic substances, sugar alcohols and all of the amino acids found in plants. These leachates can be reabsorbed by the same plant or by adjacent plants. The ability of microorganisms present on leaves to produce auxins has been demonstrated for bacteria by Klincare, Kreslina and Mishke (1971), for yeasts by Diem (1971) and for filamentous fungi by Buckley and Pugh (1971). The ability of these phylloplane organisms to produce auxins from substances known to be present in leaf leachates could have profound implications throughout the ecosystem.

In his study of fungi on living leaves of *Halimione portulacoides*, Dickinson (1965) found that they could be grouped into three categories. Some were present only as propagules on the leaf surface, and comprised the transients which were deposited as casual contaminants from the atmosphere, as well as yeasts which could easily be washed from the leaf surface. The second group consisted of those species such as *Cladosporium herbarum* which were actively growing and sporulating on the leaf surface. His third group consisted of fungi such as *Ascochytula obiones* which grew vegetatively on the leaf surface, but which subsequently produced pycnidia inside the moribund or dead leaves.

Lindsey and Pugh (1976a, b) working with the leaves of *Hippophae rhamnoides* showed that many of the fungi were associated with the mid-rib groove, and were particularly abundant on the upper surface. On the leaf lamina the phylloplane fungi were mainly found associated with the trichomes which cover the leaf surface. The most frequently occurring fungi were *Aureobasidium pullulans*, *Cephalosporium acremonium*, *Cladosporium herbarum*, *Epicoccum purpurascens*, *Phoma* sp. and *Sporobolomyces* spp. together with sterile mycelia.

During this work, a study of *Sporobolomyces roseus* on leaves (Pugh & Lindsey 1975) showed that the numbers of this fungus were affected by salt concentrations on *Hippophae* leaves, and by the washing of *Halimione* leaves during coverage by the higher spring tides. There was evidence of selection or adaptation to maritime conditions, as isolates obtained from coastal areas grew and respired more quickly at higher concentrations of salts than did other isolates obtained from inland sites.

The importance of such studies on phylloplane fungi is that these organisms, which are present from the first unfolding of the leaf, are ideally positioned to begin the decomposition of the leaf even before leaf fall. They are the primary colonists which are subsequently replaced by the ascomycetes studied by Apinis and Chesters (1964) and eventually by the soil fungi.

DISCUSSION

The regularly occurring patterns of distribution of fungi in coastal soils indicate that these organisms are a natural part of the ecosystem. Unfortunately the level of refinement of the isolation techniques used does not normally allow an estimate to be made of the activities of the fungi. Soil dilution plates have been used to count fungi, but such counts are heavily biased in favour of the heavily sporulating species, and beg the question of what is meant by the numbers. Soil crumb plates reduce the bias towards the spore producers, but neither technique can differentiate between active hyphae, inactive hyphae and dormant propagules. Techniques such as Brown's (1958b) modified impression slides have been used to estimate the amount of mycelium in dune sands; Jones and Mollison's (1948) method can be used to measure hyphal lengths, but has not apparently been used in coastal soils. The activities of micro-organisms in the soil can be further elucidated by the use of isotopes, when a labelled substrate can be buried and its breakdown products traced; by respirometric studies, and by the estimation of enzymes in the soil. These methods applied to a natural soil will produce results relating to the total activity of the organisms present, including members of the microfauna. By selectively eliminating different groups of organisms, some estimate can be made of those which remain, but it must be appreciated that the treatment may have altered the soil, while the killed organisms will provide additional substrates for the survivors to utilize. There is thus a 'Microbial Principle of Uncertainty' in which the activity cannot be measured accurately without disturbing the substrate: but by disturbing the substrate the level of activity may well be altered. It is known, however, that actinomycetes, bacteria and fungal hyphae grow on dead plant remains. Actinomycetes and bacteria help decompose fungal hyphae, and all three groups of organisms are utilized as food material by protozoa, arthropods and other animals.

In the decomposition of plant remains, the fungi, as the main organisms which can decompose cellulose, must be an important element: and during that decomposition some fungi—the mycorrhizal species—are aiding the growth of other plants.

Until the part played by all decomposer groups is more fully known it is premature to describe the role of fungi in coastal soils, or elsewhere, in quantitative terms. It is safe, however, to say that the role is vital in the decomposition process upon which higher plant growth is dependent.

REFERENCES

Apinis A.E. & Chesters C.G.C. (1964) Ascomycetes of some salt marshes and sand dunes. *Trans. Br. mycol. Soc.* **47**, 419–35.

Bayliss-Elliott J.S. (1930) The soil fungi of the Dovey Salt Marshes. *Ann. appl. Biol.* **17**, 284–305.

Borut S.Y. & Johnson T.W. (1962) Some biological observations on fungi in estuarine sediments. *Mycologia* **54**, 181–93.

Boullard B. (1958) Les Mycorrhizes des especes de contact marin et de contact salin. *Rev. de Mycologie* **23**, 282–317.

Brown J.C. (1958a) Soil fungi of some British sand dunes in relation to soil type and succession. *J. Ecol.* **46**, 641–64.

Brown J.C. (1958b) Fungal mycelium in dune soils estimated by a modified impression slide technique. *Trans. Br. mycol. Soc.* **41**, 81–88.

Buckley N.G. & Pugh G.J.F. (1971) Auxin production by phylloplane fungi. *Nature, Lond.* **231**, 332.

Codnor R.C. & Platt B.C. (1959) Light induced production of carotenoid pigments by Cephalosporia. *Nature, Lond.* **184**, 171.

de Bary A. (1887) *Comparative Morphology and Biology of the Fungi, Mycetozoa and Bacteria.* Clarendon Press, Oxford.

Dickinson C.H. (1965) The Mycoflora associated with *Halimione portulacoides.* III. Fungi on green and moribund leaves. *Trans. Br. mycol. Soc.* **48**, 603–10.

Dickinson C.H. & Kent J.W. (1972) Critical analysis of fungi in two sand-dune soils. *Trans. Br. mycol. Soc.* **58**, 269–80.

Diem H.G. (1971) Production de l'acide indolyl-3-acetique par certaines levures epiphylles. *C. r. hebd Seanc Acad. Sci.* **272**, 941–43.

Feldmann G. (1959) Une Ustilaginale marine, parasite du *Ruppia maritima* L. *Rev. gen. de Botanique* **66**, 35–40.

Fries N. (1944) Beohachtungen uber die thamniscophage mykorrhiza einiger Halophyten. *Bot. Not.* **2**, 255–64.

Fries N. (1974) Contribution to a discussion in *Veroff. Inst. Meeresforsch. Bremerhaven Suppl.* **5**, p. 416.

Grove W.B. (1913) *The British Rust Fungi*. Cambridge University Press.

Jones P.C.T. & Mollison J.E. (1948) A technique for the quantitative estimation of soil micro-organisms. *J. gen. Microbiol.* **2**, 54–69.

Klecka A. & Vukolov V. (1937) Comparative study of the mycorrhizae of meadow halophytes. *Rev. appl. Mycol.* **16**, 768.

Klincare A.A., Kreslina D.J. & Mishke I.V. (1971) Composition and activity of the epiphytic microflora of some agricultural plants. *Ecology of Leaf-Surface Micro-organisms* (Ed. by T.F. Preece & C.H. Dickinson), pp. 191–201. Academic Press, London and New York.

Leach C.M. (1965) Ultraviolet absorbing substances associated with light induced sporulation in fungi. *Can. J. Bot.* **43**, 185–200.

Lindsey B.I. & Pugh G.J.F. (1976a) Succession of microfungi on attached leaves of *Hippophae rhamnoides*. *Trans. Br. mycol. Soc.* **67**, 61–67.

Lindsey B.I. & Pugh G.J.F. (1976b) Distribution of microfungi over the surfaces of attached leaves of *Hippophae rhamnoides*. *Trans. Br. mycol. Soc.* **67**, 427–33.

Mason E. (1928) Note on the presence of mycorrhiza on the roots of salt marsh plants. *New Phytol.* **27**, 193–95.

Meyers S.P. (1974) Contribution of fungi to biodegradation of *Spartina* and other brackish marshland vegetation. *Veroff. Inst. Meeresforsch. Bremerhaven Suppl.* **5**, 357–75.

Nicolson T.H. (1959) Mycorrhiza in the Gramineae. I. *Trans. Br. mycol. Soc.* **42**, 421–38.

Nicolson T.H. (1960) Mycorrhiza in the Gramineae. II. *Trans. Br. mycol. Soc.* **43**, 132–45.

Nicot J. (1960) Some characteristics of the microflora of desert sands. *The Ecology of Soil Fungi* (Ed. by D. Parkinson & J.S. Waid), pp. 94–97. Liverpool University Press.

Pugh G.J.F. (1960) The fungal flora of tidal mud flats. *The Ecology of Soil Fungi* (Ed. by D. Parkinson & J.S. Waid), pp. 202–8. Liverpool University Press.

Pugh G.J.F. (1962) Studies on fungi in coastal soils. II. *Trans. Br. mycol. Soc.* **45**, 560–66.

Pugh G.J.F. (1966) Cellulose-decomposing fungi isolated from soils near Madras. *The Journal of the Indian Bot. Soc.* **45**, 232–41.

Pugh G.J.F. (1974) Fungi in intertidal regions. *Veroff. Inst. Meeresforsch. Bremerhaven Suppl.* **5**, 403–18.

Pugh G.J.F., Blakeman J.P., Morgan-Jones G. & Eggins H.O.W. (1963) Studies on fungi in coastal soils. IV. *Trans. Br. mycol. Soc.* **46**, 565–71.

Pugh G.J.F. & Dickinson C.H. (1965) The mycoflora associated with *Halimione portulacoides*. I. *Trans. Br. mycol. Soc.* **48**, 381–90.

Pugh G.J.F. & Hughes G.C. (1975) Epistolae mycologicae. V. Keratinophilic fungi from British Columbia coastal habitats. *Syesis* **8**, 297–300.

Pugh G.J.F. & Lindsey B.I. (1975) Studies of *Sporobolomyces* in a Maritime Habitat. *Trans. Br. mycol. Soc.* **65**, 201–9.

Pugh G.J.F. & Mathison G.E. (1962) Studies on fungi in coastal soils. III. *Trans. Br. mycol. Soc.* **45**, 567–72.

Pugh G.J.F. & van Emden J.H. (1969) Cellulose-decomposing fungi in polder soils and their possible influence on pathogenic fungi. *Neth. J. Pl. Path.* **75**, 287–95.

Pugh G.J.F. & Williams G.M. (1968) Fungi associated with *Salsola kali*. *Trans. Br. mycol. Soc.* **51**, 389–96.

Saito T. (1952) The soil fungi of a salt marsh and its neighbourhood. *Ecological Review* **13**, 111–19.

Siepmann R. (1959) Ein Beitrag zur Saprophytischer Pilzflora des Wattes der Weser mundung. I. *Veroff. Inst. Meeresforsch. Bremerhaven* **6**, 213–81.

Swart H.J. (1963) Further investigations of the mycoflora in the soil of some mangrove swamps. *Botanica Neerlandica* **12**, 98–111.

Tukey H.B. Jnr. (1971) Leaching of substances from plants. *Ecology of Leaf-surface micro-organisms* (Ed. by T.F. Preece & C.H. Dickinson), pp. 67–80. Academic Press, London and New York.

Wagner D.T. (1969) Ecological studies on *Leptosphaeria discors*, a graminicolous fungus of salt marshes. *Nova Hedwigia* **18**, 383–96.

Webley D.M., Eastwood D.J. & Gimingham C.H. (1952) Development of a soil microflora in relation to plant succession on sand-dunes, including the 'rhizosphere' flora associated with colonising species. *J. Ecol.* **40**, 168–78.

27. SECONDARY PRODUCTION OF THE BENTHOS IN AN ESTUARINE ENVIRONMENT

R.M. WARWICK, I.R. JOINT AND P.J. RADFORD

NERC Institute for Marine Environmental Research, Plymouth, U.K.

SUMMARY

Several projects relating directly or indirectly to energy flow through the benthic community on a mud-flat in the Lynher estuary, Cornwall, U.K. have been integrated by using two methods, firstly a steady-state energy-flow diagram and secondly a dynamic simulation model, in order to provide a better understanding of ecosystem function and as an aid to research planning.

For each macrofauna species and meiofauna group an energy budget has been constructed in the form $C = P + R + F + U$ (IBP terminology). This information, together with values for primary inputs of carbon, has enabled us to construct a quantitative diagram representing the flow of carbon between the faunal components of 1 m^2 of mud over 1 year. The net annual production of macrofauna is 5·46 g C m^{-2}, and of meiofauna 20·17 g C m^{-2}. Of the meiofauna production, 3·34 g is utilized within the system, so that 16·83 g remains available to mobile carnivores. The macrofauna ingest 55·74 g of primary carbon annually, and the meiofauna 107·09 g. However, the meiofauna standing crop is only 0·49 times that of the macrofauna. Nematodes and copepods are energetically the most important meiofauna groups.

Having achieved realistic simulations of secondary production of deposit-feeders, filter-feeders, *Nephtys* and meiofauna, it has been possible to investigate the effect on the system as a whole of a variety of hypothetical trophic relationships which are poorly understood, particularly the interactions between meiofauna and macrofauna. The growth of *Nephtys* on different diets is given as an example.

RÉSUMÉ

Afin de fournir une meilleure compréhension de la fonction d'écosystème ainsi qu'une contribution à la planification de recherches nous avons réuni plusieurs études concernant, directement ou indirectement, le flux énergétique à travers une communauté benthique d'une zone vaseuse de l'estuaire de Lynher, Cornwall, U.K., ceci en utilisant deux méthodes, un diagrame de flux énergétique à l'équilibre et un modèle dynamique.

Nous avons construit un profil énérgetique de la forme $C = P + R + F + U$ (terminologie IBP) pour chaque espèce de macro et meiofaune. Cette information ajoutée aux valeurs d'absorption primaire de carbone nous a permis d'établir un diagrame quantitatif du flux de carbone entre les différents composantes faunistiques de 1 m^2 de vase sur une période d'un an. La production nette annuelle de la macrofaune est de 5,46 g C m^{-2}, celle de la meiofaune 20,17 g C m^{-2}, dont 3,34 g sont utilisés par le système lui-mème. Il reste donc 16,83 g de carbone disponibles pour les carnivores mobiles. La macrofaune assimile 55,74 g de carbone primaire par an et la meiofaune 107,09 g. Cependant, le rapport meiofaune/macrofaune est seulement 0,49. Nématodes et copépodes sont les groupes de la meiofaune les plus importants du point de vue énergétique.

Grâce à des simulations proches de la réalité de production secondaire par les mangeurs de dépôts, filtreurs, *Nephtys* et meiofaune, nous avons pu étudier les effets sur l'ensemble du système de quelques relations trophiques, hypothétiques mal comprises et en particulier les rapports meiofaune-macrofaune. La croissance de *Nephtys* sous différents régimes est donnée en example.

INTRODUCTION

Over the last few years several projects relating directly or indirectly to energy flow through the benthic community of a mud-flat in the estuary of the river Lynher, Cornwall, U.K. have been completed, and a number are still continuing. The location and physical characteristics of the site are described by Warwick and Price (1975). An attempt is made here to integrate these projects by constructing two kinds of model, a steady-state energy budget and a dynamic simulation, in order to provide a better understanding of ecosystem function and as an aid to research planning. Based on the best available data, original or from the literature, potentially important taxa have been identified by the steady-state model and potentially important processes by the dynamic simulation model. This has enabled priorities to be set for future research in order that the models can be refined.

With such a study it was imperative that the complete spectrum of secondary producers be covered. Examination of fresh sediment indicated a paucity of microfauna in the form of ciliates, as is usual with very fine 'non-capillary' sediments (Fenchel 1969). These have therefore been ignored, and we have concentrated on the faunal components traditionally categorized as macrofauna and meiofauna. We have attempted to avoid the pitfall highlighted by Gray (1976), who points out that the small annelids often escape attention since they pass through a 0·5-mm sieve, yet are often ignored by meiofauna workers as being too large.

METHODS

Practical details of methods have been given, or will be given, in accounts of the individual projects. Suffice it to say that three sets of faunal samples were collected at monthly intervals throughout the year: 0·1 m^2 box cores sieved at 0·5 mm for macrofauna, 0·74 cm^2 cores sieved at 45 μm for the true meiofauna and 15·9 cm^2 cores sieved at 125 μm for the intermediate sized annelids. The biomass of species from the large and intermediate sized samples was determined as ash-free dry weight by direct weighing, and from the meiofauna cores by volume determinations. Respiration rates for macrofauna are based on data obtained seasonally at field temperatures in a Gilson respirometer, and for meiofauna by cartesian diver respirometry at 20°C.

FORMULATION AND ASSUMPTIONS OF THE MODELS

The steady-state model

This model represents the flow of energy, in units of carbon, between the faunal components of 1 m^2 of mud over one year. It was demonstrably the case that there was no net gain or loss in biomass of any component between one winter and the next. The model is based on 'real' field data and not on the simulated values discussed below. In all calculations carbon = 40% of dry weight (Steele 1974). For the meiofauna, whose biomass was estimated from volume, a density of 1·13 and a wet:dry weight ratio of 4:1 were assumed (Weiser 1960, for nematodes). The Q_{10} for respiration of all meiofauna groups was taken as 2·05 over the range of monthly temperatures experienced in the field, an average value for temperate marine poikilotherms (Miller & Mann 1973), since there is no evidence for temperature compensation in any meiofauna group except for two species of oligochaete (Lasserre 1971). In all calculations 1 ml O_2 respired = 0·4 mg carbon metabolized (Crisp 1971). The energy budget for each species can be written as:

$$C = P + R + F + U$$

where C = consumption, P = production, R = respiration, F = faeces and U = excretion. (IBP terminology, Petrusewicz 1967). Nitrogenous excretion, mucus production, etc. are here considered negligible in terms of the total carbon flux of the animals.

For all macrofauna species, production and respiration have been determined independently, except for the respiration rates of *Mya* and *Scrobicularia* which rely on literature values. However, for most of the smaller species in the community, production could not be estimated by analysis of the growth of age-classes, since age-cohorts could not be separated in the samples. Reproduction in most species appeared to be continuous, or nearly so, resulting in complete overlapping of generations. For example, among the polychaetes, eggs or developing embryos were present in the adult tubes of *Manayunkia aestuarina* in all months except November and December whilst in *Streblospio schrubsolii* eggs were present in the coleomic cavity between March and October, and embryos were still present in the adult tubes in November. Most nematode species had a relatively stable age structure with gravid females and juveniles present throughout the year, whilst females of the dominant copepod *Tachidius discipes* were found with egg sacs in all months and nauplii were present throughout the year.

For some meiofauna groups (nematodes, copepods and kinorynchs) we have been able to extrapolate from our original respiration data to production. For annelids and nematodes Teal (1962) assumed that K_2, the net growth efficiency or net production efficiency, $P/(P + R)$, was 25%. However, Welch (1968) has indicated that this value may vary between 20 and 90% and Marchant and Nicholas (1974) have demonstrated a value of 38% for a bacterial-feeding nematode in an experimental situation. Extrapolating from the empirical relationship between annual production and respiration, McNeill and Lawton (1970) have shown that, for short-lived poikilotherms, the net production efficiency increases markedly at lower production levels. Therefore, in estimating production and respiration where these values have not been independently determined, we have not assumed an efficiency but have used Engelmann's (1966) P/R relationship as updated by McNeill and Lawton (1970). For some species, particularly the small annelids, we have no data at present on either production or respiration. In these cases it has been necessary to assume the same P/B ratio as that of related species of similar habit from the mud-flat or reported in the literature, and then to determine annual respiration via the McNeill and Lawton equation.

To calculate faecal production and complete the energy budgets we have relied on literature values of assimilation efficiencies for the same or related species. For filter feeding bivalves, *Mya* is assumed to have an efficiency of 75% (Tenore & Dunstan 1973, for *Mercenaria mercenaria*) and *Cardium* an efficiency of 65% (R.I.E. Newell, personal communication); the small suspension-feeding polychaetes *Manayunkia* and *Fabricia* are taken as 70%. *Nephtys* is assumed to have the same efficiency as *Neanthes* (=*Nereis*) *virens*, 77·3–85% when feeding on animal material (Kay & Brafield 1973; Tenore & Gopalan 1974), but a lower efficiency of 28·8% when feeding on detritus (Tenore & Gopalan 1974). Since animal material probably forms a small proportion of their diet (see below) the overall efficiency is taken as 30%. Deposit-feeding polychaetes are also assumed to have this efficiency (*Ampharete*, *Pygospio*, *Streblospio*) as are both oligochaete species. Among deposit-feeding bivalves the assimilation efficiency of *Scrobicularia* is 60% (Hughes 1970) and *Macoma* is assumed to be the same. The less important macrofauna species, of which *Hydrobia ulvae* comprises the major part of the biomass, are taken as 30% because they are mainly deposit feeders. Bacterial-feeding nematodes have efficiencies of about 60% (Marchant & Nicholas 1974, for *Pelodera*), but this value may fall to as low as 12% with excess feeding (Duncan *et al.* 1974, for *Plectus*). The higher value is assumed here. Copepods fed on natural detritus have efficiencies around 75% (Harris 1973; for *Tigriopus*); ostracods and kinorhynchs are assumed to be similar. For *Protohydra* an average value for carnivores (83%) has been assumed (Winberg 1956; Kay & Brafield 1973).

Energy-flow models of this kind are subject to a number of sources of error, not only in the assumption of empirical P/R and P/B relationships (Miller, Mann & Scarratt 1971) but also because of imprecise understanding of the trophic relationships between species. For modelling purposes we have categorized the species according to their dominant feeding activity. *Scrobicularia* and *Macoma* are regarded as deposit feeders although both are known to indulge in some suspension feeding (Hughes 1969; de Wilde 1975). On the other hand the suspension feeders *Mya* and *Cardium* almost certainly utilize some resuspended deposits, which for budgeting purposes will tend to balance out the suspension feeding of *Scrobicularia* and *Macoma*. The spionid worms (*Pygospio*, *Streblospio*) and ampharetid worms (*Ampharete*) are surface deposit feeders, whereas the sabellids (*Manayunkia*, *Fabricia*) are suspension feeders (Day 1967). The oligochaetes are deposit feeders (Brinkhurst & Jamieson 1971). The trophic position of *Nephtys* is enigmatic. It has traditionally been regarded as a carnivore (Clay 1967); Clark (1962) and Warwick and Price (1975) found some invertebrate remains in the gut, but because it was the top macrofauna producer in the Lynher mud-flat, Warwick and Price suggested that it may feed on meiofauna since it could not be supported by other macrofauna. However, although the meiofauna production is sufficient to support the *Nephtys* population (see below), we have found that animals collected fresh from the field produce faecal pellets within a few days which contain exclusively algal cells. The complement of gut enzymes in *Nephtys* is also capable of digesting plant material (M.N. Moore, unpublished). We have therefore assumed that *Nephtys* is a broad-spectrum omnivore, taking food in proportion to its availability (i.e. annual production), about 10% animal and 90% plant.

Widdows (in preparation) has found that about 10% of the non-living particulate organic carbon (POC) in suspension in the water column at this site is available as a food source for filter feeders. We have assumed the same value for deposited POC as a source for deposit feeders. There appears to be no net deposition of POC, so that the only input is via the autotrophs and heterotrophic bacteria, either by death or recycling through invertebrate faeces (assuming that all animal mortality is due to predation). However, only 10% of this input to the POC pool is biologically available to the consumers, which implies a buildup of biologically refractory POC. In the model, input to the available POC pool is taken as 10% of the total POC production, and output is assumed to equal input (i.e. there is no net flux). We have not attempted to estimate the rate of food production in the water column because this is modified by transfers which are not known.

Whether or not the meiofauna is passed up the food chain to the higher macrofauna trophic levels, particularly deposit feeders, is questionable. McIntyre (1969) reviews evidence suggesting that this is not the case, and in the model we have accepted this view, with the exception of *Nephtys*. Wieser (1953) has divided nematodes into four feeding groups based on the structure of the buccal cavity, and these divisions have been adhered to with some reservations by numerous subsequent authors: Group 1A—selective deposit feeders with no buccal cavity, Group 1B—non-selective deposit feeders with an unarmed buccal cavity, Group 2A—epigrowth feeders with a weakly armed buccal cavity and Group 2B—carnivores and omnivores with a large heavily armed buccal cavity. On the Lynher mud-flat 27% of the nematode population (13 species) are in Group 1A, 27% (13 species) in Group 1B, 44% (12 species) in Group 2A and 1% (2 species) in Group 2B. Perkins (1958) found diatoms in the gut of all groups (except 1A which he did not examine) from intertidal sediments, although diatoms do not appear to comprise such an important dietary item in the sublittoral (Boucher 1972). In the blackened sulphide layer Perkins found that the nematodes fed exclusively on sulphide bacteria. In the Lynher the sulphide layer is 1–2 cm deep depending on season but is not strongly developed and, because only about 10% of the nematode population live in this layer, sulphide bacteria have been ignored as a food source. We have assumed that the nematodes depend entirely on the autotrophic production of diatoms, flagellates etc. on the mud surface and on the production of aerobic heterotrophic microbes which, however, appear to be unimportant in the turnover of carbon on this mud-flat (Joint, in preparation). Nematode species of feeding Group 2A are known to be able to pierce diatoms with their buccal armature and suck out the contents (von Thun 1968). We have also observed that even in several Group 1B species with unarmed buccal cavities (particularly *Sabatieria pulchra* and *Axonolaimus paraspinosus*) the gut was green with chlorophyll in freshly extracted specimens, though no structural details of algal cells were visible. One of the Group 2B species, *Sphaerolaimus hirsutus*, has been observed feeding on other nematodes but, since this is such a rare species, carnivory amongst the nematodes has been ignored in the model.

Except for the hydroid *Protohydra leuckarti*, which is a carnivore feeding on nematodes and harpacticoids, the other meiofauna groups are assumed to feed on the autotrophic and heterotrophic microbes. This is confirmed by the observations of Perkins (1958).

The dynamic simulation model

The process flow diagram on which the model is based is shown in Fig. 27.1. This model simulates the production of the secondary producers (meiofauna, deposit feeders, suspension feeders and *Nephtys*) of 1 m² of mud-flat for one year. The majority of assumptions made for the steady-state model also apply to this model. The small annelids are included with the deposit feeders or suspension feeders because they feed in a similar way to their macrofauna counterparts, unlike the meiofauna proper. Individual species are not simulated and each state variable (e.g. meiofauna etc.) is considered as a unit of carbon which is added to by production at the expense of other state variables, and is subtracted from by respiration, excretion of dissolved and particulate organic carbon, and predation by other state variables. The food of the

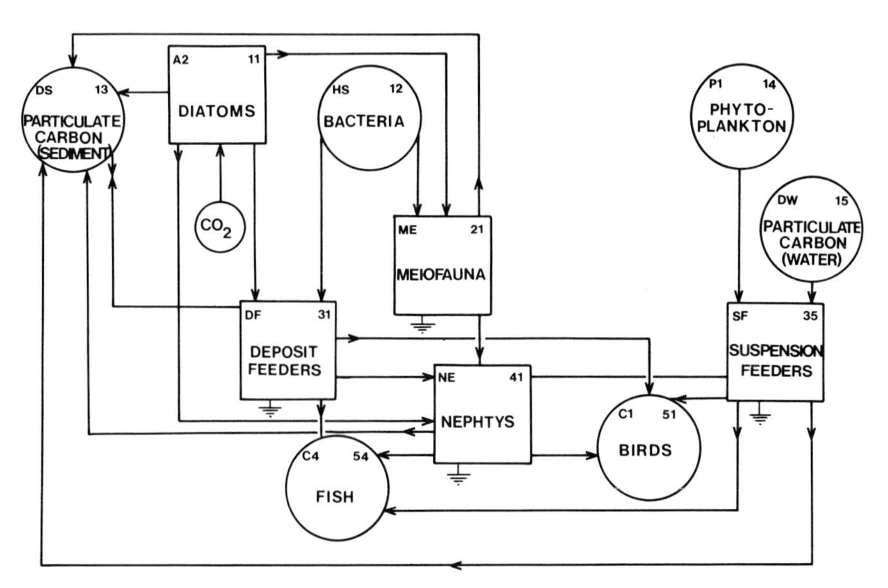

FIG. 27.1. The process flow diagram for the dynamic simulation model which indicates the transfer of carbon between state variables. Those state variables which are simulated are in boxes and those which are exogenous variables are in circles. Each state variable is designated by a number and a lettered abbreviation (e.g. deposit feeders: 31; DF). The 'earth' symbol denotes loss of carbon by respiration and the arrows represent rates of flow of carbon through the system.

secondary producers, with the exception of diatoms, are exogenous variables, since we have insufficient data to model processes such as deposition or resuspension of particulate organic carbon or horizontal inhomogeneities in phytoplankton distribution over the mud-flat. We therefore use monthly values for phytoplankton (P1), suspended organic carbon (DW), benthic bacteria (HS) and deposited organic carbon which are derived from our field data. Phytobenthos (A2) is simulated using the monthly rates of primary production measured in the field; primary production of phytobenthos occurs only when the mud-flat exposed in daylight hours, and is negligible when the mud-flat is flooded (Joint, in preparation). Other exogenous variables are the higher carnivores, birds (C1) and fish (C4), which are impossible to model because of migration, and we use the mean monthly biomass values taken from field data. Water and mud-flat surface temperature cycles are forcing functions for the model. Several of the state variables have more than one component of their diet and we have assumed that each food source is consumed in proportion to its abundance, e.g. deposit feeders will ingest phytobenthos, bacteria and particulate organic carbon and the relative proportions of each will vary with season; ingestion rate is limited by the total food available, and not by one component of the diet.

The meiofauna ingestion rates used were derived from the ingestion rates for the nematode *Pelodera*, feeding on bacteria (Marchant & Nicholas 1974) and we assume that this rate is dependent on the mud surface temperature, with a Q_{10} value of 2·05; phytobenthos and bacteria are the two components of meiofauna diet. For deposit feeders we have assumed that all animals are of the mean body size and feed at the same rate as *Scrobicularia* of the same size (Hughes 1969). We also assume that the gut of a deposit feeder is always full but that the rate of passage of material through the gut is dependent on the mud surface temperature: the gut volume, clearance rate and temperature function are those derived for *Scrobicularia* (Hughes 1969). The food of deposit feeders, phytobenthos, bacteria and particulate organic carbon, is taken in proportion to the relative abundance. The feeding rate of suspension feeders is derived from that for *Mytilus* (B.L. Bayne, personal communication); the filtration rate and the rate of production of pseudofaeces are dependent on the concentration of total particulate matter in suspension in the water column. Monthly values of suspended particulate matter for the Lynher river (J. Widdows, personal communication) are used as a forcing function; the food of suspension feeders is phytoplankton and suspended particulate organic matter, both of which are exogenous variables. Respiration rate is dependent on water temperature, with a Q_{10} value of 2·05. We have assumed that *Nephtys* is an omnivore which will feed on phytobenthos, meiofauna and on the smaller deposit feeding and suspension feeding annelids; the ingestion rate is assumed to be dependent on total food concentration and to obey Michaelis-Menten kinetics. The parameters for the Michaelis-Menten expres-

```
****CONTINUOUS SYSTEM MODELING PROGKAM****

*** VERSION 1.3 ***

TITLE   MUD-FLAT DYNAMICS LYNHER ESTUARY

FIXED NN
INITIAL
***     STATE VARIABLES INITIAL CONDITIONS - JANUARY 1974
INCON IA2=6.0,IDS=10.4,IME=0.9,IDF=2.5,ISF=3.3,INE=1.1
      KEEP=1
      NN=-1
DYNAMIC
NOSORT
      NN=NN+KEEP
      XN=FLOAT(NN)
SORT
***     EXOGENOUS VARIABLES
      P1=AFGEN(TP1,TIME)
FUNCTION TP1=0.,8.E-3,15.,8.E-3,45.,2.5E-2,74.,2.5E-2,105.,6.E-2,...
135.,2.3E-1,166.,1.5E-1,196.,5.6E-1,227.,9.E-2,258.,6.E-2,288.,3.E-2,...
315.,1.5E-2,345.,7.5E-3,365.,7.5E-3
      DW=AFGEN(TDW,TIME)
FUNCTION TDW=0.,0.289,15.,0.289,45.,0.455,74.,0.246,105.,0.237,...
135.,0.108,166.,0.194,196.,0.157,227.,0.229,258.,0.224,288.,0.237,...
315.,0.284,345.,0.360,365.,0.360
      HS=AFGEN(THS,TIME)
FUNCTION THS=0.,4.E-4,15.,4.E-4,45.,1.E-3,74.,2.E-3,105.,3.6E-3,...
135.,1.2E-2,166.,.2,196.,.36,227.,.6,258.,6.E-2,288.,1.2E-2,...
315.,3.6E-3,345.,2.E-3,365.,2.E-3
      C1=AFGEN(TC1,TIME)
FUNCTION TC1=0.,0.02,15.,0.02,45.,0.02,74.,0.,288.,0.,315.,0.04,...
345.,0.04,365.,0.04
*     DEMERSAL FISH HAVE CONSTANT BIOMASS
PARAM C4=0.05
      PARTOT=AFGEN(TPART,TIME)
FUNCTION TPART=0.,17.,15.,17.,45.,35.,74.,17.6,105.,13.2,135.,4.,...
166.,4.4,196.,3.2,227.,5.6,258.,11.2,288.,13.2,315.,25.6,345.,24.,...
365.,24.
*     RA2PR=HOURLY PRODUCTION RATE OF DIATOMS
      RA2PR=AFGEN(TA2PR,TIME)*6.E-3
FUNCTION TA2PR=0.,.6,15.,.6,45.,.9,74.,1.5,105.,5.5,135.,3.9,...
166.,6.3,196.,6.2,227.,5.6,258.,4.2,288.,2.8,315.,2.3,...
345.,1.6,365.,1.6
*     NHDAY=HOURS EXPOSED IN DAYLIGHT EACH DAY
      NHDAY=AFGEN(TNHDA,TIME)
FUNCTION TNHDA=0.,4.16,15.,4.16,45.,5.29,74.,5.58,105.,6.70,...
135.,7.68,166.,8.10,196.,7.68,227.,7.,258.,6.27,288.,5.32,...
315.,4.70,345.,5.06,365.,5.06
*     T=MEAN MONTHLY WATER TEMPERATURE
      T=AFGEN(TTW,TIME)
FUNCTION TTW=0.,8.7,15.,8.7,45.,9.1,74.,9.,105.,9.9,135.,13.,...
166.,15.,196.,15.,227.,16.6,258.,15.3,288.,12.,315.,9.4,...
345.,9.,365.,9.
*     TAIR=MEAN MONTHLY AIR TEMPERATURE
      TAIR=AFGEN(TTA1,TIME)
FUNCTION TTA1=0.,8.3,15.,8.3,45.,7.1,74.,7.,105.,9.5,135.,10.8,...
166.,14.1,196.,14.9,227.,15.,258.,12.5,288.,9.2,315.,9.1,345.,9.4,...
365.,9.4
      TAVE=(T+TAIR)/2.
*     DIATOMS(A2)
A2=INTGRL(IA2,RCOA2-RA2ME-RA2DF-RA2NE-RA2DS)
*1101 PRODUCTION OF DIATOMS
      RCOA2=RA2PR*NHDAY
```

FIG. 27.2. Listing of the program for the dynamic simulation model.

sion and assimilation efficiencies are derived from data on *Nereis* (Tenore & Gopalan 1974; Kay & Brafield 1973). Field data for monthly values of higher carnivore biomass are used as exogenous variables and we assume that the predation rates on secondary producers obey Michaelis-Menten kinetics (A.R. Longhurst & J.M. Gee, personal communication). The simulation

```
**1121 PREDATION DIATOMS BY MEIOFAUNA
PARAM Q10FME=2.,QFM010=.0075,QFM110=.075
      QFME10=INSW(Q1-.216,QFM010,QFM110)
      FME=EXP(ALOG(QFME10)+(TAVE-10.)*ALOG(Q10FME)/10.)
      RA2ME=A2*FME/Q1
      Q1=A2+HS
*1131 PREDATION DIATOMS BY DEPOSIT FEEDERS
      RA2DF=A2*FDF*(1.-TSUB)/Q2
*     TSUB=TIME EACH DAY THAT MUDFLAT IS SUBMERGED
PARAM TSUB=.55
      Q2=A2+DS+HS+ME
      FDF=(0.006+.008*TAIR)*DF
*1141 PREDATION DIATOMS BY NEPHTYS
      RA2NE=A2*FNE/Q3
      Q3=A2+ME+DF*PADF+SF*PASF
      FNE=NE*AMAX1(0.,Q3-QT3)*GNE/(QH3+Q3)
PARAM GNE=.039,QT3=.066,QH3=7.8,PADF=0.35,PASF=0.22
*1113 MORTALITY OF DIATOMS
      RA2DS=A2*MA2
PARAM MA2=0.0
*13   PARTICULATE CARBON(SEDIMENT)
*DS=INTGRL(IDS,RMEDS+RDFDS+RSFDS+RNEDS-RDSDF)
PARAM DS=10.4
*2113 MEIFAUNA PRODUCTION OF DS (E=EXCRETION)
      RMEDS=FME*EME
PARAM EME=0.325,EDF=.68,ESF=.3,E1NE=.71,E2NE=.23
*3113 DEPOSIT FEEDERS PRODUCTION OF DS
      RDFDS=FDF*EDF*(1.-TSUB)
*3513 SUSPENSION FEEDERS PRODUCTION OF DS
      RSFDSP=FOODSF*ESF
      RSFDS=(RSFDSP+TRAPSF-FOODSF)*TSUB
*4113 NEPHTYS PRODUCTION OF DS
      RNEDS=ENE*FNE
      ENE=(E1NE-E2NE)*A2/Q3+E2NE
*1331 UTILIZATION OF DS BY DEPOSIT FEEDERS
      RDSDF=DS*FDF*(1.-TSUB)/Q2
*21   MEIOFAUNA
ME=INTGRL(IME,FME-RME00-RMEDS-RMEDF-RMENE)
*1221 PREDATION BACTERIA BY MEIOFAUNA
      RHSME=HS*FME/Q1
*2100 RESPIRATION OF MEIOFAUNA
PARAM SME10=0.048,Q10SME=2.05
      SME=EXP(ALOG(SME10)+(TAVE-10.)*ALOG(Q10SME)/10.)
      RME00=ME*SME
*2131 PREDATION MEIOFAUNA BY DEPOSIT FEEDERS
      RMEDF=ME*FDF*(1.-TSUB)/Q2
*2141 PREDATION MEIOFAUNA BY NEPHTYS
      RMENE=ME*FNE/Q3
*31   DEPOSIT FEEDERS
DF=INTGRL(IDF,FDF*(1.-TSUB)-RDF00-RDFDS-RDFNE-RDFC1-RDFC4)
*1231 PREDATION BACTERIA BY DEPOSIT FEEDERS
      RHSDF=HS*FDF*(1.-TSUB)/Q2
*3100 RESPIRATION DEPOSIT FEEDERS
PARAM SDF10=.011,Q10SDF=2.05,SSF10=.016,Q10SSF=2.05
      SDF=EXP(ALOG(SDF10)+(TAIR-10.)*ALOG(Q10SDF)/10.)
      RDF00=DF*SDF
*3141 PREDATION DEPOSIT FEEDERS BY NEPHTYS
      RDFNE=DF*PADF*FNE/Q3
*3151 PREDATION DEPOSIT FEEDERS BY BIRDS
PARAM ZT2=1.,INSC1=.43,GRSC1=.033,GMC1=.6
      RDFC1=DF*FC1*(1.-TSUB)/Z2
      Z2=DF+NE+SF
      FC1=AMAX1(0.,AMIN1(C1*GC1F,(Z2-ZT2)))
      GC1F=AMIN1(GMC1,INSC1-GRSC1*TAIR)
*3154 PREDATION DEPOSIT FEEDERS BY FISH
PARAM ZT1=0.,ZH1=1.5,GC4=.175
      RDFC4=DF*FC4*TSUB/Z1
      Z1=DF+NE+SF
      FC4=C4*AMAX1(0.,Z1-ZT1)*GC4/(ZH1+Z1)
*35   SUSPENSION FEEDERS
SF=INTGRL(ISF,FOODSF-RSF00-RSFDS-RSFC1-RSFC4-RSFNE)
*1435 INTAKE PHYTOPLANKTON BY SUSPENSION FEEDERS
      INGSF=AFGEN(TINGSF,PARTOT)
```

FIG. 27.2—*continued*

```
  FUNCTION TINGSF=0.,1.,4.5,1.,10.,.40,50.,.15,100.,.1,...
  300.,0.,1000.,0.
        FOODSF=TRAPSF*INGSF
        TRAPSF=SF*FRSF*Q4*TSUB
        FRSF=AMAX1(0.,-0.64E-3*PARTOT+.192)
        RP1SF=P1*FOODSF/Q4
        Q4=P1+DW
  *1535 INTAKE DW BY SUSPENSION FEEDERS
        RDWSF=DW*FOODSF*TSUB/Q4
  *3500 RESPIRATION SUSPENSION FEEDERS
        SSF=EXP(ALOG(SSF10)+(T-10.)*ALOG(Q10SSF)/10.)
        RSF00=SF*SSF*TSUB
  *3551 PREDATION SUSPENSION FEEDERS BY BIRDS
        RSFC1=SF*FC1*(1.-TSUB)/Z2
  *3554 PREDATION SUSPENSION FEEDERS BY FISH
        RSFC4=SF*FC4*TSUB/Z1
  *3541 PREDATION SUSPENSION FEEDERS BY NEPHTYS
        RSFNE=SF*PASF*FNE/Q3
  *41   NEPHTYS
  NE=INTGRL(INE,FNE-RNE00-RNEDS-RNEC4-RNEC1)
  *4100 RESPIRATION NEPHTYS
  PARAM SNE10=6.56E-3,Q10SNE=2.05
        SNE=EXP(ALOG(SNE10)+(TAVE-10.)*ALOG(Q10SNE)/10.)
        RNE00=NE*SNE
  *4151 PREDATION NEPHTYS BY BIRDS
        RNEC1=NE*FC1*(1.-TSUB)/Z2
  *4154 PREDATION NEPHTYS BY FISH
        RNEC4=NE*FC4*TSUB/Z1
  METHOD RKS
  TIMER DELT=1.0,FINTIM=365.,PRDEL=1.,OUTDEL=1.
  PREPAR A2,DS,ME,DF,SF,NE,P1,DW,MS,C1,C4,XN
  NOSORT
        CALL DEBUG(1,0.0)
        CALL DEBUG(1,20.0)
        CALL DEBUG(1,60.0)
        CALL DEBUG(1,70.0)
        CALL DEBUG(1,110.0)
        CALL DEBUG(1,140.0)
        CALL DEBUG(1,270.0)
        CALL DEBUG(1,300.0)
  TERMINAL
  END
STOP
ENDJOB
```

FIG. 27.2—*continued*

language used is Continuous System Modelling Program, CMSP (see Radford 1971); Fig. 27.2 is a listing of the program for the dynamic simulation model.

RESULTS

The steady-state model

The energy budgets for all taxa are given in Table 27.1, and the carbon flow diagram in Fig. 27.3. This diagram is largely self-explanatory, but a few statistics can be extracted from it. The total annual production of macrofauna species is 5·46 g C m^{-2}, all of which is available to mobile predators such as birds, fish and crabs because there is no predation within the community. The total annual production of the meiofauna which feed on primary carbon sources is 20·17 g C m^{-2}, but 3·34 g of this is utilized within the system by predators (*Nephtys*, *Protohydra*), so that 'output' is only 16·83 g, three times

Table 27.1. *Population density and energy budgets.*

	No. m^{-2}	Biomass (g C m^{-2})	(g C m^{-2} yr^{-1}) Production	Respiration	Assimilation	Consumption
Macrofauna						
Mya arenaria	47·1	2·215	1·064*	10·799[a]	11·863	15·817
Nephtys hombergi	855·1	1·579	2·934*	4·723*	7·657	25·523
Scrobicularia plana	66·3	0·858	0·199*	1·763[b]	1·962	3·270
Cardium edule	28·7	0·339	0·082*	2·197*	2·279	3·506
Ampharete acutifrons	382·0	0·170	0·929*	1·351*	2·280	7·600
Macoma balthica	48·7	0·135	0·123*	0·287*	0·410	0·683
Others (6 spp.)	2,421	0·132	0·133[c]	0·527[c]	0·660	2·200
Total	3,849	5·428	5·464	21·647	27·111	58·599
'Meiofaunal' polychaetes						
Manayunkia aestuarina	346,500	0·680	3·740[d]	9·898[e]	13·638	19·483
Pygospio elegans	11,789	0·174	0·957[d]	1·998[e]	2·955	9·850
Streblospio shrubsolii	7,860	0·132	0·726[d]	1·444[e]	2·170	7·233
Fabricia sabella	27,000	0·058	0·319[d]	0·550[e]	0·869	1·241
Total	393,149	1·044	5·742	13·890	19·632	37·807

'Meiofaunal' oligochaetes						
Peloscolex benedeni	9,292	0·261	0·783[f]	1·579[e]	2·362	7·873
Tubificidae	17,535	0·128	0·385[f]	0·686[e]	1·071	3·570
Total	26,827	0·389	1·168	2·265	3·433	11·443
True meiofauna						
Nematoda (40 spp.)	12,460,000	0·788	6·623[g]	11·200*	17·823	29·705
Copepoda (8 spp.)	279,000	0·317	5·697[g]	12·794*	18·491	24·655
Ostracoda (2 spp.)	96,188	0·080	0·889[h]	1·624[e]	2·513	3·351
Protohydra leuckarti	42,188	0·020	0·222[h]	0·360[e]	0·582	0·701
Kinorhyncha (2 spp.)	25,313	0·008	0·050[g]	0·050*	0·100	0·133
Total	12,902,689	1·213	13·481	26·028	39·509	58·545

Derivation: Data on numbers and biomass original.
(*) Original.
(a) Gillfillan *et al.* (1976), assuming weight exponent of 0·75 and Q_{10} of 2·05.
(b) Hughes (1970).
(c) Based on average P/B and P/R ratios for other macrofauna species.
(d) Short-lived polychaetes assumed to have same P/B as *Ampharete* (5·5).
(e) From $\log R = 1{\cdot}1740 \log P + 0{\cdot}1352$ where R and P are in K cal m^{-2} yr^{-1} (McNeill & Lawton 1970, for short-lived poikilotherms). 1 g carbon = 12 K cal.
(f) Assumed P/B for oligochaeta = 3 (Haka *et al.* 1974).
(g) From $\log P = 0{\cdot}8262 \log R - 0{\cdot}0948$ (Ref. as e). Individual species calculated separately.
(h) Assume P/B is the mean of nematode/copepod/kinorhynch fraction of the meiofauna (11·114).

In all cases Assimilation = $P + R$ and Consumption = Assimilation × 100/Assimilation efficiency (see text).

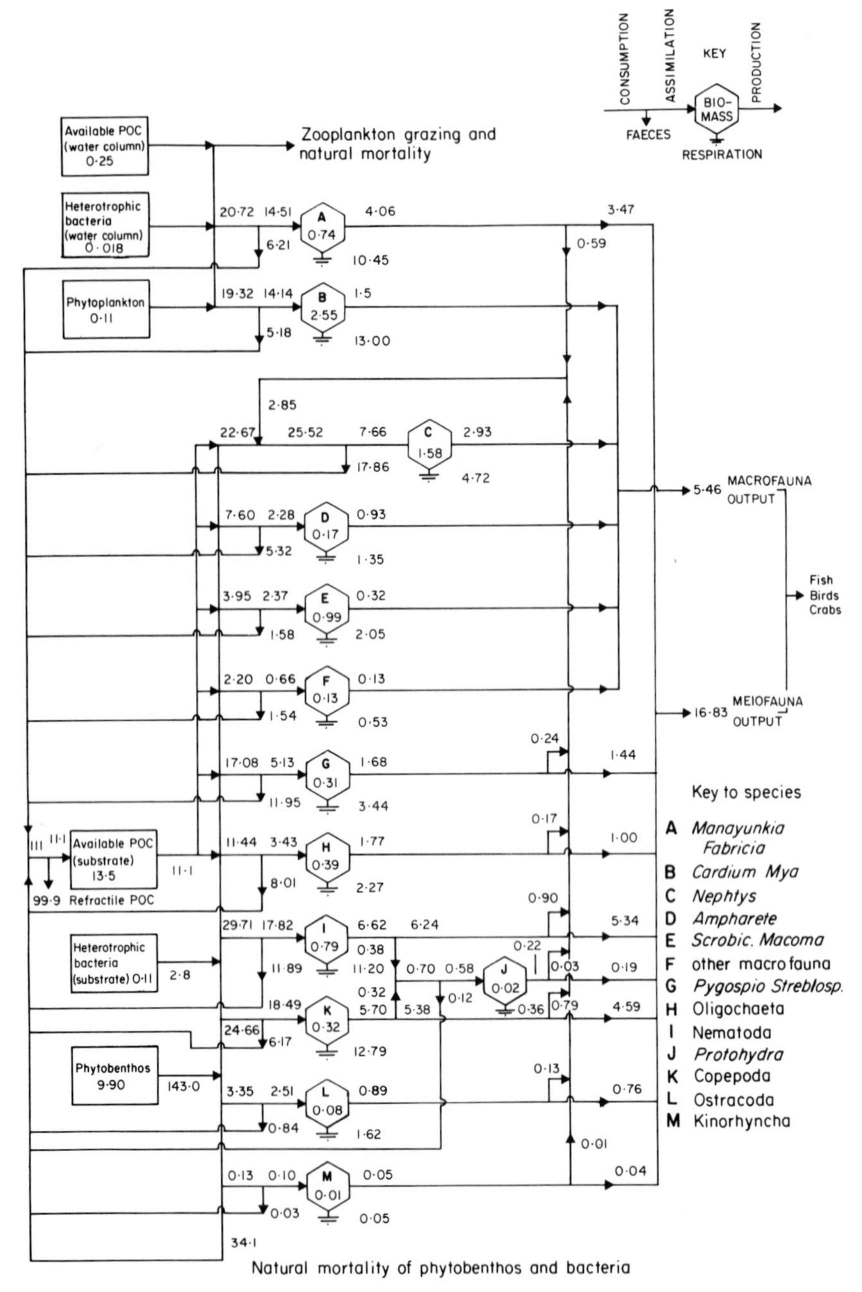

FIG. 27.3. Steady-state energy flow diagram for the Lynher mud-flat. Standing stock biomass in g C m^{-2}, rates in g C m^{-2} yr^{-1}.

greater than the macrofauna output. However, if *Nephtys* fed exclusively on meiofauna, with an assimilation efficiency of 80%, it would consume 9·57 g C of meiofauna, reducing the output by 9·86 g C. The macrofauna utilize 55·74 g of primary carbon, 35% from the water column and 65% from the sediment. Of this, 29·49 g is recycled as faeces and 21·12 g is respired. The meiofauna utilize 107·09 g of primary carbon, 19% from the water column and 81% from the sediment. Of this, 45·10 g is recycled as faeces and 41·82 g is respired. Thus the net production of meiofauna is 3·7 times that of the macrofauna, the total carbon turnover of the meiofauna is 1·9 times that of the macrofauna, whereas the meiofauna biomass is only 0·49 times that of the macrofauna.

The relative importance of individual macrofauna species has been discussed elsewhere (Warwick & Price 1975). Among the meiofauna the nematodes and copepods are dominant both in terms of production and of carbon turnover. However, the small polychaetes (particularly *Manayunkia aestuarina*) and oligochaetes are also potentially important.

The net production efficiency of the meiofauna is 32·5%. For the nematodes alone this value is 37·2%, very close to the experimental value of 38% obtained by Marchant and Nicholas (1974). For nematodes the P/B is 8·4 and for copepods 18·0. For the total meiofauna the P/B is 7·7 or, if the small annelids are excluded, 11·1. This compares with a value of 1·0 for the macrofauna.

The dynamic simulation model

The results of the dynamic simulation model for one year are given in Fig. 27.4. The simulated growth of the phytobenthos population (A2, Fig. 27.4a) shows a maximum biomass later than expected from field data, at about day 240 (Table 27.2); thereafter, the population declines as a result of reduced production and grazing. The simulated growth of *Nephtys* (NE), deposit feeders (DF) and meiofauna (ME) is shown in Fig. 27.4b. Meiofauna have a very stable biomass varying between 0·9 and 1·02 g C m^{-2}. The minimum and maximum biomass of deposit feeders are slightly later than expected but this is a result of the late increase in biomass of the phytobenthos; clearly, if the phytobenthos simulation had maximum biomass at the correct time, the simulations for the fauna would have maxima which would occur earlier. The biomass of deposit feeders varies from 2·04 to 2·5 g C m^{-2}. *Nephtys* also has minimum and maximum biomass later than expected with values from 1·07 to 1·35 g C m^{-2} (Table 27.2).

Fig. 27.4a also shows the biomass of phytoplankton in the water column, supplied as an exogenous variable, and the resulting simulation of suspension feeders (SF). There is a greater range in biomass than expected, with a minimum of 0·69 and a maximum of 8·69 g C m^{-2}. Growth occurs for a short period in the summer when the particulate load in the water column is low

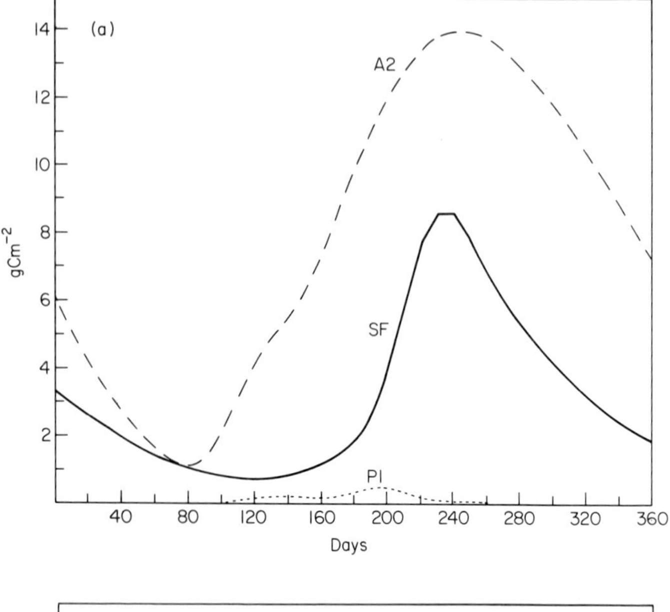

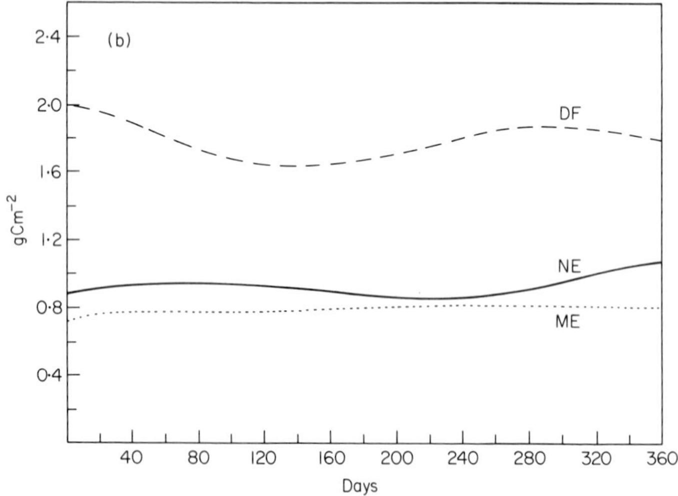

FIG. 27.4. The results of the dynamic simulation model; (*a*) for phytobenthos (– – – A2), suspension feeders (—— SF) and the values of phytoplankton (· · · · Pl) supplied as an exogenous variable and (*b*) for meiofauna (· · · · ME), deposit feeders (– – – DF) and *Nephtys* (—— NE). (*c*) the results of a simulation where *Nephtys* fed only on meiofauna and deposit feeders and not on phytobenthos and (*d*) where *Nephtys* fed exclusively on phytobenthos, and not on meiofauna or deposit feeders.

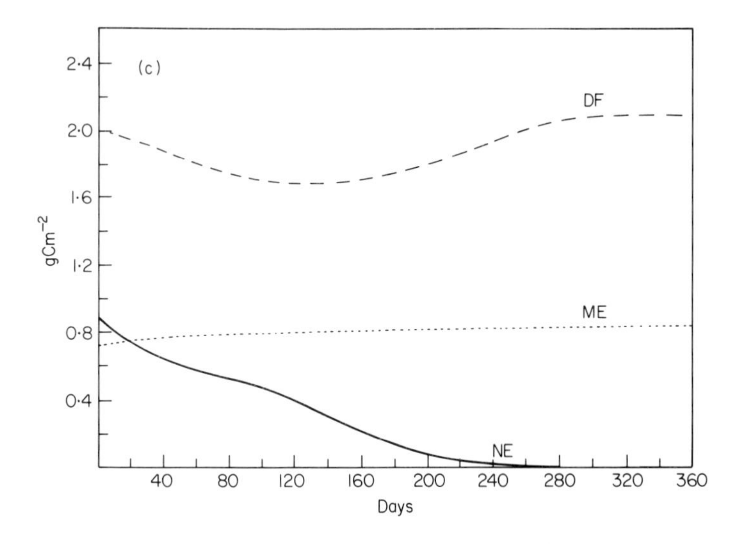

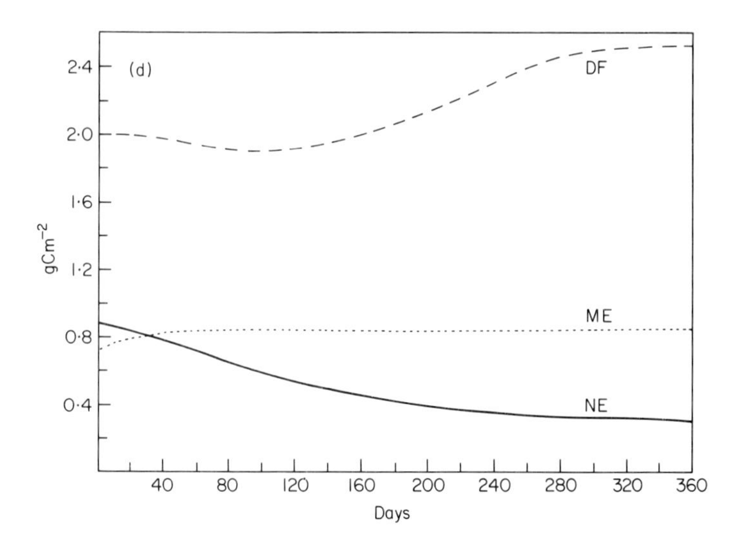

FIG. 27.4—*continued*

Table 27.2. *Comparison of simulated and field data for the state variables. Suspension feeders are omitted because the field data are insufficient to provide monthly values.*

	Simulation model				Field data			
	Max.	(Month)	Min.	(Month)	Max.	(Month)	Min.	(Month)
Phytobenthos	14·03	(Aug.)	1·08	(Mar.)	14·8	(May)	4·0	(Jan.)
Meiofauna	1·02	(Sept.)	0·9	(Jan.)	2·6	(May)	0·8	(Oct.)
Deposit feeders	2·5	(Jan.)	2·04	(May)	2·41	(Jan.)	1·49	(Feb.)
Nephtys	1·35	(Dec.)	1·07	(Jul.)	2·08	(Jun.)	1.11	(Jan.)

enough to allow assimilation of all material filtered with no production of pseudofaeces. The relationship between pseudofaeces production and particulate load in the water column is critical in this simulation, with small changes in particulate load having large changes in the biomass of suspension feeders.

The dynamic simulation model has been used to study a variety of hypothetical trophic relationships, particularly between macrofauna and meiofauna. As an example we will discuss the effect of different diets on the growth of *Nephtys*. Fig. 27.4c shows that the biomass of *Nephtys* falls rapidly when feeding as a carnivore on meiofauna and a proportion of the small deposit and suspension feeders; phytobenthos was excluded from the diet for this simulation. The biomass of deposit feeders is higher at the end of this simulation than in Fig. 27.4b because, at that time, there is no *Nephtys* left to feed on the deposit feeders; the biomass of meiofauna is very similar to that in Fig. 27.4b.

When *Nephtys* is allowed to feed exclusively on phytobenthos (Fig. 27.4d) it survives but at a lower standing stock than in the initial simulation. Small changes in the parameter values used in the *Nephtys* equations would enable *Nephtys* to maintain a higher biomass; since the parameter values used were derived from data for *Nereis*, it would not be unreasonable to make small changes in these values. However, rather large changes in these parameter values would be required before *Nephtys* could survive as a carnivore, feeding only on meiofauna and some deposit and filter feeders. In this simulation, the biomass of deposit feeders is even greater than in the previous simulation (Fig. 27.4c) because *Nephtys* does not feed on deposit feeders at any time; again, meiofauna have a similar biomass to that in other simulations.

In the model the meiofauna seem to be surprisingly insensitive to changes in the diet of *Nephtys*, with similar biomass when grazed by *Nephtys* or not grazed. This is a result of using parameter values derived from data on a nematode feeding on a bacterial suspension; the rapid assimilation of food by the meiofauna is almost balanced by respiration and exudation. With such a fast turnover of carbon, predation accounts for only a small proportion of the production and the removal of grazing by *Nephtys* has a minimal effect on meiofauna biomass.

DISCUSSION

The steady-state model

The P/B ratio of 11·1 for the true meiofauna is in accord with the estimates of McIntyre (1969) and Gerlach (1971), who suggest that this value is about 10. The close agreement in production estimates for nematodes derived from respiration data using either the McNeill and Lawton (1970) equation or assuming a net production efficiency of 38% (Marchant & Nicholas 1974) also add confidence to the production estimate. For copepods the calculations imply a net growth efficiency of 30·8%, which is above the values obtained experimentally for harpacticoids, 17·2% for *Tigriopus* (Harris 1973) and 14·1% for *Asellopsis* (Lasker *et al.* 1970), but below the values for some planktonic copepods, e.g. 38·4% for *Calanus finmarchicus* (Corner *et al.* 1967).

Meiofauna densities as high as those found in the Lynher are not exceptional for intertidal areas. Comparable densities have been found, for example, on a mud-flat in the Bristol Channel, U.K. (Rees 1940) and a salt marsh in Georgia, U.S.A. (Teal & Wieser 1966). However, the importance of meiofauna in the carbon flow may be much less in areas where the standing crop is much lower. Stripp (1969) has found that meiofauna comprise between 1 and 4% of the macrofauna biomass in a number of macrofauna communities in Kiel Bay. Of course this ratio will depend on how the meiofauna and macrofauna are categorized but, even if the small annelids are included with the macrofauna, this ratio is 17·7% in the Lynher.

It is clear that, in order to increase the reality of the model, more precise information is required on carbon flow through the meiofauna, and in particular the small annelids which appear to be potentially important but about which our data are weakest. Also much more information is required about meiofauna/macrofauna interactions, particularly the availability of meiofauna to macrofauna deposit feeders and predators. Here *Nephtys* occupies a central role, because if it fed exclusively on meiofauna it would consume about half of the meiofauna production, leaving much less available for mobile predators.

The dynamic simulation model

A realistic simulation has been obtained for meiofauna, deposit feeders and *Nephtys*. It is clear that *Nephtys* is incapable of surviving as a carnivore although it could survive solely on phytobenthos with small changes in the parameter values for the *Nephtys* equations; the most realistic simulation has been achieved by allowing *Nephtys* to feed on phytobenthos and meiofauna and some small deposit feeders and suspension feeders. Clearly experiments are required to confirm this result from the simulation model.

The simulation for deposit feeders is realistic, but there is perhaps a greater variation in biomass of suspension feeders than might be expected. The relationship between particulate load and pseudofaeces production is critical to the success of suspension feeders with small changes in particulate load resulting in lower growth rates. A less sensitive model may result from basing the feeding rate of suspension feeders on the rate of passage of material through the gut of the animal, as was used for the deposit feeders. However, these data are not available for suspension feeders.

As with the steady-state model, the simulation model has highlighted certain lacunae, especially in our understanding of the kinetics of feeding of meiofauna and macrofauna; more data are required on the food of macrofauna, especially *Nephtys* and, more importantly, on the relationships between varying food concentration and assimilation rate.

ACKNOWLEDGMENTS

We are grateful to the following members of the Institute for Marine Environmental Research who generously allowed us to quote from their unpublished data or who offered assistance in other ways: B.L. Bayne, J.M. Gee, A.R. Longhurst, R.I.E. Newell, A.J. Pomroy, R. Price, C.J. Scullard, M.J. Teare and J. Widdows. We especially thank Mrs K. Young, who ran most of the computer simulations. This work forms part of the estuarine ecology programme of the Institute for Marine Environmental Research, a component institute of the Natural Environment Research Council.

REFERENCES

Boucher G. (1972) Premières données écologiques sur les nématodes libres marins d'une station de vase côtière de Banyuls. *Vie Milieu*, ser. B, **23**, 69–100.

Brinkhurst R.O. & Jamieson B.G.M. (1971) *Aquatic Oligochaeta of the World*. University of Toronto Press.

Clay E. (1967) Literature survey of the common fauna of estuaries. 6. *Nephthys hombergi* Lamarck. ICI Paints Division, Marine Research Station, Brixham, Devon. 1–17.

Corner E.D.S., Cowey C.B. & Marshall S.M. (1967) On the nutrition and metabolism of zooplankton. V. Feeding efficiency of *Calanus finmarchicus*. *J. mar. biol. Ass. U.K.* **47**, 259–70.

Crisp D.J. (1971) Energy flow measurements. *Methods for the Study of Marine Benthos*. IBP Handbook **16**, 197–279. Blackwell Scientific Publications, Oxford.

Day J.H. (1967) A monograph on the Polychaeta of Southern Africa. Part 2. Sedentaria. *Brit. Mus. nat. Hist. Publs.* **656**, 459–878.

Duncan A., Schiemer F. & Klekowski R.Z. (1974) A preliminary study of feeding rates on bacterial food by adult females of a benthic nematode, *Plectus palustris* de Man 1880. *Pol. Arch. Hydrobiol.* **21**, 249–59.

Engelmann M.D. (1966) Energetics, terrestrial field studies and animal productivity. *Advan. Ecol. Res.* **3**, 73–115.

Fenchel T. (1969) The ecology of marine microbenthos. IV. Structure and function of the benthic ecosystem, its chemical and physical factors and the microfauna communities with special reference to the ciliated protozoa. *Ophelia* **6**, 1–182.

Gerlach S.A. (1971) On the importance of marine meiofauna for benthos communities. *Oecologia* **6**, 176–90.

Gilfillan E.S., Mayo D., Hanson S., Donovan D. & Jaing L.C. (1976) Reduction in carbon flux in *Mya arenaria* caused by a spill of No. 6 fuel oil. *Mar. Biol.* **37**, 115–23.

Gray J.S. (1976) The fauna of the polluted river Tees estuary. *Est. Coastal Mar. Sci.* **4**, 653–76.

Haka P., Holopainen I.J., Ikonen E., Leisma A., Paasivirta L., Saaristo P., Sarvala J. & Sarvala M. (1974) Pääjärven pohjaeläimistö. *Luonnon Tutkija* **78**, 157–73.

Harris R.P. (1973) Feeding, growth, reproduction and nitrogen utilisation by the harpacticoid copepod, *Tigriopus brevicornis*. *J. mar. biol. Ass. U.K.* **53**, 785–800.

Hughes R.N. (1969) A study of feeding in *Scrobicularia plana*. *J. mar. biol. Ass. U.K.* **49**, 805–23.

Hughes R.N. (1970) An energy budget for a tidal flat population of the bivalve *Scrobicularia plana* (da Costa). *J. anim. Ecol.* **39**, 357–81.

Kay D.G. & Brafield A.E. (1973) The energy relations of the polychaete *Neanthes* (=*Nereis*) *virens* (Sars). *J. anim. Ecol.* **42**, 673–92.

Lasker R., Wells J.B.J. & McIntyre A.D. (1970) Growth, reproduction, respiration and carbon utilization of the sand dwelling harpacticoid copepod, *Asellopsis intermedia*. *J. mar. biol. Ass. U.K.* **50**, 147–60.

Lasserre P. (1971) Données écophysiologiques sur la répartition des oligochètes marins méiobenthiques. Incidence des paramètres salinté-température sur le métabolisme respiratoire de deux espèces euryhalines du genre *Marionina* Michaelsen 1889. (Enchytraeidae, Oligochaeta). *Vie Mileu.* **22**, 523–40.

Marchant R. & Nicholas W.L. (1974) An energy budget for the free-living nematode *Pelodera* (Rhabditidae). *Oecologia* **16**, 237–52.

McIntyre A.D. (1969) Ecology of marine meiobenthos. *Biol. Revs.* **44**, 245–90.

McNeill S. & Lawton J.H. (1970) Annual production and respiration in animal populations. *Nature* **225**, 472–74.

Miller R.J. & Mann K.H. (1973) Ecological energetics of the seaweed zone in a marine bay on the Atlantic coast of Canada. III. Energy transformations by sea urchins. *Mar. Biol.* **18**, 99–114.

Miller R.J., Mann K.H. & Scarratt D.J. (1971) Production potential of a seaweed-lobster community in Eastern Canada. *J. Fish. Res. Bd. Can.* **28**, 1733–38.

Perkins E.J. (1958) The food relationships of the microbenthos, with particular reference to that found at Whitstable, Kent. *Ann. Mag. nat. Hist.* ser. 13, **1**, 64–77.

Petrusewicz K. (1967) Suggested list of more important concepts in productivity studies (definitions and symbols). *Secondary Productivity of Terrestrial Ecosystems*, Vol. 1, pp. 51–82. Warsaw and Cracow.

Radford P.J. (1971) The simulation language as an aid to ecological modelling. *Mathematical Models in Ecology* (Ed. by J.N.R. Jeffers), pp. 277–95. Blackwell Scientific Publications, Oxford.

Rees C.B. (1940) A preliminary study of the ecology of a mud-flat. *J. mar. biol. Ass. U.K.* **24**, 185–99.

Steele J.H. (1974) *The Structure of Marine Ecosystems.* 128 pp. Blackwell Scientific Publications, Oxford.

Stripp K. (1969) Das Vehältnis von Makrofauna und Meiofauna in den Sedimenten der Helgoländer Bucht. *Veröff. Inst. Meeresforsch. Bremerh.* **12**, 143–48.

Teal J.M. (1962) Energy flow in the salt marsh ecosystem of Georgia. *Ecology* **43**, 614–24.

Teal J.M. & Wieser W. (1966) The distribution and ecology of nematodes in a Georgia salt marsh. *Limnol. Oceanogr.* **11**, 217–22.

Tenore K.R. & Dunstan W.M. (1973) Comparison of feeding and biodeposition of three bivalves at different food levels. *Mar. Biol.* **21**, 190–95.

Tenore K.R. & Gopalan U.K. (1974) Feeding efficiencies of the polychaete *Nereis virens* cultured on hard clam tissue and oyster detritis. *J. Fish. Res. Bd. Can.* **31**, 1675–82.

Tenore K.R., Tietjen J.H. & Lee J.J. (1977) Effect of meiofauna on incorporation of aged eelgrass, *Zostera marina*, detritus by the polychaete *Nephthys incisa*. *J. Fish. Res. Bd. Can.* **34**, 563–67.

von Thun W. (1968) *Autökologische Untersuchungen an freilebenden Nematoden des Brackwassers.* thesis. Kiel University.

Warwick R.M. & Price R. (1975) Macrofauna production in an estuarine mud-flat. *J. mar. biol. Ass. U.K.* **55**, 1–18.

Welch H.E. (1968) Relationships between assimilation efficiencies and growth efficiencies for aquatic consumers. *Ecology* **49**, 755–59.

Wieser W. (1953) Die Beziehung Zwischen Mundhöhlengestalt, Ernährungsweise und Vorkommen bei freilebenden marinen Nematoden. *Ark. Zool.* Ser. 2, **4**, 439–84.

Wieser W. (1960) Benthic studies in Buzzards Bay. II. The meiofauna. *Limnol. Oceanogr.* **5**, 121–37.

de Wilde P.A.W.J. (1975) Influence of temperature on behaviour, energy metabolism and growth of *Macoma balthica* (L). *Proc. 9th Europ. mar. biol. Symp.* 239–56.

Winberg G.G. (1956) Rates of metabolism and food requirements of fishes. *Fish. Res. Bd. Can. Translation Series* **194**, 1–202.

28. FEEDING ECOLOGY AND ENERGETICS OF THE DARK-BELLIED BRENT GOOSE (*BRANTA BERNICLA BERNICLA*) IN ESSEX AND KENT

K. CHARMAN*

Institute of Terrestrial Ecology,
Colney Research Station,
Colney Lane, Norwich, U.K.

SUMMARY

1. This paper describes studies of the feeding ecology of the dark-bellied brent goose carried out during the winters 1972/75 on the Essex and Kent coasts. It discusses food preference, feeding behaviour, food consumption and examines the ability of estuaries to support feeding geese.
2. Brent geese showed a consistent pattern of food utilization from *Zostera*, to *Enteromorpha*, to salt marsh vegetation and agricultural crops.
3. The behaviour of geese feeding on *Zostera* was largely independent of food density above approximately 15% leaf cover. Below this density dispersion, consumption and the time spent feeding changed markedly.
4. Estimates of daily food consumption based on exclosures and the production of droppings, are compared with theoretical requirements. *Zostera* appeared to be the only food which provided the daily energy requirement.
5. The ability of estuaries to support feeding geese is calculated using estimates of the daily food consumption and standing crops of intertidal foods. The consequences of reduced food resources or increased goose populations are predicted.

RÉSUMÉ

1. Cette publication décrit des études sur l'écologie de nutrition des bernaches cravants (phase de coloration foncée), des côtes d'Essex et du Kent durant les hivers 1972/75. Nous discutons du choix de la nourriture, des propriétés de

* Present address: Nature Conservancy Council, 70 Castlegate, Grantham, Lincolnshire.

l'alimentation, de sa consommation et examinons les ressources de l'estuaire pour la nutrition de ces oies.
2. Les bernaches montrent une logique d'utilisation des aliments à partir de *Zostera* vers *Entomorpha*, la végétation des marais salés et les cultures agricoles.
3. La façon dont les oies se nourrissent de *Zostera* est largement indépendante de la densité de cet aliment au dessus d'une couverture de feuille d'environ 15%. Au dessous de cette densité de dispersion, la consommation et le temps passé à l'alimentation changent de façon nette.
4. Nous comparons la consommation alimentaire journalière basée sur les enclos et la production de la fiente avec la demande théorique. *Zostera* semble être le seul aliment répondant aux besoins énergétiques journaliers.
5. Nous calculons la capacité des estuaires à suppléer à l'alimentation des oies en estimant la consommation alimentaire journalière et les récoltes sur pied soumis aux balancements des marées. Nous prédisons les conséquences d'une réduction des ressources alimentaires ou de l'augmentation de la population des oies.

INTRODUCTION

As a result of the proposal to site a third London Airport in Essex on an area of intertidal sandflats known as Maplin Sands the Department of the Environment commissioned an ecological study. The objective was to predict the possible impact of the proposed development on wildlife and to suggest ways in which this impact might be minimized. The area concerned provides winter food for about 20% of the world population and most of the British population of the dark-bellied race of the brent goose (*Branta bernicla bernicla*). The purpose of this part of the work was to evaluate the effects of the loss of this area on the brent goose.

In order to predict if any displaced geese could be accommodated elsewhere it was necessary to see if areas existed which were under-utilized. This paper summarizes findings presented elsewhere (Charman 1975) on (i) the sequential utilization of food resources, (ii) the effect of food density on feeding behaviour, (iii) the estimation of food consumption and (iv) the ability of estuaries to support feeding geese. These are then used in predicting the effect of reduced food supplies on the geese.

THE STUDY AREA

In Britain, brent geese winter on the eastern and southern coasts from Lincolnshire to Devon, the majority being found between the estuaries of the

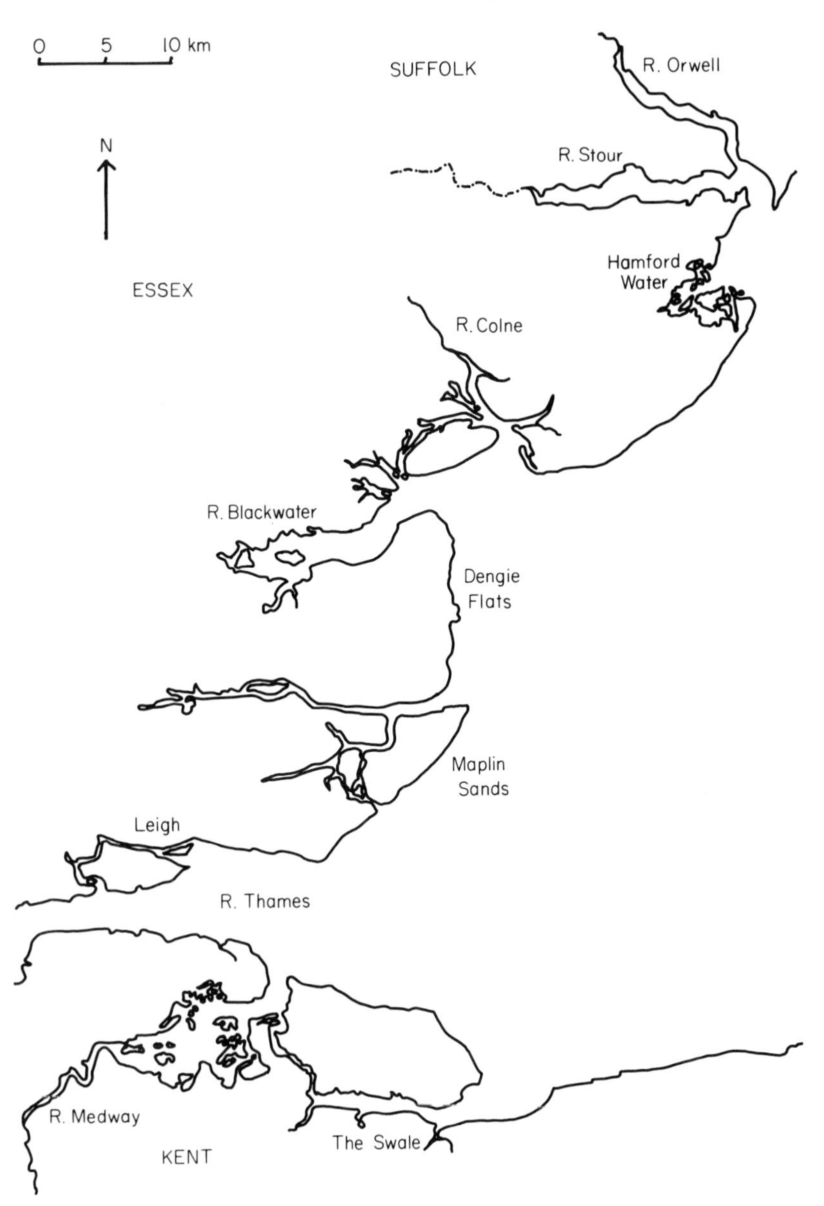

FIG. 28.1. Map showing study area.

Stour (Essex) and the Swale (Kent). Fieldwork for this study was carried out on estuaries and mudflats on the Essex and Kent coasts (Fig. 28.1) during the three winters 1972/75. The work concentrated on areas selected to represent the major food resources used by brent geese. These were:

(1) Foulness/Leigh: the stronghold of the brent goose in Essex and Kent and its major *Zostera noltii* feeding area.
(2) Dengie/Blackwater: an area which from December onwards contains large numbers of geese feeding on *Enteromorpha* spp.
(3) Colne/Fingringhoe: the only site in Essex where considerable numbers of brent geese feed on high saltings. In addition brent geese feed on *Enteromorpha* in the estuary.

All the three major study areas included sites where feeding on agricultural land occurred.

METHODS

One person could not monitor the seasonal fluctuations in numbers and distribution of brent geese in the whole of Essex and north Kent. Much of the detailed information on distribution and numbers has therefore been drawn from the 'Birds of Estuaries Enquiry' run by the British Trust for Ornithology in conjunction with the Royal Society for the Protection of Birds and the Wildfowl Trust, and organized locally by R.M. Blindell (Blindell 1975). I made direct observations on diet throughout the area.

Zostera abundance was estimated by eye as percentage leaf cover within 1 m^2 quadrats placed at intervals along a series of transects crossing the *Zostera* beds. This method gives good estimates of biomass (Wyer & Waters 1975). The pacing rate of geese was measured as the time required to make ten paces. The spacing of geese was measured by estimating, in bird lengths, the distance between individual geese and their nearest neighbour. Although at the ranges at which observations were made it was not possible to discern individual grazing or swallowing movements, the amount of time geese spent with their heads lowered and their bills in contact with the food source was measured with two stopwatches. This is termed the 'percentage of time spent feeding'.

It is possible to calculate the theoretical energy requirements for basal metabolism in birds using the relationship given by King and Farner (1961). From a series of studies Ebbinge *et al.* (1975) derived figures for the relation between basal metabolic requirements and the metabolized energy required for normal maintenance. The metabolized energy requirement can be converted into an estimate of required consumption using values for digestive efficiency and percentage assimilation.

The actual food consumption of *Zostera* by geese was estimated by comparing the biomass within a series of exclosures and grazed plots and counting the geese feeding within the area. No other species excluded took significant quantities of *Zostera*. The total weight of droppings produced per bird per day was calculated from the average weight of droppings and the frequency with which they were produced. The digestive efficiency on a weight basis was derived from the ash content of food and faeces. The daily food consumption necessary to produce a given weight of droppings could then be calculated.

RESULTS AND DISCUSSION

Seasonal pattern of food utilization

Brent geese returning to the coast of south-east England each winter did not occupy all the estuaries simultaneously but showed a sequential pattern of occupation which was similar from year to year.

Fig. 28.2 shows the proportion of the geese found on each of the estuaries and mudflats in Essex and north Kent during the winter 1973/74. Maplin Sands was occupied first and held a very high proportion until December when increasing numbers occurred on the Blackwater and other parts of the coast. The counts also indicate that some geese left the study area (i.e. Essex and Kent) completely at this time and moved to the harbours on the south coast of England and also into France (A.K.M. St Joseph, personal communication).

A clearer picture emerges if the proportion of geese on each of the food sources, rather than individual estuaries, is plotted (Fig. 28.3). Almost all the geese returning early in the winter fed on *Zostera* beds. Only when numbers had reached their maximum and much of the *Zostera* had been eaten (see below) did significant numbers of geese turn to *Enteromorpha*. The proportion of geese feeding on *Zostera* decreased throughout the winter although a small proportion remained even when leaves were almost gone because they were able to supplement their diet with *Zostera* roots.

In contrast, the proportion of geese feeding on *Enteromorpha* increased throughout the winter. In December small but significant numbers of geese began feeding on high saltings, grazing land and autumn sown cereals. Farmland became important after 1973 when, due to a series of good breeding seasons, the world population of the dark-bellied brent goose exceeded 84,000 and the numbers in Essex and Kent reached a peak of over 21,000 (Ogilvie & St Joseph 1976). This feeding behaviour may be a reaction to a shortage of their preferred intertidal food.

Other patterns emerge when the occupation of *Zostera* and *Enteromorpha* are examined in greater detail (Fig. 28.4). Maplin Sands and Leigh contain the largest quantities of *Zostera* in Essex and Kent. Geese frequently moved

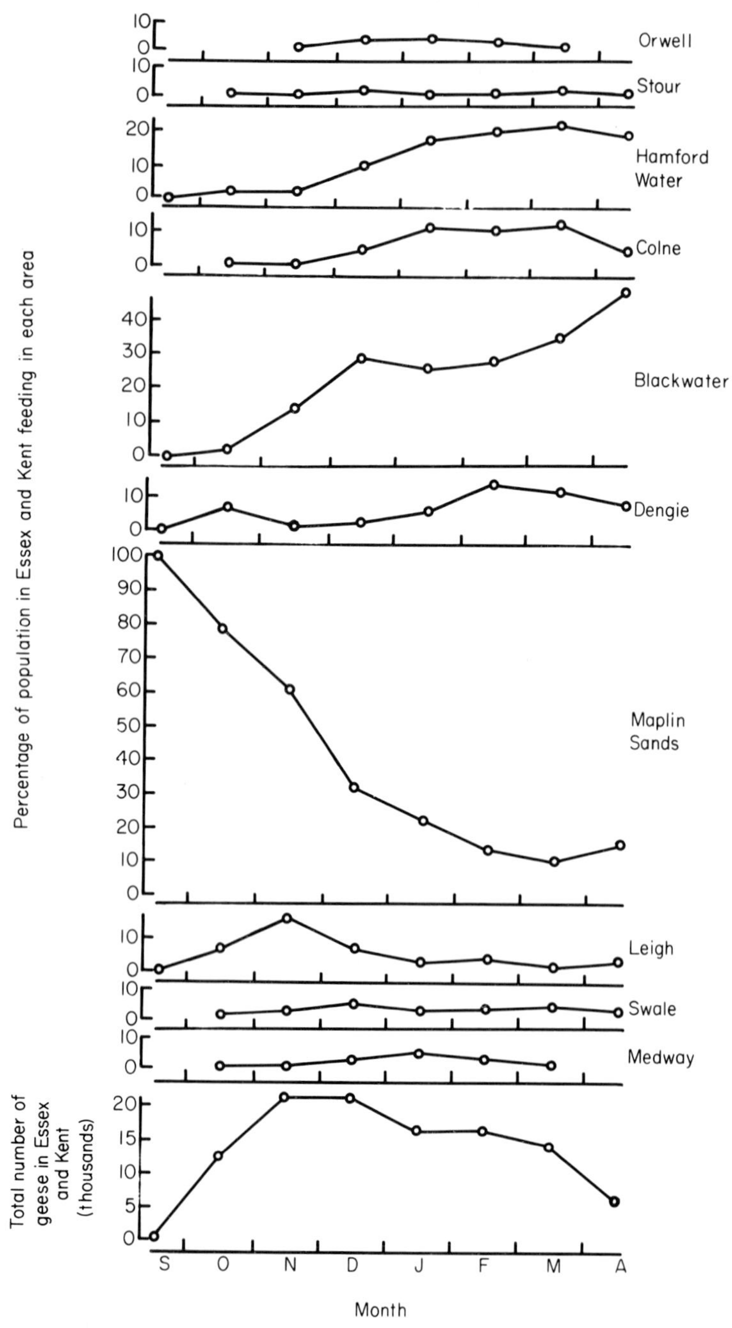

FIG. 28.2. The total number of brent geese counted in Essex and Kent for each month of the winter 1973/74 together with the percentage of these monthly figures found in each of ten areas.

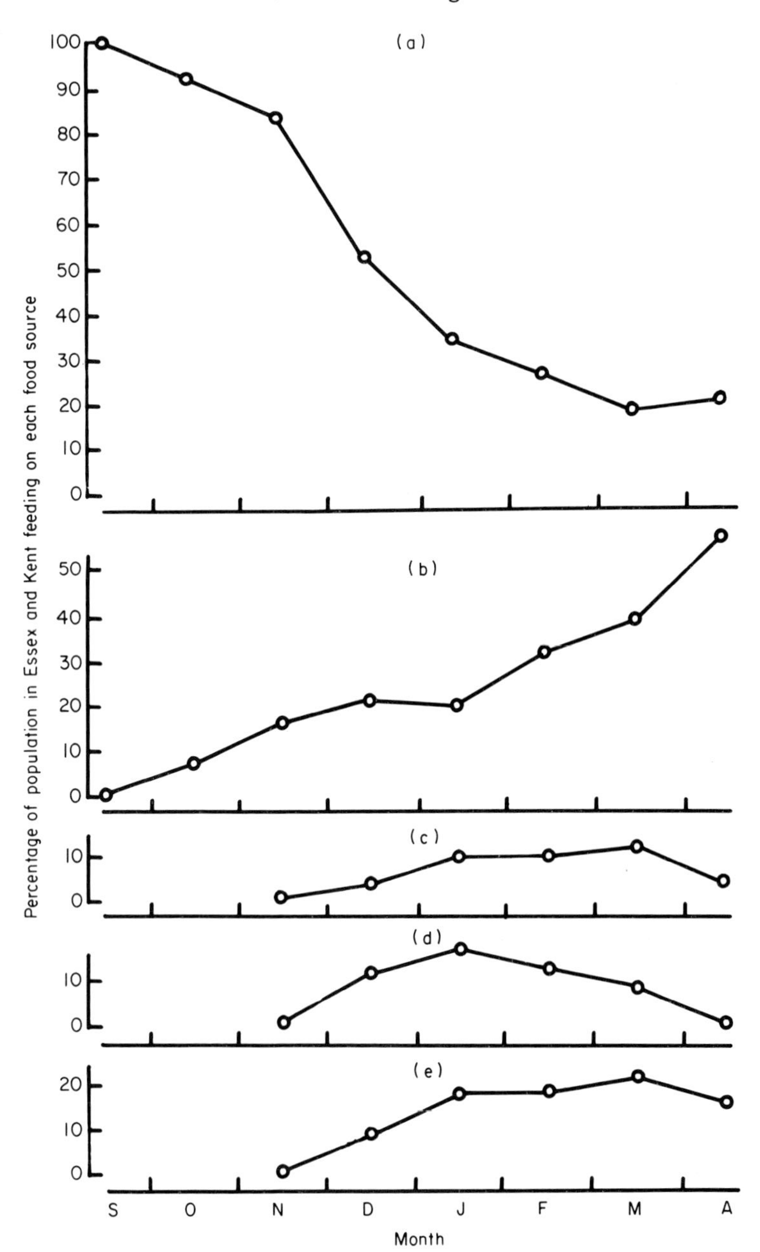

FIG. 28.3. The percentage of brent geese feeding in Essex and Kent on the five major food types for each month of the winter 1973/74. (a) *Zostera*; (b) *Enteromorpha*; (c) salt marsh; (d) cereals; (e) rough grazing.

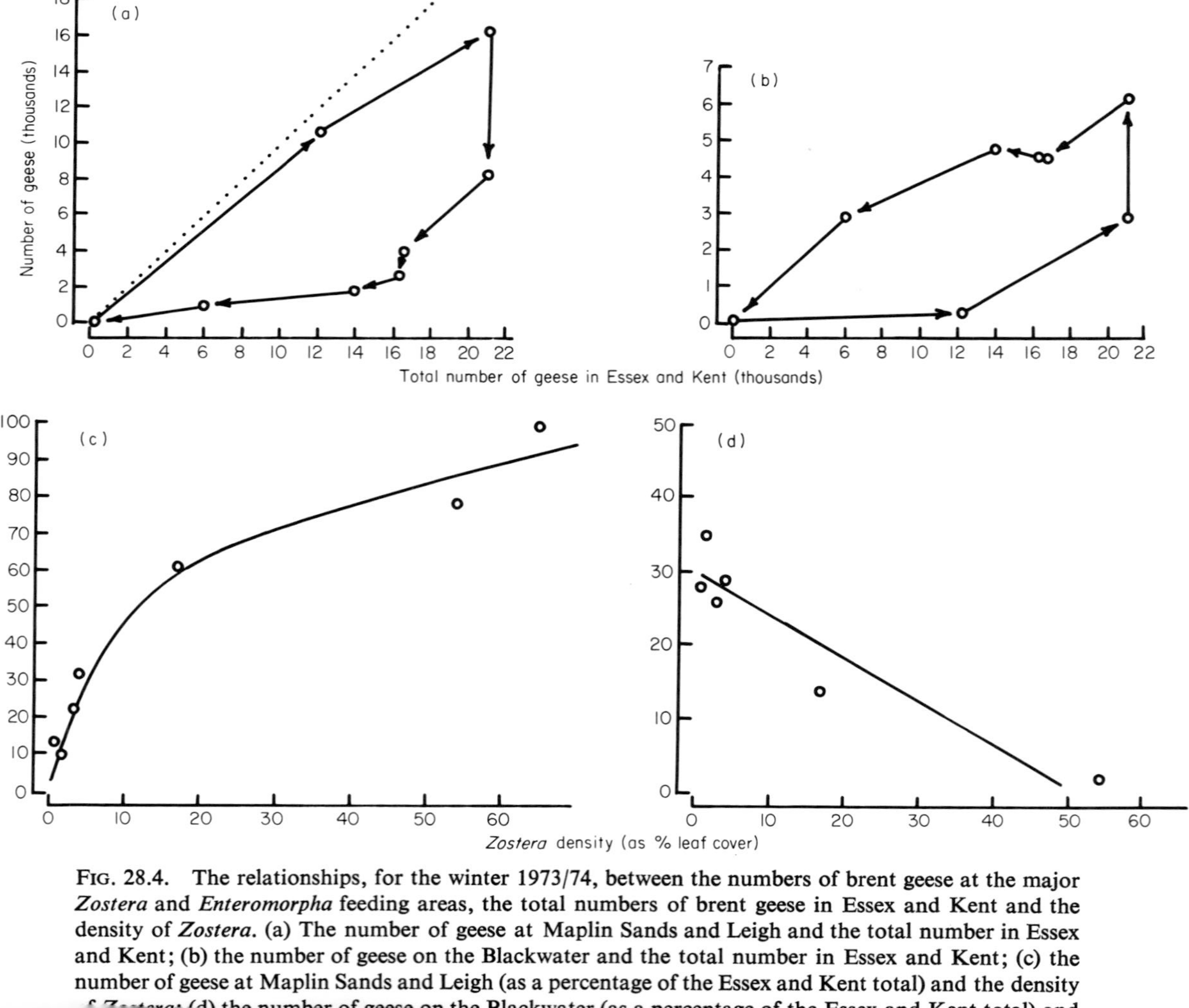

FIG. 28.4. The relationships, for the winter 1973/74, between the numbers of brent geese at the major *Zostera* and *Enteromorpha* feeding areas, the total numbers of brent geese in Essex and Kent and the density of *Zostera*. (a) The number of geese at Maplin Sands and Leigh and the total number in Essex and Kent; (b) the number of geese on the Blackwater and the total number in Essex and Kent; (c) the number of geese at Maplin Sands and Leigh (as a percentage of the Essex and Kent total) and the density of *Zostera*; (d) the number of geese on the Blackwater (as a percentage of the Essex and Kent total) and

between the two areas and initially they provided food for almost all the birds (Fig. 28.4a). As the numbers of geese in Essex and Kent increased the relative usage of Maplin and Leigh declined. Although the sites were able to provide food for additional birds, there was an increased resistance to newcomers. When the impact of grazing on the *Zostera* became marked, the numbers of geese at Maplin and Leigh rapidly declined. In contrast, the relative importance of the Blackwater (Fig. 28.4b), the major *Enteromorpha* feeding area, increased as the winter progressed.

The proportion of geese feeding on *Zostera* (Fig. 28.4c) at Maplin was closely related to the leaf cover of *Zostera* present, with a rapid decline in usage below 15% leaf cover. In contrast to this, the proportion of birds feeding on the Blackwater (Fig. 28.4d) was inversely related to the density of *Zostera.*

Feeding behaviour and food density

The density of food affected the feeding behaviour of the brent goose. When *Zostera* leaf densities were high the average daily food consumption per goose was between 100 and 150 g dry weight of leaves (Fig. 28.5a). The consumption of leaves declined rapidly when *Zostera* densities dropped below 15% leaf cover. Because the geese continued to feed at very low leaf densities of *Zostera*, the whole above-ground biomass can be considered as available for consumption.

Direct observation (Fig. 28.5b) showed that juveniles spent over 10% more time feeding than adults at high *Zostera* densities but fed for a similar amount of time when leaf cover fell below 15%. This was due to adults increasing their feeding time; juveniles apparently fed at the maximum rate possible. The remaining time was required for walking across the *Zostera* bed and for vigilance and maintenance activities.

As *Zostera* stocks declined through the winter, flocks spread out over the *Zostera* beds so that nearest neighbour distances increased (Fig. 5c). Again a food density of approximately 15% leaf cover appeared to be the critical value below which individual spacing increased markedly. Pacing rate also increased as *Zostera* stocks declined and reached a ceiling of about 25 paces per minute (Fig. 5d). Pacing rate appeared to respond to higher food densities than other aspects of feeding behaviour.

Food consumption

Exclosures gave estimates of daily food consumption of 121·6 and 99·9 g dry weight of *Zostera* leaves for 1973/74 and 1974/75 respectively. This difference may be due to the preponderance of juveniles in 1973/74; 48·5% compared with 0·04% in 1974/75 (Ogilvie & St Joseph 1976).

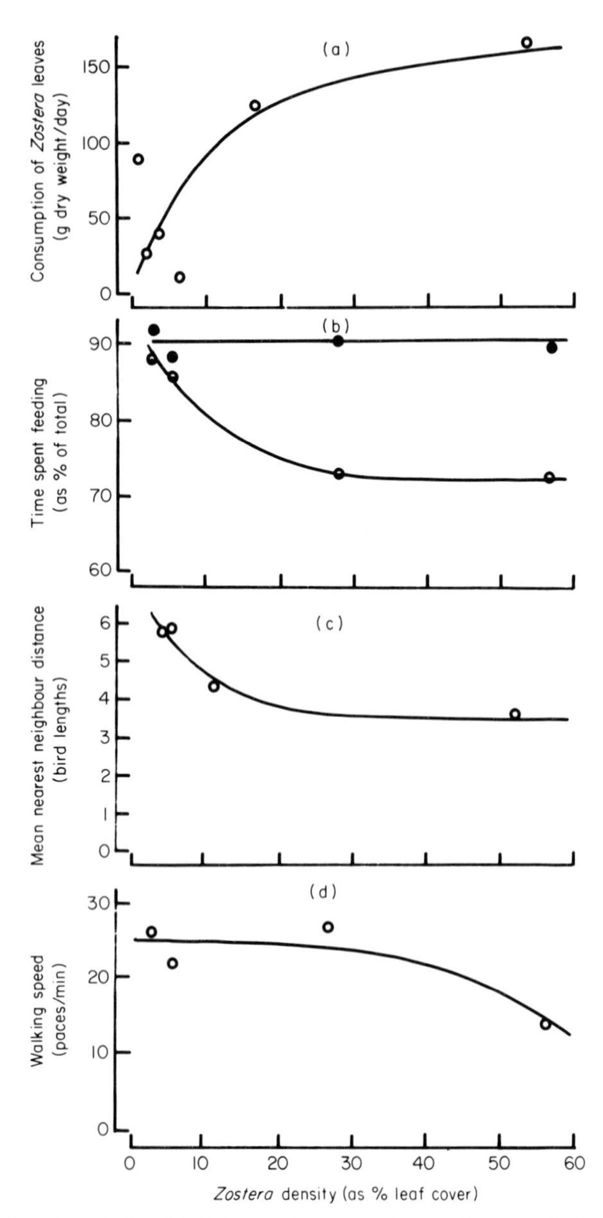

FIG. 28.5. The relationship between the percentage leaf cover of *Zostera* and four aspects of feeding behaviour in the brent goose. (a) Consumption of *Zostera* leaves; (b) percentage of time spent feeding; (c) mean nearest neighbour distance; (d) walking speed. Adults and juveniles, ○; adults, ◒; juveniles, ●.

Estimates of *Zostera* consumption based on the production of droppings showed a similar difference between 1973/74 and 1974/75 although the values of daily food consumption were considerably higher (207·6 and 152·3 g dry weight respectively) than those obtained using exclosures. These values were however, calculated from measurements made during the early half of the winter when *Zostera* densities were high and more geese were present. In contrast, consumption estimates from exclosures were based on the whole winter when, according to observations and to the functional response (Fig. 28.5a), other foods were eventually taken. An estimate of *Enteromorpha* consumption using dropping production gave a value 106·8 g dry weight per day for the winter 1974/75.

These values can be compared with the theoretical energy requirement of a bird of this size (Table 28.1). The theoretical figures are based on a stable adult weight of 1,300 g, but juveniles are likely to have higher food requirements in order to complete growth.

A theoretical requirement of 88·1 g dry weight of *Zostera* per day is not dis-similar to the 1974/75 exclosure value of 99·9 g dry weight per day. The theoretical requirement of 172·1 g dry weight per day of *Enteromorpha* is considerably higher than the estimate of daily food consumption based on dropping production (106·8 g dry weight). However, there is evidence that whilst feeding on *Enteromorpha* brent geese lose weight (A.K.M. St Joseph, personal communication) indicating that their actual consumption falls below

Table 28.1. *A comparison of the theoretical daily food requirement of an adult brent goose and the estimates of daily food consumption based on exclosures and the production of droppings. The ability of brent geese to assimilate energy from the different food sources and the energy content of each food is also given.*

	Zostera	*Enteromorpha*	Salt marsh	Cereals
Energy content kJ g^{-1} dry weight	20·1	17·2	20·9	21·8
Percentage assimilation	67·2	40·7	75·4	25·8
Theoretical daily food requirement g dry weight	88·1	172·1	75·0	211·7
Estimated daily food consumption based on exclosures g dry weight	121·6[a] 99·9[b]	— —	— —	— —
Estimated daily food consumption based on dropping production g dry weight	207·6[a] 152·3[b]	— 106·8[b]	— —	— —

a, 1973/74; b, 1974/75.

their theoretical requirements. Despite this, geese feeding on *Enteromorpha* spend long periods resting, perhaps because *Enteromorpha* is difficult to digest. *Enteromorpha* acts very much like a sponge and its bulk relative to its dry weight is very high. Consequently it is possible that the geese become gorged with food and require periods of rest between bouts of feeding in order to digest the food.

Estimates of theoretical food requirements of 75·0 and 211·7 g dry weight per day of salt marsh vegetation and cereals respectively have also been derived. The value for cereals was surprisingly high but the assimilation from this food source was relatively low. In comparison with the grass species found on grazing land (Owen 1976) *Zostera*, *Enteromorpha* and the salt marsh plants eaten by geese are fleshy and succulent and contain relatively little fibrous material (Ranwell & Downing 1959).

Ability of estuaries to support geese

Large areas of salt marsh fringe the Essex and Kent coasts, but brent geese feed consistently only on one area of high saltings. Surveys showed that no other areas were of this particular salt marsh type. Salt marsh feeding can therefore be considered a finite resource at one place which can support a limited number of geese. On the other hand winter cereals are grown in large quantities around the Essex and Kent coast and are at present a largely under-utilized source of food. But they are the least preferred food of brent geese and are only used when the intertidal foods cannot support the goose population.

Estimates of the standing crops of *Zostera* and *Enteromorpha* were produced by Wyer and Waters (1975) as part of the larger Maplin Research Programme. From these figures and estimates of daily food consumption it has been possible to calculate the ability of intertidal food to support feeding geese.

These calculations are meaningful because brent geese can feed on very low densities of food. They are able to reduce both *Zostera* and *Enteromorpha* stocks to almost zero because they also take alternative less preferred foods. For example *Zostera* roots are eaten together with leaves when the density of the latter is very low. Similarly winter cereals inside the sea wall are consumed adjacent to low density *Enteromorpha* beds.

Secondly, the measurements of the standing crops of both *Zostera* and *Enteromorpha* represent food of which all is available other than at high tide. The considerable problems encountered when trying to estimate the quantity of food available to carnivores are therefore avoided (Goss-Custard & Charman 1976). The different portions of the food can be obtained with varying degrees of difficulty and could thus be considered to have different relative availabilities. However, the food is present at all times, except at high water, within the reach of the geese and is therefore available.

Table 28.2. *A summary of the actual use by brent geese of areas in Essex and Kent relative to the potential grazing provided by intertidal food stocks together with the probable reason for any differences.*

	Potential grazing on intertidal foods in thousands of goose days	Actual goose usage as percentage of potential of intertidal foods		
		1972/73	1973/74	1974/75
Maplin Sands	692	111	168[abc]	118
Leigh	157	133[a]	144[a]	117
Hamford Water	90	120	460[c]	358[c]
Colne	74	190[d]	290[bcd]	249[d]
Blackwater	428	66[e]	186[bc]	148[bc]
Dengie	160	111	118[b]	164[b]
Orwell	234	2[e]	25[e]	11[e]
Stour	323	13[e]	13[e]	14[e]
Crouch	186	—	19[e]	27[e]
Swale	635	5[e]	11[e]	8[e]
Medway	1,671	2[e]	10[e]	5[e]
Total number of goose days in thousands		1,790	3,280	2,655

Reasons for over or under use: a, feeding on *Zostera* roots; b, feeding on cereals; c, feeding on rough grazing; d, feeding on salt marsh; e, disturbance.

The third justification for this approach is that it appears to produce a realistic model of the brent goose food system. The model is able to predict with reasonable accuracy the consequences of increased brent goose numbers on intertidal food sources of finite size (Table 28.2).

Increasing populations of brent geese would be expected to turn more to areas of their preferred food which were under-utilized and yet free from disturbance, and also to turn to less-preferred foods. In addition, if they were able to tolerate or habituate to disturbance, they would be expected to utilize disturbed areas. A further consequence of increasing numbers would be expected to be the departure of large numbers of geese completely from the area and their probable movement south to previously under-used food sources. This is precisely the pattern which was observed during the three years of the study.

CONCLUSIONS

The sequence of occupation of the different food sources used by brent geese, from *Zostera*, to *Enteromorpha*, to salt marsh vegetation and agricultural crops, was consistent from year to year and was associated with a sequential progression down a chain of food preference. The less-preferred foods, although available when the geese arrived in this country were only used when *Zostera* stocks had been largely eaten out.

Feeding behaviour and food consumption in the brent goose was largely independent of food density above 15% leaf cover of *Zostera*. Below this food density changes in behaviour were marked. Comparable measurements made on the pale-bellied race of the brent goose (*Branta bernicla hrota*) in north-east England showed similar responses to food density (P.C. Smith, personal communication).

Zostera appeared to be the only food from which brent geese obtained their daily energy requirements. Thus the stepwise progression from food to food through the winter not only represented movement to less preferred foods but also a reduction in energy intake.

The direct approach adopted in this study to calculate the number of goose days an area of intertidal food could support gave reasonably accurate predictions. This method was acceptable for two reasons. First because brent geese could feed on very low densities of food and secondly because, except at high tide, their intertidal foods were always in a position where they could be eaten.

The effect of any reduction in the quantity of food present on the Essex and Kent coasts would be to advance the sequential progression of the geese to less preferred foods. The consequences of such a change in the winter feeding conditions on the subsequent survival and breeding of the brent goose are unknown. Similar changes in winter feeding conditions are also likely to be brought about by increases in the goose population. Any subsequent changes in mortality and natality will be slow to show themselves and difficult to measure. However a knowledge of such factors is of importance if the dark-bellied brent goose is to be conserved and managed effectively.

ACKNOWLEDGMENTS

The studies described in this paper were financed by the Department of the Environment under Research Contract DGR 205/2. I am grateful to Drs L.A. Boorman, J.D. Goss-Custard, D. Jenkins, R. Mitchell, M. Owen and D.S. Ranwell for useful comments on an earlier draft.

REFERENCES

Blindell R.M. (1975) The numbers of waders and wildfowl in the Essex estuaries. *Report of the Maplin Ecological Research Programme*, Part II 2b, pp. 194–255. Department of Environment (unpublished), London.

Charman K. (1975) The feeding ecology of the brent goose. *Report of the Maplin Ecological Research Programme*, Part II 3b, pp. 259–89. Department of Environment (unpublished), London.

Ebbinge B., Canters K. & Drent R. (1975) Foraging routines and estimated daily food intake in barnacle geese wintering in the northern Netherlands. *Wildfowl* **26**, 5–19.

Goss-Custard J.D. & Charman K. (1976) Predicting how many wintering waterfowl an area can support. *Wildfowl* **27**, 157–58.

King J.R. & Farner D.S. (1961) Energy metabolism, thermoregulation and body temperature. *Biology and Comparative Physiology of Birds* (Ed. by A.J. Marshall), pp. 215–88. Academic Press, New York.

Ogilvie M.A. & St Joseph A.K.M. (1976) The dark-bellied brent goose in Britain and Europe, 1955–76. *Br. Birds* **65**, 422–39.

Owen M. (1976) The selection of winter food by white-fronted geese. *J. appl. Ecol.* **13**, 715–29.

Ranwell D.S. & Downing B.M. (1959) Brent goose winter feeding pattern and *Zostera* resources at Scolt Head Island, Norfolk. *Anim. Behav.* **7**, 42–56.

Wyer D.W. & Waters R.J. (1975) Tidal flat *Zostera* and algal vegetation. *Report of the Maplin Ecological Research Programme*, Part II 1b, pp. 26–73. Department of Environment (unpublished), London.

29. ECOLOGICAL PROCESSES CHARACTERISTIC OF COASTAL *SPARTINA* MARSHES OF THE SOUTH-EASTERN U.S.A.*

RICHARD G. WIEGERT
Department of Zoology, University of Georgia, Athens, Georgia, U.S.A.

SUMMARY

The ecological processes characteristic of the extensive *Spartina* marshes along the south-eastern coast of the United States fall into three categories: (i) those operating in the water that inundates the marsh twice each day; (ii) those active in, or on, the sediments, including the interstitial sediment water retained during low tide; (iii) those characteristic of the terrestrial species found on the vegetation or on the sediments at low tide. All of these are constrained by the physical factors of temperature, salinity, inorganic nutrient load and the actions of the two daily tidal cycles. Of the three types of processes, the last, dealing exclusively with the terrestrial species (salt marsh grasshoppers, ephydrid flies, long-billed marsh wrens, wading birds, etc.), is least important in determining overall marsh productivity and the fate of this fixed carbon. The marsh produces annually some 1,000 g C m^{-2}, that is not utilized (changed to gaseous carbon) in the marsh. The processes of category (i) are vital in determining the relative proportions degraded in the water, the amount leaving the marsh and/or the estuary proper via tidal flushing, and the growth and migration of motile species. Similarly, the productivity of *Spartina* is dominant in the system but limited largely by processes operating in the sediments. The sediment is also a site both of temporary storage of much of the surplus organic carbon and of its anaerobic degradation. Thus category (ii) is vital in regulating the dynamics of the marsh ecosystem. Tidal action, both the daily ebb and flow, and the periodic catastrophic coincidence of spring tides, heavy rain and wind, are the primary determinants of the rate at which the surplus organic carbon (and inorganic nutrients) leave the marsh. This paper will present first a coherent overview of the most important examples of

* Contribution No. 344 from the University of Georgia Marine Institute.

each of these processes. It discusses the role of a simulation model in guiding and refining the research effort devoted to an understanding of this extensive, economically valuable and aesthetically pleasing ecosystem.

RÉSUMÉ

Les processus écologiques caractéristiques des marais de l'extensive *Spartina* le long de la côte sud-est des états-unis peuvent être classés en trois catégories: (i) ceux opérant dans les eaux inondant le marais deux fois par jour. (ii) Ceux actifs dans ou sur les sédiments, incluant l'eau du sédiment retenue pendant la marée basse. (iii) Ceux caractéristiques des espèces terrestres trouvées sur la végétation ou le sédiment à marée basse. Tous subissent les contraintes des facteurs physiques de température, salinité, charge de nutriments organiques et les actions des cycles des deux marées journalières. De ces trois types de processus, le dernier concernant seulement les espèces terrestres (sauterrelles, mouches ephedrines, troglodytes des marais au long bec, les échassiers ...) est le moins important dans la détermination de la productivité globale et le sort du carbone fixé. Le marais produit annuellement quelques 1 000 g C m^{-2} qui ne sont pas utilisés (changés en gaz carbonique) dans le marais. Le processus (i) est ainsi vital dans la détermination des proportions relatives dégradées dans l'eau, la quantité quittant le marais et/ou l'estuaire à cause du reflux de la marée et la croissance et la migration des espèces migratrices. Similairement la productivité de *Spartina* est dominante dans le système, mais est limitée largement par les processus opérant dans le sédiment. Le sédiment est aussi un site pour l'accumulation temporaire de la plupart de l'excédent de carbone organique et de sa dégradation anaérobie. Ainsi la catégorie (ii) est vitale dans la régulation des dynamiques de l'écosystème du marais. L'action de la marée, à la fois le flux et le reflux journaliers et la coïncidence périodique catastrophique des marées de printemps avec les pluies importantes et le vent sont les déterminants primaires de la vitesse à laquelle l'excédent de carbone organique (et des nutriments inorganiques) quitte le marais. Dans ce papier, nous présentons premièrement une vue générale, cohérente des exemples les plus importants de chacun de ces processus. Nous discutons du rôle d'un modèle de simulation pour guider et redéfinir les efforts de la recherche pour la compréhension de cet important écosystème, économiquement valable et esthétiquement plaisant.

GENERAL CHARACTERISTICS OF SALT MARSHES

Wet coastal ecosystems of the world comprise both salt marsh and mangrove (mangal) associations, up to the limit of extreme high water (Chapman 1977). Although both salt marsh and mangal are found in similar types of physical habitat, the salt marsh is characterized by the absence of dominant arboreal vegetation. The topography slopes gently to provide a large surface within the intertidal zone. Vegetation characteristic of salt marshes usually grows on low energy coasts, behind barrier islands or in estuaries where protection from the direct action of storm waves is provided. However, this preference may simply result from the effect of waves on seedling establishment and/or unstable sediments. For example, Frey and Basan (in press) report mature *Spartina alterniflora* reducing the height of wind-generated waves by 71% and wave energy by 92%. They also report *S.alterniflora* marshes in firm peat withstanding the full force of waves along the exposed shores of Cape Cod. In the Florida Panhandle, a low energy coast, marshes also front the open sea (Tanner 1960).

The vegetation of salt marshes consists of a relatively small number of genera distributed world-wide. Some, such as *Salicornia*, dominate as annual species in some localities and as perennial species in others. Most genera of salt marsh plants are perennial (Chapman 1977). The transition zone between the marsh proper and the adjacent upland is often dominated by shrubs, although in the extensive *Spartina* marshes in south-eastern U.S.A. much of this zone may be occupied by the needle rush, *Juncus romerianus*, which grades rather abruptly into pine or hardwood forest. Vegetational zonation in the salt marshes is common, but there is disagreement about whether this zonation is to be regarded as successional or simply a reflection of the effect of elevation, salinity, tidal inundation and sediment type on the distribution of the different species (see Chapman 1977).

Because low energy coasts with the requisite physiographic characteristics for the development of salt marshes are relatively rare, the major extensive marsh systems of the world are found in semi-protected situations, such as in estuaries or behind barrier islands.

Along the Atlantic coast of south-eastern U.S.A. extensive barrier island development has provided conditions for the development of huge expanses of salt marsh dominated by *Spartina alterniflora*. In particular, the portion of Atlantic coast within the state of Georgia is ideal for the development of this type of ecosystem and, although only about 100 miles long, contains more than 31% of all salt marsh habitat on the Atlantic coast of the U.S.A. South Carolina and Georgia together contain more than 62% of these salt marshes (Reimold 1977).

BARRIER ISLAND MARSHES OF SOUTH-EASTERN U.S.A.

A number of theories have been put forward to explain the formation of the barrier islands so common along the Georgia and South Carolina coastline.

The prevailing consensus was summarized by J. Hoyt in Johnson *et al.* (1974). In the Pleistocene and Recent (Holocene) periods, large dunes were thrown up along the shore by tidal and wave action. During periods of rising sea level (glacial retreat) these dunes were partially submerged and cut off from shoreward contact with the formation of *Spartina* salt marsh between island and mainland.

A number of distinct 'belts' of old barrier islands, now reduced to ridges, can be traced inland along the Georgia coast (Hoyt, in Johnson *et al.* 1974). All of these, including the present extant barrier islands, are of Pleistocene age. However there are a few small seaward islands or extensions of islands, which are of more recent (Holocene) origin.

The present islands are all located several miles offshore. Their long axis is located in a N.N.E. to S.S.W. direction parallel to the coast (Fig. 29.1).

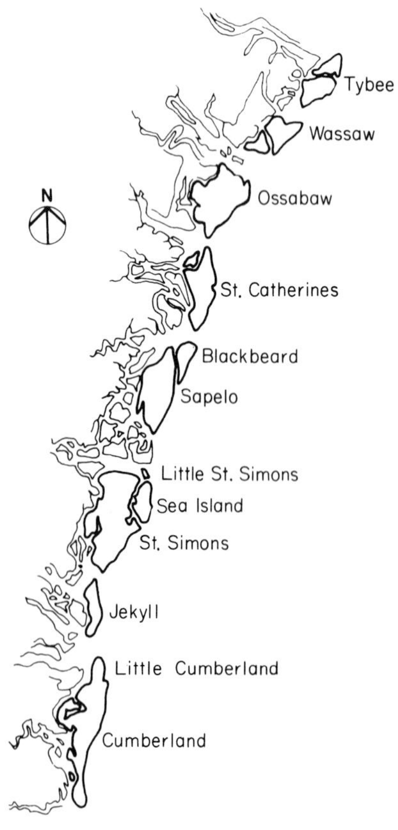

FIG. 29.1. Plan of barrier islands along the Georgia coast. (From State of Georgia, Budget in Brief, Fiscal Year 1977–78. By permission of Department of Natural Resources, State of Georgia.)

The areas between islands form the estuaries. A diagrammatic cross-section of a typical island shows the relative position of ocean beach, protective dunes, stabilized dunes, forest and tidal creek-marsh areas (Fig. 29.2). Sediment deposited in this area between island and mainland has formed the extensive flat or gently sloping topography essential for the development of salt marsh dominated by smooth cord grass, *Spartina alterniflora.* Many descriptions of this kind of salt marsh have been published, the reader should consult recent reviews by Johnson *et al.* (1974), Reimold (1977) and Frey and Basan (1978).

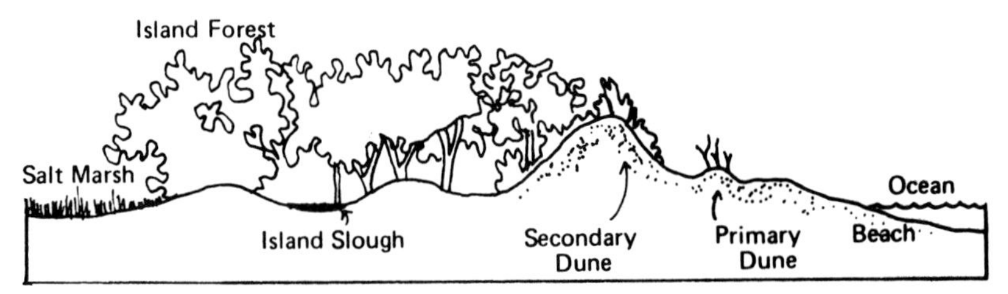

FIG. 29.2. Diagrammatic cross-section of a typical barrier island showing ocean, dune, forest and tidal creek-marsh areas. (From Clement C.D. (1971) Recreation on the Georgia Coast: An Ecological Approach. *Georgia Business* **30**, 1–24, adapted by J.R. Richardson. By permission of Department of Natural Resources, State of Georgia.)

Although the Georgia coast has a moderate to low energy surf, it has a large (2–3 m) tidal amplitude, larger than any other coastal area to the immediate north or south. This large daily fluctuation in height of the water profoundly affects the kind of marsh that develops, e.g. levée formation is much less pronounced or almost absent in South Carolina marshes. Odum (1973) spoke of the tidal flux as a 'tidal energy subsidy', but this term has unfortunately been misinterpreted by some to imply some mysterious process whereby *Spartina* 'incorporates' tidal energy instead of simply relying on the tide to supply energy to move nutrients and water onto and through the marsh.

Behind the Georgia barrier islands the *Spartina alterniflora* marshes form on the fine clay-silt sediments deposited by tidal action, and river flow in areas such as the estuaries of the Altamalia and Savannah rivers. Frey and Basan (1978) prefer the term sediment, as opposed to soil, because the former, whether allocthonous or autochthonous, carries no connotation of a parent material. As sediment deposition abetted by the flow-slowing action of the *Spartina* grass deposits yet more sediment, drainage channels develop in the hitherto smooth contours of the marsh. Headward erosion extends the drainage channels into the marsh and the effects of oyster and mussel reefs cause erosional changes in the twists and turns of the creeks (Fig. 29.3) (Frey & Basan 1978).

As water is pushed into the tidal creeks and spills over the bank to run onto the marsh proper, the sudden cessation of flow causes increased deposition of particulate material and the formation of a levée (Fig. 29.4). Because of both its increased elevation and the hydrostatic head formed when water drains below the top of the levée on an ebb tide, the levée and the marsh

FIG. 29.3. Aerial view of the Sapelo Island salt marshes showing the dendritic patterns of the drainage creeks. (Courtesy of E.P. Odum.)

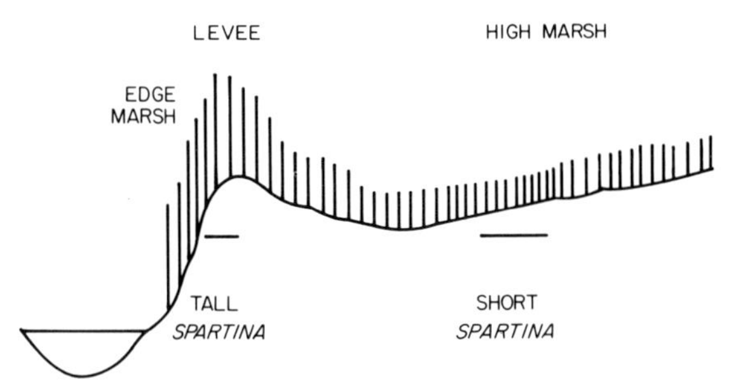

FIG. 29.4. Diagrammatic cross-section of a Sapelo Island marsh showing the relationship of low marsh and high marsh, tall *Spartina* and short *Spartina* to the tidal creek and the levée (Courtesy of W. Wiebe).

adjacent to the creek have a greater through-flow than areas of high marsh (Riedeburg 1975). This affects the salinity and nutrient content of interstitial water in the sediment of the creek banks.

CHARACTERISTIC ECOLOGICAL PROCESSES

Carbon fluxes in the Spartina *marshes*

The dynamics of the salt marsh ecosystem may be discussed in terms of any conserved unit, i.e. element or energy. Fig. 29.5 illustrates the three major physical divisions of the ecosystem: (1) air and emergent plants, in which terrestrial-type processes and organisms predominate; (2) water, both tidal creek and marsh water at high tide, in which only aquatic organisms and predominantly aerobic processes are found, and (3) sediment, in which aquatic organisms and predominantly anaerobic processes are active. These divisions emphasize an important distinction between the intertidal ecosystem and most others, namely that the air–water or sediment–water interfaces, which normally form the boundaries between distinct ecosystems, are compressed here into such a dynamically interacting system that they cannot be uncoupled easily, even for purposes of study and description. Fig. 29.5, from Wiegert and Wetzel (1978) shows a diagrammatic carbon flux web. The important processes operating in the marsh remove abiotic carbon from the air to the biotic carbon components of the air, water and sediment (*Spartina* shoots, rhizomes and roots plus the heterotrophic populations), these in turn are degraded to abiotic carbon compounds in water and sediment. The latter exchange with the water as a result of erosion, deposition and diffusion. The former in turn exchange with the estuary water through tidal action or with the air by diffusion. The magnitude of these fluxes is conditioned by the kind of ecological interactions involved, the levels of resources (nutrients, energy, etc.) available and the influence of physical factors (light, temperature, tidal regime, etc.). The following sections summarize briefly the most important of these ecological processes. An ecological process is defined, for purposes of this paper, as any transformation of carbon initiated by living organisms and mediated in some way by interactions either with other organisms or abiotic materials. The different processes are strongly influenced by environmental conditions. Thus,

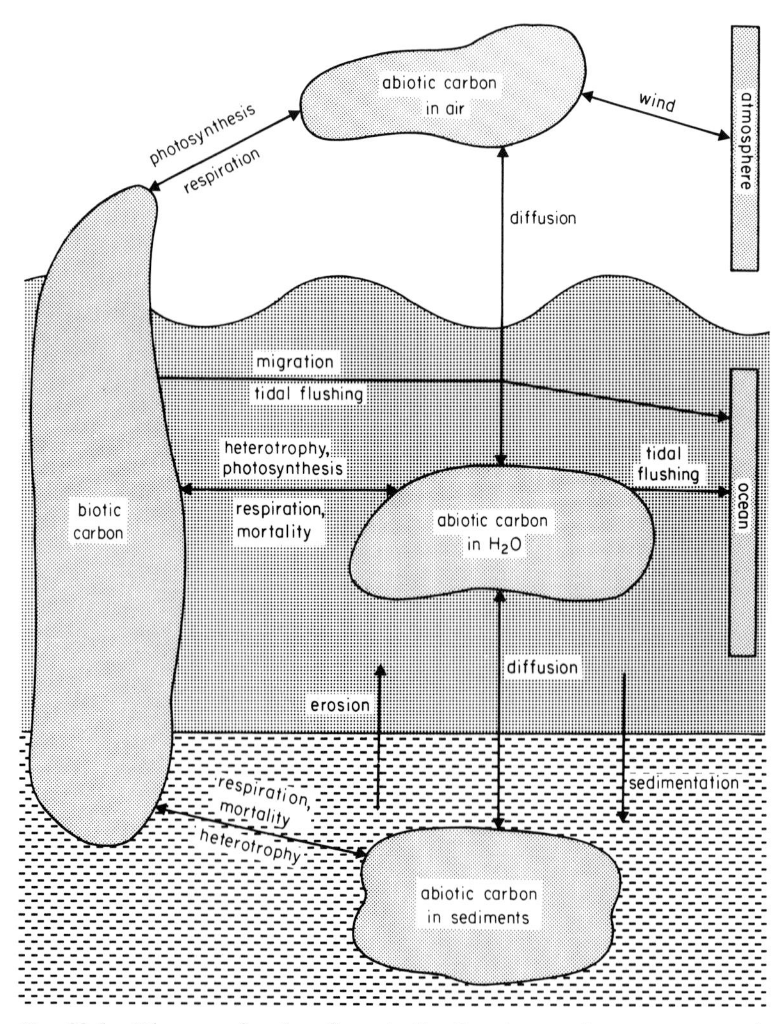

FIG. 29.5. Diagram of carbon fluxes in the *Spartina* marsh ecosystem associated with air, water and sediment. (From Wiegert & Wetzel (1978) by permission of University of South Carolina Press.)

production (primary and secondary), respiratory catabolism, predation and seasonal mortality are all examples of ecological processes important to the functioning of a *Spartina* marsh ecosystem.

The importance of the marsh system discussed in this paper will centre on the overall annual carbon balance achieved by the fluxes shown in Fig. 29.5. Two questions are posed and discussed in this paper. Does the marsh system as a whole operate as a producer or consumer of organic carbon? If it is the former, what are the fates of this organic material; and if it is the latter, what are the organic carbon sources? The importance of an ecological process will be defined in terms of the magnitude of its effect on organic carbon production or degradation. The reader interested in details of the movement of carbon in the salt marsh beyond the brief treatment offered in this paper should consult Teal (1962), Wiegert *et al.* (1975), Wiegert and Wetzel (1978) and the references given in these papers as well as those at the end of the present paper.

Ecological processes in air

The input of organic carbon to the marsh surface is associated with the annual growth of the above-ground parts of smooth cord grass, *Spartina alterniflora.* The dominance of this one species is so complete that the Georgia barrier island salt marshes seem to be a homogenous waving sea of grass. On closer inspection of the system, as Fig. 29.6 shows, this apparent homogeneity gives way to a spatially heterogenous pattern of tidal creeks, mudbanks, oyster reefs, dead stems of *Spartina,* surface detritus, animal burrows, etc. Nevertheless, *Spartina alterniflora* growth constitutes the vast majority of the net primary productivity of the marsh, and a large part of this appears as shoot growth.

Spartina productivity varies greatly with position in the marsh. Two distinct kinds, 'tall' and 'short' *Spartina*, can be distinguished (see Fig. 29.4). The tall *Spartina* area, occupying 6–7% of the total marsh surface (Reimold *et al.* 1973), includes the creek banks, the levées and the marsh surface immediately behind the levées. Tall *Spartina* is more productive and has a higher standing crop than does short *Spartina* (Gallagher *et al.* 1979; Giurgevich & Dunn pers. comm.). The 'short *Spartina*' area which, with the tidal creeks, forms the remainder of the marsh surface, consists of *Spartina* grading in height from medium, to very short near the upper, landward boundary of the marsh. At this point, *Spartina* often gives way to the black needle rush, *Juncus romerianus* wherever fresh water imput lowers the sediment salinity somewhat or it may be replaced by *Salicornia* wherever the salinity of the sediment is high. Neither of these two species occupies significantly large areas of the Sapelo Island marshes.

The productivity differences between the tall and short *Spartina* stands is indeed impressive. The harvest data of Gallagher *et al.* (1979), and the data of Giurgevich and Dunn, who used gas analysis, are contrasted in Table 29.1.

FIG. 29.6. View of the marsh showing aerial shoots of *Spartina*, exposed mud flats (sediment) and tidal creek (water). (Courtesy of R.L. Wetzel.)

The productivity difference is consistent and seems definitely to be a positional and not an ecotypic effect. Transplanting and other manipulations can result in each form being transformed into the other, although the time needed for this to be achieved may be several years. A concise summary of the current evidence on this subject can be found in Chalmers (1977).

Although the results from the use of the two methods (harvest method and the CO_2 gas analysis technique) agree exactly in the case of the short *Spartina*, the use of the latter technique produces an exceptionally high estimate of net primary production for tall *Spartina*. Giurgevich and Dunn (pers. comm.) argue that this is a physiological maximum and does not represent the field value. This is because the size of the plants necessitated an orientation of leaves within the chambers that produced a greater exposure to light than is normal in the field. Tall *Spartina* grows in sediments which have a lower salinity and higher nutrient medium than elsewhere, because of the greater tidal flushing of sediments caused by the levée formation. Mutual shading in the dense stands is a major mediator of field net production, thus the values derived from the CO_2 gas analysis data overestimate productivity. In addition the plants were so large that they could not be entirely enclosed by the field chambers, as could the short *Spartina* plants, thus the measurements of net primary production in the tall *Spartina* were conducted on younger, more metabolically active tissue, again increasing the estimate.

For the marsh as a whole, allowing for the open water and mud bank areas, the net input of fixed organic carbon derived from *Spartina alterniflora* is 1,573 g C m^{-2} yr^{-1}. Of this total 40% is estimated on the average to remain as shoot production and 60% appears in the sediments as production of roots and rhizomes.

Table 29.1. *Comparison of estimates of net annual primary productivity (g C m^{-2} y^{-1}) of tall and short* Spartina *areas on Sapelo Island, Ga. Data based on the harvest method are from Gallagher* et al. *(1979) and the estimates based on measurements of CO_2 exchange are from Giurgevich and Dunn (pers. comm.). The discrepancy in the two estimates for tall* Spartina *is discussed in the text.*

	Harvest method	CO_2 exchange method
Tall *Spartina*		
Shoot	1,658	
Root	938	
Total	2,596	7,020
Short *Spartina*		
Shoot	593	
Root	895	
Total	1,488	1,491

Control of this rate of input is largely through nutrient limitation or the effect of high interstitial salinities as shown by the presence of short *Spartina* in much of the marsh. There is abundant phosphate in the marsh sediment water, but nitrogen is scarce and nitrogen enrichment will stimulate production (Chalmers *et al.* 1976; Chalmers 1977).

Grazing of *Spartina* shoots by terrestrial organisms drains away very little of this net primary production of carbon compounds. Teal (1962) summarized the existing data on energy flow through the grazers of *Spartina* shoots. His values, converted to carbon equivalents give 31 g C m^{-2} yr^{-1} as the loss to these herbivores. Current investigations of the populations of plant-hoppers suggest this to be an overestimate. Thus, the direct removal of carbon from

Spartina shoots by terrestrial grazers cannot be an important factor in the overall carbon flux of the marsh ecosystem. The effect of grazers, if it exists, must lie in some as yet unknown control function.

Abiotic carbon balance in the air over the marsh (mainly CO_2) is maintained constant by air movement. There is, of course, diffusion exchange with the marsh water, but the convective exchange of air through wind is so rapid that the CO_2 content can be assumed to remain constant for purposes of modelling the carbon fluxes within the marsh ecosystem.

Ecological processes in water

The water in the larger tidal creeks, which do not drain entirely at low tide, and the water that covers the entire marsh at high tide, contain many aquatic organisms and a complement of organic carbon in gaseous (CO_2, CH_4), dissolved (DOC) and particulate form (POC). In contrast to the aerial component, organic carbon may enter this system by either photosynthetic fixation of gaseous inorganic carbon (CO_2) or by the release of compounds from *Spartina* shoots (the secretion of dissolved organic carbon (DOC) into the water) or by mortality and disintegration of the plant into abiotic particulate organic carbon (POC).

Two groups of organisms contribute to the photosynthetic input of organic carbon to the water, the phytoplankton and the benthic algae. Although Ragotskie (1959) found phytoplankton net production in the estuarine waters of Sapelo Island to be near zero, there are often quite dense blooms of dinoflagellates (Ragotskie & Pomeroy 1957) which are heavily grazed by zooplankton, so that the phytoplankton must result in a certain amount of organic carbon input to the system. The absolute amount per unit area, however, prorated over the entire marsh is undoubtedly very small. A major factor restricting the importance of production by phytoplankton seems to be the constant turbidity of the water, limiting the light, as nitrogen and phosphorus, both seem abundant (Pomeroy *et al.* 1969).

Net production by the diatoms and other benthic algae represents an appreciable contribution to the total primary production of the marsh. The early study by Pomeroy (1959) reported a gross production by the benthic algae of 200 g C m^{-2} yr^{-1}, with a net production of 90% or more of this value. Thus, the benthic algae, despite their small standing stock, would be contributing 10% as much as *Spartina*. Current studies by Darley (personal communication) suggest that this may be a maximum, but the benthic algal contribution is certainly substantial. Benthic net photosynthesis is related to the amount of time the marsh surface is exposed to light at low tide, but within the period of illumination predation on the algae, not nutrient scarcity, appears to be a major factor on the mud banks (Pace 1977). Within the *Spartina* itself, shading of the algae, even during exposure at low tide, is probably the major controlling factor of the production (Darley, personal communication).

By far the major input of organic abiotic carbon to the water of the marsh is through the mortality and disintegration of the *Spartina* shoots and not via photosynthesis and the subsequent death of phytoplankton. Although seasonal in nature, this ecological process goes on all year, eventually committing virtually the total shoot biomass to the water. Most of this is in the form of POC, but secretion by living shoots and the micro-organisms of dead shoots contributes a small amount of DOC. Because these processes occur primarily in response to the internal physiology of the *Spartina* and to seasonal changes in environmental conditions, there is little, if any, feedback control. Catastrophic events such as storms probably play a role in breaking up the large *Spartina* stems, but heterotrophic activity would seem to be the likely agent of particle formation. Because there is no ice in winter in these southern marshes, shearing and packing of dead *Spartina* which initiate peat formation seldom occurs as it does in more northerly marshes (Frey & Basan 1978).

Numerous transformations of organic carbon take place within the aquatic system of the marsh. Following the scheme of Fig. 29.6, all fixed carbon that is initially part of the standing stock of benthic algae must be transferred to the abiotic component by mortality and secretion, either as POC or DOC. These processes, like the corresponding ones for living *Spartina*, are both inevitable and predictable. The fate of this abiotic carbon is of more fundamental interest; is it again transformed into biotic carbon through assimilation by detritivores, or do the physical processes of sedimentation and tidal flushing remove it?

The three qualitatively different degradative ecological processes operating in the water of the salt marsh system are filter feeding, particle feeding and microbial assimilation.

The filter feeding organisms in the marsh comprise the zooplankton in the water and the mussels, oysters and polychaete worms on and in the sediments. Based on the data of Ragotskie (1959) it is assumed that the zooplankton can have no effect on the degradation of the net carbon fixed in the marsh because they utilize only the phytoplankton production and the latter contributes nothing to the 1,573 g C m^{-2} yr^{-1}, the ultimate fate of which is being investigated. Bahr (1976) found that oyster (*Crassostrea virginica*) reefs occupy 0·06% of the total marsh-water surface and degrade (respire as CO_2) an amount equivalent to 1% of the net annual primary production in the marsh (16 g C m^{-2} yr^{-1}). The energetics of the marsh mussels (*Modeolus demissus*) in the marsh were studied by Kuenzler (1961) and his data on respiration provide an estimate of 4 g C m^{-2} yr^{-1} degraded by this filter feeder. The fourth group of filter feeders, the bottom-dwelling polychaete worms, degrade only a small amount of carbon, 3 g C m^{-2} yr^{-1}, a value computed from the energy flow studies of Teal (1962). Thus, all filter feeding organisms taken together are estimated to account for only 23 g C m^{-2} yr^{-1}.

The major particle feeders in the marsh are the small mud and fiddler crabs of the genera *Uca*, *Sesarma*, *Eurytium* and *Panopeus*. Teal (1962) considered these as a group to be 'detritus eating' and computed their respiratory energy loss, which I have converted to a mean of 17 g C m^{-2} yr^{-1}. Although a more recent study shows that the most abundant genus (*Uca*) derives at least part of its energy from benthic algal instead of *Spartina* detritus (Montague, personal communication), the loss is still part of the overall degradation of the net primary production of the marsh. The mud snail (*Nassarius*) also assimilates mostly benthic algae (Wetzel 1977).

Burkholder and Bornside (1957) reported that 11% of the net production of shoots was available for rapid conversion to bacterial biomass upon death and submersion of the *Spartina*. This represents 0·11 × 1,573 × 0·50 or 69 g C m^{-2} yr^{-1}. Thus, the minimum total estimated degradation of carbon resulting from ecological processes active in the water is 209 g C m^{-2} yr^{-1} (Table 29.2). To this must be added the (currently unknown) amount degraded by the slower-acting organisms which utilize cellulose.

These ecological processes in water are all influenced by both nutrient scarcity and direct interference of competing organisms. Although we are only now beginning to understand how some of these control mechanisms operate in the Georgia coastal salt marshes, the discussion above shows clearly that a

Table 29.2. *Estimated annual net production and degradation of carbon by ecological processes active in the air, water and sediments. All values are g C m^{-2} yr^{-1}. Conversions from Kcal, where necessary, were made assuming 5 kcal g^{-1} for detritus and a carbon content of 50%. Thus g C = kcal × 0·1.*

Category	g C m^{-2} yr^{-1}
Net primary production of *Spartina*	1,573
Net primary production of algae	180
Total net primary production	1,653
Degradation in air	31
Degradation in water	209
Degradation in sediments	533
Total degraded	773
% of *Spartina* net production	49
% of total net production	47
Surplus carbon	880

relatively small portion of the total net primary production, or the net production by shoots, is degraded within the water itself.

Carbon in the water may be exchanged with the sediments via sedimentation and erosion of particulate (POC) and diffusion of dissolved organic (DOC) and gaseous carbon, both organic and inorganic (CH_4 and CO_2). Exchanges with the air are by diffusion of CH_4 and CO_2. Exchanges with the estuarine water are mediated by the net tidal water mixing and by net movements of organisms into or out of the marsh water. These latter exchanges (tidal and migrational) are discussed in detail in a subsequent section.

Ecological processes in the sediments

About 60% of the total annual net primary productivity of *Spartina alterniflora* is devoted to growth of roots and rhizomes. Upon the death of these plant parts, this carbon is available for degradation by the organisms present in this segment of the marsh. Although the death and disintegration of the roots and rhizomes into POC and DOC are modulated by season, growth stage and to some degree the action of detrivores, the standing crop of living roots and rhizomes shows little long-term change and thus the annual rate of abiotic organic carbon input to the sediment saprophages is relatively constant.

The majority of the sediment in a *Spartina alterniflora* ecosystem is anaerobic, with highly reduced compounds prevailing. In fact, the immediate chemical uptake of oxygen upon aeration of the sediment is some 25 times the steady-state biotic uptake (Teal & Kanwisher 1961). Nevertheless, there are aerobic oxidized zones deep in the sediments surrounding the *Spartina* roots. These are caused by the excess oxygen diffusing down through the stems of *Spartina* and out into the sediments (Teal & Kanwisher 1966). Furthermore, there is a thin but active aerobic zone at the surface of the sediments. But the methane (CH_4) diffusing from the deeper sediments is not oxidized in this zone, although oxidation in the aerobic rhizosphere has not been studied (King & Wiebe 1978). Thus, a reasonable measure of the total carbon degraded by the sediments should be obtained by measurement of either the oxygen supplied to the sediments, or the CO_2 and CH_4 evolved from the sediments.

Teal and Kanwisher (1961) measured the oxygen uptake by the sediment of three sites in the Sapelo marshes. Computed as $g\ C\ m^{-2}\ yr^{-1}$ their results gave an estimate of 353. However, to obtain this figure they subtracted that portion of the oxygen consumption that entered the cut-off stems of the *Spartina* plants contained inside their respiration chamber. Teal and Kanwisher (1966) later found the oxygen diffusing via the *Spartina* stems and rhizomes to the roots to be in excess of the needs of the roots by 30 to 200%. Thus, from $\frac{1}{3}$ to $\frac{2}{3}$ of the oxygen entering the cut stems in the respiration studies was actually leaving the roots and being used in aerobic processes in the sediments.

Upon the initial growth of a root into a previously anaerobic part of the sediment, chemical oxygen demand would be expected to use up the excess oxygen and it would be unavailable for oxidation of organic C by microbial degradation. Once an oxidized zone has been established around the root, however, the rhizophere organisms could be expected to use further oxygen to degrade secreted DOC to CO_2. Thus, most, if not all, of the excess oxygen should have been added to the oxygen demand found by Teal and Kanwisher (1961) when the conversion to the amount of carbon degraded was made. Teal and Kanwisher (1961) subtracted an average oxygen uptake rate of 10 $mm^3\ cm^{-2}\ hr^{-1}$ to account for root-rhizome respiration. If this was 30% too high, the correction should have been 7 and if 200% too high the correction should have been 3·3. Thus, their estimate of total sediment oxygen demand would change from 8·9 to within the range of 11·9 to 15·1 $mm^3\ cm^{-2}\ hr^{-1}$. The corresponding estimate of carbon degraded would be 479–635 $g\ C\ m^{-2}\ yr^{-1}$.

John Hall (unpublished data) measured CO_2 evolution at many sites in the Sapelo marshes over different seasons and estimated the degradation of carbon to be 533 $g\ C\ m^{-2}\ yr^{-1}$ (including 27 $g\ C\ m^{-2}\ yr^{-1}$ as methane). These two independent studies show a satisfactory agreement, although much work needs to be done to reduce the variance in the data.

Accepting Hall's figure for the present, Table 29.2 shows the total estimated degradation of carbonaceous material by all ecological processes in the marsh to be 773 $g\ C\ m^{-2}\ yr^{-1}$ or 99% of the total net primary production of *Spartina*. Below ground, the *Spartina* input is 944 $g\ C\ m^{-2}\ yr^{-1}$ of which an estimated 533 g (or 56%) is degraded as against the 240 g (38%) of net shoot production degraded in the water and air. However, the total carbon input to the water is increased by the 180 g C net production of the algae (Pomeroy 1959), so the overall percentage of marsh carbon input degraded is reduced to 47%, and 30% of the organic carbon input to water is degraded.

Knowing the possible fate and significance of this large excess carbon production is vital to a complete understanding of the impact of the marshes on the estuarine and offshore marine communities. By coupling simulation modeling with field research, researchers at Sapelo Island are attempting to obtain this information.

TIDAL TRANSPORT OF ORGANIC MATTER

A simulation model of carbon flux

Very early in the current cooperative study of the ecology of the Sapelo Island salt marsh ecosystem a simulation model was constructed and used as a guide for the continuing research program (Wiegert *et al.* 1975). Sensitive parameters singled out by this model were reinvestigated and the model evolved through three versions. An important early change was the incorporation of a tidal exchange parameter that simulated the manner in which the tidal excursions were thought to result in water exchange and thus promote the export of materials carried in the marsh water at high tide. These included algae, dissolved organic carbon and particulate organic carbon. Comparison of the output with field data on standing crop and flux rates showed them to be in reasonably good agreement (Wiegert & Wetzel 1978). Fig. 29.7 shows

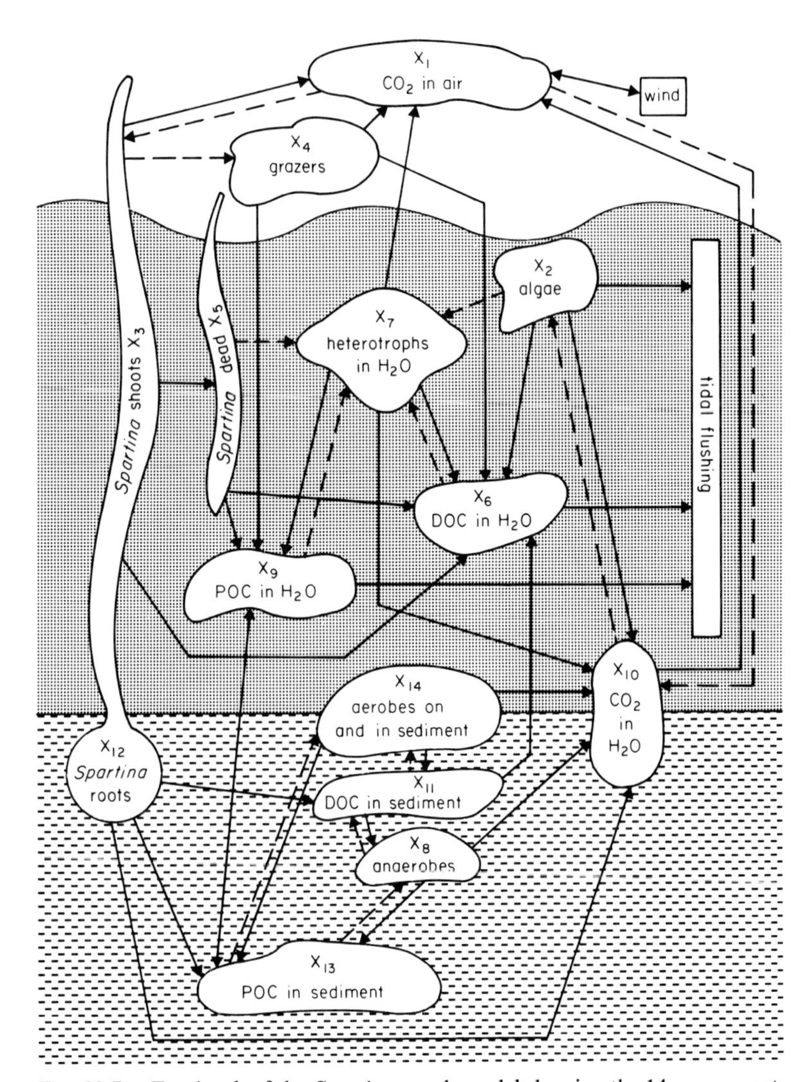

FIG. 29.7. Food web of the *Spartina* marsh model showing the 14 components and their association with the air, water and sediment flows of Fig. 29.5. (From Wiegert & Wetzel (1978) by permission of the University of South Carolina Press.)

the food web and the 14 compartments of the model with their relationships to the flows of carbon between air, water and sediment of Fig. 29.6. The dissolved organic carbon compartments (DOC) and particulate organic carbon compartments (POC) represent abiotic carbon, as do the two gaseous compartments (CO_2). In general, dead *Spartina* is an abiotic component, but is being decomposed *in situ* and so does have a CO_2 production associated with it. The dotted lines representing the biotic ingestions are controlled by resources (donors), consumers (recipients) or both. Solid lines (fluxes to abiotic components) are controlled by factors acting on the biotic donor alone. Abiotic to abiotic fluxes, e.g. CO_2 in water to CO_2 in air, are special cases treated individually. A postulated daily loss of 25% of the POC and DOC in the marsh water and 12·5% of all suspended algae produced a total annual export of 1,075 g C m^{-2} and an increment to the marsh of 26 g C m^{-2} (Wiegert & Wetzel 1978).

However, two independent lines of evidence suggested caution in embracing this explanation of carbon export. Some preliminary measurements of the $^{12}C/^{13}C$ ratios of the carbon in particulate organic matter (Haines 1976) failed to support the assumption that most or all of the particulate carbon in the water of the Duplin River was derived from C_4 plant (*Spartina*) detritus. Rather, the material appeared to derive in appreciable quantities from terrestrial plant detritus of the C_3 type, presumably originating from the water of the Altamaha River to the south. Secondly, a reconsideration of the classic paper by Odum and de la Cruz (1967) which is commonly cited as the best evidence for tidal transport of organic matter out of the marsh showed that their data, on the Study Creek drainage, when expressed on a per m^2 of watershed basis, amounted to only 118 g C m^{-2} yr^{-1}, little more than 12% of the surplus of annual net primary production. These findings could hardly be reconciled with a large daily transport of *Spartina* detritus, particulate or dissolved, off the marsh, out of the tidal creeks and into the estuaries.

Tidal exchanges between estuary and marsh

At this point it was decided to mount an intensive, interdisciplinary study of the hydrology of the Duplin River to determine just what was the daily tidally-caused exchange between the estuary, river and marsh waters. Salinity was to be used as a tracer in this effort and changes in salinity following a temporary slight lowering in the marsh water due to rain were to be related to changes in the concentrations of other substances in the waters, among them chlorophyll, O_2, POC, DOC, silicon and nitrate. The results of this study are to be published in detail elsewhere (Imberger *et al.*, in preparation). A finding of major import for the model was that daily convective exchange between marsh water and other Duplin river water was much smaller than 25%. Only following a 'catastrophe' consisting of very heavy rains (10 cm or more delivered in 1–2 hours, at low tide) would a quick (1–2 tide cycles) overturn or displacement of the marsh water downstream occur.

The central question now posed was: Is the surplus carbon, which is produced but not degraded within the marsh, exported as a result of a small daily tidal exchange together with a series of infrequent catastrophic events, of which the intense rainfall experienced is one example?

In order to determine whether such an export could be postulated without drastically altering the standing crops and fluxes rates of other components of the marsh ecosystem, the simulation model of the marsh was changed. The changes included an increase in the sedimentation rate of POC in water, a decrease in the daily tidal export of DOC, POC and algae from 25% (12·5% in case of algae to allow for the presence of benthic forms) to 1%, a major loss of water borne POC, DOC and algae in the event of a catastrophic storm and a scouring of the marsh sediments to remove any buildup above a constant fixed standing crop of organic carbon. Four simulations were made with the model in which the seasonal changes in standing crops of algae (X2), heterotrophs in H_2O (X7), dissolved organic C in H_2O (X6) and particulate organic carbon in H_2O (X9) were examined. The change involved using the low value (1%) for daily export of material mentioned above and assumed four 'catastrophic' events per year, 91 or 92 days apart. Each one resulted in the removal of 80% of the standing stocks of algae (X2), heterotrophs (X7), DOC in water (X6) and POC in water (X9) and the export of this material from the marsh. In addition, the 'catastrophe' was allowed to scour the marsh and creek bottoms, such that all of the particulate carbon of the sediment (X13) above 17,637 g C m^{-2}, the measured average standing stock was removed and exported.

The four named components of the marsh model were the only ones affected by the changes in the model structure and parameters. All other components and fluxes were indistinguishable from those predicted by the nominal version of the model. The results of these experiments with the model are being published in detail elsewhere (Imberger *et al.*, in preparation). Here I merely summarize the most important results.

The seasonal standing crop changes and annual fluxes predicted by the nominal model agreed closely with the *independently measured* standing crops and fluxes in the marsh itself (Wiegert & Wetzel 1978), thus the degree of damage done to the validity or predictive power of the model by the changes described above can be assessed from a comparison of the nominal model run and the 'catastrophic' model run for each of the four compartments.

Apart from the transient changes caused by the catastrophe itself, changes so short-lived they would be unlikely to be picked up in the course field sampling anyway, the predictions of the 'catastrophe' model differed from the nominal model in only two major respects.

First, the catastrophic flushing regime predicts an algal standing crop and productivity in winter and early spring as high as that of summer and autumn, a prediction greatly at variance with field experience. The reason is, of course, the great reduction in daily loss (export) of algae in the new version of the model coupled with the ability of the algae to recover quickly from a single catastrophic removal of 80% of its standing crop. Further studies of the photosynthetic rates of algae in winter and spring are needed, together with an investigation of the degree of zooplankton grazing.

Second, the average annual standing stock of particulate organic C in the water predicted by the catastrophic export model (annual average of 25 g C m^{-2}) is several times higher than the mean of 5 g C m^{-2} reported by Odum and de la Cruz (1967) or the 10 g C m^{-2} found by E. Haines in the upper Duplin river in July 1977 (Imberger *et al.*, in preparation). Again, as with the algae, the large decrease in the daily export of POC from the marsh to the estuarine water causes an increase in the standing stock. Further studies of sedimentation rate and the rates of POC production are certainly indicated.

Table 29.3. *Summary of data computed from Odum and de la Cruz (1967) showing daily average export of carbon (per m^{-2} of watershed) moving in and out of Study Creek, Sapelo Island, Georgia. Net annual export was g C m^{-2}.*

	g C m^{-2} day^{-1}	
Season	Flood	Ebb
Summer	0·285	0·676
Fall	0·334	0·734
Winter	0·186	0·342
Mean	0·268	0·584

Annual export = (0·584 − 0·268) × 365
= 115 g C m^{-2} day^{-1}

Interestingly, however, the daily tidal export of POC (as opposed to the 4 times yearly catastrophic removal) calculated from the 'catastrophic' model, at 91 g C m^{-2} yr^{-1}, is close to the 115 g C m^{-2} yr^{-1} (Table 29.3) measured by Odum and de la Cruz (1967). The exports in each of the three categories predicted by the new model are (in g C m^{-2} yr^{-1}) algae 4, DOC 247 and POC 778 for a total export from the marsh ecosystem of 1,029 g C m^{-2} yr^{-1}. The total carbon input from all sources is 1,573 + 180 or 1,753 g C m^{-2} yr^{-1} and the estimate of total degradation on the marsh (Table 29.2) is 746 g C m^{-2} yr^{-1} for a net surplus of 1,007. Thus, although the revised (catastrophic) model fails to represent adequately the seasonal behaviour of algae and POC in water, its simulated fluxes, particularly exports, seem to be in reasonable balance with imports.

Marsh management

The results of the dynamics of the Sapelo Island marshes suggests a system whose net primary productivity is far greater than either the utilization by heterotrophs in the system or the amount of organic deposition. The discussion in this paper has centred on the tidally mediated transport of this excess organic carbon out of the marsh into the estuarine waters. There remains the possibility that a significant portion of the excess carbon is ingested by populations of the young of commercially important estuarine and offshore fish and shellfish and thus the carbon moves out of the estuary via migratory movements of these animals. Whatever the mechanism of movement, however, these coastal Georgia marshes are clearly important producers of energy rich, organic carbon compounds. Just as clearly, the current studies do not support the hypothesis of the convective mixing of estuarine with marsh water on *each* tide as the major mechanism of export. Indeed, the marsh water (that which is pushed onto the marsh with each high tide) appears to be a very distinct mass of water which gives up or exchanges its organic content with the estuarine water only very slowly except under the pressure of catastrophic events.

Any kind of 'management' of the marshes for the benefit of man would seem to be necessarily directed at leaving the marsh as it is, so as not to damage its productive capacity and concentrate on improving the utilization of the exported organic matter by commercially valuable species of marine life. In this context, however, must be mentioned the proposal to simultaneously enrich the marsh (increase its productivity) and solve a perennial pollution problem by dumping sewage sludge in marshes and thus using the marsh as a tertiary treatment plant. Apart from the potential problems of conservative toxins, such as heavy metals, which may be present in waste sludge and subsequently appear in food chains leading to humans, the data discussed in this paper suggest a very cautious approach to such proposals for two reasons. One is the very slow exchange of water masses, particularly in the upper reaches of the tidal rivers leading to the estuary. Second is the very high natural productivity of the Georgia marshes. The theory is, of course, always that more is better, but in the case of a detritus-based ecosystem, where a large oxygen demand is generated in the water, an extra input to the tidal water, coupled with a slow exchange with offshore areas, could well cause oxygen levels which are lower than could be tolerated by the populations of the very commercial organisms that the procedure is intended to benefit. Indeed, the productivity of the *Spartina* marshes is already so high, even in the short *Spartina*, that fertilization with nitrogenous waste may cause the vegetation to be limited by other factors (light, interstitial salinity, etc.). Some fertilization studies conducted by Chalmers (1977) and Sherr (1977), summarized by Chalmers *et al.* (1976), have shown that less than 5% of the N added as sewage sludge was actually incorporated by the *Spartina* plants, and a large proportion (44%) of the sludge was moved off the study plots (into the marsh water?). The remainder (51%) remained as organic N in sludge on the plot. These results and their bearing on sludge amendments to marshes are discussed in detail by Chalmers (1977). They certainly suggest the marsh to be less than the ideal treatment plant. Some of the added sludge would appear to be washed into the marsh and tidal creek water, there to reside for some time as nutrients and organic matter in an already eutrophic system. Some is retained as increased productivity by *Spartina*, but this material ultimately is exported, and some remains in the high marsh sediment, perhaps resulting in changes in the vegetation of the marsh.

Spartina salt marshes, even along the same coastline, are quite different in their productivity, nutrient cycling and general dynamic behaviour, as a quick perusal of this paper and those of Woodwell *et al.* and Valiela and Teal (this volume) show. Any schemes for management (or other disturbance) of these aesthetically beautiful and commercially valuable ecosystems should be based on a sound ecological knowledge of the particular system involved.

ACKNOWLEDGMENTS

An overview report of this kind owes its entire existence to the data collected by a large number of research workers. These are, of course, acknowledged wherever their published or unpublished data have been employed. I would, in addition, like to thank Dr J. Imberger and Dr W. Wiebe for many helpful comments and suggestions. The studies reported here have all been supported by grant DES 72-10605-A02 from the National Science Foundation. Travel support for presentation of the paper was provided by the British Ecological Society and by grant DES 72-01605-A02. Space and facilities for completion of the manuscript were provided by Imperial College Field Station of the University of London through the courtesy of Professor T.R.E. Southwood.

REFERENCES

Bahr L.M. Jr. (1976) Energetic aspects of the intertidal system reef community at Sapelo Island, Georgia (U.S.A.). *Ecology* **57**, 121–31.

Burkholder P.R. & Bornside G.H. (1975) Decomposition of marsh grass by aerobic marine bacteria. *Bull. Torrey Bot. Club* **84**, 366–83.

Chalmers A.G. (1977) *Pools of nitrogen in a Georgia salt marsh.* Ph.D. thesis, University of Georgia, Athens.

Chalmers A.G., Haines E.B. & Sherr B.F. (1976) Capacity of a *Spartina* salt marsh to assimilate nitrogen from secondarily treated sewage. *Tech. Completion Report, USDI/OWRT Project* no. A-057-GA.

Chapman V.J. (1977) Introduction. *Ecosystem of the World: Vol. 1, Wet Coastal Ecosystems* (Ed. by V.J. Chapman), pp. 1–29. Elsevier, New York.

Frey R.W. & Basan P.B. (1978) North American Coastal Salt Marshes. *Coastal Sedimentary Environments* (Ed. by R.A. Davis), pp. 101–69. Springer-Verlag, Berlin.

Gallagher J.L., Reimold R.J., Linthurst R.A. & Pfeiffer W.J. (1979) Aerial production mortality and mineral accumulation dynamics in *Spartina alterniflora* and *Juncus roemerianus* in a Georgia salt marsh. *Ecology* (in press).

Haines E.B. (1976) Stable carbon isotope ratios in the biota, soils and tidal water of a Georgia salt marsh. *Estuarine Coast. Mar. Sci.* **4**, 609–19.

Johnson A.S., Hellestad H.O., Shanholtzer S.F. & Shanholtzer G.F. (1974) An ecological survey of the coastal region of Georgia. *Nat. Park Ser. Sci. Monog. Ser. No. 3* **15**, 13–16.

King G.M. & Wiebe W.J. (1978) Methanogenesis in a Georgia salt marsh and some factors controlling its production. *Geochim. Cosmochim Acta.* **42**, 343–48.

Kuenzler E.J. (1961) Structure and energy flow of a mussel population in a Georgia salt marsh. *Limnol. and Oceanogr.* **60**, 191–204.

Odum E.P. (1974) Halophytes, energetics and ecosystems. *Ecology of Halophytes* (Ed. by R.J. Reimold & W.H. Queen), pp. 599–602. Academic Press, New York and London.

Odum E.P. & de la Cruz A.A. (1967) Particulate organic detritus in a Georgia salt marsh–estuarine ecosystem. *Estuaries* (Ed. by G.H. Lautt), pp. 383–88. Am. Assoc. Adv. Sci. Publ. 83.

Pace M.L. (1977) *The effect of macroconsumer grazing on the benthic microbial community of a salt marsh mudflat.* M.S. thesis, University of Georgia, Athens, Georgia.

Pomeroy L.R. (1959) Algal productivity in salt marshes of Georgia. *Limnol. and Oceanogr.* **4**, 386–97.

Pomeroy L.R., Johannes R.E., Odum E.P. & Roffman B. (1969) The phosphorus and zinc cycles and productivity of a salt marsh. *Proc. 2nd Symp. Radioecol. U.S.At. Energy Comm. TID4500* (Ed. by D.J. Nelson & F.C. Evans), pp. 412–19.

Ragotzkie R.A. (1959) Plankton productivity in estuarine waters of Georgia. *Institute of Marine Sci.* **6**, 146–58.

Ragotzkie R.A. & Pomeroy L.R. (1957) Life history of a dinoflagellate bloom. *Limnol. and Oceanogr.* **2**, 62–69.

Riedburg C. (1976) *A dye study of interstitial water flow in tidal marsh sediments.* M.S. thesis, University of Georgia, Athens, Georgia.

Reimold R.J. (1977) Mangals and salt marshes of eastern United States. *Wet Coastal Ecosystems* (Ed. by V.J. Chapman), pp. 157–66. Elsevier, New York.

Sherr B.F. (1977) *The ecology of denitrifying bacteria in salt marsh soils—an experimental approach.* Ph.D. dissertation, University of Georgia, Athens, Georgia.

Tanner W.F. (1960) Florida coastal classification. *Trans. Gulf Coast Assoc. Geol. Soc.* **10**, 259–66.

Teal J.M. (1962) Energy flow in the salt marsh ecosystem of Georgia. *Ecol. Monog.* **43**, 614–24.

Teal J.M. & Kanwisher J. (1961) Gas exchange in a Georgia salt marsh. *Limnol. and Oceanogr.* **6**, 388–99.

Teal J.M. & Kanwisher J.W. (1966) Gas transport in the marsh grass, *Spartina alterniflora. J. exp. Bot.* **17**, 355–61.

Wetzel R.L. (1977) Carbon resources of a benthic salt marsh invertebrate *Nassaruis absoletus* Say (Mollusca: Nassariidae). *Estuarine Processes Vol. II. Circulation Sediments and Transfer of Material in the Estuary* (Ed. by M.L. Wiley), pp. 293–308. Academic Press, New York.

Wetzel R.G., Christian R.R., Gallagher J.L., Hall J.R., Jones R.D.H. & Wetzel R.L. (1975) A preliminary ecosystem model of coastal Georgia *Spartina* marsh. *Estuarine Research* Vol. 1 (Ed. by E. Cronin), pp. 583–601. Academic Press, New York.

Wiegert R.J. & Wetzel R.L. (1978) Simulation experiments with a fourteen-compartment model of a *Spartina* salt marsh. *Marsh–Estuarine Systems Simulation* (Ed. by R. Dame), (in press). University of S. Carolina Press, Columbia.

30. THE FLAX POND ECOSYSTEM STUDY: THE ANNUAL METABOLISM AND NUTRIENT BUDGETS OF A SALT MARSH

G.M. WOODWELL,[1] R.A. HOUGHTON,[1] C.A.S. HALL,[1]
D.E. WHITNEY,[2] R.A. MOLL[3] AND D.W. JUERS[4]

[1] *The Ecosystems Center, Marine Biological Laboratory, Woods Hole, Mass. U.S.A.*

[2] *University of Georgia, Marine Laboratory, Sapelo Island, Georgia, U.S.A.*

[3] *Great Lakes Research Division, University of Michigan, Ann Arbor, Michigan, U.S.A.*

[4] *Marine Biological Laboratory, Woods Hole, Mass. U.S.A.*

SUMMARY

Flax Pond, a 57 ha tidal marsh on the north shore of Long Island, New York, was used to examine the function of a marsh in interaction with other units of the biosphere. Inputs of carbon and other nutrient elements were negligible from the hinterland, but large exchanges occurred between the marsh and the atmosphere, marsh sediments, and Long Island Sound. About 130 g C m^{-2} yr^{-1} was exported as large fragments of *Spartina*, while 60 g C m^{-2} yr^{-1} was imported as fine particulate organic carbon. The net flow of dissolved organic carbon was into the marsh in winter and out in summer, yielding a net loss of dissolved organic carbon of about 10 g C m^{-2} yr^{-1}. The annual exchange of carbon between Flax Pond and the coastal waters was a net loss of about 80 g C m^{-2}.

Total net primary production was 535 g C m^{-2} yr^{-1}. About 400 g C m^{-2} yr^{-1} was from *Spartina alterniflora*, the dominant plant. Almost half of this production (255 g C m^{-2} yr^{-1}) was consumed by heterotrophs in the marsh, and more than a third (200 g C m^{-2} yr^{-1}) was buried in the sediments.

The marsh as a unit was slightly autotrophic over the year as a whole in that the net flux of organic carbon to the coastal waters was 15% of marsh production. During the summer, however, the marsh was a reducing environment, clearly heterotrophic. It consumed fine particulate organic carbon and chlorophyll, removed oxygen from tidal waters, and released ammonium ions and dissolved organic carbon. The phosphorus budget over the course of the year was approximately in balance, while nitrogen fixation on the marsh surface contributed toward making the marsh a small net source of inorganic nitrogen to coastal waters.

RÉSUMÉ

Nous avons utilisé Flax Pond, un marais soumis aux marées, de 57 ha, sur le rivage nord de Long Island, New York, pour examiner la fonction d'un marais en action réciproque avec d'autres éléments de la biosphère. Les flux entrants de carbone et d'autres éléments nutritifs des hinterlands étaient négligeables, mais il y arrivaient de grandes échanges entre le marais et l'atmosphère, les sédiments du marais, et Long Island Sound. Environ 130 g C m^{-2} yr^{-1} avaient été exporté sous la forme de grandes pièces de *Spartina*, tandis que 60 g C m^{-2} yr^{-1} avaient été importé comme de fines particules de carbone organique. Le flux net de carbone organique dissous entrait le marais en hivers et sortait en été, rendant une perte nette de carbone organique dissous d'environ 10 g C m^{-2} yr^{-1}. L'échange annuelle de carbone entre Flax Pond et les eaux côtières était une perte nette de 10 g C m^{-2} yr^{-1} environ.

Le net de la production primaire entier était 535 g C m^{-2} yr^{-1}. Environ 400 g C m^{-2} yr^{-1} étaient de *Spartina alterniflora*, la plante dominante. Presqu'une moitié de cette production (255 g C m^{-2} yr^{-1}) a été consummé par des hétérotrophes dans le marais, et plus d'un tiers (200 g C m^{-2} yr^{-1}) a été stocké dans les sédiments.

Le marais entier, pendant l'année dernière, a été un peu autotrophique, où le flux net de carbone organique aux eaux côtières était 15% de la production du marais. Pendant l'été, cependant, le marais a été un environnement réduisant, évidemment hétérotrophique. Il a été consommateur des fines particules de carbone organique et de chlorophylle, a relevé l'oxygène des eaux à marée, et a dégagé des ions d'ammoniaque et le carbone organique dissous. Le bilan annuel du phosphore est approximativement équilibré, bienque la fixation d'azote en surface du marais assiste le marais à se faire une petite source nette d'azote inorganique aux eaux côtières.

INTRODUCTION

Objectives in analysis of the biosphere as a set of interacting ecosystems

Remarkable advances have been made in the last decade in interpreting the function of the earth's biota in maintenance of the biosphere. Many of these advances have been aided substantially by interpretation of the biosphere as a set of interacting segments of the earth's surface that have been called, somewhat loosely, 'ecosystems'. The ecosystems include forests, lakes, streams, estuaries, oceans and the concept is frequently extended now to include agricultural landscapes and other man-dominated areas. The functions and interactions of such ecosystems are predictable, can be measured, and in some instances, controlled. The advantage of the approach which involves a study of basic units of environment is that it enables the results of the specific technical analyses that are popular in science to be interpreted with precision in a comprehensive context. Most science has 'ecological implications'; the difficulty in interpretation and application in human affairs has been the lack of a coherent theory of environment that allows effective sorting among the diverse partial analyses that are currently abundant. While interpretation of the biosphere as a set of interacting ecosystems may seem oversimplified, it focuses attention on the fundamental question of what ecosystems do as units of the earth's surface.

The largest advances of recent years have been made in instances where the question of the function of ecosystems as units has been addressed directly. A clear answer requires an especially appropriate site where measurements of major exchanges can be made conveniently and accurately. On land a single drainage basin underlain by impermeable rock offers unusual opportunity for studies of nutrient cycling in relationship to the hydrological cycle (Likens *et al.* 1977). Similar opportunities are offered by level land with well-drained soils (Woodwell *et al.* 1974, 1976, 1977). Streams and lakes with their drainage basins offer parallel opportunities; so do estuaries and marshes, although the problems here may be more complicated.

Interest in the function of estuaries and marshes has been especially intense because of the importance of these systems in support of the biotic resources of coastal areas. Stroud (1971) has estimated that an acre of salt marsh contributes vitally to the production of as much as 500 lb of fish in coastal fisheries annually. What other functions do these intensively active centres of life perform as basic units of the biosphere? This question has been addressed in several ways and in several places over the past years. Evidence is accumulating that the function of such communities is important, not only in the maintenance of coastal fisheries, but in the control of the quality of coastal waters.

The exchanges between estuaries or salt marshes and other segments of the biosphere are indicated diagrammatically in Fig. 30.1. All of these exchanges can be measured, some by several different techniques. The techniques are complex, expensive and demanding of skilled technicians but the data are necessary to interpret both the internal function of ecosystems and the interactions among ecosystems.

Such an appraisal is most appropriately addressed first to the budgets for energy, organic matter or carbon, all of which are interchangeable. Is the system a net source or sink? How is energy used within the system? The answers determine whether the ecosystem is heterotrophic or autotrophic and define its interactions with the rest of the biosphere.

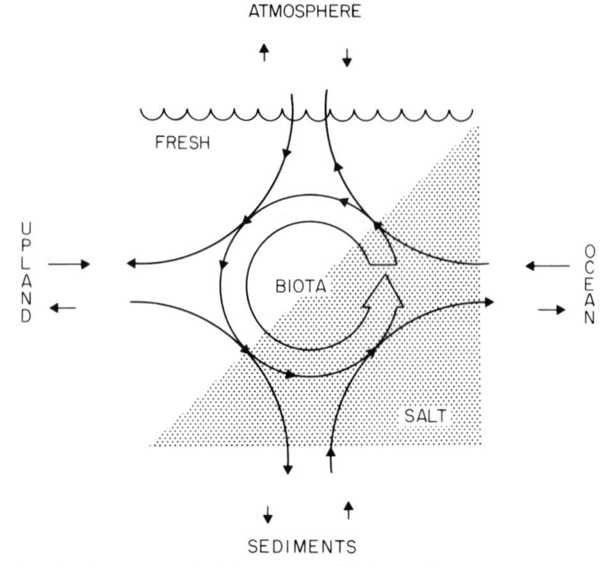

FIG. 30.1. A diagrammatic illustration of the exchanges between an estuary and its adjacent ecosystems, the coastal ocean, the sediments, the uplands, and the atmosphere.

In making such an appraisal we have the advantage of some simple models that help set forth the major elements in the exchanges. The models are the production equations, first applied in this context by Woodwell and Whittaker (1968) in analysis of a forest.

$$NPP = GP - R_A \tag{1}$$

$$NEP = GP - (R_A + R_H) \pm NP_x \tag{2}$$

The production equations provide a simplified framework for analysing the production, consumption and exchanges of energy, organic matter, or carbon in ecosystems. The first equation applies only to the community of green plants. The excess of gross production or total photosynthesis (GP) over the respiration required to support the plant community (R_A) is net primary production (NPP) and is available to consumers at higher trophic levels and to decomposers. The second equation applies to the ecosystem as a unit, including not only the green plants but also the heterotrophs, whose respiration is indicated by R_H. The difference between gross production and total respiration of the ecosystem (R_E), the sum of autotrophic (R_A) and heterotrophic respiration (R_H), is net ecosystem production (NEP), which may be positive or negative depending on whether energy is being stored or released by the entire system. In an ecosystem that has the potential for a significant exchange of net production with others, another term may be appropriate, NP_x, to account for this exchange. The information in these simple models is basic to analysis of the function of the ecosystem, its nutrient budgets, and to policies in management.

We sought an opportunity to apply this type of analysis to an estuarine marsh and selected Flax Pond on the north shore of Long Island, New York, in the town of Old Field about 50 km east of New York City (Woodwell & Pecan 1973).

FLAX POND: AN ESTUARINE MARSH

Flax Pond is a New England salt marsh (Ayers 1959), formed within a coastal cove cut off from Long Island Sound by a sand bar that is now penetrated by a single tidal channel. It is young. The oldest peat has been dated by ^{14}C as about 1,700 years. The marsh is tidal with a range of 1·7 metres and a twice daily exchange of about 80% of the water held at mean high tide. Salinity varies seasonally between 24 and 28 parts per thousand. Freshwater dilution is limited to surface runoff and is small. A two-year series of measurements of chlorinity showed that incoming tides had 0·018‰ ± 0·028‰ higher chlorinity than the water of the ebb tide. The relationship showed a tendency towards dilution but the difference was not statistically significant.

The entire marsh, including the open water, covers 57 hectares and drains a total of 170 hectares. The area of open water or bare mud is 27 hectares. The emergent vegetation is *Spartina alterniflora* over 26 ha and *S.patens—Distichlis spicata* over about 4 ha.

In any effort to determine the details of function of a unit of the earth's surface it is most important that the site selected for the study offer a sufficiently simple sampling problem that the answers to the basic questions can be obtained. Flax Pond was selected as such a site. The exchanges with other ecosystems (Fig. 30.1) are with the uplands, the coastal ocean, the atmosphere and the sediments. The exchanges with the uplands are limited in that the drainage basin is small and the flow of fresh water into the marsh is small. The exchanges with the coastal oceans are entirely through a single channel where the flow is sufficiently turbulent to mix the water thoroughly. The turbulence assures that the sampling of waters is simplified (Woodwell *et al.* 1977). There is the further advantage in use of Flax Pond in that the marsh is state-owned and will probably be preserved intact.

The immediate objective was an evaluation of the production equations with respect to the carbon budget of the marsh. Certain terms of the production equations can be measured directly by at least one technique available now. They are NPP, NEP, $(R_A + R_H) = R_E$ and NP_x. The other terms GP, R_A and R_H, must be determined indirectly by subtraction. An evaluation of the production equations sets the context within which the estuary as a unit operates, either as a heterotrophic or an autotrophic system.

In this paper we offer a summary and discussion of the research in Flax

Pond with special emphasis on the marsh as an ecosystem in interaction with other segments of the biosphere. Additional details of methods used in appraising various aspects of function of the estuary are offered in a series of special papers published elsewhere and cited in the text.

AN EVALUATION OF THE PRODUCTION EQUATIONS FOR FLAX POND

Net production exchanges (NP_x)

The exchanges between Flax Pond and the coastal waters are large and important. The tidal waters that flush the marsh twice daily have the potential for transferring a large amount of fixed carbon to or from the coastal waters. An average concentration difference of 1 mg organic matter with a continuous tidal range between flood and ebb tides of 1 m could remove as much C as the net primary production of many plant communities. To appraise this potential exchange with the coastal waters a sampling plan was devised in which the basic sampling unit was the entire tidal cycle. The tides were selected sequentially on the basis of the time of high tide: 0000, 0400, 0800, 1200, 1600 and 2000 hrs. One tidal cycle was sampled approximately weekly over about 18 months. Four water samples were taken in the well-mixed water of the channel during both the flood and ebb tides. The net fluxes of carbon as total CO_2, dissolved organic matter, and particulate organic matter were determined (Woodwell *et al.* 1977). These fluxes are shown in Fig. 30.2. The net flux has been summarized in Table 30.1 drawn from Woodwell *et al.* (1977).

Table 30.1. *Exchanges of carbon between Flax Pond and Long Island Sound. Negative values are export from marsh. Values in parentheses are standard errors.*

Form	Flux per tide (kg)	Annual flux: Whole marsh (kg × 10³)	Annual flux: Per m² (g)
Total CO_2	−1·5 (34·5)	−1·1 (24·1)	−1·9 (42·0)
DOC			
1/4–30/9	−24·8 (7·6)		
1/10–31/3	11·1 (12·6)		
Full year	−6·8 (7·7)	−4·8 (5·4)	−8·4 (9·4)
POC	50·3 (11·4)	35·2 (7·9)	61·3 (13·8)
Total organic C	43·5 (13·8)	30·4 (9·6)	53·0 (16·7)

DOC = dissolved organic carbon.
POC = particulate organic carbon.

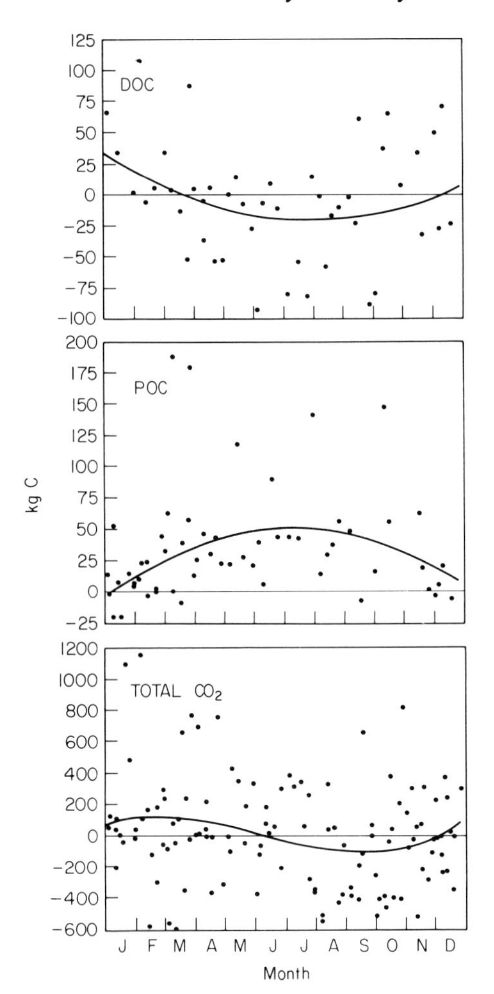

FIG. 30.2. Net tidal exchange of carbon between Flax Pond and Long Island Sound over the course of the year. Negative values are losses from the marsh; positive values, net gains. Each point is the difference between transport on the flood and transport on the ebb tide. Curves are least-squares polynomial regressions of net carbon flux on day of year. DOC, dissolved organic carbon; POC, particulate organic carbon.

It shows that although the annual total flux is approximately in balance, there are seasonal differences in the net flows. For instance, there appears to be a net loss of dissolved organic matter from the marsh in late summer and a net influx of particulate matter throughout the year. The net loss of dissolved carbon in a year was about $-8{\cdot}4$ g m^{-2}; the net input of particulate carbon was $61{\cdot}3$ g m^{-2}. The net exchange as fine particulate carbon, dissolved and total CO_2 in the carbonate-bicarbonate system was an import of about 50 g m^{-2} yr^{-1}.

These detailed studies were supplemented by less detailed analyses of the movement of large particulate organic matter in the water through the channel. The large particulate matter was in the form of algal fronds or stipes, and mats of *Spartina* stems. A special, floating net was used weekly to collect such material. These data showed no net movement of carbon in this form (Woodwell *et al.* 1977). Houghton (in preparation), however, in a separate analysis has shown by consideration of the accumulation and decay of *Spartina* stems that there is probably a net loss of *Spartina* stems from the marsh at a rate that is between 85–170 g C m^{-2} yr^{-1}. Houghton measured with litter-bag experiments the rate of decay of the *Spartina* stems that accumulate in windrows. The combined biomass of live and dead grasses, both standing on the marsh and accumulated in windrows, was measured every two months. Once the growing season was over, the total biomass decreased continuously until the following summer. But the decrease in the biomass was greater than could be explained by decay. His conclusion was that there is an irregular export of this type of organic matter from the marsh on spring and storm tides. The conclusion was strengthened by the observation that *Spartina* debris is washed up regularly along the beaches of Long Island Sound.

The exchanges of carbon between Flax Pond and Long Island Sound appear to be a more or less regular inward flux of small particulate carbon throughout the year, a late summer loss of dissolved carbon, a variable but balanced flux of total CO_2, and an irregular loss of large particulate carbon. The net exchange for all forms is a loss of 35–120 g C m^{-2} yr^{-1}, with a median value of about 80 g C m^{-2} yr^{-1}.

Net primary production (*NPP*)

The techniques for measurement are obviously complicated. The largest contribution to net primary production of the marsh is that of the *Spartina* mat. There are three approaches to measurement of the net production of the *Spartina* (Woodwell & Whittaker 1968): the harvest technique, measurement of carbon dioxide exchange rates using small chambers and the CO_2-flux technique. All of these techniques were used in this work. Above-ground production was measured with a combination of inventory and harvest techniques (Houghton, in preparation). Fifteen stations were sampled in duplicate each month for 22 months. Below-ground production was estimated from the *in situ* decay rates of root and rhizome material and by measuring the growth rates of below-ground tissue into barren sediments (Houghton, in preparation). The results indicated that *Spartina* has a net primary production of 780 g C m^{-2} yr^{-1}, including both shoots and roots. The *Spartina* mat covers 51% of the total marsh area and the net primary production for the two species of

Table 30.2. *Net primary production of the plant populations in Flax Pond (averaged over the entire marsh area).*

	g dry wt m^{-2} y^{-1}	g C m^{-2} y^{-1}
Grasses		
Above-ground	750	292
Below-ground	300	108
Total	1,050	400
Fucoid algae	150	75
Epibenthic algae	60	30
Epiphytic algae	40	20
Phytoplankton	23·4	11·7
Total	1,325	535

A factor of 0·5 was used to convert dry weights to carbon weights except for the shoots and leaves (0·39) and the roots and rhizomes (0·36) of *Spartina*, for which the conversion factors were determined by loss of weight on ignition.

Spartina averaged over the entire marsh was therefore about 400 g C m^{-2} yr^{-1} (Table 30.2). C^{14} analysis of phytoplankton production in the waters of the marsh indicated a yearly net primary production of about 23 g of dry organic matter per square metre over the entire marsh according to an elaborate series of measurements taken over 17 months by Moll (1977). From monthly sampling of macroscopic algal biomass (Brinkhuis 1976; Houghton, in preparation), we have estimated that the annual productivity of fucoids is about 75 g C m^{-2}. There may be additional contributions to the net primary production of the marsh that have not been measured in this series of studies. These include the possibility that benthic algae in the intertidal zone fix appreciable amounts of carbon when exposed. Gallagher and Daiber (1974) and Van Raalte *et al.* (1976) found epibenthic algal production to be one quarter to one third of above-ground grass production. In both studies algal production was less where grass biomass was greater. The grass biomass in Flax Pond is greater than in many other marshes (average end-of-season live above-ground biomass is 900–1,000 g dry wt m^{-2} for the vegetated surface) (Houghton, in preparation), and epibenthic production would be expected to be correspondingly lower there. On the other hand, the large biomass of fucoid algae has not been reported in other studies. Our best estimate of total algal production, including fucoid, epibenthic and epiphytic algae, is one third of the above-ground grass production, or 250 g dry wt m^{-2} yr^{-1}. Annual production by fucoid algae was assumed to be its maximum biomass, 150 g dry wt m^{-2} yr^{-1}. Epiphytic production is greatest adjacent to creeks (Day *et al.* 1973) and may be 5% of the above-ground grass production there (Seneca *et al.* 1976). This percentage gives a value of about 40 g dry wt m^{-2} yr^{-1}

for epiphytic net primary production in Flax Pond. Epibenthic production is estimated to be 60 g dry wt m^{-2} yr^{-1}. It seems unlikely to us that these quantities could be large in proportion to the net production of the *Spartina* mat. Data available at present indicate that the net primary production of the marsh, including all of the sources measured, is probably about 535 g C m^{-2} yr^{-1} or 1,325 g dry organic matter.

Net ecosystem production (*NEP*)

The net increase or loss of organic matter by the ecosystem is net ecosystem production. In Flax Pond the pools of carbon that might change in size are the *Spartina* mat, the animal communities and the sediments.

The slow rise of sea level relative to the land suggests that there is a gradual accumulation of sediments in the anaerobic zones of the marsh at a rate that maintains the level of the marsh against the sea level rise. This accumulation of carbon in sediments is the major contribution to *NEP*, but the problem is more complicated.

There are indications from historical records and from the configuration of the Flax Pond basin that Flax Pond is a young marsh; its most recent connection with the coastal waters has existed for a century or so (Woodwell & Pecan 1973). About 80% of the vegetation is *Spartina alterniflora*, and about 75% of that is the tall form. The large proportion of tall *S.alterniflora* in comparison with other marshes indicates that a large fraction of the marsh is intertidal, and hence immature. The marsh sediment is accumulating (Armentano & Woodwell 1975), possibly faster than the measured rise in sea level, and the marsh may be increasing in biomass as well. Measurements of biomass during three years did not show a significant increase but variability in such measurements is high. If the biomass were increasing it would probably be in the horizontal extension of vegetated surface rather than in the vertical accumulation of root or shoot biomass. Isolated patches of mud have been observed to become vegetated over the years, but in other patches vegetation is killed by ice abrasion or the accumulation of drifts of thatch. An estimate of increasing biomass would require a long-term study, probably from aerial photographs. Such a study is not available.

The annual deposition of carbon in sediments is probably the major contribution to *NEP* and has been measured. Armentano and Woodwell (1975) measured the age of the sediment in two locations with ^{210}Pb and determined the accretion rate over recent years to be 0·47 to 0·63 cm yr^{-1}. The authors calculated that between 146 and 196 g C m^{-2} were being deposited annually. For the evaluation of the equations we assume that the net ecosystem production is probably between 150 and 250 g C m^{-2} yr^{-1} with an average of 200 g C m^{-2} yr^{-1}.

Total respiration of the ecosystem $(R_A + R_H) = R_E$

The possibility of devising techniques for separating autotrophic from heterotrophic respiration in the field seems very small indeed, but the possibility for measuring the sum of both (or total respiration of an ecosystem) is realistic enough. The challenge in an estuary such as Flax Pond is to measure both the exchanges with the atmosphere and the tidal exchanges with the coastal oceans.

Atmospheric exchanges were measured by the CO_2 flux technique using a portable meteorological tower and automatic recording equipment to obtain a continuous record of the CO_2 and wind gradients in air over the marsh. An attempt was made initially to use the temperature-inversion modification of the flux technique (Woodwell & Dykman 1966) but inversions in the marsh proved rare. Based on 84 days of accumulated data the monthly, seasonal and annual CO_2 fluxes were calculated instead using standard equations of the aerodynamic method (Houghton & Woodwell, in preparation).

The night-time flux of CO_2 to the atmosphere was about 500 g C m^{-2} yr^{-1} (Houghton & Woodwell, in preparation). If the same rate of respiration were assumed to occur during the day, then 1,150 g C m^{-2} yr^{-1} were released from Flax Pond to the atmosphere. This value is more than twice the night-time value because the higher rates of respiration occurred in summer when the ratio of hours of darkness to hours of light was less. Because dark respiration is temperature dependent, the higher temperatures during daylight would increase the estimate of total exchange further. The recognition of photorespiration recently (Zelitch 1971) would add a further upward adjustment of unknown magnitude.

There was in addition, of course, an exchange of CO_2 from respiration in the tidal waters exchanged with Long Island Sound. This exchange was measured by the systematic sampling programme carried out in the channel to Long Island Sound. In general, Flax Pond appeared to release CO_2 to Long Island Sound during the night and to absorb it during the day.

The exchanges of dissolved oxygen offered a second method of analysis. Those tidal cycles with hours of darkness brought oxygen into the Pond; the daylight tides had a net export from the Pond. The average annual night-time sink for oxygen was 400 g oxygen m^{-2} yr^{-1}. This amount of oxygen is equivalent to 150 g C m^{-2} yr^{-1} if one mole of oxygen is assumed to produce one mole of CO_2 in respiration. The respiratory quotient (*RQ*) for aerobic respiration is 0·85, however, and an unknown fraction of the respiration in Flax Pond is anaerobic, giving most likely a still lower quotient. We assumed that a coefficient of 0·75 was appropriate and calculated the dark consumption of oxygen to be equivalent to 112·5 g C m^{-2} yr^{-1}. The annual respiration for the aquatic environment of Flax Pond was, therefore, about 225 g C m^{-2} yr^{-1}. This value, estimated from the net loss of dissolved oxygen during night tides, is an underestimate because it does not include the diffusion of atmospheric oxygen into underesaturated waters. Unfortunately, a better estimate of respiration in the aquatic phase of Flax Pond is not available. The net annual tidal flux of oxygen into the Pond (198 g O_2 m^{-2} yr^{-1}) was not balanced by an equivalent efflux of carbon dioxide in the water. Some of the oxygen left the Pond in other oxidized forms, such as nitrate and sulphate, and some was exchanged with the atmosphere as either CO_2 or O_2.

The total respiration of Flax Pond is the sum of both aquatic (225 g C m^{-2} yr^{-1}) and aerial (1,200 g C m^{-2} yr^{-1}) exchanges, or about 1,425 g C m^{-2} yr^{-1}.

Gross production and the separation of autotrophic and heterotrophic respiration

This series of appraisals of the basic segments of the estuarine carbon budget offers the opportunity to calculate by difference the remaining terms of the production equations. From Equation (2) we can determine gross production inasmuch as we have measured the other terms. Inserting the gross production in Equation (1) gives R_A and allows calculation of R_H as well. The production equations, evaluated, appear as follows:

$$\begin{array}{ccccccc} NPP & = & GP & - & R_A & & \\ 535 & & 1705 & & 1170 & & \\ NEP & = & GP & - & (R_A + R_H) & - & NP_x \\ 200 & & 1705 & & 1425 & & 80 \end{array}$$

Heterotrophic respiration (R_H) was 255 g C m^{-2} yr^{-1} (1425–1170). The carbon fixed by the green plants (gross production) was divided among various marsh processes in the following fractions:

$$\left.\begin{array}{r} R_A = 69\% \\ R_H = 15\% \end{array}\right\} R_E = 84\%$$
$$NEP = 12\%$$
$$NP_x = 5\%$$

The primary producers themselves were clearly the greatest consumers of fixed carbon. They respired more than four times the amount that all of the heterotrophs together respired. Total respiration (R_E) in the marsh accounted for 84% of the carbon fixed. The total exported to the waters of another ecosystem was about 5%. The remainder, about 12%, was the accumulation of organic sediments and an increment in biomass as well, as the *Spartina* mat expanded.

Since the export of *Spartina* rafts was not measured directly in this study, the possibility exists that the *Spartina* missing from the windrows did not leave the Pond but collected either in a location or a form that was not sampled. Small fragments may have been washed into creek and pond bottoms, for example, and may have been consumed there in respiration.

CO_2 released into the water from benthic respiration was small enough in comparison to the total CO_2 of sea water to obscure net annual exchanges on the order of ± 40 g m^{-2} (Table 30.1). The results of benthic metabolism studies (Hall, in preparation) are not yet available, but the respiration of sediments in air was approximately 152 g C m^{-2} per summer (Houghton & Woodwell, in preparation), and might have been 200 g C m^{-2} yr^{-1}. Since the total heterotrophic respiration for the marsh was 255 g C m^{-2} yr^{-1} (from the production equations), the respiration of submerged bottoms (only 15% of the marsh surface) would have to be on the order of 900 g C m^{-2} yr^{-1} to respire not only the 55 g C m^{-2} yr^{-1} left for benthic metabolism but the additional 80 g C m^{-2} yr^{-1} attributed to the export of *Spartina* rafts. If the net export of 80 g C m^{-2} yr^{-1} to Long Island Sound is correct, then the rate of submerged benthic metabolism necessary to respire 55 g C m^{-2} yr^{-1} is 367 g C m^{-2} yr^{-1}. The number is approximately the same as reported for a Louisiana estuary, 343 g C m^{-2} yr^{-1} (Day *et al.* 1973), but the uncertainties of these rates are great enough that no clear resolution of the question in Flax Pond is possible.

The amount of carbon accumulating in Flax Pond each year, net ecosystem production, should equal the net exchanges with the atmosphere and Long Island Sound. That is,

$$NEP = \text{Net Exchange}_{\text{atmosphere}} \pm \text{Net Exchange}_{\text{coastal waters}}$$
$$200 \cong 302 - 80$$

In fact net ecosystem production approximates to the sum of the inputs and outputs within the errors of this analysis.

Of the 535 g C m^{-2} yr^{-1} of net primary production in Flax Pond, approximately 255 g (48%) is respired heterotrophically in the Pond, 200 g (37%) is buried in the sediments and 80 g (15%) is either unaccounted for within the estuary or exported to coastal waters.

One of the surprises in this study is the observation that the loss of organic matter through export was small and irregular throughout the year. The total loss was probably about 80 g C m^{-2} yr^{-1}, 5% of the gross production or 15% of the net production. Losses reported in other studies have been considerably higher. Teal (1962) assumed a loss of 45% of the net production and Day *et al.* (1973), a loss of 50% from southern marshes. Because of heterotrophic activity in the Louisiana embayment (Day *et al.* 1973) only 30% of the total net primary production was exported from the marsh-estuary system, but both studies supported the idea that salt marshes export large amounts of organic matter to coastal waters. In Flax Pond about half (49%) of the net primary production was consumed by heterotrophs. Of the remaining 51% about three-quarters (200 g C m^{-2} yr^{-1}) remained within the salt marsh as sedimented organics. The evidence is strong that the marsh is heterotrophic throughout part of the year, when there is a net influx of particulate carbon and when the estuarine system is strongly reducing. The evidence becomes clearer when the cycles of other essential elements are examined with special emphasis on exchanges with Long Island Sound.

OTHER EXCHANGES WITH COASTAL WATERS

Nitrogen

The nitrogen budget of the estuary is no less complex than the carbon budget. There is a potential series of exchanges with the atmosphere, there are potential inputs from the uplands, and there are exchanges with the coastal waters (Fig. 30.1). The forms of nitrogen range from the molecular through the various ionic forms to organic forms. Our estimates were restricted to an appraisal of the inorganic exchanges in water between Flax Pond and Long Island Sound and to a limited survey of N-fixation at different sites in the estuary.

The exchanges of N in water were measured in the channel using water samples taken at the time and in the same way that water samples were taken for the studies of carbon (see above and Woodwell *et al.* 1977). The fluxes of nitrogen, calculated from the data on concentration and water flux, for the various inorganic forms are shown in Fig. 30.3. The net exchange of total N was about a 500 kg loss from the marsh, equal to $0{\cdot}9 \pm 1{\cdot}2$ g N m^{-2} yr^{-1} (Table 30.3). This net loss was the result of a complex of interactions that are apparent from the graphs of Fig. 30.3. The dominant flux by far was the winter–summer exchange of N as the ammonium ion. During winter there was an accumulation of N as ammonium in the marsh; during summer there was a systematic loss that reached a peak in September and continued through November. There was a net flow of nitrate and nitrite into the marsh as well (Table 30.3).

A survey of N-fixation in the estuary revealed that fixation occurs most commonly at the surface of the sediments (Whitney *et al.* 1975). About 50% of the total fixation occurred in the top 4 cm of the sediment. None was observed at 20 cm or below. The highest rates on the marsh occurred in blue-green algal mats and tidal pools. The rates observed in the sediments of the *Spartina* mat, on mudflats and pannes, however, which covered about 90% of the marsh, were more than high enough to account for the tidal losses of ammonium ion.

The organic nitrogen flux was not examined in this study, but it may be as large or larger than the inorganic exchanges (see Valiela & Teal, this volume). Much of the net flux of fine particulate organic matter (average about 50 kg C $tide^{-1}$), for example, was of living phytoplankton. If we assume a C:N ratio of 5·7:1 (by weight), tidal exchanges of particulate organic nitrogen would be in the same general range as the inorganic nitrogen exchanges. It is difficult to

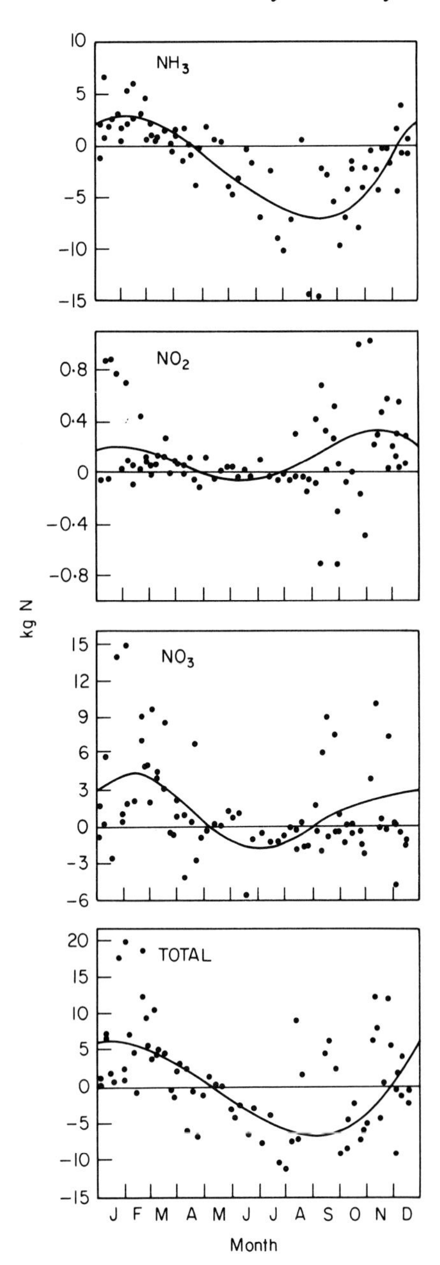

FIG. 30.3. Net tidal exchange of inorganic nitrogen between Flax Pond and Long Island Sound over the course of the year. Negative values are losses from the marsh; positive values, net gains. Each point is the difference between transport on the flood and transport on the ebb tide. Curves are least-squares polynomial regressions of net nitrogen flux on day of year.

Table 30.3. *Exchanges of inorganic nitrogen between Flax Pond and Long Island Sound. Negative values are export from marsh. Values in parentheses are standard errors.*

Form	Flux per tide (kg N)	Annual flux	
		Whole marsh (kg N)	Per m² (g N)
Ammonium			
16 April–31 Nov	−4·1 (0·78)		
1 Dec–15 April	2·5 (0·72)		
Full Year	−1·6 (0·72)	−1,140 (506)	−2·1 (0·9)
Nitrite			
16 April–31 July	0·0013 (0·0169)		
1 Aug–15 April	0·1312 (0·0459)		
Full Year	0·0906 (0·0330)	63 (23)	0·1 (0·04)
Nitrate			
1 May–31 Aug	−0·5 (0·40)		
1 Sept–30 April	1·5 (0·49)		
Full Year	0·8 (0·38)	582 (263)	1·1 (0·5)
Total inorganic			
1 May–15 Nov	−4·1 (1·1)		
16 Nov–30 April	2·9 (1·0)		
Full Year	−0·7 (0·9)	−499 (638)	−0·9 (1·2)

estimate the quantities of nitrogen associated with the transport of *Spartina* stems. The C:N ratio for decaying *Spartina* varies over time, and the windrows of thatch contained stems of different ages and different degrees of decay. Similarly, we have no good estimate of the magnitude of organic nitrogen exchanges in the dissolved form; the annual net exchange of dissolved organic carbon was close to zero and we would guess that the N-exchanges would also be very small.

Table 30.4. *Exchanges of phosphorus between Flax Pond and Long Island Sound. Negative values are export from marsh. Values in parentheses are standard errors.*

	Average tide (g)	Annual exchange	
		kg	gm^{-2}
$PO_4 - P$			
1 May–30 Dec	−1,740 (345)		
1 Jan–31 April	331 (278)		
All year	−1,000 (282)	−704 (197)	−1.41 (0.39)
Organic-P	807 (215)	565 (150)	1.13 (0.30)
Total P			
1 May–30 Sept	−1,290 (499)		
1 Oct–30 April	802 (280)		
All year	−201 (317)	−141 (222)	−0.28 (0.44)

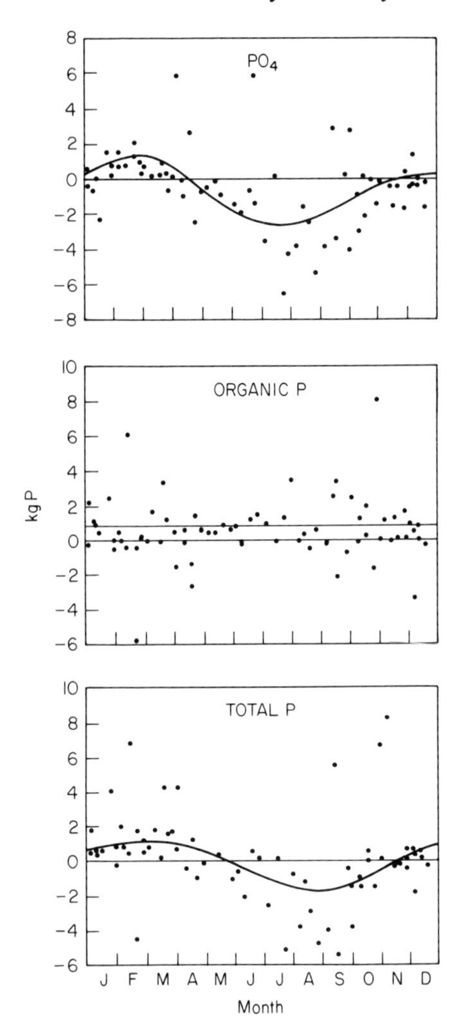

FIG. 30.4. Net tidal exchange of phosphorus between Flax Pond and Long Island Sound over the course of the year. Negative values are losses from the marsh; positive values, net gains. Each point is the difference between transport on the flood and transport on the ebb tide. Curves are least-squares polynomial regressions of net phosphorus flux on day of year.

Phosphorus

The annual flux of phosphorus through the channel appears in the graphs of Fig. 30.4 and the annual budget appears in Table 30.4 (Woodwell & Whitney 1977). Over the entire year, as might be expected, the phosphorus budget was approximately in balance with as much leaving the marsh as entered. There

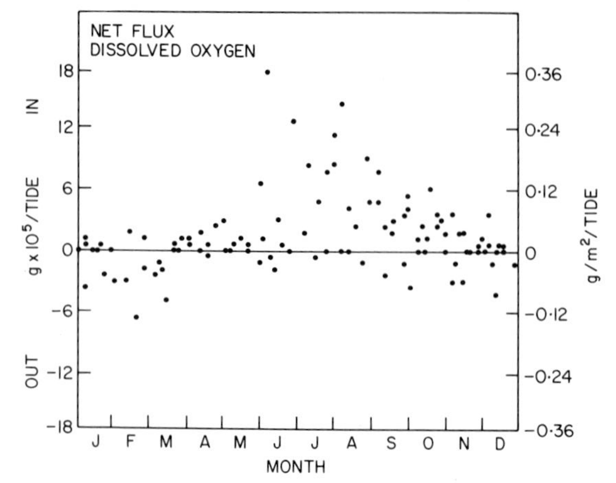

FIG. 30.5. Net tidal exchange of dissolved oxygen between Flax Pond and Long Island Sound over the course of the year. Negative values are losses from the marsh; positive values, net gains. Each point is the difference between transport on the flood and transport on the ebb tide.

was a pronounced seasonality, however, for phosphate-phosphorus with an influx from Long Island Sound in winter and an efflux in summer (Fig. 30.4). There was a net influx of phosphorus in organic form throughout the year (Woodwell & Whitney 1977).

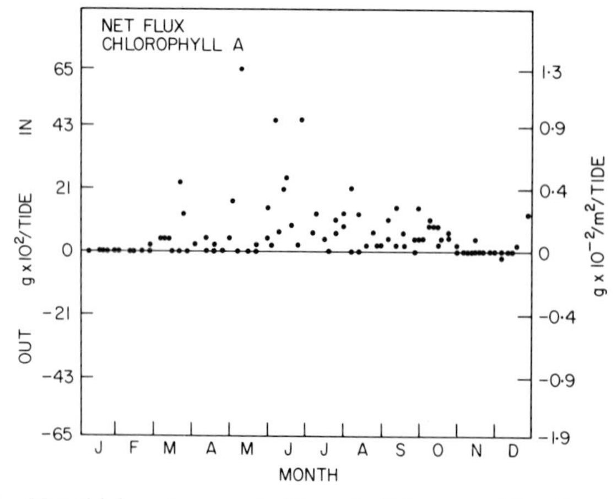

FIG. 30.6. Net tidal exchange of chlorophyll between Flax Pond and Long Island Sound over the course of the year. Negative values are losses from the marsh; positive values, net gains. Each point is the difference between transport on the flood and transport on the ebb tide. [Note: An earlier version of this figure was in error (i.e. Figure 6 in Moll 1977). The units are correct as published here.]

Oxygen and chlorophyll

On the basis of the information above one might guess that the marsh is commonly a reducing system and a net consumer of oxygen and chlorophyll. The results of measurements of oxygen and chlorophyll in the channel, taken in the same sampling programme as the data on carbon, appear in Figs. 30.5 and 30.6. There was a strong net flux of O_2 into the marsh in summer when respiration was high and a weaker loss of O_2 in winter. The flux of chlorophyll was consistently inward throughout the year (Fig. 30.6) (Moll 1977). These data simply confirm the general pattern of function set forth above.

THE FLAX POND ECOSYSTEM

What is the effect of an estuarine marsh on the rest of the world? The question has more meaning now that we have been able to sort and measure the major exchanges shown in Fig. 30.1 for a small estuary. The marsh is indeed important as a net source of fixed carbon for the biosphere, but most of that carbon is used directly in the marsh and not exported to other systems. The amount of carbon fixed is considerably less than that apparently measured by Valiela *et al.* (1976) on Cape Cod, but there is no reason to believe that the measurements reported here have been biased towards the low side. Marshes apparently vary greatly in primary productivity (Woodwell *et al.* 1973); they probably also vary in other basic functions.

Flax Pond is overall weakly autotrophic with respect to Long Island Sound; it is a small net source of fixed carbon released into those waters. Its most important functions, however, may well be related to its tendency to be heterotrophic during long periods. In summer, when rates of metabolism are high, Flax Pond removes small particulate carbon from tidal water, presumably phytoplankton because the patterns of chlorophyll and silica movement are consistent (Moll 1977), and releases the ammonium ion and dissolved organic matter. The exchange is probably less important to the coastal waters, because the total quantities of substances are small, than it is to us as evidence of details of function of the estuary. The estuary tends to be chemically reducing and heterotrophic in summer and autotrophic, releasing fixed carbon and more free oxygen in winter.

The fate of carbon within the estuary is most revealing. There is a steady and important flux into long-term storage in sediments, a part of net ecosystem production. The flux into sediments is probably controlled by the rise of water against the land, a process that appears to be occurring at a rate of millimetres per year along the eastern coast of North America. In Flax Pond the topic has been examined in detail by Armentano and Woodwell (1975). Presumably the rise of sea level determines the rate of storage of carbon in sediments up to some maximum fraction of the net primary production after which the marsh loses its race against inundation and burial beneath the coastal waters.

The importance of this physical control over the function of the marsh can hardly be exaggerated because of the extent to which sedimentation rates can control the amount of fixed carbon available for metabolism or export and because of the possibility that changes in water level will affect primary production itself and the potential for the marsh to maintain itself. The impression left by the experience with Flax Pond is that the estuary is a rapidly changing place, subject to strong physical forces that include changing coastal currents, shifting patterns of water flow, devastating storm tides, and other periodic disturbances that may alter drastically in hours a situation that has appeared to be stable for years. We can expect therefore great variability in our studies of estuaries, a variability that appears now to be apparent in the growing experience with studies like this one in which the estuary is considered as a unit and is found sometimes to be a net source of fixed carbon, sometimes a sink, and usually changeable in a short time.

REFERENCES

Armentano T.V. & Woodwell G.M. (1975) Sedimentation rates in a Long Island marsh determined by ^{210}Pb dating. *Limnol. Oceanogr.* **20**, 452–56.

Ayers J.G. (1959) The hydrography of Barnstable Harbor, Massachusetts. *Limnol. Oceanogr.* **4**, 448–62.

Brinkhuis B.H. (1976) The ecology of temperate salt-marsh fucoids. I. Occurrence and distribution of *Ascophyllum nodosum* ecads. *Marine Biology* **34**, 325–38.

Day J.W. Jr., Smith W.G., Wagner P.R. & Stowe W.C. (1973) *Community structure and carbon budget of a salt marsh and shallow bay estuarine system in Louisiana.* Publ. No. LSU-SG-72-04, Louisiana State Univ., Baton Rouge.

Gallagher J.L. & Daiber F.C. (1974) Primary production of edaphic algal communities in a Delaware salt marsh. *Limnol. Oceanogr.* **19**, 390–95.

Likens G.E., Bormann F.H., Pierce R.S., Eaton J.S. & Johnson N.M. (1977) *Biogeochemistry of a Forested Ecosystem.* Springer-Verlag, New York.

Moll R.A. (1977) Phytoplankton in a temperate-zone salt marsh: Net production and exchanges with coastal waters. *Marine Biology* **42**, 109–18.

Seneca E.D., Stroud L.M., Blum U. & Noggle G.R. (1976) An analysis of the effects of the Brunswick Nuclear Power Plant on the productivity of *Spartina alterniflora* (smooth cordgrass) in the Dutchman Creek, Oak Island, Snow's Marsh and Walden Creek Marshes, Brunswick County, North Carolina, 1975–76. *Third Annual Report to Carolina Power and Light Company, Raleigh, N.C.*

Stroud R.H. (1971) Introduction to Symposium. *Symposium on the Biological Significance of Estuaries* (Ed. by P.A. Douglas & R.H. Stroud), pp. 3–8. Sport Fishing Institute, Washington.

Teal J.M. (1962) Energy flow in the salt marsh ecosystem of Georgia. *Ecology* **43**, 614–24.

Valiela I., Teal J.M. & Perrson N.Y. (1976) Production and dynamics of experimentally enriched salt marsh vegetation: Belowground biomass. *Limnol. Oceanogr.* **21**, 245–52.

Van Raalte C.D., Valiela I. & Teal J.M. (1976) Production of epibenthic salt marsh algae: Light and nutrient limitation. *Limnol. Oceanogr.* **21**, 862–72.

Whitney D.E., Woodwell G.M. & Howarth R.W. (1975) Nitrogen fixation in Flax Pond: A Long Island salt marsh. *Limnol. Oceanogr.* **20**, 640–43.

Woodwell G.M. (1977) Recycling sewage through plant communities. *American Scientist* **65**, 556–62.

Woodwell G.M., Ballard J.T., Clinton J. & Pecan E.V. (1976) Nutrients, toxins and water in terrestrial and aquatic ecosystems treated with sewage plant effluents. *Final Report of the Upland Recharge Program.* BNL 50513, Brookhaven National Laboratory, Upton, N.Y.

Woodwell G.M., Ballard J.T., Small M., Pecan E.V., Clinton J., Wetzler R., German F. & Hennessy J. (1974) Experimental eutrophication of terrestrial and aquatic ecosystems. *First Annual Report of the Upland Recharge Project.* BNL 50420, Brookhaven National Laboratory, Upton, N.Y.

Woodwell G.M. & Dykeman W.R. (1966) Respiration of a forest measured by carbon dioxide accumulation during temperature inversions. *Science* **154**, 1031–34.

Woodwell G.M. & Pecan E.V. (1973) Flax Pond: An estuarine marsh. BNL 50397, Brookhaven National Laboratory, Upton, New York.

Woodwell G.M., Rich P.H. & Hall C.A.S. (1973) Carbon in estuaries. *Carbon and the Biosphere, Brookhaven Symposium in Biology* No. 24 (Ed. by G.M. Woodwell & E.V. Pecan), pp. 221–39. U.S. Atomic Energy Commission, Division of Technical Information, Oak Ridge, Tenn.

Woodwell G.M. & Whitney D.E. (1977) Flax Pond ecosystem study: Exchanges of phosphorus between a salt marsh and the coastal waters of Long Island Sound. *Marine Biology* **41**, 1–6.

Woodwell G.M., Whitney D.E., Hall C.A.S. & Houghton R.A. (1977) The Flax Pond ecosystem study: Exchanges of carbon in water between a salt marsh and Long Island Sound. *Limnol. Oceanogr.* **22**, 833–38.

Woodwell G.M. & Whittaker R.H. (1968) Primary production in terrestrial ecosystems. *American Zoologist* **8**, 19–30.

Zelitch I. (1971) *Photosynthesis, Photorespiration and Plant Productivity.* Academic Press, New York and London.

VI

APPLIED COASTAL ECOLOGY

31. STRATEGIES FOR THE MANAGEMENT OF COASTAL SYSTEMS

D.S. RANWELL

Institute of Terrestrial Ecology,
Colney Research Station,
Colney Lane, Norwich, U.K.

SUMMARY

The land to water boundary at the coast has abundant water supplies useful to industry, beaches attractive for recreation, natural food and sea defence resources liable to oil pollution from the sea or drainage pollution from the land, and essentially dynamic landscapes. Effective management strategies demand understanding of these special features. At the Individual site level (examples are given) strategies tend to be conservative and frequently do not take account of known trends such as the rate of silting in estuaries or the dune development cycle on open coasts. At the Regional level, studies at Morecambe Bay, the Wash and on the Essex and North Kent coast have added to knowledge of resources and functions. They point to the need for extensive survey, computer data handling and more studies on how coastal systems work. National management strategies require knowledge of orders of size of coastal habitat resources and this can be obtained, quite cheaply if recent map and air photographs exist. Detailed knowledge of wildlife resources requires extensive surveys. Process studies, essential for planning management strategies, compete with surveys for funds. Individual ecologists can contribute to national management strategies by learning to recognize invasive species, studying them, and advising on their management. A dune shrub (*Hippophaë*), a prostrate herb (*Acaena*) and a newly-arrived seaweed (*Sargassum*) are discussed as examples.

RÉSUMÉ

Sur la côte, la frontière terre mer produit d'importantes ressources utiles pour l'industrie, des plages attractives pour les loisirs, l'alimentation naturelle et les moyens de défense de la mer susceptibles à la pollution du pétrole venant de la mer ou la pollution des égoûts venant de la terre, et des paysages essentiellement dynamiques. Les stratégies de direction efficace demandent une compréhension de ces propriétés spécifiques. Au niveau d'un site individuel (des exemples sont donnés) les stratégies tendent au conservatisme et souvent ne tiennent pas compte des tendances connues, tel l'envasement des estuaires, ou le cycle de développement de la dune sur les côtes découvertes. A un niveau régional, les études de la baie de Morecambe, le Wash et la côte de l'Essex et du Kent nord ont amélioré la connaissance des ressources et des fonctions. Elles mettent en évidence le besoin d'un examen approfondi, d'un traitement des expériences par ordinateur et de plus d'études sur le travail des systèmes côtiers. L'administration nationale nécessite la connaissance des ordres de grandeur des ressources de l'habitat côtier, et ceci peut être obtenu, relativement bon marché, si des cartes récentes et des photographies aériennes existent. La connaissance précise des ressources de la vie sauvage demande des examens extensifs. Les études de processus, essentiellement pour les stratégies de planification rivalisent avec les études pour les fonds. Individuellement, les écologistes peuvent contribuer aux stratégies d'administration nationale, en apprenant à reconnaître les espèces envahissantes, en les étudiant et en conseillant sur leur gestion. Un arbrisseau des dunes (*Hippophaë*) une herbe prosternée (*Acaena*) et une herbe marine nouvellement arrivée (*Sargassum*), sont les exemples étudiés.

INTRODUCTION

Our laboratory at Colney has been fortunate in being associated with a series of quite large scale research studies on coastal systems, and some experience relevant to the management of coastal sites has been built up.

This paper attempts to put together some ideas relevant to management strategies at local, regional and national levels. It is a personal point of view based on one set of experiences, but I hope it will generate discussion on management strategies at the coast. The paper concludes with a brief reference to some of the problems posed by invasive species at the coast.

McCarthy (1976) found that common problems relating to the management of coastal systems both in Europe and North America were:

1. The conflict between economic and environmental needs.
2. Establishing common ground between administrators and ecologists.
3. Gathering and handling large amounts of scientific and planning data.
4. Translating results into field action.

These are all effectively problems of communication and have to be overcome before effective management strategies for coastal systems can be put into operation.

Special conditions at the coast relating to the conflict between economic and environmental needs include (a) the abundant supplies of cooling water for power stations and their visual impact in areas of wide horizons; (b) strandlines where oil from off-shore sources can accumulate and damage inter-tidal communities; (c) emergence of land drainage water rich in nitrates resulting in excess algal growth and oxygen demand in estuarine benthic communities and (d) the natural recreational attractions of the coast resulting in a concentration of tourist activities potentially damaging the environment.

Coasts present special problems too for communication between administrators and ecologists since administrative boundaries are rarely designed to take into account the cross boundary relationships (such as the nitrate seepage) which makes coastal habitats of such special interest to the ecologist. Tidal activities present special risks and problems in data gathering exercises, and the rapidity of change in unstable coastal habitats demands collection and analysis of big data banks if the complex processes at work are to be understood. Translation of results into field action is beset with difficulties arising from legal problems relating to ownerships in the coastal zone and the fact that management action in one part of the coast (e.g. groyning) may have deleterious effects in another (e.g. reduction in beach sediment supply).

Most sites on European coasts are not managed to an agreed strategy other than in the broadest sea defence or zonal planning use senses. The case studies reported in other papers show to some extent what has been done in different countries. Now that we have the common problem of oil spillage in the North Sea there is an outstanding need to bring together ideas on how management strategy could be better co-ordinated on European coasts.

In selecting a management strategy it is useful to consider three questions: (a) What is the aim? (b) What is the full historical and environmental context? (c) What are the options? Management options common to inland and coastal situations involve restoring, protecting, modifying, adding, subtracting or diversifying species or parts of the environment. Coastal examples of these have been reviewed elsewhere (Ranwell 1975).

Strategies for the management of coastal systems have to take account of the highly dynamic nature of these systems and their special vulnerability to disturbance from erosive forces. For example, the temporal and spatial cycles related to dune building and mobility are frequently not known, let alone assimilated into planning by all people who have responsibility for managing dune coasts. This cycle is a phenomenon that may span the lifetime of man in the order of 50–100 years and I think it is not unconnected that cyclic phenomena of this order of time are not readily appreciated from one generation to another.

Even though coastal systems are subject to periodic and often rapid change the slate is rarely wiped clean. More often than not, events that have occurred over several millenia stamp their impress on the environment even today. The effects of lesser events become overlaid to form the patina on which the current dusting of plant and animal life forms an ever changing pattern. For example, the 6,000 acre mudflat at Bridgwater Bay, Somerset fringed by 50-year-old salt marsh today is a submerged platform dating back 5 or 6,000 years when it was afforested. Mud holes from which clay was extracted a century or more ago to make bricks to build London are still recognizable in the Medway estuary, Kent (Kirby 1969). In Hebridean dune systems, ancient land surfaces 4,000 years old or more, may be re-exposed by wind erosion alongside contemporary surfaces (Crawford & Switzur 1977).

Most good planners are strongly imbued with a sense of history so far as buildings are concerned, few have the ecologist's sense of history relating to landscapes. I believe ecologists and archaeologists could work together to communicate this sense of landscape history to planners, though it would be unrealistic to suppose that management strategies stressing historical preservation alone could be extensively pursued.

LEVELS OF OPERATION

Three levels of operation have been recognized: (1) the individual site level, (2) regional level and (3) national level. These are defined as follows. Individual sites in the British coastal context would be up to 500 ha or so in size (often much less) and characterized usually by one dominant coastal habitat type, e.g. a sand dune or mud flat/salt marsh system. They are essentially of local interest and their use is likely to be governed by local socio-economic and political factors. At the regional level extensive tracts of often quite diverse types of coastline are involved and the socio-economic and political factors are governed by overall land use and the regional administrative structure with its legal, financial and political tie-lines to central government. The national level is self-evident and relates to the entire coastal resources of the country and both the internal and external socio-economic and political factors governing their use.

Individual site

Management strategies for local sites tend to be heavily weighted towards the options concerned with restoration and protection at the present time. This is largely because of the increasing pressures for development of European coastlines, but also from reluctance to increase the amount of change. Sometimes it is necessary to take drastic action to restore, as at Camber in Sussex (Pizzey 1974) where sand, mobilized by trampling, threatened property and had to be contoured before replanting. There is a strong case for deliberately recreating diversity in such a system, but such a strategy has yet to be adopted.

Visible evidence of coastal erosion leads to strong pressure for a management strategy aimed at stabilization. This could be not merely costly, but unnecessary. For example there is little point in stabilizing a mature eroding coast dune when no roads or buildings behind it are threatened (as at Morfa

Dyffran, Merioneth, Wales). The logical strategy is to let it erode and make way for the regeneration of a new dune at the strandline. Such a strategy recognizes the dynamic nature of this type of coastal habitat and allows for the fact that certain types of ephemeral communities (e.g. those of strandlines, fore dunes and primary dune slacks) can only exist at specific times in the cycle of development. It is a strategy that allows therefore for the protection of a greater range of wildlife than is present on the site at any one moment in time.

Once one moves from local site strategies to those for whole coastal systems communication tends to expand from individuals to institutions with particular interests to safeguard and divergent viewpoints.

Clarke (1976) has pointed out, 'Much of the conflict in planning and management study, at both lay and professional levels, and at individual and institutional scales, results from disparities between different perceptions of problems'.

To a coastal land owner the salt marsh grass *Spartina anglica* is a re-claimer of mud flats, to a yachtsman it reduces sailing areas. Sea Buckthorn (*Hippophaë rhamnoides*) is a dune stabilizer to a coastal engineer, a source of bird food in winter to an ornithologist, a shrub providing shelter for a picnic to those concerned with recreation, and a case for eradication to a Nature reserve warden charged with maintaining a diverse sand dune flora.

Vegetated shingle complexes accumulate at a level only just above high water mark. Over-riding considerations of sea defence make it necessary to bull-doze shingle into breaches to make good storm damage.

The special value of extensive and relatively undisturbed vegetated shingle beaches such as Dungeness, Kent is therefore accentuated, and management strategy would normally lay particular stress on the preservation of such areas from disturbance so far as possible. However, recent conflicts of interest in connection with power generation and gravel and water extraction have demonstrated that a realistic management strategy must accept that preser-vation will not be a primary consideration. The ecological emphasis of management strategy might therefore shift to a more dynamic approach associated with design of new habitats in worked-out wet shingle pits, as the Royal Society for the Protection of Birds is in fact doing at Dungeness.

In the case of croppable coastal systems like salt marshes, management strategy can be based on spatial and temporal separation of activities. In the Bridgwater Bay National Nature Reserve, Somerset for example it was decided over 20 years ago to allow sheep grazing in the terminal salt marshes and prevent it in the central section. This has resulted in two quite distinct marsh communities, grazed salting pasture and reed marsh. However one of the current problems is that *Agropyron pungens* (a grass of low palatability) is spreading into the grazing marshes by seed produced in the ungrazed section. It has been shown possible to keep the salting pasture below the level at which

Agropyron invades by turf cutting, but mowing is likely to prove a more practical way of controlling its spread once it is in the salting pasture. Sheep grazing in summer alternates with wildfowl grazing in winter on this high level salting pasture and it has been shown experimentally that a crop of turf could be taken once every 5 years or so without more than local temporary interruption to the grazing regime. Here then is an example of a quite sophisticated management strategy involving modified farming and wildlife protection operated simultaneously by zoning in space and time.

The sand dunes of St Ouen's Bay, Jersey, Channel Islands, are among the ten largest single dune systems in the British Isles and the fourth richest in vascular plants. This site has had a highly varied history of land use yet parts have remained relatively undisturbed for centuries though much of the island is intensively cultivated. The stimulus for the development of a rational management strategy for this area has derived from a fruitful collaboration between planners and those concerned for the wildlife. Here the management strategy proposed attempts to embrace exploitation of mineral resources, recreational development, rubbish disposal, grazing, tree plantation and wildlife preservation by zoning in space and phasing in time. The success of such strategies depends on respect for historical landscapes, and a willingness to compromise on the part of all concerned.

Regional level

Ecological studies on whole regions of the British Isles such as Morecambe Bay, Lancashire (Corlett 1970), the Wash (Anon 1976), and the Essex and North Kent coast (Boorman & Ranwell 1977) have been completed in recent years. Each was funded as a result of political decisions relating to water resources or transport requirements (airport development). None has resulted in an accepted overall management strategy, but the information base is there on which to build strategy providing this is done soon, before major changes occur.

What do we learn from such large scale studies and what are the implications for management strategy? The studies in both Morecambe Bay and the Wash were concerned with predicting the effects on wildlife of building fresh water reservoirs in the intertidal zone of large coastal inlets into each of which several rivers drain. Studies at Morecambe Bay investigated sediment circulation patterns (Kestner 1970, 1972), and the distribution of intertidal plants (e.g. Gray 1972; Gray & Bunce 1972) and animals (e.g. Anderson 1972). They demonstrated three aspects relevant to planning effective management strategies:

1. The need for extensive survey to provide inventory information so that future changes can be measured.

2. The need to develop computer techniques to handle the large body of data involved.
3. The need for process studies on how the coastal systems work.

The Wash studies organized by Dr Gray included work on the interface between mud flats and salt marsh and this brought out the distinctive structure and function of this boundary zone involving mud mounds, micro-algae and accretion as Dr Coles (this volume) has already described. Boundaries between major habitat types are not merely a gradation from one to another, they have distinctive physical and biological features and special functional relationships with the habitats on either side of them. Such boundaries are important sites of change just as certain phases of development at the soil surface within habitats become at certain times important sites of change. At such sites and such times habitats are transformed naturally, and probably most effectively manipulated by management.

Salt marsh studies developed by Dr Randerson on the Norfolk coast and the Wash led to a useful predictive model which is currently being tested on data from Bridgwater Bay. The significance of the model lies less in what it will predict than in its ability to account for the observed facts and comprehend valid relationships and, through its deficiencies, indicate where further research is needed. The basic components of the model include such physical factors as standard accretion per thousand hours, tidal submergence, proportion of sand in tideborne sediment (which has nutritional connotations) and rate of accretion of windborne sediment. Biological components include zonation (species tidal limits for growth), biomass (root and shoot separately), growth (maximal and seasonal) and species grazing factors.

The Wash salt marsh studies also helped to confirm the conclusions of Kestner (1975), that given no interference salt marsh seaward extension is self-limiting. Increasing velocity of ebb flow dependent on increasing volume of tidal water to be drained off the marsh as its area increases results eventually in the erosive effects of ebb flow balancing out accreting effects of flood flow so that further seaward extension of marsh ceases until the hydraulic conditions change or are altered by reclamation.

The Essex and North Kent coastal habitat and wildlife population surveys were organized by Dr Boorman to discover the probable effects of developing an airport at the mouth of the Thames estuary (Maplin) and what might be done to mitigate the effects on the coast and its wildlife. Dr Wyer's component of this study was focussed on quantitative estimates of the populations of *Zostera* and *Enteromorpha*. This enabled a quantitative estimate of these food resources and their calorific value for the Brent goose to be calculated which Dr Charman showed was near to limiting for the population. The possibility of transplanting *Zostera* (a principal food of Brent geese) was demonstrated (Ranwell *et al.* 1974), but it seems that suitable sites for this are lacking, possibly as a consequence of past reclamation activities on this coast.

Both the Wash and the Maplin studies provided inventory information on a large scale and drew attention to the importance of bird behavioural features (Goss-Custard *et al.* 1977). The Maplin airport study, as discussed by Dr Charman (this volume), also demonstrated the remarkably rapid behavioural adjustment in the feeding of Brent geese when a threshold population level exceeded the normal food capacity.

Strategies for management at the regional level require not only some sort of inventory for comparative purposes, but a keen awareness of the functional importance of boundaries, thresholds, and behavioural characteristics of both physical and biological elements of the environment. The use of conceptual and mathematical models in exploring regional strategies seems a sensible approach. Also the concept of functional environmental units such as the coastal cell (Dolan & Hayden 1973) seems a useful basis for modelling.

The recent re-organization of regional government in this country has opened the way for development of regional management strategy. It might for example prove a better strategy to designate certain eroding parts of the East Anglian coast as feeder zones for accretion elsewhere rather than the present strategy which results in increasing engineered defences, so costly to erect and maintain (Clayton 1976). Such a strategy might go hand in hand with attempts to boost the regeneration of coastal plants capable of holding sediment in place where accretion does occur.

I believe it is true to say it is no one's job at the present time to even consider such regional strategies, let alone implement them.

National level

It is not yet possible to give examples of national coastal management strategies. Instead two aspects relevant to such strategies, national survey of habitat resources and the problem of countrywide invasive species at the coast are singled out for discussion.

Habitat resources

If we are to get our priorities right in the management of coastal systems, we need to know orders of size in which the different systems are represented in a defined area (Fig. 31.1). In this country for example we estimate that tidal flats, salt marsh, sand dune, and vegetated shingle beach, occur in the ratios 5:1:1:0·5 approximately, with tidal flat occupying some 500,000 acres and vegetated shingle beach only some 40,000 acres. To have reached these figures we have clearly had to sum individual units and so we know in addition that more than half the dunes in Great Britain occur in Scotland, over a quarter of the salt marshes in south-east England, and where the biggest unit areas

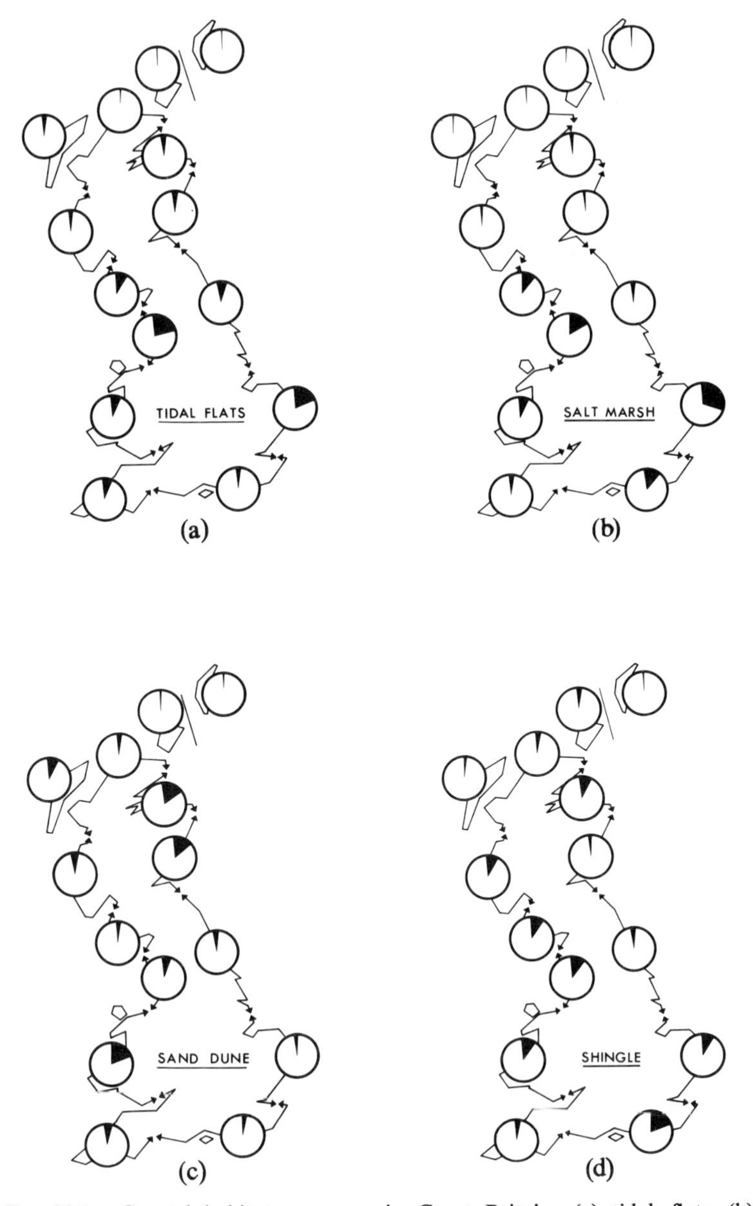

FIG. 31.1. Coastal habitat resources in Great Britain: (a) tidal flat; (b) salt marsh; (c) sand dune; (d) vegetated shingle. The distributions given are very approximate, but best available estimates derived from map and documentary evidence dating in general over the last decade.

of each type of habitat are to be found. That the Maplin mud flats at the mouth of the Thames Estuary are the biggest continuous tidal flats in Britain is not unconnected with the fact that it supports the biggest continuous population of *Zostera noltii* a staple diet for one fifth of the world population of Brent geese. Also we know that Dungeness, Kent alone accounts for one fifth of all the vegetated shingle in Britain. These sorts of facts are, or should be, highly relevant to planning management strategies at the national level and therefore should be part of the stock in trade of anyone concerned with coastal management on a countrywide basis. How much I wonder is such information readily available for the European coast as a whole? It is not costly to obtain provided appropriate maps and air photographs are available.

The detailed distribution of individual species is very much more time consuming and expensive information to obtain. It is one thing to list the flora and fauna and the communities of coastal systems, but quite another to know exactly where and in what amount individual species and communities are to be found. The Institute of Terrestrial Ecology is currently engaged on a survey of the flora of 94 (mainly sand dune) sites in Scotland for the Nature Conservancy Council. This includes not only flowering plants, but terrestrial bryophytes and lichens as well. In addition a wide spectrum of environmental features is being recorded. The aim of this survey is to record an unbiased sample at a particular time of what is present in these sites, so that following computer analysis objective comparisons can be made between sites for planning purposes in relation to human impacts on the coastline of Scotland. Knowledge of the flora and human impacts is also amplified by subjective assessment to aid in interpretation of results. Surveys on this scale are expensive, but they do contribute to both inventory and distribution information on a reliable and repeatable quantitative basis, and help to found a factual background on which to base choices of management strategies. Nevertheless, the problems of interpreting data collected at an instant in time on phenomena which fluctuate in time are very real. For example there will be records of the numbers of rabbit burrows in some 4,000 objectively selected 200 × 200 m sample squares. One can foresee that these may quite possibly be a better indicator of potentially burrowable ground rather than an indicator of rabbit population density.

Armed with the sort of information now being generated for Scottish dune systems it should be possible to plan a comprehensive national management strategy in which the needs for example, of sand mining, agriculture, archaeology, wildlife protection, recreation and industrial or urban use so far as dune systems are concerned can be rationalized. The actual strategy would be a compromise because existing use at many sites would continue to have a dominating influence at least for some time. Ultimately perhaps we shall learn to switch uses smoothly and with minimum damage to the environment so that the maximum potential use of sites can be achieved.

Invasive species
I want to conclude with something a little more practical, to show how individual ecologists can influence national management strategy, at least on a small scale.

An example is the study group set up by the Nature Conservancy to recommend a national management policy for Sea Buckthorn (*Hippophaë rhamnoides*) on sand dunes in Great Britain (Ranwell 1972). After a thorough review of existing information by a small group with differing interests and viewpoints, recommendations for the management of this species in 45 sites (mainly nature reserves), were made and have since been at least partially implemented at many of these sites. The strategy contained options designed on a national basis ranging from preventing establishment, eradication, partial control, to no control.

To prevent establishment requires some expenditure of management resources on regular inspections and the will to persist with these. It is difficult to convey the need for this to a dune manager who has not seen for himself the effects of *Hippophaë* on a dune system (almost complete elimination of the flora beneath dense *Hippophaë*), so visits to heavily infested areas should be arranged. Sites are particularly susceptible to *Hippophaë* invasion when exceptionally cold winters encourage birds such as Fieldfares (*Turdus pilaris*) to feed on their fruits and distribute viable seeds. Special vigilance is needed when such conditions are combined with low rabbit populations due to myxomatosis.

Eradication is only feasible where *Hippophaë* is present in relatively small amount (e.g. less than 2 ha) as at Whiteford Burrows, South Wales where combined attack by uprooting seedlings, and cutting and spraying mature growths, has effectively eliminated the shrub. Hesitation in deciding on an eradication policy early enough can result in populations very difficult and expensive to control in a decade or two. The population at Spurn Head, Yorkshire more than doubled in seven years and converted open *Ammophila arenaria* dune with a characteristically varied flora to impenetrable *Hippophaë* thicket beneath which few associated species could survive.

Partial control requires a continued maintenance input to maintain areas free of *Hippophaë*, but where rides are cut through dense growths as at Gibraltar Point, Lincolnshire, their width could be such that rabbit grazing from the two shrub boundary limits can help to maintain the ride in an open condition. However a further problem has been encountered at this site where long established growths have been cut. *Hippophaë* has root nodules with nitrogen-fixing bacteria and its growth so enhances nutrient levels that persistent tall weedy growth of *Urtica dioica* and *Chamaenerion angustifolium* replaced the natural dune communities in cut areas.

The *Hippophaë* report with its practical proposals has influenced management strategies towards this species in some sites outside those for which specific recommendations were made.

We lack effective management strategy for recognizing, monitoring the behaviour of, and effectively controlling potential problem species. This seems to me an area in which ecologists have a powerful, but as yet largely undeclared role to play.

For example it should be possible to recognize and predict (even before they arrive), the special characteristics of species likely to become invasive, and likely to thrive in a particular country. Indeed this was actually done in the case of the seaweed *Sargassum muticum* (Druehl 1973), one of the most recent and invasive arrivals on European shores.

Once such species have arrived a management strategy needs to be devised and put into action at once. There is no official procedure in this country (let alone in Europe) at present to do this. In the case of *Sargassum* a purely voluntary group of scientists from Universities, Government and Industrial organizations came together, agreed on an attempt to eradicate the species, and lobbied for funds to study its biology. Eradication, not surprisingly, did not prove feasible and an alternative management strategy has yet to be devised. One essential piece of information for this is knowledge of the potentially colonizable limits on European coasts. In my opinion carefully controlled transplants to determine these limits are justified in the case of *Sargassum*, and necessary to develop a co-ordinated strategy which might well eventually include cropping for alginates on European coasts.

The New Zealand Pirri-pirri burr (*Acaena anserinifolia*) has been introduced as a wool alien to most parts of the British Isles and locally is steadily building up extensive populations, for example on the Holy Island dunes in Northumberland. This is one of the most widespread weeds in the Southern Hemisphere, its hooked burrs diminish the value of wool in sheep country and can so impede nestling ground-nesting birds as to result in their starving to death. The species is highly polymorphic, fecund, readily dispersed, and favoured by trampling and burning—all hallmarks of a problem plant for the next generation.

Yet at the present time only a handful of botanists take any interest in this plant and there is no clearly-defined strategy relating to it. As a first step, the Nature Conservancy Council has had the Holy Island population surveyed and I have mapped the limits of all known populations in four English counties, as a basis from which to measure future change.

I think we need a list of these 'super-species' that are actively radiating out from their centres of origin and, perhaps along with the Red data book for rare species, the World Wildlife Organization might consider a Green data book for those potential problem species which at present have an evolutionary 'green light for go'.

REFERENCES

Anderson S.S. (1972) The Ecology of Morecambe Bay. II. Intertidal invertebrates and factors affecting their distribution. *J. appl. Ecol.* **9**, 161–78.

Anon (1976) *The Wash water storage feasibility study.* A report on the ecological studies. N.E.R.C. Pubn. Series C. No. 15, London.

Boorman L.A. & Ranwell D.S. (1977) *Ecology of Maplin Sands and the coastal zones of Suffolk, Essex and North Kent.* Institute of Terrestrial Ecology (N.E.R.C.) Pubn., Cambridge.

Clarke M.J. (1976) *The Relationship between Coastal Zone Management and Offshore Economic Development.* Southampton University Report.

Clayton K. (1976) *Research on the East Anglian Coast: a progress report.* University of East Anglia.

Corlett J. (1970) *Morecambe Bay Barrage Feasibility Study.* Report to Natural Environment Research Council on biological aspects. N.E.R.C., London.

Crawford I.A. & Switzur R. (1977) Sandscaping and C_{14}: The Udal, North Uist. *Antiquity* **51**, 124–36.

Dolan R. & Hayden B. (1973) *Classification of Coastal Environments, Procedures and Guidelines.* Tech. Rep. 2, U.S. Office of Naval Research Geography Programmes. Task No. NR 389-158.

Druehl L.D. (1973) Marine transplantations. *Science* **179**, 12.

Goss-Custard J.D., Kay D.G. & Blindell R.M. (1977) The density of migratory and overwintering Redshank, *Tringa totanus* (L.) and Curlew, *Numenius arquata* (L.), in relation to the density of their prey in south-east England. *Estuarine and Coastal Marine Science* **5**, 497–510.

Gray A.J. (1972) The ecology of Morecambe Bay. V. The salt marshes of Morecambe Bay. *J. appl. Ecol.* **9**, 207–20.

Gray A.J. & Bunce G.H. (1972) The Ecology of Morecambe Bay. VI. Soils and vegetation of the salt marshes: a multivariate approach. *J. appl. Ecol.* **9**, 221–34.

Kestner F.J.T. (1972) The effects of water conservation works on the regime of Morecambe Bay. *Geogr. J.* **138**, 178–208.

Kestner F.J.T. (1975) The loose boundary regime of the Wash. *Geogr. J.* **141**, 388–414.

Kirby R. (1969) *Sedimentary environments, sedimentary processes and river history in the Lower Medway estuary, Kent.* Ph.D. thesis, University of London.

McCarthy J. (1976) *Coastal Conservation in U.S.A., Some Experience from the Eastern States.* Winston Churchill Trust Fellowship Report.

Pizzey J.M. (1974) Assessment of dune stabilisation of Camber, Sussex, using air photographs. *Biological Conservation* **7**, 275–88.

Ranwell D.S. (Ed.) (1972) *The Management of Sea Buckthorn* (Hippophaë rhamnoides) *on selected sites in Great Britain.* Report of the Hippophaë Study Group. Nature Conservancy (N.E.R.C.) Pubn., London.

Ranwell D.S., Wyer D.W., Boorman L.A., Pizzey J.M. & Waters R.J. (1974) *Zostera* transplants in Norfolk and Suffolk, Great Britain. *Aquaculture* **4**, 185–98.

Ranwell D.S. (1975) Management of salt marsh and coastal dune vegetation. *Estuarine Research II* (Ed. by L.E. Cronin), 471–83. Academic Press, London.

32. RESPONSES OF SALT MARSH VEGETATION TO OIL SPILLS AND REFINERY EFFLUENTS

JENIFER M. BAKER
Oil Pollution Research Unit,
Orielton Field Centre,
Pembroke, U.K.

SUMMARY

1. Salt marsh plants show a wide range of susceptibility to typical crude oils, ranging from annuals such as *Salicornia* spp. which are killed by one oiling, to *Oenanthe lachenalii* which withstands 12 successive monthly applications of fresh Kuwait crude oil.
2. *Spartina anglica* survives most single oil spillages well but does not tolerate chronic pollution. Experimental evidence suggests that oil interferes with the normal oxygen diffusion process down the plants into the roots.
3. Competitive advantages are gained by tolerant species following oiling. This is particularly striking in the upper salt marsh community studied, where the susceptible *Juncus maritimus* was replaced by *Oenanthe* and by *Agrostis stolonifera.*
4. Marsh cleaning methods (burning, cutting or treatment with chemical dispersants) do not appear to decrease the damage done by oil pollution, and often increase it.
5. Different crude oils and products vary considerably in their toxicity. Light aromatics appear to be the greatest contributor to acute toxicity. Some heavy or weathered oils stimulate the growth of salt marsh plants.

RÉSUMÉ

Réponses de la végétation des marais salants à la pollution par le pétrole et les rejets des raffineries.

1. Les végétaux des marais salés présentent une gamme étendue de sensibilité aux pétroles classiques; ainsi les végétaux annuels tels *Salicornia* spp. sont morts à la suite d'un seul épandage, alors que *Oenanthe lachenalii* a résisté pendant douze mois à l'application de pétrole frais du Kowait.

2. *Spartina anglica* survit relativement bien à une seule application de pétrole mais ne tolère pas une pollution chronique. Des données expérimentales suggèrent une interférence du pétrole avec le processus de diffusion de l'oxygène descendant vers les racines.
3. Après une pollution, l'espèce tolérante est avantagée pour la compétition. Ceci est particulièrement frappant dans la communauté des marais salés supérieurs où *Oenanthe* et *Agrostis* ont remplacé *Juncus maritimus*, plus sensible.
4. Les méthodes de nettoyage des marais (par le feu, la coupe ou le traitement chimique) ne semblent pas diminuer le dommage causé par la pollution du pétrole; souvent elles l'augmentent.
5. Les différents pétroles et déchets varient considérablement dans leur toxicité. Les composés aromatiques semblent y contribuer pour la plus grande part. Quelques pétroles lourds ou patinés stimulent la croissance des végétaux des marais salés.

INTRODUCTION

Salt marsh vegetation often acts as an 'oil trap' for stranded spills, and in some locations marshes may be subjected to several successive oilings within a few years. In addition, some British coastal refinery effluents are discharged near or into salt marshes. The effects of these influences have been studied over a period of nine years by the Oil Pollution Research Unit of the Field Studies Council, using field surveys and monitoring, field experiments and laboratory experiments. Details of much of the work are available (Baker 1971a, b; Dicks 1976), and this paper is an updated summary. Nomenclature follows Clapham, Tutin and Warburg (1962), except for the case of *Spartina* × *townsendii* H. & J. Groves which is here called *Spartina anglica* following Hubbard (1968).

EFFECTS OF A SINGLE OIL SPILLAGE

Numerous observations, both previously published and original, are brought together in Tables 32.1 and 32.2. The overall impression is that salt marsh vegetation often recovers well from a single spillage, but there is a large range of effects on vegetation and sediments according to oil type, amount of oil, time of year, cleaning treatment and species of plant. Oils with the greatest immediate toxicity are the lighter crude oils and products, such as Nigerian crude oil and No. 2 fuel oil. This is in agreement with numerous published observations on pesticidal oils (reviewed by Baker 1970) which show that within each series of hydrocarbons, the smaller molecules are more toxic than the larger. Toxicity also increases along the series alkanes–alkenes–aromatics. Heavy oils, if present in sufficient quantities, cause death by smothering, and recolonization of areas so affected is likely to be slow. Long-term effects are also associated with penetration of oil into salt marsh sediments, which is most likely to occur with light oils or in the presence of dispersants. Cleaning by means of dispersants, cutting or burning does not appear to decrease mortality of oiled salt marsh plants, and the effects of trampling of cleaning contractors are likely to add to the damage of the vegetation.

EXPERIMENTAL SUCCESSIVE SPILLAGES

Three experimental sites were chosen on the north Gower coast, south Wales in areas dominated respectively by *Spartina anglica, Puccinellia maritima* and *Juncus maritimus.* The *Spartina* site is flooded nearly every day, the *Juncus* site is flooded only during high spring tides and the *Puccinellia* site occupies an intermediate position. In each case a large area of vegetation was chosen, as free as possible from patchiness and on level ground lacking creeks and pans. The experimental layout at each site consists of two randomized blocks of five plots, each plot being 2 × 5 m. Each block contains an untreated plot and plots which have received two, four, eight or twelve successive monthly sprayings of 4·5 litre fresh Kuwait crude oil. This amount, applied using a knapsack sprayer, is sufficient to cover all the vegetation with a thin oil film. The vegetation was recorded before oiling and at intervals afterwards using point quadrats, usually 100 per plot. All species touched by each pin were recorded, but not all the contacts per species per pin. In the case of *Spartina anglica,* recording was by counts of live shoots within ten 25 × 25 cm quadrats per plot chosen at random.

In all cases the initial effect of the oil films on the plants was yellowing and death of leaves. Recovery took place from protected growing points in or under the soil surface. Dead oily leaves remained in the plots for varying lengths of time. They had largely disappeared from the *Spartina* plots a year following cessation of treatment, but oily *Juncus maritimus* was beaten down by winter weather to form a dense mat, visible traces of which persisted for five years. No oiled plots showed a decrease of level compared with the controls, presumably because sediments were protected from erosion by mats of oily vegetation or by the quickly recovering populations of filamentous algae.

A crucial number of light successive monthly oilings appears to be four for the survival of most species. Up to this number the major species *Spartina anglica, Puccinellia maritima* and *Festuca rubra* show good recovery within

Table 32.1. *Effects of single accidental oil spillages.*

Date of incident	Locality	Oil	Observations	References
Many occasions	Gulf of Mexico, Barataria Bay area	Louisiana crude	Plants killed by 25 ml oil per 0·09 m^2 of water surface. Lush recolonizing growth.	Mackin 1950a, b
Jan. 1961	Poole Harbour, Dorset	Fuel oil	Good recovery	Ranwell & Hewett 1964
Sept. 1966	Medway, Kent	Nigerian light crude 1,700 tonne spill, 'Gamlen' dispersant used for cleaning	Vegetation recovered well by the following May	Buck & Harrison 1967 Harrison 1967 Harrison & Buck 1967
Jan. 1967	Milford Haven	Arabian light crude 250–500 tonne spill from 'Chryssi P. Goulandris'. Dispersants used for cleaning.	Severe immediate effects, generally good recovery. Dispersant use on one marsh increased damage to *Spartina*, and oil is still visible in the mud (1977).	Cowell 1969 Cowell & Baker 1969 Dalby 1969 Baker unpublished material
Mar. 1967	Cornwall	Kuwait crude from 'Torrey Canyon'	Immediate damage less than 'Chryssi P. Goulandris' spill, possibly due to loss of volatile fractions while the oil was weathering at sea	Cowell 1969 Ranwell 1968 Ranwell & Stebbings 1967 Smith 1968
Mar. 1967	Britanny	Kuwait crude from 'Torrey Canyon'	New growth observed in most areas 16 months after heavy pollution. Some apparent growth stimulation. Anaerobic conditions evident below thick oil layers.	Ranwell 1968 Ranwell & Stebbings 1967 Stebbings 1968
Feb. 1968	Milford Haven (Sandy Haven)	Heavy fuel oil	In general, vigorous new growth observed four months after pollution, but oiled *Halimione* branch ends dead	Baker 1971b

Nov. 1968	Milford Haven (Hazelbeach)	Fresh crude oil	Oily *Spartina* shoots cut and removed by cleaning contractors. A few dead patches persisted during 1970.	Baker 1971b
Feb. 1969	Milford Haven (Martinshaven)	Heavy fuel oil, in places 5 cm thick, but covering only a small part of the marsh	Rapid recovery of *Schoenoplectus tabernaemontani* and *Spartina anglica*. Thickly oiled *Juncus maritimus* died, and recolonization is still (1977) incomplete.	Baker 1971b
Sept. 1969	Buzzards Bay, Massachusetts	No. 2 fuel oil	Small area of *Spartina alterniflora* had not recovered by 1976. Oil persists in some sediments.	Hampson & Sanders (personal communications)
Feb. 1970	Chedabucto Bay	Bunker C	*Spartina alterniflora* grew well the year following the spill, then declined, and has since shown recovery. Oil penetrated sediments.	Thomas 1977, 1978
Aug. 1974	Eastern Straits of Magellan	Light Arabian crude and Bunker 'C'. Large quantity of 'chocolate mousse' emulsion formed because of the rough water conditions.	*Salicornia ambigua*, *Senecio patagonicus* and *Lepidophyllum cupressiforme* affected. Some recovery, but effects persist in two small, heavily polluted inlets.	Baker *et al.* 1976, and subsequent observations by the Institute of Patagonia
Oct. 1974	Winsor Cove, Bourne, Massachusetts	No. 2 fuel oil	Death of *Spartina alterniflora* and penetration of oil into sediments	Hampson & Moul 1978
Feb. 1976	Chesapeake Bay	No. 6 fuel oil	*Spartina alterniflora* showed increased net productivity as measured by standing crop; increased density, decreased mean height, and increased flowering success	Hershner & Moore 1977

Table 32.2. *Effects of single experimental oil spillages.*

Date and location of experiment	Experiment	Observations	References
1968–71 Milford Haven	Randomized block of 7 transects (3 controls, + Feb., May, Aug, and Nov. treatments with fresh Kuwait crude at 0·5 litre per m^2)	Cover measurements made using point quadrats showed that according to season of oiling, reduction in germination or in flowering can occur. Recovery of all transects was complete within 3 years.	Baker 1971a, b
1968–71 Milford Haven	*Spartina anglica* plots oiled with fresh Kuwait crude (0·5 and 1·0 litre per m^2), and cleaned by burning, cutting, or dispersant application	Recovery was measured using shoot density. None of the cleaning treatments significantly decreased mortality of oiled *Spartina*	Baker 1971a, b
1969–71 University College, Swansea	Comparative toxicity expts. using 30 different crude oils, products and dispersants on *Puccinellia*/*Festuca* turves. Usual application rate 0·8 litre per m^2	Fresh Kuwait crude was much more toxic than 50% Kuwait residue. The residue and some heavy crudes stimulated growth. Light crudes, e.g. Nigerian, and low-boiling distillates caused severe damage. All undiluted dispersants tested were toxic, but none caused permanent damage if applied as sea water solutions containing less than 10% dispersant	Baker 1971a, b
July 1973 Gulf Coast Research Lab., Mississippi	Approx. 70 litre of Empire Mix crude poured into intertidal pond	*Spartina alterniflora* shoots appeared dead after 10 days, but new shoots appeared within 3 weeks. Recovery good within 2 months	Lytle 1975
Sept. 1975 York River, Virginia	Fresh and weathered Louisiana crude released in salt marsh enclosures	Both fresh and weathered crude reduced production of *Spartina alterniflora*	Bender *et al.* 1977
1975 Virginia	No. 2 fuel oil released in marsh	Death of patches of *Spartina alterniflora*, and penetration of oil into sediments	Herschner (personal communication)

one year. Further successive oilings result in a rapid decline. *Juncus maritimus* is severely affected by two successive oilings, so the upper salt marsh community is likely to change considerably in appearance after relatively light pollution. Large areas of mud are exposed in the *Spartina* and *Puccinellia* communities with eight or twelve sprayings, but this number does not affect the upper salt marsh community in the same way due to the growth of the resistant species *Agrostis stolonifera* and *Oenanthe lachenalii.* Plots treated eight or twelve times in 1968 still (1977) show large differences from the controls in the case of the *Puccinellia* and *Juncus* dominated communities. In the case of *Spartina*, recovery is complete. The bare plots were recolonized by rhizomes growing into them from unoiled areas outside the plots. Some long term effects on the upper salt marsh community are shown in Figs. 32.1 and 32.2, which illustrate the reactions of a susceptible species (*Juncus maritimus*) and one which gains a competitive advantage (*Agrostis stolonifera*). Early results from this series of experiments are described in more detail in Baker (1971a).

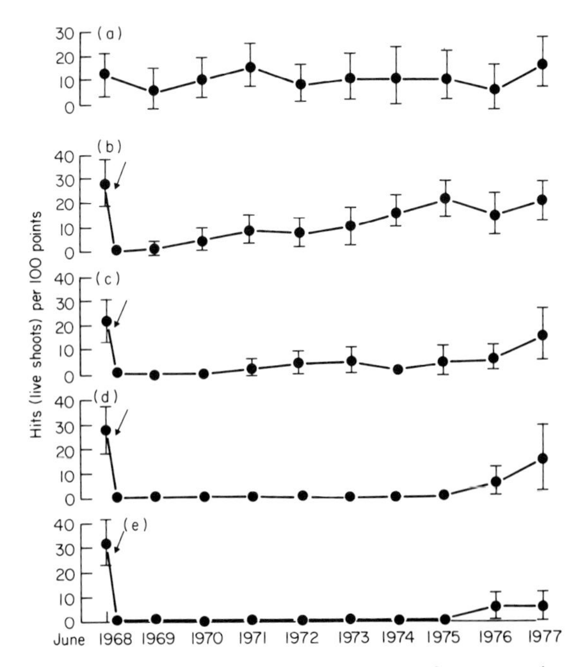

FIG. 32.1. Effects of successive monthly oilings with fresh Kuwait crude oil on *Juncus maritimus*. Results expressed as percentage cover of live shoots; vertical bars represent 95% confidence intervals. (a) untreated; (b) 2 oilings; (c) 4 oilings; (d) 8 oilings; (e) 12 oilings. Arrows indicate start of treatments.

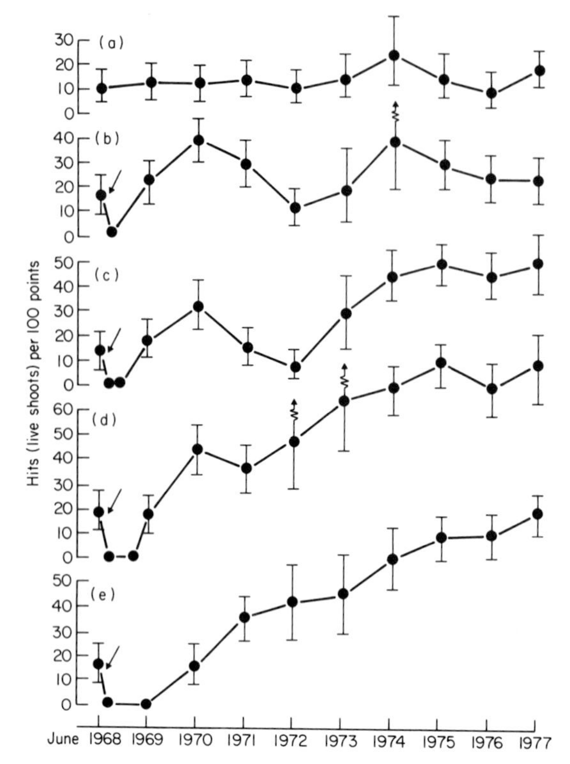

FIG. 32.2. Effects of successive monthly oilings with fresh Kuwait crude oil on *Agrostis stolonifera*. Results expressed as percentage cover of live shoots; vertical bars represent 95% confidence intervals. (a) untreated; (b) 2 oilings; (c) 4 oilings; (d) 8 oilings; (e) 12 oilings. Arrows indicate start of treatments.

THE SIGNIFICANCE OF SEASON

Another series of experiments, also detailed in Baker 1971a, was designed to find the importance of season as a factor determining extent of damage. A series of transects was sprayed with fresh Kuwait crude oil at different times of year, and recovery of the different species measured using point quadrats as described above. There was little long-term vegetative damage to most perennial species but populations of annuals (*Suaeda maritima* and *Salicornia* spp.) were severely reduced, except in the case of a transect oiled in November. With this transect, it was concluded that seed was set before oiling, that seeds were not damaged by the oil, and that the oil weathered sufficiently during the winter to allow the penetration of water for germination in the spring. Marked reduction of flowering of *Juncus gerardii*, *Festuca rubra*, *Plantago maritima* and *Spartina anglica* occurred if plants were oiled when the flower buds were developing (May in the case of the first three species and June and July in the case of the *Spartina*). This effect lasted one season only.

TOLERANCE OF DIFFERENT SPECIES TO OIL SPILLAGES

The following grouping is based on observations made during the experiments described above, and data from other field experiments and oil spillage incidents as summarized in Tables 32.1 and 32.2.

In terms of Raunkiaer's life form classification (Raunkiaer 1934), therophytes (e.g. *Salicornia* spp.) and chamaephytes (e.g. *Halimione portulacoides*) are more susceptible than hemicryptophytes and geophytes. The majority of British salt marsh plants are hemicryptophytes, but this cannot be assumed for marshes elsewhere.

Group 1—very susceptible

This group comprises shallow rooting, usually annual, plants with no underground storage organs; they are quickly killed by a single oil spillage. Examples: *Suaeda maritima*, *Salicornia* spp. and seedlings of all species.

Group 2—susceptible

This is a mixed group containing *Halimione portulacoides*, a shrubby perennial with exposed branch ends which are easily damaged by oil; filamentous green algae, which are killed but which recolonize quickly; and *Juncus maritimus*.

Group 3—intermediate

This group includes perennials which usually recover from a spillage or up to four light experimental oilings, but decline rapidly if further oiled. Examples: *Spartina anglica*, *Puccinellia maritima* and *Festuca rubra*.

Group 4—resistant

This group includes two kinds of species. The first kind are perennials which have a competitive advantage in vegetation recovering from oil, due to fast growth rate and mat-forming habit. Examples: *Agrostis stolonifera* and *Agropyron pungens*.

The second kind are perennials, usually of rosette habit, with robust underground storage organs (e.g. tap roots). Examples: *Armeria maritima*, *Plantago maritima* and *Triglochin maritima*.

Group 5—very resistant

In a group of its own is *Oenanthe lachenalii*. This species is not usually an important component of salt marsh vegetation, and it does not take over oiled areas in the manner of *Agrostis stolonifera*. Nevertheless, nearly all the *Oenanthe* plants in the successive spillages experiment responded to the twelfth oiling by rapidly producing yet more healthy shoots. Van Overbeek and Blondeau (1954) found that the plasma membranes of leaf and root cells of Umbellifers are particularly resistant to petroleum alkanes and alkenes, but the reasons are not known.

EFFECTS OF A REFINERY EFFLUENT

A *Spartina anglica* dominated salt marsh in Southampton Water, which has had refinery effluents discharged through its creek system since 1951, was surveyed in 1969 and 1970 to assess the extent of ecological damage. The salt marsh has been re-surveyed twice a year since 1972 to monitor any changes in the distribution of plant species in association with an effluent improvement programme started by the refinery (Dicks 1976).

The main pollutants that occur in refinery effluents are oil, sulphides, phenols, nitrogen compounds such as ammonia, and sludge (Blokker 1970). The Southampton Water effluent, in 1970, averaged 31 ppm oil content with a total discharge volume of approximately 515,000 litre min^{-1} through outfalls 1 and 2 (see Fig. 32.3). These figures have been substantially improved with in 1972 an average oil content of 25 ppm and volume of 470,000 litre min^{-1}, and in 1974 an average oil content of 14 ppm and volume of 390,000 litre min^{-1} (Dicks 1976). The effluent is discharged at high tide level to two large creeks. As well as the oil normally dispersed in the effluent, oil from other sources occasionally enters the creek system. These sources include:

a. accidental spillage within the refinery, which might produce unusually high oil contents in the effluent;
b. oil spillage from the refinery jetty installations and tankers;
c. oil spillages from other jetty installations in the area;
d. oil spillage from the considerable ship traffic using Southampton Water.

Full details of the marsh history in association with refinery developments are given by Baker (1971a, b), Dicks (1976) and subsequent unpublished reports of the Oil Pollution Research Unit. These can be summarized as follows:

1950 Aerial photograph taken by RAF shows the whole marsh area to be covered by healthy *Spartina anglica.*

1951 Outfall No. 1 starts operation.

1953 Outfall No. 2 starts operation.

1954 Aerial photograph taken by RAF shows what can be interpreted as oily vegetation along the edges of the creeks through which the discharges pass.

1962 Photographs of the marsh taken by Dr D.S. Ranwell show that large areas of *Spartina anglica* have died and decomposed around the two outfalls.

1966 Ordnance survey map the area and produce sheets no. SU 4404 and SU 4504 showing large areas of bare mud in the effluent discharge area.

1969 A series of transects across the marsh shows extensive areas of bare mud with the remains of *Spartina* stems and roots. The mud level in the denuded area was 15–25 cm lower than the nearby healthy marsh, indicating differential erosion and/or deposition (Baker 1971a, b).

1970 Re-survey shows little change from 1969. Conclusions reached at this stage (Baker 1971a, b) were that the main damage to the marsh was due to repeated light oilings from films of oil coming partly from the effluent and partly from spillages.

1972 to the present time, and continuing. Six-monthly vegetation mapping surveys show that the formerly denuded marsh has been progressively re-colonized, *Salicornia* spp. occupying the greatest areas so far. *Spartina anglica* is growing back, but its progress is slow compared with the opportunist *Salicornia.* Some areas devoid of higher plant life remain near the effluent discharge points. The formerly denuded marsh is still lower than the healthy marsh.

A reasonable interim conclusion is that the effluent improvement scheme has resulted in a decreased area of effect on the salt marsh, though reduced numbers of oil spills from nearby jetties may also be an important factor.

Fig. 32.3 shows the marsh with its creek system, the two discharge points, and the area now re-colonized by *Salicornia* spp.

EFFECTS OF OIL ON THE OXYGEN DIFFUSION PATHWAY OF *SPARTINA ANGLICA*

Using cylindrical platinum electrodes, Armstrong (1967) found that oxygen diffused from the roots of ten species of bog and marsh plants, including *Spartina anglica.* This result for *Spartina* has been repeated (Baker 1971a, b), using *Spartina* plants embedded in de-oxygenated agar jelly with electrodes round their roots. A reduction in oxygen diffusion rate was measured following the oiling of the leaves with slightly weathered Kuwait crude oil.

There is some evidence that *Spartina anglica* can eventually succumb to

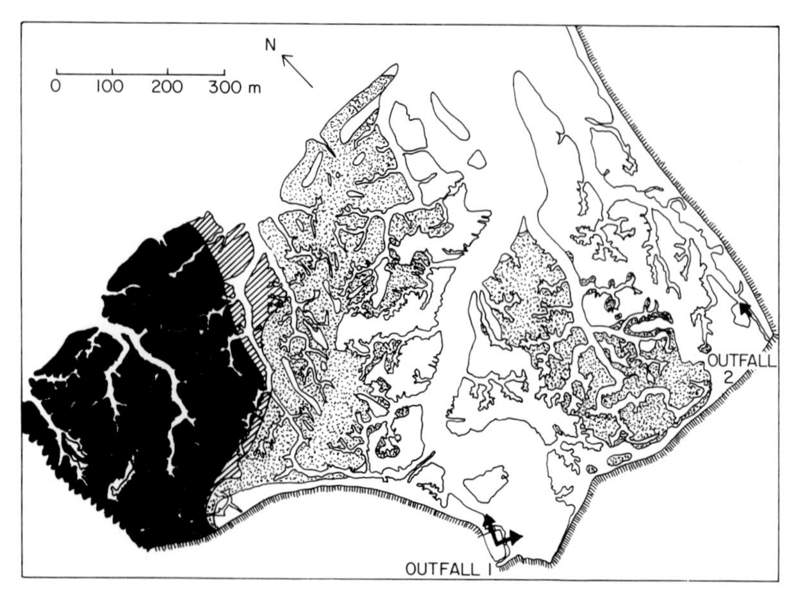

FIG. 32.3. Major vegetation changes from 1972 to 1977 on a Southampton Water salt marsh, in relation to two oil refinery effluent outfalls. Black area, normal marsh; cross-hatched area, *Spartina anglica* recolonization 1972–77; stippled area, *Saliconia* spp. recolonization 1972–77.

reducing conditions—Goodman and Williams (1961) decided that 'die-back' of *Spartina* was due to a toxic reduced ion, possibly sulphide, in the substrate. As oil on the leaves will promote reducing conditions by reducing the oxygen diffusion out of the roots, it is possible that oil polluted *Spartina* dies from both ends at once.

GROWTH STIMULATION FOLLOWING OIL POLLUTION

Growth stimulation of a variety of plants following oil pollution seems to be a not uncommon phenomenon. Mackin (1950a, b) observed that crude oil rapidly killed the grass *Distichlis spicata* but that later the plants completely repopulated the area and produced lush growth which he thought might be due to fertilization from oil decomposition products. Following 'Torrey Canyon' oil pollution in Brittany, Stebbings (1968) made the following observations: 'oil pollution is always considered to be entirely detrimental to the higher plants, but at St Anne's inlet stands of *Agropyron pungens*, *Juncus maritimus*, *Scirpus maritimus* and *Festuca rubra* were extremely vigorous. All except the latter were over 100 cm high and were dark green in colour'.

Growth stimulation was also noticed after experimental oil spraying originally designed to investigate seasonal and chronic oil pollution effects. Some separate work was therefore carried out to measure the stimulation and if possible find reasons for it. Shoot length and dry weight measurements made on *Puccinellia/Festuca* turves treated with Kuwait residue showed that this oil produced a highly significant stimulation (Baker 1971a, b). The reasons are still not clear and many factors may be involved. Two interesting points are that some crude oils contain compounds related to auxins and gibberellins (Gudin 1973), and that some bacteria are known to be able to utilize hydrocarbon substrates and also fix atmospheric nitrogen. One such organism is *Pseudomonas methanitrificans* (Davis *et al.* 1964). Apart from this, nutrients could be supplied by oil-killed animals, plant material and bacteria, or liberated from the oil itself.

ACKNOWLEDGMENTS

I would like to thank the staff of the Oil Pollution Research Unit for their assistance with field work, the oil companies who have provided information, oil samples and access to effluents, and the Gower Commoners' Association for permission to use experimental sites. I much appreciate the helpful continuing work. The financial support of the World Wildlife Fund and the oil industry is acknowledged with thanks.

REFERENCES

Armstrong W. (1967) The oxidising activity of roots in waterlogged soils. *Physiologia Plantarum* **20**, 920–26.

Baker J.M. (1970) The effects of oils on plants. *Environ. Pollut.* **1**, 27–44.

Baker J.M. (1971a) Studies on saltmarsh communities. *The Ecological Effects of Oil Pollution on Littoral Communities* (Ed. by E.B. Cowell), pp. 16–101. Applied Science Publishers, Barking.

Baker J.M. (1971b) *The effects of oil pollution and cleaning on the ecology of salt marshes.* Ph.D. thesis. University of Wales.

Baker J.M., Campodonico I., Guzman L., Texera J.J., Texera W., Venegas C. & Sanhueza A. (1976) An oil spill in the Straits of Magellan. *Marine Ecology and Oil Pollution* (Ed. by J.M. Baker), pp. 441–71. Applied Science Publishers, Barking.

Bender M.E., Shearls E.A., Ayres R.P., Hershner C.H. & Huggett R.J. (1977) Ecological effects of experimental oil spills on eastern coastal plain estuarine ecosystems. *1977 Oil Spill Conference Proceedings*, pp. 505–10. EPA/API/USCG.

Blokker P.C. (1970) Prevention of water pollution from refineries. *Water Pollution by Oil* (Ed. by P. Hepple), pp. 21–36. Applied Science Publishers, Barking.

Buck W. & Harrison J.G. (1967) Some prolonged effects of oil pollution on the Medway estuary. *Ann. Rep. Wildfowl Ass.* for 1967, 32–33.

Clapham A.R., Tutin T.G. & Warburg E.F. (1962) *Flora of the British Isles* 2nd edn. Cambridge University Press, London.

Cowell E.B. (1969) The effects of oil pollution on salt marsh communities in Pembrokeshire and Cornwall. *J. appl. Ecol.* **6**, 133–42.

Cowell E.B. & Baker J.M. (1969) Recovery of a salt marsh in Pembrokeshire, S. Wales, from pollution by crude oil. *Biological Conservation* **1**, 291–95.

Dalby D.H. (1969) Some observations on oil pollution of salt marshes in Milford Haven. *Biological Conservation* **1**, 295–96.

Davis J.B., Coty V.F. & Stanley J.P. (1964) Atmospheric nitrogen fixation by methane oxidising bacteria. *J. Bacteriol.* **88**, 468–72.

Dicks B. (1976) The effects of refinery effluents: the case history of a saltmarsh. *Marine Ecology and Oil Pollution* (Ed. by J.M. Baker), pp. 227–45. Applied Science Publishers, Barking.

Goodman P.J. & Williams W.T. (1961) Investigations into 'die-back' in *Spartina townsendii* agg. III. *J. Ecol.* **49**, 391–98.

Gudin C. (1973) *Mise en évidence dans le pétrole de régulateurs de la croissance des plantes.* Doctoral thesis, University of Paris.

Hampson G.R. & Moul E.T. (1978) No. 2 fuel oil spill in Bourne, Massachusetts: Immediate assessment of the effects on marine invertebrates and a 3-year study of growth and recovery of a salt marsh. *J. Fish. Res. Board Canada* **35**, 731–44.

Harrison J.G. (1967) Oil pollution fiasco on the Medway estuary. *Birds* **1**, 134–36.

Harrison J.G. & Buck W.F.A. (1967) An account of the Medway estuary oil pollution of September 1966. Supplement to *Kent Bird Report* **16**.

Hershner C.H. & Moore K. (1977) Effects of the Chesapeake Bay oil spill of February 2, 1976 on salt marshes of the lower Bay. *1977 Oil Spill Conference Proceedings*, pp. 529–34. EPA/API/USCG.

Hubbard C.E. (1968) *Grasses* 2nd edn. Penguin Books Ltd., Harmondsworth.

Lytle J.S. (1975) Fate and effects of crude oil on an estuarine pond. *1975 Oil Spill Conference Proceedings*, pp. 595–600. EPA/API/USCG.

Mackin J.G. (1950a) Report on a study of the effect of application of crude petroleum on salt grass *Distichlis spicata* (L.) Greene. Texas A. and M. Res. Found. Project 9 report.

Mackin J.G. (1950b) A comparison of the effect of application of crude petroleum to marsh plants and to oysters. Texas A. and M. Res. Found. Project 9 report.

Ranwell D.S. (1968) Extent of damage to coastal habitats due to the Torrey Canyon incident. *The Biological Effects of Oil Pollution on Littoral Communities* (Ed. by J.D. Carthy & D.R. Arthur), pp. 39–47. Supplement to *Field Studies* 2.

Ranwell D.S. & Hewett D. (1964) Oil pollution in Poole Harbour and its effect on birds. *Bird Notes* **31**, 192–97.

Ranwell D.S. & Stebbings R.E. (1967) Coastal ecology section report on the effects of Torrey Canyon oil pollution and decontamination methods on coastal habitats in Cornwall and Brittany. The Nature Conservancy, London.

Raunkiaer C. (1934) *The Life Forms of Plants and Plant Geography*. Oxford University Press, London.

Smith J.E. (Ed.) (1968) *Torrey Canyon Pollution and Marine Life*. Marine Biological Ass. of the U.K., Cambridge University Press, London.

Stebbings R.E. (1968) Torrey Canyon oil pollution on salt marshes and a shingle beach in Brittany 16 months after. The Nature Conservancy, Furzebrook Research Station.

Thomas M. (1977) Longterm biological effects of Bunker C oil in the intertidal zone. *Fate and Effects of Petroleum Hydrocarbons in Marine Organisms* (Ed. by D.A. Wolfe), pp. 238–45. Pergamon Press, New York.

Thomas M. (1978) Comparison of oiled and unoiled intertidal communities in Chedabucto Bay, Nova Scotia. *J. Fish. Res. Board Canada* **35**, 707–16.

Van Overbeek J. & Blondeau R. (1954) Mode of action of phytotoxic oils. *Weeds* **3**, 55–65.

33. ENVIRONMENTAL MANAGEMENT OF COASTAL DUNES IN THE NETHERLANDS

E. VAN DER MAAREL

Division of Geobotany,
University of Nijmegen,
The Netherlands

SUMMARY

Coastal dunes in the Netherlands have always been utilized by man for many purposes. However, unlike many other dunes of the world they are still largely in a semi-natural or near-natural state. This is the result of regular management brought about by governmental action since the 14th century. Initially, laws were introduced as a reaction to previous mismanagement, viz. extensive clearing of forests in the 12th century and overstocking with rabbits after their introduction (c. 1300). A marine transgression, which culminated c. 1350, and human disturbance have led to the formation of a new landscape, the 'Younger Dunes', over most of the former landscape. This landscape, the 'Older Dunes', which were formed between 3,000 BC and 0 AD, comprises a system of low ridges parallel to the coast separated by large wet flats.

The interior of the Older Dunes have been partly built upon (e.g. parts of The Hague and Haarlem) or taken into cultivation. A considerable part, however, remains as a beautiful 'park landscape' as the Dutch call it, which was created from the 16th to the 19th century when many country-seats were established.

The Younger Dunes may be divided into three major regions: the Holland region (H; provinces of North and South Holland, between Hoek van Holland and Alkmaar), Delta region (D) and Wadden region (W). Their main environmental features are extensive beaches with primary dunes (W), high and broad dunes (esp. H), both rich in lime (H, D) and poor in lime (W), pronounced zonation in vegetation and soil development (H, D), extensive primary slacks (D, W), secondary (blow-out) slacks (H, W), extensive transition between dunes and salt marshes (W, D).

The main functions of the coastal dunes are (in chronological order of their recognition and preservation): protection of the hinterland, water reservoir, scientific interest and biotic resources, landscape recreation.

Environmental management started as a direct single-function preservation and now includes the rigorous raising of seaward dunes, afforestation mainly with exotics, water catchment (including canals and infiltration works), creation of extensive recreation facilities and establishment of nature reserves.

Main environmental impacts include lowering of the ground water table, infiltration with polluted water, recreational pressures, disturbances from sand extraction, raising of seaward dunes, production of gas and military training.

Present-day ecological management is based upon three strategies, viz. internal management, external management and environmental zoning. These strategies are discussed and exemplified, with emphasis on the possible combinations of functions each coastal dune area could exert.

RÉSUMÉ

L'homme a toujours utilisé de nombreuses façons, les dunes côtières des Pays-Bas. Cependant, à la différence de beaucoup d'autres dunes, elles sont encore pour la plupart dans un état semi-naturel ou même quasi-naturel. Ceci est le résultat du contrôle gouvernemental régulier depuis le quatorzième siècle. Initialement on introduisit des lois en réaction à une mauvaise exploitation, en particulier contre l'abattage extensif de la forêt au douzième siècle, et la surpopulation des lapins après leur introduction en 1300. Une transgression marine culminant en 1350 et l'action de l'homme ont formé un nouveau paysage—les «Dunes récentes»—remplaçant presque partout l'ancien paysage. Ce dernier—les «Dunes anciennes»—qui s'etait formé entre 3000 avant J.-C. et l'an zéro, comprend des cordons parallèles à la côte, séparés par de larges plaines humides.

L'intérieur des anciennes dunes était partiellement construit (La Hague, Haarlem) ou cultivé. Une partie considérable cependant, reste le «Parklandschap» (ainsi appelé par les hollandais) qui a été créé à partir du seizième siècle jusqu'au dix-neuvième siècle, quand des dizaines de maisons de campagne ont été construites.

Les dunes plus récentes peuvent être divisées en trois grandes régions : La Hollande (H; province du sud et du nord entre Hoek van Holland et Alkmaar), le Delta (D) et le Wadden (W). Leurs principales caractéristiques environnementales sont : Larges plages avec dunes primaires (W), hautes et larges dunes (esp. H) calcaires (H, D) pauvres en calcaire (W), zonation prononcée de la végétation et du développement du sol (H, D), larges vallées humides primaires (D, W) ou secondaires (H, W), larges régions de transition entre les dunes et les vases salées (W, D).

Les principales fonctions (dans l'ordre chronologique de reconnaissance et de préservation) sont : Protection de l'arrière pays, réserves d'eau, intérêt scientifique et ressources biotiques, récréation.

L'aménagement de l'environnement qui était au début une simple préservation, comprend le renforcement des cordons dunaires, boisement, surtout avec d'arbres exotiques, la captation d'eaux (canaux et travaux d'infiltration), la création de larges approvisionnements pour la récréation et l'établissement de réserves naturelles.

Les principaux problèmes de l'environnement comprennent l'abaissement du niveau phréatique, l'infiltration d'eau polluée, de la pression de la récréation, les perturbations par l'extraction du sable, le renforcement des cordons dunaires, la production de gaz et l'entraînement militaire.

L'aménagement actuelle est basé sur trois stratégies, aménagement interne, externe et zonation de l'environnement. Nous discutons ces stratégies et donnons des exemples en mettant l'accent sur les possibles combinaisons de fonctions de chaque région dunaire.

INTRODUCTION

Coastal dunes in the Netherlands have always been utilized by man for many purposes. However, unlike many other dunes of the world they are still largely in a semi-natural or near-natural state, which is the result of continual governmental action since the 14th century. Inevitably Public Authorities and private organizations have been concerned mainly with the protection of the hinterland, the 'low countries', from flooding as well as from shifting sand.

This concern is largely a reaction to the serious mismanagement of most of the dunes during the 11th, 12th and 13th centuries, to which introduction of laws in the 14th century protecting the dunes was a response of the government of the time. Mismanagement, mainly extensive clearing of forests, contributed to the formation of a new and still developing dune system, the 'Younger Dunes' over most of the former dune system, the 'Older Dunes'. Environmental management of the present dunes should be understood in the light of historical events. Geological and historical evidence on the development of the Dutch coastal dunes will be summarized in the next section. This survey also will make clear that at least three major dune regions must be distinguished: the Holland region, i.e. the dune region in the provinces of North and South Holland between Hoek van Holland and Alkmaar, the Delta region and the Wadden region, each with its own history and management practices.

HISTORY

Holland region

A detailed account of the geological and vegetational history of the coastal dunes exists for the mid-western Netherlands (Pons *et al.* 1963; Jelgersma &

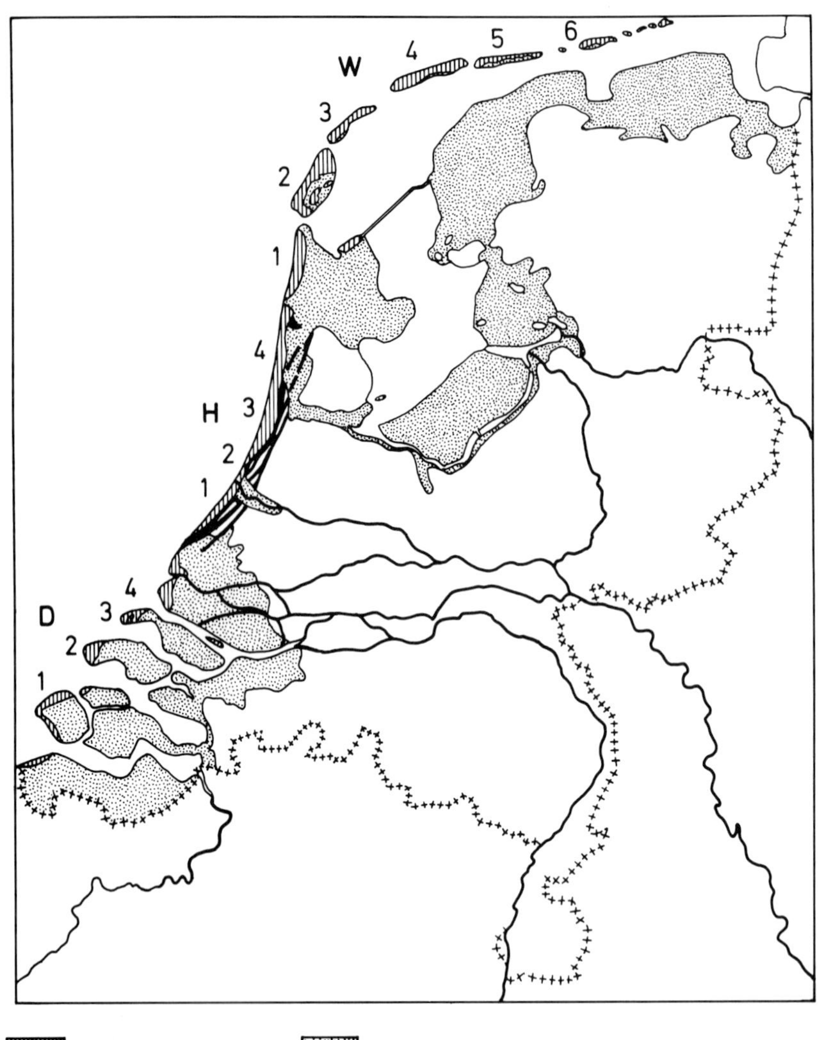

FIG. 33.1. Generalized geological map of the Netherlands showing the three major coastal regions, with dune areas viz. the Delta region south of Hook of Holland (D1–4), provinces of South and North Holland between Hook of Holland and Bergen (near Alkmaar) (H1–4) and the Wadden region north of Bergen (W1–6). The names of the areas are listed in Table 33.3. The figure also shows the position of the Older and Younger Dunes. Simplified from Jelgersma *et al.* (1970).

van Regteren Altena 1969; Jelgersma *et al.* 1970; Boerboom & Zagwijn 1966; Doing 1963; Doing & Doing-Huis in't Veld 1971; Doing 1974; Lambert 1971). The review articles of Zagwijn (1971) and Doing (1975) summarize the findings (see Fig. 33.1).

The continuous history of the present dunes goes back to 5,000 years ago when, after a period of quickly rising sea levels, the coastline was stabilized as a result of the formation of a number of sand barrier ridges ('strandwallen') separated by large wet flats ('strandvlakten'). On the ridges relatively low dunes were formed which became largely covered by forests. Here Bronze Age and Roman settlements grew up and the land nearby these settlements was grazed by stock.

About 700 AD this system, called the Older Dunes, began to be cleared of forests by the Carolinian people. A marine transgression started which caused a regression of the coastline. By 1200 AD the deforestation of the drier parts was almost complete and possibly as a result of this the older dunes were overblown with large quantities of fresh, carbonate-rich sand.

Up till 1600 AD the so-called Younger Dunes were formed, largely in two 'waves'. The first one reached the inner part of the Older Dunes in some areas (see Fig. 33.1). A new vegetation cover must have been established but this must have been kept open by rabbits which had been introduced c. 1300 AD and soon the dunes became overstocked. Moreover woodcutting, hunting and grazing took place during this period.

In the same period the population of the direct hinterland, mainly living in the cities of The Hague, Leiden, Haarlem and Alkmaar, grew considerably. Much of this urban development occurred on the remaining ridges of the Older Dunes. Another part of the Older Dunes became a dune heath under the influence of grazing and the flats were turned into meadows and fields. Thus both the Older Dunes and the hinterland of the State of Holland, by then already the economic and cultural heart of the country, were threatened by the effects of devastation associated with the movement of the Younger Dunes. These developments indicate the reasons underlying the introduction of the legislation governing the planting of marram (*Ammophila arenaria*), hunting and wood-cutting in the 14th century. Until the 19th century the coastline further regressed and the seaward dunes were rejuvenated. In the Older Dunes a new landscape arose as a result of the lay-out and maintenance of country-seats for the patricians of the nearby cities.

In the 19th century the seaward dunes became stabilized and attempts to reclaim the older dune slacks were started. These failed, although old fields with hedgerows and woodlands still remain. Most of the remaining dune heath was either built upon (The Hague, Haarlem) or dug to supply sand for housing on peaty sites and to prepare fields for horticulture (bulbgrowing). In the second half of this century the first extensive industrial activities occurred, notably the construction of the Rotterdam and Amsterdam waterways, connecting the harbours of these cities with the sea. In both cases further dune areas were covered by harbour installations.

By the end of the 19th century locally extensive planting of trees especially *Pinus nigra* took place; also water extraction from the dunes began. The establishment of the first nature reserves in the dunes dates from the early 20th century.

Finally recreation began to develop. In the last 20 years the use of the dunes for recreational purposes has been particularly evident. A new development in the water catchment of the dunes is infiltration with Rhine water. The regression of the coastline is increasing again. Storms and storm surges of which the one of 31 January 1953 is the most notorious, have damaged the seaward ridge and caused its retraction over tens of metres, whilst the beach has become smaller.

Delta region

Much less information is available on the history of the south-western coastal dunes (van der Maarel 1966; Adriani & van der Maarel 1968; Anon. 1972; Beeftink 1975; Pons *et al.* 1963; Jelgersma *et al.* 1970). The Older Dunes must once have formed a more or less continuous chain except for inlets, from Calais to Denmark, but in the south-western part of the Netherlands they were largely destroyed by the sea by 0 AD. The oldest dunes known there probably date from the early Middle Ages. The top soil is poor in lime and the land has been used for extensive grazing for many centuries. Most of the present dunes which probably were formed between 1200 and 1600 AD lie on peat and clay instead of sand.

The dunes of Voorne differ from other Delta dunes in that a new system of ridges and wet flats has been formed during the present century in a manner similar to the formation of the Older Dunes.

A special feature of the Delta region is the formation of so-called green beaches ('groene stranden'), large elevated beaches mostly situated along the estuaries, which are partly covered by the tide from the estuarine side. Such beaches are usually sandy, but at lower elevations silt has been deposited on the beaches. Unlike the 'yellow beach' (the foreshore facing the sea) the green beaches are densely covered with vegetation, and consist of a mosaic of salt marsh communities and wet and dry dune communities. Usually also embryo dunes with *Elytrigia* (*Agropyron*) *junceiformis** develop here (Westhoff *et al.* 1961).

The Delta Plan (which dates from 1938 but was approved by parliament

* Nomenclature of plant species follows Henkels & van Ooststroom (1975); nomenclature of plant communities follows Westhoff & den Held (1969).

only shortly after the catastrophic storm surge of 31 January 1953) is a major issue in the history of the Delta area. It aims at closing most of the estuaries and raising the seaward dunes and dikes to prevent any large-scale breaching by the sea. The major loss in ecosystems resulting from this plan is the green beaches. Parts of them were destroyed by new dikes, parts were cut from the sea and the remainder are changing as a result of increasing salinity.

On the other hand there are two developments of ecological interest:

1. the creation of sand flats in estuaries as a result of the lowering of the water table in some of the enclosed sea-arms;
2. the development of new salt marsh and green beach areas in front of the present coastline, the so-called fore-delta.

Wadden region

This region includes all the Westfrisian islands and the dunes of north Holland, north of Bergen which generally contain less than 0·1% $CaCO_3$ (Westhoff 1947). The difference in $CaCO_3$ content between northern and southern dunes is linked also with considerable differences in the mineralogical composition of the sands (Eisma 1968).

The history of the Wadden region is both more dynamic and less well known than that of the other regions (Pons *et al.* 1963; Jelgersma *et al.* 1970; Abrahamse & Veenstra 1976; Waterbolk 1976). Small dune systems of the 'Older Dunes' type have always existed but the number and the position of islands have changed, with a strong tendency towards 'wandering' to the east and to a lesser extent to the south. This movement of dunes increased during the medieval marine transgression and was accompanied by the formation of 'Younger Dunes'.

The Wadden dunes always have been connected with extensive salt marshes; the latter developed both along the Waddenzee coastline and on the large sand flats on the extremities of many islands.

The history of intensive human use of the Wadden region is well known from the Middle Ages onwards. Grazing started on the higher salt marshes but after dikes were constructed the polders directly behind the dunes became farm land. Because of the low $CaCO_3$ content of the dunes, dune scrub characteristic of southern dunes has only developed as a low growth of *Hippophaë rhamnoides* in the outer dunes. Most of the dunes were covered with open dune herb vegetation, dune grassland and extensive dune heath, all of which were grazed by sheep and cattle. Grazing also occurred on the salt marshes. The main occupation of the islanders was fishing, so that the demands on the resources of the dunes were not as strong as in Holland.

Afforestation, mainly with *Pinus nigra* has occurred on all islands. Together with drainage to the adjacent polders this has resulted in a considerable decrease in ground water level, although many small lakes and extensive wet slacks still occur in the area.

Urbanization has never been extensive. However, during the last 20 years there has been a rapid increase in recreation activities, including the development of camping sites and second homes. Finally, military training occurs on some of the sand flats.

Despite the low population density on the islands, dunes always have been regarded as an important coastal defence against incursions of the sea. During the present century, sand dikes ('stuifdijken') have been constructed on the large sand flats to the east and west of the islands and these have been stabilized with wood, *Ammophila arenaria* and × *Ammocalamagrostis baltica.* Such sand bars have enabled the development of sandy salt marshes and green beaches to occur on the Waddenzee side of the bars, which has increased greatly the variety of plant and animal communities on the flats.

PRESENT LAND USE

Main types of land use

From the historical description, the continuous human use of the coastal dunes, within or beyond their natural capacities, has become clear. At present we may distinguish four major and many minor uses of coastal dunes. These major uses are, in chronological order: coastal defence, water catchment, recreation and nature preservation. Most dune areas are managed primarily for one of these main uses. Table 33.1 presents some relevant statistics. The total dune area amounts to 73,000 ha of which 22,000 ha are more or less definitely transformed into urban-industrial and agricultural areas. Although the remaining more or less natural dune area is less than 1·5% of the Dutch territory, I agree with Doing (1975) that with respect to size and variation, the Dutch dunes as a whole can be considered one of the best developed and best preserved dune systems of the world.

Admission to most dune areas is either free or very cheap. Usually access is restricted to roads and footpaths. A few hundred hectares are 'strict reserve' and usually only accessible for research purposes. 80% of the reserves in the Wadden area are not accessible to the public during the breeding season. Altogether almost 90% of the present dune area is more or less open to the public. This, of course, is of considerable social benefit, but it also presents a growing problem for environmental management of the dune system as a whole.

Table 33.1. *Land use of all Dutch coastal dunes expressed as percentages. Figures are based on estimations of area from maps of the Netherlands and data from a handbook of nature reserves and recreation areas, Anon. (1976). Parentheses indicate a value of less than one per cent.*

	Dune region (Site)			
	Delta	Holland	Wadden	Total
Total dune area	10	56	34	100
Beaches, mobile seaward dunes	2	1	13	16
Slacks, scrubs, heaths, grasslands	5	17	16	38
Park landscape Older Dunes	( )	4	( )	4
Plantations (pine wood)	1	7	4	12
Agricultural land (Older Dunes)	( )	16	( )	16
Urban-industrial areas	2	11	1	14
Coastal defence areas	2	1	6	9
Water catchment areas	1	13	( )	14
Recreation areas; state-owned	( )	2	9	11
Ibid., society-owned or managed	( )	3	( )	3
Ibid. accessible municipal and private areas	( )	4	1	5
Nature reserves, state-owned	2	4	15	21
Ibid., society-owned or managed	2	1	2	5
Remainder (not accessible areas)	1	1	( )	2

Functions of the natural environment

The concept of function of the natural environment was developed for ecological planning of the physical environment of the Netherlands (van der Maarel & Vellema 1975; van der Maarel 1976b, 1978; Anon. 1977). A function is defined here as a possible land use.

Fig. 33.2 presents a general system of functions suggested for physical planning. The main division is based on exchanges of matter and energy as well as the flow of information between the natural environment and society. Production functions imply thc flow of matter and energy from natural environment to society, i.e. from abiotic, biotic and agricultural resources. Carrier functions are connected with a flow in the opposite direction, i.e. with the provision of space for human activities, artificial elements and waste. Information functions imply a flow of information towards society for orientation, scientific research, education and signalling of environmental changes through indicator organisms. Regulation functions are concerned with those processes in the natural environment which enable the continuous supply of matter and energy as well as information, i.e. processes of waste assimilation

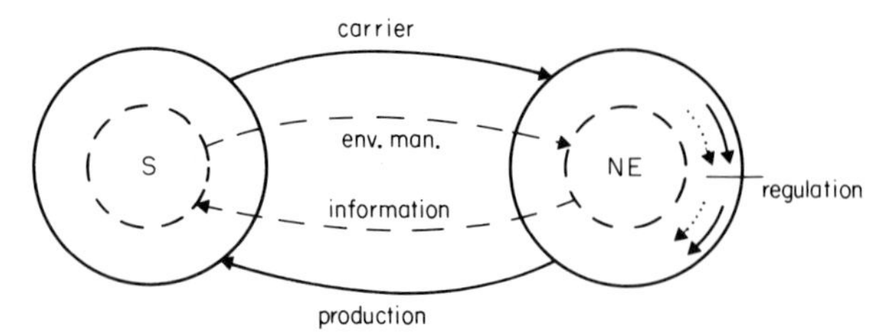

FIG. 33.2. Functions of the natural environment in relation to exchange of matter-and-energy and information between natural environment and society. S, Society; NE, Natural environment; env. man., environmental management. From Anon. (1977).

and environmental stabilization. The fourth flow in this scheme, i.e. the flow of information from society to natural environment, typically is the implementation of environmental management programmes.

Functions of the coastal dunes

The main types of land use mentioned above, as well as the various minor forms of land use, will now be listed and discussed in terms of the functions and activities involved.

Coastal defence (carrier function). The protecting seaward dunes are largely managed through the planting of marram, the erection of fences and recently also through mechanical displacement of sand and fixation with binders like bitumen (Adriani & Terwindt 1974).

Water catchment (abiotic production function). The water is pumped up and carried away through canals. Recently this has been accompanied by infiltration through pipelines and canals, with artificial dune lakes as reservoirs.

Recreation (carrier function). This can be divided into substrate-, landscape- and nature-recreation. In the dunes the latter types predominate and include walking, hunting, as well as plant and bird watching. Substrate recreations such as playing, swimming and picnicking occur on the beach and are of relevance because, under conditions of changeable weather, many of these activities may be transferred to the dunes where special facilities, including artificial bathing-lakes have been developed.

Nature preservation. This is the activity concerned with the maintenance of information and regulation functions. In the case of the coastal dunes we may mention the supply of information for nature-recreation, scientific research and education, indicators of air pollution and soil conservation.

Production of petroleum and natural gas (abiotic production function). Drilling for these fuels takes place in the dunes of North and South Holland and also in the Waddenzee area. Test borings on a much wider scale are planned for the latter area and even on some of the islands, which would of course have big repercussions for the area as a whole (Abrahamse *et al.* 1976).

Extraction of sand (abiotic production function). This took place mainly at the landward side of the Older Dunes and it still takes place in the Waddenzee where it affects currents and deposition of materials in the sea (Abrahamse *et al.* 1976).

Collection of fruits (biotic production function). A peculiar form of commercial fruit picking occurs on Terschelling where a population of the cranberry, *Oxycoccus macrocarpos*, has become established in some dune slacks, probably after an American ship was wrecked there in the 18th century (Westhoff *et al.* 1970). Various fruits of shrubs, notably *Hippophaë rhamnoides* are commercially collected. Picking of dewberries, *Rubus caesius*, especially in the outer dunes of North and South Holland is nowadays mainly a recreational activity.

Forestry (agricultural production function). The extensive plantations, mainly of *Pinus nigra* (Table 33.1), were originally planted as a timber resource and as a means of stopping soil erosion, but nowadays they are managed largely for recreational purposes.

Military training (carrier function). Except for a small artillery range near The Hague there is no space reserved for military training in the Delta and Holland dunes. However, the south-western sand flats of three Wadden islands are used for target practice.

Environmental management, the concept and the application of management practices to coastal dunes

In Fig. 33.2 environmental management is defined as similar to, but wider in scope than what Usher (1973) called the preferable concept of biological conservation. Environmental management is essentially Man's concern for an optimal fulfilment of the functions of the natural environment, on the basis of those properties of the environment which determine the regulation and information functions. Such functions are bound to the more or less natural landscape types (Westhoff 1971; van der Maarel 1975b, 1978) and may be grouped as *natural functions*.

The natural environment's capacity or suitability to fulfil these functions may be estimated through the process of 'evaluation' (van der Maarel 1975b,

1976b, 1977; van der Maarel & Vellema 1975; Kalkhoven *et al.* 1976; Seibert 1975; cf. Wright 1977 for a recent survey of English and American work). This approach yields a more quantitative assessment of a need for conservation which had been recognized earlier by Mörzer Bruyns (1967) and Westhoff (1971b) for the Netherlands and Helliwell (1973) for England. Our own studies distinguish four evaluation indices as representing the various information and regulation functions: a diversity index and a scarcity index are used to estimate the complex of information functions, a structual differentiation index (based on stratification and life-form diversity) and an index of naturalness (based on originality of structure and proportion of spontaneously occurring species) are used to estimate the complex of regulation functions, particularly the resistance-stability (Holling 1973) of the systems.

STRUCTURE AND FUNCTIONING

Main dune landscape types and their plant communities

Landscape types are defined here as relatively homogeneous recognizable units of the earth's surface with a characteristic land form, soil type and complex of plant and animal communities. They may also be called geotopes or ecotopes (van der Maarel 1978). Tables 33.2 and 33.3 give a comprehensive survey of the main dune landscape types and their principal plant communities, as well as their occurrences in 14 coastal dune areas (see also Fig. 33.1).

Beaches with embryo dunes occur most extensively on the large sand flats of the Wadden islands of Terschelling, Ameland and Schiermonnikoog (Westhoff 1951; Westhoff *et al.* 1970; Abrahamse *et al.* 1976). Green beaches, one of the very typical Dutch landscape types, have largely disappeared in the Delta region but still occur in the Wadden area (Westhoff 1951; Westhoff *et al.* 1961). Mobile dunes with a luxuriant growth of *Ammophila arenaria* are virtually extinct now, except on the Wadden islands.

Outstanding dune slacks have developed on Voorne (van der Maarel & Westhoff 1964; Adriani & van der Maarel 1968) but others occur elsewhere, mainly on the Wadden islands. Secondary slacks are well differentiated according to different amounts of calcium carbonate in their top soils. The secondary slacks of the islands of Terschelling and Schiermonnikoog show the largest ecological variation. The biotic variation of dune slacks as a whole is determined mainly by the differences in ground water regime (Ranwell 1972; van der Maarel 1971).

There are only three dune lakes of any particular size, viz. Quackjeswater and Breede Water in the Voorne dunes and the Zwanewater near Callantsoog (area H1, Fig. 33.1), and a larger number of smaller ones (Leentvaar 1963, 1967).

Table 33.2. *Main coastal dune landscape types in the Netherlands and the principal characteristic plant communities and their occurrence in fourteen coastal dune areas.*

	Region													
	Delta				Holland				Wadden					
Dune landscape type	1	2	3	4	1	2	3	4	1	2	3	4	5	6
A. Beach with embryo dunes		·		·							·	+	+	+
B. Green beach			+	+						+	·	+	+	+
C. Mobile seaward dune		·	·	·						·	+	+	+	+
D. Primary dune slack		·	·	+						·	·	+	+	·
E. Secondary dune slack of Younger Dunes, rich in lime				+	·			·						
F. Dune lake				+	·				+					
G. Dune scrub mosaic of Younger Dunes, rich in lime	·	·	·	+	+	+	+	+		·				·
H. Dune grassland mosaic of Older Dunes, inner zone Younger Dunes	·	+	+	+	·	·	·	+						
I. Dune forest, park landscape of Older Dunes	·			·	+	+	+	·						
J. Secondary dune slack of Younger Dunes, poor in lime									·	·	·	+	·	+
K. Dune scrub mosaic of Younger Dunes, poor in lime									·	+	+	+	+	+
L. Dune heath										·	·	+	+	·
M. Dune pine woods		·		·	·	·	·	·	·	·	·	·	·	·

Table 33.3. *Plant communities associated with 13 landscape types given in Table 33.2.*

A. Soc. *Cakile maritima*, soc. *Honkenya peploides*, Agropyretum boreoatlanticum
B. Ibid. + Sagino maritimae-Cochlearietum danicae, Junco-Caricetum extensae, comm. of all. Agropyro-Rumicion crispi, transitions to salt marsh communities
C. Elymo-Ammophiletum
D. Centaurio-Saginetum moniliformis, ass with *Parnassia palustris* and *Schoenus nigricans* of all. Caricion davallianae, Caricetum trinervis nigrae
E. Ibid. + comm. of order Molinietalia
F. Samolo-Littorelletum, Scirpo-Phragmitetum, willow comm. of all. Salicion cinereae, Macrophorbio-Alnetum
G. Hippophao-Ligustretum, tall scrubs with *Crataegus monogyna* and/or *Betula verrucosa* of all. Berberidion, Convallario-Quercetum dunense, therophyte and grassland ass, of all. Galio-Koelerion
H. Grassland ass. of all. Galio-Koelerion and Thero-Airion, Calluna comm. of all. Calluno-Genistion, comm. of all. Agropyro-Rumicion
I. Mixed woodland comm. of suball. Ulmion carpinifoliae, Convallario-Quercetum dunense, Macrophorbio-Alnetum
J. Samolo-Littorelletum, ass. with *Schoenus nigricans* of all. Caricio davallianae, Caricetum trinervi-nigrae, ass. with *Salix repens* and *Salix pentandra* of all. Salicion cinereae, Pyrolo-Salicetum, Empetro-Ericetum
K. *Hippophaë rhamnoides* comm. of all. Salicion arenariae; Polypodio-Empetretum, therophyte and grassland ass. of all. Galio-Koelerion and all. Thero-Airion
L. Empetro-Genistetum tinctoriae, Festuco-Galietum
M. Leucobryo-Pinetum, Convallario-Quercetum dunense

soc. = sociation; ass. = several associations; all. = alliance; comm. = several communities.

The *Hippophaë-Ligustrum-Crataegus* dune scrub, probably the most conspicuous vegetation of the Dutch dunes occurs almost everywhere, but is especially well developed on Voorne and between The Hague and Haarlem (Sloet van Oldruitenborgh 1976). This scrub can be found in a large number of communities, each with its own place in the zonation from the seaward to the landward dunes, and in the range from dry to moist habitats. In the Wadden area this dune scrub is limited to the outer dunes, where it usually does not develop beyond the *Hippophaë*-stage because of the rapid acidification of the lime-poor sand. Instead *Empetrum nigrum* and *Salix repens* form a low scrub. In the landward zones of the broader dunes, notably those near The Hague and Haarlem, the scrub reaches heights of 6–8 m and forms a mosaic with woodland and grassland.

Planted woods of deciduous trees, mainly consisting of *Populus* species and *Quercus robur*, may form a more or less natural element, but the same cannot be said of most of coniferous woods which consist largely of *Pinus nigra*. Only in some of the Wadden regions, is a natural development of a moss and dwarf shrub layer found, with typical boreal species such as *Listera cordata* and *Goodyera repens* (Westhoff 1952, 1959). At many places in the lime-rich dunes, the *Pinus* woods are declining (Doing 1974).

The semi-natural park woodland which occurs in the Older Dunes, contains a number of different woodland communities with a large number of both native and naturalized herbs (Westhoff 1952; Doing 1963).

The dune grasslands still cover extensive areas in the inner parts of the Delta dunes, and contain a large variety of different plant communities and plant species (Westhoff *et al.* 1961; van der Maarel 1966, 1975). These grasslands much resemble the Scottish machair (Gimingham 1964; Ranwell 1974).

Dune heath, which usually alternates with grassland, is well developed in the Wadden area (de Smidt 1975). Estimations of the area occupied by these different types of vegetation are given in Table 33.1.

Flora and fauna

The flora and fauna of coastal dunes are very rich, but generally not well known (Ranwell 1972). Only two Dutch dune areas have been described in some detail: the water catchment dunes of The Hague (Croin Michielsen 1974) and Voorne (Adriani & van der Maarel 1968). The Voorne dunes are a representative dune system as far as biotic diversity is concerned. Table 33.4 gives some data on this area. There are 29 groups of plants and animals distributed over nine types of dune landscape. By 1968 over 3,000 different species had been identified. The Voorne dunes, although covering only 1,600 ha, or less than 0·5% of the Dutch territory, provide so much variation that 50–70% of the species and communities found in the Netherlands occur there.

Table 33.4. *Some biological characteristics of the different landscape types of Voorne (after Adriani & van der Maarel 1968).*

Landscape type	A	B	C	D	E
Beaches and flats	20	5	4	1	9
Green beaches and salt marshes	80	0	12	2	18
Mobile dunes	60	0	3	1	11
Dune grasslands	220	20	13	12	20
Dune scrubs	250	0	7	11	19
Dune slacks	280	30	20	13	25
Dune lakes	80	20	12	9	23
Dune forests	150	40	5	11	17
Park landscape inner dunes	120	25	6	8	16
Total	700	111	56	22	29
% of total for the Netherlands	52	66	63	54	70

A. Number of characteristic vascular plant species.
B. Number of characteristic bird species.
C. Number of plant sociological orders.
D. Number of plant and animal groups with many representatives.
E. Number of plant and animal groups with at least some representatives.

Diversity of plant species also has been calculated in relation to average species numbers for areas of different size, based on Preston's species-area relationship (Adriani & van der Maarel 1968; van der Maarel 1971). The richness of plant species for different dune areas was approximately 2 to 3 times greater than the average values for areas of comparable size in the Netherlands as a whole. The value for the Voorne dunes was 4·5 times greater.

The total number of plant species for the Wadden islands is estimated as 863 (Joenje *et al.* 1976).

The total number of plant species occurring in the coastal dunes is estimated to be as high as 1,100. Ranwell (1972) mentioned 900 species found in 43 of the more important British dune systems, and later reported a total list of nearly 1,100 species (Ranwell 1974). Of course, not all of these species are native, but they are at least naturalized. Of a total of 1,400 plant species present in the Netherlands, 80% occur in coastal dunes. Approximately 140 out of 170 regularly breeding bird species, which is over 80% of the national total, have been recorded for the coastal dunes.

It goes almost without saying that such extraordinarily high biological diversity is linked with the presence of rarities. Indeed, this has been demonstrated for Voorne (Adriani & van der Maarel 1968) and the Wadden islands (Joenje *et al.* 1976). Another aspect of the large diversity is the special floristic (and possibly faunistic) character of each of the dune areas. This was recognized early for the Wadden islands (Westhoff 1947) and recently stressed again by Joenje *et al.* (1976); it has been documented also for the Delta dunes by van der Maarel (1966), who characterized them as 'a chain of isolates'. Indeed the relatively recent isolation of both Delta and Wadden islands is one of the most interesting features of the Dutch dune system. So far no signs of endemism at the species level have been recognized, but there may be some endemic varieties present in the dunes (Westhoff 1947).

Gradients within coastal dunes

Westhoff *et al.* (1961), van der Maarel and Westhoff (1964), van Leeuwen (1966), Adriani and van der Maarel (1968), van der Maarel (1971, 1976) have stated repeatedly that environmental gradients in dunes are of decisive importance for the development of both within-community and between-community

Table 33.5. *Main environmental gradients and processes occurring in coastal dunes (adapted from van Leeuwen & van der Maarel 1971, van der Maarel 1976).*

Environmental factor	Topographical gradient	Process
Moisture	dry—moist	desiccation—inundation
Texture	sand—clay	sand accretion—silt accretion
Carbonate content	poor in lime—rich in lime	acidification—calcification
pH	acid—base rich	
Organic matter	humic—mineral	humification—mineralisation
Nutrients	poor in nutrients (N, P) —rich in nutrients	oligotrophication—eutrophication
Chloride	fresh—salt	desalination—salination
Level of rabbit grazing	low—high	
Environmental dynamics	low—high	

diversity. Table 33.5 indicates the main environmental gradients, most of which occur in different forms and in different combinations in coastal dunes. As a result of subtle complex gradients high levels of biotic diversity may develop in dunes, both at the α and β level of diversity (see van der Maarel 1971 for examples).

Most of these environmental gradients reflect dynamic processes within dunes, i.e. they are continuously changing (van Leeuwen & van der Maarel 1971).

The gradients discussed above are small-scale gradients, but in addition large-scale gradients, notably the one from outer to inner dunes, exist in which the age of the different dunes has been correlated with the vegetational zonation (Doing 1974, 1975) and the decreasing influence of air-borne salt

(Adriani & van der Maarel 1968; Sloet van Oldruitenborgh 1976). A gradient on a still larger scale occurs along the coastline of north-west and west Europe as a whole, where the mediterranean–atlantic influences of southern Europe are replaced gradually by the boreal–atlantic influences of the north (Westhoff, this volume).

ENVIRONMENTAL IMPACTS

Survey of influences

The different land uses of coastal dunes result in a variety of primary and secondary effects on the natural environment. In accordance with recent terminology these effects are examples of different impacts on the environment. Each impact may be considered as the result of an interaction between an influence exerted on the environment through any human activity, and the reaction of the environment. The latter reflects both the resistance and resilience of the environment to the perturbation. Usually influence and reaction are considered at the ecosystem level and the ecosystem's reaction to human influences (mainly an adverse one) is expressed in terms like sensitivity and vulnerability.

The main influences are listed in Table 33.6, together with the type of land use with which they are associated. This scheme should be interpreted in a general way, as a framework for detailed case studies.

Water catchment and recreation may be considered the two most important forms of land use in coastal dunes in relation to their impact on dunes and the possibility of developing guidelines for a better control of these forms of land use. They will be discussed now as examples.

Impacts of water catchment

The main impact is the gradual lowering of the ground water table. In the principal catchment areas this amounts to well over 5 metres as very large quantities of water have been extracted. For the water catchment dunes of The Hague, the yearly amount of extracted water increased from c. 5×10^6 m^3 around 1900 to over 40×10^6 m^3 in 1970. Until 1930 the extracted water came from the upper sand layers. From that time onwards water from deeper layers also was extracted and since 1955 the bulk of the total extracted water comes from infiltrated water. The impact of phreatic water extraction can be inferred from the fact that the yearly net influx of water is c. 5×10^6 m^3 and between 1890 and 1955 the average yearly extraction is estimated to be $8{\cdot}5 \times 10^6$ m^3. Thus the total excess extraction is c. 200×10^6 m^3 over that

Table 33.6. *Human influences present in coastal areas arranged according to the ecosystem component they are affecting.*

Ecosystem	Human influence	Coastal defence	Water catchment—extraction	Water catchment—infiltration	Recreation—substrate	Recreation—landscape	Recreation—nature, research	Production of petroleum, gas	Extraction of sand	Collection of fruits	Forestry	Military training
Substrate	Appearance of artifacts	×		×	×			×				×
	Accretion of sand	×										
	Removal of sand	×			×				×			
	Mobilisation of sand				×				×			×
	Fixation of sand	×									×	
Soil structure	Ploughing up							×	×			×
	Compaction				×	×	×					
Soil water	Lowering phreatic water table		×									
	Raising phreatic water table			×								
	Inundation			×								
Nutrition	Eutrophication		×	×	×	×						
	Calcification				×				×			×
	Mineralisation		×		×							
Plants	Removal	×			×	×				×		
	Introduction	×				×					×	
Vegetation	Removal	×			×				×			
	Treading				×	×	×					
	Planting	×									×	
Animals	Removal					×						
	Introduction										×	
	Disturbance				×	×		×				×

period (Tuinzaad 1974). As a consequence, the original ecosystems of dune slacks and shallow dune lakes (Tables 33.2–33.3) in the catchment areas have been virtually eliminated.

Infiltration has compensated for the loss of water and indeed new wet dune slacks and lakes have appeared in the infiltration areas of The Hague and Haarlem, and at first sight this would appear to be an ecological compensation. However two side-effects have been noticed which make infiltration a doubtful practice in relation to ecological objectives.

The first effect, noticed in the dunes near The Hague, was that a dune grassland community Taraxaco-Galietum maritimi occurring in drier valleys and lower parts of dune slopes, largely disappeared as a result of inundation. This also led to the decline of various rare species including some probably endemic *Taraxacum* species.

A second effect is the occurrence of eutrophication. The infiltration water is much richer in nutrients than the dune soil. London (1975) and van der Werf (1970, 1974) followed vegetational developments on permanent plots and demonstrated undesired changes as a result of the presence of infiltration water. The most conspicuous change was the decrease in number of species on most plots. Van der Werf gives the following example:

year	1959	1963	1968	1972
number of species	41	47	24	4

Infiltration started in 1956 and it first caused a slight increase, but later a strong decrease, in the number of species at the site. The few remaining species are nitrophilous ruderals.

Impact of recreation

Recreation in coastal dunes consists of many activities and requires the presence of many facilities (Anon. 1958, 1964, 1965, 1974). These include the construction of special foot, cycle and bridle paths, picnic and parking places, camping and second home sites and even artificial lakes (Roderkerk 1961; Anon. 1965). Chief recreational activities are walking, picnicking, cycling, riding, hunting and fruit and flower collecting. The impacts of these activities are clear (Table 33.6) but difficult to assess in quantitative terms.

Studies have been performed, both on the Wadden islands and in some dunes in the Holland region. We may summarize the effects as follows:

1. Walking causes the decrease and subsequent disappearance of erect plants and the promotion of creepers and rosette plants. Subsequent disturbance of the soil causes an increase in nitrophilous species.
2. New bare paths are formed, at least in the drier dunes. Simple measurements of total path length from air photographs have shown quite considerable increases in only a few years.

3. Bare spots may become blow outs.
4. Bird populations in heavily visited areas decrease (cf. Westhoff 1967; van der Werf 1970, 1974; Anon. 1974).

From work in the dunes near The Hague it has become clear that various types of vegetation differ in their vulnerability to recreational pressures; grasslands on steep slopes show maximal vulnerability and tall scrub and woodland on flat sites are least affected.

A general problem here is that ecologists can indicate the dangers, but it needs considerably more time to demonstrate such changes based on the results of long-term studies. Authorities are often reluctant to accept the ecologist's view without adequate proof.

STRATEGIES FOR MANAGEMENT

Three strategies

Three different strategies of environmental management may be distinguished. The first strategy relates to the direct management of an area, usually a nature reserve, with the aim to restore, maintain or increase its ecological significance, i.e. its information and regulatory capacity. It is usually called *internal management* in the Netherlands and it includes the counteracting of adverse effects of influences acting upon the area.

The second strategy, which we call *external management* is concerned with the influences themselves and includes attempts to counteract adverse effects outside the area concerned. The first and second strategy together make up environmental management *sensu* Duffey and Watt (1971).

The third strategy relates to the management of the area under concern in a much wider context in which a balance between different types of land use including the conservation of certain areas is aimed at. This strategy may be called *environmental zoning.*

Internal management

Internal management is first of all concerned with the regulation of influences from human activities, both external influences *sensu* Ranwell (1972), like catchment of water, and internal influences, like cropping or grazing. Possible measures are regulation of the intensity of grazing, removal of biomass to cope with eutrophication and diversion of footpaths to protect sites which are vulnerable to trampling.

Where the ground water table has been lowered, the soil can be removed as far as the phreatic level. This was done in the Kennemer duinen, but as

Londo has explained (1970) the measure did not result in the re-establishment of dune slack communities because the newly created slopes were too steep.

It should be realized that in cases of semi-natural ecosystems, proper internal management implies the maintenance of practices like grazing or mowing. Further it may be necessary to take measures against effects of human influences which have ceased in the recent past.

Finally internal management may be concerned with situations without any current adverse effects from human influences. In such cases measures may be devised to increase abiotic and biotic diversity.

Dune ecosystems which respond to internal management are dune scrub, dune slacks and dune grasslands; these are the most biologically diverse habitats. The management of dune scrub will be discussed here, based on the recent study of Sloet van Oldruitenborgh (1976).

In her view, scrub development is a response to a lowering in the intensity of some environmental factor, such as the removal of grazing, the building of a new dune ridge, or the stabilization of older sand dunes. Such a development can take place too rapidly. The spread of *Hippophaë rhamnoides* in the British dunes is a good example of rapid colonization by a species which has resulted in a loss of biological diversity in dune systems (Ranwell 1972b). In such an extreme case, cutting or even burning of the scrub might be an appropriate measure to conserve the diversity. Sloet considers the introduction of grazing a more suitable measure for the Dutch dunes. An alternative to grazing is mowing, which is carried out at several places particularly in moist slacks. Another measure is the removal of exotic trees or planted woodland in scrub zones, because they enhance the stabilization of the dune system and this may not be considered desirable for the scrub zone as a whole.

External management

External management in its simplest form is simply attempting to stop or reduce activities which nullify the objectives of management. Local sand extraction, high population densities of rabbits or people are examples of undesirable factors. If simple control measures are not possible because of arrangements for the multiple-use of the area, the setting up and proper management of buffer zones should be considered. In relation to recreational pressures, providing alternative routes, location of attractive sites away from the vulnerable reserves, location of obstacles to prevent access to areas and restriction of certain activities around reserves (e.g. hunting) are all methods which may be used to limit damage. Where eutrophication of an area which borders an oligotrophic site has occurred, a buffer zone should be developed which would restrict nutrient movement into the oligotrophic habitat (e.g. planting of trees, or cropping of vegetation in order to reduce nutrient load).

The adverse impact of infiltration with polluted water has been discussed already. Purification of infiltration water would be necessary prior to its discharge, or else less polluted water should be used. The most effective external management strategy at present is preventing the catchment of water in areas which still have a high phreatic water table.

Environmental zoning

External management can only solve part of the problem of overall environmental management and we have to think of the third and most general of the three strategies: environmental zoning. First we need to distinguish between the compartmental approach and zoning in a strict sense. The meaning of the term environmental compartments is similar to that defined by Odum (1969), who distinguished between a protective (natural), a productive (agricultural), an urban–industrial and a compromise (multiple use) compartment. A compartment may be defined as a part of the environment in which a specific combination of functions occurs. Doxiadis (1975) distinguished 12 'ecological types of space' or 'basic zones' according to the degree of human impact, ranging from virginal 'naturareas' to densely industrialized 'industrareas'.

In cooperation with the Dutch Planning Agency we distinguished eight compartments, each of which is characterized by a combination of functions: information + regulation, information + biotic production, information + recreation, rural, agricultural, substrate-bound recreation, urban and industrial + abiotic production (van der Maarel 1978; Anon. 1977). The next step was the study of the incompatibility of environmental compartments. Some compartments could hardly be combined in space, but others can be and here the idea of zoning comes in.

The main environmental compartments we derive for the dunes are:

Strict nature reserve
Nature reserve, national park area
General recreation area
Coastal defence area
Water catchment area
Urban–industrial area
Agricultural area

It is easy to see which compartments are largely incompatible, e.g. urban–industrial with water catchment or recreation with information-regulation. In such cases buffer compartments should be planned. Fig. 33.3 presents a zoning model with a similar compartment system devised for the Wadden area. Seven activities are distinguished and their possible ecological acceptance in the seven zones is indicated, based on an ecological assessment of their likely impact (Joenje *et al.* 1976b).

Obviously the most vulnerable compartment is the strict nature reserve.

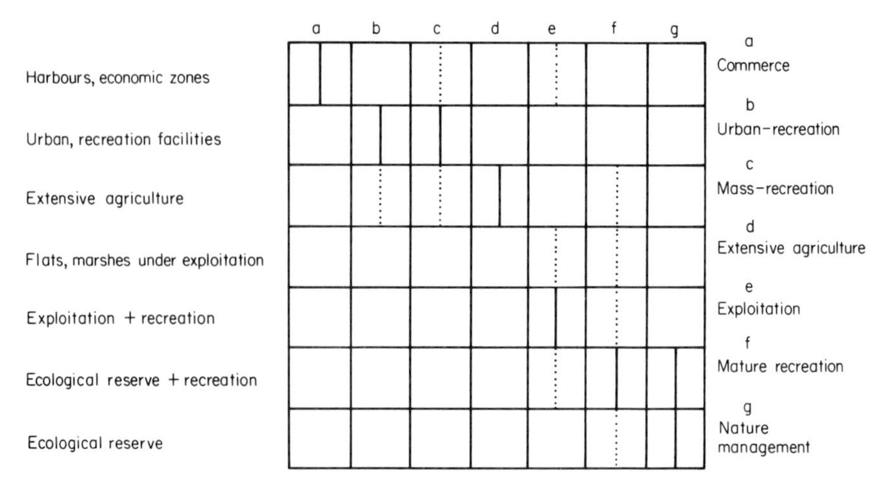

FIG. 33.3. Relation between function-compartments and forms of land use in the Wadden area. a = commerce; b = urban and recreational activities; c = mass recreation; d = extensive agriculture; e = nature exploitation; f = nature recreation; g = nature management. From Joenje *et al.* (1976b).

Consequently the choice and location of land designated as a nature reserve and its subsequent management is of prime importance in any zoning system (Westhoff 1971, 1971b; Helliwell 1976).

Special attention in zoning should be given to the water catchment areas which serve many functions but there is a growing awareness that management must include the ecological implications of land use (Croin Michielsen 1974).

EPILOGUE

In conclusion environmental management of the Dutch coastal dunes benefits from earlier management practices which have provided our generation with a largely semi-natural and an extremely varied dune landscape. However, management is hampered by our inadequate knowledge of ecosystem changes which are taking place in response to former drastic environmental changes. This indicates the need for continuous ecological research, especially the analysis of permanent plots and repeated vegetation mapping.

Finally, we are confronted with ever growing demands from the large population of the Netherlands and much of W. Germany. The deleterious effects of water catchment and recreational use of dunes could in the end destroy much of the ecological interest of the dunes. Here we may touch upon a fourth aspect in the strategy: the promotion of changes in society, which go beyond an environmental plan. Society should accept an ecological view-point and find alternative water supplies, accept a more rigid regulation of recreation activities and give up petroleum and gas exploration in and around the coastal dunes. The production and sale of many splendid books notably Westhoff *et al.* (1970), Croin Michielsen (1974) and Abrahamse *et al.* (1976) and the activities of large societies like the Waddenzee Society possibly represent a welcome growing awareness of the ecological view-point.

ACKNOWLEDGMENTS

The considerable assistance of Liesbeth Diemel in preparing Table 33.1 and parts of the text and the comments on the manuscript by Dr Clara Sloet van Oldruitenborgh and Dr Henk Doing are gratefully acknowledged.

REFERENCES

Review articles in English are marked with an asterisk.

Anon. (1958) *Recreatie en beplanting in de Haagse duinen* (English summary). Instituut voor Toegepast Biologisch Onderzoek in de Natuur, Meded. 39, Arnhem.

Anon. (1964) *Recreatie en natuurbescherming in het Noordhollands Duinreservaat* (English summary). Instituut voor Toegepast Biologisch Onderzoek in de Natuur, Meded. 69, Arnhem.

Anon. (1965) *Strand. en duinrecreatie.* Koninklijke Nederlandse Toeristenbond A.N.W.B., Den Haag.

Anon. (1972) *De kleuren van Zuidwest-Nederland, visie op milieu en ruimte.* Contact-commissie voor Natuur- en Landschapsbescherming, Amsterdam.

Anon. (1974) *Recreatie en natuurbehoud in het Waddengebied.* Koninklijke Nederlandse Toeristenbond A.N.W.B., Den Haag.

Anon. (1976) *Handboek van natuurreservaten en wandelterreinen in Nederland.* Vereniging tot Behoud van Natuurmonumenten in Nederland, 's-Graveland.

***Anon. (1977)** *Summary General Ecological Model.* Series 'General Physical Planning Outline', part 3. Ministry of Housing and Physical Planning, The Hague.

Abrahamse J., Joenje W. & van Leeuwen-Seelt N. (Eds.) (1976) *Waddenzee, natuurgebied van Nederland, Duitsland en Denemarken.* Vereniging tot Behoud van de Waddenzee, Vereniging tot Behoud van Natuurmonumenten in Nederland, Harlingen & 's-Graveland (also in German & Danish).

Abrahamse J. & Veenstra H. (1976) Waddengebied van Nederland. In: Abrahamse J. *et al.* (1976), pp. 47–61.

Adriani M.J. & van der Maarel E. (1968) *Voorne in de branding.* Stichting Wetenschappelijk Duinonderzoek, Oostvoorne.

Adriani M.J. & Terwindt J.N.J. (1974) *Sand stabilization and dune building.* Rijkswaterstaat Comm. 19, The Hague.

Beeftink W.G. (1966) Vegetation and habitat of the salt-marshes and beach plains in the south-western part of the Netherlands. *Wentia* **15**, 83–108.

Beeftink W.G. (1975) The ecological significance of embankment and drainage with respect to the vegetation of the South-West Netherlands. *J. Ecol.* **63**, 423–58.

Croin Michielsen N. (Ed.) (1974) *Meyendel, Duin-Water-Leven.* W. ter Hoeve B.V., Den Haag & Baarn.

Doing H. (1963) De buitenplaatsen en bossen langs de binnenduinrand van Noord- en Zuid-Holland (English summary). *Natuur Landschap* **16**, 261–81.

Doing H. (1974) Landschapsoecologie van de duinstreek tussen Wassenaar en IJmuiden (English summary). *Meded. Landbouwhogeschool Wageningen* 74-12.

Doing H. (1975) Beobachtungen und historische Tatsachen über die Sukzession von Dünen-Ökosystemen in den Niederlanden. *Sukzessionsforschung* (Ed. by W. Schmidt), pp. 107–18, Cramer, Vaduz.

Doing H. & Doing-Huis in 't Veld C.J. (1971) History of landscape and vegetation of coastal dune areas in the province of North-Holland. *Acta Bot. Neerl.* **20**, 183–90.

Doxiadis C.A. (1975) The ecological types of space that we need. *Envir. Conserv.* **2**, 3–13.

Duffey E. & Watt A.S. (Eds.) (1971) *The Scientific Management of Animal and Plant Communities for Conservation.* Blackwell Scientific Publications, Oxford.

Eisma D. (1968) Composition, origin and distribution of Dutch coastal sands between Hoek van Holland and the island of Vlieland. *Neth. J. Sea Res.* **4**, 123–267.

Ellenberg H. (1972) Belastung und Belastbarkeit von Oekosystemen. *Tagungsber. Ges. Oekologie Giessen* 19–26.

Gimingham C.H. (1964) Maritime and Submaritime communities. *The Vegetation of Scotland* (Ed. by J.H. Burnett), pp. 67–142. Oliver & Boyd, Edinburgh and London.

Helliwell D.R. (1973) Priorities and values in nature conservation. *J. Envir. Management* **1**, 85–127.

Helliwell D.R. (1976) The extent and location of nature conservation areas. *Envir. Conserv.* **3**, 255–58.

Heukels H. & van Ooststroom S.J. (1975) *Flora van Nederland*, 18e druk. Wolters-Noordhoff, Groningen.

Holling C.S. (1973) Resilience and stability of ecological systems. *Ann. Rev. Ecol. Syst.* **4**, 1–23.

***Jelgersma S., Jong J. de, Zagwijn W.H. & van Regteren Altena J.F. (1970)** The coastal dunes of the western Netherlands; geology, vegetational history and archaeology. *Meded. Rijks Geol. Dienst* N.S. **21**, 93–167.

***Jelgersma S. & van Regteren Altena J.F. (1969)** An outline of the geological history of the coastal dunes in the Western Netherlands. *Geologie Mijnbouw* **48**, 335–42.

Joenje W. & Thalen D. (1968) Het Groene Strand van Schiermonnikoog (English summary). *Levende Natuur* **71**, 97–107.

Joenje W., Westhoff V. & van der Maarel E. (1976) Plantengroei. In Abrahamse J. *et al.* 1976, pp. 177–95.

Joenje W., Wolff W. & van der Maarel E. (1976b). Waardering van het Waddengebied. In Abrahamse J. *et al.* 1976, pp. 333–42.

Kalkhoven J.T.R., Stumpel A.H.P. & Stumpel-Rienks S.E. (1976) *Landelijke milieukartering. Een landschapsecologische kartering van het natuurlijk milieu in Nederland ten behoeve van de ruimtelijke planning op nationaal niveau* (English summary). Staatsuitgeverij, Den Haag.

***Lambert A.M. (1971)** *The Making of the Dutch Landscape, an Historical Geography of the Netherlands.* Seminar Press, London and New York.

Leentvaar P. (1963) Dune waters in the Netherlands. I. Quackjeswater, Breede Water and Vogelmeer. *Acta Bot. Neerl.* **12**, 498–520.

Leentvaar P. (1967) Duinmeren II: Zwanewater, Muy, Oerd en van Hunenplak (English summary). Biol. Jb. Dodonaea **35**, 228–66.

Leeuwen C.G. van (1966) A relation theoretical approach to pattern and process in vegetation. *Wentia* **15**, 25–46.

Leeuwen C.G. van & van der Maarel E. (1971) Pattern and process in coastal dune vegetation. *Acta Bot. Neerl.* **20**, 191–98.

Londo G. (1971) *Patroon en proces in duinvalleivegetaties langs een gegraven meer in de Kennemerduinen* (English summary). Thesis, Nijmegen.

Londo G. (1975) Infiltreren is nivelleren (English summary). *Levende Natuur* **78**, 74–79.

***Maarel E. van der (1966)** Dutch studies on coastal sand dune vegetation, especially in the Delta region. *Wentia* **15**, 47–82.

Maarel E. van der (1971) Plant species diversity in relation to management. *The Scientific Management of Animal and Plant Communities for Conservation* (Ed. by E. Duffey & A.S. Watt), pp. 45–63. Blackwell Scientific Publications, Oxford.

Maarel E. van der (1975) Observations sur la structure et la dynamique de la végétation des dunes de Voorne (English summary). *La végétation des dunes maritimes* (Ed. by J.-M. Géhu), pp. 167–83. Cramer, Vaduz.

Maarel E. van der (1975b) Man-made natural ecosystems in environmental management and planning. *Unifying concepts in ecology* (Ed. by W.H. van Dobben & R.H. Lowe-McConnell), pp. 263–74. Junk & Pudoc, The Hague and Wageningen.

Maarel E. van der (1976) On the establishment of plant community boundaries. *Ber. Deutsch. Bot. Ges.* **89**, 415–43.

Maarel E. van der (1976b) De winning en aanvulling van grondwater: ecologische gevolgen (English summary). *H_2O* **9**, 533–42.

***Maarel E. van der (1978)** Ecological principles for physical planning. *The Rehabilitation of Severely Damaged Land and Freshwater Ecosystems in Temperate Zones* (Ed. by M.W. Holdgate & M.J. Woodman), pp. 413–50. Plenum, New York.

Maarel E. van der & Vellema K. (1975) Towards an ecological model for physical planning in the Netherlands. *Ecological Aspects of Economic Development Planning*, 128–43. Economic Commission for Europe, U.N., Geneva.

Maarel E. van der & Westhoff V. (1964) The vegetation of the dunes near Oostvoorne. *Wentia* **12**, 1–61.

Mörzer Bruijns M.F. (1967) Value and significance of nature conservation. *Nature Man* 1967, 37–47.

Odum E.P. (1969) The strategy of ecosystem development. *Science* **164**, 262–70.

***Pons L.J., Jelgersma S., Wiggers J. & Jong J.D. de (1963)** Evolution of the Netherlands coastal area during the Holocene. *Verh. Kon. Ned. Geol. Mijnbouwk. Gen. Geol. Ser.* 21–2, 197–208.

***Ranwell D.S. (1972)** *Ecology of Salt Marshes and Sand Dunes.* Chapman & Hall, London.

Ranwell D.S. (Ed.) (1972b) *The Management of the Sea Buckthorn,* Hippophaë rhamnoides *on Selected Sites in Great Britain.* Rep. Nature Conservancy Norwich.

Ranwell D.S. (Ed.) (1974) *Sand Dune Machair.* Report Seminar Colney Research Station, Norwich.

Roderkerk E.C.M. (1961) *Recreatie, recreatieverzorging en natuurbescherming in de Kennemerduinen* (English summary). Thesis, Wageningen.

Seibert P. (1975) Versuch einer synoptischen Eignungsbewertung von Oekosystemen und Landschaftseinheiten (English summary). *Forstarchiv* **46** (5), 89–97.

Sloet van Oldruitenborgh C.J.M. (1976) *Duinstruwelen in het Deltagebied* (English summary). Thesis, Wageningen.

Smidt J.T. de (1975) *Nederlandse heidevegetaties* (English summary). Thesis, Utrecht.

Tuinzaad H. (1974) Duin voor het water, water voor het duin. In Croin Michielsen N. (1974), pp. 211–15.

Usher M.B. (1973) *Biological Management and Conservation.* Chapman & Hall, London.

Veenstra H. (1976) Getijdenlandschap: struktuur en dynamiek. In Abrahamse J. *et al.* (1976), pp. 19–45.

Waterbolk H.T. (1976) Oude bewoning in het Waddengebied. In Abrahamse J. *et al.* (1976), pp. 211–21.

Werf S. van der (1970) Recreatie-invloeden in Meyendel (English summary). *Meded. Landbouwhogeschool Wageningen* 70-17.

Werf S. van der (1974) Infiltratie: met water meer plant. In Croin Michielsen N. (1974), pp. 226–30.

***Westhoff V. (1947)** *The vegetation of dunes and salt marshes on the Dutch islands of Terschelling, Vlieland and Texel.* Thesis, Utrecht.

Westhoff V. (1951) De Boschplaat op Terschelling (English summary). *Natuur Landschap* **5**, 15–32.

Westhoff V. (1952) Gezelschappen met houtige gewassen langs de binnenduinrand (English summary). *Dendr. Jb.* 9–49.

Westhoff V. (1959) The vegetation of Scottish pine woodlands and Dutch artificial coastal pine forests; with some remarks on the ecology of *Listera cordata. Acta Bot. Neerl.* **8**, 422–48.

Westhoff V. (1961) Die Dünenbepflanzung in den Niederlanden. *Angew. Pfl. Soziol.* **17**, 14–21.

Westhoff V. (1967) The ecological impact of pedestrian, equestrian and vehicular traffic on vegetation. *I.U.C.N. Publ.* N.S. **7**, 218–23.

Westhoff V. (1971) The dynamic structure of plant communities in relation of the objectives of conservation. *The Scientific Management of Animal and Plant Communities for Conservation* (Ed. by E. Duffey & A.S. Watt), pp. 3–14. Blackwell Scientific Publications, Oxford.

Westhoff V. (1971b) Choice and management of nature reserves in the Netherlands. *Bull. Jard. Bot. Nat. Belgique* **41**, 231–45.

Westhoff V., Bakker P.A., Leeuwen C.G. van & Voo E.E. van der (1970) *Wilde planten. Flora en vegetatie van onze natuurgebieden*, deel I. Verenging tot Behoud van Natuurmonumenten in Nederland, Amsterdam.

Westhoff V. & Held A.J. den (1969) *Plantengemeenschappen in Nederland.* Thieme, Zutphen.

Westhoff V., Leeuwen C.G. van, Adriani M.J. & Voo E.E. van der (1961) Enkele aspecten van vegetatie en bodem der duinen van Goeree, in het bijzonder de contactgordels tussen zout en zoet milieu (English summary). *Jb. Wet. Gen. Goeree-Overflakkee*, 47–92.

Wright D.F. (1977) A site evaluation scheme for use in the assessment of potential nature reserves. *Biol. Conserv.* **11**, 293–305.

***Zagwijn W.H. (1971)** Vegetational history of the coastal dunes in the Western Netherlands. *Acta Bot. Neerl.* **20**, 174–82.

34. THE ECOLOGY OF VEGETATION OF THE DUNES IN DOÑANA NATIONAL PARK (SOUTH-WEST SPAIN)

FRANCISCO GARCIA NOVO

Department of Ecology,
University of Seville,
Seville, Spain

SUMMARY

In this paper the ecology of the vegetation of stabilized and mobile sand dunes in the National Park of Doñana is described, with emphasis on the successional processes taking place in the area.

The vegetation was studied using numerical methods (factor analysis) in order to detect gradients of variation as well as interpretable vegetation types. Environmental measurements included chemical analyses of soil, monitoring of water table fluctuations and the study of the patterns of movement of the dune fronts.

The vegetation of the area is dominated by shrubs, with a few tree species in specific localities. As shown in the analysis, the vegetation on the mobile dunes contrasts with that of the stabilized sands. Within both of these groups a strong contrast appears between vegetation of lower and elevated ground. In the stabilized sands, the predominant environmental factor is the depth of the soil water table. In the mobile dunes the predominant environmental factor is substrate stability but in the more stable areas (slacks) it is the distance to soil water table.

Within this framework, the precise distribution of some shrub species is discussed in relation to water table depth and to the fluctuations of their leaf water potential from winter to summer. The main series of shrub succession are also discussed.

The distribution of trees both in the stabilized sands and in the mobile dunes is discussed in terms of water availability, fire frequency and substrate stability.

The management of the Park and some of the problems that are facing it are discussed as well as the early measures that have been taken for the conservation of Doñana area.

RÉSUMÉ

Nous décrivons l'écologie de la végétation des dunes de sable, stabilisées et mobiles, du parc national de Doñana, et en particulier les processus de succession qui s'y déroulent.

La méthodologie comprend l'étude de la végétation par des méthodes numériques (analyse factorielle) afin d'interpréter les directions de variation, ainsi que les types de végétation.

Les facteurs de l'environnement comprennent l'analyse chimique du sol, le contrôle des fluctuations du niveau de l'eau dans le sol, et l'étude des modèles du mouvement des fronts de dunes. La végétation dominante comprend un petit nombre d'espèces d'arbustes, dans une situation bien déterminée.

Les analyses montrent une opposition entre les types de végétation des dunes mobiles et ceux des sables stabilisés. Dans chaque groupe il y a une forte opposition entre les terrains supérieurs et inférieurs. Dans les sables stabilisés le facteur prédominant de l'environnement est la distance du sol par rapport au niveau de l'eau. Dans les dunes mobiles le facteur prédominant de l'environnement est la stabilité du sable et pour les zones stables (les dépressions humides des dunes) c'est la distance du sol par rapport au niveau de l'eau.

Dans cette optique, nous discutons de la distribution précise de quelques espèces d'arbustes en relation avec le niveau d'eau, par la mesure des fluctuations du potentiel de l'eau des feuilles, de l'hiver à l'été. Nous discutons aussi des principales séries de successions de ces arbustes.

Nous discutons de la distribution des arbres dans les dunes de sable stabilisées et également dans les dunes mobiles, en terme de disponibilité de l'eau, fréquence du feu, et stabilité du sable.

Nous examinons l'administration du parc et quelques problèmes qui en découlent, ainsi que les premières mesures prises pour la préservation de la région de Doñana.

INTRODUCTION

The National Park of Doñana was set up in 1969 close to the mouth of River Guadalquivir to preserve a sector of the south-west coast of Spain well known for its rich fauna. Since that time, research on the ecology of the Park has proved that the flora and fauna as well as the geomorphology are unique. Extensive destruction of coastal landscapes of Spain as a result of recreational

---

Nomenclature follows *Flora Europaea*. For those families not yet published, as well as for a few critical taxa, Galiano and Cabezudo (1976) have been followed.

pressures has created further interest in the preservation of the Park and it was enlarged in 1978 from 37,000 to 49,000 ha.

The ecology of vegetation of dunes within the National Park of Doñana will be described in this paper, which summarizes work done by the Department of Ecology of the University of Seville since 1970. Emphasis will be given to the ecological processes which occur in the area as a whole, rather than to detailed descriptions of ecosystems which are already published.

DESCRIPTION OF THE AREA

The National Park of Doñana includes two distinct geological structures: an ancient depression of tectonic origin now silted up with deltaic deposits of Guadalquivir River, that forms 'la marisma' (marsh) and a large coastal plain locally covered with aeolian sands, that is called 'las arenas' (sands). A wide sand barrier on which there are mobile dunes, separates the marsh and the plain from the sea (Fig. 34.1).

The marsh

The National Park of Doñana includes but a fraction of the vast marsh area that once existed over the region. Although extensively reclaimed for agricultural developments in recent years (Grande Covian 1975) unspoilt marshes still form a vast flat area on the west bank of the Guadalquivir River; some 20,000 ha of those belong to the Park. According to Vanney (1970) and Menanteau and Pou (1977) the marsh is in an advanced stage of evolution and is largely filled with clay sediments which are rich in calcium and magnesium. Rivers entering the marsh circulate through sinuous channels with raised banks (levées) that eventually drain into the Guadalquivir River. The combination of different topography, sediment composition and water origin (rivers, rain, infiltration, sea) provides a high diversity of biotopes for aquatic insects (Soler Andres *et al.* 1976), fishes (Hernando 1978) and vegetation (Gonzalez Bernaldez *et al.* 1977; Allier *et al.* 1977). According to the latter authors, plant cover is dominated by *Scirpus maritimus* and *S.lacustris* with *Ranunculus baudotii* and *Typha latifolia* in deeper waters. Shallow waters and increased salinity favour species of *Salicornia*, *Artrocnemum* and *Suaeda*. On sandy substrates, where the waters may be acidic (pH 4–5·5), rush stands develop (*Juncus maritimus*, *J.acutus*, *J.effusus*, *Holoschoenus vulgaris*) with *Trifolium maritimum* and other annual species. The high productivity of the marsh is strongly seasonal with a spring maximum after the winter rains.

FIG. 34.1. An infra-red picture showing the main features of Doñana taken by the satellite Landstat-1 on 13 March 1973. The dark area in the bottom left is the Atlantic Ocean (Gulf of Cadiz), and the large river to the right is the Guadalquivir River. The wide black area to the left of the river is the marshes; they appear black because they were covered with water at the time the picture was taken. Separating the marsh from the sea, the mobile dunes extend for over 50 km in the picture; they appear light in colour due to high reflection. Darker areas correspond to aforested slacks. Grey areas to the left of the marsh correspond with shrub vegetation.

The marsh is very rich in wildfowl (ducks and geese) with colonies of herons, spoonbills and egrets (Valverde 1960, 1967).

The sands

The second large geological unit of Doñana corresponds to a coastal plain that originated in the Pliocene (I.G.M.E. 1975). Within the boundaries of the Park, it is covered with sand dunes in which seven successive dune phases have been recognized (Pou 1976). They all originated under the influence of west or south-west winds and the dunes are parabolic (Chapman 1976) or transgressive (Davies 1972). Present-day topography (Fig. 34.1) shows a series of arched ridges (old dunes) 0·5–2·0 km long and 100–200 m wide, separated by flat depressions (the old slacks). A fairly shallow water table results in frequent flooding of depressions to give temporary lagoons and ponds. The drainage network is poorly developed and functions only after heavy rains. All the dunes were fixed long ago and no sand movement takes place at present.

The mobile dunes

The mobile dunes system corresponds with a sand barrier separating the marsh from the sea. It also extends over a sector of coastal plain, advancing inland from the beach. The sand is carried to the beach by marine currents parallel to the coast and then it is blown inland by south-west winds. The large size of this sand barrier (25 km long and some 5 km wide at its maximum) and the presence of underlying consolidated marine deposits suggests a complex origin. The sand barrier has been interpreted as originating from coastal drift (Vanney 1970), coastal uplift (I.G.M.E. 1972), sedimentation (Gavala 1952) or a combination of them.

The mobile dunes of the system are of the transgressive type (Davies 1972) with fairly straight fronts 200–500 m long parallel to the shoreline. Dune slopes are 30–35° on the active fronts and 1–2° on the retreating tail. There are four or five successive dunes between the beach and the inner boundary of the system, each separated by a flat depression (slack) from the next (Fig. 34.1). Dune size increases from the beach (5–10 m height) to the innermost dune front which is up to 30 m high. To the west only two or three successive dune fronts remain; to the east dune movement is interferred with by underlying landforms, because the dunes progress over the ancient banks of meanders near to the mouth of the Guadalquivir river.

The slacks associated with the dunes are elongated and run parallel to the dune fronts. Their length varies between 200 and 500 m, although a few exceed one kilometre. Their width fluctuates around 200 m. The bottoms of the slacks are flat but contain a number of small ridges (0·5–2 m high) parallel to the dune front which are called 'contradunas' (Allier *et al.* 1974). The water table in the soil usually lies within 2 m of the surface and shows an annual fluctuation of approximately 1 metre. Phreatic water finally discharges into shallow lagoons aligned along the main dune ridges. Their precise location slowly shifts as the dune fronts progress; displacements of up to 0·5 m per year have been observed (Valverde, personal communication). Lagoon size ranges from 1 to 50 ha, in winter flooding. During rainy winters lagoons discharge into one another from west to east; however they do differ in the chemical composition of their waters (Hernando 1978), phytoplankton (Margalef 1976) and zooplankton (Armengol 1976).

Climate

Annual precipitation averages 550 mm with a winter peak of 90 mm during November or December and a secondary peak of 85 mm during February or March. Summer drought is severe with less than 0·5 mm during July and August and an average of 24 mm for June. Summer drought is coupled with high temperatures. Average mean temperatures are 20·5°C for June, 23·4°C for July and 23·6°C for August. Winters are comparatively mild with average mean temperatures of 9·3°C for the coldest months (December and January). The oceanic influence results in mild winters and relatively cool summers compared with other areas of mediterranean climate.

THE ECOLOGY OF DUNE VEGETATION

The two approaches used in the study of the ecology of vegetation on the stabilized sands and mobile dunes of the Doñana National Park are:

1. The description of the vegetation and the environmental factors influencing the distribution of plants.
2. The interpretation of the patterns of plant succession.

Two sampling procedures were used. In stratified sampling, plots were recorded at random within each apparently homogeneous unit of vegetation. On other occasions, transects were laid out along environmental gradients and the vegetation was recorded at appropriate intervals as described later.

Methods

Only general methods used in the study of the vegetation and environmental factors are described here; those methods applied to specific problems are described in the appropriate references.

Sampling of vegetation

For the stratified sampling 20 × 20 m plots were placed at random. Data recorded included either the presence or absence of each species, or in the case of perennials, the cover of each species. Cover was measured on each plot by the line-intercept method along five 20 m lines parallel to one another but placed at 5 m intervals. Along the transects, 5 × 5 m plots with interception lines of 5 m set 1 m apart, were used. Frequency data for annuals was collected using 10 quadrats of size 20 × 20 cm, randomly distributed over each plot.

Detailed descriptions of the methods are given in Ramirez Diaz (1973), Torres Martinez (1975), Gonzalez Bernaldez *et al.* (1975a), Ramirez Diaz and Torres Martinez (1977).

Other vegetation studies

Water relations of shrub plants were studied along a gradient of water availability in the dunes. Water potentials of the leaves of all shrub species were measured along the transect under conditions of maximum water stress (late summer) and minimum water stress (early spring). Field measurements of water potential of leaves were carried out using the Scholander pressure chamber and the Chardakov solution equilibrium method. Details of the methods are given by Merino *et al.* (1976).

Water table depth measurements

Plastic pipes (PVC, 2·5 cm diameter) which had plastic mesh across the end to prevent clogging, were buried in the ground to a depth between 1·5 and 3·5 m. Water levels were recorded using a graduated rod provided with an electrical sensor at the tip.

Dune movement

Three methods were used to measure dune movement. The first method involved measurements of the movement of dune fronts in relation to fixed points (e.g. trees and posts) during a five year period. The second method was based on the comparison of aerial photographs taken over a 20-year span from 1956 onwards. The third method was based on the demography of umbrella pine (*Pinus pinea*) populations in the slack. The basis of this method is that a slack is laid bare by the tail of a dune as it moves inland. As the dune recedes, the ground is available for plant growth and pine seedlings become established. The age of a particular pine stand is thus a measurement of the time that has elapsed since the dune moved from the site. In order to apply the method, a transect is laid down across the slack in the direction of the movement of the dune. Along the transect, the age of oldest pine per plot is recorded together with its distance from the tail of the dune. Plotting the distance against the age, a rectilinear relationship is obtained. The slope of the line of regression has been taken as the average rate of dune movement along the transect. A detailed discussion may be seen in Garcia Novo *et al.* (1975), Torres Martinez (1975) and Allier *et al.* (1975).

Succession studies

Short-term changes in the composition of the vegetation have been monitored in permanent quadrats and along permanent transects from 1974 until 1977. Two of the transects were laid across an area by the 'Casa de Santa Olalla' which suffered a fire in the summer of 1974. Plant cover, plant density and plant frequency for both annuals and perennials have been recorded late in spring and again in autumn in the 5 × 5 m plots. Another 16 areas also have been studied where fires or disturbances are known to have taken place during the last 20 years. In each area the cover of shrubs was recorded on plots of 20 × 20 m by the line interception method. Following flooding, plant succession also has been monitored since 1975 at the 'Laguna del Brezo', a transient lagoon with extreme fluctuations in water level. Here, four transects placed radially to the lagoon and 25 other permanent quadrats of 2 × 2 m were laid out. Plant cover and frequency have been recorded. Individual perennials have been identified and monitored as they become established on the dried bottom of the lagoon. Details of the methods may be found in Martin (1978).

Numerical analysis

Analysis of some of the data has been carried out using ordination techniques. Principal component analysis (Gittins 1969) with 'varimax' rotation was used for the quantitative vegetational data presented here. A full description of the numerical methods employed in this study is given by Ramirez Diaz (1973), Torres Martinez (1975) and Gonzalez Bernaldez *et al.* (1975a, b). A discussion on the interpretation of factor analysis in environmental gradient studies based on Doñana data may be seen in Ramirez Diaz *et al.* (1976).

The vegetation of the sands

A principal component analysis of the vegetation of the sands was performed with cover data of 31 perennial species on 59 plots, including all the types of vegetation which had been initially recognized (Garcia Novo *et al.* 1971). The main components of variation are represented in Fig. 34.2. The vegetation of mobile dunes is separated from that of stabilized sands. The former may be further subdivided into that present in the dune and that characteristic of the slack. Vegetation of stabilized sands is subdivided between juniper (*Juniperus phoenicea*) forest, shrubs of elevated ground (naves) and shrubs of the low ground. The existence of several scrub types in the area was apparent, although their relationships were obscured in the analysis due to the heterogeneity of the data. A new analysis of the 22 scrub plots (same cover data as above but now reduced to 20 species) produced clearer results. Three groups of samples were neatly defined (Fig. 34.3). In order to explore the relationships between shrub types a transect was laid down so that it included the maximum variation in shrub vegetation from that present in a dry ridge of an old dune to that in a temporary pond in the bottom of the slack. Principal component analysis of 31 plots produced a detailed picture of shrub composition showing four definite vegetation types (Gonzalez Bernaldez *et al.* 1975b). Data of fluctuations in water table at the different types of site along the transect are

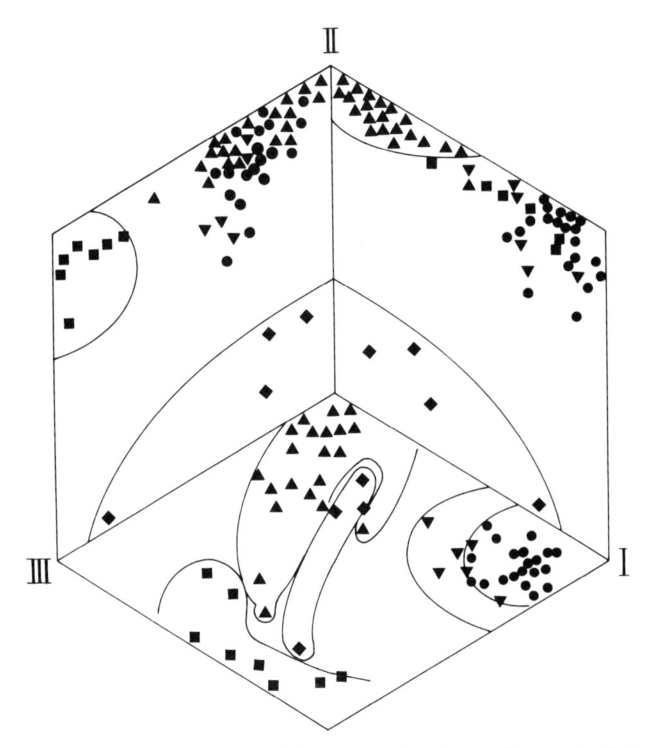

FIG. 34.2. Ordination analysis of the vegetation from the sand of Doñana National Park; 59 stands are represented on the planes formed by the first three components of a principal components analysis. On the plane I-III (bottom) five vegetation types are defined; these include mobile dunes (triangles), slacks (squares), Juniper forest (inverted triangles), xerophytic shrub (circles) and hygrophytic shrub (diamonds).

given in Table 34.1 together with species associated with the main vegetation types.

A geomorphological interpretation of the landforms was used in order to map the scrub types and to produce an ecological map of an area 7,500 ha within the Park (Allier *et al.* 1974).

Several components of variation of the vegetation were associated with a single environmental gradient, the depth of water table in the soil. In order to study the relationships between the depth of the water table and the vegetation, the water potential of shrub species along the transect was measured (Merino *et al.* 1976). Differences in the annual fluctuation of water potential in different species were related to position on the transect. The Atlantic species, characteristic of the hygrophytic type of shrub vegetation (*Erica ciliaris*, *E.scoparia*) showed little fluctuation in water potential from winter to

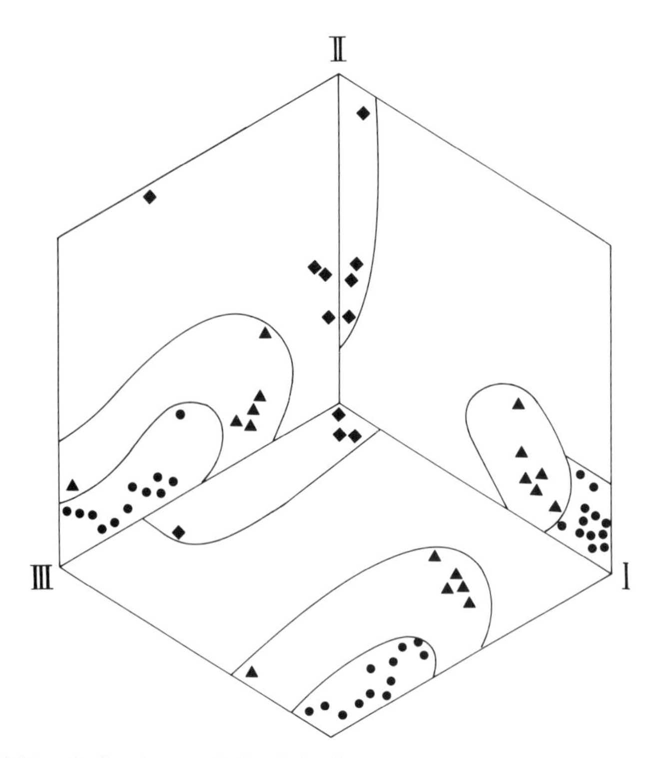

FIG. 34.3. Ordination analysis of shrub vegetation; 22 stands were analysed using principal components analysis. The ordination of samples on the first three components of the analysis is represented here. A good discrimination of samples into three groups (vegetation types) is achieved. Xerophytic shrub (circles), *Halimium halimifolium* shrub, 'monte blanco', (triangles), hygrophytic shrub (diamonds).

summer in their leaves. The tissues of these species do not exhibit potentials below -10 to -15 bars. This suggests that these species are restricted to those areas where soil water is easily available during summer drought, such as the bottom of the slacks or the vicinity of temporary lagoons. The tissues of species of xerophytic shrubs growing on dry elevated ground (*Rosmarinus officinalis*, *Cistus libanotis*, *Lavandula stoechas*) can withstand very low water potentials (-70 bars in summer). If the annual fluctuation in water potential for each species is plotted against the depth of the water table where it grows, a linear relationship is obtained (Fig. 34.4). *Halimium halimifolium* is a noted exception showing little fluctuation of water potential during the year, apparently because of transpiration control. It is also the most abundant species in the shrub vegetation and there is some evidence that it has extended its area as a consequence of fire (Garcia Novo 1977).

Plant performance measured as plant cover would be expected to decrease

Table 34.1. *Scheme of vegetation composition changes as a function of two factors: average water table depth (in columns) and succession stages (in rows) in Doñana area.*

Depth of water table		Types of vegetation		
Winter	Summer	Grassland	Shrub	Forest
2 to 3 m	over 3 m	*Malcolmia lacera* *Brassica barrellieri* *Linaria viscosa* *Loeflingia baetica* *Tuberaria guttata*	*Cistus libanotis* *Lavandula stoechas* *Halimium commutatum* *Thymus mastichina* *Helichrysum angustifolium* *Halimium halimifolium*	*Juniperus phoenicea* *Pistacia lentiscus* *Osyris quadripartita* *Quercus coccifera*
1 to 2 m	1·5 to 3 m	*Agrostis stolonifera* *Holoschoenus vulgaris* *Hypericum tomentosum*	*Halimium halifolium* *Stauracanthus genistoidis* *Calluna vulgaris* *Erica scoparia*	*Quercus suber* *Arbutus unedo* *Olea europaea* *Phillyrea angustifolia*
0 to 0·5 m	0·5 to 1·5 m	*Mentha pulegium* *Hypericum tomentosum* *Anagallis crassifolia* *Illecebrum verticillatum* *Armeria gaditana* *Agrostis stolonifera* *Eleocharis palastris*	*Ulex minor* *Erica ciliaris* *Erica scoparia* *Erianthus ravennae* *Imperata cilindrica*	*Quercus suber* *Arbutus unedo* *Pyrus bourgeana* *Fraxinus angustifolia* *Myrtus communis* *Phillyrea angustifolia*
−1 to 0 m	0 to 0·5 m	*Echinodorus ranunculoides* *Hydrocotyle vulgaris* *Ranunculus* sec. *Batrachium* *Carex* spp.	*Juncus* spp. *Typha latifolia*	*Populus alba* *Tamarix africana* *Tamarix canariensis*

Succession →

Increased depth of water table in the soil ↑

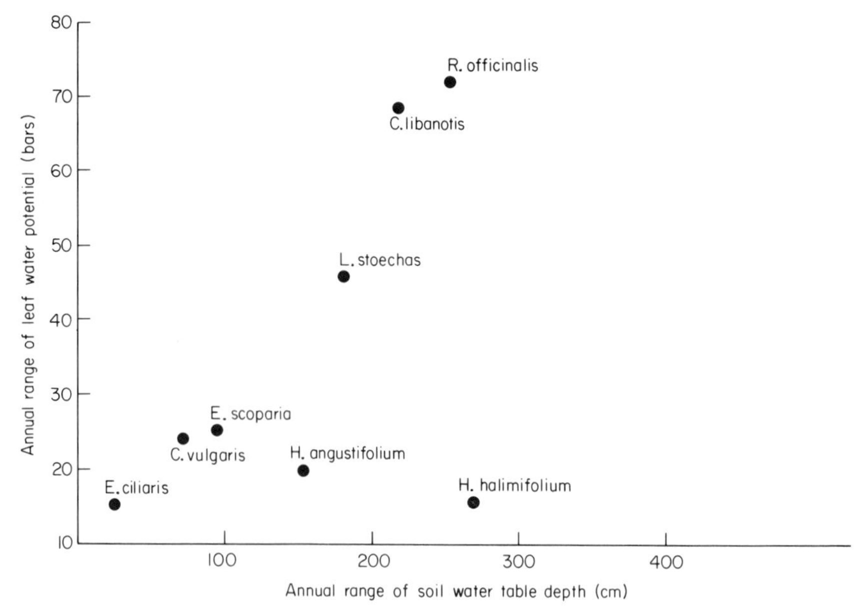

FIG. 34.4. The relationship between the annual range in leaf water potential and the annual range of water table depth for species in the shrub vegetation of Doñana. See text for full names of species.

as summer water potential becomes more negative. However, this predicted relationship is exhibited only by *Erica* species (Fig. 35.5), while *Calluna vulgaris*, *Cistus libanotis*, *Helichrysum angustifolium* and *Halimium halimifolium* show bell-shaped relationships. This has been interpreted by Merino *et al.* (1976) as a demonstration of a difference between ecological and physiological optima (Ellenberg 1963).

Shrub succession

Shrub vegetation development occurs in areas following the destruction of vegetation by flooding or fire. Shifting cultivation probably was the predominant environmental factor which led to the presence of open areas until two decades ago. Preliminary work by Martin and Haeger (unpublished) has shown that shrub development in depressions following flooding is remarkably fast: after a 4-year period the cover of *Cistus salvefolius* and *C.monspeliensis* reached 100% with an average height of plants of 60–80 cm. Variability in plant responses to flooding (Crawford 1972) results in intricate patterns of plant species around the edges of the lagoons.

Although secondary succession after fire is not well understood, it appears that the rate of succession at a site is dependent largely on water table depth. Maximum rates are recorded for the hygrophytic scrub with *Erica australis* and *Ulex minor* because of their rapid sprouting of new shoots after fire and

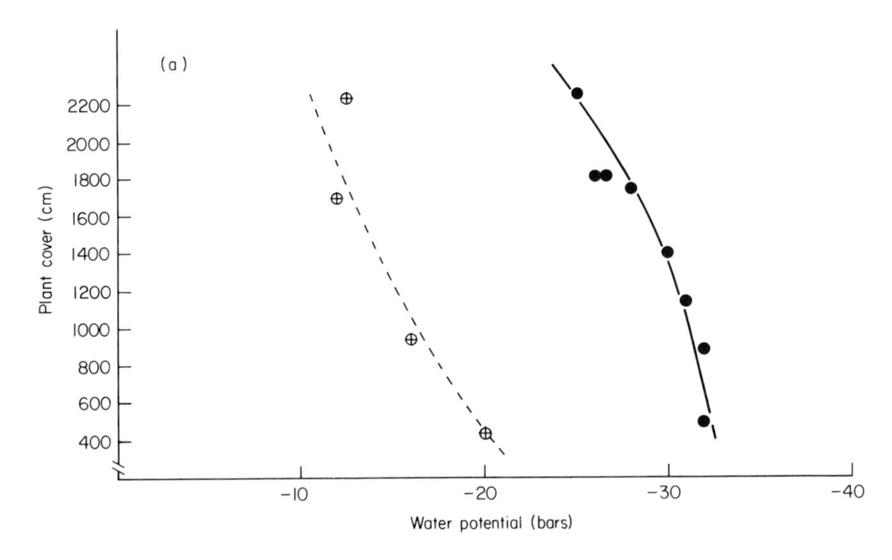

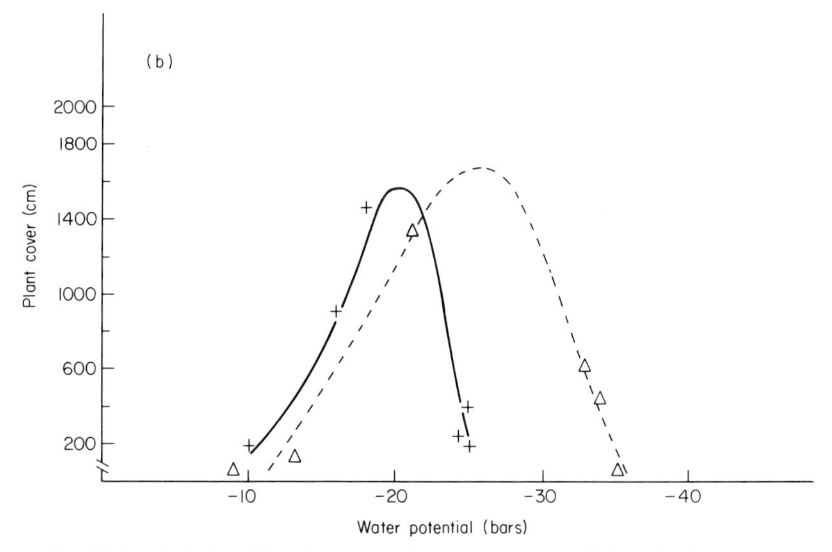

FIG. 34.5. Relationships between plant water potential and plant cover along a gradient of water availability. Two response patterns are shown by the shrubs: (a) *Erica scoparia* (●) and *Erica ciliaris* (⊕) show a decrease in plant cover at more negative water potentials; (b) *Calluna vulgaris* (+) and *Halimium halifolium* (△) show a bell-shaped response curve. Cover is expressed as the total distance (cm) where the species was present along a 2·5-m transect for each plot.

enhanced plant growth when water is available. On drier soils plants of *Stauracanthus genistoidis* grow very rapidly after fire, sprouting from their base. Plants of *Halimium halimifolium* and *Cistus salvifolius* fail to survive fire but seeds of both species germinate rapidly after fire. The combined recovery of *S.genistoidis*, *H.halimifolium* and *C.salvifolius* restores scrub vegetation after about 5 years. After a period of 10–12 years the species composition of the shrub vegetation approaches that of the original.

The succession of herbaceous species after fire, as described by Martin (1978) and Garcia Novo (1977), shows an early phase of 1–2 years in which large, nutrient-demanding species (*Senecio jacobaea*, *Carlina corymbosa*) predominate. Smaller species, many of them tolerant to grazing (*Ornithopus pinnatus*, *O.sativus*, *Anthoxanthum ovatus*, *Vulpia* spp., *Briza maxima*, *Brassica barrellieri*, *Loeflingia baetica*, *Evax pygmaea*), increase from the 2nd to the 5th year after a fire and only decrease as the shrub perennials become dominant. Subsequently only a few herb species are able to grow in the scrub, such as *Erodium cicutarium*, *Tuberaria guttata*, *Plantago psyllum* and *Malcolmia lacera*.

This succession occurs at sites of intermediate elevation. On lower ground close to the water table, very dense growth of *Holoschoenus vulgaris* with *Juncus* spp. and *Carex* spp. takes place immediately after fire. The hygrophytic shrub community dominated by *Erica* species and *Calluna vulgaris* which develops later, resembles the original vegetation in less than ten years after fire. Xerophytic shrub communities of drier sites may take over 25 years to regain the original composition, but even then, plant density and biomass will be below the original values. In summary, the succession of the scrub on the stabilized sands is completed in a short interval (10–30 years). Subsequently no further changes are apparent and no tree regeneration occurs with the exception of *Juniperus phoenicea* in dry areas and planted stands of *Pinus pinea*.

The significance of shrub vegetation

In order to discuss the significance of shrub vegetation of the stabilized sands, it is important initially, to describe the distribution of tree species. Apart from the artificial pine plantations, the only forest-forming species is juniper (*Juniperus phoenicea*) which grows on the drier dune ridges to the west and south of the Park. Nevertheless a number of tree species occur; cork oak (*Quercus suber*) with some 400 individuals present is the most common. The oaks appear as isolated old individuals, associated with depressions or temporary lagoons; a group of some 100 trees forms an open parkland close to the ecotone with the marsh. Populations of other tree species are small, comprising about 10 individuals of kermess oak (*Quercus coccifera*), 25 wild olives (*Olea europaea* var. *sylvestris*) and about 10 strawberry trees (*Arbutus unedo*). This suggests the earlier existence of a mediterranean type of forest in the area now occupied by shrub communities. Probably it was composed largely of cork oak with the strawberry tree present in lower parts subject to oceanic influence. On higher ground a drier type of forest dominated by juniper (*Juniperus phoenicea*) existed, with *Juniperus oxycedrus* ssp. *macrocarpa* on those sites open to oceanic influences. At very dry hot locations, kermess oak may have appeared mixed with wild olive and cork oak. *Pinus pinea* may have been restricted to unstable areas. A mediterranean forest composed of species with such different ecological requirements is unusual (Tomaselli 1976; Ruiz de la Torre 1971). However the environmental conditions in Doñana allow the coexistence of species with markedly different requirements. These conditions include poor soils in which a wide range of water regimes exist and a mediterranean climate which is ameliorated by strong local oceanic influences. This effect has been pointed out in relation to other taxonomic groups such as vertebrates (Valverde 1960, 1967) and lepidoptera (Fernandez Haeger *et al.* 1976).

The destruction of this original forest may have been caused by fire and shifting agriculture. Although the forest species mentioned above withstand fires to some degree, the effects of fire in Doñana are unusually important due to the poor sandy soil. The nutrients released from organic matter mineralization, as well as those of ashes, are easily leached and lost from the ecosystem. This repeated process considerably impoverishes the soil. With fewer seed producing trees due to fire and poor soil due to leaching, tree regeneration becomes difficult; if pressure from grazers and browsers is high, the survival of seedlings is very low. The overall outcome is the disappearance of the forest, which according to early records, has taken place during the last 300 years (Valverde, unpublished).

Thus the shrub vegetation of the stabilized sand of Doñana appears to be a secondary type of vegetation. It is maintained as a result of periodical fire, to which it shows extremely good adaptation. Apparently, the present-day composition of the shrub vegetation differs from the understorey shrub of the original forest. Fast-growing species and pyrophytes appear to be more common than they were previously. The only sites where fragments of the original forest have been preserved are near to drainage channels and temporary lagoons and within mobile dunes: these are all situations where fires are unlikely to spread and cultivation has not occurred.

The mobile dunes

As the establishment and growth of vegetation is largely controlled by substrate stability in the mobile dunes, the patterns of dune movement were investigated prior to the study of vegetation.

Dune movement

The movement of dune fronts for the last 40 years was evaluated. A remarkable result was the steady movement of dune fronts all over the system advancing inland at a rate of 5–6 m yr^{-1}. Considering the large size of the fronts this figure is unusually high (Ranwell 1972). When dune movement is analysed over shorter intervals of time and over shorter segments of dune front, differences in the rates of movement are apparent (Garcia Novo *et al.* 1975); very fast movements (over 30 m yr^{-1}) occur at some places. Secondary dune fronts oriented NE–SW along wind direction, show very low rates of movement (around 0·1 m yr^{-1}).

The slacks show substrate stability, although because of the obliteration of the drainage system associated with the movement of the dunes, the slacks are often flooded after heavy rain.

Distribution of trees in the mobile dunes

Pinus pinea is the most abundant tree species of the mobile dunes. Its origin in the area has been questioned, although pollen grains of the species have been identified in peat deposits in the base of the sediments found under the dune system which correspond to the Atlantic period (Ceratini & Viguier 1973).

Pines grow both in dunes and slacks. Dune populations are composed of scattered old individuals (over 35 years) isolated by dune movement. In contrast, pines in the slacks are very common. The populations show an age distribution with marked fluctuations which are probably related to the periodicity of flooding (Figueroa & Merino abstract, this volume). Pine distribution in the slacks is strongly contagious with trees clustered on small ridges (contradunas) and almost no trees or seedlings over the flat depressions in between these ridges. The pattern has been shown by Figueroa (1976) to be the effect of a fluctuation of water table in the soil of the slack.

Other tree species present in the mobile dunes include *Juniperus phoenicea*, *J.oxycedrus* ssp. *macrocarpa*, *Populus alba* and *Tamarix africana*. *J.phoenicea* is restricted to less active fronts or to stabilized sectors of the dunes and it also occurs around stable depressions and along dry ridges. In the more active parts of the dune system it is very scarce either appearing as seedlings in the slacks or else as old individuals associated with old landforms. *J.oxycedrus* is represented by a few large individuals in an area west of the dunes and more dense stands to the east. It is associated with stable substrates and, commonly, with old landforms and buried soil profiles. This evidence suggests (Garcia Novo 1977) that a large *J.oxycedrus* forest once developed over the sands. Under stable conditions, *J.oxycedrus* probably was the dominant species in the more stable parts with *J.phoenicea* occurring in the less stable drier parts. Probably *Pinus pinea* was restricted to unstable sands or mobile dunes where disturbances occurred.

In mediterranean regions (Harant & Jarry 1967) the pine is replaced by juniper (*J.phoenicea*) during succession. However, fast dune movement favours pine over juniper in Doñana slacks. Slacks are constantly invaded by the oncoming dunes at the same time as ground is cleared by the retreating face. In order for a perennial to survive within a slack it must germinate, grow, produce seeds and those seeds must spread and germinate to keep pace with the rate of dune movement. Vegetative reproduction does not allow a fast enough rate of spread. Apparently, only pines can keep pace with the advance of the front.

A few individuals of *Populus alba* and *Tamarix africana* also occur on mobile dunes. These appear to be associated with ancient landforms and are represented by large individuals which are either isolated or in small groups. Their presence on the dunes is another strong indication of a previous stage of stability in the mobile dune system when these plants were able to germinate on the banks of a channel or around a temporary lagoon, now buried under the sands.

PAST AND PRESENT VEGETATION OF THE DUNES OF DOÑANA NATIONAL PARK

The evidence so far presented reveals the existence in Doñana area of an ancient dune system that has undergone several dune building phases separated by intervals of stability. Present-day vegetation of the area corresponds to two different physiognomic entities: a forest vegetation associated with fire exclusion and substrate stability and a shrub vegetation associated to substrate instability, disturbance and fire.

The original forest vegetation of the area has been suppressed by fire, shifting agriculture, grazing pressure, timber exploitation and soil disturbance followed by sand mobility. Some of the processes responsible are still recognizable (fire, instability); others can only be inferred, such as ancient culture and timber exploitation. Although the ecological history of the area is not completely understood, it appears that fire has played an important rôle in the replacement of the original mediterranean forest by the type of shrub that is presently found in the area. Fire resistent cork oak is the only tree species that has survived; all other species have suffered stringent reductions in their numbers.

Evidence from dune morphology and vegetation indicates a recent destruction of the original forest followed by unstable conditions and widespread dune building up and dune movement. The original juniper forest of the dunes has been replaced by pine forest. Lagoons and channels between dunes have been obliterated but there are still relict populations of poplars and tamarisks which were growing round their borders.

Whether dune mobility and a higher frequency of fires are a result of climatic shift or because of human destruction of the forests cannot be ascertained. However it is likely that a single cause was not responsible for all these changes.

THE CONSERVATION OF DOÑANA NATIONAL PARK

Early steps

For centuries Doñana has been praised in Spain as a magnificent game reserve. By the end of 19th century naturalists were attracted by the rich fauna of Doñana. Reports spread on the interest of the site and the 'coto de Doñana' expeditions of 1952, 1956 and 1957 were organized. Many noted zoologists of the time congregated in Doñana, producing overwhelming evidence of the importance of the area and its conservation interest. An excellent account of these expeditions is given by Mountfort (1968). The growing interest in the area helped to set up the World Wildlife Fund to bring support to conservation movement and specially to Doñana. With funds from the WWF and the Spanish Government, an area of 7,500 ha was purchased to form the Biological Reserve of Doñana. It was officially opened on 1 April 1966 as a protected area, open to scientists and visitors. The Reserve, under the leadership of its director, Professor Valverde, catalysed the interest in the area, so that in 1969 the National Park of Doñana was created. With an area of 37,000 ha, it included the former Biological Reserve as well as other private and state-owned lands. In 1978 the Park was further enlarged to cover 49,000 ha, including 25 km of atlantic coast.

The management of the Park

The National Park of Doñana is under the authority of ICONA (Instituto para la Conservacion de la Naturaleza) a branch of the Spanish Ministry of Agriculture which is concerned with forest management and the conservation of natural areas.

The lands of the Park are largely privately owned. This means that the Park authority exerts a limited control within large sectors of Doñana; access by visitors or scientists is usually discouraged in the private lands thus concentrating work in the State-owned Biological Reserve.

Since the creation of the Park in 1969, management policies have aimed at the consolidation of the Park, developing adequate centres for scientists and visitors, and trying to secure a zone of protection round the Park. Traditional activities such as cattle raising and game have been maintained (under close control), although others (shifting cultivation, cork harvesting) have been suppressed altogether. At the same time steps to control the composition of waters entering the Park from neighbouring areas were taken; also two deep wells were drilled to secure a permanent supply of fresh water to the marsh in summer.

An intense effort has been made over these years to try and enlarge the Park so as to bring it up to 'natural' boundaries. The mobile dune system has been specially threatened as a scheme has been proposed to build a new road through the dunes 1 km inland from the coast running parallel to it. The purpose of this was to set up an enormous tourist resort in the strip of land between the Park and the sea. In 1978 the Park was finally enlarged to include this disputed strip of land.

An important agricultural development including drainage, desalinization and intense cultivation of the marshes (Grande Covian 1975) threatens the Park from the north. Increasing salinity and eutrophication of surface waters entering the Park is feared. Also, the misuse of pesticides in an area so close to the Park may result in serious damage. It should be recalled that in the summer of 1973 over 45,000 birds died in the Park due to poisoning by pesticides and this in turn gave rise to an epidemic of botulism.

Future policies

Now the Park has been consolidated and enlarged to a safe area, new management programmes have started. The original mediterranean forest will be regenerated in two degraded areas of shrub. Also, a policy of repeated burning on a limited scale is carried out to maintain species and ecosystems characteristic of the pioneer stages of succession. The aim of both policies is to increase the range of biotopes in the Park. However the overall idea in the management is not to venture into new policies until the research programmes now under way clearly point the way to follow.

Visitors

Early in the development of the Park, visitors were few (mostly scientists) and they were assisted by personnel of the Biological Reserve. Only 100 visitors per day are now allowed to the Park, although there is a growing pressure to increase this figure substantially. However Doñana ecosystems are very fragile: sandy soils are poor and unstable and plant recovery after injury is very slow. Vehicle tracks are still recognizable after 10 years both in the scrub and in the marsh. The flat topography of Doñana means that the presence of visitors disturbs vertebrates for a very long distance.

In order to alleviate visitor pressure on the Park, two Reception Centres are under construction on the boundaries. Together with the standard facilities for accommodation, services, lecture theatre etc., a series of artificial ponds for waterfowl will give the opportunity to observe many species at a close range. With appropriate scheduling of visits and a careful distribution of visitor pressure over the Park it has been estimated that 500,000 visitors per year could be accommodated without permanent damage to the Park.

In the foreseeable future the importance of the Doñana area will increase, as other wet areas shrink and peoples' concern for nature increases. One of the important rôles of Doñana will be to help people understand nature and to appreciate it, particularly in the rich Andalusian region where it is located. Consequently, a programme to bring school children from all over the region to the Park in a special series of visits has already started. Also a series of nature trails within the Park and in neighbouring areas was started in 1976.

Research

A constant policy has been to promote research on the species and ecosystems of Doñana. The Spanish Higher Research Council (CSIC) initiated such activities in 1965 within the Biological Reserve, concentrating on the zoology of the area. The initial centre has now evolved into a fairly large research centre with several laboratories. Research activities have expanded, with many research teams sponsored by Universities (especially the University of Seville), the ICONA and other institutions from all over the world.

ACKNOWLEDGMENTS

I thank A. Martin for unpublished data on the succession of vegetation. For valuable information and comments on Doñana ecology I thank J.A. Valverde. Also I thank J. Castroviejo for his support to our research programmes within the Biological Reserve. The research summarized in this paper was carried out in part by F. Gonzalez Bernaldez, J. Merino, L. Ramirez Diaz, A. Torres, E. Figueroa, J. Fernandez Haeger, F. Sancho Royo, F. Pineda and A. Pou at the Department of Ecology of the University of Seville.

Much of the research has been supported by grants from the Spanish Instituto para la Conservacion de la Naturaleza (ICONA).

REFERENCES

Allier C., Gonzalez Bernaldez F. & Ramirez Diaz L. (1974) *Reserva Biológica de Doñana. Ecological map*. Estación Biológica de Doñana. C.S.I.C., Sevilla.

Allier C., Garcia Novo F., Ramirez Diaz L. & Torres Martinez A. (1975) Dynamique actuelle et végétation du systeme dunaire littoral de Doñana (Golfe de Cadix). *C.R. Soc. Biogeographie* **440–42**, 95–111.

Allier C. & Bresset V. (1977) Etude phytosociologique de la marisma et de sa bordure (Reserve Biologique de Doñana). In *ICONA. Monografia No. 18*. Ministerio de Agricultura, Madrid.

Armengol J. (1976) Crustáceos acuáticos del Coto de Doñana. *Oecol. Aquatica* **2**, 93–97.

Caratini C. & Viguier C. (1973) Etude palynologique et sedimentologique des sables halogenes de la falaise littorale d'El Asperillo (Province de Huelva). *Estud. Geol. Esp*. **29**, 325–28.

Crawford R.M.M. (1972) Some metabolic aspects of ecology. *Trans. Edin. bot. Soc.* **41**, 309–16.

Chapman V.J. (1976) *Coastal Vegetation.* Pergamon, Oxford and London.

Davies J.L. (1972) *Geographical Variation in Coastal Development.* Oliver and Boyd, Edinburgh.

Ellenberg H. (1963) *Vegetation Mitteleuropas.* Eugen Ulmer, Stuttgart.

Fernandez Haeger J., Garcia Garcia Isabel & Aguilar Amat J. (1976) Guia de las mariposas de Doñana. *Naturalia* No. 6, ICONA, Ministerio de Agricultura, Madrid.

Figueroa E. (1976) Ecologia del *Pinus pinea* L. en el Parque Nacional de Doñana. *Tesina de Licenciatura.* Universidad de Sevilla.

Galiano E.F. & Cabezudo B. (1976) Plantas de la Reserva Biologica de Doñana (Huelva) *Lagascalia* **6**, 117–76.

Garcia Novo F. (1977) Fire effect on the vegetation of Doñana National Park (S.W. Spain). *Symposium on the Environmental Consequences of Fire and Fuel Management in Mediterranean Ecosystems* U.S.D.A. Technical Report. WD–3. pp. 318–25.

Garcia Novo F., Ramirez Diaz L. & Torres Martinez A. (1975) El sistema de dunas de Doñana. *Naturalia Hispanica* No. 5. ICONA, Ministerio de Agricultura, Madrid.

Gavala F. & Laborde J. (1952) Memoria explicativa de la hoja 1033 (La Marismilla). *Mapa Geologico de España a escala 1:50,000.* IGME, Madrid.

Gittins R. (1969) The application of ordination techniques. *Ecological Aspects of The Mineral Nutrition of Plants* (Ed. by I.H. Rorison), pp. 37–66. Blackwell Scientific Publications, Oxford.

Gonzalez Bernaldez F., Garcia Novo F. & Ramirez Diaz L. (1975a) Analyse factorielle de la vegetation des dunes de la Reserve Biologique de Doñana (Espagne). I. Analyse numerique des donnes floristiques. *Isr. J. Bot.* **24**, 106–17.

Gonzalez Bernaldez F., Garcia Novo F. & Ramirez Diaz L. (1975b) Analyse factorielle de la vegetation des dunes de la Reserve Biologique de Doñana (Espagne). II. Analyse d'un gradient du milieu. Etude spéciale du probleme de la non-linearité. *Isr. J. Bot.* **24**, 173–82.

Gonzalez Bernaldez F., Ramirez Diaz L., Torres Martinez, A. & Diaz Pineda (1977a) Estructura de la vegetación de Marisma de la Reserva Biológica de Doñana. I. Análisis factorial de datos cualitativos. *Anales Edafologia* **36**, 989–1003.

Grande Covian R. (1975) *Las marismas del Guadalquivir.* IRYDA, Ministerio de Agricultura, Madrid.

Harant H. & Jarry D. (1967) *Guide du Naturaliste dans le Midi de la France.* Delachaux et Niestle, Neuchatel.

Hernando J. (1978) *Tesis Doctoral.* Universidad de Sevilla.

I.G.M.E. (1972) Hoja 80-81. *Mapa Geológico de España 1:200,000.* Madrid.

I.G.M.E. (1975) Hoja 1.033. *Mapa Geológico de España 1:50,000.* Madrid.

Margalef R. (1976) Algas de agua dulce de Doñana. *Oecol. aquatica* **2**, 79–91.

Martin A. (1978) *Tesis Doctoral.* Universidad de Sevilla.

Menanteau L. & Pou A. (1977) Les Marismas du Guadalquivir apport de la teledeteccion et de l'archeologie a la reconstitution du paysage. Colloque '*Pur une Archeologie du paysage*'. Universite de Tours.

Merino J., Garcia Novo F. & Sanchez Diaz M. (1976) Annual fluctuation of water potential in the xerophytic shrub of the Doñana Biological Reserve (Spain). *Oecol. Plant.* **11**, 1–11.

Mountfort G. (1968) *Portrait of a Wilderness.* David and Charles, Newton Abbot.

Pou A. (1976) Implicaciones paleoclimáticas de los sistemas dunares de Doñana. *V Reunion de Climatologia Agricola.* Universidad de Santiago de Compostela.

Ramirez Diaz L. (1973) *Tesis Doctoral.* Universidad de Sevilla.

Ramirez Diaz L., Garcia Novo F. & Merino Ortega J. (1976) On the ecological interpretation of principal components in factor analysis. *Oecol. Plant.* **11**, 2–9.

Ranwell D.S. (1972) *Ecology of Salt Marshes and Sand Dunes.* Chapman and Hall, London.

Ruiz de la Torre J. (1971) *Árboles y arbustos de la España Peninsular.* Instituto Forestal de Investigaciones y Experiencias, Madrid.

Soler Andres A.G., Montes Del Olmo C. & Ramirez Diaz L. (1976) Analyse factorielle des biocenoses de coleopteres aquatiques des marais (Marisma) du bas Guadalquivir (Espagne). *Annls. Limnol.* **12**, 89–103.

Tomaselli R. (1976) Forets et maquis méditerranéens: écologie, conservation et amenagement. *Notes techniques du MAB* **2**, 35–76. Unesco, Paris.

Torres Martinez A. (1975) *Tesis Doctoral.* Universidad de Sevilla.

Valverde J.A. (1960) Vertebrados de las marismas del Guadalquivir. *Archivos Inst. Aclimatacion Almeria* **9**, 1–168.

Valverde J.A. (1967) *Estructura de una comunidad mediterranea de vertebrados terrestres.* Monografias de Ciencia Moderna No. 76. C.S.I.C., Madrid.

Valverde J.A. (1975) Doñana y las marismas del Guadalquivir: su rescate y sus problemas presentes y futuros. *Ardeola* **21**, 25–58.

Vanney J.R. (1970) *L'hidrologie du bas Guadalquivir.* Casa de Velazquez. Instituto de Geografia Aplicada Alonso de Herrera. C.S.I.C., Madrid.

35. THE COASTAL LAGOONS OF ITALY

CESARE F. SACCHI
Istituto di Ecologià animale ed Etologia,
Università di Pavia, Italy

SUMMARY

The numerous lagoons round the coasts of Italy and its larger islands, which are what remains of a larger number now reduced by natural silting or by artificial drainage for agriculture, can be classified into two main categories.

The North Adriatic lagoons (Fig. 35.1), which are large and deep, with significant tides, often show ecological and biocoenotic characteristics of a type different from that of the rest of the Mediterranean (Adriatic 'sub-atlanticism'; Fig. 35.2). The best-known example of this is provided by the Lagoon of Venice.

The lagoons of the peninsula and island coasts (Fig. 35.3). These are smaller, not closely connected with the sea, and very variable in their chemical content and populations. They are less threatened by industrial pollution of the Venetian type, but are much affected by agriculture in their vicinity, and by tourism. The small meromictic pool of Faro, near Messina in Sicily, is exceptional in that its deeper waters have a high content of H_2S.

A sensible and rational management policy for these lagoons would therefore involve effective protection against alterations in the environment, and only limited exploitation of their resources of the traditional type.

RÉSUMÉ

Les nombreuses lagunes qui bordent les côtes de l'Italie et de ses grandes îles, restes d'une série plus nombreuse qui a été réduite par comblement naturel ou pour les besoins de l'agriculture, peuvent être classées en deux grandes catégories.

Les lagunes nord-adriatiques (Fig. 35.1) plus vastes et profondes, intéressées par une remarquable marée, présentent souvent des caractères écologiques et biocénotiques non typiquement méditerranéens («subatlantisme» adriatique; Fig. 35.2) : l'exemple le plus connu est constitué par la lagune de Venise.

Les lagunes des côtes péninsulaires et insulaires (Fig. 35.3) plus petites, mal reliées à la mer, très variables dans leur chimisme et dans leurs peuplements; elles sont moins menacées par des pollutions industrielles de type venitien mais très influencées par l'agriculture environnante et par le tourisme.

Une place à part merite le petit étang méromictique du Faro (Messine) très riche en H_2S dans ses eaux profondes.

Un aménagement prudent et rationnel de ces lagunes doit par conséquent se proposer une efficace défense contre les altérations de l'environnement et une exploitation limitée et traditionnelle des ressources locales.

INTRODUCTION

Only a few of the littoral lagoons which in historical times bordered much of the coast of Italy still survive. Many of them have silted naturally, or have been drained artificially, either for agriculture or as a precaution against malaria. In a country so deficient in good arable land, the claims of agriculture have been given precedence over the less productive fish and shellfish cultures formerly practised in these waters. There remain about thirty lagoons and brackish ponds (*stagni*) available for study, which can be categorized into the North Adriatic lagoons and those of the Italian peninsula and islands.

THE NORTH ADRIATIC LAGOONS

The North Adriatic lagoons (Fig. 35.1) are very extensive; the Venetian Lagoon for example covers an area of 55,000 hectares (210 sq. miles), and sometimes they are deep. They represent what remains of a series which, in the early Middle Ages, once extended from the mouth of the Timavo, near Trieste, as far south as Ravenna, and even beyond. They formed an internal waterway of great economic and strategic importance, linked by rivers and canals across the plain of the Po to the foothills of the Alps. They are supposed to have originated from ancient estuaries which became cut off from the sea by their own sedimentation, assisted by eustatic submersion and other local effects.

At present there remain three bodies of water: the Friulian lagoon of Grado-Marano in the north, the Venetian Lagoon which is separated from it by the recently drained former Caorle Lagoon, and finally, south of the Po, the 'Valli di Comacchio', now much reduced by draining in the 1950s. There is also a series of minor lagoons, called *sacche*, which have been formed over the past few centuries at the very edge of the Po delta, between the Adige and Comacchio (Barbujani 1975).

Great ports and trading centres were sited on the North Adriatic lagoons in historical times, from the Etruscan-Greek city of Spina (near Comacchio) to Aquileia (near Grado), of which Venice survives as an important commercial, industrial and cultural metropolis. This economic development was favoured by the sheltered geographical position, the self-scouring of openings to the sea, due to the relatively large tidal range of up to 1·5 m (unusual in the Mediterranean), and the highly developed inland navigation system in the

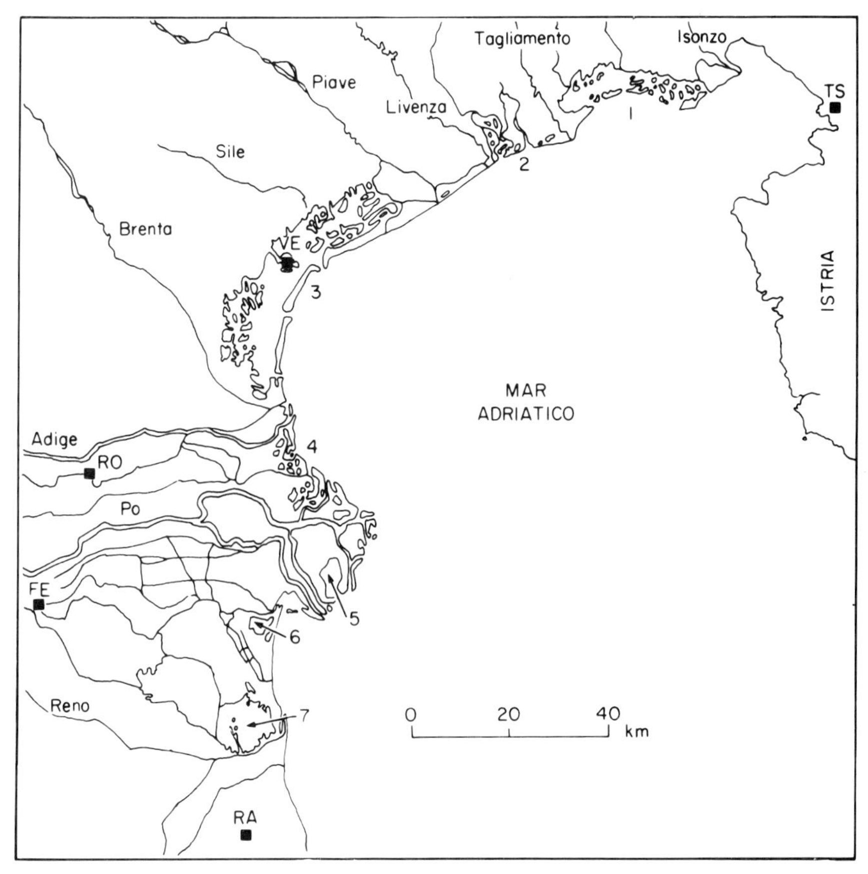

FIG. 35.1. *The North Adriatic Lagoons:* (1) Grado-Marano, (2) the remains of Caorle, drained for agriculture, (3) Venice, (4) '*Valli*' of the Po delta, (5) Sacca Scardovari, (6) Valle Bertuzzi, (7) remains of the Valli di Comacchio. TS = Trieste; VE = Venice; RO = Rovigo; FE = Ferrara; RA = Ravenna.

hinterland. Notwithstanding periodical flooding, and exceptionally high tides ('*acque alte*') associated with meteorological events (Giordani-Soika 1976), the maintenance of a free tidal circulation was essential for both the biological and the ecological life of the lagoons, and it was carefully preserved by the Venetian Republic for 1,000 years. This was achieved by the construction of defence works against the sea (dykes, and stone walls called *murazzi*), by channel diversions designed to prevent the rivers silting up, and by strict legislation.

The peculiar climate of the coastal regions of the Po plain (Mennella 1972), with a cold winter and spring, and damp summers, results in the fauna of these lagoons showing a number of 'sub-atlantic' characteristics. These environmental conditions are reinforced by others, such as the large volume of cold water flowing down rivers from the Alps, which reduces the salinity of the area and results in the waters of the larger rivers (Adige and Po) containing a relatively high nutrient content. This 'sub-atlanticism' may be interpreted (Sacchi 1977) as an example of reduced ecological differentiation of the fauna between the Atlantic and the Adriatic, compared with the rest of the Mediterranean (Fig. 35.2). It is even possible to find an Atlantic type of zonation of communities in the North Adriatic lagoons, with a 'Schorre' (called *barene* in Italian) and a 'Slikke' (locally known as *velme*). The former is an upper intertidal zone which includes Spartinetum and Salicornietum, whereas the latter is a lower zone of almost abiotic muds.

The commercial and industrial activities of Venice represent a permanent threat to the biological and cultural well-being of the area, although the recently constructed 'Canale dei petroli' (Oil Canal) has at least diverted tankers and other shipping to the highly polluted suburb of Mestre-Marghera.

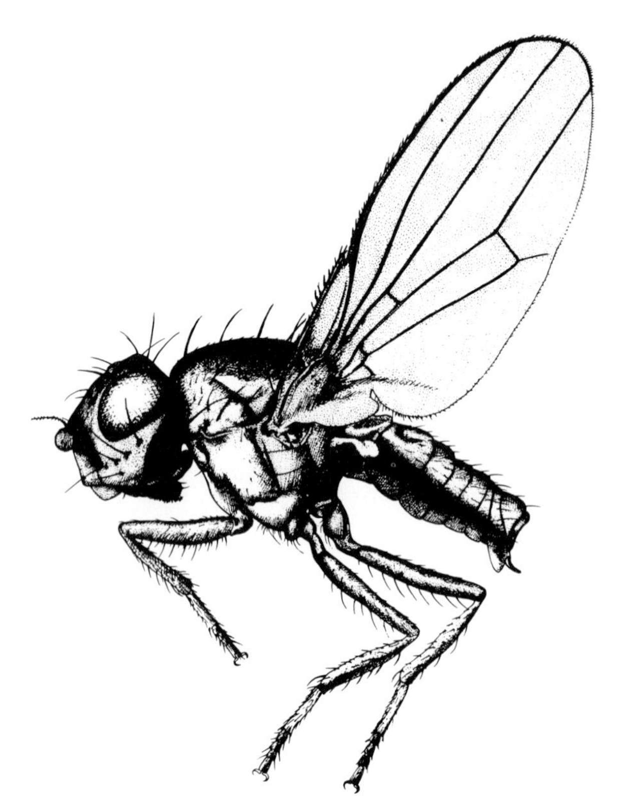

FIG. 35.2. The dipterous fly *Canace nasica* Hal., with an aquatic larva, quite common on lagoon shores up the Po delta, represents a 'sub-atlantic' element in the fauna of the North Adriatic lagoons, being replaced elsewhere by its Mediterranean sibling species, *C.salonitana* Str. (G. d'Este *del.*).

This has resulted in an increased rate of 'marine vivification'. 'Vivification' is a term commonly used in Mediterranean lagoon studies, which indicates a complex of physical, chemical and biological effects of the sea upon the lagoons or brackish ponds, which keeps them *alive*, not allowing their transformation into freshwater ponds or marshes. It involves thermic and water-level regulation, the introduction of salt, colonization by euryhaline organisms, and the introduction of biological material of marine origin.

The spread of marine organisms, which were previously confined to the neighbourhood of the three *porti* (entrances) of Lido, Malamocco and Chioggia (Giordani-Soika & Perin 1974) has occurred into the lagoon itself.

A large section of the Venetian Lagoon (principally the southern part, called the *laguna viva*, or 'living lagoon' in contrast with the northern 'dead' *laguna morta*, which is less affected by tides and subject to a greater inflow of fresh water) is occupied by mussel beds, and in addition there are fish farms in the artificially dammed *valli*. The Grado-Marano Lagoon, which is more open to the sea, is similarly exploited. The Comacchio is also used for fish farming (mainly eels), now that the Ferrarese ports have declined, as a result of the extension of the Po delta, and the construction of *tagli* (diversions) by the Republic of Venice, in an attempt to prevent the silting up of the Venetian Lagoon.

In the Ferrarese lagoons the tide is less important, the climate slightly more continental, marine 'vivification' less complete, and the fauna and flora show more Mediterranean features (Cognetti *et al.* 1975). This system can be regarded as transitional towards the next type, which is described below.

THE LAGOONS AND PONDS OF THE ITALIAN PENINSULA AND ITS LARGER ISLANDS

These are smaller and shallower than the North Adriatic lagoons (Fig. 35.3). There are many of them on the Tyrrhenian coast, some on the west coast of Sardinia, and a few in the southern Adriatic: all have a truly Mediterranean climate, with mild winters, hot summers and the intermediate seasons short, with the rains mainly in the winter, unlike the Po plain where there is more rain in the spring and autumn. The tidal range is reduced to a few centimetres, and 'vivification' is often poor, being by way of *foci* (mouths) through sand spits, which are easily shut off by wave action. There are no important rivers which bring in large volumes of fresh water. Although the streams and springs contain the limited volume of water, on occasions additions of rain water may reduce the salinity to extremely low levels, corresponding to the '*climax préléthal*' of Petit (1962), which results in a marked reduction of species including euryhaline species that are often represented by very large numbers of individuals. When the *foci* are reopened, either by natural overflowing of

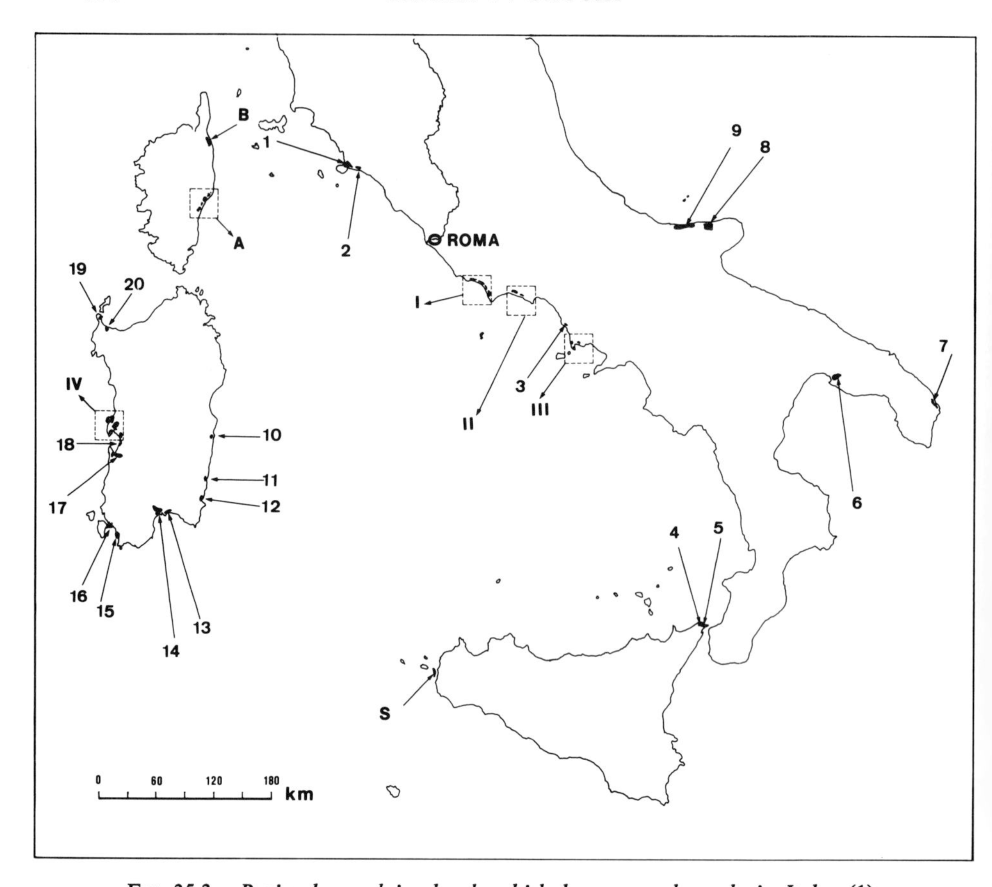

FIG. 35.3. *Peninsular and insular brackish lagoons and ponds in Italy*: (1) Orbetello, (2) Burano, (3) Patria, (4) Faro, (5) Ganzirri, (6) Mar Piccolo, (7) Alimini, (8) Varano, (9) Lesina, (10) Tortolí, (11) Colostrai, (12) Santa Giusta, (13) Quartu, (14) Cagliari, (15) Maestrale-Brebeis, (16) Santa Caterina—Botte, (17) Marceddi—S. Giovanni, (18) Santa Giusta, near Cabras, and the ponds around it, (19) Casaraccio, (20) Pilo. I = remains of the Pontine marshes (Fogliano, Monaci, Caprolace, Sabaudia). II = remains of the Fondi marshes (Fondi and San Puoto). III = the 'phlegrean' brackish lakes (Fusaro, Miseno, Lucrino). IV = the Arborea Plain group (Is Benas, Cabras, Mistras). *Corsican ponds*: A = Aleria Plain group, B = Biguglia. S = the 'lagoon' sea of Stagnone (Genovese 1967). There are also several lesser brackish (sometimes astatic) environments, especially near the mouths of the torrent-like intermittent rivers from the Appenine mountains which flood at certain times of the year.

the sandy obstructions, or by human intervention as in the Lago di Patria (Merola, Sacchi & Troncone 1965), the stock of marine organisms can increase dramatically (Sacchi 1961). Where the freshwater inflow is low, these lagoons may constitute true arms of the open sea, with high salinities, low nutrient contents and biocoenoses like those of harbours, since the bottoms are usually muddy. The Lago Fusaro, west of Naples (Sacchi & Renzoni 1962) and other Phlegrean ponds are examples of this type of system. Most of the Pontine lagoons have much in common compared with the Lago di Patria, but more typical conditions (low salinity and eutrophic conditions) are found in the Lago di Sabaudia. These are all that remains of what was once an almost continuous series of lagoons between Rome and Naples, planned in antiquity to form an inner waterway, though only the 'Fossa Neronis' (the former Lago di Licola) was actually completed.

In Tuscany there are important remnants of the large and shallow lagoons which the Etruscans used as sheltered harbours, but which turned later into unhealthy marshes. The small hypohaline Lago di Burano is now a bird sanctuary, while the larger Lago di Orbetello is exploited for fish farming, as are several of the lagoons and brackish ponds of Sardinia, although these are less well protected against pollution and over-fishing than is Orbetello (Cottiglia *et al.* 1968).

In Apulia there are two lagoons north of Gargano (the shallower and hypohaline Lesina and the better 'vivified' Varano), and the Laghi Alimini south of Brindisi. The smaller and relatively deep Lago di Faro, north-west of Messina, is quite different, constituting a miniature 'Black Sea' (Genovese 1953). Its surface waters are normally oxygenated, and used for fishing and mussel culture, but the bottom of the basin, which is 30 metres deep, is strongly anaerobic and contains a high concentration of H_2S. The chemocline of this sharply meromictic lagoon fluctuates according to the season and the weather, and between the aerobic and anaerobic layers there is an intermediate strongly micro-aerobic zone, allowing the formation of bacterial 'red water' (*acqua rossa*). The Mar Piccolo ('Little Sea') of Taranto is also unusual, since it is partly occupied by oyster and mussel storage beds and also serves as a Naval base (Vatova 1963).

A small surface area, the presence of shallow and calm water, precarious marine 'vivification', a limited wet season, a poor connection to the sea and water deficient in nutrients, characterize this class of lagoons (to which the ponds along the east coast of Corsica may be added, Casabianca 1967). The ecological characteristics of these lagoons have been summarized below (Sacchi 1967, 1973).

1. Large ranges of temperature, salinity and oxygen concentration. These fluctuate according to nychthemeral and annual cycles, which can lead to anoxia in the summer and anaerobic decomposition of the biomass.

2. Chemical fertility depends mainly on freshwater inflow, either natural or modified by agriculture. This may result in the introduction of water rich in phosphorus leading to eutrophication. Autochthonous nutrients may also show cyclical variation (Carrada & Rigillo 1975).
3. Reduced tidal range, the effects of which are limited to those parts of the lagoons nearest to the sea, which may still have tidal cycles of the Adriatic type, on a smaller scale. The North Adriatic lagoons, for their part, show seasonal and nychthemeral cycles only in their less open sections.
4. Unstable connections with the sea which often depend upon human intervention.
5. Biocoenoses, frequently deficient in species, though often with a significant biomass represented by a few species. The biomass rapidly increases during the spring but there is a dystrophic crisis in the summer, followed by a slight recovery in the rate of productivity in the autumn, and a limited decline in the rate during the winter.

CONCLUSIONS

A sensible and rational management policy for conservation (Sacchi 1973) should be based on the following points.

(a) Limiting industrial pollution, including diverting it away from the lagoons.
(b) Protection against agricultural and (more often) recreational pressures on the environment, which would involve a different planning policy for the use of the narrow littoral plains in Italy than that which exists at present.
(c) Careful maintenance of the connections between lagoons and open sea, to allow free tidal exchange.
(d) Scientific planning of fish farming etc. Attempts should be made to concentrate on the harvesting of species of eels, mugilidae, mussels and *Tapes* (*Venerupis*) which are commonly present on the sandy bottoms and for which a tradition of collection exists.
(e) The creation, where possible, of wildlife sanctuaries and nature reserves, the exploitation of the resources of which would be confined to local people.

ACKNOWLEDGMENTS

I am very grateful to Dr C.B. Goodhart, of the Museum of Zoology at Cambridge, for his help with the English translation of the text.

This work was partly supported by the Italian Research Council (P.F. contr. 76.00964.90).

REFERENCES

Barbujani R. (1974) *Nascita e sviluppo del delta padano.* Padova.

Carrada G.C. & Rigillo Troncone M. (1975) Nychthemeral cycle of nutrients in a meromictic brackish lagoon (Lago Lungo). *Rapp. Comm. int. Mer Médit.* **23**, 81–84.

Casabianca M.L. (1967) Etude écologique des étangs de la côte orientale corse. *Bull. Soc. Hist. nat. Corse* **1**, 41–74.

Cognetti G., De Angilis C.M. & Orlando E. (1975) Attuale situazione ecologica delle valli di Comacchio e proposte per la loro salvaguardia. *Quad. di Italia nostra* **12**, 1–70.

Cottiglia M., Manca G. & Mascia C. (1968) Fenomeni d'inquinamento nelle acque della Sardegna. *Acqua industriale* **56**, 1–17.

Genovese S. (1963) The distribution of the H_2S in the Lake of the Faro (Messina) with particular regard to the presence of 'red water'. *Symposium on Marine Microbiology* (Ed. by C.H. Oppenheimer), pp. 194–204.

Genovese S. (1969) Données écologiques sur le 'Stagnone' de Marsala (Sicile occidentale). *Rapp. Comm. int. Mer Médit.* **19**, 823–26.

Giordani-Soika A. (1976) *Venezia e il problema delle acqua alte.* Venezia.

Giordani-Soika A. & Perin G. (1974) L'inquinamento della Laguna di Venezia: studio delle modificazioni chimiche e del popolamento sottobasale dei sedimenti lagunari negli ultimi vent'anni. *Boll. Mus. civ. St. nat. Venezia* **26**, 25–68.

Mennella C. (1972) *Il clima d'Italia II* (Conte Ed. Napoli).

Merola A., Sacchi C.F. & Rigillo Troncone M. (1965) Gli ambienti studiati ed i fattori ecologici. In *Ricerche ecologiche sul lago salmastro litoraneo di Patria.* Delpinoa (2) 5 suppl., 9-240.

Petit G. (1962) Quelques considérations sur la biologie des eaux saumâtres méditerranéennes. *Pubbl. Staz. zool. Napoli* **32**, suppl., 205–18.

Sacchi C.F. (1961) L'évolution récente du milieu dans l'étang saumâtre dit 'Lago di Patria' (Naples) analysée par sa macrofaune invertébrée. *Vie et Milieu* **12**, 37–65.

Sacchi C.F. (1967) Rythmes des facteurs physico-chimiques du milieu saumâtre et leur emploi comme indice de production. *Problèmes de productivité biologique* (Ed. by A. Bourlière & M. Lamotte). Masson, Paris.

Sacchi C.F. (1973) Les milieux saumâtres méditerranéens: dangers et problèmes de productivité et d'aménagement. *Archo Ocean. Limnol. Ven.* **18**, suppl., 23–58.

Sacchi C.F. (1977) Il delta del Po come elemento disgiuntore nell'Ecologia delle spiagge alto-adriatiche. *Boll. Mus. cov. St. Nat. Venezia* **29**, 43–72.

Sacchi C.F. & Renzoni A. (1962) L'écologie de *Mytilus galloprovincialis* (Lam.) dans l'étang du Fusaro et les rythmes saisonniers et nycthéméraux des facteurs environnants. *Pubbl. Staz. zool. Napoli* **32**, suppl., 255–93.

Vatova A. (1960) Condizioni ecologiche e fasi di marea nell'alta laguna veneta. *Nova Thalassia* **2** (9), 1–61.

Vatova A. (1963) Conditions hydrographiques de la Mer Grande et de la Mer Piccolo de Tarente. Rapp. Comm. int. Mer Médit. **17**, 749–51.

36. FACTORS AFFECTING THE VEGETATION OF FRESH WATER RESERVOIRS ON THE GERMAN COAST

KUNO BREHM

Landesstelle für Vegetationskunde Neue Universität Kiel, B.R.D.

SUMMARY

Areas of the Waddensea off the west coast of Schleswig-Holstein have been reclaimed by diking for use as agricultural land and reservoirs. One of these, the Hauke-Haien-Koog, enclosed in 1959, includes two fresh water reservoirs with a total area of 7 km^2. In these, the water level is allowed to rise to 0·5 m above mean sea level (MSL) in March; throughout the summer the level gradually falls back to MSL in September, when the level is artificially lowered to 1·25 m below MSL by opening the sluices on the ebb-tides. Under this regime, of the area of 540 ha investigated, 250 ha is subject to annual flooding and a further 135 ha is permanently inundated. Apart from water storage the reservoirs are also used as a wildfowl sanctuary and for cattle and sheep grazing.

The development of the vegetation has been followed from 1959 to the present day. During the first two years a rapid succession, beginning with *Salicornia stricta* and *S.europaea*, took place. Areas above the inundation level now bear species-rich meadows dominated by *Agrostis stolonifera* or *Festuca rubra*, or else typical Lolio-Cynosuretum. The inundated zones are dominated by *Phragmites australis* and *Bolboschoenus maritimus*. Certain areas still support halophytic vegetation, where sea water percolates through underground sandy layers. The influence of hydrology, soil salinity, grazing, wildfowl foraging and reed-cutting on the various types of vegetation is discussed.

It is concluded that although further loss of the Waddensea is regrettable, the fresh water reservoirs produced are of considerable ecological interest.

RÉSUMÉ

Grâce à l'endiguement, des régions du Waddensee, à distance de la côte ouest de Schleswig-Holstein, ont été utilisées pour l'agriculture et les réservoirs.

L'une d'elles le Hauke-Haien-Koog, enclos en 1959, comprend deux réservoirs d'eau douce, avec une surface totale de 7 km^2. Ici l'eau peut monter en mars jusqu'à 0·5 m au dessus du viveau moyen de la mer (MSL). Au cours de l'été le niveau décroît progressivement pour atteindre le niveau de la mer. En septembre le niveau est artificiellement abaissé jusqu'à 1·25 m au dessous du MSL par l'ouverture des écluses à marée basse. Sur un total de 540 ha, 250 ha sont submergés annuellement et 135 nouveaux hectares sont inondés en permanence. Outre le stockage de l'eau, les réservoirs servent également de refuge aux gibiers d'eau et de pâture pour le bétail et les moutons.

Nous avons suivi le développement de la végétation depuis 1959 jusqu'a nos jours. Durant les deux premières années il y a eu une rapide succession commençant avec *Salicornia stricta* et *S.europaea.* Les terrains au dessus du niveau d'innondation sont maintenant des prairies riches en espèces dominées par *Agrostis stolonifera* ou *Festuca rubra* ou d'autres plantes typiques comme Lolio-Cynosuretum. Les zones inondées sont dominées par *Phragmites australis* et *Bolboschoenus maritimus.* Dans les régions où la mer s'infiltre à travers les couches sableuses du sous-sol il y a encore une végétation halophyte. Nous discutons de l'influence de l'hydrologie, de la salinité du sol, de la pâture, du fourrage du gibier d'eau et de la coupe des roseaux sur les differents types de végétation.

Bien que cette nouvelle perte du Waddensee soit regrettable, nous concluons que ces réservoirs d'eau douce sont d'un intérêt écologique considérable.

INTRODUCTION

Some nine hundred years ago there were large swamps along the North Sea coast of Germany, including about 4,000 km^2 of marshes, swamps and bogs north of the Elbe estuary. In 1362 this area was nearly completely inundated and eroded during heavy storm tides: the Waddensea was reduced to about 3,600 km^2. During the following centuries 1,100 km^2 have been diked, and today 2,500 km^2 are left.

One of the most recent diking projects was completed in 1959; this is the Hauke-Haien-Koog which has a total area of 12 km^2. The area is used for agriculture (5 km^2) on the land that once formed the upper foreshore, and for two water reservoirs (7 km^2), which cover those parts lying below mean sea level (MSL). The catchment area of the Bongsiel River, which runs through this Koog, consists of about 550 km^2 of the sandy *Geest* and 170 km^2 of the region of the *Marsch*. The Bongsiel River can be dammed by sluices in the sea-dike to stop the influx of sea water into the river during storms. Up to 7,000,000 m^3 of water can be collected in these fresh water reservoirs.

It should be emphasized that the necessity for establishing a fresh water reservoir has its origin in the general amelioration campaign in the catchment area of the Bongsiel River during the past few decades. Therefore more rainfall is brought down from the Geest into the low Marsch than before.

Building a dike on the bare Wadden was a pioneer experiment of international significance. New developments have brought further dangers for the Waddensea. Nowadays large dikes may be built independent of the state of the foreshore. This can be seen in further diking projects that have been completed or are being planned, e.g. the Meldorfer Bucht (bight), the Eidersperrwerk (damming of the estuary of the River Eider), the Sperrwerke of the Rivers Stör, Krückau and Pinnau, the Nordstrander Bucht and the Rodenäs Vorland.

The fresh water reservoirs of the Hauke-Haien-Koog have multiple functions and have to fulfil several management objectives, the most important of which are discussed below.

Hydrological factors in the fresh water reservoirs

The fresh water reservoir in the Hauke-Haien-Koog is divided into three sections (Fig. 36.1), each of which may be flooded separately to a height of 1·25 m above MSL. The full storage capacity has not yet been used. Since 1961, fresh water in the northern and southern reservoirs has been dammed on about the 15th of March each year and the water level allowed to rise to 0·5 m above MSL. Depending on the amount of rainfall from March to September each year the water level decreases gradually to a level not lower than MSL because at this point a dynamic equilibrium is established between the salt water penetrating from the sea through the sandy layers and fresh water from rainfall. There is no replenishment with water from Bongsiel River during this time. From the end of September each year the water level is artificially lowered to 1·25 m below MSL by opening the sluices to the Bongsiel River which allows water to drain through the sluices in the sea-dike during ebb-tides.

The areas which have been investigated in the fresh water reservoirs comprise about 540 ha, located in the northern and in the southern reservoirs. In springtime about 385 ha of this area is flooded with water to a depth of 0·5 m above MSL, but in late summer only about 135 ha are covered with water. Thus about 250 ha are flooded annually with fresh water; the remaining 135 ha are covered with deeper water resulting from the removal of sand for the dike and only these areas are waterlogged throughout the winter.

The development of the vegetation cover and plant succession are influenced by:

1. The absence of tidal inundation of salt water.
2. An annual damming of fresh water where the level of water in the reservoirs decreases during the summer and is low throughout the autumn and winter.
3. The infiltration of the underground sandy layers with sea water.

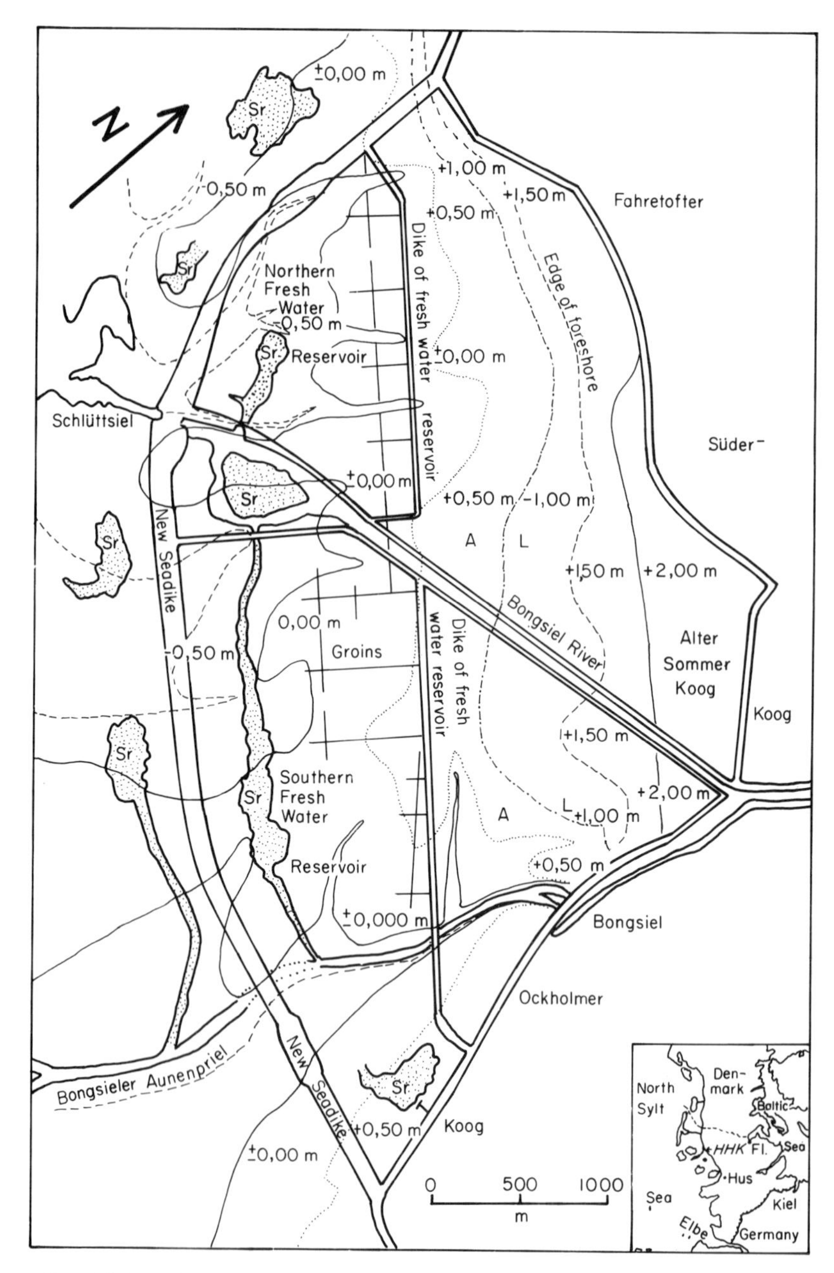

FIG. 36.1. The geographical and topographical position of the Fresh Water Reservoirs in the Hauke-Haien-Koog in Nordfriesland in the FRG. (S.r., Sites of sand removal for the dike, i.e. deep water areas.)

4. The changing salt content of the soil solution in the rhizosphere which is related to the rainfall, temperature and transpiration-evaporation regime.

Biological factors in the fresh water reservoirs

There are further ecologically relevant factors influencing the vegetation besides the hydrological factors outlined above.

1. Grazing by sheep and cattle. The density cannot be given exactly; grazing pressure depends on the mode of rotation of the herds. Up to five sheep per hectare or two cattle per hectare are approximate densities.
2. The area serves as a wildlife sanctuary; wildfowl foraging may have effects on some of the vegetation such as the grazing of shoots and rhizomes and the dissemination of propagules of several species. 'Wildfowl pressure' is lowest in October and November.
3. *Phragmites australis* is harvested in winter time in suitable places (about 20–40 ha). This hardly has an effect on further spread of *Phragmites* over the fresh water reservoirs because only dense *Phragmites* stands are utilized.

THE VEGETATION AND SUCCESSION

The vegetation on the foreshore and of the wadden in 1957–59 before diking took place has been documented by Heydemann (1963). There was a series of communities from the Salicornietum to the Armerietum in which the community of *Festuca rubra littoralis* dominated. The greater part of the foreshore of the former Hauke-Haien area was not covered by phanerogams.

Vegetational changes in several permanent plots have been observed from 1961 to 1965 by Jahns (unpublished). In 1961 values of vegetation cover in the newly reclaimed areas showed a gradient from 25% in areas of the former foreshore to 5% in the lower Wadden. Regrettably the corresponding measurements of salt contents in the soil profiles have not been available. Immediately after reclamation the composition of species was nearly the same as described by Christiansen (1955) for a typical Salicornietum with a transition from *Salicornia stricta* to *Salicornia europaea*, which was evident until about 1964. *Salicornia stricta* showed its highest frequency in 1961 but was replaced rapidly by *S.europaea*, and in 1964 the former species was restricted to the lowest areas of the fresh water reservoirs.

In 1961 *Salicornia stricta* and *Suaeda maritima* grew to a maximum height of 0·7 m under conditions of intensified drainage, partial desalination, aeration and nitrogen mineralization of the mostly sandy soils. In 1962 growth of the above mentioned species decreased. *Suaeda maritima* and *Atriplex hastata* var. *salina* especially decreased in growth after 1963 suggesting a fall in available nitrogen in the soil.

Similar phenomena have been reported after reclamation of other sites, for example in the Finkhaushalligkoog (Wohlenberg & Plath 1953), in the Veerse Meer in the Netherlands (Beeftink *et al.* 1971), in the Lauwerszeepolder in the Netherlands (Joenje 1972, 1974; Meijer 1973), and in the Meldorfer Bucht in Schleswig-Holstein.

From 1963 onwards with the decrease in *Salicornia stricta* a number of species colonized the sites, including *Salicornia europaea*, *Spergularia salina*, *S.marginata* and *Puccinellia distans*. The salt content of the soil solution showed a wide amplitude at sites where these species grow (Jahns, personal communication).

The vegetation would have been likely to change gradually into a Juncetum gerardii during the following years, with further desalination. But this likely development has been modified as a consequence of the annual inundation of fresh water to the level of 0·5 m above the MSL.

Between 1969 and 1971 a new vegetation map was prepared by Brehm and Eggers (1974) (Fig. 36.2). The annual inundation of fresh water and the continuous grazing appear to have led to a certain stabilization or at least to a decreased rate of change in the diversity of species and in the plant communities. This trend was evident some ten years after diking but now, 18 years after diking, only rather slow changes are taking place. Communities now present are described below.

Salicornia *communities*

Communities of *Salicornia europaea* are present still at certain sites in the fresh water reservoirs (Fig. 36.3, S). These sites remain uncovered during inundation and they are grazed intensively thus restricting the growth of higher vegetation. At some places salt water may percolate through the sandy layers from the wadden area beyond the dike. Species of such sites include: *Salicornia europaea*, *Spergularia marginata*, *Puccinellia distans*, *P.maritima* and *Suaeda maritima*.

Puccinellia distans *communities*

Puccinellia distans has covered sites lying several centimetres higher than those occupied by plants of *Salicornia*. A certain similarity of this community to that of the Puccinellietum distantis juncetosum (Westhoff 1947) can be recognized, and *Juncus ranarius*, *J.gerardii*, *Aster tripolium* as well as *Agrostis stolonifera* and *Alopecurus geniculatus* are the common species.

Although *Puccinellia distans* spread most rapidly during 1964–65, since that time it has decreased continuously. It is now restricted to a few small patches where it may be favoured by salinity and the prevailing grazing regime.

Puccinellia maritima *communities*

Swards of *Puccinellia maritima* were extensive on the former foreshore where they were usually grazed by sheep. In 1969–71 several patches were still in existence but these have further decreased in size (Fig. 36.3).

Agrostis stolonifera/Juncus gerardii *inundation meadows*

Those areas artificially inundated with fresh water each year during March to a depth of 0·5 m above MSL are inundation meadows in which *Agrostis*

FIG. 36.3. A view across the southern Fresh Water Reservoir (SW → NE) of the Hauke-Haien-Koog (2 October 1977). Bo *Bolboschoenus maritimus*; LC, *Lolium/Cynosurus*-meadows; Phr, *Phragmites australis*; Pm, *Puccinellia maritima* swards; S, *Salicornia* swards; sh, sheep grazing area; w, water at about MSL.

stolonifera and *Juncus gerardii* dominate. The meadows have a poor species-diversity. These inundation meadows are influenced by several ecological factors which have been investigated by Tschach (1977).

Agrostis stolonifera *meadow*

Hardly differing in height from the species-poor *Agrostis stolonifera* inundation meadow, this community contains a larger number of species which includes *Trifolium repens*, *Plantago major*, *Leontodon autumnalis*, *Taraxacum officinale*, *Potentilla anserina*, *Rumex crispus*. Grazing by sheep and cattle produces drastic changes in the vegetation of these meadows.

Festuca rubra *meadows*

At a slightly higher elevation than the *Agrostis* meadows is a species-rich meadow in which *Festuca rubra* is dominant. Bearing in mind the extremely small differences in height between the *Agrostis* meadow and the *Festuca rubra*-meadow, it seems difficult to accept that differences in the hydrology account for the differences in the vegetation. Perhaps competitive interactions between the species may be responsible for their different distributions.

Lolium/Cynosurus *meadows*

The vegetation cover of the slopes of the dikes results from the artificial sowing of a seed-mixture. This community contains species characteristic of fertilized meadows. *Trifolium repens* is the dominant dicotyledon, and *Cynosurus cristatus*, *Festuca rubra* and *F.pratensis* are common grasses. There has been no further fertilization of these meadows which are intensively grazed (Fig. 36.3, LC).

Communities of Phragmites australis *and* Bolboschoenus maritimus

Between the levels of 0·1 m and 0·4 m above MSL there has been an extensive development of reed beds from 1965 onwards. These two species are still colonizing sites beyond the elevations indicated above.

Phragmites occurs as physiognomically very different types and forms a heterogeneous mosaic. Originally, widely separated *Phragmites* sods had been planted in the fresh water reservoirs. Heterogeneity may partly be a result of vegetative propagation of rhizomes. At present we do not know enough about the population biology of *Phragmites* to make any predictions about its further spread (cf. Björk 1967; van der Toorn 1972).

During recent years *Phragmites* has invaded the *Agrostis* meadows and *Phragmites* is now in competition with *Agrostis stolonifera*. Tschach (1977) demonstrated that conditions varied at the different sites and that these differences may affect the outcome of competition at these sites.

1. The level of the soil surface *per se* has no significant effect on the distribution of different species.
2. The level of the ground water at different sites is not the same. In springtime the *Phragmites*-site is slightly more inundated with fresh water than the *Agrostis*-meadows. This difference remains during much of the summer when the ground water level decreases to about 0·5 m below the soil surface in the *Agrostis*-meadow and to about 0·2 m below the soil surface in the *Phragmites* area.
3. The acidity of the surface soil layer (0–5 cm) is slightly lower in the *Agrostis*-meadow than in *Phragmites*-soils. The values of pH (c. 6·5) are similar to the *Phragmites*-soils of the Rantum-Becken (Isle of Sylt), which are muddy soils (Müller-Suur 1972). The Rantum-Becken is like a fresh water reservoir. It was diked in 1938 but its function as a sewage plant has led to extreme eutrophication at that site.
4. There are only slight differences between the pore volumes of sediments under stands of *Phragmites* and *Agrostis*. In 1974 the volumes were 45% and 55% respectively to a depth of 0·3 m below the surface. The *Agrostis* soils become dry earlier in the season than the sediments in the area where *Phragmites* grows.
5. The salinity of the soil water in the rhizosphere of *Phragmites* was found to lie between 2·4 and 4·0‰. Investigations from the Rantum-Becken show that salinities of 6·4‰ may bring about reduced growth of *Phragmites* (Müller-Suur 1972). Soil water from *Agrostis* meadows has less than half the salinity of water from the sediments where *Phragmites* grows. Thus salinity may play a great part in differentiating between stands of *Phragmites* and *Agrostis*.
6. Grazing by domestic animals may be a major factor which favours *Agrostis stolonifera* in competition with *Phragmites*, at least in those parts where cattle are grazed. No well-developed *Phragmites* exists where cattle grazing is extensive (Fig. 36.3, Phr). Sheep grazing only slightly affects *Phragmites* (Fig. 36.3, sh). Sheep may graze the young shoots of *Phragmites* but do not touch older plants. Thus the effects of sheep grazing favour *Phragmites* in the *Agrostis*/*Juncus gerardii*-meadows and the spread of *Phragmites* is likely to continue where the *Agrostis* meadows are grazed by sheep.
7. Mowing of *Phragmites* has little effect on the outcome of plant competition because it is done in wintertime. Mowing early in the growing season would be a suitable method of biotope management for the breeding of birds, e.g. waders, shoveler or garganey.

Bolboschoenus maritimus seems to be dependent on similar factors to *Phragmites* in order to compete successfully with *Agrostis stolonifera*. Cattle grazing severely damages *Bolboschoenus* (Fig. 36.3, Bo). Sheep grazing in contrast has little effect on *Bolboschoenus* because of irregular rotation of the flocks. It is probable that sheep grazing results in more damage to *Bolboschoenus* than to *Phragmites* under otherwise identical circumstances.

Mowing of *Bolboschoenus* early in the season might be a suitable method for biotope management because almost only coot breed in dense *Bolboschoenus* reeds. According to proposals made by A. Koridon (personal communication), mowing must take place beneath the water level. A mowing in November may lead to little regrowth in the subsequent year; the regrowth must then be cut once or twice in the subsequent year. Mowing in June and July decreases the amount of regrowth even more and it needs to be cut only once in the subsequent year. The control of *Phragmites* reeds in drier habitats is much more troublesome, and repeated mowing is necessary over a period of at least three years (J. van der Toorn, personal communication).

CONCLUSION

Van Duin (1969, 1976) describes the management of a similar meadow sanctuary 'Kievitslanden' (98 ha) in East Flevoland in the Netherlands. The water level is gradually lowered from 15 March to the 15 September and then allowed to rise again (0·5 m difference). Forty three ha serve as pasture for 1–1·5 cattle per ha, and 55 ha are mown. Kievitslanden has been colonized by several meadow birds in the course of the years, such as by *Limosa limosa*, *Vanellus vanellus*, *Tringa totanus*, *Philomachus pugnax*, *Anas platyrhynchos*, *A.querquedula*, *A.strepera* and *A.clypeata* (de Jong 1972; van Duin 1976).

Damming of salt water instead of fresh water would be of great interest. Maybe some experience can be gained in future when new parts of the Waddensea are diked on the North Sea coast of Schleswig-Holstein.

There are some older areas which resemble the fresh water reservoirs in the Hauke-Haien-Koog. These are the marshes of the Isle of Fehmarn in the Baltic Sea (Sulsdorfer Wiek, Wallnau), which were diked at the end of the last century and have for long been managed as brackish fishponds. Partly comparable is the Rantum-Becken on the Isle of Sylt which was diked in 1938 and now serves as a sewage plant.

There are certain similarities in the results of this study with other diking projects, as for example in the IJsselpolder (Feekes 1936; Feekes & Bakker 1954; Westhoff 1969; van der Toorn *et al.* 1969), or with areas in the Delta Region in the Netherlands (Beeftink 1962, 1966; Beeftink & Daane 1973; van der Toorn & Mook 1973), in the Veerse Meer (Beeftink 1967; Beeftink *et al.* 1971), in the Ringkøbing-Fjord in Denmark (Gravesen 1972) or at the French coast (Corillion 1953) or in the British salt marshes (Tansley 1949; Gillham 1957; Brereton 1971; Gray 1976, 1977).

Further diking is taking place in the Schleswig-Holstein Waddensea. The Meldorfer Bucht consists of 50 km^2 of which 3·8 km^2 are being planned as fresh water reservoirs or perhaps partly as a salt water reservoir for nature conservation purposes. There are plans for further diking projects. The Nordstrander Bucht (about 56 km^2) will comprise finally 8 km^2 of water reservoirs of similar character, and the recent Rodenäs diking of about 20 km^2 will leave perhaps 2·8 km^2 as fresh or salt water reservoirs.

This diking will lead to a further decrease of the ecological value of the unique Waddensea (Abrahamse *et al.* 1977). Although they may be very interesting ecological units, fresh water reservoirs cannot serve as compensation for the Waddensea, because more than ninety percent of the newly reclaimed land is changed into agricultural land. As Beeftink (1975) has stated, 'A plea is made for an evaluation of the real economic advantage of such works against the social advantages of the present state of the coast's natural features'.

Nevertheless, if further diking is sponsored by the government, regretable though it is, water reservoirs will reward the attention of ecologists as well as of nature conservationists.

REFERENCES

Abrahamse J., Joenje W. & van Leeuwen-Seelte N., (Eds.) (1977) *Wattenmeer* Lendelijke Vereniging tot Behoud van de Waddenzee, Harlingen and Vereniging tot Behoud van Natuurmonumenten in Nederland, 's-Graveland. Neumünster/FRG.

Beeftink W.G. (1962) Conspectus of the phanerogamic salt plant communities in the Netherlands. *Biol. Jaarb. Antwerpen*, 1962, 325–62.

Beeftink W.G. (1966) Vegetation and habitat of the salt marshes and beach plains in the south-western part of the Netherlands. *Wentia* **15**, 83–108.

Beeftink W.G. (1967) Veranderingen in bodem en vegetatie van de voormalige slikken en schorren langs het Veerse Meer. *Driemaandelijks Bericht Deltawerken* **41**, 1–8.

Beeftink W.G., Daane M.C. & De Munck W. (1971) Tien jaar botanisch-oecologische verkenningen langs het Veerse Meer. *Natuur en Landschap* **25**, 50–65.

Beeftink W.G. & Daane M.C. (1973) Ontwikkelingen in het plantendek op het Groene Strand, Oostvoorne. *Delta Landschaap* 7/8, 23–27.

Beeftink W.G. (1975) The ecological significance of embankment and drainage with respect to the vegetation of the south-west Netherlands. *J. Ecol.* **63**, 423–58.

Björk S. (1967) Ecologic investigations of *Phragmites communis*—Studies in theoretic and applied limnology. *Folia limnol. scand.* **14**, 1–248.

Brehm K. (1971) *Seevogelschutzgebiet Hauke-Haien-Koog*. Barmstedt (Holst.), 1971.

Brehm K. & Eggers T. (1974) Die Entwicklung der Vegetation in den Speicherbecken des Hauke-Haien-Kooges (Nordfriesland) von 1959 bis 1974. *Schr. Naturwiss. Verein Schleswig-Holstein* **44**, 27–36.

Brereton A.J. (1971) The structure of the species populations in the initial stages of salt marsh succession. *J. Ecol.* **59**, 321–38.

Christiansen W. (1955) Solicornietum. *Mitt. Florist.-soziol. Arbeitsgem., NF* **5**, 64–65.

Corillion R. (1953) Les halipèdes du nord de la Bretagne (Finistère, Côtes-du-Nord. Ille-et-Vilaine). Etude phytosociologique et phytogéographique. *Revue Générale de Botanique* **60**, 609–58 and 705–75.

Dittmer E. (1956) Die Versalzung des Grundwassers an der schleswig-holsteinischen Westküste. *Die Küste* **5**, 87–102.

van Duin R.H.A. (1969) Natuurbouw in Flevoland. *Rijksdienst voor de Ijsselmeerpolders*, 1–8.

van Duin R.H.A. (1976) Broedterreinen voor Weidevogels. *Rijksdienst voor de Ijsselmeerpolders, werk document*, 1976–87 B, 15–26.

Eggers T. (1969) Über die Vegetation im Gotteskoog (Nordfriesland) nach der Melioration. *Mitt. der Arbeitsgem. Geobotanik Schleswig-Holstein* **17**, 1–103.

Feekes W. (1936) *De ontwikkeling van de natuurlijke vegetatie in de Wieringermeerpolder, de erste groote droogmakerij van de Zuiderzee.* Proefschrift, Amsterdam.

Feekes W. & Bakker D. (1954) De ontwikkeling van de natuurlijke vegetatie in de Nordoostpolder. *Van Zee tot Land* **6**, 1–92.

Gillham M.E. (1957) Vegetation of the Exe estuary in relation to water salinity. *J. Ecol.* **45**, 735–56.

Glue D.E. (1971) Salt marsh reclamation stages and their associated bird-life. *Bird Study* **18**, 187–98.

Gravesen P. (1972) Plant communities of salt-marsh origin at Tipperne, Western Jutland. *Bot. Tidsskr.* **67**, 1–32.

Gray A.J. (1976) The Ouse Washes and the Wash. *Nature in Norfolk: a Heritage in Trust*, pp. 123–29. Jarrold, Norwich.

Gray A.J. (1977) Reclaimed land. *The Coastline* (Ed. by R.S.K. Barnes), pp. 253–70. Wiley, London.

Heydemann B. (1963) Deiche der Nordseeküste als besonderer Lebensraum. Ökologische Untersuchungen über die Arthropoden-Besiedlung. *Die Küste* **11**, 90–130.

Hughes R.E., Milner C. & Dale J. (1964) Selectivity in grazing. *Grazing in Terrestrial and Marine Environments* (Ed. by D.J. Crisp), pp. 189–202. Blackwell Scientific Publications, Oxford.

Joenje W. (1972) De ontwikkeling van natuurlijke begroeningen in de Lauwerszee polder. *Natura* **69**, 172–75.

Joenje W. (1974) Production and structure in the early stages of vegetation development in the Lauwerszee-polder. *Vegetatio* **29**, 101–8.

de Jong H. (1972) Het Weidevogelreservaat in Oostelijk Flevoland. *Limosa* **45**, 49–57.

Meijer J. (1973) Die Besiedlung des neuen Lauwerszeepolders durch Laufkäfer (*Carabidae*) und Spinnen (*Araneae*). *Faunist.-ökolog. Mitteilungen* **4**, 169–84.

Müller-Suur A. (1972) *Vegetations- und Standortsuntersuchungen im Rantum-Becken auf Sylt.* Dissertation, Göttingen.

Schreitling K.-Th. (1959) Beiträge zur Erklärung der Salz-vegetation in den nordfriesischen Kögen. *Mitt. d. Arbeitsgem. Geobotanik Schleswig-Holstein* **8**, 1–98.

Tansley A.G. (1949) *The British Islands and their Vegetation.* Cambridge University Press, London.

Tschach E. (1977) Untersuchungen an *Agrostis stolonifera*—Beständen und ihren Ausgangsgesellschaften. Dissertation, Kiel.

van der Toorn J. (1972) *Variability of* Phragmites australis (Cav.) *Trin. ex Steudel in relation to the environment.* Proefschrift, Rijksuniversiteit Groningen.

van der Toorn J., Brandsma M., Bates W.B. & Penny M.G. (1969) De vegetatie van Zuidelijk Flevoland in 1968. *De Levende Natuur* **72**, 56–61.

van der Toorn J. & Mook J.H. (1973) Waarnemingen verricht in 1972 en 1973 op het rietproefveld N 17 in Zuidelijk Flevoland. *Jaarversl. Inst. Oecologisch Onderzoek*, 1973.

Westhoff V. (1969) Langjährige Beobachtungen an Aussüßungs-Dauerprobeflächen beweideter und unbeweideter Vegetation an der ehemaligen Zuiderzee. *Exp. Pfl. soziol.*, Symp. 1965, Den Haag, pp. 246–53.

Wohlenberg E. & M. Plath (1953) Produktionsbiologische Untersuchungen auf eingedeichten Wattflächen. *Die Küste* **2**, 5–23.

37. PLANT SUCCESSION AND NATURE CONSERVATION OF NEWLY EMBANKED TIDAL FLATS IN THE LAUWERSZEEPOLDER

W. JOENJE

Rijks Universiteit Groningen, Biological Centre, Department of Plant Ecology, Haren (Gn), The Netherlands

SUMMARY

Environmental characteristics and processes during the colonization by plants of recently exposed marine sand flats are exemplified by a case study in the Lauwerszeepolder. A major part of this former inlet of about 9,100 ha of the Dutch Waddenzee is destined to be a nature reserve. Soil factors, water and salt movements and nutrient status are described. The migration and colonization of the principal species are discussed.

Aspects of ecosystem development such as the increase in the annual net primary production are discussed together with data on the population dynamics of two species of *Salicornia**, the principal colonists. Within a few years after drainage the availability of nitrogen limits primary production. Abiotic and biotic phenomena lead to changes in the composition of species present in the different plant communities and these changes are accompanied by accumulation of organic matter and development of a well-aerated topsoil layer, with abundant plant roots. The main lines of succession on the sand and silty sand are summarized and some probable successional trends are discussed in relation to soil formation.

Finally the present management practices and future objectives are discussed. It is concluded that sufficient data or expertise do not exist to predict long term changes in these new ecosystems. The need for ecological research into the effects of different management practices is stressed.

RÉSUMÉ

A partir d'un exemple dans le Lauwerszeepolder nous étudions les caractéristiques de l'environnement et les processus pendant la colonisation par des

* Nomenclature of *Salicornia* follows Tutin *et al.* (1964); for other taxa it follows Heukels and Van Ooststroom (1970).

plantes, de surfaces sableuses marines récemment exposées. Une grande part de cette ancienne crique d'environ 9 100 ha du Waddenzee hollandais est destinée à une réserve naturelle. Nous décrivons les facteurs du sol, les mouvements du sel marin et l'état des nutriments. Nous discutons de la migration et de la colonisation des principales espèces.

Nous discutons des aspects du développement de l'écosystème, tel l'augmentation de la production primaire nette annuelle, avec les expériences sur les dynamiques de population de deux espèces de *Salicornia*, les principaux colonisateurs. Peu d'années après le drainage, les ressources en azote limitent la production primaire. Des phénomènes biotiques et abiotiques conduisent à des changements de la composition des espèces présentes dans les différentes communautés végétales et ces changements sont accompagnés d'une accumulation de matière organique et du développement d'une couche bien aérée de racines abondantes, à la surface du sol. Nous résumons les principales lignes de succession sur le sable et la vase et discutons de quelques directions probables de succession en relation avec la formation du sol.

Finalement nous discutons des pratiques de la présente administration et des objectifs futurs. Nous concluons à l'insuffisance d'expériences ou d'expertises pour la prédiction de changements à long terme de ces nouveaux écosystèmes sur des zones sableuses primitivement soumises aux marées. Nous insistons sur le besoin de recherches écologiques sur les effets des pratiques de différentes directions.

INTRODUCTION

Soft coastal ecosystems along the European coast have been modified by man for a long time. The silty coasts of the delta region of the Rhine and Meuse, the coast of the Wadden area and much of the east coast of England have been reclaimed and used as pasture and meadow land. New marshes, as they developed, have been embanked regularly for agricultural purposes (Beeftink 1975; Gray 1976; Verhoeven 1976). Most scientific knowledge of the development of vegetation and soil on reclaimed land has been gained from a series of studies in the Netherlands in the large embankments in the former Zuiderzee, which was cut off from the Waddenzee in 1932 (Feekes 1936, 1943; Feekes & Bakker 1954; van Schreven 1965; van der Toorn *et al.* 1969; Zuur 1961; and many reports of the Zuiderzee Commissie in the *Nederlands Kruidkundig Archief* 1927–45). Many of the results obtained in these studies are applicable to other reclaimed areas.

After the storm surge of 1953 many dikes have been reconstructed and inlets and estuaries closed to the open sea. In order to solve drainage problems in the low holocene coastal plains, these engineering works have been combined with the construction of fresh water lakes on the former seabed, such as the Hauke-Haien-Koog (Schleswig-Holstein) and the Lauwerszeepolder. The new land created as a result of the construction of recent embankments consists of tidal flats and channels and sands which have little or no agricultural value. Hence large areas have been left for the natural development of plant communities to occur. In contrast to the earlier polders, little is known about the development of vegetation and soil on these newly emergent marine sands. They seem to represent a new type of environment and offer excellent opportunities for the study of colonizing species and ecosystem development.

In spite of their recent origin the sandflats and reservoir lakes already have developed interesting plant and animal communities and thus they have considerable potential as sites for nature conservation, although this is no compensation for the lost marine wetland.

In this paper some results are summarized of a study carried out from 1969 to 1975 on the emergent sand flats in the Lauwerszeepolder which became dry in May 1969. The study consisted of an ecological investigation of the colonization and production of vascular plants and bryophytes in the Lauwerszeepolder and an examination of the abiotic factors affecting these processes. Some remarks are made on future developments, in relation to management objectives for nature conservation.

ENVIRONMENTAL CHARACTERISTICS

Soil

Sedimentation and transport phenomena in the Waddenzee, as influenced by tidal streams and waves, result in the formation of a typical system of gullies and sand flats, which run dry at low tide. The Lauwerszeepolder (Fig. 37.1) may be considered to represent such a system.

The extensive sand flats lie at about 95 cm above the reservoir level (NAP* −90 cm) and gradients in elevations and soil texture are characteristic features of the area. The sediments in the areas at low elevation generally contain more silt. The soil profile often consists of layers of sediment which differ slightly in silt content. The main sediment constituents consist of fine sand fractions up to 200 μm in size (Fig. 37.2). The percentage of silt, clay (<2 μm), organic matter and carbonate content in the sediments are low. Because of these and other features, the sand flats are characterized by a low nutrient content.

---

* NAP = Nieuw Amsterdams Peil = Amsterdam Ordnance Datum ≈ mean sea level.

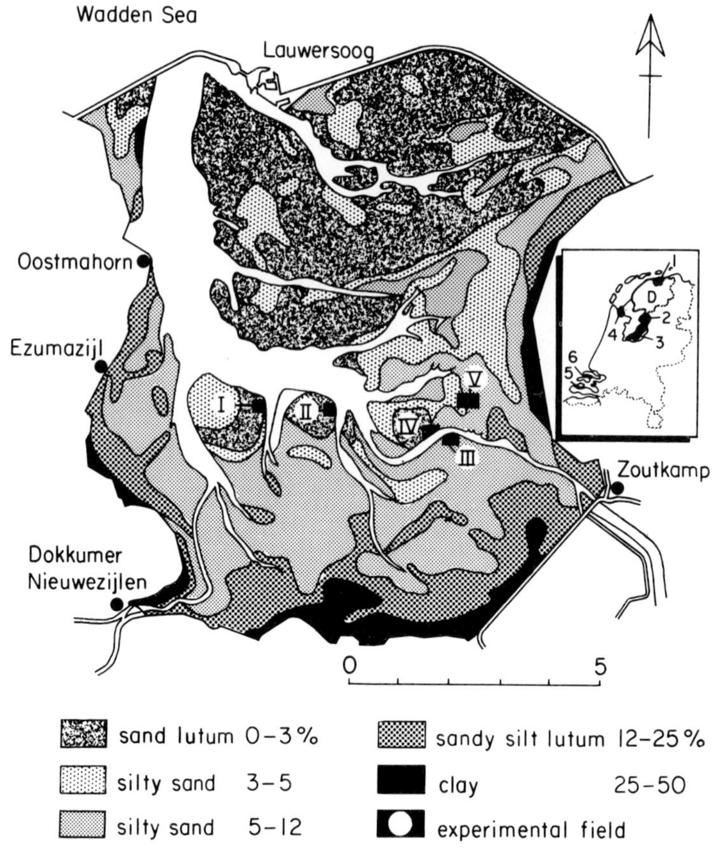

FIG. 37.1. Main soil types in the Lauwerszeepolder and location of the experimental fields (numbered I to V) Inset: location of the Lauwerszeepolder (1), the discharge area (D) and some other embankments in the Netherlands: North-Eastern Polder (2), Flevo Polders (3) Wieringermeer (4), Veerse Meer (5) and Grevelingen (6). The scale indicates 0–5 km (after RIJP, Lelystad).

Hydrology

Whenever possible the level in the reservoir is kept at NAP −85 cm in summer and NAP −95 cm in winter. However depending on the supply of water from the hinterland and the amount of water which is allowed to discharge into the Waddenzee at low tide, the level of water in the reservoir fluctuates. During short periods in winter when the sand flats are soaked with rainwater, maximum water levels of about NAP −20 cm are reached and a very large area is flooded.

The fluctuations in soil water conditions (Fig. 37.3) follow a characteristic pattern. The differences in silt content and grain size are accompanied by

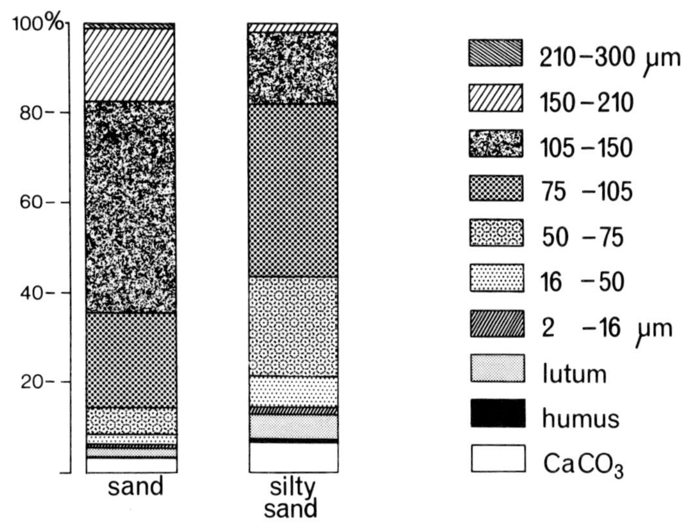

FIG. 37.2. Components of sand and silty sand as a percentage of the dry weight of the 0–20 cm layer of sediment.

differences in pore space which affect micro water movements and the precipitation–evaporation balance. Fig. 37.4 gives an example of the fluctuations in soil water tables caused by two showers of 12 and 14 mm respectively which occurred after a long dry period. The rise of the water table in silty sand is higher than that at other sites, probably due to the higher moisture content and better capillary rise. Capillary rise generally occurs up to 50 cm above the phreatic level. Lateral water movements are very slow, resulting in convex water tables in summer (Fig. 37.4) and waterlogged conditions in winter (Fig. 37.3). After a short period of heavy rain the water-saturated surface layer traps air in the lower sediments and excess water runs off superficially. These phenomena provide an explanation why the soil water table at a site is independant of height of the sand above O.D. and the distance from the reservoir. The silt content of sediments markedly affects the level of the water table; the largest fluctuations and the lowest levels of the water table occur in silty sands (Fig. 37.3). These differences between the hydrology of sand and silty sands probably will increase in future when a more extensive cover of vegetation on the silty sands than exists at present, leads to an increase in evapotranspiration. On the highest sands, which are poorly vegetated, the water table falls in dry periods so that the upper layers of sand are beyond the reach of capillary water and consequently loss of water as a result of evapotranspiration is limited.

Desalination

The movement of salt is related to the movement of water and in principle the passage of water through the soil can be calculated from salinity figures

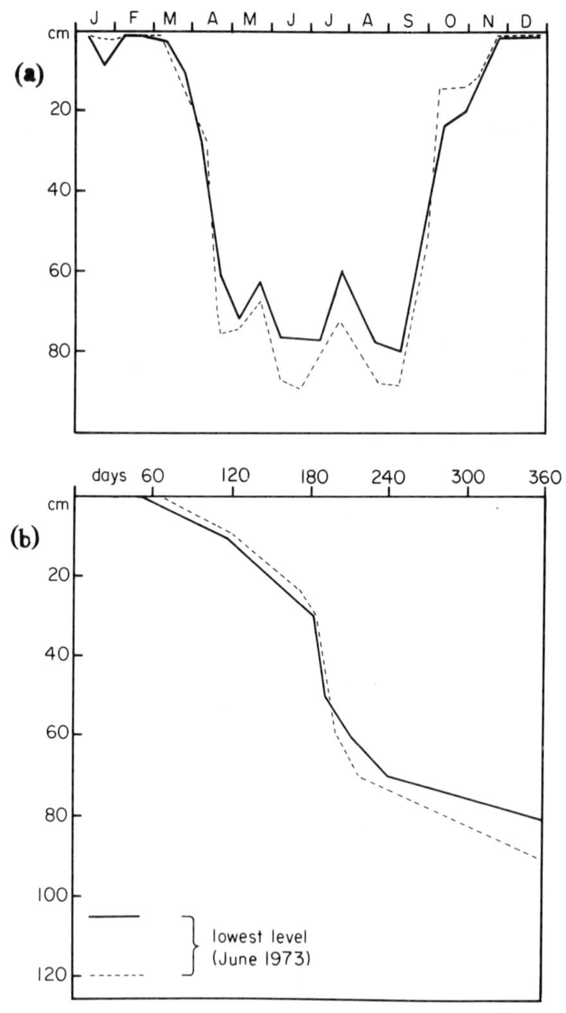

FIG. 37.3. (a) Mean depth of the water table for each month (1970–74) on sand (——) and silty sand (– – –). (b) The time in days during which the water table has been at a certain level.

(Verhoeven 1953; Zuur 1938). However because of differences in hydrology mentioned earlier, the processes of desalination on the sand and silty sand are dissimilar. In Fig. 37.5 the general processes are depicted. However the actual values of salinity, especially in the top centimetre of the soil were often much higher than indicated in the figure and these concentrations in the range of 50–150 g NaCl l^{-1} soil moisture are lethal to most plant species. Salt rising with the capillary water in the silty habitats was and still is the dominating

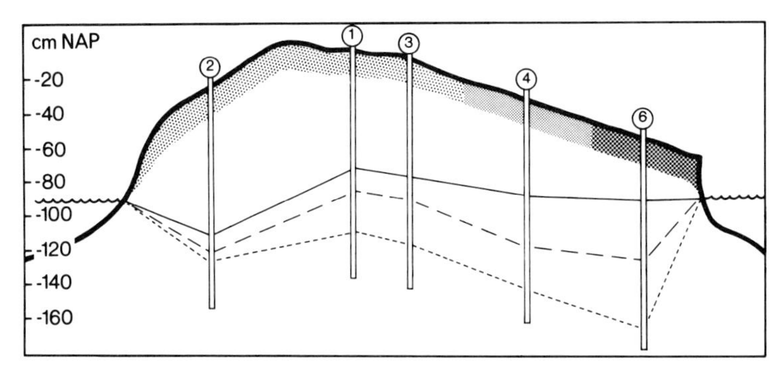

FIG. 37.4. The effects of 12 and 14 mm precipitation on the water table of a sandflat as measured in 5 tubes along a 400 m transect from sand (left) to silty sand (right): (· · · · · ·), level 17 June 1970, after a dry period; (– – – –), level 24 June 1970, after 12 mm; (———), level 1 July 1970, after 14 mm; (〰〰), reservoir level.

environmental factor which markedly affects the establishment of new species. From the data on desalination it has been estimated that in the most areas only 10–20% of the precipitation moves through the soil. However some of the salt is lost by diffusion into superficial surface water.

Initial soil formation

Pedogenetic development of soils of reclaimed land involves many physical, chemical and microbiological changes (Smits *et al.* 1962). These include increased aeration of the sediments and the subsequent oxidation of the dark

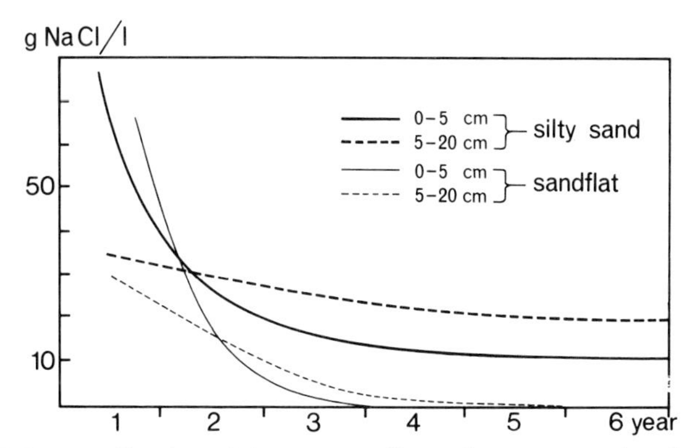

FIG. 37.5. Desalination of the upper sediment layers on sand and silty sand over a period of 6 years after drainage of the tidal flats. The salinity is given in g NaCl l^{-1} of the interstitial water.

sulphides, mainly FeS and FeS_2. Movement of oxygen from the roots of the pioneer vegetation also results in the oxidation of these sulphides as shown by the brown colour of the ferric hydroxides in the vicinity of the roots. Within a year of the draining of the Lauwerszeepolder the evidence of oxidation was visible to a depth of about 50 cm, especially around the old burrows of the marine fauna and around plant roots.

A component of the small organic fraction present in the sediments was mineralized quickly, resulting in a luxurious growth of pioneer plants in the first two years after drainage occurred. The subsequent growth of higher plants however suffered from nitrogen deficiency. Important, though less visible, changes took place in the ion-exchange properties of the sediments; calcium replaced sodium, and potassium was lost gradually from the sands. The low cation exchange capacity of about 100–600 meq per 100 g dry soil and the low nutrient content will not increase, until after several years an organic residue has been built up.

Decalcification occurs slowly, but as there is a low lime content (1–2%), a significant drop in pH, now about 7·5, in the top centimetre of sediment is predicted within a few decades. However this process will be retarded by the activities of the soil fauna (e.g. *Bledius*, *Heterocirrus* and larvae of *Tipulidae* spp.) which result in the mixing of materials within the soil profile. The initial soil formation is completed only after the alkali metals have been leached from the soil (Zuur 1961) and further pedogenesis results in the development of a podzol, a well-known phenomenon in the pleistocene sands on the adjacent mainland.

Invading flora

The surrounding coastal landscapes, consisting of arable and pasture land within polders of different ages, do not offer much environmental variation. Within 2 km of the Lauwerszeepolder, nearly 400 species of vascular plants were found, but only a few of these became established on the new land. In Table 37.1 several species are listed, which were grown in soil taken prior to the enclosure of the polder and watered with fresh water. The extreme salinity of the flats only permitted the establishment and growth of halo-tolerant species. A few localities such as the former mussel-banks, lost salt rapidly and were quickly colonized by plants. At this site there were 37 and 176 species, respectively in the first and second years including both halophytes and glycophytes (Joenje 1978a, b).

In the saline habitats eu-hydatochorous dispersal of plant propagules played a dominant role in the initial stages of colonization by plants. Colonization of the flats started with a sparse halophytic vegetation mainly consisting of species of the *Chenopodiaceae*, notably *Salicornia dolichostachya*, *S.europaea*, together with *Suaeda maritima* and *Atriplex hastata*. In the first year the plant

Table 37.1. *Number of living propagules of plant species present in soil samples of 20 quadrats (0·5 m²) taken from the sea-bed prior to the enclosure.*

Species	Number	Species	Number
Scirpus maritimus	54	*Suaeda maritima*	1
Atriplex hastata	36	*Triglochin maritima*	1
Aster tripolium	33	*Cochlearia anglica*	1
Spartina townsendii	18	*Rorippa islandica*	1
Ranunculus sceleratus	11	*Lolium perenne*	1
Puccinellia maritima	2	*Senecio vulgaris*	1
Phragmites australis	2	*Epilobium hirsutum*	1
Agrostis stolonifera	2	*Chenopodium rubrum*	1
Salicornia europaea	1	Total (17)	167

densities ranged from zero to more than 300 ha^{-1}. These different densities reflected the silt content of the sediments and the distance of the site from nearby salt marshes. *Salicornia* spp. had the highest densities on silty sand. *Suaeda*, *Atriplex* as well as *Spartina* became established in sandy areas, a probable outcome of differences in hydatochorous dispersal (Joenje 1974). At a later stage of development an interesting difference between the two species of *Salicornia* emerged. The tetraploid *S.dolichostachya* was more prevalent on the central sands than *S.europaea* and this could be another example of the effects of differences in dispersal between species. *Aster tripolium* appeared on the partially desalinated sands from the first year after drainage, but colonized the flats only after several years. The high salinity at this site together with the low nutrient level immediately after drainage probably account for the delayed establishment of this species rather than the inefficient dispersal of seed. *Spergularia marina* an annual or sometimes biennial species characteristic of unstable open habitats in salt marshes (Sterk 1968), was a successful invader from the third year after drainage and became co-dominant within a few years, at the expense of *Salicornia* species. *Spergularia media*, a perennial species with winged seeds, characteristic of more stable salt marshes, only became dominant in the saline silty habitats after 5 years, together with *Puccinellia maritima*. For all of the abundant species as well as for many others the dispersal of the propagules on or in sheets of rainwater blown across the saturated surface of the sparsely vegetated flats was most effective. This was true for species such as *Puccinellia capillaris*, *P.distans*, *Poa trivialis*, *P.annua*, *Juncus bufonius*, *Sagina maritima* and for several Bryophytes such as *Funaria hygrometrica* and *Pottia heimii* (Joenje & During 1977).

On the shallow desalinated sandy habitats the first glycophytes appeared in the second year after drainage. Here the rapidly increasing abundance of plants of species of *Taraxacum*, *Epilobium*, *Cirsium* and *Sonchus* together with

Erigeron canadensis and *Senecio vulgaris* illustrates the effectiveness of anemochorous dispersal in non-saline open habitats, as was shown earlier by Bakker and van der Zweep (1957). In Table 37.2 changes in the dispersal spectra of the flora from 1969 to 1974 on sandy and silty habitats are shown.

Table 37.2. *Dispersal spectra of the flora on sand and silty sand in the first, second and sixth year after drainage of the tidal flats.*

		1969				1970				1974			
		A	H	O	R	A	H	O	R	A	H	O	R
Sand	species	1	6	0	0	6	15	7	0	30	25	17	54
	percent	14	86	0	0	21	54	25	0	24	20	13	43
Silty sand	species	0	6	0	0	0	8	1	0	1	14	1	7
	percent	0	100	0	0	0	89	11	0	4	61	4	31

Dispersal types (partly after Westhoff 1947).
A. Eu-anemochores: propagules dispersed by wind over some hundred metres to many kilometres.
H. Eu-hydatochores: propagules dispersed by water, floating for some hours to years.
O. Combination of type A and H, typical in the bare environments: dispersal in and on sheets of rain water, blown across the saturated soil surface.
R. other dispersal types: propagules dispersed by various agents, mostly over shorter distances.

Given the fact that there exist only very few halo-tolerant species in which the propagules are wind dispersed, a general conclusion from this study is that differences in colonizing ability of species are related to differences in the rate of desalination (viz. Fig. 37.5).

ASPECTS OF ECOSYSTEM DEVELOPMENT

After drainage of the area the high salinity and the presence of their seeds on the seabed resulted in a rapid colonization of plants of species of *Salicornia*, *Suaeda* and *Atriplex*. A wide range of morphological variation was shown by individuals of each of the species in the first two years of colonization, a phenomenon reported earlier by Feekes (1936) (Fig. 37.6).

In the third and subsequent years this morphological variation was much reduced, probably due to selection or plastic response of the plants to the high densities. The significance of this intraspecific variation in relation to ecosystem development deserves more attention. Aggregation, another characteristic phenomenon in plant colonization (Weaver & Clements 1938) was clearly observed in the monospecific *Salicornia* stands. Seedling aggre-

FIG. 37.6. Intraspecific variation in *Atriplex hastata* on silty sand in the second year. Two plants of this species and several plants of *Salicornia* spp. are shown.

gates of luxuriously thriving plants were widespread especially in the second and third year. They disappeared in subsequent years when dense *Salicornia* stands were replaced by *Spergularia marina*, *Aster tripolium* and several grasses such as *Puccinellia capillaris*, *P.distans*, *Poa trivialis* and *P.annua*. In relation to environmental differences mentioned above, the plant successional changes on sand and silty sand were not the same. The quantitative changes in the contribution each group of principal species makes to the annual primary production are summarized in Fig. 37.7. The estimated annual production of up to 400 g d wt. m^{-2} (above and below ground) is low, but this estimate is based on measurements of standing crop at the end of the growing season. Observations indicate that as the salinity drops and less halo-tolerant and glycophyte species which are mainly perennial, invade the area, annual production increases in both the sandy and silty sand habitats.

On silty sand the salinity of the soil solution remained well above 5 g NaCl l^{-1} (Fig. 37.5), a toxic level for most glycophyte species (Feekes 1936). After 8 years the vegetation consisted of Puccinellietum maritimae with a local dominance of *Spergularia media* and several species of the lower salt marsh such as *Juncus gerardii*, *Triglochin maritima*, *Artemisia maritima* and *Plantago maritima*. However the presence of *Agrostis stolonifera* and *Alopecurus geniculatus* indicates a probable trend towards decreasing salinity, caused by incidental inundations from the eutrophic fresh-water reservoir. This also allows a large surface of silty sands to gain additional nutrients.

On the sandy flats leaching of nutrients prevails. Here *Salicornia* only colonized in the second year while in the third and fourth year *Atriplex hastata*

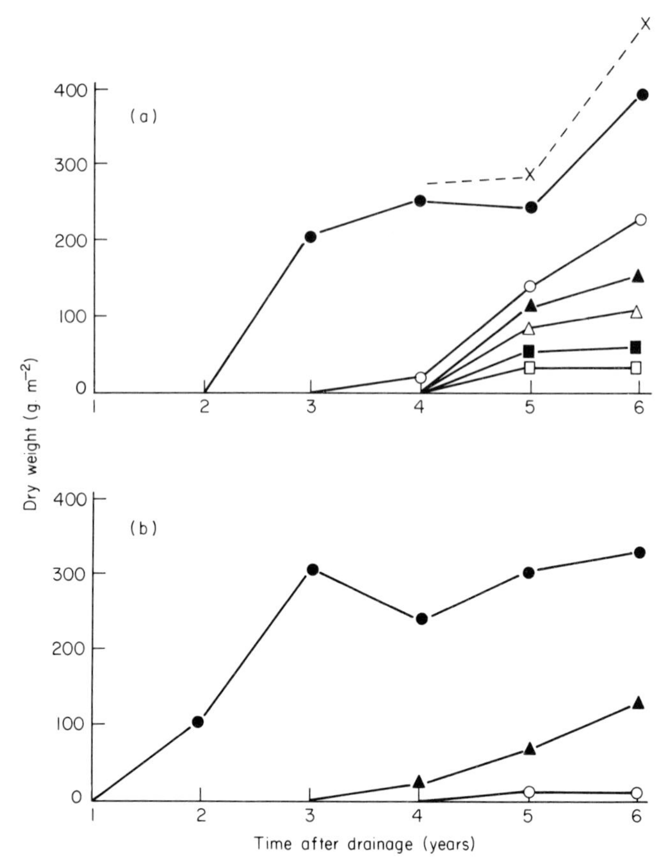

FIG. 37.7. Quantitative changes in total biomass and the contribution of the principal species groups in g d wt. m^{-2} (above and below ground biomass) on sand (a) and silty sand (b) over a period of 6 years after drainage of the Lauwerszeepolder. (– – –), Mosses (*Funaria hygrometrica, Bryum bicolor, Pottia heimia et al.*); (●), Annual halophytes (*Salicornia europaea, Salicornia dolichostachya, Atriplex hastata, Suaeda maritima*); (□) Forbs (many species e.g. *Epilobium hirsutum, E. parviflorum, E. adnatum, Senecio vulgaris, Sonchus asper, Cirsium vulgare, Rumex crispus*); (○), *Aster tripolium;* (▲) *Spergularia marina and S.media;* (△), *Poa annua and P.trivialis;* (■), *Puccinellia distans and P.capillaris.*

was a co-dominant with it. In the following years, the effect of desalination permitted the establishment of many species, several of which (notably perennials) made significant contributions to the biomass (Fig. 37.7).

Mosses became an important component of the vegetation from the fifth year onwards and their establishment is further evidence of the diversification of the system. The mosses modified the surface of the sand and increased the overall biomass particularly in early spring.

Table 37.3. *Lifeform spectra (after Raunkiaer 1934) of the flora on sand and silty sand in the first, second and sixth year after drainage of the tidal flats.*

	1969		1970		1974	
	Sand	Silty sand	Sand	Silty sand	Sand	Silty sand
T	6	4	13	6	42	7
He	2	2	11	4	67	16
G	0	0	2	0	10	0
C	0	0	0	0	2	0
P	0	0	0	0	10	0
Total species	8	6	26	10	131	23

(T), Therophytes; (He), Hemicryptophytes; (G), Geophytes; (C), Chamaephytes; (P), Phanerophytes.

In both habitats the number of perennial species increased, which is reflected in the lifeform spectra (Table 37.3), but their contribution to the biomass is only apparent on the sand where the accumulation of organic litter occurred. Here the presence of mosses resulted in the production of an organic layer, the importance of which in relation to cation-exchange capacity and soil formation is now being investigated. The accumulation of organic matter on silty sand is less obvious. Mosses are absent here and the major part of the standing crop of annuals is eaten by geese and waterfowl.

Population dynamics of Salicornia *spp.*

A closer examination of the establishment and growth of *Salicornia* was carried out, because it was possible to harvest the two species *S.dolichostachya* (tetraploid, 2n = 36) and *S.europaea* (diploid, 2n = 18) separately. In 1969 the number of *S.europaea* colonizing the area was about twice that of *S.dolichostachya.* In the following years plants of *S.europaea* were the more abundant and persisted. In the sandy habitats as mentioned above *Salicornia* colonization took place one year later but here plants of *S.dolichostachya* out-numbered plants of *S.europaea* in the ratio of 2:1. In both habitats however *S.dolichostachya* is replaced by *S.europaea* within a few years (Fig. 37.8). This succession reflects the situation normally found in the halosere, where the tetraploid plants again are pioneer at the lowest levels.

Biotic factors

The early stages of vegetation development are subject to grazing by geese, ducks, hares, rabbits and mice. Grazing pressure is maximal at the end of the growing season, when the major part of the annual production notably of

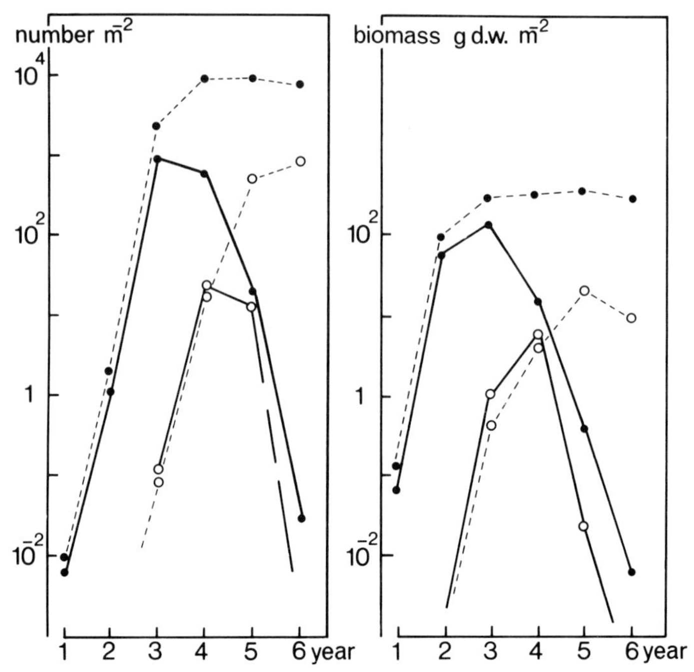

FIG. 37.8. Density and standing crop (above and below ground) at the end of the growing season of two species of *Salicornia*, on sand (○) and silty sand (●) over 6 years after drainage of the Lauwerszeepolder. (———), *Salicornia dolichostachya*, (– – –), *Salicornia europaea*.

Salicornia and *Spergularia* is removed by waterfowl. This could tend to perpetuate the early stages; such an effect was also observed in mowing experiments.

SUCCESSIONAL TRENDS

Although other studies of comparable situations are limited and the process of plant succession has only just begun, it is tempting to predict the likely vegetational changes during the next 20 to 30 years at these sites. The main evidence available at present is based on the water and nutrient conditions prevailing in the soil and the existing vegetation. Additional evidence is derived from the results of similar studies in the Veerse Meer (Beeftink *et al.* 1971), on sandy flats in the IJsselmeer (Feekes 1943) and in the Hauke-Haien-Koog (Brehm & Eggers 1974). The succession at Tipperne, Jutland, studied by Gravesen (1971) also may provide relevant data for a period of 40–50 years.

Salt present in the sand will be likely to dominate local conditions for some decades after drainage, while changes in nutrient status of the sand, pH and the lime content will influence the rate of plant succession in the long run.

The sands will be characterized by summer drought, low nutrient level, accumulating organic matter and a slowly decreasing pH. Species and community diversity is expected to increase with time as succession proceeds, and already species of *Salix*, *Crataegus monogyna*, *Hippophaë rhamnoides* and *Rubus caesius* have appeared. The well-drained areas are covered by mosses and here the invasion of lichens very likely will occur. Over a longer period decalcification of the sand will probably permit the invasion of heathland species such as *Erica tetralix*, *Nardus stricta*, *Festuca ovina*. On somewhat lower areas of sand *Calamagrostis epigejos* is likely to become dominant for a long period. In general it is predicted that the sites will resemble in time the older and partly decalcified dunes in the Wadden district.

The silty sand will probably remain saline for another 10–20 years at least, and may become enriched with nutrients as a result of incidental winter inundations. The invading plants of *Agrostis stolonifera*, *Alopecurus geniculatus* and *Juncus gerardii* suggest that large areas eventually will be covered with these species, and perhaps a few others. The area close to the lake which is likely to be flooded periodically, probably will be invaded by *Phragmites australis*.

The intermediate zones with gradients in soil texture and elevation of the *limes divergens* type (*sensu* van Leeuwen 1966) already have and probably will maintain the highest diversity of plant species.

NATURE CONSERVATION AND MANAGEMENT

The tidal flats (7,800 ha) have been recognized as a nature conservation area and an environmental plan prepared accordingly. It has been recognized that 'laisser faire' management for this whole area would be unsatisfactory and that some deliberate management would be required to conserve the outstanding ecological features of the site. The general objectives of management are difficult to define in both theoretical and practical terms and it is only when nature management plans are worked out for local areas, that the objectives can be clearly stated. Only from the results of field studies together with an understanding of natural developments and the effects of human influences, can one decide on the best approach to the management of a site (viz. also Gray 1977).

The value of the Lauwerszeepolder for nature conservation lies in its role as a foraging and wintering ground for large numbers of migratory birds, notably geese (esp. *Branta leucopsis*) and waterfowl (*Anas crecca*, *Anas penelope*). It is also an increasingly important breeding area for many species, such as lapwing (*Vanellus vanellus*), avocet (*Avocetta recurvirostra*), black-tailed godwit (*Limosa limosa*), redshank (*Tringa totanus*), kestrel (*Falco tinnunculus*), short-eared owl (*Asio flammeus*) and marsh harrier (*Circus aeruginosus*). The value of the site as a grazing area however is due to the presence of a large crop of annual halophytes on the silty sand, and this could change as plant succession proceeds. The numbers of migratory birds probably will diminish, but the number of species present will be likely to increase.

Management plans are being considered to keep some of the area in a young, productive stage as a feeding ground for geese and waterfowl by using them as agricultural grassland, thus deflecting the process of plant succession. The difficulty at present is knowing whether the geese and waterfowl will use these areas for grazing. The importance of sites such as the Lauwerszeepolder as bird refuges has been recognized from similar studies of other embankments (Brehm 1974, Lebret 1971).

The botanical value of the young and changing plant communities is their uniqueness as representative of the early stages of succession; as yet, only a few rare species are involved. The most interesting plant communities occur on the sandy habitats where there are gradients in elevation and soil conditions. Among the 320 plant species found in the polder many are regarded as valuable, such as *Parnassia palustris*, *Centaurium littorale*, *C.pulchellum* and *Odontites verna* ssp. *serotina*. However they may be members of transient plant communities. Here the objectives of management should be to increase community structure and diversity, rather than maximize suitable conditions for a few species.

Management of the ecologically different sand and silty sand areas should include the following activities:

(1) grazing and eventually hay making on silty sand, combined with local reed cutting along the reservoir lake.
(2) 'laisser faire' management eventually followed by very extensive grazing on sand.

Experimental sheep grazing was started in summer 1977 on an area of about 300 ha comprising sand and silty sand habitats. The animals left most of the halophytic vegetation to the waterfowl and preferred the forbs present at the sandy site. In view of the low primary productivity, the plant species involved and the considerable natural grazing pressure, the sheep density has to be kept low.

CONCLUSION

A general conclusion may be that some of the interesting aspects of the young stages of plant succession are difficult or impossible to maintain. Grazing and mowing will be useful as management techniques, at least on the silty sands. The large areas which are managed offer good opportunities for the experimental use of different grazing regimes with different types of cattle. Management plans for more than 10 different areas in the Lauwerszeepolder are being made by the Authority of the Rijksdienst voor de IJsselmeerpolders and these will be adjusted annually. This has established the need for a close liason between ecological field studies and work on the effects of different management practices.

ACKNOWLEDGMENTS

This study was supported financially by the Netherlands Organization for Advancement of Pure Research (Z.W.O.).

REFERENCES

Bakker D. & van der Zweep W. (1957) Plant-migration studies near the former island of Urk in The Netherlands. *Acta Bot. Neerl.* **6**, 60–73.

Beeftink W.G. (1975) The ecological significance of embankment and drainage with respect to the vegetation of the South-West Netherlands. *J. Ecol.* **63**, 423–58.

Beeftink W.G., Daane M.C. & De Munck W. (1971) Tien jaar botanisch-oecologische verkenningen langs het Veerse Meer. *Natuur en Landschap* **25**, 50–63.

Brehm K. (1974) Landschaft-Mensch-Vogel. Zur Entstehung der 'Totalen Kulturlandschaft' in Schleswig-Holstein. *Vogelleben zwischen Nord- und Ostsee* (Ed. by G.A.J. Schmidt & K. Brehm), pp. 183–241. Karl Wachholtz Verlag, Neumünster.

Brehm K. & Eggers T. (1974) Die Entwicklung der Vegetation in den Speicherbecken des Hauken-Haien-Kooges (Nord-friesland) von 1959 bis 1974. *Schr. Naturwiss. Ver. Schleswig Holstein* **44**, 27–36.

Feekes W. (1936) *De ontwikkeling van de natuurlijke vegetatie in de Wieringermeer-polder, de eerste groote droogmakerij van de Zuiderzee.* Thesis, Amsterdam.

Feekes W. (1943) De Piamer Kooiwaard en de Makkumerwaard. *Ned. Kruidk. Arch.* **53**, 288–331.

Feekes W. & Bakker D. (1954) De ontwikkeling van de natuurlijke vegetatie in de Noordoostpolder. *Van Zee tot Land* **6**, N.V. Uitgeversmij. Tjeenk Willink, Zwolle.

Gravesen P. (1971) Plant communities of salt-marsh origin at Tipperne, Western Jutland. *Bot. Tidsskrift* **67**, 1–32.

Gray A.J. (1976) The Ouse Washes and the Wash. *Nature in Norfolk—a Heritage in Trust*, pp. 123–29. Jarrolds, Norwich.

Gray A.J. (1977) Reclaimed land. *The coastline* (Ed. by R.S.K. Barnes), pp. 253–70. John Wiley & Sons, London.

Heukels H. & Van Ooststroom S.J. (1970) *Flora van Nederland.* Wolters-Noordhoff N.V., Groningen.

Joenje W. (1974) Production and structure in the early stages of vegetation development in the Lauwerszeepolder. *Vegetatio* **29**, 101–8.

Joenje W. & During H.J. (1977) Bryophytes colonizing a desalinating Waddenpolder. *Vegetatio* **35**, 177–85.

Joenje W. (1978a) *Plant colonization and succession on embanked sandflats.* Thesis, Groningen.

Joenje W. (1978b) Migration and colonization by vascular plants into a new polder. *Vegetatio* (in press).

Lebret T. (1972) Vogels van de Natuurreservaten in het Veerse Meer na de afsluiting 1961–1970. *Limosa* **45**, 1–24.

Van Leeuwen Chr. G. (1966) A relation theoretical approach to pattern and process in vegetation. *Wentia* **15**, 25–46.

Raunkiaer C. (1934) *The life forms of plants and statistical plant geography.* Oxford University Press, New York.

Van Schreven D.A. (1965) Stikstofomzettingen in jonge IJsselmeerpoldergronden. *Van Zee tot Land* **41**. N.V. Uitgeversmij, Tjeenk Willink, Zwolle.

Smits H., Zuur A.J., van Schreven D.A. & Bosma W.A. (1962) De fysische, chemische en micro-biologische rijping der gronden in de IJsselmeerpolders. *Van Zee tot Land* **32**. N.V. Uitgeversmij, Tjeenk Willing, Zwolle.

Sterk A.A. (1968) *Een studie van de variabiliteit van* Spergularia media *and* Spergularia marina *van Nederland.* Thesis, Utrecht.

Sterk A.A. (1968) *Een studie van de variabiliteit van Spergularia media and Spergularia marina van Nederland.* Thesis, Utrecht.

Van der Toorn J., Brandsma M., Bates W.B. & Penny M.G. (1969) De vegetatie van Zuidelijk Flevoland in 1968. *De levende Natuur* **72**, 56–61.

Tutin T.G., Heywood V.H., Burges N.A., Valentine D.H., Walters S.M. & Webb D.A. (Eds.) (1964) Flora Europaea 1, Cambridge University Press, London.

Verhoeven B. (1953) *Over de zout- en vochthuishouding van geïnundeerde gronden.* Thesis, Wageningen.

Verhoeven B. (1976) Landaanwinning, dijkbouw en waterbeheersing. *Waddenzee* (Ed. by Abrahamse J., Joenje W. & Van Leeuwen-Seelt N.), pp. 249–60. Land. Ver. Behoud Waddenzee, Harlingen and Ver. Beh. Natuurmonumenten in Ned.'s Graveland.

Weaver J.E. & Clements F.E. (1938) *Plant Ecology.* McGraw-Hill, New York.

Westhoff V. (1947) *The vegetation of dunes and salt-marshes on the Dutch islands of Terschelling, Vlieland and Texel.* Thesis, Utrecht.

Zuur A.J. (1938) *Over de ontzilting van de bodem in de Wieringermeer.* Thesis, Wageningen.

Zuur A.J. (1961) Initiele bodemborming bij mariene gronden. *Med. van de Landbouwhogeschool en de opzoekingsstations van de Staat te Gent* ?4, 7–33.

VII

ABSTRACTS OF POSTER PAPERS

The population dynamics of *Ammophila arenaria*

A.H.L. Huiskes *School of Plant Biology, University College of North Wales, Bangor, Gwynedd, U.K. Present address: Delta Institute for Hydrobiological Research, Yerseke, The Netherlands.*

Populations of tillers of *Ammophila arenaria* (L.) Link were very stable in density during two years of observation. Higher densities occurred in the mobile areas of the dunes, as compared with more stabilized areas. The birth rates and death rates of cohorts of leaves showed a strong seasonal fluctuation. High birth rates and death rates were found in the higher areas of the dunes, as compared with lower lying dune grasslands and slacks. Fertilizer application increased both the population density and the population flux, but only if the nutrients could reach the root system of *A.arenaria*, i.e. in areas where there was little root competition from other species.

REFERENCE

Huiskes A.H.L. (1977) *The population dynamics of* Ammophila arenaria *(L.) Link*. Ph.D. thesis, University of Wales.

Nitrogen fertilizers and the establishment and growth of *Ammophila arenaria* on eroded sand dunes

Paul Johnson *Department of Botany, University of Liverpool. Present address: Land Capability Consultants Ltd.*, 52 *Station Road, Fulbourn, Cambridge, U.K.*

The supply of nitrogen from rainfall, sea-spray and N-fixing micro-organisms was found to be adequate to maintain established *Ammophila arenaria* (L.) Link at a low level of productivity on nitrogen deficient dunes. Comparisons of various N-fertilizers added to promote growth of *Ammophila* which had been planted to stabilize eroding dunes indicated that there is no significant benefit to be gained from using expensive slow release N-fertilizers rather than cheap fast release forms such as ammonium nitrate. Under normal conditions, actively growing roots of planted *Ammophila* are capable of rapidly taking up large quantities of available nitrogen before it is lost from the system due to leaching.

The contrasting adaptations of two biennial species to survival in the dune habitat

L.A. Boorman and R.M. Fuller *I.T.E., Colney Research Station, Colney Lane, Norwich, U.K.*

Lactuca virosa and *Cynoglossum officinale* both occur on the dunes at Holkham, north Norfolk, as scattered individuals. *Cynoglossum* is more often found on sites where heavy rabbit grazing has reduced the vegetation height while *Lactuca* occurs mostly on sites where there is little rabbit grazing and the vegetation is tall. *Lactuca* produces very large numbers of small wind-dispersed seeds (seed wt. <1 mg) that germinate readily; while *Cynoglossum* produces smaller numbers of animal-dispersed seeds that are relatively large (seed wt. c. 30 mg) and have a considerable degree of dormancy. The significance of the points of similarity and difference are considered in relation to the overall strategies for survival of these species.

The ecophysiology of the sand sedge (*Carex arenaria*): the response to local mineral sources

T. TIETEMA *Botanical Laboratory, State University of Utrecht, Lange Nieuwstraat* 106, *Utrecht, The Netherlands.*

The distribution of mineral nutrients in *Carex arenaria* L. indicates that a local mineral source does not supply the plant as a whole. There appears to be no long distance water and mineral nutrient translocation along the rhizome. Redistribution via the phloem does not seem to occur to such an extent that the plant could use this supply to maintain a high growth rate in those parts of the plant not fed with an external source of mineral nutrients.

The population dynamics of *Vulpia fasciculata* (Forskål) Samp.

A.R. WATKINSON *School of Biological Sciences, University of East Anglia, Norwich, U.K.*

The behaviour of populations of *Vulpia fasciculata* on two Anglesey dune systems was remarkably similar. In early summer each plant produced on average 1·7 seeds. Loss of seed viability (5%), seed predation (0–15%) and the failure of seedlings to establish (12–28%) due principally to wind-drag were the major factors involved in the loss of individuals prior to seedling establishment in late autumn. More than 99·5% of the seeds germinated in the year of their production. The probability of a seedling surviving to maturity was high (0·69), but the potential seed rain on some plots was considerably reduced by rabbit grazing. There was a negatively density-dependent relationship between the reproductive output and density of flowering plants. Combining the field estimates of density-dependent control of reproduction with the measured density-independent mortality was sufficient to account for the observed range of densities found in natural populations.

REFERENCE

Watkinson A.R. & Harper J.L. (1978) The demography of a sand dune annual: *Vulpia fasciculata*. I. The natural regulation of populations. *J. Ecol.* **66**, 15–33.

Braunton Burrows: *Pyrola* and shade

RODERICK HUNT *Department of Botany, The University, Bristol. Present address: Unit of Comparative Plant Ecology (NERC), Department of Botany, The University, Sheffield.*

Pyrola rotundifolia ssp. *maritima* has, in recent years, become a frequent occurrence in the dune-slack vegetation at Braunton Burrows, N. Devon. Newly-arrived seedlings have given rise to a great proliferation of clonal rosettes. These occur almost exclusively in situations which are quite deeply shaded in summer by a canopy of *Salix repens* ssp. *argentea*, but, both in the field and in the glasshouse, it has been shown that *Pyrola*'s adaptability to shade is low. The considerable success of *Pyrola* in shaded situations is thus considered to be due to factors other than a simple shade requirement for vegetative growth.

Structure and development of dune scrub communities in south-west Netherlands

CLARA J.M. SLOET VAN OLDRUITENBORGH *Department of Nature Conservation, Agricultural University, Wageningen, The Netherlands.*

The structure, species composition and ecology of dune scrub communities in the Dune-district of The Netherlands are described. Their positions in space and time are illustrated by means of transect diagrams and ordination models. Their development must be interpreted, in particular, against the background of nearly complete devastation of the vegetation at the beginning of the 20th century, by cutting, heavy grazing and cultivation, and also more recent activities in the dunes such as water-catchment and recreation. The different types of succession and the scrub species characteristic for each of these types are described, and guidelines are given for appropriate management.

REFERENCE

Sloet van Oldruitenborgh C.J.M. (1976) Duinstruwelen in het Deltagebied. Thesis, Wageningen.

Environmental control of a population of *Pinus pinea* in the coastal sand dune system of Doñana National Park (S.W. Spain)

M.E. FIGUEROA AND J. MERINO *Department of Ecology, University of Sevilla, Sevilla, Spain.*

Environmental control of the distribution and age-structure of umbrella pine (*Pinus pinea* L.) has been studied. Three different scales of pattern (approximately, 80,20 and 10 m) were associated with topography. Populations showed different survival rates on dry and wet sites and furthermore the age structure of the pine populations showed a markedly anomalous distribution.

Intense precipitation causes a rise in the soil water table, and tree seedlings growing on lower ground tend to be killed by flooding. Repeated killing results in aggregation of trees on higher ground. Thus it is hypothesized that aggregation, differential mortality rates and anomalous age distribution all represent responses of pine populations to a single environmental factor, the phreatic water level oscillation.

REFERENCE

Figueroa M.E. (1976) *Ecología de* Pinus pinea *L. en el Parque Nacional de Doñana.* Tesis de Licenciatura, Universidad de Sevilla.

Coastal sand dunes of Oregon and Washington, U.S.A.

A.M. WIEDEMANN *The Evergreen State College, Olympia, Washington, U.S.A.*

Sand dunes occur along 300 km of Pacific coast in the states of Oregon and Washington, U.S.A. There are three distinct dune systems: parallel ridges, parabola dunes and a long, continuous sand plain with complex dune topography. Herbaceous plant communities develop on bare sand surfaces created by deflation (wet sites) or by deposition-erosion (moving sand sites). Succession to shrub and tree communities is rapid. *Pinus contorta* is the predominant pioneer tree species on all sites. The introduced grass, *Ammophila arenaria*, has brought about significant changes in the dune landscape.

REFERENCE

Wiedemann A.M., Dennis L.J. & Smith F.H. (1969) *Plants of the Oregon coastal dunes.* O.S.U. Book Stores Inc., Corvallis, Oregon, U.S.A.

Ecological processes in a subtropical Australian dune forest

W.E. WESTMAN AND R.W. ROGERS *Geography Department, University of California, Los Angeles, U.S.A. and Botany Department, University of Queensland, Brisbane, Australia.*

A coastal *Eucalyptus*-heath forest in Queensland was the subject of studies on seasonal productivity, nutrient stocks and flows, forest biomass, litter turnover and phenology for a 26-month period. Significant differences in seasonality of growth patterns, ratios of nutrient stocks to biomass and responses to fertilization suggest separate biogeographic origins for the shrub and tree synusiae. Fossil dates and modern biogeographic distributions for members of the two synusiae further support this hypothesis.

The dominant tree species, *Eucalyptus signata* and *E.umbra* ssp. *umbra*, are capable of interbreeding. No hybrids occur in the field, however. Our investigations suggest that the two species have differentiated and specialized in niche behaviour (flowering times, pollinator use, nutrient partitioning, time and arrangement of growth of plant parts) sufficiently to put hybrid offspring at a competitive disadvantage.

Calculation of the nutrient budget indicated that the ecosystem is dependent on periodic release of potassium and phosphorus by fire to sustain forest growth. Gradual leaching of nutrients over the past 500,000 years has resulted in dwarfing of the vegetation on older soils.

REFERENCES

Westman W.E. & Rogers R.W. (1978) Nutrient stocks in a subtropical eucalypt forest, N. Stradbroke Island. *Aust. J. Ecol.* **2**, 447–60.

Westman W.E. (1978) Inputs and cycling of mineral nutrients in a coastal subtropical eucalypt forest. *J. Ecol.* **66**, 513–31.

The sand-dune vegetation of south-eastern Australia and problems of management

D.M. CALDER AND A.J.E. PARK *Botany School, University of Melbourne, Australia.*

Thirty-seven percent of the Victorian coast (total length about 2,000 km) is sandy, and in places dunes extend several kilometres inland. These sandy shorelines are generally receding, often resulting in dunes cliffed at their seaward margin with limited development of a primary foredune. Zonation of dune vegetation proceeds from foredune grasses to dune shrubs and eventually to either woodland or heath. The shrub zone is eroded as the shoreline recedes, and the key to dune management and protection of the hinterland is the control of this erosion. Control is most difficult in areas of intensive human use. Various techniques of management are being assessed, including the planting of native sand-binding species.

Response of coastal ecosystems to the mechanical stress of off-road vehicles

PAUL J. GODFREY *National Park Service Cooperative Research Unit and Department of Botany, University of Massachusetts, Amherst, Massachusetts, U.S.A.*

Experiments have been done for the past three years in Cape Cod National Seashore, Mass., to determine the response of coastal ecosystems to vehicle impact. Research involved controlled impacts of each ecosystem; beaches, dunes, salt marshes and tidal flats. Salt marshes and tidal flats were found to be the most sensitive. Low levels of impact severely affected dune vegetation and developing dune lines; recovery was most rapid where *Ammophila* grew near fresh sources of sand. *Hudsonia* was most sensitive. Intertidal ocean beaches showed least stress of systems studied, but require more research. Shorebirds habituated to passing vehicles if nesting sites were protected.

REFERENCES

Godfrey P.J., Leatherman S.A. & Buckley P.A. (1978) Impact of off-road vehicles on coastal ecosystems. In *Coastal Zone '78 Proc.*, pp. 581–600. Amer. Soc. of Civil Eng., New York.

Brodhead J.M.B. & Godfrey P.J. (1977) Off-road vehicle impact in Cape Cod National Seashore. *Internat. Jour. Biometeor.* **21**, 299–306.

The role of littoral bacteria in the solution of calcareous sand grains

ZIMA DARTEVELLE *Institut royal des Sciences naturelles, Bruxelles, Belgium.*

Bacteria able to dissolve calcium carbonate have been isolated from marine littoral sand grains. Even after twenty consecutive rinsings with sterile water, the grains contained bacteria which hydrolysed calcium carbonate, when this was incorporated in the medium. The bacteria probably adhered to the grain surface or developed inside cavities in the grains. When these bacteria were identified, some of the genera coincided with those we have found in littoral interstitial water in Belgium, and especially with the new *Bacillus* ssp. we have described. However, most of them belong to the genus *Staphylococcus*.

REFERENCE

Brisou J., Dartevelle Z. & de Barjac H. (1976) Sur un *Bacillus* halophile préférentiel isolé du sable littoral belge. *XXVe Congrès-Assemblée plénière de la C.I.E.S.M.*, Split.

The adaptability of estuarine bacteria to high concentration of heavy metals

ZIMA DARTEVELLE AND MICHEL MERTENS *Institut royal des Sciences naturelles, Bruxelles, Belgium.*

We have tested the ability of estuarine bacteria to adapt themselves to increasing concentrations of chromium and zinc. Freshly-sampled water was enriched with peptone and incubated for 48 hr. Then 1 ml of the culture was used to seed a new medium supplemented with a few ppm of metal. We repeated this process 48 hr later, increasing the concentration of metal. This procedure was repeated every 2 or 3 days.

The increasing metal concentration led to progressive selection of the bacteria. The experiment was terminated when the medium seemed to contain only one bacterial species. Medium supplemented with 1,280 ppm of Cr^{6+} led to selection of *Staphylococcus* and 280 ppm of Zn^{2+} to selection of *Pseudomonas fluorescens*.

Neuronal adaptations to osmotic stress in an extreme estuarine osmoconformer

J.E. TREHERNE, J.A. BENSON AND H. LE B. SKAER *Department of Zoology, University of Cambridge, U.K.*

The giant axon of the serpulid worm, *Mercierella enigmatica* Fauvel, can adapt to massive variations in the ionic and osmotic concentration of the body fluids (between 84 and 2,304 m OsM). At the unusually high potassium concentration (30 mM) of the blood of sea-water adapted animals the resting membrane potential ($53{\cdot}6 \pm 1{\cdot}4$ mV) lies on a steep portion of the curve relating the external potassium concentration to membrane potential. At this membrane potential a relatively large proportion of the sodium channels (which carry the inward current of the action potential) appear to be inactivated. During dilution of the bathing medium the increase in resting potential (which results from dilution of external potassium ions), reduces the inactivation of the sodium channels and, also, enables the active membrane to approach the sodium equilibrium potential and still maintain an action potential of appreciable amplitude even at dilutions as low as 5% ($[Na^+]_0 =$ 24·0 mM).

REFERENCE

Treherne J.E., Benson J.A. & Skaer H. le B. (1977) Axonal accessibility and adaptation to osmotic stress in an extreme osmoconformer. *Nature* **269**, 430–31.

Natural geo-electric fields and electro-orientation in fish

N. PALS *Laboratory of Comparative Physiology, Jan van Galenstraat* 40, *Utrecht, The Netherlands.*

In the last two decades it has been established that several non-electric fish (Elasmobranchs, Siluriforms, Polypteriforms and Dipnoians) are equipped with a special sensory system for weak electric currents. Although it is known that the electric fields of animals are used in prey-detection, the significance of inanimate electric fields to electro-sensitive fish is not completely understood. It is proven that uniform electric fields can be used for orientation in some species.

The main objective of our present study is the measurement of electric fields in the sea with special regard to local phenomena near the bottom, which might be important in some kind of local orientation in electro-sensitive fish.

REFERENCE

Pals N. (1977) The measurement of electric fields in the sea in relation to the electric sense of some marine fishes. *Proc. Int. Un. Physiol. Sc.* **13**, 576.

Quelques aspects du rôle biogéochimique des *Nereis* dans deux facies estuariens (Baie de Somme—France)

M. LOQUET ET J.P. DUPONT *Laboratoire de Biologie végétale et d'Ecologie et Laboratoire de Géologie, Faculté des Sciences et des Techniques de Rouen, Université de Rouen, France.*

L'existence de populations denses de *Nereis* nous a amené à rechercher l'influence biogéochimique exercée par ces vers sur leur biotope.

Deux biofaciès ont été distingués *in situ* :

1. l'un en milieu réducteur de slikke (3 500 terriers/m^2);
2. l'autre, subfossile, au niveau de la microfalaise du schorre (2 800 terriers/m^2) présentant des gaines rouillées.

L'étude sédimentologique, géochimique et microbiologique des deux zones a été réalisée en dissociant terriers et sols témoins.

L'activité des vers en développant des interfaces eau/sédiment ou air/sédiment entraîne localement des modifications physicochimiques. Ces dernières influent sur la composition, la répartition et l'activité de la microflore.

Une activité biologique solubilisatrice du Fer a été démontrée.

REFERENCE

Loquet M. & Dupont J.P. (1976) Etude morphologique et microbiologique des terriers à *Nereis* dans la slikke en baie de Somme (Cap Hornu)—Essais préliminaires. ***Bull. Fr. de Pisciculture*** **261**, 170–86.

Iranian oil as germination inhibitor and stress factor for two halophytes

PETER JANIESCH AND WERNER MATHYS *Inst. f. Angew. Botanik and Hygiene Institut, Universität Münster, D-44 Münster, B.R.D.*

The influence of increasing concentrations of Iranian oil on germination, peroxidase activity and organic acid concentrations in *Aster tripolium* and *Triglochin maritima* was tested.

(1) The range of concentration tested was between 0–4 g litre^{-1}. Even the

lowest concentration of 0·08 g litre^{-1} caused slight inhibition. Germination was reduced to 50% by 0·8 g litre^{-1} and 1·2 g litre^{-1} in *Triglochin* and *Aster* respectively.
(2) Peroxidase activity increased at high oil concentrations. At 4 g litre^{-1} the activity reached 143% and 189% respectively of unpolluted controls in *Aster* and *Triglochin*.
(3) Production of malate and citrate was diminished in both species above concentrations of 1·5 g litre^{-1}, whereas the amount of succinate seemed to be augmented.
(4) In all experiments the pollution effects were more marked in soil culture than in water cultures using only water soluble extracts of the oil.

The differential sensitivity of the two species indicates that oil may play an increasing role in affecting the ecological balance of halophyte communities.

The response of halophytes to short-term fluctuations in salinity and water stress

M.J. Bailey,[1] F.B. Thompson[2] and S.R.J. Woodell[3] *Botany School, Oxford University. Present address:* [1] *Oxford Scientific Films Ltd., Long Hanborough, Oxford,* [2] *Department of Forestry, Oxford University and* [3] *Botany School, Oxford University.*

The response of *Limonium vulgare* Mill. and *Aster tripolium* L. to short term fluctuations in salinity was investigated. The roots of plants in liquid culture were flushed with sea water solutions of varying strengths while transpiration, net carbon dioxide uptake, stomatal and mesophyll resistances were monitored.

An immediate stimulation of net carbon dioxide uptake and transpiration was followed within a few minutes by a steady but sharp decline in both rates, accompanied by a rapid fall in leaf water potential. After about four hours, there was a rather abrupt change to a much slower rate of decline; finally, after about sixteen hours, steady rates of net carbon dioxide uptake and transpiration were attained. Thus the response could be analysed into four distinct stages, which showed different susceptibilities to environmental factors. Experiments with *Phaseolus vulgaris* L. indicated that this was not a specifically halophytic response. Experiments using mannitol solutions showed that the response was primarily due to water stress rather than to salinity.

Longer term adaptations were also noted: saline pretreatment of the plants reduced the water loss during the period of rapid stomatal closure.

REFERENCE

Bailey M.J. (1976) *Physiological ecology of salt marsh plants.* M.Sc. thesis, Oxford University.

Salt balance of *Avicennia marina*

Y. WAISEL *Department of Botany, Tel-Aviv University, Tel-Aviv, Israel.*

The coast of the Gulf of Eilath is an extremely dry and salty habitat and plants must have a well adapted salt and water balance in order to survive there. *Avicennia marina* is the only species of mangrove which grows there. It is a salt-excreting species and the trees' leaves are continuously covered with salt crystals. Transpiration rates, salt secretion and xylem sap concentrations were measured and the overall efficiency of such a salt removing system was investigated. Diurnal rhythms were observed in all three processes: highest transpiration rates were found during midday and lowest during the night. Highest sap concentrations were found at midnight whereas the lowest concentrations occurred at noon. Two peaks in salt secretion, at noon and midnight, were observed.

REFERENCE

Waisel Y. (1972) *Biology of Halophytes.* Academic Press, New York and London.

Geographical variation in British salt marsh vegetation

PAUL ADAM *The Botany School, Downing Street, Cambridge.*

Data on the occurrence of plant communities on 133 salt marsh sites around the British coast were subjected to minimum-variance cluster analysis which suggested the possibility of distinguishing three major types of salt marsh in Britain. One type is largely restricted to sites in south-east England, a second type is represented by the majority of marshes in north-west England and Wales, while the third type is characteristic of the head of lochs in western Scotland. The differences between the three types of marsh originate from the action of various factors of which grazing, substrate and climate are the most important.

REFERENCE

Adam P. (1978) Geographical variation in British salt marsh vegetation. *J. Ecol.* **66,** 339–66.

The pattern of plant distribution on an Essex salt marsh

S.B. OTHMAN AND S.P. LONG *Department of Biology, University of Essex, U.K.*

The pattern of distribution of plant species on a saltings area at Colne Point, Essex, has been examined by the Bray and Curtis Ordination technique. This analysis suggests two distinct association of plant species. The first association contains mainly *Puccinellia maritima* and *Halimione portulacoides*, and the other is more species rich, the most abundant species being: *Limonium vulgare*, *Plantago maritima*, *Armeria maritima* and *Triglochin maritima*.

Soil salinity, structure, drainage and nutrient status have been studied over a one year period at the sites used in constructing the ordination pattern. Cover of *Puccinellia*, *Halimione* and *Suaeda maritima* had significant negative correlation coefficient with exchangeable sodium. *Limonium* cover had a significant positive correlation coefficient with exchangeable magnesium. However, only *Puccinellia* had a significant correlation coefficient with ammonium nitrogen.

Vegetation of a subarctic salt marsh in James Bay (Ontario)

WALTER A. GLOOSCHENKO *Canada Centre for Inland Waters, P.O. Box* 5050, *Burlington, Ontario, Canada.*

The vegetation of Hudson and James Bay salt marshes is quite similar to that of salt marshes in Alaska, the Canadian Arctic, Greenland, the northern British Isles and Scandinavia. In contrast to Atlantic and Gulf coast salt marshes, *Spartina alterniflora* does not occur, and marshes are dominated by such species as *Puccinellia phryganodes*, *P.lucida*, *Salicornia europaea*, *Glaux maritima*, *Scirpus maritimus*, *Triglochin maritima*, *Potentilla egredii*, *Plantago maritima*, *Juncus balticus*, *Cicuta maculata*, *Carex salina*, *Carex paleacea*, *Hordeum jubatum* and *Atriplex patula*. Primary productivity has been estimated at 500 g m^{-2} yr^{-1} which is less than that of temperate marshes. These salt marshes grade into fresh-water marshes, fens and old beach ridge communities.

REFERENCE

Glooschenko W.A. (1978) Above-ground biomass of vascular plants in a subarctic James Bay (Ontario) salt marsh. *Can. Field Nat.* **92,** 30–37.

Vegetation associations and primary productivity of a salt marsh on Puget Sound, Washington, U.S.A.

M.E. BURG, D.R. TRIPP AND E.S. ROSENBERG *Nisqually Delta Laboratory*, 9131 *D'Milluhr Road, Olympia, Washington, U.S.A.*

A vegetation and primary productivity study was conducted on the salt marsh of the Nisqually River delta located at the southern end of Puget Sound in Washington, U.S.A. Twelve plant associations were defined and a vegetation map was prepared showing their extent and location. Productivity values were estimated for eight of the associations. Average annual net productivity was 750 g dwt. m^{-2} with a total of 1,380 dry weight metric tons produced over the 193·1 ha sampled. The *Carex lyngbyei* association is the most productive with an annual net productivity of 1,390 g dwt. m^{-2}.

Changes in species composition in fertilized salt marsh plots

J.M. TEAL AND I. VALIELA *Woods Hole Oceanographic Institution and Boston University Marine Program, Woods Hole, Massachusetts, U.S.A.*

Increased nutrients, particularly nitrogen, lead to changes in the flora of salt marshes. In our experiments, enrichment led to an initial dominance by opportunistic species (*Salicornia virginica* and *Distichlis spicata*), followed by replacement by the usual dominants (*Spartina alterniflora* and *S.patens*). After several growing seasons *Salicornia europaea*, an annual species, is aggressively invading the fertilized plots, while the controls remain as they were before the treatments. The diversity of the vegetation was reduced in all the fertilized plots.

REFERENCE

Valiela I., Teal J.M. & Sass W.J. (1975) Production and dynamics of salt marsh vegetation and the effects of experimental treatment with sewage sludge. *J. appl. Ecol.* **12**, 973–82.

Methods in determining the below-ground live biomass in salt marsh ecosystems

R. DUNN, A. HUSSEY AND S.P. LONG *Department of Biology, University of Essex, Wivenhoe Park, Colchester, Essex, U.K.*

Extraction and determination of live below-ground biomass (BGM) presents many difficulties in accessible dry terrestrial sites. On remote inter-tidal mudflats these problems are amplified. The estimation of live BGM has been attempted on two marshes in Essex, one on soft sediments dominated by *Spartina townsendii*, the other on consolidated sediments dominated by *Puccinellia maritima*. On the latter a simple aluminium tube was used to obtain cores, whilst on the former a cover with a closing flap system had to be developed before satisfactory cores could be obtained. The practicability and accuracy of eight methods of distinguishing live biomass from dead was assessed, these ranged in complexity from flotation to ^{86}Rb uptake. On both marshes, live BGM constituted on average 40% of the total live plant biomass, but occasionally it was as much as 70%, emphasizing the importance of considering below-ground material in studies of primary productivity in salt marshes.

Tidal flows in salt marsh creeks

T.P. Bayliss-Smith, R. Healey, R. Lailey, T. Spencer and D.R. Stoddart *Department of Geography, University of Cambridge, U.K.*

Data from north Norfolk, England indicate a three-fold division of tidal flow regimes. On upper marshes the majority of tides are 'below-marsh' and, apart from a small initial pulse of velocity on the flood, they generate only modest flows in the creeks. The level of the marsh surface seems to provide a threshold at which, with higher 'marshful' tides, velocities can increase to reach peaks shortly before and shortly after high tide. These velocity pulses are greater on the ebb than on the flood, but the flood pulse occurs at a higher stage. The even higher but infrequent 'over-marsh' tides, usually associated with storm surges, have a substantial flood maximum in both velocity and discharge. It is hypothesized that only these extreme tides might be capable, on upper marshes, of achieving significant erosion and deposition, and their much higher frequency on lower marshes can account for the observed rapidity of geomorphic change at this earlier stage in salt marsh development.

Simulation of successional processes on salt marshes

P.F. Randerson *Department of Applied Biology, UWIST, Cardiff, U.K.*

Relationships between edaphic and floristic factors of the salt marsh ecosystem were used to formulate a mathematical model representing the dynamic processes of the halosere, and its behaviour explored using the simulation language CSMP. The structure of the model was determined by successional processes assumed to be generally typical of haloseres, but its calibration differed in accordance with the widely different sedimentation regimes of the Norfolk coast and the Wash. The models were extended to include the effects of windborne sand, stochastic variation in plant invasion and, in the case of the Wash, the effects of grazing. Comparisons of simulation results with field data indicate the utility of the model in re-appraising the ecological assumptions on which it is based.

REFERENCES

Randerson P.F. (1979) The ecology of the Wash V: A simulation model of salt marsh succession. *J. appl. Ecol.* (in press).

Randerson P.F. (1975) *An ecological model of succession on a Norfolk salt marsh.* Ph.D. thesis. University of London.

Ecology and production of eelgrass in the former Grevelingen estuary, an evolving saline lake in the S.W. Netherlands

P.H. NIENHUIS *Delta Institute for Hydrobiological Research, Vierstraat* 28, *Yerseke, The Netherlands.*

The Grevelingen estuary was cut off from the North Sea and from the influences of the river Rhine by a dam in 1971, and became a stagnant salt-water lake. The production and the ecology of *Zostera marina* L. were studied in 1968 and in 1973–76. Standing crop estimations were made and biomass changes in permanent quadrats recorded. Correlations of the distribution patterns of this species with ecological factors also were attempted.

After the closure of the estuary the area occupied and the density of the eelgrass beds increased strongly. Eelgrass annual above-ground production, based on doubled maximum standing crop values in July–August, was estimated at 50 g C m^{-2} in 1968, 121 g C m^{-2} in 1973 and 91 g C m^{-2} in 1975 in *Zostera* beds.

REFERENCE

Nienhuis P.H. & Bree B.H.H. de (1977) Production and ecology of eelgrass (*Zostera marina* L.) in the Grevelingen estuary, The Netherlands, before and after the closure. *Hydrobiol.* **52**, 55–66.

The annual cycle of a coastal lagoon in the Ebro Delta

F.A. COMÍN AND X. FERRER *Departamento de Ecología, Facultad de Biología, Universidad de Barcelona, Spain.*

The seasonal cycle of a coastal lagoon, the Encañizada (Ebro Delta, north-eastern Spain), was studied from January 1976 until May 1977. Water level, temperature, conductivity, pH, oxygen, alkalinity, Cl^-, nutrients, phytoplankton numbers, pigments, primary production and macrophyte cover, were determinated at monthly intervals.

Fresh water inflow to the lagoon from irrigation channels is artificially regulated. Water level and Cl^- fluctuate in accordance with this human influence. The lagoon is characteristically eutrophic, with high values of alkalinity, pH, nutrients and mineral salts. However, nitrogen is limiting towards the end of the summer at the time of a cyanophyceae bloom. Phytoplankton also show important differences between winter and the rest of the year. A clear succession was observed in the composition of the major algal groups and in the cover and flowering of macrophytes.

REFERENCE

Comin F.A. & Ferrer X. (1978) Lacunes costeres. Caracteristiques i problemâtica limnológica. *Cuad. Ecol. Ap.* **4** (in press).

The vegetation of the Gippsland Lakes, south-eastern Australia

D.M. CALDER AND S.C. DUCKER *Botany School, University of Melbourne, Australia.*

The vegetation of the Gippsland Lakes has been surveyed. The lakes are interconnected, and are separated from Bass Strait by a sand barrier through which an entrance was cut in 1889. Salinity in the lakes varies from fresh water to hypersaline. The aquatic vegetation is composed of seagrasses (*Zostera, Lepilaena, Ruppia*), benthic algae (*Cladophora, Lamprothamnium, Gracilaria,* diatoms) and phytoplankton. The land vegetation includes reed swamps, salt marshes, as well as heath, scrub and forest communities.

In contrast to the land flora, which has a high degree of endemism, the aquatic flora has a generic composition fairly typical of estuaries throughout the world.

Pressing ecological problems include the die-back of *Phragmites* and the subsequent erosion of the shoreline. This may be related to increasing salinity or to the population explosion of the introduced European carp.

Port development and marsh integrity at the Squamish River estuary, British Columbia

C.D. LEVINGS *Fisheries and Environment Canada, Resources Services Branch,* 4160 *Marine Drive, West Vancouver, B.C., Canada.*

Port development at the Squamish River estuary (south-western British Columbia) led to modification of fresh-water flow across the delta and land filling. The results of disruption on the sedge marsh ecosystem at the estuary was of interest since an amphipod (*Anisogammarus confervicolus*) important in the diet of juvenile salmon utilizes *Carex lyngbyei* for structural and functional reasons. Increased salinities have resulted from dyke construction, but the marsh has maintained its integrity over five years of observation. In 1974, net primary production of *C.lyngbyei* was approximately 1,300 g dwt. m^{-2} season^{-1}. The recovery of a portion of the marsh from silt spillage was incomplete four years after disruption.

REFERENCE

Levings C.D. & Moody A.I. (1976) Studies on intertidal vascular plants, especially sedge (*Carex lyngbyei*) at the disrupted Squamish River estuary, B.C. Fisheries and Marine Service (Canada). Research and Resources Services Directorate, *Tech. Rep.* No. **606**.

The role of *Spartina capensis* in the reclamation of eroding banks of the Kowie river (Eastern Cape, South Africa)

R.A. LUBKE AND BARBARA A. CURTIS *Department of Plant Sciences, Rhodes University, South Africa.*

In conjunction with the other halophytic species *Spartina capensis* Nees ex Trin. is shown to be an important stabilizer of eroding banks on the Kowie estuary. A study of soil from four different sites revealed no reason why *Spartina* plants should not grow in uncolonized areas below the eroding banks. Subsequent transplant experiments proved successful, time of planting and position in the intertidal region being critical factors affecting productivity. It is calculated that clumps planted at 1×1 m intervals should achieve complete cover in about two years. Attempts at transplanting other species have not been very successful.

Ecological strategies of plants on retreating barrier beaches in northern and southern climates

PAUL J. GODFREY *National Park Service Cooperative Research Unit and Department of Botany, University of Massachusetts, Amherst, Massachusetts, U.S.A.*

Responses of plants to forces of barrier beach recession (dune migration, overwash, inlet dynamics) vary significantly between northern and southern environments of the U.S. Grassland species of southern barriers are well adapted to overwash flooding and burial. Vegetative recovery occurs by growth up through overwash deposits. On northern barriers, no species are adapted to overwash. Re-vegetation occurs by *Ammophila* growth starting on top of overwash deposits. Northern populations of *Spartina patens* are not adapted to overwash while populations in the south are well adapted and respond quickly. These ecological responses may play a major role in the geomorphology of barrier beaches.

REFERENCE

Godfrey P.J. (1977) Climate, plant response and development of dunes on barrier beaches along the U.S. East Coast. *Intern. Journ. Biometeor.* **21**, 203–15.

The population dynamics of *Erigone arctica* (*Araneae, Linyphiidae*) in a coastal plain

WALTER K.R.E. VAN WINGERDEN *Institute for the Education of Teachers, Free University, sect. Biology, Amsterdam, The Netherlands.*

Fluctuations in density of the spider *E.arctica* (White) are caused mainly by fluctuations in density of the prey (*Hypogastrura viatica: Collembola*), as differences in prey-density cause differences in egg production, and it appeared from experiments that mortality was higher under low prey-densities. Furthermore drought lowers the accessibility of the prey.

REFERENCE

Wingerden W.K.R.E. van (1978) Population dynamics of *Erigone arctica* (White) (*Araneae*, *Linyphiidae*), part II: Further analysis. *Symp. zool. Soc. Lond.* **42**, 195–202.

A survey of coastal landscapes in the Netherlands

H. DOING *Department of Plant Ecology, Agricultural University, Wageningen, The Netherlands.*

Landscape maps (scale 1:25,000) are expected to be available in 1978 for all sand dune areas (including adjacent salt marshes) in the Netherlands, in order to provide a general framework for research, conservation and management of coastal areas. Mapping units are mainly based on vegetation, but also on soil complexes and geomorphology.

Various examples were shown, e.g. a schematic map of a hypothetical Wadden island.

REFERENCES

Doing H. (1974) Landschapsoecologie van de duinstreek tussen Wassenaar en IJmuiden (with maps and English summary: Landscape ecology of sand dune areas between Wassenaar en IJmuiden, Netherlands). *Meded. Landbouwhogeschool Wageningen* **74-12**, 1–111.

Doing H. (1975) Beobachtungen und historische Tatsachen über die Sukzession von Dünen-Ökosystemen in den Niederlanden. *Sukzessionsforschung* (Ed. by W. Schmidt), pp. 107–22. J. Cramer, Vaduz.

The Fleet Study Group

J.M. FITZPATRICK *Dorset Institute of Higher Education, Weymouth, U.K.*

The Fleet Study Group, formed in 1975, is an *ad hoc* body representing landowning interests of the Dorset Naturalists Trust, environmental authorities and individual scientists. Its aim is to evaluate the state of scientific knowledge of the Fleet Water and the Chesil Beach, Dorset, England. It is also attempting to coordinate research done in the area. The Fleet Water itself is likely to be sensitive to changes made by man as it is a shallow brackish lagoon some 13 km long with a limited exchange of water through a narrow channel into Portland Harbour. The Group is based at the Weymouth Centre of the Dorset Institute. Information is being collected under separate sections—Geomorphology, Hydrology and Ecology.

AUTHOR INDEX

Figures in italics refer to pages where full references appear.

SUBJECT INDEX